Advanced Molecular
Solutions for Cancer Therapy

Advanced Molecular Solutions for Cancer Therapy

Editor

Dumitru A. Iacobas

Basel • Beijing • Wuhan • Barcelona • Belgrade • Novi Sad • Cluj • Manchester

About the Editor

Dumitru A. Iacobas

Dumitru A. Iacobas is an expert in both experimental and computational genomics. He has a Ph.D. in Biophysics from the University of Bucharest (Romania) and is currently a Research Professor of the Undergraduate Medical Academy, Prairie View A&M University, TX, U.S.A. Dr. Iacobas introduced the Principle of Transcriptomic Stoichiometry and developed advanced mathematical algorithms to fully characterize the transcriptome and quantify its remodeling in disease and remodeling in response to treatments.

Preface

"Advanced Molecular Solutions for Cancer Therapy" is a collection of 25 articles presenting clinical investigations, bioinformatics analyses, and experimental studies in support of novel molecular anti-cancer approaches. It continues *"Molecules at Play in Cancer"*, which examined the pro- and anti-carcinogenic roles of several important molecules. In addition to the Editorial that discusses the limits of the biomarker approach, the reprint is organized into 15 cutting-edge research articles, 9 well-documented reviews, and 1 promising case report.

The reprint covers a wide range of malignant tissues and organs, i.e., blood, breast, cervix, head and neck, kidney, liver, lung, nervous system, ovary, prostate, skin, and thyroid gland. The research studies used advanced experimental methods, including DNA, RNA, exome and protein sequencing, gene and microRNA expression microarrays, qPCR, CRISPR, and ultra-high-performance liquid chromatography. Along with analyzing the involved molecular mechanisms and proposing improvements to existing immunotherapy, chemotherapy, and gene therapy, the Special Issue also assessed the benefits of repurposing as anticancer agents existing drugs like aspirin and tezosentan used to treat other diseases.

We were honored by the contributions of 145 scientists affiliated with academic and clinical institutions from Belgium, Canada, China, Egypt, France, Germany, India, Japan, Korea, Lebanon, Lithuania, Mexico, Morocco, the Netherlands, Poland, Portugal, Romania, Russia, South Africa, the U.K., and the U.S.A. The number of coauthors and the geographical distribution of their institutions indicate both the actuality of the proposed theme and the level of international cooperation to fight cancer.

Although our efforts were mainly addressed to the respected community of scientists involved in cancer research and oncology clinicians devoted to curing their patients, we do hope that the general public will also find the presented book both interesting and informative.

We would like to acknowledge that this book would not have been possible without the excellent editorial work of the team led by Ms. Norah Tang, *CIMB* Managing Editor, and the rigorous evaluations provided by the invited reviewers.

Dumitru A. Iacobas
Editor

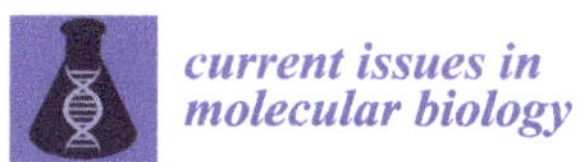

Editorial

Advanced Molecular Solutions for Cancer Therapy—The Good, the Bad, and the Ugly of the Biomarker Paradigm

Dumitru Andrei Iacobas

Laboratory of Personalized Genomics, Undergraduate Medical Academy, Prairie View A&M University, Prairie View, TX 77446, USA; daiacobas@pvamu.edu; Tel.: +1-936-261-3086

Citation: Iacobas, D.A. Advanced Molecular Solutions for Cancer Therapy—The Good, the Bad, and the Ugly of the Biomarker Paradigm. *Curr. Issues Mol. Biol.* **2024**, 46, 1694–1699. https://doi.org/10.3390/cimb46030109

Received: 9 January 2024
Accepted: 17 February 2024
Published: 22 February 2024

Identifying the most effective actionable molecules whose "smart" manipulation might selectively kill/slow down/stop the proliferation of cancer cells, with few side effects on the normal cells of the tissue, was for decades the single major objective of countless investigators. This Special Issue (SI), a continuation of the *previous Current Issues in Molecular Biology* SI "Molecules at Play in Cancer" [1], aimed to present the latest developments in the molecular solutions for cancer therapy, and the ways of personalizing the treatment to the individual characteristics of the patient. The authors have contributed their best efforts, either by performing accurate experiments, reanalyzing publicly accessible genomic datasets (including The Cancer Genome Atlas (TCGA) [2]), or writing comprehensive reviews of the literature. Among the interesting proposed solutions, of note are: the targeted delivery of chimeric antigen receptor into T cells via CRISPR [3], the metabolic silencing via methionine-based amino acid restriction [4], the inhibition of ERK5 [5], the activation of *ADRA2A* [6], and the aspirin treatment of breast cancer [7].

Most articles dealing with molecular mechanisms that should be activated or inhibited to destroy the cancer cells either directly, (e.g., [5]), or by increasing the efficacy of immuno- [3,8], chemo- [6], and radiotherapy were aligned with the main stream "biomarker paradigm" (BMP). For a long time, the vast majority of investigators and clinical oncologists have believed that the mutation and/or altered expression of certain genes, called "biomarkers", are responsible for triggering cancerization. Moreover, it was hoped that restoring the normal status of such gene biomarkers would provide the natural anti-cancer therapy. The efficacy of various biomarker-oriented gene therapies was tested on both standard human cancer cell cultures and animal models.

However, let us have a candid discussion about how cancer biomarkers have been discovered, what their real values are for diagnostic and therapy, and how reliable their testing is on animal models and human cell cultures. Decades of work at almost all stages of both experimental and theoretical genomic research using numerous types of platforms entitle and oblige me to take the risky enterprise of discussing "the good, the bad and the ugly" of this paradigm. Studies of my lab-profiled cell cultures and tissues from surgically removed human tumors, as well as a wide variety of cells and tissues from animal (mouse, rat, rabbit, dog, chicken) models of human diseases. In addition to optimizing the wet protocols, we have introduced the Genomic Fabric Paradigm and developed advanced mathematical algorithms and computer software to analyze the genomic data.

The GOOD (BMP promises). Cancer is a multi-factorial disease regulated by an enormous number of widely diverse favoring conditioners, understanding of which is far from being complete. Therefore, relying on a few gene biomarkers (even with each of them presenting several variants [9]) is a considerable simplification of the diagnostic process. One of the greatest advantages of BMP is that (in theory) it provides molecular explanations of the origin of various cancer forms.

Several test kits (e.g., [10–12]) have been developed and are currently in use for the genomic detection of an existing, particular cancer form or the perspective of a tissue that will undergo sooner or later a malignant transformation. For instance, the "Invitae Multi-Cancer Panel" [10] of 70 genes specifically designed for heritable germline mutations in blood, saliva, or buccal swab specimens sequences *BRCA1* and *BRCA2* to detect hereditary breast and ovarian cancer syndrome. The "nCounter PanCancer IO 360™ Panel" [11], a unique 770 gene expression assay looking for the transcriptomic signatures of various cancer forms, which provides a number of cancer risk scores. "TissueScan™ Cancer and Normal Tissue cDNA Arrays" [12] were developed for differential gene expression analysis by comparing patient samples with pathologist-verified cancer and normal tissues.

The biomarkers are also considered legitimate targets for cancer gene therapy, offering significant economic advantages in the large-scale production of biomarker-manipulating medicines that should be prescribed to all persons affected by the same cancer form. In contrast, the economic incentives of the precision medicine that tries to tailor the treatment to the patient's characteristics are still limited, especially for low-income countries [13,14]. Owing to these (mostly desired rather than real) benefits, BMP serves to standardize the oncological recommendations and procedures (e.g., [15,16]).

To resume, the very important BMP (believed) GOODs are that:

(1) it provides a simple molecular mechanistic explanation of cancerization in any human and animal regardless of race/strain, sex, age, or any other personal and environmental characteristic;

(2) it is the theory behind organizing the genes in functional pathways by the very popular software: Ingenuity Pathway Analysis [17], DAVID [18], and KEGG [19];

(3) it provides the reason for developing universal assays to detect the existing cancer of a particular form and/or estimating the chances of future cancerization for any person;

(4) it stimulated the development of animal models and engineered cell cultures to mimic various forms of human cancer, validate their genetic etiology, and test gene therapy;

(5) it is the basis of designing therapeutic solutions that target the molecular mechanisms of cancerization;

(6) it supports the standardization of the oncological procedures by the National Comprehensive Cancer Network (e.g., [20]);

(7) it has been adopted by the vast majority of genomic researchers (as of 6 January 2024, PubMed [21] listed 421,759 "cancer biomarker" and 957,483 "cancer genetic etiology" publications);

(8) it benefits from the most generous research funding by public and private agencies and is of major interest for the pharma industry.

The BAD (BMP reality). According to the 39.0 release (12 April 2023) of the NIH-NCI Harmonized Cancer Datasets [22], almost every single gene was found as mutated in at least one case of almost every form of cancer and every form of cancer exhibited mutations in almost all genes. So, what about the specificity of the biomarkers?

Nonetheless, together with the blamed biomarker(s), hundreds of other genes appear as mutated and/or regulated when comparing tissues from cancer stricken and healthy persons; however, their potential contributions to the cancer phenotype are mostly ignored by the BMP users. One reason for disregarding the other altered genes might be that their combination is practically never exactly repeated among patients and changes (slowly but steadily) in time for the same person.

It is legitimate to ask how the developers of the cancer test kits determined and validated the predictive values of the transcriptomic signatures since not 770 but only 30 genes that can be up-/down-/not regulated form over 2×10^{14} distinct combinations (many more than living humans).

Most biomarkers were identified through meta-analyses that compared DNA sequences and/or RNA expression profiles from tissues of cancer patients and healthy counterparts. In order to increase the statistical significance of the biomarkers, numerous analyses covered data collected by several laboratories (using sometimes different wet protocols and/or equipment) from large and in many cases heterogeneous (as in race, sex, age) populations. Thus, beyond possible human errors, protocol differences, and the normal technical noise of the platform, the data were biased by the distinct cancer prevalence among races, sexes, and age groups (to name just a few favoring factors) [23]. Owing to the dispersions of the gene expression levels among cancer-stricken and healthy persons, the difference between the mean values of the two distributions is most likely smaller than between the extreme values within each distribution. Thus, two healthy persons may have larger transcriptomic differences than a healthy and a cancer-stricken person (Figure 1).

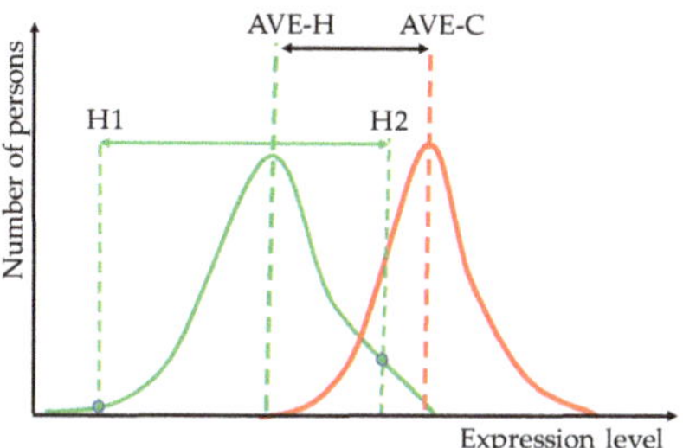

Figure 1. Potential distributions of the expression levels of a hypothetical gene biomarker within healthy (H) and cancer-stricken (C) persons. Note that the difference between the expression levels in two healthy individuals (H1 and H2) may be larger than that between the average healthy (AVE-H) person and the average cancer-stricken (C) person.

Nonetheless, the tumors are heterogeneous, harboring cell subpopulations with distinct genotypes and phenotypes, and it is a very slim probability that in the large repository, the profiled samples have been collected from exactly the same type of clones and/or region of the tissue (affected by the microenvironment) from different individuals. Intra-tumor genetic and transcriptomic heterogeneity was reported by many groups (e.g., [24–28]). In our gene expression studies (e.g., [29,30]), we found large differences even between equally graded cancer nodules from the same tumor. So, what do "genetic etiology" and "transcriptomic signature" (both deduced by comparing the average cancer patient with the average healthy counterpart) stand for when genes from different locations within the same tumor exhibit different mutations and expression regulations? Therefore, the best comparison is not between the tissues of the average cancer-stricken person and average healthy counterpart, but between the cancer nodule(s) and surrounding cancer-free tissue from the same tumor. This strategy has already been adopted by several laboratories (e.g., [29–34]).

About the BMP-based therapy: as selected from the most frequently altered genes in large populations of people harboring similar cancer forms, biomarkers appear as the least protected by the homeostatic mechanisms, an indication of their low importance. Therefore, restoring their normal status might be of little consequence.

In summary, the BMP BADs are:

(1) low diagnostic specificity owing to the large number of cancer forms harboring the same biomarker;

(2) insufficient diagnostic sensitivity, numerous cases missing the alleged biomarker (e.g., [35]);

(3) disconsiders major personal favoring factors of the patient, including race, sex, age, diet, and environmental factors, such as exposure to radiation, toxins and stress;

(4) differences between healthy persons might be larger than between the average healthy and the average cancer-stricken persons;

(5) disregard of the contributions of the many other genes whose sequences and/or expression levels are altered (even in not repeatable combination) in the cancer of each individual;

(6) BMP-based functional pathways do not discriminate with regard to race, sex, and age, do not change with the cancer progression and/or in response to external stimuli and treatments, and are reduced to unique gene networking;

(7)   it selects low cell players for gene therapy.

The UGLY (BMP justification). Whilst even a germline mutation is supposed to be present in all cells, only a part of them evolves into a cancer phenotype, indicating the importance of the local environment. Moreover, most cancers were not inherited but occur and disappear spontaneously in some cells of otherwise homo-cellular tissues.

A very important issue is related to the experimental resolutions of both sequencing and expression studies. Although the single-cell sequencing technology discriminates the genomic characteristics of cell subpopulations with distinct phenotypes, it is still unable to quantify gene alterations within a single cell. Not only does it take a critical number of neighboring cells harboring synchronously the same mutation/expression regulation to start developing a tumor but this is also necessary to be detected by the experimental platform.

Despite the strict control exerted by the cellular homeostatic mechanisms, both DNA replication and transcription are affected by errors caused by the stochastic nature of the involved chemical reactions. The average one of each of a 1000 nucleotides being mutated at any time (i.e., over 3 mil spontaneous mutations in every cell genome) makes it practically impossible to find an unmutated gene in any cell of the human body. The expected number of mutations in a gene equals the number of its composing kilo bases but there is a negligible overlapping of the mutations among the cells of the tissue, making most of them impossible to detect.

Moreover, the expression level of each gene fluctuates around the cell-cycle dependent value to provide the needed rate of protein synthesis and, although correlated, fluctuations are not synchronized among all cells of the tissue. Therefore, the transcriptome is not homogeneous across the tissue and the detection of significantly regulated genes (and implicitly the transcriptomic signature) depends on the selection of the profiled regions in the compared tissues. Not to mention that the arbitrarily introduced absolute fold-change cut-off (e.g., 1.5×) for a gene to be considered as significantly up-/down-regulated is too stringent for stably expressed genes across biological replicas and low-noise platforms and too lax for variably expressed genes and noisier platforms.

The experimental validation of both BMP-based diagnostic and targeted therapy is disputable owing that manipulating the sequence and/or the expression level of one gene has ripple effects on hundreds of other genes. There is no way of dissecting the contribution of the biomarker from those of the other genes in any genetically engineered animal model or cell line.

Therefore, the BMP UGLY aspects are related to the limited resolution and high technical noise of the actual genomic platforms, together with the impossibility of experimentally validating the functional roles of the biomarkers and the benefits of the biomarker-targeting therapy.

The continuing development of the next generation sequencing and the unlimited power of AI algorithms will soon decrease the need of the oversimplified BMP looking for genes universally responsible for a particular cancer form in any human. The clinical oncologists are already moving from the "fit-for-all" model to personalized gene therapies. The spectacular advancements in gene-editing technologies and the economic incentives are expected to convince the pharma industry to start producing shelf-ready constructs to manipulate the master regulators identified for each cancer patient (e.g., [29,30]).

Conflicts of Interest: The author declares no conflict of interest.

References

1.   Iacobas, D.A. Molecules at Play in Cancer. *Curr. Issues Mol. Biol.* **2023**, *45*, 2182–2185. [CrossRef]
2.   The Cancer Genome Atlas (TCGA) Research Network. Available online: https://www.cancer.gov/tcga (accessed on 4 January 2024).
3.   Moço, P.D.; Farnós, O.; Sharon, D.; Kamen, A.A. Targeted Delivery of Chimeric Antigen Receptor into T Cells via CRISPR-Mediated Homology-Directed Repair with a Dual-AAV6 Transduction System. *Curr. Issues Mol. Biol.* **2023**, *45*, 7705–7720. [CrossRef]

4. Wünsch, A.C.; Ries, E.; Heinzelmann, S.; Frabschka, A.; Wagner, P.C.; Rauch, T.; Koderer, C.; El-Mesery, M.; Volland, J.M.; Kübler, A.C.; et al. Metabolic Silencing via Methionine-Based Amino Acid Restriction in Head and Neck Cancer. *Curr. Issues Mol. Biol.* **2023**, *45*, 4557–4573. [CrossRef]

5. Hwang, J.; Moon, H.; Kim, H.; Kim, K.-Y. Identification of a Novel ERK5 (MAPK7) Inhibitor, MHJ-627, and Verification of Its Potent Anticancer Efficacy in Cervical Cancer HeLa Cells. *Curr. Issues Mol. Biol.* **2023**, *45*, 6154–6169. [CrossRef]

6. Albanna, H.; Gjoni, A.; Robinette, D.; Rodriguez, G.; Djambov, L.; Olson, M.E.; Hart, P.C. Activation of Adrenoceptor Alpha-2 (ADRA2A) Promotes Chemosensitization to Carboplatin in Ovarian Cancer Cell Lines. *Curr. Issues Mol. Biol.* **2023**, *45*, 9566–9578. [CrossRef]

7. Savukaitytė, A.; Bartnykaitė, A.; Bekampytė, J.; Ugenskienė, R.; Juozaitytė, E. DDIT4 Downregulation by siRNA Approach Increases the Activity of Proteins Regulating Fatty Acid Metabolism upon Aspirin Treatment in Human Breast Cancer Cells. *Curr. Issues Mol. Biol.* **2023**, *45*, 4665–4674. [CrossRef]

8. Filin, I.Y.; Mayasin, Y.P.; Kharisova, C.B.; Gorodilova, A.V.; Chulpanova, D.S.; Kitaeva, K.V.; Rizvanov, A.A.; Solovyeva, V.V. T-Lymphocytes Activated by Dendritic Cells Loaded by Tumor-Derived Vesicles Decrease Viability of Melanoma Cells In Vitro. *Curr. Issues Mol. Biol.* **2023**, *45*, 7827–7841. [CrossRef] [PubMed]

9. Chatrath, A.; Przanowska, R.; Kiran, S.; Su, Z.; Saha, S.; Wilson, B.; Tsunematsu, T.; Ahn, J.H.; Lee, K.Y.; Paulsen, T.; et al. The pan-cancer landscape of prognostic germline variants in 10,582 patients. *Genome Med.* **2020**, *12*, 15. [CrossRef] [PubMed]

10. Invitae Multi-Cancer Panel. Available online: https://www.invitae.com/us/providers/test-catalog/test-01101 (accessed on 11 December 2023).

11. nCounter®PanCancer IO 360TM Panel. Available online: https://nanostring.com/products/ncounter-assays-panels/oncology/pancancer-io-360/ (accessed on 11 December 2023).

12. TissueScan™ Cancer and Normal Tissue cDNA Arrays. Available online: https://www.origene.com/products/tissues/tissuescan (accessed on 12 November 2023).

13. Vellekoop, H.; Huygens, S.; Versteegh, M.; Szilberhorn, L.; Zelei, T.; Nagy, B.; Koleva-Kolarova, R.; Tsiachristas, A.; Wordsworth, S.; Rutten-van Mölken, M. Guidance for the harmonisation and improvement of economic evaluations of personalised medicine. *Pharmacoeconomics* **2021**, *39*, 771–788. [CrossRef] [PubMed]

14. Chen, W.; Wong, N.C.B.; Wang, Y.; Zemlyanska, Y.; Butani, D.; Virabhak, S.; Matchar, D.B.; Prapinvanich, T.; Teerawattananon, Y. Mapping the value for money of precision medicine: A systematic literature review and meta-analysis. *Front. Public Health* **2023**, *11*, 1151504. [CrossRef] [PubMed]

15. Kato, T.; Casarini, I.; Cobo, M.; Faivre-Finn, C.; Hegi-Johnson, F.; Lu, S.; Özgüroğlu, M.; Ramalingam, S.S. Targeted treatment for unresectable EGFR mutation-positive stage III non-small cell lung cancer: Emerging evidence and future perspectives. *Lung Cancer* **2023**, *187*, 107414. [CrossRef] [PubMed]

16. Lutgendorf, S.K.; Telles, R.M.; Whitney, B.; Thaker, P.H.; Slavich, G.M.; Goodheart, M.J.; Penedo, F.J.; Noble, A.E.; Cole, S.W.; Sood, A.K.; et al. The biology of hope: Inflammatory and neuroendocrine profiles in ovarian cancer patients. *Brain Behav. Immun.* **2023**, *116*, 362–369. [CrossRef]

17. QIAGEN Ingenuity Pathway Analysis (QIAGEN IPA). Available online: https://digitalinsights.qiagen.com/products-overview/discovery-insights-portfolio/analysis-and-visualization/qiagen-ipa/ (accessed on 7 January 2024).

18. Database for Annotation, Visualization and Integrated Discovery (DAVID). Available online: https://david.ncifcrf.gov (accessed on 7 January 2024).

19. Kyoto Encyclopedia of Genes and Genomes. Wiring Diagrams of Molecular Interactions, Reactions and Relations. Available online: https://www.genome.jp/kegg/pathway.html (accessed on 7 January 2024).

20. Schaeffer, E.M.; Srinivas, S.; Adra, N.; An, Y.; Barocas, D.; Bitting, R.; Bryce, A.; Chapin, B.; Cheng, H.H.; D'Amico, A.V.; et al. Prostate Cancer, Version 4.2023, NCCN Clinical Practice Guidelines in Oncology. *J. Natl. Compr. Cancer Netw.* **2023**, *21*, 1067–1096. [CrossRef]

21. NIH National Library of Medicine/PubMed. Available online: https://pubmed.ncbi.nlm.nih.gov/?term=cancer+biomarker&sort=date (accessed on 3 January 2024).

22. NIH-National Cancer Institute Genomic Data Commons Data Portal. Available online: https://portal.gdc.cancer.gov (accessed on 29 November 2023).

23. Ren, A.H.; Fiala, C.A.; Diamandis, E.P.; Kulasingam, V. Pitfalls in Cancer Biomarker Discovery and Validation with Emphasis on Circulating Tumor DNA. *Cancer Epidemiol Biomark. Prev.* **2020**, *29*, 2568–2574. [CrossRef]

24. Kulac, I.; Roudier, M.P.; Haffner, M.C. Molecular Pathology of Prostate Cancer. *Surg. Pathol. Clin.* **2021**, *14*, 387–401. [CrossRef] [PubMed]

25. Tolkach, Y.; Kristiansen, G. The Heterogeneity of Prostate Cancer: A Practical Approach. *Pathobiology* **2018**, *85*, 108–116. [CrossRef] [PubMed]

26. Tu, S.-M.; Zhang, M.; Wood, C.G.; Pisters, L.L. Stem Cell Theory of Cancer: Origin of Tumor Heterogeneity and Plasticity. *Cancers* **2021**, *13*, 4006. [CrossRef]

27. Berglund, E.; Maaskola, J.; Schultz, N.; Friedrich, S.; Marklund, M.; Bergenstråhle, J.; Tarish, F.; Tanoglidi, A.; Vickovic, S.; Larsson, L.; et al. Spatial maps of prostate cancer transcriptomes reveal an unexplored landscape of heterogeneity. *Nat. Commun.* **2018**, *9*, 2419. [CrossRef]

28. Brady, L.; Kriner, M.; Coleman, I.; Morrissey, C.; Roudier, M.; True, L.D.; Gulati, R.; Plymate, S.R.; Zhou, Z.; Birditt, B.; et al. Inter- and intra-tumor heterogeneity of metastatic prostate cancer determined by digital spatial gene expression profiling. *Nat. Commun.* **2021**, *12*, 1426. [CrossRef]
29. Iacobas, S.; Iacobas, D.A. A Personalized Genomics Approach of the Prostate Cancer. *Cells* **2021**, *10*, 1644. [CrossRef]
30. Iacobas, D.A.; Obiomon, E.A.; Iacobas, S. Genomic Fabrics of the Excretory System's Functional Pathways Remodeled in Clear Cell Renal Cell Carcinoma. *Curr. Issues Mol. Biol.* **2023**, *45*, 9471–9499. [CrossRef] [PubMed]
31. Liu, Y.; Weber, Z.; San Lucas, F.A.; Deshpande, A.; Jakubek, Y.A.; Sulaiman, R.; Fagerness, M.; Flier, N.; Sulaiman, J.; Davis, C.M.; et al. Assessing inter-component heterogeneity of biphasic uterine carcinosarcomas. *Gynecol Oncol.* **2018**, *151*, 243–249. [CrossRef]
32. Fujimoto, H.; Saito, Y.; Ohuchida, K.; Kawakami, E.; Fujiki, S.; Watanabe, T.; Ono, R.; Kaneko, A.; Takagi, S.; Najima, Y.; et al. Deregulated Mucosal Immune Surveillance through Gut-Associated Regulatory T Cells and PD-1+ T Cells in Human Colorectal Cancer. *J. Immunol.* **2018**, *200*, 3291–3303. [CrossRef]
33. Yang, C.; Gong, J.; Xu, W.; Liu, Z.; Cui, D. Next-generation sequencing identified somatic alterations that may underlie the etiology of Chinese papillary thyroid carcinoma. *Eur. J. Cancer Prev.* **2023**, *32*, 264–274. [CrossRef]
34. Clark, D.J.; Dhanasekaran, S.M.; Petralia, F.; Pan, J.; Song, X.; Hu, Y.; da Veiga Leprevost, F.; Reva, B.; Lih, T.M.; Chang, H.Y.; et al. Clinical Proteomic Tumor Analysis Consortium. Integrated Proteogenomic Characterization of Clear Cell Renal Cell Carcinoma. *Cell* **2019**, *179*, 964–983.e31. [CrossRef] [PubMed]
35. Zhao, H.; Wang, L.; Fang, C.; Li, C.; Zhang, L. Factors influencing the diagnostic and prognostic values of circulating tumor cells in breast cancer: A meta-analysis of 8,935 patients. *Front Oncol.* **2023**, *13*, 1272788. [CrossRef] [PubMed]

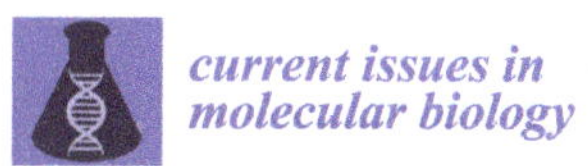
current issues in molecular biology

Article

2-Methoxyestradiol as an Antiproliferative Agent for Long-Term Estrogen-Deprived Breast Cancer Cells

Masayo Hirao-Suzuki [1], Koki Kanameda [2], Masufumi Takiguchi [1], Narumi Sugihara [2] and Shuso Takeda [2,*]

[1] Laboratory of Xenobiotic Metabolism and Environmental Toxicology, Faculty of Pharmaceutical Sciences, Hiroshima International University, 5-1-1 Hiro-koshingai, Kure-shi 737-0112, Hiroshima, Japan; m-hirao@hirokoku-u.ac.jp (M.H.-S.); m-takigu@hirokoku-u.ac.jp (M.T.)

[2] Laboratory of Molecular Life Sciences, Faculty of Pharmacy and Pharmaceutical Sciences, Fukuyama University, Sanzou 1, Gakuen-cho, Fukuyama-shi 729-0292, Hiroshima, Japan; p7118025@fukuyama-u.ac.jp (K.K.); sugihara@fukuyama-u.ac.jp (N.S.)

* Correspondence: s.takeda@fukuyama-u.ac.jp; Tel.: +81-84-936-2112

Abstract: To identify effective treatment modalities for breast cancer with acquired resistance, we first compared the responsiveness of estrogen receptor-positive breast cancer MCF-7 cells and long-term estrogen-deprived (LTED) cells (a cell model of endocrine therapy-resistant breast cancer) derived from MCF-7 cells to G-1 and 2-methoxyestradiol (2-MeO-E2), which are microtubule-destabilizing agents and agonists of the G protein-coupled estrogen receptor 1 (GPER1). The expression of GPER1 in LTED cells was low (~0.44-fold), and LTED cells displayed approximately 1.5-fold faster proliferation than MCF-7 cells. Although G-1 induced comparable antiproliferative effects on both MCF-7 and LTED cells (IC_{50} values of >10 µM), 2-MeO-E2 exerted antiproliferative effects selective for LTED cells with an IC_{50} value of 0.93 µM (vs. 6.79 µM for MCF-7 cells) and induced G2/M cell cycle arrest. Moreover, we detected higher amounts of β-tubulin proteins in LTED cells than in MCF-7 cells. Among the *β-tubulin* (*TUBB*) isotype genes, the highest expression of *TUBB2B* (~3.2-fold) was detected in LTED cells compared to that in MCF-7 cells. Additionally, siTUBB2B restores 2-MeO-E2-mediated inhibition of LTED cell proliferation. Other microtubule-targeting agents, i.e., paclitaxel, nocodazole, and colchicine, were not selective for LTED cells. Therefore, 2-MeO-E2 can be an antiproliferative agent to suppress LTED cell proliferation.

Keywords: 2-methoxyestradiol; 2-MeO-E2; LTED cells; tubulin; G-1; breast cancer

Citation: Hirao-Suzuki, M.; Kanameda, K.; Takiguchi, M.; Sugihara, N.; Takeda, S. 2-Methoxyestradiol as an Antiproliferative Agent for Long-Term Estrogen-Deprived Breast Cancer Cells. *Curr. Issues Mol. Biol.* **2023**, *45*, 7336–7351. https://doi.org/10.3390/cimb45090464

Academic Editors: Sung-Kun (Sean) Kim and Dumitru A. Iacobas

Received: 11 July 2023
Revised: 28 August 2023
Accepted: 7 September 2023
Published: 9 September 2023

1. Introduction

Estrogen receptors (ERs) are pivotal in breast cancer development and progression, and their expression and activity are tightly regulated. Among patients with breast cancer, >70% are ER-positive at diagnosis. Clinically, ER-positive patients are treated with agents such as aromatase inhibitors and fulvestrant (also known as ICI 182780, an antagonist of ERα), which exert estrogen-lowering effects by suppressing estrogen signaling [1]. Although >50% of patients are sensitive to the medications, most patients exhibit relapse [1]. Thus, resistance to endocrine therapy remains a major clinical concern in breast cancer, and understanding the molecular mechanism(s) responsible for the acquired resistance to endocrine therapy is important. An experimental cell model (i.e., long-term estrogen-deprived [LTED] cells) was established by maintaining parental human breast cancer MCF-7 cells (ER-positive) under estrogen-deprived conditions for approximately six months [2–4]. In addition to the previously reported feature that ERs expressed in LTED cells display ligand-independent activity [2–6], we recently reported that LTED cells express very low levels of G protein-coupled estrogen receptor 1 (GPER1, formerly known as GPR30), such that its expression is comparable to that of GPER1 produced by human breast cancer MDA-MB-231 cells, a GPER1-negative (or very low), and an ER-negative cell line [6]. GPER1 is a membrane-type receptor discovered in 1996 in breast cancer tissue [7]. G-1 (also known as

LNS-8801) (Figure 1A), a GPER1-selective agonist, was identified in 2006 by Bologa et al. and has been shown to exert no significant effects on other protein-coupled receptors or ERα/β [8]. The biological impact of G-1 treatment on breast cancer cell proliferation and survival is controversial, although G-1-mediated stimulation of breast cancer cells has been reported [9–13]. However, we and others demonstrated that G-1 abrogates the proliferation of both MCF-7 (ER/GPER1-positive) and MDA-MB-231 (ER/GPER1-negative) breast cancer cells, with more substantial effects on MDA-MB-231 cells than on MCF-7 cells [6,14].

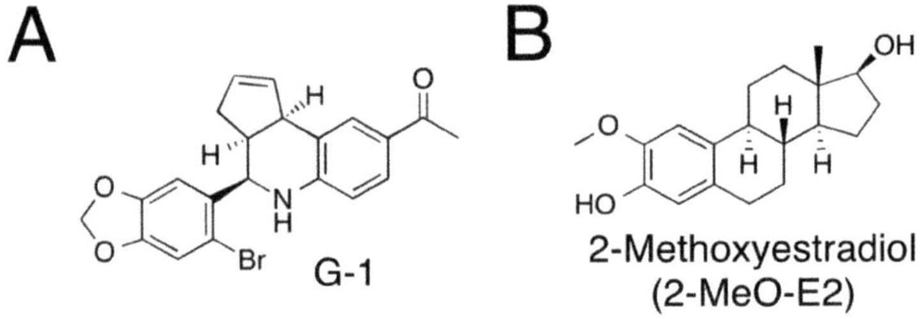

Figure 1. The structures of G-1 (**A**) and 2-methoxyestradiol (2-MeO-E2) (**B**).

Wang's research group has reported that G-1 can also act as a microtubule-destabilizing agent and target β-tubulins to kill MDA-MB-231 and MCF-7 cells [14,15]. Similar to G-1, 2-methoxyestradiol (2-MeO-E2) (Figure 1B), a metabolite of 17β-estradiol, has been shown to bind to the colchicine-binding site on β-tubulins [14,16] and has the potential to activate GPER1 as an agonist [17]. Although E2 metabolites are known to either stimulate or abrogate breast cancer cell proliferation, 2-MeO-E2 can act as an antiproliferative agent in both ER-positive and -negative breast cancer cells [18–20]. Microtubules comprise two types of tubulin proteins, α- and β-tubulin, that heterodimerize [21]. α/β-Tubulins comprise multiple isotypes, each encoded by a different gene. Microtubules are targeted for anti-cancer therapies, and a large majority of the United States Food and Drug Administration-approved tubulin inhibitory drugs target β-tubulins. The fluctuating expression of β-tubulin isotypes is profoundly related to resistance to microtubule-targeting agents [22,23]. We and others have reported that compared with parental MCF-7 cells, LTED cells exhibit stronger resistance to several antiproliferative agents, including etoposide (a topoisomerase IIα inhibitor), LY2835219 (a cyclin-dependent kinase 4/6 inhibitor), and trichostatin A (a histone deacetylase inhibitor) [24]. Based on the above-mentioned findings, we analyzed the expression level of β-tubulin and its isotype involved in the aggressive behavior of LTED cells. We sought to investigate whether G-1 and 2-MeO-E2 are effective antiproliferative agents for LTED cells.

2. Materials and Methods

2.1. Reagents

G-1 (Chemical Abstract Service [CAS] number 881639-98-1, purity $\geq$ 98%) and 2-MeO-E2 (CAS number 362-07-2, purity $\geq$ 95%) were purchased from Cayman Chemicals (Ann Arbor, MI, USA). Paclitaxel (CAS number 33069-62-4, purity $\geq$ 98%) was obtained from FUJIFILM Wako Pure Chemical Corporation (Osaka, Japan). Nocodazole (CAS number 31430-18-9, purity $\geq$ 99%) was supplied by Sigma-Aldrich (St. Louis, MO, USA), and colchicine (CAS number 64-86-8, purity $\geq$ 97%) was purchased from Tokyo Chemical Industry Co., Ltd. (Tokyo, Japan). All chemicals were dissolved in dimethyl sulfoxide (molecular biology grade).

2.2. Cell Culture, Chemical Treatments, and Cell Morphology Analysis

The cell culture methods were based on previously described procedures [6,25,26]. Briefly, human breast cancer MCF-7 cells (obtained from the American Type Culture Collection, Rockville, MD, USA) were routinely grown in phenol red-containing minimum essential medium α (MEMα) (FUJIFILM Wako Pure Chemical Corporation) supplemented with 10 mM 4-(2-hydroxyethyl)-1-piperazineethanesulfonic acid (HEPES), 5% fetal bovine serum (FBS), 100 U/mL penicillin, and 100 µg/mL streptomycin in a humidified incubator with an atmosphere of 5% CO_2 at 37 °C. LTED cells were derived from parental MCF-7 cells as previously described [2–4,6]. The LTED cell batches obtained were routinely cultured in phenol red-free MEMα (FUJIFILM Wako Pure Chemical Corporation) supplemented with 10 mM HEPES, 5% dextran-coated charcoal-treated FBS (DCC-FBS), 100 U/mL penicillin, and 100 µg/mL streptomycin. Prior to the 24 h chemical treatments, the culture medium was changed to phenol red-free MEMα supplemented with 10 mM HEPES, 5% DCC-FBS, 100 U/mL penicillin, and 100 µg/mL streptomycin. The cells were treated with G-1, paclitaxel, nocodazole, 2-MeO-E2, or colchicine in the culture medium at different concentrations and time points, as indicated in each Figure legend. Cell morphology analysis was performed as previously described [6].

2.3. Preparation of Total RNA and Real-Time Reverse Transcription Polymerase Chain Reaction (RT-PCR)

Total RNA was extracted, and RT-PCR was performed as previously described [25,26]. The total RNA concentration was determined using a NanoDrop 2000 UV spectrophotometer (Thermo Fisher Scientific, Waltham, MA, USA). The following PCR primers were used: GPER1 (sense), 5′-TTT GTG GGC AAC ATC CTG ATC-3′; GPER1 (antisense), 5′-CAC CGC CAG GTT GAT GAA GTA-3′; histone deacetylase 6 (HDAC6) (sense), 5′-TGC CTC TGG GAT GAC AGC TT-3′; HDAC6 (antisense), 5′-CCT GGA TCA GTT GCT CCT TGA-3′; the β-tubulin (TUBB) isotypes, TUBB (sense), 5′-CTG CCA CAT CAG TGT TTG AGT C -3′; TUBB (antisense), 5′-AAA AAG ATG GAG GAG GGT TCC-3′; TUBB2A (sense), 5′-GAC GAA CAA GGG GAG TTC G-3′; TUBB2A (antisense), 5′-GGA TGC ACG ATT GAT CTG AG-3′; TUBB2B (sense), 5′-AGG ACG GAC AGA CCC AGA C-3′; TUBB2B (antisense), 5′-CTG ATG ACC TCC CAA AAC TTG-3′; TUBB3 (sense), 5′-GGC CTT TGG ACA TCT CTT CA-3′; TUBB3 (antisense), 5′-CGG TCG GGA TAC TCC TCA-3′; TUBB4A (sense), 5′-TGG TAC ACG GGC GAG GGC AT-3′; TUBB4A (antisense), 5′-GTG GGA AGC GAT GGG AGC AGC-3′; TUBB4B (sense), 5′-CTG CTG CTG TTT GTC TAC TTC C-3′; TUBB4B (antisense), 5′-GCT GAT CAC CTC CCA AAA CT-3′; TUBB6 (sense), 5′-CCA GTT CCT AGC GCA GAG CCG-3′; TUBB6 (antisense), 5′-GCA CGC TGT CCA TGG TGC CT-3′; TUUBB1 (sense), 5′-GGA GAT GAT TGG TGA GGA ACA C-3′; TUBB1 (antisense), 5′-GGT TCT AGG TCC ACC AAG ACT G-3′; β-actin (sense), 5′-GGC CAC GGC TGC TTC-3′; β-actin (antisense), 5′-GTT GGC GTA CAG GTC TTT GC-3′. The primers for human *GPER1, HDAC6, TUBB, TUBB2A, TUBB2B, TUBB3, TUBB4A, TUBB4B, TUBB6, TUBB1,* and *β-actin* were determined in previous reports [6,27–31]. The mRNA levels of *GPER1, HDAC6, TUBB, TUBB2A, TUBB2B, TUBB3, TUBB4A, TUBB4B, TUBB6,* and *TUBB1* were normalized to that of *β-actin.*

2.4. Ponceau Staining and Western Blot Analysis

Antibodies specific to GPER1 (ab39742; Abcam, Cambridge, MA, USA), β-actin (sc-47778; Santa Cruz Biotechnology, Dallas, TX, USA), α-tubulin (017-25031; FUJIFILM Wako Pure Chemical Corporation), and β-tubulin (014-25044; FUJIFILM Wako Pure Chemical Corporation) were used. Whole-cell extracts were prepared, and Western blot analysis was performed as previously described [32]. The protein concentrations of whole-cell extracts were determined using the RC DC^{TM} Protein Assay (Bio-Rad, Hercules, CA, USA). The polyvinylidene difluoride membrane was stained with Ponceau S solution (Beacle Inc., Kyoto, Japan). Pink-stained blots were scanned, and the images were converted to grayscale for publication. Band intensities were quantified using the ImageJ 1.46r software "https://imagej.nih.gov/ij/ (accessed on 21 September 2022)". The values obtained for the GPER1 band were normalized to the intensity of the β-actin band (internal control).

2.5. Cell Proliferation Analysis (MTS Assay) and Cell Counting

MTS assays were performed to determine cell proliferation as previously described [25]. The cells were seeded in 96-well plates at a density of 5×10^3 cells/well. After chemical treatment, cell proliferation was analyzed using the CellTiter 96® AQ$_{ueous}$ One Solution Cell Proliferation Assay Kit (MTS Reagent; Promega, Madison, WI, USA). Cells were counted using a TC10^{TM} Automated Cell counter (Bio-Rad).

2.6. Cell Cycle Analysis Using Flow Cytometry

Cells were stained with propidium iodide as previously described [6]. Stained cells were analyzed using a FACSCalibur^{TM} flow cytometer (BD Biosciences, Franklin Lakes, NJ, USA). The obtained data were analyzed using ModFit LT^{TM} v3.0 (Verity Software House, Topsham, ME, USA).

2.7. Gene Silencing

To avoid off-target effects, we used commercially available TUBB2B siRNA (sc-105006; Santa Cruz Biotechnology) and a pool of three target-specific siRNAs. Control siRNA (sc-37007; Santa Cruz Biotechnology) was used as the negative control. TUBB2B and control siRNAs were transfected using Lipofectamine RNAiMAX reagent (Thermo Fisher Scientific). The siRNA concentration used for transfection was 15 nM.

2.8. Data Analysis

The IC$_{50}$ values were calculated by fitting the dose–response curves using the SigmaPlot 11 software (Systat Software, Inc., San Jose, CA, USA). Differences were considered statistically significant when p values were less than 0.05. The statistical significance of the difference between the two groups was determined using Student's t-test. The statistical significance of the differences between multiple groups was determined using analysis of variance (ANOVA) with Dunnett's or Tukey–Kramer's post hoc test. These calculations were performed using the StatView 5.0 J software (SAS Institute Inc., Cary, NC, USA).

3. Results

3.1. On LTED Cells, 2-MeO-E2 Exerts Selective Antiproliferation

Prior to conducting experiments to determine whether G-1 and 2-MeO-E2 negatively affected the proliferation of LTED and MCF-7 cells by targeting GPER1, we confirmed the expression profile of GPER1 in these breast cancer cells. The mRNA expression level of *GPER1* was less than that (0.23-fold, $p < 0.05$) in LTED cells compared to that in parental MCF-7 cells (Figure 2A). Consistent with the mRNA expression profile, GPER1 protein expression displayed a similar trend (0.44-fold) (Figure 2A, *inset*) [6]. As shown in Figure 2B, LTED cells exhibited more aggressive proliferative features than MCF-7 cells. The reduced expression of GPER1 favors cell aggressiveness, such as stimulated cell proliferation, particularly in LTED cells, as indicated in the reports suggesting that GPER1 can act as a tumor suppressor in different types of cancers, including breast cancer [33–36]. Therefore, the possible negative (antiproliferative) effects of GPER1 agonists (G-1 and 2-MeO-E2) [8,17] on the aggressive behavior of LTED cells were not observed or weakened. Therefore, we investigated the validity of this hypothesis. Both LTED and MCF-7 cells were exposed to G-1 and 2-MeO-E2 in varying concentrations from 1 pM to 10 µM for 48 h; 2-MeO-E2 exhibited potent/selective inhibitory effects on the proliferation of LTED cells as compared with that on MCF-7 cells (IC$_{50}$ values: 6.79 ± 0.71 and 0.93 ± 0.11 µM, respectively) (Figure 3A,B). Furthermore, the antiproliferative effects of 2-MeO-E2 on LTED cells were prolonged in a time-dependent manner for up to 96 h, with lower IC$_{50}$ values of 0.55 ± 0.02 and 0.40 ± 0.02 µM observed at 72 and 96 h, respectively (Supplemental Figure S1). To confirm the antiproliferative selectivity of 2-MeO-E2 in LTED cells, we investigated other representative microtubule-destabilizing agents (i.e., paclitaxel, nocodazole, and colchicine) belonging to the same group as G-1 and 2-MeO-E2 [24,37]. Although they exhibited stronger antiproliferative effects than 2-MeO-E2, none of the antiproliferative agents displayed selective inhibitory potential for the proliferation of LTED cells (Figure 4). Thus, among the molecules that interact with tubulins, it is strongly suggested that 2-MeO-E2 is a selective antiproliferative molecule for LTED cells.

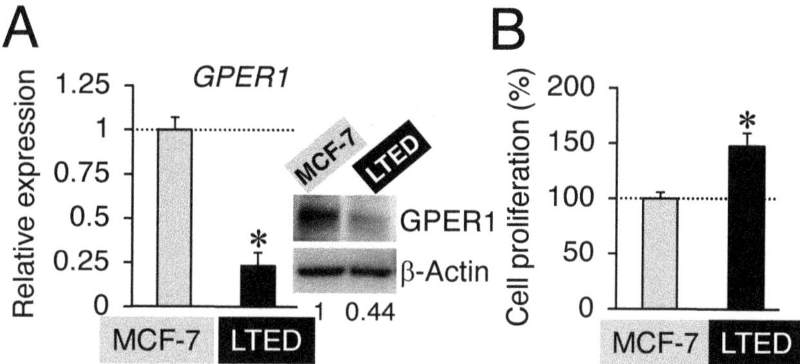

Figure 2. Highly proliferative long-term estrogen-deprived (LTED) cells display a lower expression of G protein-coupled estrogen receptor 1 (GPER1). (**A**) Basal *GPER1* mRNA and GPER1 protein (*inset*) expression in MCF-7 and LTED cells. Data are presented as the mean $\pm$ S.E. (n = 6) of the fold induction from MCF-7 cells. Significant differences (by Student's t-test) to MCF-7 cells are marked with an asterisk (* $p < 0.05$). (**A**, *inset*) Western blot analysis was performed using antibodies specific for GPER1 and β-actin, respectively. Representative images are shown. The band intensity of GPER1 (MCF-7 lane as 1) was quantified using the ImageJ 1.46r software and normalized to the band intensity of β-actin. Data are shown below the blot image. (**B**) Basal proliferation in MCF-7 and LTED cells. Data are presented as the mean $\pm$ S.E. (n = 6) percentage of MCF-7 cells. Significant differences (by Student's t-test) to MCF-7 cells are marked with an asterisk (* $p < 0.05$).

3.2. G-1 Arrests LTED Cells at the G2/M Phase with Accompanying Sub-G1

Depending on the concentration used, G-1 can evoke suppressive effects on the proliferation of MCF-7 cells, coupled with apoptosis [6,14]. Furthermore, G-1 decreased the percentage of the cell population in the G1 phase and increased the number of MCF-7 cells in the G2/M phase [14]. Based on these findings, we first performed flow cytometry analysis to study the LTED cell cycle progression after exposure to 1 μM G-1 for 48 h and used MCF-7 cells as a positive control. The flow cytometry results showed that G-1 induced G1 phase downregulation and G2/M phase stimulation in MCF-7 cells compared to that in the vehicle-only treated control (Figure 5A, upper panel; Figure 5B, left panel). We focused on the interplay between G-1 and LTED cells and found that the cell population exhibited characteristics similar to those of MCF-7 cells after exposure to G-1 (Figure 5A, lower panel; Figure 5B, right panel). G-1 induced a population of cells in the sub-G1 phase (a hallmark of apoptosis) in both breast cancer cell lines. The fold induction of the sub-G1 population after G-1 treatment was approximately 355-fold (0.06% vs. 21.3% in MCF-7 cells) and 617-fold (0.017% vs. 10.3% in LTED cells) higher than that in the controls (Figure 5A,C). Cellular morphology analysis after exposure to 1 μM G-1 for 48 h indicated membrane blebbing. One of the apoptotic cell markers [38] was detected in both MCF-7 and LTED cells (Supplementary Figure S2). Considering the significant appearance of the sub-G1 peak after G-1 exposure, G-1 treatment was presumed to decrease the proliferation of LTED cells and induce apoptotic signaling, as previously observed in MCF-7 cells [14].

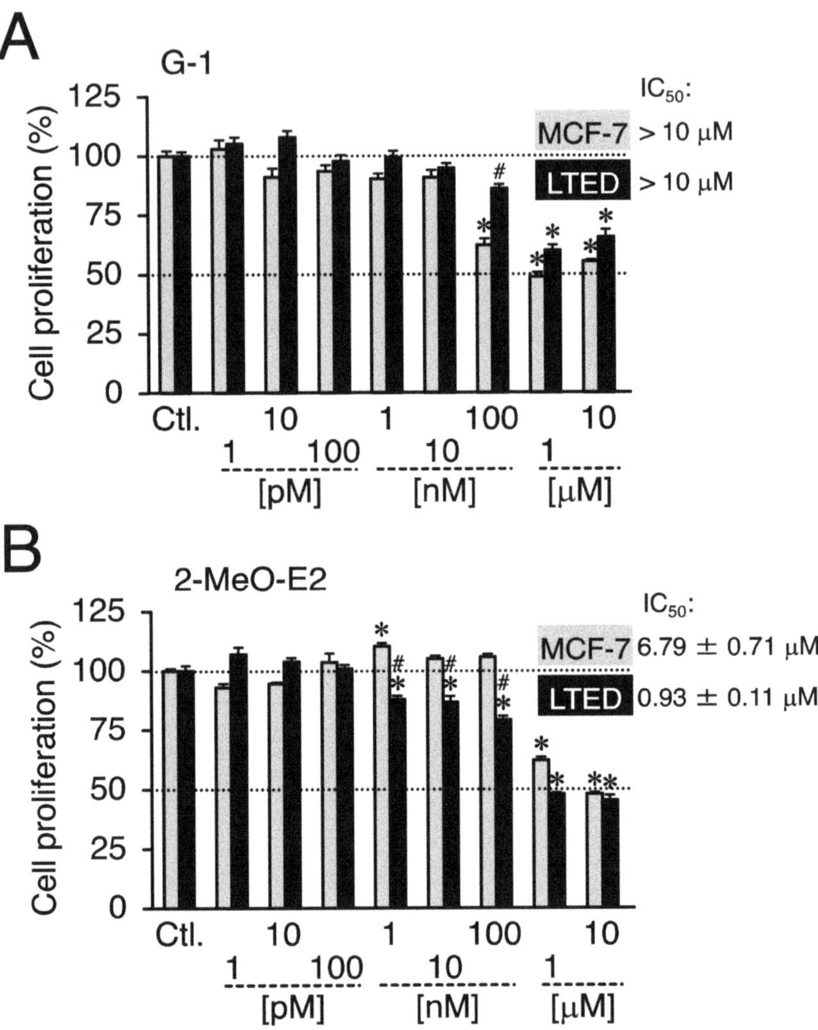

Figure 3. On LTED cells, 2-methoxyestradiol (2-MeO-E2) exerted selective antiproliferative effects. MCF-7 and LTED cells were treated with G-1 (1 pM to 10 μM) (**A**) or 2-MeO-E2 (1 pM to 10 μM) (**B**) for 48 h. The control sample (indicated as Ctl.) cells were treated with the vehicle. Data are presented as the mean ± S.E. (n = 6) percentage of the vehicle-treated control. Significant differences (by two-way ANOVA, followed by Tukey–Kramer's post hoc test) as compared with the vehicle-treated control for each cell and compared with MCF-7 cells are marked with asterisks (* $p < 0.05$) and hashes ([#] $p < 0.05$), respectively.

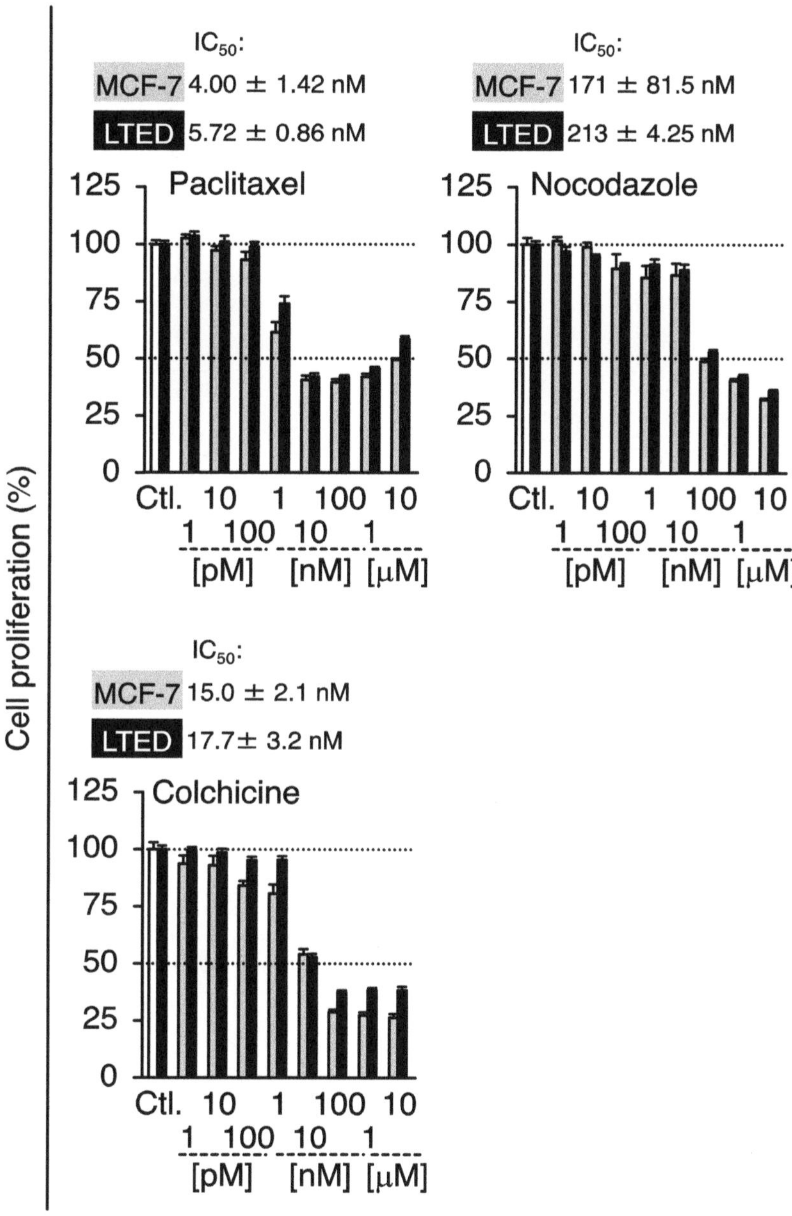

Figure 4. Effects of paclitaxel, nocodazole, and colchicine on the proliferation of MCF-7 and LTED cells. MCF-7 and LTED cells were treated with paclitaxel (**upper left panel**), nocodazole (**upper right panel**), or colchicine (**lower left panel**) (1 pM to 10 µM) for 48 h. The control sample (Ctl.) was treated with the vehicle. Data are presented as the mean ± S.E. ($n = 6$) percentage of the vehicle-treated control.

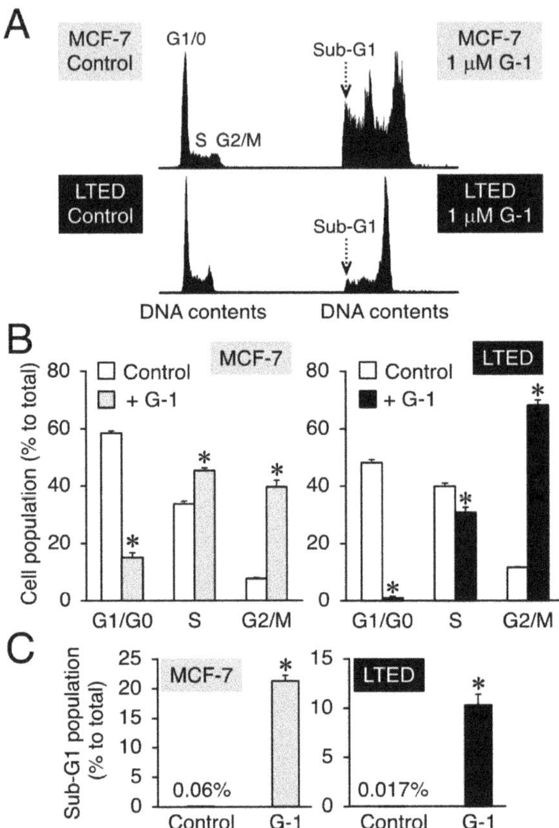

Figure 5. G-1 arrests LTED cells at the G2/M phase with accompanying sub-G1. MCF-7 and LTED cells were treated with 1 μM G-1 for 48 h. The control sample was treated with the vehicle. (**A**) Representative histograms are shown. (**B,C**) The percentages of cells in G1, S, G2/M (**B**), and sub-G1 (**C**) phases are shown. Data are presented as the mean ± S.E. (n = 3). Significant differences (by Student's *t*-test) as compared with the vehicle-treated control are marked with asterisks (* $p < 0.05$).

3.3. 2-MeO-E2 Arrests LTED Cells at the G2/M Phase without Accompanying Sub-G1

Cell morphology and cell cycle analyses of MCF-7 and LTED cells exposed to 0.1 μM or 1 μM 2-MeO-E2 revealed differences in the morphology of MCF-7/LTED cells exposed to G-1 and 2-MeO-E2 (Supplementary Figure S2 and Figure 6A). Similar to the effects of G-1, 2-MeO-E2 induced cell cycle arrest at the G2/M phase and reduced the number of MCF-7 cells in the G1/G0 phase (Figure 6B,C) [14,39]. However, unlike the effects of G-1 on MCF-7 and LTED cells (Figure 5 and Supplementary Figure S2), 2-MeO-E2 did not affect the G1/G0 phase, even at 1 μM; however, it caused a concentration-dependent increase in the G2/M cell population (Figure 6B,C). Furthermore, the sub-G1 phase did not appear in LTED or MCF-7 cells exposed to 2-MeO-E2 (Figure 6B), suggesting that 2-MeO-E2 may have an antiproliferative mechanism(s) distinct from that of G-1.

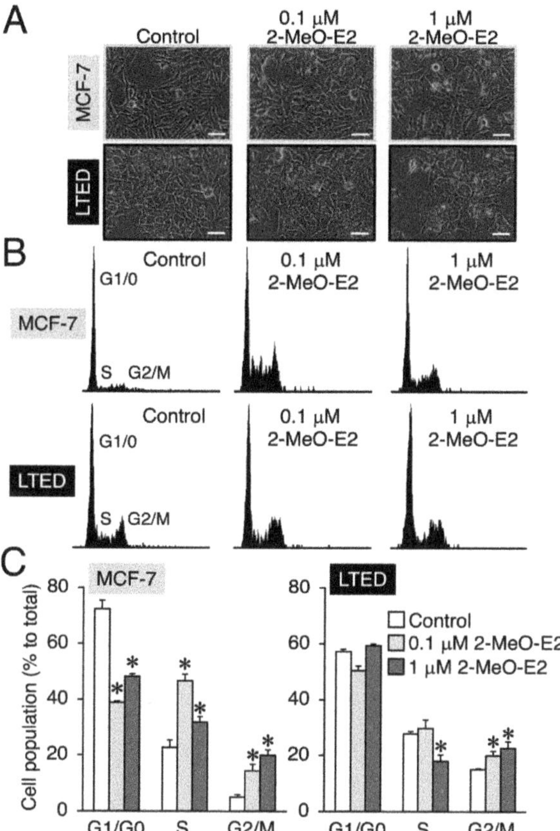

Figure 6. At the G2/M phase, 2-MeO-E2 arrests LTED cells without accompanying sub-G1. MCF-7 and LTED cells were treated with 2-MeO-E2 (0.1 and 1 µM) for 48 h. The control sample was treated with the vehicle. (**A**) Morphology of 2-MeO-E2-treated cells. Representative images are shown. The images were acquired at $400\times$ magnification. Scale bar is indicated as 50 µm. (**B**) Representative histograms are shown. (**C**) The percentage of cells in the G1, S, and G2/M phases are shown. Data are presented as the mean $\pm$ S.E. ($n = 3$). Significant differences (by one-way ANOVA, followed by Dunnett's post hoc test) as compared with the vehicle-treated control are marked with asterisks ($^{*} p < 0.05$).

3.4. Expression Profiles of β-Tubulin and Its Isotype in MCF-7 and LTED Cells

We investigated the possible targets underlying the 2-MeO-E2-mediated inhibition of LTED cell proliferation. G-1 and 2-MeO-E2 are microtubule (i.e., polymers of α/β-tubulin)-targeting agents that bind to the colchicine-binding sites of β-tubulin and exert antiproliferative activity in MCF-7 cells [14]. As shown in Figure 7A, Western blot analysis indicated that a high protein expression of α- and β-tubulins (1.62- and 1.55-fold, respectively, $p < 0.05$), but not β-actin, was detected in LTED cells compared to MCF-7 cells. The post-translational modification of α-tubulin has been suggested to be engaged with the aggressive nature of breast cancer cells [40,41]. It has been reported that in the ERα-positive MCF-7 cells, ligand-dependent activation of ERα can upregulate HDAC6, leading to deacetylation of α-tubulin mediated by HDAC6 [41,42]. When focusing on the *HDAC6* mRNA expression between MCF-7 and LTED cells, LTED cells that largely display ligand-independent ERα activation [6,32] showed reduced *HDAC6* levels (~0.79-fold, $p < 0.05$) in comparison to MCF-7 cells (Figure 7B), indicating that the involvement of ERα in the HDAC6/α-tubulin axis might not be high in LTED cells. Additionally, the altered expression of the isotype has been suggested to be involved in resistance to chemotherapy [22,23]. Hereafter, we focused on the β-tubulins and investigated the expression profile of β-tubulin isotypes expressed in MCF-7 and LTED cells. Real-time PCR analysis revealed that compared with MCF-7 cells, a low expression of *TUBB3* and *TUBB4A* ($p < 0.05$) but a significantly higher expression of *TUBB2B* (3.44-fold, $p < 0.05$) was detected in LTED cells (Figure 7C). Thus, among β-tubulin isotypes, the *TUBB2B* gene is a possible target of 2-MeO-E2-mediated antiproliferation in LTED cells.

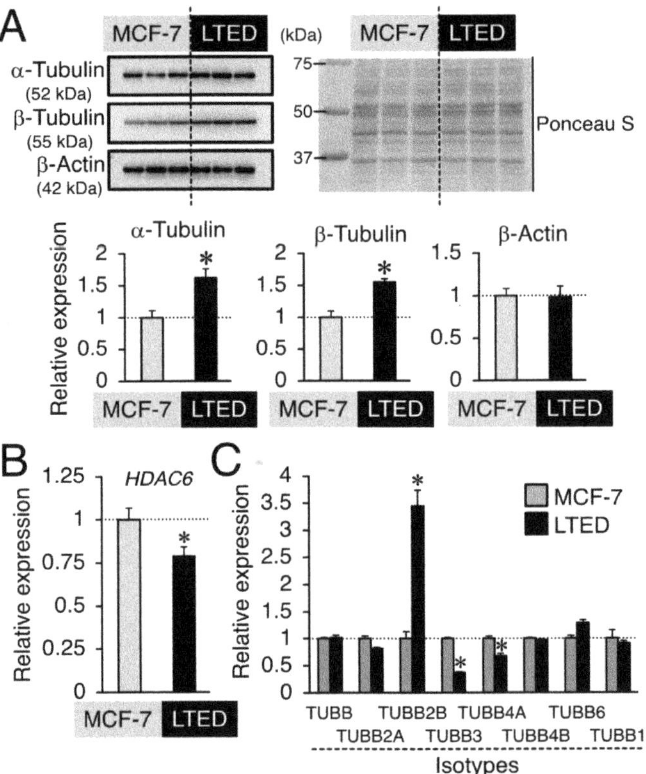

Figure 7. Expression profiles of β-tubulin and its isotype in MCF-7 and LTED cells. (**A**) Basal α-tubulin, β-tubulin, and β-actin protein expression in MCF-7 and LTED cells. (**Upper left panel**) Western blot analysis was performed using antibodies specific for α-tubulin, β-tubulin, and β-actin, respectively. (**Upper right panel**) Ponceau S staining was performed to indicate total protein loading in each lane. (**Lower panel**) The band intensity was quantified using the ImageJ 1.46r software. Data are presented as the mean ± S.E. ($n = 3$) of the fold induction from MCF-7 cells. Significant differences (by Student's *t*-test) to MCF-7 cells are marked with asterisks (* $p < 0.05$). (**B,C**) mRNA expression of Basal *histone deacetylase 6* (*HDAC6*) (**B**) and *β-tubulin* (*TUBB*) isotypes (**C**) in MCF-7 and LTED cells. Data are presented as the mean ± S.E. ($n = 6$) of the fold induction from MCF-7 cells. Significant differences (by Student's *t*-test) to MCF-7 cells are marked with asterisks (* $p < 0.05$).

3.5. Effects of siTUBB2B on the Antiproliferation by 2-MeO-E2 in LTED Cells

Given that TUBB2B is an antiproliferative target evoked by 2-MeO-E2 in LTED cells, TUBB2B siRNA (siTUBB2B) treatment would somewhat restore the antiproliferative effects of 2-MeO-E2 on LTED cells. We first investigated the effects of siTUBB2B on the basal proliferation of LTED cells. Treatment with siTUBB2B resulted in reduced proliferation (87%, $p < 0.05$) compared to the control siRNA-treated (siControl) group (Supplementary Figure S3), implying the functional role of TUBB2B in the proliferation of LTED cells. To confirm the above-mentioned possibility, we applied 2-MeO-E2 to siTUBB2B-treated LTED cells maximally at 100 nM based on the results in Figure 3B, which revealed selective antiproliferation by 2-MeO-E2 against LTED cells below 100 nM. As shown in Figure 8A, the treatment with siTUBB2B caused a 0.23-fold decrease in *TUBB2B* compared to that in the siControl group. Compared with the siControl group, siTUBB2B restored the reduction in cell viability caused by 2-MeO-E2 over almost the same concentration range (at 10 and 100 pM; $p < 0.05$), although the rescue effect of siTUBB2B was lower, particularly at higher concentrations of 2-MeO-E2 (>10 nM) (Figure 8B).

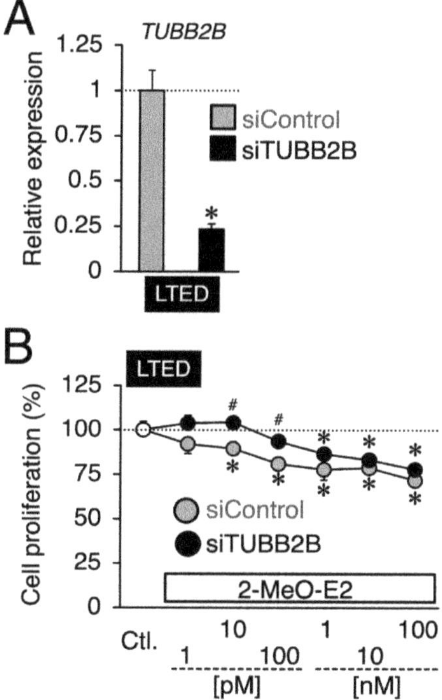

Figure 8. Effects of TUBB2B siRNA (siTUBB2B) on antiproliferation by 2-MeO-E2 against LTED cells. (**A**) mRNA expression of *TUBB2B* in LTED cells transfected with control siRNA (siControl) or siTUBB2B. Data are presented as the mean ± S.E. ($n = 3$) of the fold induction from the siControl-transfected group. Significant differences (by Student's *t*-test) to the siControl-transfected group are marked with an asterisk (* $p < 0.05$). (**B**) LTED cells were transfected with siControl or siTUBB2B, followed by treatment with 2-MeO-E2 (1 pM to 100 nM). Data are presented as the mean ± S.E. ($n = 6$) percentage of the vehicle-treated control. Significant differences (by two-way ANOVA, followed by Tukey–Kramer's post hoc test) as compared with the vehicle-treated control and compared with the siControl-transfected group are marked with asterisks (* $p < 0.05$) and hashes (# $p < 0.05$), respectively.

3.6. Effects of G-1 on the 2-MeO-E2-Mediated Antiproliferation in LTED Cells

The binding potential of G-1 to β-tubulin is stronger than that of 2-MeO-E2 [14,16]. Therefore, we investigated whether adding G-1 abrogated the 2-MeO-E2-mediated inhibition of LTED cell proliferation. Compared with the control group, 2-MeO-E2-treated cells showed reduced proliferation in a concentration-dependent manner at concentrations up to 10 μM. At 1 μM, G-1 treatment alone reduced the proliferation of LTED cells to approximately 60% of that of the control. However, an additive antiproliferative interaction between 2-MeO-E2 and G-1 was not observed, and the degree of inhibition was comparable to that of the G-1 alone system (Figure 9). Thus, G-1 and 2-MeO-E2 may share a mechanism of action with β-tubulin in LTED cells.

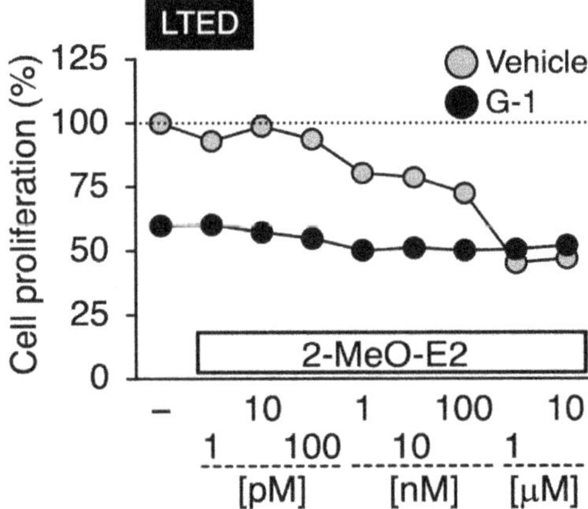

Figure 9. Effects of G-1 on the 2-MeO-E2-mediated antiproliferation in LTED cells. LTED cells were treated with 2-MeO-E2 (1 pM to 10 μM) in the absence or presence of 1 μM G-1 for 48 h. The control sample was treated with vehicle only. Data are presented as the mean ± S.E. (n = 6) percentage of the vehicle-treated control.

4. Discussion

We used two breast cancer cell lines, MCF-7 (a parental cell line for LTED) and LTED cells (a cell model resistant to endocrine therapy). Compared to MCF-7 cells, LTED cells showed a much higher proliferative potential (Figure 2B) and displayed resistance to microtubule-destabilizing agents (paclitaxel, nocodazole, and colchicine) (Figure 4). Additionally, we reported that LTED cells were tolerant to different kinds of established antiproliferative agents, etoposide, LY2835219, and trichostatin A [24]. We sought to identify chemicals that selectively kill LTED cells and focused on two antiproliferative molecules, G-1 and 2-MeO-E2, which can target both GPER1 and β-tubulin to display their antiproliferative activities. Whereas G-1 preferentially inhibited MCF-7 cell proliferation, 2-MeO-E2 exhibited antiproliferative activity in LTED cells (Figure 3). The protein expression analysis of GPER1 and β-tubulin in MCF-7 and LTED cells revealed upregulated β-tubulin but downregulated GPER1 in LTED cells compared with MCF-7 cells (Figures 2A and 7A). Among the *β-tubulin* gene isotypes, the highest expression of *TUBB2B* was detected in LTED cells compared to MCF-7 cells (Figure 7C). The introduction of siTUBB2B restored 2-MeO-E2-driven antiproliferation (Figure 8B), indicating that 2-MeO-E2 is an effective pharmacological modality; that is, the 2-MeO-E2–TUBB2B interaction is a promising candidate for the abrogation of LTED cell proliferation.

We could not detect the sub-G1 phase in LTED or MCF-7 cells exposed to 2-MeO-E2 (Figures 5 and 6), suggesting that 2-MeO-E2 may utilize mechanism(s) distinct from G-1 for the abrogation of breast cancer cell proliferation. In support of this notion, a recent study reported that 2-MeO-E2 induces centrosome declustering in breast cancer cells, resulting in mitotic arrest via multipolar spindle formation [43].

G-1 and 2-MeO-E2 can bind to the colchicine-binding site of β-tubulin, and G-1 displays a stronger binding potential to the tubulin site than 2-MeO-E2 [14,44]. Compared to the results in which LTED cells were singly treated with 2-MeO-E2, the antiproliferative effects of 2-MeO-E2 were completely blocked by the co-introduction of G-1 (1 μM) (Figure 9). Given that 2-MeO-E2 targets β-tubulin to exert its antiproliferative activity in LTED cells, the binding potency itself might not be essential for the action. Furthermore, after 2-MeO-E2 interaction with β-tubulin (TUBB2B), the downstream antiproliferative signaling pathways might differ between G-1 and 2-MeO-E2. Although little is known about the functional participation of the TUBB2B subtype in breast cancer, including endocrine therapy-resistant ER-positive breast cancer, TUBB2B can function as an oncogene in hepatocellular carcinoma and plays a role in promoting cell proliferation [45]. Among the β-tubulin isotypes, significant upregulation of *TUBB2B* was noted in LTED cells compared to parental MCF-7 cells, suggesting the possibility that TUBB2B can play a common biological role in cancer cells. Although 2-MeO-E2 was suggested to have no antiproliferative effect in vivo and in vitro (on MCF-7 cells) because of its low stability [46,47], the antiproliferative effects of 2-MeO-E2 on LTED cells were not only detected but also prolonged in a time-dependent manner for up to 96 h, with lower IC_{50} values of 0.55 ± 0.02 and 0.40 ± 0.02 μM observed at 72 and 96 h, respectively (Figure 3 and Supplemental Figure S1). Although further research is required, this study strongly suggests that 2-MeO-E2 can be used as a prototype for developing anticancer agents against endocrine therapy-resistant breast cancers.

5. Conclusions

LTED cells were used as an experimental model of endocrine therapy-resistant ER-positive breast cancer. Compared to the parental MCF-7 cells, LTED cells display highly aggressive cancerous behavior. Due to the acquisition of resistance, effective modalities for abrogating LTED cell proliferation are required. In this study, we focused on several tubulin-inhibitory drugs, including paclitaxel, nocodazole, colchicine, G-1, and 2-MeO-E2. It was demonstrated that 2-MeO-E2 selectively suppresses the proliferation of LTED cells compared to that of MCF-7 cells (i.e., IC_{50} values: 0.93 μM for LTED cells vs. 6.79 μM for MCF-7 cells). Among the β-tubulin isotypes, LTED cells display the highest upregulated expression of *TUBB2B* compared to parental MCF-7 cells. In the G2/M phase, 2-MeO-E2 arrested LTED cells, in conjunction with a reduced number of cells in the S phase, and targeted TUBB2B for antiproliferative activity. Our findings suggest that TUBB2B-targeting 2-MeO-E2 is a potential candidate molecule for preventing LTED cell proliferation. However, further studies are required to assess whether 2-MeO-E2 abrogates LTED cell-driven breast cancer progression in vivo.

Supplementary Materials: The following supporting information can be downloaded from https://www.mdpi.com/article/10.3390/cimb45090464/s1: Figure S1: Effect of 2-methoxyestradiol (2-MeO-E2) on long-term estrogen-deprived (LTED) cell proliferation; Figure S2: Effects of G-1 on MCF-7 and LTED cell morphology; and Figure S3: Effects of TUBB2B siRNA (siTUBB2B) on the proliferation of LTED cells.

Author Contributions: Conceptualization, S.T.; methodology, M.H.-S., K.K. and S.T.; validation, M.H.-S.; formal analysis, M.H.-S. and K.K.; investigation, M.H.-S. and K.K.; data curation, M.H.-S. and K.K.; writing—original draft preparation, S.T.; writing—review and editing, M.H.-S., K.K., M.T., N.S. and S.T.; visualization, M.H.-S.; supervision, S.T.; project administration, S.T.; funding acquisition, M.H.-S., N.S. and S.T. All authors have read and agreed to the published version of the manuscript.

Funding: This study was supported in part by JSPS KAKENHI (21K12261 [to S.T.] and 22K15288 [to M.H.-S.]). This study was also supported in part by a Setouchi Satoyama-Satoumi Research Project Grant from Fukuyama University (to S.T. and N.S.), and in part by the Cooperative Research Program of the Network Joint Research Center for Materials and Devices (Research No. 20231305 [to S.T.]).

Institutional Review Board Statement: Not applicable.

Informed Consent Statement: Not applicable.

Data Availability Statement: Not applicable.

Conflicts of Interest: The authors declare no conflict of interest.

References

1. Early Breast Cancer Trialists' Collaborative Group (EBCTCG). Aromatase inhibitors versus tamoxifen in early breast cancer: Patient-level meta-analysis of the randomised trials. *Lancet* **2015**, *386*, 1341–1352. [CrossRef]
2. Katzenellenbogen, B.S.; Kendra, K.L.; Norman, M.J.; Berthois, Y. Proliferation, hormonal responsiveness, and estrogen receptor content of MCF-7 human breast cancer cells grown in the short-term and long-term absence of estrogens. *Cancer Res.* **1987**, *47*, 4355–4360. [PubMed]
3. Jeng, M.H.; Shupnik, M.A.; Bender, T.P.; Westin, E.H.; Bandyopadhyay, D.; Kumar, R.; Masamura, S.; Santen, R.J. Estrogen receptor expression and function in long-term estrogen-deprived human breast cancer cells. *Endocrinology* **1998**, *139*, 4164–4174. [CrossRef] [PubMed]
4. Chan, C.M.W.; Martin, L.-A.; Johnston, S.R.D.; Ali, S.; Dowsett, M. Molecular changes associated with the acquisition of oestrogen hypersensitivity in MCF-7 breast cancer cells on long-term oestrogen deprivation. *J. Steroid Biochem. Mol. Biol.* **2002**, *81*, 333–341. [CrossRef]
5. Martin, L.A.; Ribas, R.; Simigdala, N.; Schuster, E.; Pancholi, S.; Tenev, T.; Gellert, P.; Buluwela, L.; Harrod, A.; Thornhill, A.; et al. Discovery of naturally occurring *ESR1* mutations in breast cancer cell lines modelling endocrine resistance. *Nat. Commun.* **2017**, *8*, 1865. [CrossRef]
6. Hirao-Suzuki, M.; Takeda, S.; Kodama, Y.; Takiguchi, M.; Toda, A.; Ohara, M. Metalloestrogenic effects of cadmium are absent in long-term estrogen-deprived MCF-7 cells: Evidence for the involvement of constitutively activated estrogen receptor α and very low expression of G protein-coupled estrogen receptor 1. *Toxicol. Lett.* **2020**, *319*, 22–30. [CrossRef] [PubMed]
7. Carmeci, C.; Thompson, D.A.; Ring, H.Z.; Francke, U.; Weigel, R.J. Identification of a gene (GPR30) with homology to the G-protein-coupled receptor superfamily associated with estrogen receptor expression in breast cancer. *Genomics* **1997**, *45*, 607–617. [CrossRef]
8. Bologa, C.G.; Revankar, C.M.; Young, S.M.; Edwards, B.S.; Arterburn, J.B.; Kiselyov, A.S.; Parker, M.A.; Tkachenko, S.E.; Savchuck, N.P.; Sklar, L.A.; et al. Virtual and biomolecular screening converge on a selective agonist for GPR30. *Nat. Chem. Biol.* **2006**, *2*, 207–212. [CrossRef]
9. Albanito, L.; Madeo, A.; Lappano, R.; Vivacqua, A.; Rago, V.; Carpino, A.; Oprea, T.I.; Prossnitz, E.R.; Musti, A.M.; Ando, S.; et al. G protein-coupled receptor 30 (GPR30) mediates gene expression changes and growth response to 17β-estradiol and selective GPR30 ligand G-1 in ovarian cancer cells. *Cancer Res.* **2007**, *67*, 1859–1866. [CrossRef]
10. Lucki, N.C.; Sewer, M.B. Genistein stimulates MCF-7 breast cancer cell growth by inducing acid ceramidase (*ASAH1*) gene expression. *J. Biol. Chem.* **2011**, *286*, 19399–19409. [CrossRef]
11. Vivacqua, A.; Romeo, E.; De Marco, P.; De Francesco, E.M.; Abonante, S.; Maggiolini, M. GPER mediates the egr-1 expression induced by 17β-estradiol and 4-hydroxitamoxifen in breast and endometrial cancer cells. *Breast Cancer Res. Treat.* **2012**, *133*, 1025–1035. [CrossRef] [PubMed]
12. Scaling, A.L.; Prossnitz, E.R.; Hathaway, H.J. GPER mediates estrogen-induced signaling and proliferation in human breast epithelial cells and normal and malignant breast. *Horm. Cancer* **2014**, *5*, 146–160. [CrossRef] [PubMed]
13. Santolla, M.F.; Avino, S.; Pellegrino, M.; De Francesco, E.M.; De Marco, P.; Lappano, R.; Vivacqua, A.; Cirillo, F.; Rigiracciolo, D.C.; Scarpelli, A.; et al. SIRT1 is involved in oncogenic signaling mediated by GPER in breast cancer. *Cell Death Dis.* **2015**, *6*, e1834. [CrossRef] [PubMed]
14. Lv, X.; He, C.; Huang, C.; Hua, G.; Wang, Z.; Remmenga, S.W.; Rodabough, K.J.; Karpf, A.R.; Dong, J.; Davis, J.S.; et al. G-1 inhibits breast cancer cell growth via targeting colchicine-binding site of tubulin to interfere with microtubule assembly. *Mol. Cancer Ther.* **2017**, *16*, 1080–1091. [CrossRef] [PubMed]
15. Wang, C.; Lv, X.; He, C.; Hua, G.; Tsai, M.-Y.; Davis, J.S. The G-protein-coupled estrogen receptor agonist G-1 suppresses proliferation of ovarian cancer cells by blocking tubulin polymerization. *Cell Death Dis.* **2013**, *4*, e869. [CrossRef]
16. D'Amato, R.J.; Lin, C.M.; Flynn, E.; Folkman, J.; Hamel, E. 2-Methoxyestradiol, an endogenous mammalian metabolite, inhibits tubulin polymerization by interacting at the colchicine site. *Proc. Natl. Acad. Sci. USA* **1994**, *91*, 3964–3968. [CrossRef]
17. Prossnitz, E.R.; Arterburn, J.B. International union of basic and clinical pharmacology. XCVII. G protein-coupled estrogen receptor and its pharmacologic modulators. *Pharmacol. Rev.* **2015**, *67*, 505–540. [CrossRef] [PubMed]
18. Cavalieri, E.; Chakravarti, D.; Guttenplan, J.; Hart, E.; Ingle, J.; Jankowiak, R.; Muti, P.; Rogan, E.; Russo, J.; Santen, R.; et al. Catechol estrogen quinones as initiators of breast and other human cancers: Implications for biomarkers of susceptibility and cancer prevention. *Biochim. Biophys. Acta* **2006**, *1766*, 63–78. [CrossRef]
19. Mueck, A.O.; Seeger, H.; Huober, J. Chemotherapy of breast cancer-additive anticancerogenic effects by 2-methoxyestradiol? *Life Sci.* **2004**, *75*, 1205–1210. [CrossRef]

20. Zhu, B.T.; Conney, A.H. Is 2-methoxyestradiol an endogenous estrogen metabolite that inhibits mammary carcinogenesis? *Cancer Res.* **1998**, *58*, 2269–2277.
21. Ludueña, R.F.; Shooter, E.M.; Wilson, L. Structure of the tubulin dimer. *J. Biol. Chem.* **1977**, *252*, 7006–7014. [CrossRef] [PubMed]
22. Ranganathan, S.; Benetatos, C.A.; Colarusso, P.J.; Dexter, D.W.; Hudes, G.R. Altered β-tubulin isotype expression in paclitaxel-resistant human prostate carcinoma cells. *Br. J. Cancer* **1998**, *77*, 562–566. [CrossRef] [PubMed]
23. Kavallaris, M.; Burkhart, C.A.; Horwitz, S.B. Antisense oligonucleotides to class III β-tubulin sensitize drug-resistant cells to taxol. *Br. J. Cancer* **1999**, *80*, 1020–1025. [CrossRef] [PubMed]
24. Takeda, S.; Hirao-Suzuki, M.; Yamagishi, Y.; Sugihara, T.; Kaneko, M.; Sakai, G.; Nakamura, T.; Hieda, Y.; Takiguchi, M.; Okada, M.; et al. Effects of the ethanol extract of *Neopyropia Yezoensis*, cultivated in the Seto Inland Sea (Setonaikai), on the viability of 10 human cancer cells including endocrine therapy-resistant breast cancer cells. *Fundam. Toxicol. Sci.* **2021**, *8*, 75–80. [CrossRef]
25. Sakai, G.; Hirao-Suzuki, M.; Koga, T.; Kobayashi, T.; Kamishikiryo, J.; Tanaka, M.; Fujii, K.; Takiguchi, M.; Sugihara, N.; Toda, A.; et al. Perfluorooctanoic acid (PFOA) as a stimulator of estrogen receptor-negative breast cancer MDA-MB-231 cell aggressiveness: Evidence for involvement of fatty acid 2-hydroxylase (FA2H) in the stimulated cell migration. *J. Toxicol. Sci.* **2022**, *47*, 159–168. [CrossRef]
26. Takeda, S.; Hirao-Suzuki, M.; Shindo, M.; Aramaki, H. (–)-Xanthatin as a killer of human breast cancer MCF-7 mammosphere cells: A comparative study with salinomycin. *Curr. Issues Mol. Biol.* **2022**, *44*, 3849–3858. [CrossRef]
27. Duong, V.; Bret, C.; Altucci, L.; Mai, A.; Duraffourd, C.; Loubersac, J.; Harmand, P.O.; Bonnet, S.; Valente, S.; Maudelonde, T.; et al. Specific activity of class II histone deacetylases in human breast cancer cells. *Mol. Cancer Res.* **2008**, *6*, 1908–1919. [CrossRef]
28. Narvi, E.; Jaakkola, K.; Winsel, S.; Oetken-Lindholm, C.; Halonen, P.; Kallio, L.; Kallio, M.J. Altered TUBB3 expression contributes to the epothilone response of mitotic cells. *Br. J. Cancer* **2013**, *108*, 82–90. [CrossRef]
29. Saussede-Aim, J.; Matera, E.-L.; Ferlini, C.; Dumontet, C. β3-Tubulin is induced by estradiol in human breast carcinoma cells through an estrogen-receptor dependent pathway. *Cell Motil. Cytoskelet.* **2009**, *66*, 378–388. [CrossRef]
30. Azumi, M.; Yoshie, M.; Nakachi, N.; Tsuru, A.; Kusama, K.; Tamura, K. Stathmin expression alters the antiproliferative effect of eribulin in leiomyosarcoma cells. *J. Pharmacol. Sci.* **2022**, *150*, 259–266. [CrossRef]
31. Li, W.; Zhai, B.; Zhi, H.; Li, Y.; Jia, L.; Ding, C.; Zhang, B.; You, W. Association of ABCB1, β tubulin I, and III with multidrug resistance of MCF7/DOC subline from breast cancer cell line MCF7. *Tumour Biol.* **2014**, *35*, 8883–8891. [CrossRef] [PubMed]
32. Hirao-Suzuki, M.; Takiguchi, M.; Yoshihara, S.; Takeda, S. Repeated exposure to 4-methyl-2,4-bis(4-hydroxyphenyl)pent-1-ene (MBP) accelerates ligand-independent activation of estrogen receptors in long-term estradiol-deprived MCF-7 cells. *Toxicol. Lett.* **2023**, *378*, 31–38. [CrossRef] [PubMed]
33. Wei, W.; Chen, Z.-J.; Zhang, K.-S.; Yang, X.-L.; Wu, Y.-M.; Chen, X.-H.; Huang, H.-B.; Liu, H.-L.; Cai, S.-H.; Du, J.; et al. The activation of G protein-coupled receptor 30 (GPR30) inhibits proliferation of estrogen receptor-negative breast cancer cells in vitro and in vivo. *Cell Death Dis.* **2014**, *5*, e1428. [CrossRef] [PubMed]
34. Girgert, R.; Emons, G.; Gründker, C. Inactivation of GPR30 reduces growth of triple-negative breast cancer cells: Possible application in targeted therapy. *Breast Cancer Res. Treat.* **2012**, *134*, 199–205. [CrossRef]
35. Weißenborn, C.; Ignatov, T.; Ochel, H.-J.; Costa, S.D.; Zenclussen, A.C.; Ignatova, Z.; Ignatov, A. GPER functions as a tumor suppressor in triple-negative breast cancer cells. *J. Cancer Res. Clin. Oncol.* **2014**, *140*, 713–723. [CrossRef]
36. Weißenborn, C.; Ignatov, T.; Poehlmann, A.; Wege, A.K.; Costa, S.D.; Zenclussen, A.C.; Ignatov, A. GPER functions as a tumor suppressor in MCF-7 and SK-BR-3 breast cancer cells. *J. Cancer Res. Clin. Oncol.* **2014**, *140*, 663–671. [CrossRef]
37. Lu, Y.; Chen, J.; Xiao, M.; Li, W.; Miller, D.D. An overview of tubulin inhibitors that interact with the colchicine binding site. *Pharm. Res.* **2012**, *29*, 2943–2971. [CrossRef]
38. Mills, J.C.; Stone, N.L.; Erhardt, J.; Pittman, R.N. Apoptotic membrane blebbing is regulated by myosin light vhain phosphorylation. *J. Cell Biol.* **1998**, *140*, 627–636. [CrossRef]
39. van den Brand, A.D.; Villevoye, J.; Nijmeijer, S.M.; van den Berg, M.; van Duursen, M.B.M. Anti-tumor properties of methoxylated analogues of resveratrol in malignant MCF-7 but not in non-tumorigenic MCF-10A mammary epithelial cell lines. *Toxicology* **2019**, *422*, 35–43. [CrossRef]
40. Hubbert, C.; Guardiola, A.; Shao, R.; Kawaguchi, Y.; Ito, A.; Nixon, A.; Yoshida, M.; Wang, X.-F.; Yao, T.-P. HDAC6 is a microtubule-associated deacetylase. *Nature* **2002**, *417*, 455–458. [CrossRef]
41. Azuma, K.; Urano, T.; Horie-Inoue, K.; Hayashi, S.; Sakai, R.; Ouchi, Y.; Inoue, S. Association of estrogen receptor α and histone deacetylase 6 causes rapid deacetylation of tubulin in breast cancer cells. *Cancer Res.* **2009**, *69*, 2935–2940. [CrossRef] [PubMed]
42. Yoshida, N.; Omoto, Y.; Inoue, A.; Eguchi, H.; Kobayashi, Y.; Kurosumi, M.; Saji, S.; Suemasu, K.; Okazaki, T.; Nakachi, K.; et al. Prediction of prognosis of estrogen receptor-positive breast cancer with combination of selected estrogen-regulated genes. *Cancer Sci.* **2004**, *95*, 496–502. [CrossRef] [PubMed]
43. El-Zein, R.; Thaiparambil, J.; Abdel-Rahman, S.Z. 2-Methoxyestradiol sensitizes breast cancer cells to taxanes by targeting centrosomes. *Oncotarget* **2020**, *11*, 4479–4489. [CrossRef] [PubMed]
44. Dumontet, C.; Jordan, M.A. Microtubule-binding agents: A dynamic field of cancer therapeutics. *Nat. Rev. Drug Discov.* **2010**, *9*, 790–803. [CrossRef]
45. Wang, X.; Shi, J.; Huang, M.; Chen, J.; Dan, J.; Tang, Y.; Guo, Z.; He, X.; Zhao, Q. TUBB2B facilitates progression of hepatocellular carcinoma by regulating cholesterol metabolism through targeting HNF4A/CYP27A1. *Cell Death Dis.* **2023**, *14*, 179. [CrossRef]

46. Mooberry, S.L. New insights into 2-methoxyestradiol, a promising antiangiogenic and antitumor agent. *Curr. Opin. Oncol.* **2003**, *15*, 425–430. [CrossRef]
47. Sutherland, T.E.; Schuliga, M.; Harris, T.; Eckhardt, B.L.; Anderson, R.L.; Quan, L.; Stewart, A.G. 2-Methoxyestradiol is an estrogen receptor agonist that supports tumor growth in murine xenograft models of breast cancer. *Clin. Cancer Res.* **2005**, *11*, 1722–1732. [CrossRef]

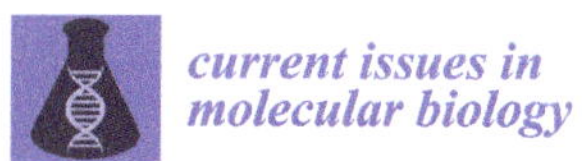

current issues in molecular biology

Article

Unveiling the Molecular Mechanism of Trastuzumab Resistance in SKBR3 and BT474 Cell Lines for HER2 Positive Breast Cancer

Anna Kokot [1,2], Sachin Gadakh [2], Indrajit Saha [2,3], Ewa Gajda [1], Michał Łaźniewski [2], Somnath Rakshit [2,†], Kaustav Sengupta [2,4,‡], Ayatullah Faruk Mollah [2,§], Michał Denkiewicz [2,‖], Katarzyna Górczak [5], Jürgen Claesen [6], Tomasz Burzykowski [1,5] and Dariusz Plewczynski [2,*]

1. Department of Clinical Molecular Biology, Medical University of Bialystok, 15-089 Bialystok, Poland; rusek.annamaria@gmail.com (A.K.); e.gajda@gmail.com (E.G.); tomasz.burzykowski@uhasselt.be (T.B.)
2. Centre of New Technologies, University of Warsaw, 02-097 Warszawa, Poland; s.gadakh@cent.uw.edu.pl (S.G.); indrajit@nitttrkol.ac.in (I.S.); m.lazniewski@cent.uw.edu.pl (M.Ł.); somnath@utexas.edu (S.R.); kaustav.sengupta@pw.edu.pl (K.S.); afmollah@aliah.ac.in (A.F.M.); m.denkiewicz@cent.uw.edu.pl (M.D.)
3. Department of Computer Science and Engineering, National Institute of Technical Teachers' Training and Research, Kolkata 700106, India
4. Faculty of Mathematics and Information Science, Warsaw University of Technology, Koszykowa 75, 00-662 Warszawa, Poland
5. Department of Mathematics and Statistics, Hasselt University, 3500 Hasselt, Belgium; katarzyna.gorczak@uhasselt.be
6. Department of Epidemiology and Data Science, Amsterdam Universitair Medische Centra, VU University, 1081 HV Amsterdam, The Netherlands; j.claesen@amsterdamumc.nl
* Correspondence: d.plewczynski@cent.uw.edu.pl
† Current address: School of Information, The University of Texas at Austin, Austin, TX 78712, USA.
‡ Current address: The Jackson Laboratory for Genomic Medicine, Farmington, CT 06032, USA.
§ Current address: Department of Computer Science and Engineering, Aliah University, IIA/27 Newtown, Kolkata 700160, India.
‖ Current address: Faculty of Mathematics and Information Science, Warsaw University of Technology, Koszykowa 75, 00-662 Warszawa, Poland.

Citation: Kokot, A.; Gadakh, S.; Saha, I.; Gajda, E.; Łaźniewski, M.; Rakshit, S.; Sengupta, K.; Mollah, A.F.; Denkiewicz, M.; Górczak, K.; et al. Unveiling the Molecular Mechanism of Trastuzumab Resistance in SKBR3 and BT474 Cell Lines for HER2 Positive Breast Cancer. *Curr. Issues Mol. Biol.* **2024**, *46*, 2713–2740. https://doi.org/10.3390/cimb46030171

Academic Editor: Dumitru A. Iacobas

Received: 25 January 2024
Revised: 15 March 2024
Accepted: 16 March 2024
Published: 21 March 2024

Abstract: HER2-positive breast cancer is one of the most prevalent forms of cancer among women worldwide. Generally, the molecular characteristics of this breast cancer include activation of human epidermal growth factor receptor-2 (HER2) and hormone receptor activation. HER2-positive is associated with a higher death rate, which led to the development of a monoclonal antibody called trastuzumab, specifically targeting HER2. The success rate of HER2-positive breast cancer treatment has been increased; however, drug resistance remains a challenge. This fact motivated us to explore the underlying molecular mechanisms of trastuzumab resistance. For this purpose, a two-fold approach was taken by considering well-known breast cancer cell lines SKBR3 and BT474. In the first fold, trastuzumab treatment doses were optimized separately for both cell lines. This was done based on the proliferation rate of cells in response to a wide variety of medication dosages. Thereafter, each cell line was cultivated with a steady dosage of herceptin for several months. During this period, six time points were selected for further in vitro analysis, ranging from the untreated cell line at the beginning to a fully resistant cell line at the end of the experiment. In the second fold, nucleic acids were extracted for further high throughput-based microarray experiments of gene and microRNA expression. Such expression data were further analyzed in order to infer the molecular mechanisms involved in the underlying development of trastuzumab resistance. In the list of differentially expressed genes and miRNAs, multiple genes (e.g., *BIRC5*, *E2F1*, *TFRC*, and *USP1*) and miRNAs (e.g., hsa miR 574 3p, hsa miR 4530, and hsa miR 197 3p) responsible for trastuzumab resistance were found. Downstream analysis showed that *TFRC*, *E2F1*, and *USP1* were also targeted by hsa-miR-8485. Moreover, it indicated that miR-4701-5p was highly expressed as compared to *TFRC* in the SKBR3 cell line. These results unveil key genes and miRNAs as molecular regulators for trastuzumab resistance.

Keywords: microRNA; microarray; HER2; breast cancer; drug resistance

1. Introduction

Cancer continues to pose a substantial global health threat despite advancements in diagnosis and treatment [1]. In a recent update in 2020, there were an estimated 19.3 million new cases and 10.0 million cancer-related deaths [2], an increase from 18.1 million cases and 9.6 million deaths in 2018 [3]. This rise can be attributed to factors such as population growth, increased exposure to risk factors like smoking and obesity, and changing reproductive patterns due to economic development and urbanization. Lung cancer is the most frequently diagnosed and deadliest, followed closely by breast cancer. Breast cancer is the most common cancer in women worldwide, with over 2.3 million new cases in 2020, significantly contributing to cancer-related mortality [4]. Breast cancer is now recognized as a diverse group of diseases with distinct clinical behaviors, molecular components, risk factors, prognostic markers, and responses to treatment. Molecular classification relies on markers like estrogen receptors (ER), progesterone receptors (PR) and HER2 and Ki67 proliferation rate [5]. As a result, breast cancer has been categorized into five subtypes: luminal A, luminal B, triple-negative, and two HER2-positive types. Breast cancers with HER2 overexpression constitute 15–25% of cases, being aggressive and challenging to treat. Trastuzumab was approved by the FDA in 1998 and demonstrated a 37% relative improvement in overall survival with an increase of about 9% in the probability of 10-year OS when combined with chemotherapy [6]. Mutations in PIK3R1, activating PI3K/Akt/mTOR, drive trastuzumab resistance, especially in HER2-overexpressing breast cancer [7]. Approximately 70% of HER2-positive breast cancer patients develop resistance to trastuzumab within a year of treatment initiation, despite initial responsiveness [8].

In this context, the present study aimed to comprehensively investigate the molecular mechanisms underlying trastuzumab resistance using a two-fold approach. Initially, well-established breast cancer cell lines, namely SKBR3 and BT474 [9], were used to mimic the in vitro conditions representative of HER2-positive breast cancer. Optimal trastuzumab treatment doses were determined for each cell line based on their respective proliferation rates in response to a wide range of medication dosages. Subsequently, both cell lines were continuously cultivated with a steady dosage of trastuzumab over several months, leading to the development of trastuzumab-resistant cell lines. To capture the dynamic changes occurring during the acquisition of resistance, six distinct time points were selected for further analysis, encompassing the progression from the untreated cell line at the outset to the fully resistant cell line at the termination of the experiment. Concurrently, a second stage of the study involved the extraction of nucleic acids from the aforementioned cell lines for subsequent high-throughput microarray experiments targeting gene and microRNA expressions. The resulting expression data were subjected to comprehensive bioinformatics analyses to unravel the intricate molecular mechanisms underpinning the development of trastuzumab resistance. Through this approach, we focused on the 25 genes and microRNAs with the most statistically significant changes in expression levels across time, implicated in the development of trastuzumab resistance, and demonstrated crucial roles in protein–protein interactions. Notably, among the identified genes and microRNAs, *BIRC5*, *E2F1*, *TFRC*, and *USP1* emerged as the top candidates influencing trastuzumab resistance, while miR-574-3p, miR-4530, miR-8485, and miR-197-3p were identified as the key microRNAs regulating this process. In-depth investigations using the prediction by the miRDB database [10] highlighted the miR-4701-5p targeting *TFRC* and the targeting of *E2F1* and *USP1* by hsa-miR-8485. Moreover, the expression revealed substantial upregulation of miR-4701-5p compared to *TFRC* in the SKBR3 cell line, providing crucial insights into the intricate regulatory mechanisms governing trastuzumab resistance. Overall, the findings of this study shed light on the critical molecular players and pathways driving trastuzumab resistance in HER2-positive breast cancer, offering valuable insights into potential therapeutic targets and strategies for overcoming treatment challenges associated with this aggressive subtype of breast cancer.

2. Materials and Methods

2.1. Breast Cancer Cell Lines

SKBR3 and BT474 human breast cancer cell lines were chosen for this study due to their HER2 receptor overexpression, trastuzumab sensitivity [11–13], and potential for trastuzumab resistance [14,15]. These certified, mycoplasma-tested cell lines, obtained from American type cell collection, were used as controls and parental lines for drug-resistant cell line development. Despite sharing HER2 overexpression, their genetic backgrounds and origins differ.

2.2. Cell Culture Conditions

Cells were cultured in DMEM/F12 with 10% FBS and antibiotics to prevent contamination. Adherent cell cultures were maintained in T75 flasks at 37 °C with 5% CO_2. Parental cell lines were cryo-preserved at 80–90% confluence with 10% DMSO/FBS. Fresh medium or trypsin-EDTA was used to maintain cell lines as needed.

2.3. Drug Resistance Development Conditions and Monitoring

A long-term, consistent dose of trastuzumab (Herceptin) was used to develop drug-resistant cell lines with a strong response. Initial experiments used a proliferation assay to test various drug doses (ranging from 0.05 µg/mL to 500 µg/mL) on two cell lines. As indicated in Figure 1, increasing the dose to more than 5 µg/mL and 10 µg/mL did not substantially decrease the proliferation rate. Consequently, doses of 5 µg/mL and 10 µg/mL were selected based on the aforementioned results and literature [16]. Control cells were cultured similarly without drug exposure. Regular proliferation assays were conducted, and cell samples were preserved for analysis. Over time, cells became more resistant, with SKBR3 cells starting to be more sensitive than BT474. Six key time points were chosen for further investigation, including a control.

2.4. RNA Extraction

Total RNA was extracted from frozen cell pellets using the MirVana TM Isolation Kit as per the manufacturer's instructions, aiming to study mRNA and microRNA changes in the same material, as recommended by Agilent for microarray experiments.

2.5. Gene Expression Microarray Experiments

Gene expression microarray experiments utilized Agilent's SurePrint G3 Human GE 8x60K v2 Microarrays, encompassing over 50,000 biological features, including long intergenic non-coding RNAs (linc RNAs). Labeled cRNA was generated from 200 ng of input, purified with the RNeasy Mini Kit, and quantified. Microarray slides were washed, scanned using an Agilent Scanner, and analyzed with Feature Extraction software.

2.6. MicroRNA Microarrays Experiments

MicroRNA expression experiments utilized Sure Print G3 Unrestricted miRNA 8x60K microarrays by Agilent Technologies, 5301 Stevens Creek Boulevard Santa Clara, CA, USA, designed based on the miRBase-microRNA database. One Color approach was employed with two replicates for each cell line and time point. RNA labeling and hybridization followed standard protocols, and microarray data were analyzed with Agilent Scanner and Feature Extraction software, all meeting quality control standards.

2.7. Gene and microRNA Expression Microarrays Data Analysis

Gene and microRNA expression data from BT474 and SKBR3 cell lines were separately analyzed. A total of 34,756 genes and 2549 microRNAs were examined, with some genes and microRNAs having multiple measurements. In particular, for each of the six time points, two separate replicates were placed on two (out of three) different slides in a balanced design to allow within-slide pairwise comparisons between the time points. Additionally, for some genes and microRNAs, there were multiple probes. Measurements resulting from all the replicates and probes were analyzed for each gene and microRNA. In particular, a linear model with slide and time-point factor (and a probe effect for genes with multiple probes) was used for genes. The model included slide, time point, and probe factors for microRNAs. The slide effect accounted for slide-to-slide variations and served as a normalization factor. ANOVA was used for genes with multiple measurements, and LIMMA [17] was employed for genes with non-replicated probes to enhance error variance estimation. In the analysis, the null hypothesis of no mean expression difference among all time points was tested first for both genes and microRNAs. The Benjamini–Hochberg procedure [18] was used to correct for multiple testing (FDR = 0.05). Upon rejection of the null hypothesis, Tukey's multiple-testing procedure α = 0.05 was applied to conduct pairwise comparisons between time points.

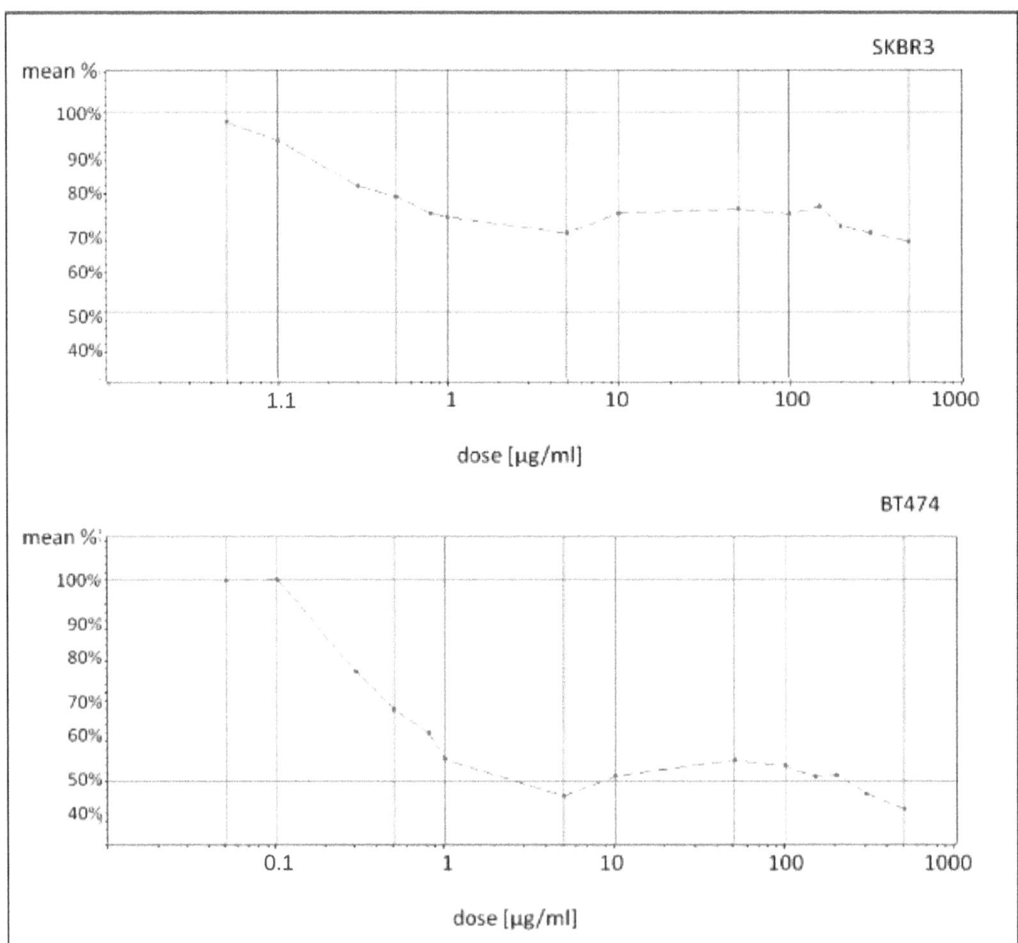

Figure 1. Plots representing the relationship between drug dose ("dose [µg/mL]"axis on the logorithmic scale) and proliferation intensity, presented as a percentage of the proliferation rate of drug-treated cells compared to the control untreated cells, based on proliferation assay measurements ("mean %" axis). Proliferation rates were measured based on six biological replicates and two technical replicates. The experiment was carried out in three independent repetitions.

2.8. Association of miRNA Target Genes

Based on the list of identified top 25 most significant genes and top 25 most significant miRNAs from both cell lines, we derived the miRNA target gene pairs using the miRDB database [10]. The miRDB database uses a Target score for target prediction, where the score is assigned by the miRNA target prediction program based on support vector machines (SVMs) and high-throughput training datasets. We then examined the relationship between miRNAs and their target genes by considering the distribution of log2FC scores for each pair.

2.9. Bioinformatics Analysis of Statistically Processed Microarray Data

Two initial gene datasets were generated by analyzing microarray data from BT474 and SKBR3 cell lines. Genes with *p*-values below 0.05 were chosen, resulting in 8874 genes for BT747 and 13,892 genes for SKBR3. A subsequent dataset was created by selecting 5674 genes that exhibited altered expression in both cell lines, meeting the same *p*-value threshold. To identify enriched Gene Ontology (GO) terms, the topGO package in R [19] was employed. It used a combination of "elim" and "weight" algorithms [20], considering the hierarchical structure of the GO database. The Fisher exact test assigned *p*-values for each GO term enrichment. Similarly, KEGG pathway enrichment analysis was conducted to identify pathways affected by trastuzumab treatment in BT747 and SKBR3 cell lines. Due to the complexity of KEGG pathways, the Fisher exact test was applied. Only genes analyzed with the SurePrint G3 Human GE 8x60K v2 Microarray were included. Additionally, protein–protein interaction analysis was performed by mapping gene names to Ensemble and Uniprot IDs, creating proteome networks based on various databases. The focus was on explaining the connections between the top 25 significant genes in each cell line via the proteomic network, seeking the shortest path within 17 hops. Eight sub-network images were generated for essential genes in both cell lines using the top four significant genes as reference points.

3. Results

In this section, we first describe the cancer cell response to trastuzumab, observing, notably, a temporal decline in the response. Six time points were selected within the studied time period for further microarray experiments. The microarray experiment results fulfill the acceptable quality control metrics for all the samples across different time points for RNA and miRNA. Next, we discuss the results of the microarray expression data, revealing statistically significant genes and miRNA, and their role in the development of trastuzumab resistance.

3.1. Identification of the Molecular Changes That Occur during the Emergence of Trastuzumab Resistance

3.1.1. Cell Culture: Drug Resistance Development Conditions and Monitoring

Cell cultures were maintained in Dulbecco's Modified Eagle Medium (DMEM) supplemented with F12 nutrient mixture, 10% fetal bovine serum (FBS), and antibiotics to prevent bacterial contamination. The cells were incubated in T75 treated flasks at 37 °C with 5% CO_2. Before any passage, the parental cell lines were cultured without drug treatment until they reached 80–90% confluence and then cryo-preserved in 10% DMSO/FBS in liquid nitrogen. Additional vials were frozen as a backup at an early passage number. A prolonged and consistent dosage of trastuzumab (Herceptin, Roche , Engelhorngasse 3, Vienna, Austria) was administered to establish drug-resistant cell lines and determine an adequate response. Initial experiments evaluated the proliferation intensity following exposure to a wide range of drug doses (0.05 μ/mL–500 μg/mL) in both cell lines (Figure 1).

The cell response to Herceptin was assessed via a proliferation assay using the BioTek Cytation TM 3 imaging reader. Based on the preliminary findings and existing literature, two treatment doses (5 μg/mL and 10 μg/mL) were chosen. The control cell lines were cultured alongside the resistant cells, following the same procedures except for drug exposure. Regular proliferation assays were performed monthly to monitor the development of resistance by observing changes in cell response to the drug. At each time point, cell samples were preserved in liquid nitrogen or frozen at $-80\,^{\circ}$C for future analysis. At the commencement of the study, both the SKBR3 and BT474 cell lines showed differing levels of sensitivity to the drug, with the BT474 primary tumor cells displaying greater sensitivity compared to the metastatic SKBR3 cells, as evidenced by a 72% and 50% proliferation rate, respectively. Despite the difference, this indicates a successful establishment of sensitivity to trastuzumab. By the study's conclusion, both cell lines exhibited increased proliferation intensity, with the SKBR3 cells reaching 92% and the BT474 cells reaching 89% as compared to untreated controls when exposed to a 100 μg/mL drug concentration. Furthermore, there was a discernible trend indicating a decrease in the cells' responsiveness to the drug over time, suggestive of the development of resistance during the duration of exposure (Figure 2). Six specific time points, including the control, were selected for subsequent microarray experiments.

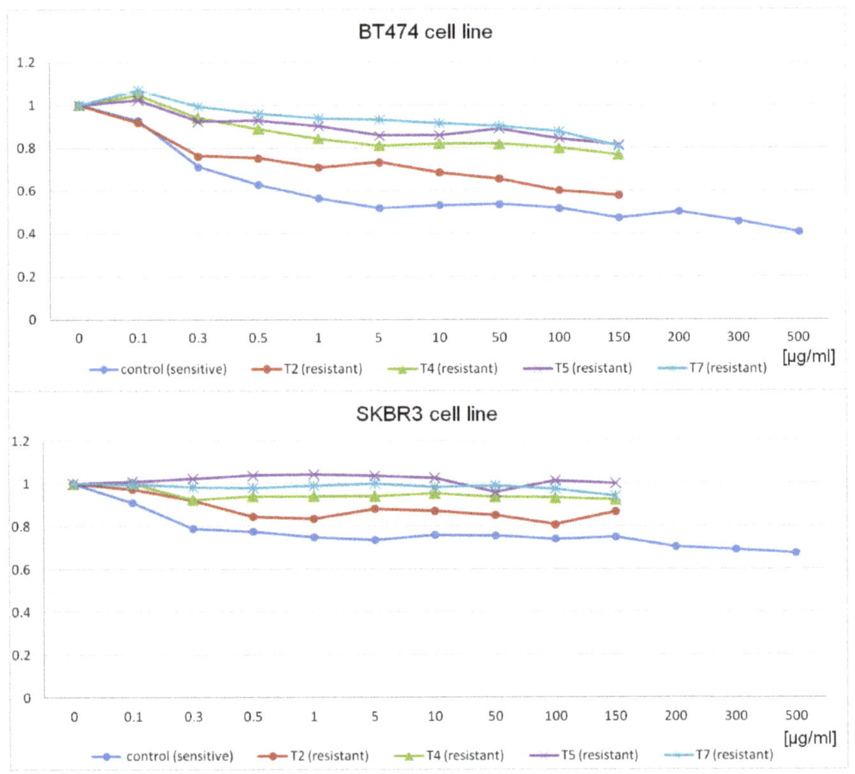

Figure 2. Diagrams presenting changes in proliferation rates at different time points of the experiment during the whole time of cell exposure to trastuzumab. The charts represent the relationship between drug dose ("[μg/mL]"axis) and proliferation intensity (y-axis) presented as a decimal fraction of the proliferation of drug-treated cells compared to the control untreated cells and are based upon proliferation assay measurements, with each curve corresponding to a particular time point. "Control" means cells purchased from ATCC and confirmed to be trastuzumab sensitive and able to develop resistance. Control cell lines were not treated with the drug at any time in the experiment and reaction for the drug was stable. T2, T4, T5, and T7 were treated at for 2, 4, 5, and 7 months, respectively. The time points were selected based on the significant increase in proliferation rates (decreased reaction for the drug) to the previous month, reflecting a gradual development of resistance to trastuzumab.

3.1.2. RNA Extraction and Quality Evaluation

For each designated time point, total RNA was isolated from frozen cell pellets using the MirVana TM Isolation Kit (Ambion, Life Technologies, Carlsbad, CA, USA) as directed in the online protocol https://assets.fishersci.com/TFS-Assets/LSG/manuals/MAN00111 31_A27828_magmax_mirvanatotalrna_manualextraction_ug.pdf, accessed on 15 March 2024. This approach aimed to enable the examination of both mRNA and microRNA in the same material, in accordance with Agilent's recommendations for microarray experimentation. Each RNA sample underwent rigorous assessment of its concentration and quality using the NanoDrop 2000 spectrophotometer (Thermo Scientific, Waltham, MA, USA) to ensure acceptable purity (260/280 nm: 1.9–2.1; 260/230 nm: 1.8–2.2). Additionally, Agilent Bioanalyzer was employed to evaluate concentration and integrity, ensuring a minimum RNA Integrity Number (RIN) of 7 for all samples (Figure S1). Repeated extractions and assessments were conducted for samples that failed the quality checks.

3.1.3. Gene Expression Microarrays Experiment

The gene expression microarray analysis was conducted using the Sure Print G3 Human GE 8x60K v2 Microarrays from Agilent Technologies, designed to detect a wide array of biological features, including long intergenic non-coding RNAs (linc RNAs). Labeled cRNA targets were prepared from 200 ng of input using the One Color Low Input Quick Amp Labeling Kit and One-Color Spike-In Kit from Agilent Technologies, followed by purification with the RNeasy Mini Kit from Qiagen. The quality of the cyanine3CTP-labeled cRNA was assessed to ensure samples had an activity exceeding 6 pmol/μg, meeting the manufacturer's standard for high-quality samples. Table 1 presents the results of labeled cRNA quality control. The hybridization procedure was performed overnight, with two replicates for each cell line at every time point. Microarray slides were washed using GE Wash Buffers and then scanned with an Agilent Scanner version C (G2505C). Feature Extraction software was utilized for image analysis, yielding raw data files, quality control PDFs for each array, and a comprehensive summary protocol. All arrays exhibited satisfactory quality control metrics, enabling a robust comparative analysis across different time points without the need for dye-swap experiments.

Table 1. The results of labeled cRNA quality control. R—repeated extraction; the number of "R" means the number of extraction repetitions.

Sample ID	RNA ng/μL	Yield μg	260/280	DyeConc. ng/μL	Activity pmol/μg
BT474 controlRR	351.3	10.539	2.19	7.5	21.34927412
SKBR3 controlRRR	426.7	12.801	2.18	8.7	20.38903211
SKBR3 T2	221.4	6.642	2.23	2.6	11.74345077
BT474 T2	269.2	8.076	2.25	3.5	13.00148588
SKBR3T5C	241.4	7.242	2.21	3.2	13.25600663
BT474 T5	378.6	11.358	2.22	6.8	17.96090861
SKBR3 T7	333.3	9.999	2.23	4.9	14.70147015
BT474 T7R	354.5	10.635	2.21	5.8	16.36107193
SKBR3T3R	374.4	11.232	2.2	5.6	14.95726496
BT474 T3RR	387.3	11.619	2.19	6.2	16.00826233
SKBR3T4R	311.6	9.348	2.2	5.2	16.68806162
BT474 T4R	443.5	13.305	2.2	8.6	19.39120631

3.1.4. MicroRNA Expression Microarray Experiment

The microRNA expression microarray experiment was conducted by utilizing the Sure Print G3 Unrestricted miRNA 8x60K microarrays (Agilent Technologies), encompassing probes for almost all known human microRNAs based on the miRBase database. Following the One Color approach, two replicates were performed for each cell line and time point.

Starting from 100 ng of input, dephosphorylated and labeled RNA was prepared using the miRNA Complete Labeling and Hybridization Kit along with the microRNA Spike In Kit, followed by desalt procedures with the Micro Bio-Spin P6 Gel Column. The hybridization mixture, including Hyb Spike In, underwent overnight hybridization. Subsequently, microarray slides were washed and scanned in an Agilent Scanner version C (G2505C). Feature Extraction software was employed for image analysis (GRID: 070156DF20141006), confirming satisfactory quality control rates across all arrays.

3.2. Analysis of the Molecular Changes That Occur during the Emergence of Trastuzumab Resistance

3.2.1. Gene Expression Analysis

It is essential to note that only about one-third (5675/17,091 = 33%) of the genes with statistically significant expression-level changes in SKBR3 or BT474 were common to both cell lines. About 19% (3199/17,091) of the genes were unique to the SKBR3 cell line, while about 48% (8217/17,091) were specific to BT474. Furthermore, a substantial fraction (5675/8874 = 64%) of the genes important for BT474's resistance development were also found to have statistically significantly altered expression levels in SKBR3, but the reverse was less true, with only 41% (5675/13,892) of genes related to SKBR3's resistance development identified in BT474 (Figure 3).

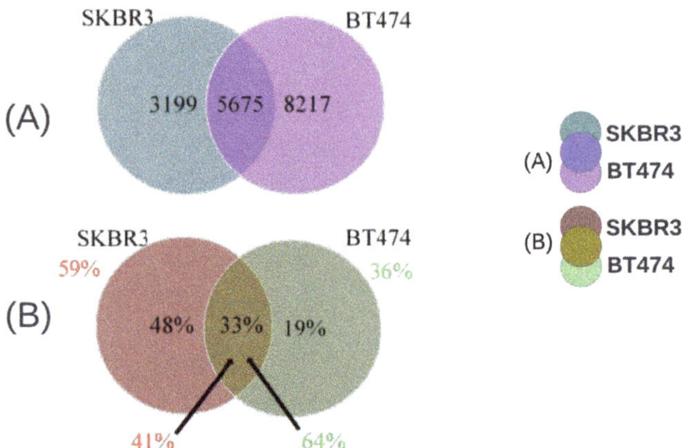

Figure 3. (**A**) Venn diagram shows the number of genes uniquely significant in SKBR3 and BT474 cell lines for trastuzumab resistance and the number of genes common to both. (**B**) Venn diagram to display the percentage of significant genes unique to SKBR3 and BT474 cell lines, the percentage of common significant genes, the percentage of genes significant in one cell line but important in the other (with an arrow), and the percentage of unique genes for each cell line among all significant genes in that cell line.

This disparity could be attributed to a significant imbalance in the number of significant transcripts found in both cell lines, with SKBR3 having more relevant genes identified as compared to BT474. In the second part of the statistical analysis, once the global null hypothesis of no change in expression levels across time was rejected for a gene, pairwise comparisons between all time points were conducted for that gene separately for each cell line (Table S1). The comparisons between T2 vs. T0 and T3 vs. T0 yielded the most statistically significant results, indicating that the most important global gene expression changes occurred at the beginning of trastuzumab treatment. This was followed by a temporary decrease in activity at T3, with more balanced gene expression modifications thereafter. Similar trends were observed in SKBR3 (Table S2), with the largest changes in gene expression occurring at the start of Herceptin exposure and a secondary increase in activity at T4. Pairwise comparisons are presented in (Figure 4).

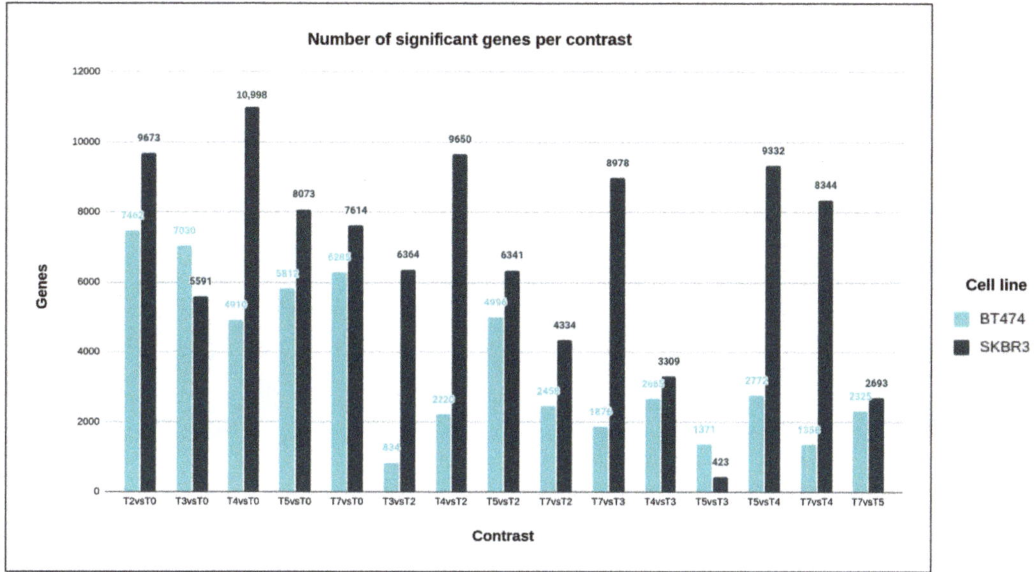

Figure 4. Bar chart presenting the number of genes (y-axis) with statistically significant differences between different pairs of time points (x-axis) for the two cell lines.

In both cell lines, the majority of genes at each time point of drug resistance development were downregulated as compared to the parental cell line (Table 2), suggesting a potential role for tumor-suppressive genes in the process.

Table 2. Number of genes for which statistically significant changes of expression levels across time were obtained and which were either up-regulated or down-regulated.

		BT474			SKBR3			
Time-Point	Up	% of Signif	Down	% of Signif	Up	% of Signif	Down	% of Signif
T2 vs. T0	174	2	7288	98	1328	14	8345	86
T3 vs. T2	614	74	220	26	4670	73	1694	27
T4 vs. T3	1790	95	86	5	648	7	8330	93
T5 vs. T4	892	32	1881	68	7788	83	1544	17
T7 vs. T5	1389	60	936	40	1645	61	1048	39

However, a different pattern emerged when considering sequential changes throughout the process. In BT474, 98% of genes were initially downregulated at the beginning of drug exposure (T2 vs. T0) (Table 3), followed by two instances of global gene overexpression (74% at T3 vs. T2 and 95% at T4 vs. T3) before more balanced changes toward the end of the process (60% vs. 40% in both T5 vs. T4 and T7 vs. T5). In SKBR3, while the global gene expression changes appeared more variable, they shared some similarities with BT474. The majority of genes (86%) were initially downregulated (T2 vs. T0) (Table 3), followed by a reverse situation (73% overexpressed at T3 vs. T2) and balanced expression patterns at the end (40% vs. 60% in T7 vs. T5) (Figure S2, Table 3).

Table 3. Number of genes for which statistically significant changes of expression levels across time were obtained and which were either up-regulated or down-regulated at a specific time point when compared with the control (parental) cell line.

Time-Point		BT474					SKBR3		
	Up	% of Signif	Down	% of Signif	Up	% of Signif	Down	% of Signif
T2 vs. T0	174	2	7288	98	1328	14	8345	86
T3 vs. T0	267	4	6763	96	500	9	5091	91
T4 vs. T0	342	7	4568	93	402	4	10,596	96
T5 vs. T0	562	10	5250	90	550	7	7523	93
T7 vs. T0	502	8	5783	92	341	4	7273	96

These results suggest that trastuzumab had an inhibitory effect at the initial stages of the experiments, but as time progressed, cells seemed to adapt to the drug treatment, resulting in increasing global gene expression and dynamic changes throughout the process. A more detailed investigation focused on the most statistically important genes in the subsequent part of the gene expression profile analysis. The top 25 significant genes (Table 4, remaining significant genes are listed in Table S3) were identified for both BT474 and SKBR3 cell lines, and it was found that 80% of the top 25 genes in BT474 were among the top 200 in SKBR3. Conversely, as much as 92% of the genes were common. These findings underscore the consistency of the results and the substantial overlap between both cell lines, suggesting shared patterns in the development of drug resistance.

Table 4. List of the 25 genes with the most statistically significant changes during the whole period of trastuzumab resistance development in both cell lines: BT474 and SKBR3.

Top 25 Most Significant in BT474 Cell Line			
Gene	**Adj *p*-Val**	**Gene**	**Adj *p*-Val**
KLK11	7.6×10^{103}	ZNF195	2.0×10^{-51}
IGF2BP1	3.4×10^{-72}	C6orf48	2.2×10^{-51}
GSTM3	1.6×10^{-65}	ADM	2.7×10^{-51}
IGFBP3	3.5×10^{-64}	RND3	2.7×10^{-51}
TNFRSF11B	1.1×10^{-63}	CBPB	1.9×10^{-50}
RASD1	3.8×10^{-61}	PLA2G16	1.5×10^{-49}
CHAF1B	3.8×10^{-58}	2F1	1.5×10^{-49}
TRIT1	1.7×10^{-56}	KIAA0586	3.4×10^{-49}
PMP22	8.1×10^{-55}	ADK	3.7×10^{-49}
PSN1	1.4×10^{-53}	TFRC	4.0×10^{-49}
PLAT	2.5×10^{-53}	CNP	8.2×10^{-48}
BIRC5	2.7×10^{-52}	RO1L	3.1×10^{-47}
USP1	2.0×10^{-51}		

Top 25 Most Significant in SKBR3 Cell Line			
Gene	**Adj *p*-Val**	**Gene**	**Adj *p*-Val**
CCL2	7.7×10^{108}	HN1	6.6×10^{-90}
F8A1	2.7×10^{104}	C20orf24	8.3×10^{-90}
E2F1	1.0×10^{102}	EXO1	1.1×10^{-88}
GINS2	6.4×10^{100}	KIAA0101	4.3×10^{-88}
YWHAH	6.4×10^{100}	TFRC	2.6×10^{-87}

Table 4. *Cont.*

Top 25 Most Significant in SKRBR3 Cell Line			
Gene	**Adj *p*-Val**	**Gene**	**Adj *p*-Val**
PRMT6	7.7×10^{100}	*NACC2*	1.5×10^{-86}
DTL	7.8×10^{-99}	*DDIT4*	2.5×10^{-86}
GSTP1	4.4×10^{-98}	*CNOT10*	2.8×10^{-86}
BIRC5	6.5×10^{-98}	*FAHD1*	7.3×10^{-84}
DOLK	6.5×10^{-98}	*USP1*	7.3×10^{-84}
SAP30	7.5×10^{-92}	*MGST2*	1.7×10^{-83}
RPP40	1.3×10^{-91}	*BRCA1*	2.5×10^{-83}
RTN4IP1	3.9×10^{-90}		

Research on the association between specific genes and the development of trastuzumab resistance in the BT474 cell line has yielded crucial insights. Among the 25 most significant genes identified (Table S4), CHAF1B stands out, with its steadily increasing expression during trastuzumab exposure, indicating a role in DNA replication and cell division, possibly as an effector. *E2F1*, a key transcription factor controlling cell cycle and apoptosis, appears linked to the molecular mechanism driving drug resistance. TRIT1, a mitochondrial tRNA modifier and potential tumor suppressor, shows expression changes suggesting complex regulation. ADM's role, despite its distinct expression pattern, remains unclear due to its multifunctional nature. Additional candidates of interest include *BIRC5* and *USP1*, both prominent in both cell lines and potentially involved in apoptosis prevention. IGFBP3's role in extending IGF's half-life, a factor in HER2 signaling pathway crosslinking with trastuzumab resistance, is noteworthy. IGF2BP1's mRNA binding and translational regulation may also play a role. *GSTM3* and *RASD1*, while having complex expression patterns, are associated with drug resistance. Genes like *KLK11, CENPE, TFRC*, and *KIAA0586*, though not fully understood, could be indirectly involved. Some genes with diverse expression profiles (*CEBPB, SNHG32, TNFRSF11B*) or unclear functions (*PMP22, PSEN1, PLAT, ZNF195, RND3,* and *ERO1L*) require further investigation into the context of drug resistance development. Table S5 presents the 25 genes with the most statistically significant differences associated with trastuzumab resistance in the SKBR3 cell line, each accompanied by a brief description of their molecular functions. However, it is worth noting that only eight of these genes (*CCL2, F8A1, PRMT6, DTL, RTN4IP1, RAB5IF, EXO1, FAHD1*) were among the top 25 genes with the most statistically significant differences, while the remaining 15 are listed in Table S17 (*PFDN6, LOXL2, MRPS23, HMOX1, WASHC5, PPIP5K2, AHSA1, DNAJA3, RAD50, SRRT, SUZ12, PSMD6, PCNA, TSFM, FAM25*). Several genes, such as *BIRC5, E2F1, TFCR, GSTP1, YWHAH, DTL, DOLK, NACC2, DDIT, BRACA1,* and *DNAJA3*, seem to play crucial roles in trastuzumab resistance development. *BIRC5* and *E2F1* were also significant in the BT474 cell line, suggesting a common mechanism. Furthermore, some genes are linked to the p53 pathway, a well-known tumor suppressor. For example, *NACC2* represses transcription and inhibits *MDM2*, stabilizing *TP53*. *DDIT4* is involved in p53-mediated apoptosis regulation and may connect to HER2 signaling through mTOR. *DNAJA3* interacts with both p53 and HER2 and stimulates Hsp70 chaperone activity. Other genes involved in DNA repair and damage response, like *BRCA2, PCNA,* and *RAD50*, are also implicated in trastuzumab resistance. While some genes have unclear associations, such as *CCL2* and *MGST2* related to immune response, they require further investigation. Additionally, Supplementary Materials (Supplementary Tables S3–S6) highlight the most statistically significant changes in gene expression during trastuzumab resistance development, including long non-coding RNAs and proteins with limited information, offering avenues for future research into the molecular mechanisms of resistance.

3.2.2. MicroRNA Expression Analysis

The primary objective of the microRNA expression analysis was to identify microRNAs that played a statistically significant role at all stages of trastuzumab resistance development. Surprisingly, for nearly all tested microRNAs, the differences in expression levels over time were statistically significant, with 99.6% in BT474 and 99.3% in SKBR3 cell lines found to yield significant results at the significance level of 0.05 (adjusted for multiple testing). In contrast, gene expression analysis showed a more limited number of significant transcripts, only 25% in BT474 and about 40% in SKBR3 (Table 5). This disparity suggests that microRNA molecules play a substantial role in trastuzumab resistance development, highlighting their involvement in complex regulatory networks.

Table 5. Expression of the genes and microRNAs undergoing statistically significant expression changes during the development of trastuzumab resistance in both quantitative and percentage terms.

Genes	BT474	SKBR3
all	34,756	34,756
significant p0.05	8874	13,892
significant p0.05 (%)	**25.50%**	**40.00%**
significant p0.01	4018	7529
significant p0.01 (%)	**11.60%**	**21.70%**
MicroRNAs		
all	2549	2549
significant p0.05	2540	2541
significant p0.05 (%)	**99.60%**	**99.70%**
significant p0.01	2532	2537
significant p0.01 (%)	**99.30%**	**99.50%**

Statistical analysis was used to examine microRNA expression changes in two cell lines, BT474 and SKBR3, during the development of drug resistance. Pairwise comparisons between different time points were conducted for each microRNA with a statistically significant test of the global null hypothesis of no change in expression across time (Figure 5). In the BT474 cell line, the T7 vs. T5 comparison showed statistically significant results for 88.2% of microRNAs, and 12.7% for T4 vs. T2. In the SKBR3 cell line, the T5 vs. T4, T4 vs. T0, and T7 vs. T5 comparisons yielded the highest significant results (96.8%, 83.5%, and 83.4%, respectively), while T7 vs. T2 yielded the lowest fraction of 11.8%. Both cell lines displayed a similar pattern of microRNA expression changes, but SKBR3 cells seemed to respond faster to changing conditions than BT474 cells (Table S6).

In the next phase of our microRNA expression profile analysis, we delved into the detailed study of the 25 microRNAs with the most statistically significant changes (Table 6, remaining significant miRNAs are in Table S8), focusing on their role in trastuzumab resistance within BT474 and SKBR3 cell lines. Surprisingly, 24 of the top 25 microRNAs in BT474 were also among the top 200 in SKBR3, suggesting substantial overlap in their resistance patterns. However, for the reverse scenario, we found less than 48% common microRNAs, indicating a stronger representation of BT474 results in SKBR3. Gene expression analysis further revealed that SKBR3 had a more complex network of genes and pathways involved in trastuzumab resistance, possibly affecting microRNA engagement and leading to this disparity in results.

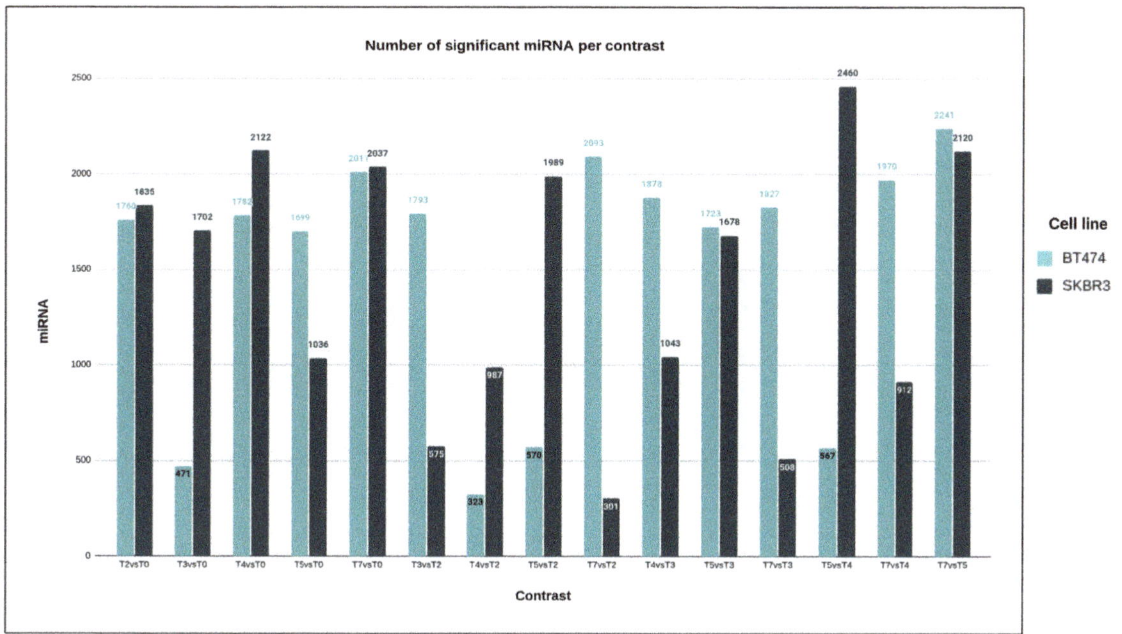

Figure 5. Bar chart presenting the number of microRNAs (y-axis) with statistically significant differences between different pairs of time points (x-axis) for the two cell lines.

Table 6. Top 25 most statistically significantly expressed microRNAs during the whole process of trastuzumab resistance development separately for both biological models: BT474 and SKBR3.

Top 25 Most Significant in BT474 Cell Line			
microRNA	**Adj p-Val**	**microRNA**	**Adj p-Val**
hsa-miR-574-3p	2.2×10^{213}	hsa-miR-6775-3p	2.4×10^{177}
hsa-miR-1207-5p	4.5×10^{204}	hsa-miR-466	2.2×10^{174}
hsa-miR-8485	4.3×10^{197}	hsa-miR-4649-3p	2.9×10^{173}
hsa-miR-6886-3p	6.5×10^{196}	hsa-miR-1281	2.4×10^{172}
hsa-miR-6088	3.4×10^{190}	hsa-miR-6743-3p	5.9×10^{171}
hsa-miR-4254	5.6×10^{190}	hsa-miR- 4436b-5p	5.5×10^{170}
hsa-miR-4701-5p	2.2×10^{185}	hsa-miR-4530	2.0×10^{166}
hsa-miR-3151-3p	4.5×10^{184}	hsa-miR-3162-3p	5.3×10^{166}
hsa-miR-197-3p	1.7×10^{182}	hsa-miR-483-3p	1.7×10^{164}
hsa-miR-1825	4.5×10^{180}	hsa-miR-4725-5p	1.9×10^{164}
hsa-miR-6834-3p	4.5×10^{180}	hsa-miR-663a	1.1×10^{163}
hsa-miR-4281	5.7×10^{180}	hsa-miR-6794-3p	2.9×10^{163}
hsa-miR-3591-3p	1.3×10^{177}		

Table 6. *Cont.*

Top 25 Most Significant in SKBR3 Cell Line			
microRNA	**Adj *p*-Val**	**microRNA**	**Adj *p*-Val**
hsa-miR-7977	1.1×10^{235}	hsa-miR-5100	4.5×10^{163}
hsa-miR-6826-5p	2.2×10^{214}	hsa-miR-4530	7.4×10^{163}
hsa-miR-7975	6.5×10^{203}	hsa-miR-6869-5p	2.1×10^{161}
hsa-miR-6165	1.7×10^{190}	hsa-miR-6085	3.0×10^{161}
hsa-miR-5739	4.6×10^{185}	hsa-miR-15b-5p	1.6×10^{156}
hsa-miR-574-3p	4.6×10^{185}	hsa-miR-20a-5p	1.6×10^{156}
hsa-let-7a-5p	3.2×10^{179}	hsa-miR-574-5p	5.1×10^{155}
hsa-miR-6749-5p	5.6×10^{176}	hsa-miR-8485	1.7×10^{153}
hsa-miR-3162-5p	8.4×10^{173}	hsa-miR-197-3p	7.0×10^{152}
hsa-miR-1202	1.5×10^{171}	hsa-miR-6090	8.0×10^{149}
hsa-miR-4284	2.6×10^{168}	hsa-miR-29c-3p	1.3×10^{148}
hsa-let-7e-5p	2.1×10^{167}	hsa-miR-4455	4.2×10^{148}
hsa-miR-24-3p	1.5×10^{165}		

Upon further analysis, we found that the eight most crucial microRNAs identified in the BT474 cell line were also among the top 50 (Table S8) microRNAs in the SKBR3 cell line. These microRNAs include hsa-miR-6886-3p, hsa-miR-4254, hsa-miR-4701-5p, hsa-miR-3151-3p, hsa-miR-6834-3p, hsa-miR-4281, hsa-miR-4649-3p, and hsa-miR-3162-3p. Additionally, two microRNAs with highly statistically significant changes discovered in the SKBR3 cell line, hsa-miR-6826-5p, and hsa-miR-6869-5p, were found within the top 50 (Table S8) important microRNAs in the BT474 cell line. These microRNA molecules may play a role in the development of trastuzumab resistance mechanisms. Four microR-NAs, including hsa-miR-574-3p, hsa-miR-4530, hsa-miR-8485, and hsa-miR-197-3p, were identified among the 25 microRNAs with the most statistically significant results of the test of the global null hypothesis of no change in expression across time in both cell lines, suggesting their crucial role in trastuzumab resistance. Three of the four microRNAs (hsa-miR-574-3p, hsa-miR-4530, hsa-miR-8485) were among the 34 and 38 microRNAs (Table S7) in BT474 and SKBR3 cell lines, respectively, for which all pairwise timepoint comparisons were statistically significant. This underscores their potential importance in trastuzumab resistance development. Noteworthily, for hsa-miR-8485, all pairwise comparisons were significant in both cell lines.

3.2.3. Analysis of miRNA Regulation of Genes

In our research, we employed a comprehensive approach by selecting the 25 genes and miRNAs with the most statistically significant changes from their respective cell lines. Our objective was to identify potential interactions between miRNAs and target genes using the miRDB database [10]. The miRDB predicted three pairs of miRNA-gene targets, as indicated in (Table 7).

Table 7. Pair of significant genes and miRNAs where miRNA target genes were predicted by miRDB database.

miRNA.Name	Gene.Symbol	Target.Score
hsa-miR-4701-5p	*TFRC*	69
hsa-miR-8485	*E2F1*	89
hsa-miR-8485	*USP1*	51

It is important to note that the miRDB predictions are assigned scores ranging from 50 to 100, with higher scores indicating greater statistical confidence in the prediction outcomes. Intriguingly, we observed that among these miRNA-gene pairs, four of the miRNAs and genes belonged to the top 25 genes and miRNAs with the most statistically significant changes identified in both cell lines. To gain deeper insights into the regulatory dynamics at play, we conducted an analysis of the association between miRNA-gene target pairs (miR-4701-5p-*TFRC* in Figure 6, figures for the remaining pairs from (Table 7) are presented in Supplementary Figure S17), and the distribution of log2 fold change (log2FC) values at their respective time points (Figure 6A) and average of time points (Figure 6B). This allowed us to investigate the extent to which miRNAs exerted control over gene expression within the context of the specific cell line under study.

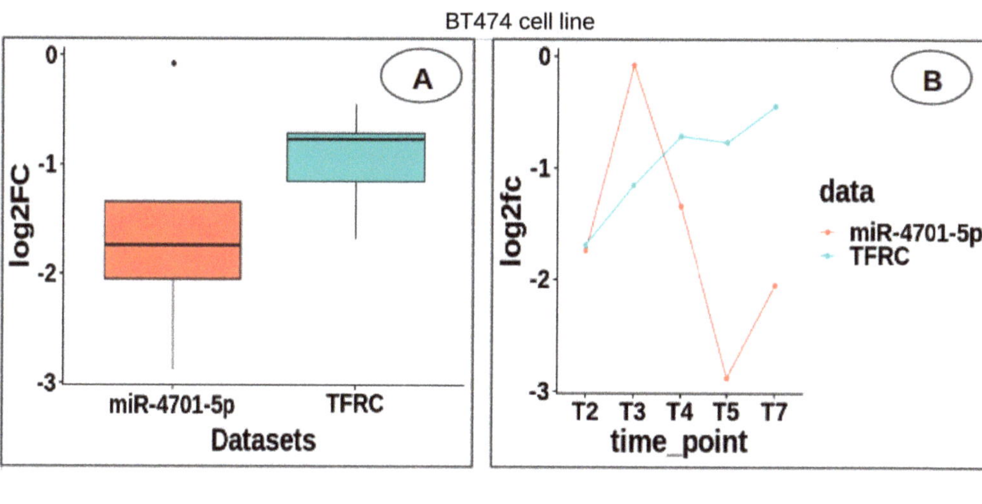

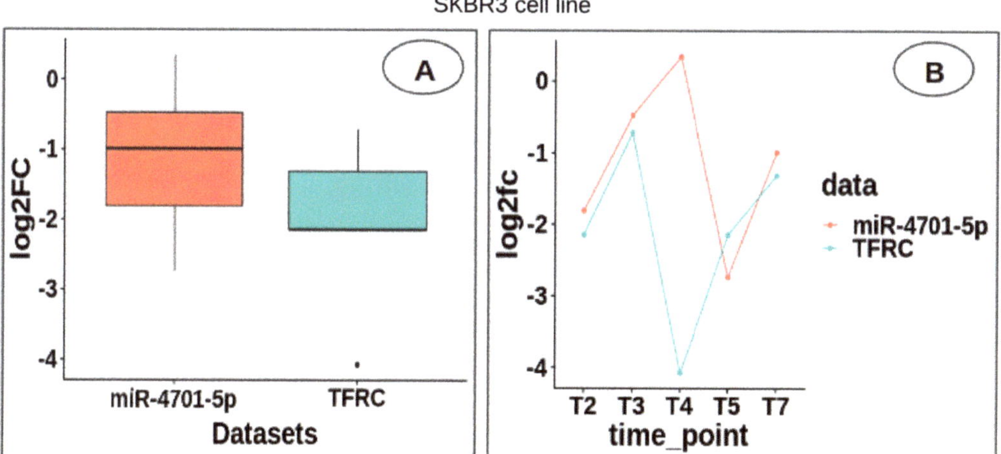

Figure 6. (**A**) Box plot and (**B**) line plot of the distribution of log2 fold change (log2FC) values of miRNA−target−gene pair, at their average and respective time points.

3.2.4. Gene Ontology Terms (GO Terms) Enrichment Analysis

This study leveraged high-throughput experiments to identify thousands of genes associated with trastuzumab resistance development in breast cancer. These genes exhibited diverse expression patterns throughout the study. The researchers conducted a Gene

Ontology (GO) enrichment analysis [21] to gain insights into the underlying biological processes. The analysis focused on molecular function (MF) and biological process (BP) GO terms [22] separately for two cell lines, SKBR3 and BT474. SKBR3 showed 193 enriched MF GO terms (Table 8) and 600 enriched BP GO terms (Table 9), while BT474 had 103 (Table 10) and 354 (Table 11), respectively. Further investigation involved in-depth scrutiny of the top 20 statistically significant and the top 20 most overrepresented MF and BP GO terms for both cell lines. Gene significance density plots were generated to visualize gene distribution by *p*-value. To ensure result consistency within each cell line, the study examined the overlap between the most significant and most enriched GO terms. A comparative analysis between SKBR3 and BT474 included checking the overlap of GO terms, verifying enriched terms, and comparing gene significance density plots. Lastly, a global analysis was performed to categorize significant GO terms by molecular functions and biological processes, shedding light on the major functional groups driving trastuzumab resistance. Figures S3 and S4 depict density plots revealing the distribution of the top 20 statistically significant MF GO terms in the BT474 cell line. Five exhibits a symmetric distribution, encompassing broad actions like DNA binding and protein homodimerization. In contrast, most GO terms display a left-sided asymmetric distribution, including specific functions like DNA replication origin binding and mRNA 5′-UTR binding. Notably, a deeper analysis reveals that 10 of these terms are common between the most significant and enriched categories, emphasizing their importance in BT474. These findings highlight the crucial roles of these molecular functions in the studied process, as evidenced by their asymmetric gene significance enrichment profiles. Figures S5 and S6 display density plots of the top 20 most significant molecular function (MF) Gene Ontology (GO) terms in the SKBR3 cell line. Most SKBR3 MF GO terms exhibit symmetric distributions, while only nine show asymmetric distributions, with NADH dehydrogenase and single-stranded DNA binding being notable. Among these terms, nucleosomal DNA binding stands out as the most influential in the SKBR3 cell line despite some data inconsistency. Comparing the most significant Molecular Function Gene Ontology (GO) terms in two different cell lines, BT474 and SKBR3, revealed limited overlap, suggesting cell line-specific mechanisms in trastuzumab resistance. Three common GO terms were found, such as single-stranded DNA binding, RNA binding, and structural constituents of ribosome. However, their distribution patterns varied. SKBR3 displayed more asymmetric plots and a higher percentage of enriched GO terms, possibly due to its larger dataset. Despite differences, a substantial proportion of shared molecular functions implies the existence of some universal mechanisms contributing to trastuzumab resistance, albeit influenced by cell line-specific factors.

Figures S7 and S8 include density plots showcasing the top 20 most significant biological process (BP) Gene Ontology (GO) terms associated with the BT474 cell line. Like the molecular function (MF) GO terms, many BP GO terms exhibit asymmetric gene significance distributions. For example, "sister chromatid cohesion", "DNA replication initiation", and "SRP-dependent cotranslational protein targeting to the membrane" display left-skewed plots. In contrast, 8 out of the top 20 BP GO terms exhibit relatively symmetric distributions, including "cell division", "cell proliferation", and "DNA repair". Interestingly, there is no direct correlation between the number of genes within a particular GO term and its gene significance distribution. However, there's a tendency for these terms to occupy higher hierarchical positions. When comparing the most significant and enriched GO terms for BT474, only 4 out of 20 BP GO terms overlap. These common terms include "DNA replication initiation", "regulation of transcription involved in G1/S transition of the mitotic cell cycle", "NLS-bearing protein import into the nucleus", and "positive regulation of pri-miRNA transcription by RNA polymerase II". Furthermore, all these terms exhibit an asymmetric profile of gene significance enrichment, suggesting their importance in trastuzumab resistance in the BT474 cell line. Moving on to the SKBR3 cell line, Figures S9 and S10 depict density plots for the top 20 most significant BP GO terms. Unlike the MF GO terms, most BP GO terms in SKBR3 show asymmetric gene significance distributions, with left-skewed plots for terms such as "mitochondrial translational elongation" and "mitotic cytokinesis". Conversely, 8 out of 20 BP GO terms display relatively symmetric gene significance distributions, including "cell division" and "DNA repair". Unlike the BT474 cell line, there is no clear correlation between gene significance distribution, the number of genes in a GO term, or its hierarchical level in SKBR3. Notably, there is no overlap between the most significant and most enriched GO terms for SKBR3, indicating the complexity of processes involved in trastuzumab resistance in this cell line. Comparing the most significant BP GO terms between SKBR3 and BT474, 7 out of 20 are shared, such as "cell division", "DNA repair", and "regulation of signal transduction by p53 class mediator". However, the gene significance density plots reveal diversity, with some terms showing asymmetric distributions in one cell line and symmetric distributions in the other. This underscores the importance of certain biological processes, like "G1/S transition of mitotic cell cycle", in drug resistance development. In summary, while commonalities exist in the main processes contributing to trastuzumab resistance between SKBR3 and BT474, specific aspects and regulatory components differ, highlighting the cell line-specific nature of these mechanisms. Nevertheless, universal biological processes are shared across both cell lines, emphasizing their significance in trastuzumab resistance.

Table 8. Top 10 most significant and enriched resp. Molecular Function Gene Ontology (GO) terms out of a total of 193 that have been identified for the SKBR3 cell line. A complete list of significant and enriched GO terms in Table S10.

MF GO Term Analysis for SKBR3 Cell Line					
Top 10 Most Significant MF GO Terms			Top 10 Most Enriched MF GO Terms		
ID	Term	Adjusted *p*-Value	ID	Term	% Signif/All
GO:0003723	RNA binding	1.0×10^{-30}	GO:0061608	nuclear import signal receptor activity	100.00
GO:0003735	structural constituent of ribosome	1.9×10^{-17}	GO:0008097	5S rRNA binding	100.00
GO:0045296	cadherin binding	2.1×10^{-13}	GO:0016884	carbon-nitrogen ligase activity, with glutamine as amido-N- donor	100.00
GO:0031625	ubiquitin protein ligase binding	5.5×10^{-13}	GO:0140142	nucleocytoplasmic carrier activity	95.00
GO:0005524	ATP binding	2.2×10^{-9}	GO:0000339	RNA cap binding	94.00
GO:0019899	enzyme binding	5.0×10^{-9}	GO:0005123	death receptor binding	94.00
GO:0031492	nucleosomal DNA binding	4.6×10^{-8}	GO:0031492	nucleosomal DNA binding	93.00
GO:0003697	singl-strandd DNA binding	8.4×10^{-8}	GO:0030515	snoRNA binding	93.00
GO:0005525	GTP binding	2.0×10^{-7}	GO:0008353	RNA polymerase II carboxy-terminal domain kinase activity	93.00
GO:0008565	protein transporter activity	2.3×10^{-7}	GO:0003688	DNA replication origin binding	93.00

Table 9. Top 10 most significant and enriched resp. BP GO terms out of a total of 600 that have been identified for the SKBR3 cell line. Top 300 list of significant and top 20 enriched GO terms in Table S12.

BP GO Term Analysis for SKBR3 Cell Line					
Top 10 Most Significant BP GO Terms			**Top 10 Most Enriched BP GO Terms**		
ID	**Term**	**Adjusted p-Value**	**ID**	**Term**	**% Signif/All**
GO:0051301	cell division	2.5×10^{-15}	GO:0000920	cell separation after cytokinesis	100
GO:0006364	rRNA processing	2.6×10^{-14}	GO:1904874	positive regulation of telomerase RNA localization to Cajal body	100
GO:0070125	mitochondrial translational elongation	2.1×10^{-12}	GO:0000054	ribosomal subunit export from nucleus	100
GO:1902036	regulation of hcmatopoietic stem cell differentiation	3.3×10^{-12}	GO:0070525	tRNA threonylcarbamoy-ladenosine metabolic process	100
GO:0070126	mitochondrial translational termination	5.2×10^{-12}	GO:0051315	attachment of mitotic spindle microtubules to kinetochore	100
GO:0043488	regulation of mRNA stability	1.0×10^{-11}	GO:0060707	trophoblast giant cell differentiation	100
GO:1901796	regulation of signal transduction by p53 class mediator	2.4×10^{-11}	GO:0090646	mitochondrial tRNA processing	100
GO:0016579	protein deubiquitination	7.7×10^{-11}	GO:0090151	establishment of protein localization to mitochondrial membrane	100
GO:0038061	NIK/NF-kappaB signaling	1.3×10^{-10}	GO:0072425	signal transduction involved in G2 DNA damage checkpoint	100
GO:0031145	anaphase-promoting complex-dependent catabolic process	2.5×10^{-10}	GO:0016024	CDP-diacylglycerol biosynthetic process	100

Table 10. Top 10 most significant and enriched resp. Molecular Function Gene Ontology (GO) terms out of a total of 103 that have been identified for the BT474 cell line. A complete list of significant and top 20 enriched GO terms in Table S9.

MF GO Term Analysis for BT474 Cell Line					
Top 10 Most Significant MF GO Terms			**Top 10 Most Enriched MF GO Terms**		
ID	**Term**	**Adjusted p-Value**	**ID**	**Term**	**% Signif/All**
GO:0005515	protin binding	1.3×10^{-12}	GO:0003688	DNA rplication origin binding	86.00
GO:0003697	single-stranded DNA binding	2.1×10^{-5}	GO:0030983	mismatched DNA binding	86.00

Table 10. *Cont.*

MF GO Term Analysis for BT474 Cell Line					
Top 10 Most Significant MF GO Terms				**Top 10 Most Enriched MF GO Terms**	
ID	**Term**	**Adjusted p-Value**	**ID**	**Term**	**% Signif/All**
GO:0003677	DNA binding	3.9×10^{-5}	GO:0097617	annealing activity	85.00
GO:0003688	DNA replication origin binding	5.3×10^{-5}	GO:1990825	sequence-specific mRNA binding	80.00
GO:0030983	mismatched DNA binding	5.3×10^{-5}	GO:0008242	omega peptidase activity	77.00
GO:0003682	chromatin binding	7.6×10^{-5}	GO:0043138	$3'$-$5'$ DNA helicase activity	75.00
GO:0003723	RNA binding	1.3×10^{-4}	GO:0031996	thioesterase binding	73.00
GO:0097617	annealing activity	1.4×10^{-4}	GO:0004303	estradiol 17-beta dehydrogenase activity activity	73.00
GO:0030331	estrogen receptor binding	9.0×10^{-4}	GO:0000400	four-way junction DNA binding	69.00
GO:0048027	mRNA $5'$-UTR binding	1.1×10^{-3}	GO:0031994	insulin-like growth factor I binding	69.00

Table 11. Top 10 most significant and enriched resp. BP GO terms out of a total of 354 that have been identified for the BT474 cell line. A complete list of significant and enriched GO terms in Table S11.

BP GO Term Analysis for BT474 Cell Line					
Top 10 Most Significant BP GO Terms				**Top 10 Most Enriched BP GO Terms**	
ID	**Term**	**Adjusted p-Value**	**ID**	**Term**	**% Signif/All**
GO:0051301	cell division	2.5×10^{-10}	GO:0090161	Golgi ribbon formation	82.00
GO:0007062	sister chromatid cohesion	1.3×10^{-9}	GO:0006744	ubiquinone biosynthetic process	79.00
GO:0000082	G1/S transition of mitotic cell cycle	2.8×10^{-8}	GO:2000651	positive regulation of sodium ion transmembrane transporter activity	77.00
GO:0008283	cell proliferation	2.1×10^{-7}	GO:0048715	negative regulation of oligodendrocyte differentiation	75.00
GO:0006270	DNA replication initiation	4.1×10^{-7}	GO:0031573	intra-S DNA damage checkpoint	73.00
GO:0006614	SRP-dependent cotranslational protein targeting to membrane	4.3×10^{-7}	GO:0000083	regulation of transcription involved in G1/S transition of mitotic cell cycle	71.00

Table 11. *Cont.*

BP GO Term Analysis for BT474 Cell Line					
Top 10 Most Significant BP GO Terms			Top 10 Most Enriched BP GO Terms		
ID	Term	Adjusted p-Value	ID	Term	% Signif/All
GO:0000184	nuclear-transcribed mRNA catabolic process, nonsense-mediated decay	1.9×10^{-6}	GO:0006607	NLS-bearing protein import into nucleus	71.00
GO:0019083	viral transcription	2.3×10^{-6}	GO:0035461	vitamin transmembrane transport	71.00
GO:0006364	rRNA processing	1.7×10^{-5}	GO:0060576	intestinal epithelial cell development	70.00
GO:0000083	regulation of transcription involved in G1/S transition of mitotic cell cycle	2.1×10^{-5}	GO:0003183	mitral valve morphogenesis	70.00

In the final phase of the Gene Ontology study, significant GO terms were analyzed separately for the BT474 and SKBR3 cell lines to uncover common molecular and cellular factors driving resistance to trastuzumab. The analysis revealed that critical molecular functions related to drug resistance included receptor binding (such as estrogen, retinoic acid, and death receptors), protein kinase activities, GTPase-related functions, and ATP/ATPase mechanisms. Additionally, significant functions were linked to transferase activities, RNA processing, DNA replication and repair, and p53 binding. Key biological processes involved cell cycle regulation, mitochondrial function, apoptosis, microRNA activity, stress response, viral infection, microtubule organization, and DNA damage repair. Several pathways, including Wnt signaling and insulin receptor pathways, were also affected during trastuzumab treatment and resistance development. For a detailed list of GO terms, refer to the Supplementary Materials.

3.2.5. KEGG Pathways Enrichment Analysis

Section 2.1 highlights the complexity of trastuzumab resistance, involving numerous genetic factors. To gain deeper insights into the underlying biological processes and functional interpretation of high-throughput data, Section 3.2.4 focuses on Gene Ontology (GO) enrichment analysis. This method helps identify significant molecular functions and biological processes driving trastuzumab resistance, shedding light on the primary mechanisms at play. Additionally, the section delves into KEGG Pathways enrichment analysis, initiated by Professor Minoru Kanehisa in 1995 as part of the Japanese Human Genome Program [23]. KEGG Pathways database offers manually curated pathway maps encompassing molecular interactions, reactions, and networks involving genes, proteins, RNAs, chemical compounds, and more. These pathways span various categories, including replication, metabolism, transcription, and cellular processes, providing a comprehensive understanding of biological processes [24]. This analysis examines genes identified as significant in trastuzumab resistance development in BT474 and SKBR3 cell lines. Notably, the study reveals that Herceptin treatment significantly affects 9 pathways in BT474 and a striking 75 pathways in SKBR3. This discrepancy aligns with findings from the GO term analysis and gene expression studies, underscoring the complexity of trastuzumab resistance. The complete list of significant KEGG Pathways for both cell lines can be found in the Supplementary Materials. A comprehensive comparative analysis was conducted on the significance of various pathways in the BT474 and SKBR3 cell lines (Table S13 and Table S14, respectively), focusing on their response to trastuzumab treatment. Three common key pathways were identified, cell cycle, cellular senescence, and DNA replication, signifying their importance in both cell lines. Examining overrepresented pathways (Table S15 and Table S16, respectively), SKBR3 displayed a notably higher percentage of genes significantly affected within KEGG pathways (ranging from 76% to 94%) compared to BT474 (ranging from 43% to 65%).

Four pathways—DNA replication, cell cycle, colorectal cancer, and bladder cancer—were prominent in both cell lines, highlighting their significance. Furthermore, in the SKBR3 cell line, the 20 most significant and 20 most overrepresented KEGG pathways shared seven pathways related to molecular activities (e.g., DNA replication, cell cycle, apoptosis) and tumor development (e.g., pancreatic cancer, colorectal cancer). Interestingly, eight pathways (Table 12) were found to be involved in drug resistance development in both cell lines, underscoring the substantial overlap between these biological cell lines. Notably, the SKBR3 cell line revealed greater complexity in molecular activities induced by Herceptin exposure compared to BT474. These findings emphasize common resistance mechanisms while acknowledging the unique aspects of each cell line's response to treatment.

Table 12. KEGG Pathways common for both BT474 and SKBR3 cell lines (order is based on lower adjusted *p*-value).

ID	Pathway
path:hsa04110	Cell cycle—Homo sapiens (human)
path:hsa03460	Fanconi anemia pathway—Homo sapiens (human)
path:hsa04218	Cellular senescence—Homo sapiens (human)
path:hsa04115	p53 signaling pathway—Homo sapiens (human)
path:hsa03030	DNA replication—Homo sapiens (human)
path:hsa05169	Epstein–Barr virus infection—Homo sapiens (human)
path:hsa05219	Bladder cancer—Homo sapiens (human)
path:hsa05210	Colorectal cancer—Homo sapiens (human)

3.2.6. PPI Network Analysis

Integrating various "omics" data is essential for understanding intricate cellular-level biological processes. In this study, a straightforward ID mapping method was employed to link genomic data to proteomic data, focusing on protein-coding genes in the complete human genome and proteome. To navigate complex biological interactions, the study favored protein–protein interaction (PPI) networks over genomic networks due to their higher density and connectivity. Unlike gene–gene interaction networks, PPI networks provided a more reliable means to identify the shortest path between essential genes associated with trastuzumab resistance. The statistical analysis of the PPI network, particularly the Gold set of interactions, is summarized in (Table 13). The study then delved into detailed PPI network analyses for the top 25 significant genes in BT474 and SKBR3 cell lines, complemented by literature-reported trastuzumab resistance-related genes. The results revealed a close and complex interplay among these significant genes, i.e., *BIRC5* (Figure 7), (Figure 8), *E2F1* (Figures S11 and S12), *USP1* (Figures S15 and S16), and TFRC (Figures S13 and S14), shedding new light on the pathways and networks involved in trastuzumab resistance development.

Table 13. Statistical parameters obtained by protein–protein interaction analysis in BT474 and SKBR3 cell lines).

PPI Statistics	SKBR3	BT474
Total Genes	6588	7009
Total Unmapped Genes	128	142
Total Mapped Genes	6460	6867
Total Proteins Mapped	5871	6020
Total Unique Proteins Mapped	5871	6013
Total Edges Mapped	20,904	20,718
Total Nodes	6730	6667
Total Edges	16,421	16,301

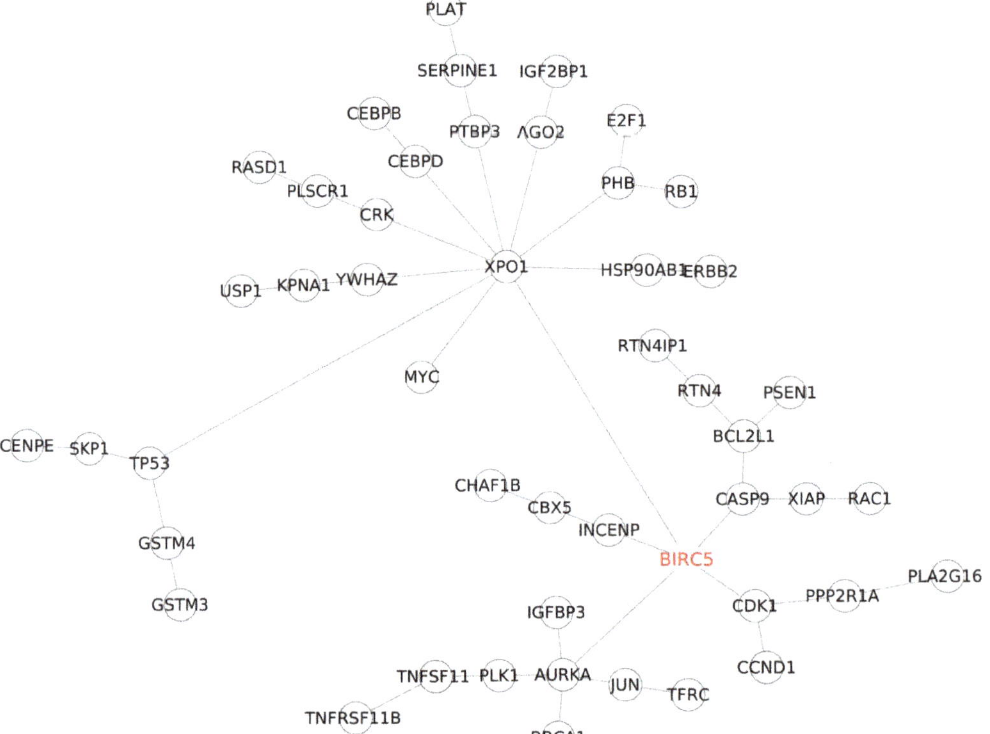

Figure 7. Protein–protein interaction (PPI) sub-network focused on the BIRC5 gene, incorporating the top 25 most significant genes in the BT474 cell line and genes associated with trastuzumab resistance development.

Remarkably, *BIRC5*, *E2F1*, and *RB1* are pivotal players at the core of these networks, engaging with numerous proteins. As depicted in Figures 7 and 8, *BIRC5* forms five direct connections shared across both cell lines: caspase 9 (*CASP9*), exportin (*XPO1*), Aurora kinase (*AURKA*), cyclin-dependent kinase 1 (CDK1), and inner centromere protein (IN-CENP) [25]. *CASP9*, linked to apoptosis, *AURKA*, involved in cell cycle regulation and tumorigenesis, and CDK1, essential for cell cycle progression, are vital contributors. *XPO1* stands out due to its multiple connections and its role in regulating protein transport. In SKBR3 cells, *BIRC5* also connects directly with *DIABLO* and *BECN1*, influencing apoptosis and autophagy [14]. Additionally, *BIRC5* interacts indirectly with *BRCA1, TFRC, IGFBP3* via *AURKA*, and *RAC1* via *CASP9*, forming numerous connections via *XPO1*, including *MYC*, *RB1*, *E2F1*, and *IGF2BP1* [14].

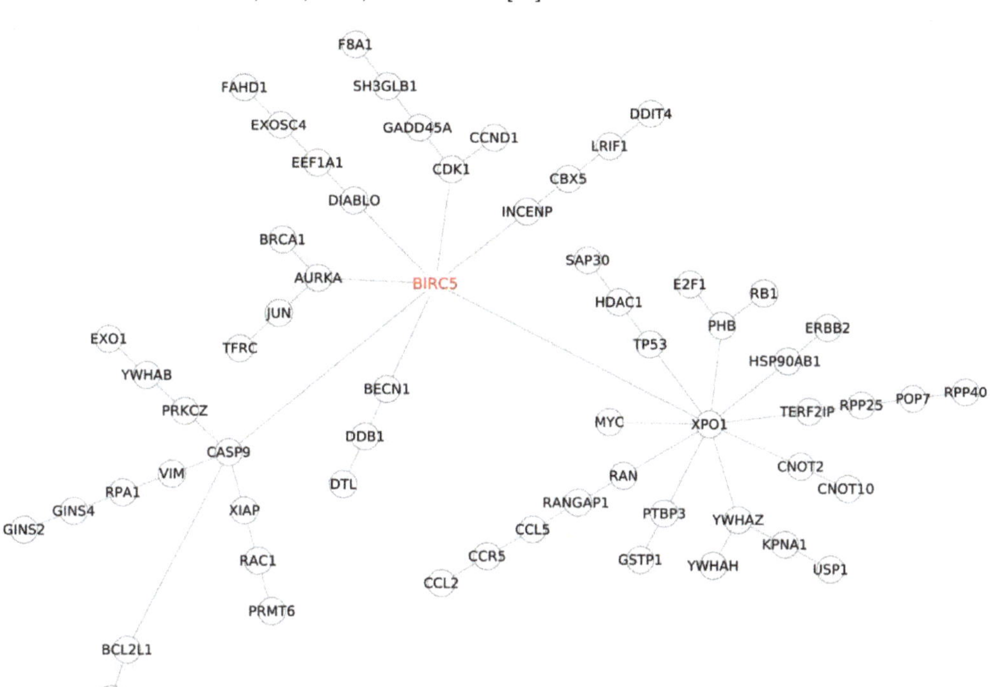

Figure 8. Protein–protein interaction (PPI) sub-network focused on the BIRC5 gene, incorporating the top 25 most significant genes in the SKBR3 cell line and genes associated with trastuzumab resistance development.

4. Discussion

The primary aim of this research was to delve into the molecular mechanisms responsible for developing resistance to trastuzumab, a drug used in breast cancer treatment. Traditional studies on Herceptin resistance mainly focused on identifying differences between resistant and sensitive cells [26–28]. However, this approach has limitations because cells can actively adapt to changing environments and treatment conditions. To address this, we proposed a novel approach combining a longitudinal investigation of in vitro resistance development with high-throughput microarray technology. Our comprehensive analysis of gene expression data encompassed over 34,000 genes and identified 8873 and 13,891 genes significantly contributing to trastuzumab resistance in two breast cancer cell lines, BT474 and SKBR3, with 5675 genes common to both cell lines. Further analysis grouped these genes into various molecular functions and biological processes using Gene Ontology terms. Additionally, we identified significant signaling pathways through KEGG Pathway analysis, with eight pathways commonly affected in both cell lines. Our study highlights the complexity of trastuzumab resistance development, emphasizing the need for a multi-dimensional understanding of the process. This research contributes valuable insights into potential targets for more effective breast cancer treatments [29,30]. In both cell lines, four genes—*BIRC5*, *E2F1*, *USP1*, and *TFRC*—emerged as highly significant transcripts within the top 25. *BIRC5* and *E2F1* exhibited a distinct expression pattern throughout the study. They initially decreased upon drug exposure and then gradually increased, mirroring the gradual development of drug resistance. This pattern aligns with existing research on *BIRC5*'s role in trastuzumab resistance, where the overactive PI3K/Akt [31] pathway leads to survivin overexpression, contributing to resistance [32]. In patients with HER2-overexpressing breast cancer, higher pretreatment survivin RNA levels correlated with poorer responses to trastuzumab, indicating *BIRC5*'s involvement in primary Herceptin resistance [32]. *E2F1*, a member of the E2F transcription factor family, shares similarities with survivin (BIRC5) in contributing to trastuzumab resistance. *E2F1* regulates cell growth and apoptosis, often activated in response to DNA damage [33]. Fanconi anemia DNA repair pathway emerged as crucial in trastuzumab resistance through KEGG pathway analysis. *E2F1* can also activate *BIRC5*, a known resistance factor. *E2F1* is linked to HER2 signaling and trastuzumab actions, with research showing its involvement in proliferative breast tumors. Interestingly, *E2F1* expression levels rise during trastuzumab exposure, indicating its potential role in drug resistance through independent HER2 pathway activation. While the roles of *BIRC5* and *E2F1* in drug resistance in HER2 breast cancer are understood, the involvement of *USP1* and *TRFC* remains unclear. *USP1* has been linked to cancer development and metastasis but not yet to trastuzumab resistance. Interestingly, *USP1* inhibitors have shown promise in leukemia treatment. Suppressing *USP1* leads to the degradation of the ID1 transcription factor, crucial for cancer progression. The high representation of *USP1* in molecular function and biological process gene ontology terms suggests its significant role in trastuzumab resistance mechanisms. Unlike *USP1*, current scientific literature does not mention *TFRC*'s role in Herceptin resistance development. Recent research has revealed *TFRC*'s involvement in various cancer-related signaling pathways, notably the endocytosis pathway, which is significant in trastuzumab resistance. While *TFRC*'s contribution to Gene Ontology terms is less pronounced than *USP1*, it plays a role in the "response to drug" biological process. This study emphasizes the importance of further investigating both genes in trastuzumab resistance development, as their correlation with drug resistance is not yet clear, and their molecular functions in the context of cancer progression and trastuzumab resistance require deeper exploration.

Understanding the molecular mechanisms behind resistance to Herceptin is critical. This study identified several genes, such as *BIRC5*, *BRCA1*, *RB1*, *ERBB2*, and others, as significant players in trastuzumab resistance, supporting previous research. New candidate genes like *E2F1*, *USP1*, and *TFRC* were also highlighted. Some genes, like *IGF2BP1* and *GSTP1*, showed potential involvement but require further confirmation. Surprisingly, this study did not confirm the involvement of certain previously reported genes in resistance. The findings suggest that primary Herceptin resistance may be linked to alterations in downstream components of HER2 signaling pathways or antiapoptotic proteins, rather than HER2 receptor activity itself. In contrast, acquired resistance may involve changes at the receptor level, such as epitope masking or upregulation of receptor components. Some intrinsic resistance mechanisms may overlap with acquired resistance, aligning with earlier research. Overall, this study underscores the complexity of trastuzumab resistance involving both HER2-dependent and independent pathways. Our study significantly advances our understanding of trastuzumab resistance by identifying new molecular contributors and genetic pathways. This knowledge informs future cancer genetics and molecular medicine research, potentially leading to effective adjuvant therapies. However, clinical validation is crucial, as preclinical and in vitro findings may not always translate. Additionally, we uncovered unexplored long non-coding RNAs and proteins, offering opportunities for further basic research.

This study unequivocally demonstrates the substantial role of microRNA molecules in developing resistance to trastuzumab. In both cell lines, more than 99.6% of the tested human microRNAs exhibited statistically significant expression changes during the development of drug resistance, even when applying stringent statistical criteria. Notably, using a more rigorous threshold for medical purposes revealed a high percentage of significant microRNAs, emphasizing the specificity of microRNA action. MicroRNA molecules play crucial roles in cellular processes such as proliferation, development, metabolism, differentiation, and apoptosis [34,35]. Their ability to target multiple mRNAs due to imperfect matching and their widespread regulatory influence on protein-coding genes further highlight their complexity [34]. Despite constituting only a small portion of the human genome, microRNAs are predicted to regulate a substantial portion of protein-coding genes. Four microRNAs, specifically hsa-miR-574-3p, hsa-miR-4530, hsa-miR-8485, and hsa-miR-197-3p, ranked among the top 25 most significant microRNAs in both cell lines, underscoring their significant involvement in the development of trastuzumab resistance. In silico analysis was conducted to identify potential targets of specific microRNAs (miRNAs) associated with Herceptin resistance in breast cancer cell lines. For hsa-miR-574-3p, TargetScan predicted potential targets such as *RAC1, BIRC5, E2F1, PMP22*, and *EGFR* [36]. Luciferase reporter assays confirmed direct regulation of *RAC1* and *EGFR* by miR-574-3p, both implicated in Herceptin resistance [37]. Similarly, miR-4530 showed potential targets, including FOXO1, MAPK4, and AKT predicted by miRDB [10] whereas Target Scan predicted *BIRC5, TFRC, HER2, ESR2*, and AURKA [36], some of which are involved in HER2 signaling and were important in Herceptin resistance. Another miRNA, hsa-miR-8485, was found to potentially target genes like *BIRC5, E2F1, USP1, RAC1, EPHA2, PTEN*, and *CCND1* [10,36], where, *E2F1* and *USP1*, being from the top four, were highly significant in both cell lines. Additionally, miR-4701-5p was identified as targeting TFRC [10], which was highly significant in both cell lines. These findings provide insights into the molecular mechanisms underlying Herceptin resistance in breast cancer. The Hsa-miR-197-3p, a microRNA, plays a crucial role in cancer progression, particularly in breast, bladder, and thyroid cancers. Multiple studies emphasize the influence of long non-coding RNAs (ncRNAs) on regulating miR-197-3p expression. LIFR-AS1 inhibits cell proliferation, migration, and invasion in breast cancer by repressing miR-197 [38]. Similar results were observed in bladder [39] and thyroid cancers [40], where miR-197-3p downregulation led to decreased cell proliferation, migration, and invasion due to the actions of specific ncRNAs. Notably, miR-197-3p is linked to the PTEN/PI3K-Akt pathway [40], which plays a key role in HER2 signaling. This suggests that miR-197-3p might be involved in an alternative pathway compensating for trastuzumab's therapeutic effects, a drug targeting HER2. Additionally, miR-197 targets *MAPK1*, a gene associated with trastuzumab resistance. Overexpressing miR-197 can reverse drug resistance by inhibiting *MAPK1*, as seen in gastric cancer cells [41]. Overall, miR-197-3p has a significant role in trastuzumab resistance development, potentially through its regulation of *MAPK1* and involvement in alternative signaling pathways [41]. Furthermore, miR-197-3p directly regulates other genes implicated in trastuzumab resistance, including *FOXJ2* [42], *MTHFD1* [43], *RAN* [44], *TUSC2* [45], and *FUS1* [46], though their specific roles in drug resistance and cancer progression require further investigation. These findings support the hypothesis that miR-197-3p is a key player in developing resistance to trastuzumab, a critical drug in breast cancer treatment.

In summary, our study suggests that nearly all known human microRNAs may play a role in developing resistance to the drug trastuzumab. We identified four microRNAs that are particularly important in both biological cell lines studied. Two of these microRNAs are confirmed to be involved in either HER2 signaling or drug resistance, supporting our findings' reliability. The other two highly significant microRNAs, which have limited existing information, were identified through computational analysis as potential regulators of genes associated with trastuzumab resistance. However, further research is needed to validate these hypotheses and understand the underlying mechanisms.

5. Conclusions

The current study conducted high-throughput microarray experiments to investigate the intricate dynamics of gene and microRNA expression changes during the development of trastuzumab resistance in two prominent breast cancer cell lines, BT474 and SKBR3. The analysis of a pool of 34,000 genes revealed distinct patterns of differential expression, with 8874 and 13,892 genes implicated in resistance development in BT474 and SKBR3, respectively. Remarkably, our findings highlighted the significant involvement of key genes, including *BIRC5, E2F1, USP1,* and *TFRC,* which are crucial players in both cell lines, particularly within the context of HER2 signaling. Moreover, the identification of novel contributors to Herceptin resistance, such as *IGF2BP1, GSTM3, RASD1, KLK11, GSTP1, YWHAH, DTL, DOLK, NACC2, DDIT,* and *DNAJA3,* among the top 25 significant genes, suggests the existence of previously unrecognized mechanisms in drug resistance. Importantly, our research also underscored the role of established genes, including *BIRC5, E2F1, BRCA1, RB1, ERBB2, EPHA2, IGFBP3, ADAM10, FOXM1, RAC1, MYC, CCND1, PTEN, TP53, MAP2K4,* and *PI3KCA* in contributing to resistance. Notably, protein–protein interaction analysis illuminated the pivotal roles of *BIRC5, E2F1,* and *RB1* as central hubs within networks linked to Herceptin resistance. Furthermore, the significant impact of long non-coding RNAs and microRNAs in this resistance context was evident, indicating their potential as vital regulators in the process. Gene Ontology analysis highlighted enriched molecular functions such as receptor binding, protein kinase activities, and DNA replication, while biological processes encompassed crucial aspects like cell cycle regulation, apoptosis, and DNA damage repair. Pathway analysis brought to light 9 and 75 affected networks in BT474 and SKBR3, respectively, with the convergence of eight common pathways, notably including cell cycle and p53 signaling. Notably, our investigation revealed HER2-dependent and independent resistance mechanisms, thereby ruling out the involvement of epitope masking and other ERBB receptors. Noteworthy complexities observed in SKBR3 possibly arose from disparities in the cancer stage, considering the primary vs. metastatic distinction. Intriguingly, our study highlighted the significant role of microRNAs in Herceptin resistance, with hsa-miR-574-3p, hsa-miR-4530, hsa-miR-8485, and hsa-miR-197-3p emerging as critical contributors, including some previously unreported microRNAs specific to each cell line. These comprehensive findings shed light on the multifaceted landscape of trastuzumab resistance in breast cancer, providing valuable insights for the development of more effective therapeutic strategies and personalized treatment approaches. Admittedly, the findings have been obtained based on an analysis of a limited amount of material (that included, due to resource constraints, only two independent cell-line replicates at each time point). Evaluation of their credibility should take this aspect of the study into account. In this respect, further validation of the findings would be important. For instance, an in vivo validation and analysis of gene expression at the protein level could be used to ensure the accuracy and comprehensiveness of the microarray-based findings. An investigation of the expression of identified DEGs in samples of patients included in clinical databases and of the association with patients' outcomes could shed light on the relevance of the study's findings to clinical practice. These extensions are left for future research.

Supplementary Materials: The following supporting information can be downloaded at: https://www.mdpi.com/article/10.3390/cimb46030171/s1.

Author Contributions: Conceptualization, performed experiments, participation in drafting paper, A.K.; performed bioinformatics analysis, miRNA-gene association analysis, key role in drafting paper, S.G.; performed cellular experiments, assistance in grant application, E.G.; performed bioinformatics analysis, gene expression analysis, M.Ł.; performed microRNA data analysis, M.D. and S.R.; built protein–protein interaction networks, K.S. and A.F.M.; performed statistical analysis, K.G. and J.C.; co-supervision, performed statistical analysis, T.B.; performed microRNA data analysis, M.D.; performed microRNA data analysis, guidance, and participation in drafting paper, I.S.; participation in bioinformatics and statistical data analysis, co-supervision, guidance, and participation in drafting paper, D.P. All authors have read and agreed to the published version of the manuscript.

Funding: This work has been supported by the Leading National Research Center, PRELUDIUM research grant funded by Polish National Science Center (2015/17/N/NZ2/01932) and a TEAM research grant from The Foundation for Polish Science (POIR.04.04.00-00-3CA6/16-00). The research was funded by the Warsaw University of Technology within the Excellence Initiative: Research University (IDUB) programme. This work has been co-supported by the Polish National Science Centre (2019/35/O/ST6/02484 and 2020/37/B/NZ2/03757). Computations were performed thanks to the Laboratory of Bioinformatics and Computational Genomics, Faculty of Mathematics and

Information Science, Warsaw University of Technology using the Artificial Intelligence HPC platform financed by the Polish Ministry of Science and Higher Education (decision no. 7054/IA/SP/2020 of 2020-08-28).

Institutional Review Board Statement: Not applicable.

Informed Consent Statement: Not applicable.

Data Availability Statement: The data presented in this study are available in Supplementary Files such as Supplementary tables in Spreadsheet and Supplementary figures in Documents.

Conflicts of Interest: The authors declare no conflicts of interest.

References

1. Ma, X.; Yu, H. Cancer issue: Global burden of cancer. *Yale J. Biol. Med.* **2006**, *79*, 85.
2. Sung, H.; Ferlay, J.; Siegel, R.L.; Laversanne, M.; Soerjomataram, I.; Jemal, A.; Bray, F. Global cancer statistics 2020: GLOBOCAN estimates of incidence and mortality worldwide for 36 cancers in 185 countries. *CA Cancer J. Clin.* **2021**, *71*, 209–249. [CrossRef] [PubMed]
3. Bray, F.; Ferlay, J.; Soerjomataram, I.; Siegel, R.L.; Torre, L.A.; Jemal, A. Global cancer statistics 2018: GLOBOCAN estimates of incidence and mortality worldwide for 36 cancers in 185 countries. *CA Cancer J. Clin.* **2018**, *68*, 394–424. [CrossRef] [PubMed]
4. Arnold, M.; Morgan, E.; Rumgay, H.; Mafra, A.; Singh, D.; Laversanne, M.; Vignat, J.; Gralow, J.R.; Cardoso, F.; Siesling, S.; et al. Current and future burden of breast cancer: Global statistics for 2020 and 2040. *Breast* **2022**, *66*, 15–23. [CrossRef] [PubMed]
5. Inoue, K.; Fry, E.A. Novel molecular markers for breast cancer. *Biomark. Cancer* **2016**, *8*, 25–42. [CrossRef] [PubMed]
6. Perez, E.A.; Romond, E.H.; Suman, V.J.; Jeong, J.H.; Sledge, G.; Geyer, C.E.; Martino, S.; Rastogi, P.; Gralow, J.; Swain, S.M.; et al. Trastuzumab Plus Adjuvant Chemotherapy for Human Epidermal Growth Factor Receptor 2—Positive Breast Cancer: Planned Joint Analysis of Overall Survival From NSABP B-31 and NCCTG N9831. *J. Clin. Oncol.* **2014**, *32*, 3744–3752. [CrossRef]
7. Bonazzoli, E.; Cocco, E.; Lopez, S.; Bellone, S.; Zammataro, L.; Bianchi, A.; Manzano, A.; Yadav, G.; Manara, P.; Perrone, E.; et al. PI3K oncogenic mutations mediate resistance to afatinib in HER2/neu overexpressing gynecological cancers. *Gynecol. Oncol.* **2019**, *153*, 158–164. [CrossRef] [PubMed]
8. Wang, Z.H.; Zheng, Z.Q.; Jia, S.; Liu, S.N.; Xiao, X.F.; Chen, G.Y.; Liang, W.Q.; Lu, X.F. Trastuzumab resistance in HER2-positive breast cancer: Mechanisms, emerging biomarkers and targeting agents. *Front. Oncol.* **2022**, *12*, 1006429. [CrossRef] [PubMed]
9. Dai, X.; Cheng, H.; Bai, Z.; Li, J. Breast cancer cell line classification and its relevance with breast tumor subtyping. *J. Cancer* **2017**, *8*, 3131. [CrossRef]
10. Chen, Y.; Wang, X. miRDB: An online database for prediction of functional microRNA targets. *Nucleic Acids Res.* **2020**, *48*, D127–D131. [CrossRef]
11. Ginestier, C.; Adelaide, J.; Goncalves, A.; Repellini, L.; Sircoulomb, F.; Letessier, A.; Finetti, P.; Geneix, J.; Charafe-Jauffret, E.; Bertucci, F.; et al. ERBB2 phosphorylation and trastuzumab sensitivity of breast cancer cell lines. *Oncogene* **2007**, *26*, 7163–7169. [CrossRef] [PubMed]
12. Kauraniemi, P.; Hautaniemi, S.; Autio, R.; Astola, J.; Monni, O.; Elkahloun, A.; Kallioniemi, A. Effects of Herceptin treatment on global gene expression patterns in HER2-amplified and nonamplified breast cancer cell lines. *Oncogene* **2003**, *23*, 1010–1013. [CrossRef] [PubMed]
13. Neve, R.M.; Chin, K.; Fridlyand, J.; Yeh, J.; Baehner, F.L.; Fevr, T.; Clark, L.; Bayani, N.; Coppe, J.P.; Tong, F.; et al. A collection of breast cancer cell lines for the study of functionally distinct cancer subtypes. *Cancer Cell* **2006**, *10*, 515–527. [CrossRef] [PubMed]
14. Yoshida, R.; Tazawa, H.; Hashimoto, Y.; Yano, S.; Onishi, T.; Sasaki, T.; Shirakawa, Y.; Kishimoto, H.; Uno, F.; Nishizaki, M.; et al. Mechanism of resistance to trastuzumab and molecular sensitization via ADCC activation by exogenous expression of HER2-extracellular domain in human cancer cells. *Cancer Immunol. Immunother.* **2012**, *61*, 1905–1916. [CrossRef] [PubMed]
15. Scaltriti, M.; Serra, V.; Normant, E.; Guzman, M.; Rodriguez, O.; Lim, A.R.; Slocum, K.L.; West, K.A.; Rodriguez, V.; Prudkin, L.; et al. Antitumor Activity of the Hsp90 Inhibitor IPI-504 in HER2-Positive Trastuzumab-Resistant Breast Cancer. *Mol. Cancer Ther.* **2011**, *10*, 817–824. [CrossRef] [PubMed]
16. Han, S.; Meng, Y.; Tong, Q.; Li, G.; Zhang, X.; Chen, Y.; Hu, S.; Zheng, L.; Tan, W.; Li, H.; et al. The ErbB2-targeting antibody trastuzumab and the small-molecule SRC inhibitor saracatinib synergistically inhibit ErbB2-overexpressing gastric cancer. *mAbs* **2013**, *6*, 403–408. [CrossRef]
17. Ritchie, M.E.; Phipson, B.; Wu, D.; Hu, Y.; Law, C.W.; Shi, W.; Smyth, G.K. limma powers differential expression analyses for RNA-sequencing and microarray studies. *Nucleic Acids Res.* **2015**, *43*, e47. [CrossRef]
18. Benjamini, Y.; Hochberg, Y. Controlling the False Discovery Rate: A Practical and Powerful Approach to Multiple Testing. *J. R. Stat. Soc. Ser. (Methodol.)* **1995**, *57*, 289–300. [CrossRef]
19. Gene Set Enrichment Analysis. Available online: https://bioconductor.org/packages/release/bioc/vignettes/topGO/inst/doc/topGO.pdf (accessed on 16 October 2023).
20. Alexa, A.; Rahnenführer, J.; Lengauer, T. Improved scoring of functional groups from gene expression data by decorrelating GO graph structure. *Bioinformatics* **2006**, *22*, 1600–1607. [CrossRef]

21. Ashburner, M.; Ball, C.A.; Blake, J.A.; Botstein, D.; Butler, H.; Cherry, J.M.; Davis, A.P.; Dolinski, K.; Dwight, S.S.; Eppig, J.T.; et al. Gene ontology: Tool for the unification of biology. *Nat. Genet.* **2000**, *25*, 25–29. [CrossRef]
22. Consortium, G.O. The gene ontology project in 2008. *Nucleic Acids Res.* **2008**, *36*, D440–D444. [CrossRef]
23. Kanehisa, M. A database for post-genome analysis. *Trends Genet. TIG* **1997**, *13*, 375–376. [CrossRef] [PubMed]
24. Kanehisa, M.; Goto, S. KEGG: Kyoto encyclopedia of genes and genomes. *Nucleic Acids Res.* **2000**, *28*, 27–30. [CrossRef] [PubMed]
25. Gene Resources. Available online: https://www.ncbi.nlm.nih.gov (accessed on 16 October 2023).
26. Merry, C.R.; McMahon, S.; Forrest, M.E.; Bartels, C.F.; Saiakhova, A.; Bartel, C.A.; Scacheri, P.C.; Thompson, C.L.; Jackson, M.W.; Harris, L.N.; et al. Transcriptome-wide identification of mRNAs and lincRNAs associated with trastuzumab-resistance in HER2-positive breast cancer. *Oncotarget* **2016**, *7*, 53230. [CrossRef] [PubMed]
27. Zazo, S.; González-Alonso, P.; Martín-Aparicio, E.; Chamizo, C.; Cristóbal, I.; Arpí, O.; Rovira, A.; Albanell, J.; Eroles, P.; Lluch, A.; et al. Generation, characterization, and maintenance of trastuzumab-resistant HER2+ breast cancer cell lines. *Am. J. Cancer Res.* **2016**, *6*, 2661.
28. Mercogliano, M.F.; De Martino, M.; Venturutti, L.; Rivas, M.A.; Proietti, C.J.; Inurrigarro, G.; Frahm, I.; Allemand, D.H.; Deza, E.G.; Ares, S.; et al. TNFα-induced mucin 4 expression elicits trastuzumab resistance in HER2-positive breast cancer. *Clin. Cancer Res.* **2017**, *23*, 636–648. [CrossRef]
29. Arteaga, C.L.; O'Neill, A.; Moulder, S.L.; Pins, M.; Sparano, J.A.; Sledge, G.W.; Davidson, N.E. A phase I-II study of combined blockade of the ErbB receptor network with trastuzumab and gefitinib in patients with HER2 (ErbB2)-overexpressing metastatic breast cancer. *Clin. Cancer Res.* **2008**, *14*, 6277–6283. [CrossRef]
30. Baselga, J.; Lewis Phillips, G.D.; Verma, S.; Ro, J.; Huober, J.; Guardino, A.E.; Samant, M.K.; Olsen, S.; de Haas, S.L.; Pegram, M.D. Relationship between tumor biomarkers and efficacy in EMILIA, a phase III study of trastuzumab emtansine in HER2-positive metastatic breast cancer. *Clin. Cancer Res.* **2016**, *22*, 3755–3763. [CrossRef] [PubMed]
31. Liu, D.; Yang, Z.; Wang, T.; Chen, H.; Hu, Y.; Hu, C.; Guo, L.; Deng, Q.; Liu, Y.; Yu, M.; et al. β2-AR signaling controls trastuzumab resistance-dependent pathway. *Oncogene* **2016**, *35*, 47–58. [CrossRef]
32. Chakrabarty, A.; Bhola, N.E.; Sutton, C.; Ghosh, R.; Kuba, M.G.; Dave, B.; Chang, J.C.; Arteaga, C.L. Trastuzumab-resistant cells rely on a HER2-PI3K-FoxO-survivin axis and are sensitive to PI3K inhibitors. *Cancer Res.* **2013**, *73*, 1190–1200. [CrossRef]
33. Stevens, C.; Smith, L.; La Thangue, N.B. Chk2 activates E2F-1 in response to DNA damage. *Nat. Cell Biol.* **2003**, *5*, 401–409. [CrossRef] [PubMed]
34. Hussain, M.U. Micro-RNAs (miRNAs): Genomic organisation, biogenesis and mode of action. *Cell Tissue Res.* **2012**, *349*, 405–413. [CrossRef] [PubMed]
35. Xu, L.; Yang, B.F.; Ai, J. MicroRNA transport: A new way in cell communication. *J. Cell. Physiol.* **2013**, *228*, 1713–1719. [CrossRef] [PubMed]
36. McGeary, S.E.; Lin, K.S.; Shi, C.Y.; Pham, T.M.; Bisaria, N.; Kelley, G.M.; Bartel, D.P. The biochemical basis of microRNA targeting efficacy. *Science* **2019**, *366*, eaav1741. [CrossRef] [PubMed]
37. Chiyomaru, T.; Yamamura, S.; Fukuhara, S.; Hidaka, H.; Majid, S.; Saini, S.; Arora, S.; Deng, G.; Shahryari, V.; Chang, I.; et al. Genistein up-regulates tumor suppressor microRNA-574-3p in prostate cancer. *PLoS ONE* **2013**, *8*, e58929. [CrossRef] [PubMed]
38. Xu, F.; Li, H.; Hu, C. LIFR-AS1 modulates Sufu to inhibit cell proliferation and migration by miR-197-3p in breast cancer. *Biosci. Rep.* **2019**, *39*, BSR20180551. [CrossRef]
39. Jiang, Y.; Wei, T.; Li, W.; Zhang, R.; Chen, M. Circular RNA hsa_circ_0002024 suppresses cell proliferation, migration, and invasion in bladder cancer by sponging miR-197-3p. *Am. J. Transl. Res.* **2019**, *11*, 1644.
40. Liu, K.; Huang, W.; Yan, D.Q.; Luo, Q.; Min, X. Overexpression of long intergenic noncoding RNA LINC00312 inhibits the invasion and migration of thyroid cancer cells by down-regulating microRNA-197-3p. *Biosci. Rep.* **2017**, *37*, BSR20170109. [CrossRef]
41. Xiong, H.L.; Zhou, S.W.; Sun, A.H.; He, Y.; Li, J.; Yuan, X. MicroRNA-197 reverses the drug resistance of fluorouracil-induced SGC7901 cells by targeting mitogen-activated protein kinase 1. *Mol. Med. Rep.* **2015**, *12*, 5019–5025. [CrossRef]
42. Dubey, R.; Saini, N. STAT6 silencing up-regulates cholesterol synthesis via miR-197/FOXJ2 axis and induces ER stress-mediated apoptosis in lung cancer cells. *Biochim. Biophys. Acta (BBA)-Gene Regul. Mech.* **2015**, *1849*, 32–43. [CrossRef]
43. Minguzzi, S.; Selcuklu, S.D.; Spillane, C.; Parle-McDermott, A. An NTD-Associated Polymorphism in the 3′ UTR of MTHFD 1 L can Affect Disease Risk by Altering mi RNA Binding. *Hum. Mutat.* **2014**, *35*, 96–104. [CrossRef] [PubMed]
44. Tang, W.F.; Huang, R.T.; Chien, K.Y.; Huang, J.Y.; Lau, K.S.; Jheng, J.R.; Chiu, C.H.; Wu, T.Y.; Chen, C.Y.; Horng, J.T. Host microRNA miR-197 plays a negative regulatory role in the enterovirus 71 infectious cycle by targeting the RAN protein. *J. Virol.* **2016**, *90*, 1424–1438. [CrossRef] [PubMed]
45. Samandari, N.; Mirza, A.H.; Nielsen, L.B.; Kaur, S.; Hougaard, P.; Fredheim, S.; Mortensen, H.B.; Pociot, F. Circulating microRNA levels predict residual beta cell function and glycaemic control in children with type 1 diabetes mellitus. *Diabetologia* **2017**, *60*, 354–363. [CrossRef] [PubMed]
46. Du, L.; Schageman, J.J.; Subauste, M.C.; Saber, B.; Hammond, S.M.; Prudkin, L.; Wistuba, I.I.; Ji, L.; Roth, J.A.; Minna, J.D.; et al. miR-93, miR-98, and miR-197 regulate expression of tumor suppressor gene FUS1. *Mol. Cancer Res.* **2009**, *7*, 1234–1243. [CrossRef]

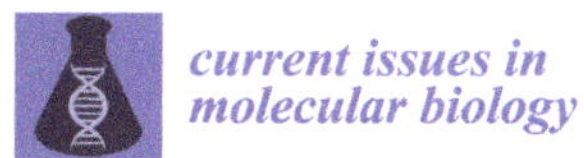

current issues in molecular biology

MDPI

Article

Significant Association of Estrogen Receptor-β Isoforms and Coactivators in Breast Cancer Subtypes

Young Choi [1],* and Simcha Pollack [2]

[1] Department of Pathology, Yale School of Medicine, 434 Pine Grove Lane, Hartsdale, NY 10530, USA
[2] Department of Statistics, St. John's University, New York, NY 11423, USA; pollacks@stjohns.edu
* Correspondence: yok9012@nyp.org

Abstract: Nuclear receptor coregulators are the principal regulators of Estrogen Receptor (ER)-mediated transcription. ERβ, an ER subtype first identified in 1996, is associated with poor outcomes in breast cancer (BCa) subtypes, and the coexpression of the ERβ1 isoform and AIB-1 and TIF-2 coactivators in BCa-associated myofibroblasts is associated with high-grade BCa. We aimed to identify the specific coactivators that are involved in the progression of ERβ-expressing BCa. ERβ isoforms, coactivators, and prognostic markers were tested using standard immunohistochemistry. AIB-1, TIF-2, NF-kB, p-c-Jun, and/or cyclin D1 were differentially correlated with ERβ isoform expression in the BCa subtypes and subgroups. The coexpression of the ERβ5 and/or ERβ1 isoforms and the coactivators were found to be correlated with a high expression of P53, Ki-67, and Her2/neu and large-sized and/or high-grade tumors in BCa. Our study supports the notion that ERβ isoforms and coactivators seemingly coregulate the proliferation and progression of BCa and may provide insight into the potential therapeutic uses of the coactivators in BCa.

Keywords: estrogen receptor β; coactivator; correlation; coregulation; prognosis; therapy

check for updates

Citation: Choi, Y.; Pollack, S. Significant Association of Estrogen Receptor-β Isoforms and Coactivators in Breast Cancer Subtypes. *Curr. Issues Mol. Biol.* **2023**, *45*, 2533–2548. https://doi.org/10.3390/cimb45030166

Academic Editor: Dumitru A. Iacobas

Received: 20 February 2023
Revised: 8 March 2023
Accepted: 10 March 2023
Published: 17 March 2023

1. Introduction

1.1. Two Estrogen Receptors

There are two estrogen receptor (ER) genes (ESR1/ERα and ESR2/ERβ). ERα and ERβ are members of the nuclear receptor superfamily of transcription factors and share some structural similarities, including a high degree of homology (96%) in their DNA-binding regions. However, they also have distinct differences in genotype, tissue distribution, and binding to pharmacological agents; they share only moderate homology in the ligand-binding region, and they have markedly distinct NH^2-terminal activation function-1 (AP-1) regions. ERα and ERβ can form heterodimers [1]; when coexpressed, ERβ acts as a transdominant inhibitor of ERα transcriptional activity. Thus, the relative levels of ERα and ERβ in breast cancer (BCa) are likely to affect cell proliferation, signaling pathways, and their response to ER ligands [2,3]. ERβ has different variant forms that interact with multiple protein partners, as well as ligands, and heterodimerize with ERα, thereby creating a highly complex labyrinth of functions. Furthermore, ERβ localizes in different cellular compartments and is susceptible to different posttranscriptional modifications (PTM) [4–6].

The exact role of ERβ in BCa has not yet been fully established. Highly variable and even opposite effects have been ascribed to the expression of ERβ isoform mRNA and protein expression in BCa, including both proliferative and growth-inhibitory actions, as well as favorable or adverse clinical outcomes [7,8]. Our recent study showed that ERβ1 protein expression is associated with poor prognostic markers [9]. ERβ2 and ERβ5mRNA expression are risk factors for OS in BCa subtypes and are associated with poor prognostic biomarkers, particularly in ERα-negative BCa and TNBC [10]. Overall, the outcome results of ERβ expression in BCa are inconsistent. Such inconsistent and controversial results may be due to the complexity of ERβ isoforms and the lack of standardized testing protocols but may also relate to various downstream signaling pathways, their PTM, and the their involvement of coregulators in its transcription.

1.2. Nuclear Receptor Coregulators

Nuclear receptor (NR) coregulators have emerged as the principal regulators of gene expression by directly interacting with and modulating the activity of NRs. ER-mediated transcriptional and biological activities require the recruitment of a diverse array of coregulator proteins to ERs. Coregulator complexes enable the ERs to respond to hormones or pharmacological ligands and communicate with the transcription apparatus at target gene promoters. Ligand-dependent and ligand-independent ERα and ERβ receptors recruit coactivators and corepressors and activate or repress ER-mediated transcription [11–15]. Alterations in ER conformation induced by binding to different estrogen response element (ERE) sequences modulate ERα and ERβ interaction with coactivators and corepressors [16].

Steroid receptor coactivator (SRC) family members, the p160 class, of coactivators are a gene family characterized as the primary coactivators for NRs and are required for NR-mediated transcription. They have been widely implicated in the regulation of steroid hormone action by mediating functions of NRs and facilitating the assembly of transcriptome complexes at target genes [14,17,18]. The SRC family consists of three members: SRC-1 (NCOA1), transcriptional intermediary factor-2 (TIF-2/SRC-2/GRIP-1/NCOA2), and amplified in breast cancer-1 (SAIB-1/SRC-3/NCOA3). The alteration or deregulation of SRC coregulators is common in BCa and enhances both ligand-independent and ligand-dependent ERα signaling to drive the proliferation, progression, and invasive capacity of neoplastic cells [13–15,19].

Among the SRC family members, SRC-3/AIB-1 is the primary coactivator for ERα and is overexpressed in BCa, and it is a crucial driver of mammary tumorigenesis [20–24]. AIB-1 mRNA and protein overexpression correlate with the expression of high Her2/neu, larger tumor size, higher tumor grade, and poor disease-free survival (DFS). AIB-1 also interacts with coactivates p65/NF-κB and plays an essential role in the NF-kB signaling pathway [17,25]. Furthermore, AIB-1 facilitates the transition of downstream genes encoding cyclin D1 and the insulin-like growth factor-1 (IGF1) pathway [14,18,19], and it promotes the epithelial–mesenchymal transition through its interaction with ERα and worse outcomes in Erα-positive BCa [19,26]. In tamoxifen (TAM)-treated patients, high AIB-1 expression is associated with TAM resistance and poorer DFS [19,27–29]. The overexpression of AIB-1 correlated with poor prognosis in TNBC patients [19,30].

TIF-2 is frequently overexpressed in various neoplasms. Recurrent prostate cancers have exhibited high expression levels of THE androgen receptor, TIF-2, and SRC-1 [31]. TIF-2 correlates significantly with lymph node (LN)-positive BCa [32].

SRC-1 frequently correlates with high Her2/neu expression, LN metastasis, disease recurrence, poor DFS, and more advanced disease stage in BCa [33,34]. SRC-1 is a coactivator that can switch BCa from a steroid-responsive to a steroid-resistant state and promote the aggressive BCa phenotype. It has been implicated in aromatase inhibitor-resistant recurrent BCa [35]. SRC-1 and its homolog transcriptional co-activators p/CIP have been shown to be the coactivators for NF-kB, CREB, and STAT-1 [36].

NF-kB is a pleiotropic transcription factor and is the key activator of genes involved in host immunity and inflammatory responses with the induction of a large number of genes that influence cellular proliferation and inflammation. NF-kB activity promotes tumor proliferation, regulates cell apoptosis, and also induces the epithelial–mesenchymal transition, which facilitates distant metastasis and transactivates the expression of cyclin D1 and c-myc [37,38].

C-Jun is a component of the transcription factor AP-1. Extra- or intracellular signals, including growth factors and transforming oncoproteins, stimulate the phosphorylation of c-Jun at serine 63/73 and activate c-Jun-dependent transcription. Activated c-Jun has been demonstrated to be associated with proliferation and angiogenesis [39], as well as epithelial stem cell expansion [40].

Cyclin D1 is frequently overexpressed in BCa and contributes to ERα activation in BCa. AIB-1 and other steroid receptor coactivators can enhance the functional interaction of ERα with the cyclin D1 promoter [41], while cyclin D1 can recruit SRC-1 and AIB-1 to ERα in the absence of a ligand [42]. High cyclin D1 expression is associated with high proliferation and a higher risk of death from BCa in ERα-positive BCa. However, no significant prognostic impact of cyclin D1 expression has been found among ERα-negative cases [43], and the reverse relationship was demonstrated for cyclin D1 overexpression in invasive ductal carcinoma [44].

Overall, ERα-coactivator proteins enhance ligand-dependent and ligand-independent ERα signaling, progression, endocrine therapy resistance, and metastasis in BCa. Suen et al. [45] demonstrated that AIB-1 selectively enhances ERα but does not enhance ERβ-dependent gene transcription. TAM-induced AIB-1 recruitment to the ER-ERE enhanced interaction between AIB-1 and ERα but not ERβ. However, Liu et al. [46] observed opposing actions of ERα and ERβ with the dominance of ERβ over ERα in the activation of cyclin D1 gene expression. Estrogens, which activate cyclin D1 gene expression with ERα, inhibit expression with ERβ. The different recruitments of AIB-1 to ERα and ERβ may, in part, explain the different associations between ERs and response to endocrine treatment [47].

On the other hand, Bai et al. [48] observed that both ERα and ERβ can interact with the coactivator receptor interaction domains (RIDs) of all three SRC isoforms in living cells. Other studies have also demonstrated that ERβ transactivation recruits members of the SRC family [49,50]. The phosphorylation of AF-1 by MAP kinase (MAPK) leads to the recruitment of SRC-1 by ERβ and provides a molecular basis for the ligand-independent activation of ERβ via the MAPK cascade [51]. ERβ expression was significantly correlated with SRC-1, TIF-2, and NCOR protein levels in BCa and the upregulation of expression levels of ERβ and cofactors during the development of intraductal carcinomas [32]. ERβ and GRIP1/TIF2 has been shown to interact in vitro in a ligand-dependent manner and the transcriptional responses to estrogen in nonsmall cell lung cancer cells [52] and colon cancer via ERβ [53].

In summary, the combinations of ligand and ER subtypes can effectively recruit the three p160 coactivators albeit with differences in the levels and dose–response for coactivator recruitment by some of the ligands, with respect to their agonist activity [49,54]. Thus, coactivators seem to play an important role in directing ERβ-regulating genes or gene sets, further contributing to the functional complexity of ERβ.

Our previous study showed that high ERβ1 protein expression in BCa-associated myofibroblasts (MFs) was significantly associated with AIB-1 and TIF-2 expression in high-grade carcinoma with desmoplastic reaction and heavy lymphocytic infiltration [55]. Furthermore, our recent studies showed that high ERβ1 protein expression in ERα-negative BCa was correlated with high Ki-67, P53, and Her2/neu expression [9], and the expression of high ERβ2 and -5 isoform mRNAs is a poor risk factor and associated with high Ki-67 expression in BCa subtypes and subgroups [10]. As Erβ has strong affinity preferences for particular coactivators, in this study, we aimed to identify specific co-activators that interact with the Erβ isoform and are involved in the progression of BCa with ERβ expression.

2. Materials and Methods

2.1. Patients

All procedures involving patients with Bca were performed according to the ethical standards of the Institutional Research Board, Bridgeport Hospital, Bridgeport, CT (IRB# 090101). This study included 65 Erα-negative (43 TNBC) and 73 ERα-positive BCa from 138 patients with a follow-up period from 2003 to 2010. The demographic and clinical characteristics of all subjects were retrieved from medical records and cancer registry reports, as well as pathology records for hormone receptor reports, histologic types, tumor grades, tumor size, and AJCC tumor stages. Histological grades were assessed according to the Bloom–Richardson classification criteria. The AJCC tumor stages consisted of 75 in stage 1, 45 in stage II, and 18 in stage III. The follow-up period ranged from 1 to

96 months (median: 60 months); 20 patients died during this period. The phenotypic BCa patterns were determined according to Erα, HER2/neu, and progesterone receptor (PR) status following consensus guidelines. The proliferation marker Ki-67 was evaluated for all tumors. The molecular types comprised 50 luminal A ($Er\alpha^+/PR^+/HER2^-$), 25 luminal B ($Er\alpha^+$ and/or $PR^+/HER2^+/Ki-67^+$), 17 HER2 ($Er\alpha^-/PR^-/HER2^+$), 17 basal-like ($ER\alpha^-/PR^-/HER2^-/CK5/6^+$), and 29 unclassified types [10].

2.2. Tissue Microarray (TMA) Preparation

Hematoxylin and eosin sections of formalin-fixed paraffin-embedded (FFPE) tumor samples were evaluated. The TMA blocks were constructed using triplicate 0.6 mm diameter cores selected from the most representative tumor cellular areas of the primary Bca.

2.3. Immunohistochemistry

Standard immunohistochemistry (IHC) was performed using 4 µm thick sections of TMA slides of BCa following antigen retrieval with a steam-heating (95 °C) system in 0.01 M citrate buffer (pH 6.0) for 20 min or 1 mmol/L Tris–EDTA buffer at pH 9.0. The slides were stained with the appropriately diluted antibodies (Table 1) using an automated immunostainer (Dako, Santa Clara, CA, USA). Different clones of ERβ isoform antibodies, prognostic markers, and coactivators (Table 1) were tested for the optimum and reproducible immunoreaction, following the standard IHC testing protocol established in our laboratory. The IHC testing was conducted on the following antibodies: Erα, Erβ1, Erβ2, ERβ5, p-c-Jun (1:100), cyclin D1 (1:50), NF-kBp65 (1:100), SRC-1 (1:100), TIF-2 (1:50), AIB-1 (1:100), Ki-67, P53, CK 56, PR, and Her2/neu. The ERβ1 (38/AR385-10R), ERβ2 (57/3), and ERβ5 (5/75) antibody clones used in our previous study [10] and in this study have been tested by many investigators [7]; the immunogens were found to be peptide specific to the ERβ2 and Erβ5 splice variants [56–61]. Under the optimum immunostaining condition, ERβ1 (385p/AR385-10R) antibody displayed a consistent immunoreaction with each IHC test. Myofibroblasts were identified by smooth muscle actin staining using the EnVision G/2 double stain system. The positive and negative tissue and reagent controls were used. The immunoreactions of nuclear staining were evaluated using a semiquantitative Allred scoring system [10], summing the proportion of positive cells (scored on a scale of 0–5) and staining intensity (scored on a scale of 0–3) to produce a cumulative score of 8. A total score of 0–2 was regarded as negative, and a total score > 3 with 1–10% immunoreactive cells as positive. For Erβ isoform protein expression, >20% nuclear positivity was taken as the cutoff of positivity for ERβ1 and 2 isoforms, while >40% was applied for Erβ5 protein expression [10]. Over 1% of ERα and PR nuclear staining was considered positive. The Her2/neu expression was interpreted following the HercepTest kit guidelines and was scored according to the ASCO/CAP guidelines and considered positive for 3+ Her2/neu staining or 2+ Her2 staining with fluorescent in situ hybridization positivity. A nuclear immune reaction of Ki-67 > 15% and p53 > 5% was considered positive. The positive nuclear reaction of AIB-1, TIF-2, SRC-1, NF-kB, cyclin D1, and p-c-Jun in BCa were compared with those of normal breast tissues.

2.4. Statistical Analysis

The associations and correlations between the Erβ isoform protein, coactivators, and clinical characteristics were assessed for the entire cohort and the subtypes and subgroups of BCa using Fisher's exact test and by Spearman's rank-order test, respectively. Overall survival (OS) was calculated from the date of BCa diagnosis to death or the last follow-up visit, and the OS outcomes were estimated using Cox univariate and multivariate proportional hazard (PH) regression models. The hazard ratios were determined with 95% confidence intervals. Results with a p-value < 0.05 were considered significant.

This study sample size was sufficient statistically to detect correlations as small as ±0.17 and to detect relationships that explain at least 3% of the variance in dependent variables. All analyses were conducted using SAS 9.4 (SAS Institute Inc., Cary, NC, USA).

Table 1. Antibodies for immunohistochemistry study.

Antibody	Antibody Clone	Supplier
ERβ1	385P/AR 385-10R	Biogenex, San Ramon, CA, USA
ERβ2	MCA2279S/57/3	Bio-rads, Hercules, CA, USA
ERβ5	MCA4676/5/25	Bio-rads, Hercules, CA, USA
AIB-1	clone 34, mouse monoclonal	BD Transductuction Labs, San Jose, CA, USA
TIF-2	clone 29, mouse monoclonal	BD Transductuction Labs, San Jose, CA, USA
NF-kB p65	ABCAM E379	Waltham, MA, USA
SRC-1	clone 128E7, rabbit monoclonal	Cell Signaling Technology, Daners, MA, USA
Cyclin D1	DCS-6	DAKO, Carpintena, CA, USA
p-c-Jun	822, KM-1	Santa Cruz Biotech, Dallas, TX, USA
Actin-SMA	clone 2A4, mouse antihuman	DAKO, Carpintena, CA, USA
Ki-67	MIB-1	DAKO, Carpintena, CA, USA
P53	D07	DAKO, Carpintena, CA, USA
HER2/neu	HerceptTest	DAKO, Carpintena, CA, USA
ERα	ID5	DAKO, Carpintena, CA, USA
PR	Pg363	DAKO, Carpintena, CA, USA

3. Results

The immunostaining of ERβ isoform 1, 2 and 5 proteins was strongly positive in the nuclei of luminal epithelial and myoepithelial cells, and stromal cells including fibroblasts, myofibroblast (MF), endothelial cells, and lymphocytes in the benign breast tissues, whereas that of Erα protein was positive only in the nuclei of luminal epithelial cells. The polyclonal ERβ1 (385p/AR385-10R) and ERβ5 (57/3) antibodies produced a stronger nuclear staining and some cytoplasmic staining than ERβ2. ERβ isoform 1, 2, or 5 protein expression was detected in 61.5%, 44.9%, and 59.5% of the entire cohort, respectively. ERβ1 protein expression showed differential expression in BCa subtypes with higher expression in well-differentiated duct carcinoma and lobular carcinoma than in poorly differentiated BCa. The ERβ1 protein expression was coexpressed with a high Her2/neu and p53 expression in the ERα-negative BCa. A. high Ki-67 positivity > 15% correlated with ERβ1, ERβ2, and/or ERβ5 protein expression in the various subtypes of BCa, as shown in our previous study [10].

A high immunoreaction, as determined by an Allred score > 3, for AIB-1, TIF-2, SRC-1, NF-kB, and p-c-Jun protein expression was consistently observed in the nuclei of neoplastic epithelial cells, as well as in some stromal cells, particularly in MF (Figure 1). The nuclear expression of ERβ1 in epithelial cells was positively correlated with that in MF. On the contrary, ERα was neither expressed in the stromal cells nor in the MF. The ERβ1 expression was significantly associated with AIB-1, TIF-2, and p-c-Jun and with high-grade carcinoma with desmoplastic reaction and heavy lymphocytic infiltration. The nuclear expression of AIB-1, TIF-2, NF-kB, and p-c-Jun in MF gradually increased from the benign proliferative disease to carcinoma. Overall, AIB-1 protein expression was exclusively present in BCa and high-grade tumors and was higher in invasive BCa than in benign proliferative breast tissues. The cyclin D1 reaction levels in ERα-positive BCa (32.9%) were higher than those of ERα-negative BCa (11.4%) and TNBC (9.4%). Overall, the positive immunoreaction levels of cyclin D1 and p-c-Jun were lower than those of AIB-1, TIF-2, and NF-kB.

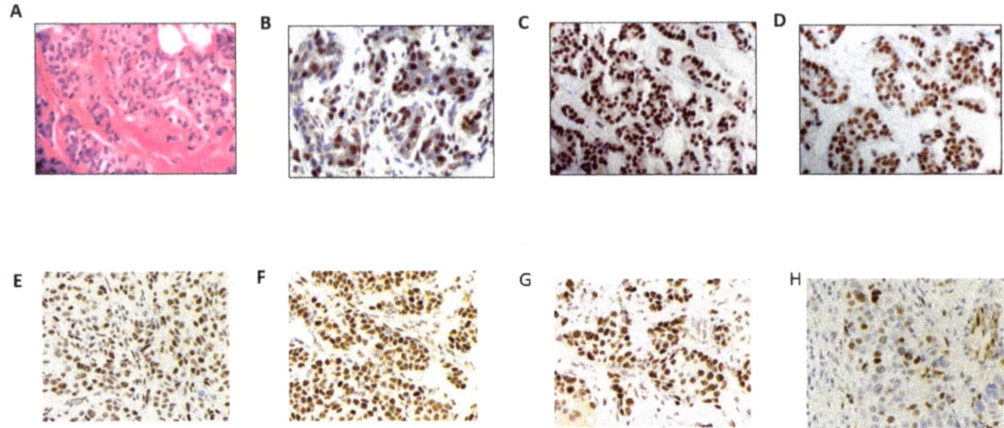

Figure 1. Immunohistochemistry stains of ERβ1 expression and coactivators in infiltrating duct carcinoma: (**A**) H & E staining of infiltrating duct carcinoma; (**B**) Erβ1 expression in the nuclei of benign epithelial cells and myoepithelial cells, stromal cells, and lymphocytes; (**C**) ERβ1 expression in the nuclei of neoplastic epithelial cells of infiltrating carcinoma and stromal cells and lymphocytes (immunohistochemistry staining using polyclonal ERβ1 385p/AR385-10R antibody); (**D**) ERα is expressed only in the nuclei of epithelial cells (original magnification 20×); immunohistochemistry stains of (**E**) AIB-1, (**F**) TIF-2, (**G**) SRC-1, and (**H**) p-c-Jun coactivators showing a positive nuclear reaction in the infiltrating carcinoma (original magnification 40×).

3.1. Association of Coactivators and Clinical Parameters in BCa Subtypes

A high cyclin D1 immunoreaction was positively associated with ERα-positive BCa, while that of TIF-2 and SRC-1 was associated with P53 > 5% positivity and that of p-c-Jun was associated with high Her2/neu expression (Table 2). However, there was an inverse association between cyclin D1 expression and luminal-B-type and TNBC (Table 3).

3.2. Spearman Rank Order Correlation between Coactivators and ERβ Isoforms

High expression levels of coactivators were significantly and differentially correlated with the expression of ERβ isoforms and clinical parameters. In the entire cohort (Table 4), high expression levels of AIB-1, TIF-2, and NF-kB were correlated with high ERβ1 and -5 expressions, while SRC-1, cyclin D1, and p-c-Jun were not associated with any of the ERβ isoforms. A high expression of ERβ5 isoform was correlated with high Ki-67, her2/neu, P53, and high-grade BCa; high ERβ1 expression was correlated with high Ki-67, high-grade and large-size BCa; and ERβ2 expression was correlated with lymph-node positive BCa and luminal-A-type BCa.

Table 2. Association between coactivators and clinical characteristics.

Variables		AIB-1 Pos	Neg	*p*-Value *	NF-kB Pos	Neg	*p*-Value	TIF-2 Pos	Neg	*p*-Value	SRC-1 Pos	Neg	*p*-Value	p-c-Jun Pos	Neg	*p*-Value	Cyclin D1 Pos	Neg	*p*-Value
ERα status	Pos	49	11	0.3	58	5	1	60	4	1	49	16	1	48	14	0.83	22	50	**0.024**
	neg	49	6		43	3		48	3		43	13		40	14		7	57	
Her-2/neu	Pos	34	6	1	35	3	1	33	1	0.67	29	11	0.49	33	4	**0.022**	14	29	0.08
	Neg	69	11		66	5		75	6		63	17		55	24		17	78	
PR	Pos	46	9	0.79	51	4	1	54	4	1	47	13	0.83	43	14	1	20	47	0.07
	Neg	52	8		50	4		54	3		45	15		45	14		11	60	
Ki-67	>15%	27	1	**0.07**	25	1	0.67	27	1	1	70	5	0.61	23	8	0.61	8	21	0.46
	<15%	71	16		76	6		81	6		22	23		68	20		23	86	
Grade	Grade 2/3	85	16	0.69	88	7	1	93	6	1	80	23	0.54	79	22	0.19	29	91	0.36
	Grade 1	13	1		13	1		15	1		12	5		9	6		22	16	
Tumor size	>2 cm	42	7	1	43	3	1	44	2	0.7	39	11	0.83	36	12	1	15	39	0.29
	<2 cm	56	10		58	5		64	5		53	17		52	16		16	68	
Nodal status	Pos	23	3	0.76	26	1	0.67	25	2	0.67	22	7	1	20	6	1	7	24	1
	Neg	75	14		76	7		83	5		70	21		68	22		24	83	
CK5/6	Pos	12	2	1	13	1	1	14	1	1	13	1	0.18	11	24	0.75	3	13	1
	Neg	86	15		88	7		84	6		79	27		77	24		28	94	
P53 > 5%	Pos	60	6	0.3	50	2	0.27	56	0	**0.013**	48	8	**0.032**	46	10	0.13	14	44	0.68
	Neg	48	11		51	6		52	7		44	20		42	18		17	63	

* All *p*-values were calculated with the Fisher's Exact test.; bold: significant *p*-value < 0.05.

Table 3. Associations between coactivators and molecular types of breast cancers.

	AIB-1			NF-kB			TIF-2			SRC-1			p-c-Jun			cyclin D1		
	Pos	Neg	*p*-Value *	Pos	Neg	*p*-Value	Pos	Neg	*p*-Value	Pos	Neg	*p*-Value	Pos	Neg	*p*-Value	Pos	Neg	*p*-Value
Types																		
Luminal A type (50)	31	8	0.26	36	5	1	40	4	0.42	35	7	0.66	30	12	0.49	14	36	0.29
	67	9		69	3		69	3		57	19		58	16		17	71	
Luminal B type (25)	21	3	1	23	2	1	21	0	0.34	17	7	0.43	19	3	0.27	10	16	**0.039**
	77	14		78	6		87	7		95	21		9	25		21	91	
Basal-like type (17)	13	2	1	14	1	1	5	1	1	13	3	0.51	11	5	0.53	2	15	0.36
	85	15		87	7		2	6		79	26		77	23		29	92	
HER2 type (17)	12	3	0.46	14	1	1	15	0	0.59	14	1	0.19	11	4	0.75	5	12	0.53
	86	14		87	7		93	7		78	27		77	24		26	95	
TNBC (43)	32	3	0.26	29	2	1	34	2	1	28	7	0.64	25	11	0.34	4	40	**0.009**
	66	14		72	6		74	5		64	21		63	17		27	67	

* All *p*-values calculated with the Fischer Exact test.; bold: significant *p*-value < 0.05.

Table 4. Spearman rank correlation of ERβ isoform protein expression with coactivators and clinical parameters in the entire cohort.

	ERβ1 Protein	ERβ2 Protein	ERβ5 Protein
AIB-1	**0.19 (0.047)**	0.066 (0.48)	**0.25(0.0064)**
NF-kB	**0.21 (0.028)**	0.13 (0.17)	**0.41 (<0.0001)**
TIF-2	0.24 (0. 008)	−0.01 (0.90)	**0.31 (0.0005)**
SRC-1	0.17 (0.07)	0.07 (0.45)	0.12 (0.17)
p-c-Jun	−0.04 (0.63)	0.08 (0.38)	0.06 (0.49)
Cyclin D1	0.11 (0.18)	0.11 (0.18)	0.13 (0.14)
Ki-67	**0.38 (<0.0001)**	0.14 (0.088)	**0.34 (<0.0001)**
P53	0.085 (0.33)	0.06 (0.49)	**0.30 (0.029)**
Grade 3	**0.17 (0.049)**	0.17 (0.071)	**0.19 (0.024)**
>2 cm	**0.17 (0.04)**	0.06 (0.49)	0.10 (0.23)
Her2/neu+	0.26 (0.89)	0.25 (0.10)	**0.31 (0.045)**
LN+	0.09 (0.29)	**0.23 (0.007)**	0.14 (0.08)
ERα+	0.01 (0.81)	0.15 (0.07)	−0.0005 (0.99)
PR+	0.005 (0.95)	0.07 (0.39)	−0.08 (0.32)
Luminal A type	0.005 (0.94)	**0.17 (0.045)**	−0.08 (0.32)
Luminal B type	−0.01 (0.87)	−0.05 (0.54)	−0.07 (0.37)
HER2 type	−0.07 (0.39)	−0.09 (0.27)	0.08 (0.36)
Basal type	0.03 (0.7)	0.05 (0.58)	**0.17 (0.038)**

Bold: significant *p*-value < 0.05.

In the subtypes and subgroups of BCa (Table 5), the coexpression of high AIB-1, NF-kB, p-c-Jun, and TIF-2 and ERβ isoforms was significantly correlated with poor clinical prognostic markers, such as high Ki-67, p53, high-grade BCa, large-size BCa, and/or positive LN and with different types of BCa and molecular types. The coexpression of cyclin D1 and ERβ5 was correlated with ERα- and PR-positive BCa and luminal-A-type BCa, while p-c-Jun and ERβ5 were correlated with luminal-B-type BCa. Furthermore, the coexpression of high ERβ1 and NF-kB, as well as TIF-2 was correlated with high-grade BCa, and the expression of high ERβ1 and cyclin D1 was correlated with high Her2/neu BCa and luminal-B-type. Coexpression of TIF-2 and both of the ERβ5 and ERβ1 isoforms in TNBC suggests that TIF-2 may coregulate the proliferation and progression of Erβ-expressing TNBCs.

Among the ERβ isoforms, the ERβ 1 and -5 isoforms, predominantly ERβ5, were significantly correlated with coactivators in BCa, while ERβ2 was not associated with coactivators. Among the coactivators, AIB-1, NF-kB, p-c-Jun, and TIF-2 were significantly associated with ERβ isoform expression, while SRC-1 was not. Thus, SRC-1 seems independent of the other coactivators.

3.3. Cox Univariate OS and Cofactors Expression in BCa Subtypes and Subgroups

Using a Cox univariate proportional hazards model (Table 6), it was found that among the entire cohort, AIB-1, TIF-2, SRC-1, and NF-kB did not show any significant association with OS. However, in the subgroups, cyclin D1 expression was the risk factor for OS in ERα-positive BCa (*p* = 0.0336), PR-positive BCa (*p* = 0.0128), and luminal-A-type BCa (*p* = 0.0320).

Table 5. Spearman rank correlation of ERβ isoforms and coactivators in subtypes and subgroups.

Correlation with	subgroups	ERβ 5 Expression r (*p*-value) *	ERβ1 Expression r (*p*-value)	ERβ2 Expression r (*p*-value)
AIB-1	ERα+	0.34 (0.0082)	0.18 (0.16)	0.15 (0.25)
	ERα-	0.18 (0.20)	0.21 (0.11)	0.012 (0.93)
	Luminal A type	0.38 (0.019)	0.22 (0.17)	0.15 (0.34)
	HER2 type	0.54 (0.035)	0.46 (0.08)	−0.09 (0.73)
	>2 cm tumor	0.29 (0.042)	0.11 (0.44)	−0.16 (0.27)
	Grade 3	0.23 (0.02)	0.14 (0.15)	−0.12 (0.21)
	Her2/neu+	0.35 (0.027)	0.19 (0.23)	0.1 (0.34)
NF-kB	ERα+	0.43 (0.0004)	0.22 (0.07)	0.24 (0.06)
	ERα-	0.39 (0.0078)	0.20 (0.17)	0.05 (0.72)
	Luminal A type	0.51 (0.0007)	0.15 (0.33)	0.11 (0.50)
	HER2 type	0.68 (0.035)	0.07 (0.08)	0.03 (0.9)
	Ki-67 > 15%	0.31 (0.019)	0.16 (0.23)	−0.07 (0.59)
	Her2/neu+	0.45 (0.0046)	0.20 (0.21)	0.15 (0.36)
	>2 cm tumor	0.29 (0.042)	0.14 (0.35)	−013 (0.39)
	Grade 3	0.42 (<0.0001)	0. 25 (0.013)	0.15 (0.16)
	PR	0.47 (0.0003)	0.023 (0.08)	−0.13 (0.33)
	LN+	0.47 (0.013)	−0.13 (0.51)	0.27 (0.16)
TIF-2	ERα+	0.35 (0.0039)	−0.21 (0.09)	−0.15 (0.23)
	ERα-	0.34 (0.043)	−0.13 (0.34)	−0.19 (0.16)
	TNBC	0.33 (0.046)	0.33 (0.05)	−0.19 (0.26)
	Luminal A type	0.35 (0.019)	0.17 (0.24)	0.28 (0.07)
	Ki-67 > 15%	0.32 (0.01)	0.15 (0.21)	−0.12 (0.33)
	p53 > 5%	0.28 (0.029)	0.12 (0.33)	0.12 (0.38)
	Grade 3	0.30 (0.0023)	0.25 (0.012)	0.05 (0.61)
	PR	0.28 (0.03)	0.16 (0.23)	0.19 (0.16)
Cyclin D1	ERα+	0.29 (0.011)	0.19 (0.11)	0.16 (0.18)
	Her2/neu+	0.09 (0.53)	0.3 (0.049)	0.12 (0.45)
	Luminal A type	0.25 (0.004)	0.11 (0.45)	0.25 (0.07)
	Luminal B type	0.1 (0.49)	0.45 (0.02)	0.02 (0.94)
	PR	0.24 (0.046)	0.22 (0.07)	0.09 (0.43)
p-c-Jun	Luminal B type	0.44 (0.039)	0.007 (0.97)	−0.008 (0.71)
SRC-1	ERα+	0.18 (0.17)	0.09 (0.49)	0.01 (0.93)
	ERα-	0.08 (0.48)	0.22 (0.07)	0.08 (0.52)
	Her2/neu+	0.53 (0.34)	0.10 (0.53)	−0.09 (0.54)
	PR	0.12 (0.34)	0.20 (0.11)	0.19/0.13

* bold: significant *p*-value < 0.05.

Table 6. Cox univariate overall survival analysis of coactivators in breast cancer subtypes and subgroups.

Subgroups (Case#)	AIB-1		NF-kB		TIF-2		SRC-1		p-c-Jun		Cyclin D1	
	p-Value	HR (CI)	*p*-Value	HR (CI)	*p*-Value	HR (CI)	*p*-Value	HR (CI)	*p*-Value	HR (CI)	*p*-Value	HR (CI)
ERα- positive BCa (73)	0.25	0.99 (0.96–1.011)	0.58	0.733 (0.25–2.17)	0.103	0.98 (0.96–1.004)	0.53	0.99 (0.97–1.02)	0.64	0.94 (0.77–1.02)	0.034	1.02 (1.001–1.037)
ERα -negative BCa (65)	0.23	1.01 (0.99–1.03)	0.76	1.1 (0.52–2.45)	0.22	1.01 (0.99–1.03)	0.79	0.99 (0.98–1.02)	0.24	1.01 (0.99–1.03)	0.68	0.99 (0.91–1.06)
TNBC (43)	0.5	1.008 (0.98–1.03)	0.54	0.71 (0.23–2.2)	0.17	1.02 (0.99–1.05)	0.43	0.99 (0.97–1.02)	0.49	1.008 (0.99–1.03)	0.99	0.06 (0.000–1.0000)
Her2/neu+ (39)	0.35	0.98 (0.97–1.013)	0.72	1.2 (0.44–3.29)	0.67	0.99 (0.97–1.022)	0.76	0.99 (0.96–1.03)	0.79	1.004 (0.98–1.03)	0.59	1.006 (0.98- 1.03)
PR+ (54)	0.54	0.99 (0.97–1.02)	0.59	0.73 (0.23–2.34	0.19	0.98 (0.96–1.007)	0.53	0.99 (0.97–1.016)	0.66	0.99 (0.97–1.021)	0.0128	1.03 (1.005–1.05)
Luminal A type (50)	0.87	1.003 (0.97–1.04)	0.97	1.04 (0.206–204)	0.1	0.9 (0.94–1.006)	0.72	0.99 (0.97–1.024)	0.59	0.99 (0.959–1.03)	0.032	1.028 (1.002–1.055)
Luminal B type (25)	0.12	0.95 (0.89–1.01)	0.48	0.56 (0.12–2.71)	0.34	1.1 (0.8–1.41)	0.61	0.99 (0.95–1.03)	0.87	1.004 (0.96–1.05)	0.61	1.007 (0.98–1.034)
HER2 type (17)	**		**		**		**		**		**	
Basal type (17)	0.12	1.03 (0.99–1.07)	0.21	3.1 (0.54–17.38)	0.34	1.03 (0.97–1.097)	0.93	1.001 (0.97–1.03)	0.96	1.001 (0.97–1.03)	0.99	0.46 (0.000–3.87)
Grade 2/3 (115)	0.97	1.000 (0.98–1.013)	0.93	0.97 (0.49–1.9)	0.78	0.99 (0.99–1.011)	0.2	0.99 (0.97–1.006)	0.71	1.003 (0.99–1.012)	0.23	1.009 (0.99–1.023)
Grade 1 (23)	**		**		**		**		**		**	
>2 cm tumor (51)	0.91	1.001 (0.98–1.02)	0.88	1.05 (0.54–2.05)	0.72	0.99 (0.98–1.013)	0.99	1.000 (0.98–1.02)	0.56	1.005 (0.98–1.023)	0.65	1.004 (0.99–1.02)
<2 cm tumor (87)	0.82	1.004 (0.97–1.03)	0.84	0.85 (0.18–3.9)	0.77	0.99 (0.97–1.02)	0.25	0.98 (0.94–1.02)	0.58	1.009 (0.978–1.04)	0.31	1.019 (0.98–1.04)
>15% Ki-67 (63)	0.85	0.99 (0.98–1.014)	0.9	1.05 (0.50–2.19)	0.27	0.99 (0.98–1.006)	0.19	0.99 (0.97–1.007	0.73	1.003 (0.98–1.020)	0.11	1.013 (0.99–1.03)
LN positive (34)	0.83	1.002 (0.98–1.03)	0.85	1.16 (0.25–5.33)	0.82	0.99 (0.98–1.02)	0.83	1.00 (0.9–1.02)	0.25	1.02 (0.99–1.06)	0.28	1.011 (0.99–1.03)
p53 > 5% (57)	0.59	0.99 (0.98–1.013)	0.92	0.95 (0.37–2.5)	0.097	0.98 (0.97–1.003)	0.31	0.99 (0.96–1.013)	0.31	1.011 (0.9–1.032)	0.57	0.99 (0.9–1.02)

** sample too small to reliably calculate COX; HR: Hazard Ratio, CI: Confidence Interval.

4. Discussion

Studies on the role of coactivators in BCa have largely been investigated in ERα-positive BCa. In ERα-positive BCa, AIB-1 amplification has been associated with worse outcomes [26], progression of these tumors [62], resistance to TAM, early relapse during treatment, and distant recurrences. Moreover, high AIB1 expression in patients with Her2/neu-overexpressing tumors has been associated with an increased risk of relapse on tamoxifen [63] and, along with poor prognostic factors, with poorer DFS and OS in ERα-positive and -negative BCa [24]. This supports the notion of crosstalk between ERα and growth factor receptor pathways through specific coactivator proteins. Furthermore, high expression levels of cyclin D1 were significantly correlated to ERα positivity and with luminal A type [64], as well as high proliferation and a higher risk of death in ERα-positive BCa [43].

Studies on the role of ERβ isoforms and cofactors in BCa are limited. In this study, the most pertinent findings are the significant association and correlation between the expression of the ErβXXX5 and/or Erβ1 isoforms and AIB-1, NF-kB, TIF-2, p-c-Jun, and cyclin D1 coactivators in the BCa subtypes and subgroups. However, ERβ2 was not associated with coactivators and SRC-1 was not associated with ERβ expression. The coactivators were found to be differentially correlated with ERβ5 and/or ERβ1 expression in ERα-positive and ERα-negative BCa, as well as with TNBC and different molecular types of BCa. Their coexpression is associated and correlated with high-grade and large-sized tumors and high Her2/neu, p53, and Ki-67 positive BCa. High Ki-67 expression in BCa with high NF-kB, TIF-2, and AIB-1 expression suggests that the coactivators may be involved in the proliferation and growth of BCa. The coexpression of both ERβ5 and ERβ1 and TIF-2 in TNBC suggests that both the TIF-2 and ERβ isoforms may be implicated in poor prognosis in TNBC. The coexpression of high ERβ expression and AIB-1 and TIF-2 in MF in high-grade carcinoma with desmoplastic reaction and heavy lymphocytic infiltration suggests that the activation of AIB-1 and TIF-2 signal transductions in the MF may be involved in the initiation and progression of ERβ1-expressing BCa [65], as MF are the predominant cells in the cancer microenvironment that orchestrate the epithelial–mesenchymal crosstalk [66]. Tzelepi et al. [67] also reported that AIB-1 was more frequently expressed in the MF of dysplastic or cancer-associated mucosa stroma compared with normal mucosa. Enhanced nuclear ERβ1 expression and elevated nuclear AIB-1 expression were more frequently noted in the MF of carcinomas of an advanced stage, supporting the notion of the possible role of these coactivators in the initiation and progression of colorectal carcinomas through paracrine actions [22].

Although ERβ2 expression in this study was not associated with the coactivators, others have reported that ERβ2 mRNA levels are correlated with AIB-1 mRNA levels [68], and ERβ2 protein expression was found to be strongly associated with p-c-Jun and NF-kBp65 in ERα-negative BCa [69].

Furthermore, SRC-1 in our study was not correlated with any ERβ isoforms in BCa. However, others [70] have observed that patients with high expression levels of Her2/neu in combination with SRC-1 have a greater probability of recurrence on endocrine treatment compared with those who are Her2/neu positive but SRC-1 negative. SRC-1 was associated with nodal positivity and resistance to endocrine treatment. Fleming et al. [34] reported that SRC-1 was inversely associated with ERβ, negatively associated with DFS, and positively correlated with Her2/neu.

There was no significant association between OS and AIB-1, TIF-2, SRC-1, and NF-kB. However, among the subtypes, cyclin D1 was a significant risk factor for OS in ERα-positive BCa ($p = 0.0336$), PR-positive BCa ($p = 0.0128$), and luminal-A-type BCa ($p = 0.0320$). Others reported that among the ERα-negative subgroup, strong AIB-1 protein expression correlated with poorer DFS and overall survival and correlated with the amplification of the Her2/neu gene [24]. AIB-1 was found to enhance the estrogen-dependent induction of cyclin D1 expression by ERα [41].

5. Conclusions

Our study is the first comprehensive simultaneous investigation of the correlation and association of the ERβ1, ERβ2, and ERβ5 isoforms with multiple coactivators, including AIB-1, NF-kB, cyclin D1, SRC-1, p-c-Jun, and TIF-2, in the entire cohort, as well as in the subtypes and subgroups of BCa. AIB-1, NF-kB, p-c-Jun, and TIF-2 were found to be associated and correlated with ERβ5 and ERβ1 expression, as well as with poor clinical parameters, and were differently associated with the subtypes of BCa, including different molecular types. ERβ5 was determined to be the predominant ERβ isoform associated and correlated with coactivators in the subtypes and subgroups of BCa, while ERβ2 did not demonstrate the relationship. High Ki-67 expression with the coexpression of coactivators and ERβ5 suggests a potential involvement of the coactivators in the proliferation of ERβ-expressing BCa. SRC-1 is not associated with any ERβ expression. Cyclin D1 was the risk factor for OS only in the BCa subtypes.

In summary, although this study was limited by its relatively small sample size with respect to the subtypes and groups, we firmly believe that the sample size sufficiently supported both our positive and negative results.

ERβ interacts with the members of the SRC family and other coactivators and coregulate the development and growth of BCa [49–51,54]. As ERβ isoforms were found to be the risk factors and associated with unfavorable clinical outcomes in BCa in our previous study [10], the significant correlation between ERβ isoforms and the coactivators in the present study supports the notion that the coactivators are co-implicated in the proliferation of BCa and the risk factors of ERβ-expressing BCa.

Previous studies [71–74] have demonstrated that the activity of ERs depends on the coordinated activity of ligand binding, PTM, and interaction with their partner coregulators and that distinct receptor subtype-specific coregulators are recruited at the transcription sites and factors, such as ERα or ERβ. Thus, further studies with other coregulators and large cohorts of BCa subgroups and subtypes, including the BRCA-1-associated TNBC [75], are needed to determine the involvement of specific coactivators in ERβ-expressing BCa.

This may provide insights into the potential usefulness of the coactivators as therapeutic targets in BCa in the adjuvant setting. The blocking of coactivators may slow disease progression and potentially play an important role in the adjuvant setting to prevent disease recurrence and the development of metastases in the subtypes of BCa [16,37,38,76–78].

Author Contributions: All authors contributed to the study conception and design; Y.C. designed the study, performed patient data collection, supervised the TMA, immunohistochemistry, and final interpretation of the data, and prepared the entire manuscript. The statistical analysis was conducted by S.P. Both authors critically reviewed and revised final version of the manuscript. All authors have read and agreed to the published version of the manuscript.

Funding: This work was supported in part by a grant-in-aid from Dong-A ST Research Institute 21, Geumhwa-ro, 105 Beon-Gil, Giheung-gu, Yongin-si, Gyeonggi-do, Republic of Korea, and Zip: 17073.

Institutional Review Board Statement: All procedures involving human subjects were performed according to the ethical standards of the Institutional Research Board, The Bridgeport Hospital, Bridgeport, CT (IRB# 090101). Owing to the retrospective nature of the study, informed patient consent was not required.

Informed Consent Statement: All procedures involving patients with BCa were performed according to the ethical standards of the Institutional Research Board, Bridgeport Hospital, Bridgeport, CT (IRB# 090101).

Data Availability Statement: The datasets generated during and/or analyzed during the current study are included in this manuscript. Any additional request is available from the corresponding author upon reasonable request.

Conflicts of Interest: The authors have no conflict of interest to declare.

References

1. Powell, E.; Shanle, E.; Brinkman, A.; Keles, S.; Wisinski, K.; Huang, W.; Xu, W. Identification of estrogen receptor dimer selective ligands reveals growth-inhibitory effects on cells that co-express ERα and ERβ. *PLoS ONE* **2012**, *7*, e30993. [CrossRef] [PubMed]
2. Hall, J.; McDonnell, D. The estrogen receptor β isoform (ERβ) of the human estrogen receptor modulates ERα transcriptional activity and is a key regulator of the cellular response to estrogens and antiestrogens. *Endocrinology* **1999**, *40*, 5566–5578. [CrossRef] [PubMed]
3. Borgquist, S.; Zholm, C.; Stendahl, M.; Anagnostaki, L.; Landberg, G.; Jirstrom, K. Oestrogen receptors α and β show different associations to clinicopathological parameters and their co-expression might predict a better response to endocrine treatment in breast cancer. *J. Clin. Pathol.* **2008**, *61*, 197–203. [CrossRef] [PubMed]
4. Leung, Y.; Lee, M.; Lam, H.; Tarapore, P.; Ho, S. Estrogen receptor-beta ad breast cancer: Translating biology into clinical practice. *Steroids* **2012**, *77*, 727–737. [CrossRef] [PubMed]
5. Bozovic, A.; Mandusic, V.; Todorovic, L.; Krajnovic, M. Estrogen receptor beta: The promising biomarker and potential target in metastasis. *Int. J. Mol. Sci.* **2021**, *22*, 1656. [CrossRef]
6. O'Malley, B.W.; Qin, J.; Lanz, R.B. Cracking the coregulatory codes. *Curr. Opin. Cell Biol.* **2008**, *20*, 310–315. [CrossRef]
7. Choi, Y. Estrogen receptor β expression and its clinical implication in breast cancers: Favorable or unfavorable? *J. Breast Cancer* **2022**, *25*, e9. [CrossRef]
8. Zhou, Y.; Liu, X. The role of estrogen receptor beta in breast cancer. *Biomark. Res.* **2010**, *8*, 39. [CrossRef]
9. Choi, Y.; Pinto, M. Estrogen receptor β in breast cancer: Associations between ERβ hormonal receptors, and other prognostic biomarkers. *App. Immunohistochem. Mol. Morphol.* **2005**, *13*, 19–25. [CrossRef]
10. Choi, Y.; Kim, H.; Pollack, S. ERβ isoforms have differential clinical significance in breast cancer subtypes and subgroups. *Curr. Issues Mol. Biol.* **2022**, *44*, 1564–1586. [CrossRef]
11. Hall, J.M.; McDonnell, D.P. Coregulators in nuclear estrogen receptor action: From concept to therapeutic targeting. *Mol. Interv.* **2005**, *5*, 343–357. [CrossRef] [PubMed]
12. Lonard, D.M.; Lanz, R.B.; O'Malley, B.W. Nuclear receptor coregulators and human disease. *Endocr. Rev.* **2007**, *28*, 575–587. [CrossRef] [PubMed]
13. York, B.; O'Malley, B.W. Steroid receptor coactivator (SRC) family: Master of Systems Biology. *J. Biol. Chem.* **2010**, *285*, 38743–38750. [CrossRef]
14. Dasgupta, S.; Lonard, D.; O'Malley, B.W. Nuclear receptor coactivators: Master regulators of human health and disease. *Annu. Rev. Med.* **2014**, *65*, 279–292. [CrossRef]
15. Altwegg, K.; Vadlamudi, R. The role of estrogen receptor correlators in endocrine resistant breast cancer. *Explor. Target Antitumor Ther.* **2021**, *2*, 385–400. [CrossRef]
16. Klinge, C.M.; Jernigan, S.C.; Mattingly, K.A.; Risinger, K.E.; Zhang, J. Estrogen response element-dependent regulation of transcriptional activation of estrogen receptors alpha and beta by coactivators and corepressors. *J. Mol. Endocrinol.* **2004**, *33*, 387–410. [CrossRef] [PubMed]
17. Wu, R.C.; Qin, J.; Hashimoto, Y.; Wong, J.; Xu, J.; Tsai, S.Y.; Tsai, M.-J.; O'Malley, B.W. Regulation of SRC-3 (pCIP/ACTR/AIB-1/RAC-3/TRAM-1) coactivator activity by I kappa B kinase. *Mol. Cell. Biol.* **2002**, *22*, 3549–3561. [CrossRef]
18. Xu, J.; Wu, R.C.; O'Malley, B.W. Normal and cancer-related functions of the p160 steroid receptor co-activator (SRC) family. *Nat. Rev. Cancer* **2009**, *9*, 615–630. [CrossRef]
19. Li, L.; Deng, C.; Chen, Q. SRC-3, a Steroid receptor coactivator: Implication in cancer. *Int. J. Mol. Sci.* **2021**, *22*, 4760. [CrossRef]
20. List, H.J.; Reiter, R.; Singh, B.; Wellstein, A.; Riegel, A.T. Expression of the nuclear coactivator AIB1 in normal and malignant breast tissue. *Breast Cancer Res. Treat.* **2001**, *68*, 21–28. [CrossRef]
21. Torres-Arzayus, M.I.; Zhao, J.; Bronson, R.; Brown, M. Estrogen-dependent and estrogen-independent mechanisms contribute to AIB-1-mediated tumor formation. *Cancer Res.* **2010**, *70*, 4102–4111. [CrossRef] [PubMed]
22. Yan, J.; Tsai, S.Y.; Tsai, M.J. SRC-3/AIB1: Transcriptional coactivator in oncogenesis. *Acta Pharmacol. Sin.* **2006**, *27*, 387–394. [CrossRef] [PubMed]
23. Gojis, O.; Rudraraju, B.; Gudi, M.; Hogben, K.; Sousha, S.; Coombes, R.; Cleator, S.; Palmieri, C. The role of SRC-3 in human breast cancer. *Nat. Rev. Clin. Oncol.* **2010**, *7*, 83–89. [CrossRef] [PubMed]
24. Lee, K.; Lee, A.; Song, B.; Kang, C. Expression of AIB1 protein as a prognostic factor in breast cancer. *World J. Surg. Oncol.* **2011**, *9*, 139. [CrossRef]
25. Dasgupta, S.; O'Malley, B.W. Transcriptional coregulators: Emerging roles of SRC family of coactivators in disease pathology. *J. Mol. Endocrinol.* **2014**, *53*, R47–R59. [CrossRef]
26. Burandt, E.; Jens, G.; Holst, F.; Jänicke, F.; Müller, V.; Quaas, A.; Choschzick, M.; Wilczak, W.; Terracciano, L.; Simon, R.; et al. Prognostic relevance of AIB1 (NCoA3) amplification and overexpression in breast cancer. *Breast Cancer Res. Treat.* **2013**, *137*, 745–753. [CrossRef]
27. Zhao, W.; Zhang, Q.; Kang, X.; Jin, S.; Lou, C. AIB1 is required for the acquisition of epithelial growth factor receptor-mediated tamoxifen resistance in breast cancer cells. *Biochem. Biophys. Res. Comm.* **2009**, *380*, 699–704. [CrossRef] [PubMed]
28. Osborne, C.K.; Bardou, V.; Hopp, T.A.; Chamness, G.C.; Hilsenbeck, S.G.; Fuqua, S.A.; Wong, J.; Allred, C.; Clark, G.; Schiff, R. Role of the estrogen receptor coactivator AIB1(SRC-3) and HER-2/neu in tamoxifen resistance in breast cancer. *J. Natl. Cancer Inst.* **2003**, *95*, 353–361. [CrossRef]

29. Shou, J.; Massarweh, S.; Osborne, C.K.; Wakeling, A.E.; Ali, S.; Weiss, H.; Schiff, R. Mechanisms of tamoxifen resistance: Increased estrogen receptor-HER2/neu crosstalk in ER/HER2-positive breast cancer. *J. Natl. Cancer Inst.* **2004**, *96*, 926–935. [CrossRef]

30. Song, X.; Zhang, C.; Zhao, M.; Chen, H.; Liu, X.; Chen, J.; Lonard, D.M.; Qin, L.; Xu, J.; Wang, X.; et al. Steroid receptor coactivator-3 (SRC-3/AIB1) as a novel therapeutic target in triple-negative breast cancer and its inhibition with a phospho-bufalin prodrug. *PLoS ONE* **2015**, *10*, e0140011. [CrossRef]

31. Gregory, C.; He, B.; Johnson, R.; Ford, H.; Mohler, J.; French, F.; Wilson, E. A mechanism for androgen receptor–mediated prostate cancer recurrence after androgen deprivation therapy. *Cancer Res.* **2001**, *61*, 4315–4319.

32. Hudelist, G.; Czerwenka, K.; Kubista, E.; Marton, E.; Pischinger, K.; Singer, C.F. Expression of sex steroid receptors and their cofactors in normal and malignant breast tissue: AIB1 is a carcinoma-specific co-activator. *Breast Cancer Res. Treat.* **2003**, *78*, 193–204. [CrossRef] [PubMed]

33. Fleming, F.J.; Hill, A.D.; McDermott, E.W.; O'Higgins, N.J.; Young, L.S. Differential recruitment of coregulator proteins steroid receptor coactivator-1 and silencing mediator for retinoid and thyroid receptors to the estrogen receptor-estrogen response element by beta-estradiol and 4- hydroxytamoxifen in human breast cancer. *J. Clin. Endocrinol. Metab.* **2004**, *89*, 375–383. [CrossRef]

34. Myers, E.; Fleming, F.; Crotty, T.; Kelly, G.; McDermott, E.; O'Higgins, N.; Hill1, A.; Young, L. Inverse relationship between ER-b and SRC-1 predicts outcome in endocrine-resistant breast cancer. *Br. J. Cancer* **2004**, *91*, 1687–1693. [CrossRef] [PubMed]

35. McBryan, J.; Theissen, S.M.; Byrne, C.; Hughes, E.; Cocchiglia, S.; Sande, S.; O'Hara, J.; Tibbitts, P.; Hill, A.D.; Young, L.S. Metastatic progression with resistance to aromatase inhibitors is driven by the steroid receptor coactivator SRC-1. *Cancer Res.* **2012**, *72*, 548–559. [CrossRef] [PubMed]

36. Lee, S.-K.; Kim, H.-J.; Na, S.-Y.; Kim, T.S.; Choi, H.-S.; Im, S.-Y.; Lee, J.W. Steroid receptor coactivator-1 coactivates activating protein-1-mediated transactivations through interaction with the c-Jun and c-Fos subunits. *J. Biol. Chem.* **1998**, *273*, 16651–16654. [CrossRef]

37. Xia, Y.; Shen, S.; Verma, I. NF-kB, an active player in human cancers. *Cancer Immunol. Res.* **2014**, *2*, 823–830. [CrossRef]

38. Lin, Y.; Bai, L.; Chen, W.; Xu, S. The NF-κB activation pathways, emerging molecular targets for cancer prevention and therapy. *Expert Opin. Ther. Targets* **2010**, *14*, 45–55. [CrossRef]

39. Vleugel, M.; Wall, E.; Van Diest, P. C-jun activation is associated with proliferation and angiogenesis in invasive breast cancer. *Hum. Pathol.* **2006**, *37*, 668–674. [CrossRef]

40. Jiao, X.; Katiyar, S.; Willmarth, N.; Liu, M.; Ma, X.; Flomenberg, N.; Lisanti, M.; Pestell, R. c-Jun induces mammary epithelial cellular invasion and breast cancer stem cell expansion. *J. Biol. Chem.* **2010**, *285*, 82188226. [CrossRef]

41. Planas-Silva, M.D.; Shang, Y.; Donaher, J.L.; Brown, M.; Weinberg, R.A. AIB-1 enhances estrogen-dependent induction of cyclin D1 expression. *Cancer Res.* **2001**, *61*, 3858–3862. [PubMed]

42. Zwijsen, R.; Buckle, R.E.; Hijmans, M.; Loomans, C.; Bernards, R. Ligand-independent recruitment of steroid receptor coactivators to estrogen receptor by cyclin D1. *Genes Dev.* **1998**, *12*, 3488–3498. [CrossRef] [PubMed]

43. Ahlin, C.; Lundgren, C.; Embretse'n-Varro, E.; Jirstro¨m, K.; Blomqvist, C.; Fja¨llskog, M. High expression of cyclin D1 is associated to high proliferation rate and increased risk of mortality in women with ER-positive but not in ER-negative breast cancers. *Breast Cancer Res. Treat.* **2017**, *164*, 667–678. [CrossRef] [PubMed]

44. Mohammadizadeh, F.; Hani, M.; Ranaee, M.; Bagheri, M. Role of cyclin D1 in breast cancer. *J. Res. Med. Sci.* **2013**, *18*, 1021–1025. [PubMed]

45. Suen, C.S.; Berrodin, T.J.; Mastroeni, R.; Cheskis, B.J.; Lyttle, C.R.; Frail, D.E. A transcriptional co-activator, steroid co-activator-3 selectively augments steroid receptor transcriptional activity. *J. Biol. Chem.* **1998**, *273*, 27645–27653. [CrossRef] [PubMed]

46. Liu, M.; Albanese, C.; Anderson, C.; Hilty, K.; Webb, P.; Uht, R.; Price, R.; Pestell, R.; Kushner, P. Opposing action of estrogen receptors α and β on cyclin D1 gene expression. *J. Biol. Chem.* **2002**, *277*, 24353–24380. [CrossRef]

47. McIlroy, M.; Fleming, F.; Buggy, Y.; Hill, A.; Young, L. Tamoxifen-induced ER-α–SRC-3 interaction in HER2 positive human breast cancer; a possible mechanism for ER isoform specific recurrence. *Endocr.-Relat. Cancer* **2006**, *13*, 1135–1145. [CrossRef]

48. Bai, Y.; Giguerère, V. Isoform-Selective Interactions between Estrogen Receptors and Steroid Receptor Coactivators Promoted by Estradiol and ErbB-2 Signaling in Living Cells. *Mol. Endocrinol.* **2000**, *17*, 589–599. [CrossRef]

49. Wong, C.; Boris, B.; Cheskis, J. Structure-function evaluation of ER α and β interplay with SRC family coactivators. *Biochemistry* **2001**, *40*, 6756–6765. [CrossRef]

50. Leung, Y.K.; Ho, S.M. Estrogen receptor beta: Switching to a new partner and escaping from estrogen. *Sci. Signal.* **2011**, *4*, e19. [CrossRef]

51. Tremblay, A.; Tremblay, G.; Labrie, F.; Giguere, V. Ligand-Independent Recruitment of SRC-1 to Estrogen Receptor β through Phosphorylation of Activation Function AF-1. *Mol. Cell* **1999**, *3*, 513–519. [CrossRef] [PubMed]

52. Herschberger, P.; Vasquez, A.; Kanterewicz, B.; Land, S.; Siegfried, J.; Nichols, M. Regulation of endogenous gene expression in human non-small cell lung cancer cells by estrogen receptor ligands. *Cancer Res.* **2005**, *65*, 1598–1605. [CrossRef] [PubMed]

53. Giannini, R.; Cavallini, A. Expression analysis of a subset of coregulators and three nuclear receptors in human colorectal carcinoma. *Anticancer Res.* **2005**, *25*, 4287–4292.

54. Kraichely, D.; Sun, J.; Katzenellenbogen, J.; Katzenellenbogen, B. Conformational changes and coactivator recruitment by novel ligands for estrogen receptor-α and estrogen receptor-β: Correlations with biological character and distinct differences among SRC coactivator family members. *Endocrinology* **2000**, *121*, 3534–3545. [CrossRef]

55. Choi, Y. ERβ, AIB-1 and TIF-2 expression in breast cancer-associated myofibroblasts. *Mod. Pathol.* **2010**, *23*, 40A.
56. Shaaban, A.; Green, A.; Karthik, S.; Alizadeh, Y.; Hughes, T.; Harkins, L.; Ellis, I.; Robertson, J.; Paish, E.; Saunders, P.; et al. Nuclear and Cytoplasmic Expression of ERB1, ERB2, and ERB5 Identifies Distinct Prognostic Outcome for Breast Cancer Patients. *Clin. Cancer Res.* **2008**, *14*, 16. [CrossRef]
57. Collins, F.; Itani, N.; Esnal-Zufiaurre, A.; Gibson, D.; Fitzgerald, C.; Saunders, P. The ERβ5 splice variant increases oestrogen responsiveness of ERα^{pos} Ishikawa cells. *Endocr.-Relat. Cancer* **2020**, *27*, 55–66. [CrossRef]
58. Collins, F.; MacPherson, S.; Brown, P.; Bombail, V.; Williams, A.; Anderson, R.; Jabbour, H.; Saunders, P. Expression of oestrogen receptors, ERα, ERβ, and ERβ variants, in endometrial cancers and evidence that prostaglandin F may play a role in regulating expression of ERα. *BMC Cancer* **2009**, *9*, 330. [CrossRef]
59. Liu, J.; Sareddy, G.; Zhou, M.; Viswanadhapalli, S.; Li, X.; Lai, Z.; Tekmal, R.; Brenner, A.; Vadlamudi, R. Differential effects of estrogen receptor b isoforms on glioblastoma progression. *Cancer Res.* **2018**, *78*, 3176–3189. [CrossRef]
60. Li, W.; Wintersa, A.; Poteeta, E.; Ryoua, M.; Lin, S.; Haob, S.; Wu, Z.; Yuan, F.; Hatanpaa, K.J.; Simpkins, J.W. Involvement of estrogen receptor b5 in the progression of glioma. *Brain Res.* **2013**, *29*, 97–107. [CrossRef]
61. Wimberly, H.; Han, G.; Pinnaduwage, D.; Murphy, L.C.; Yang, X.R.; Andrulis, I.L.; Sherman, M.; Figueroa, J.; Rimm, D.L. ERβ splice variant expression in four large cohorts of human breast cancer patient tumors. *Breast Cancer Res. Treat.* **2014**, *146*, 657–667. [CrossRef]
62. Azorsa, D.O.; Cunliffe, H.E.; Meltzer, P.S. Association of steroid receptor coactivator AIB1 with estrogen receptor-alpha in breast cancer cells. *Breast Cancer Res. Treat.* **2001**, *70*, 89–101. [CrossRef]
63. Kirkegaard, T.; McGlynn, L.; Campbell, F.; Muller, S.; Tovey, S.; Dunne, B.; Nielsen, K.; Cooke, T.; Barlett, J. Amplified in Breast Cancer 1 in human epidermal growth factor receptor-positive tumors of tamoxifen—Treated breast cancer patients. *Clin. Cancer Res.* **2007**, *13*, 1405–1411. [CrossRef]
64. Guo, L.; Liu, S.; Yilamu, D.; Wang, B.; Yan, J. Positive expression of cyclin D1 is an indicator for the evaluation of the prognosis of breast cancer. *Int. J. Clin. Exp. Med.* **2015**, *8*, 18656–18664.
65. Grivas, P.; Tzelepi, V.; Otiropoulou-Bonikou, G.; Kefalopoulou, Z.; Papavassiliou, A.; Kalofonos, H. Estrogen receptor α/β, AIB1, and TIF2 in colorectal carcinogenesis: Do coregulators have prognostic significance? *Int. J. Color. Dis.* **2009**, *24*, 613–622. [CrossRef] [PubMed]
66. Nascimento, C.; Ferreira, F. Tumor microenvironment of human breast cancer, and feline mammary carcinoma as potential study model. *BBA-Rev. Cancer* **2021**, *1876*, 188587. [CrossRef]
67. Tzelepi, V.; Grivas, P.; Kefalopoulou, Z.; Kalofonos, H.; Varakis, J.; Melachrinou, M.; Sotiropoulou-Bonikou, G. Estrogen signaling in colorectal carcinoma microenvironment: Expression of ERβ1, AIB-1, and TIF-2 is upregulated in cancer-associated myofibroblasts and correlates with disease progression. *Virchows Arch.* **2009**, *454*, 389–399. [CrossRef] [PubMed]
68. Girault, I.; Lerebours, F.; Amarir, S.; Tozlu, S.; Tubiana-Hulin, M.; Lidereau, R.; Bieche, I. Expression analysis of estrogen receptor alpha coregulators in breast carcinoma: Evidence that NCOR1 expression is predictive of the response to tamoxifen. *Clin. Cancer Res.* **2003**, *9*, 1259–1266. [PubMed]
69. Skliris, G.; Leygue, E.; Watson, P.; Murphy, L. Estrogen receptor alpha negative breast cancer patients: Estrogen receptor beta as a therapeutic target. *J. Steroid Biochem. Mol. Biol.* **2008**, *109*, 1–10. [CrossRef]
70. Fleming, F.; Myers, E.; Kelly, G.; Crotty, T.; McDermott, E.; O'Higgins, N.; Hill, A.; Young, L. Expression of SRC-1, AIB1, and PEA3 in HER2 mediated endocrine resistant breast cancer; a predictive role for SRC-1. *J. Clin. Pathol.* **2004**, *57*, 1069–1074. [CrossRef]
71. Paramanik, V.; Krishnan, H.; Thakur, M. Estrogen receptor α -and β interacting proteins contain consensus secretory structures: An Insilico study. *Ann. Neurosci.* **2018**, *25*, 1–10. [CrossRef]
72. Mananathi, B.; SAmanthapudi, V.; Gajulapalli, N. Estrogen receptor coargulators and pioneer factors: The orchestrators of mammay gland cell fate and development. *Front. Cell Dev. Biol.* **2014**, *2*, 1–13. [CrossRef]
73. Madak-Erdogan, Z.; Charn, T.; Edison, Y.; Katzenellenbogen, J. Integrative genomics of gene and metabolic regulation by estrogen receptors α -and β, and their coregulators. *Mol. Syst. Biol.* **2013**, *9*, 67. [CrossRef] [PubMed]
74. Heldring, N.; Oike, A.; Andersson, S.; Mathews, J.; Cheng, G.; Hartman, J.; Tujague, M.; Strom, A.; Treuter, E.; Warner, M. Estrogen receptors: How do they signal and what are their targets. *Physiol. Rev.* **2007**, *87*, 905–931. [CrossRef] [PubMed]
75. Chen, J.; Russo, J. Erα negative and triple-negative breast cancer: Molecular features and potential therapeutic approaches. *Biochim. Biophys. Acta* **2009**, *1796*, 162–175. [CrossRef]
76. Hsia, E.Y.; Goodson, M.L.; Zou, J.X.; Privalsky, M.L.; Chen, H.W. Nclear receptor coregulators as a new paradigm for therapeutic targeting. *Adv. Drug Deliv. Rev.* **2010**, *62*, 1227–1237. [CrossRef] [PubMed]
77. Song, X.; Chen, J.; Zhao, M.; Zhang, C.; Yu, Y.; Lonard, D.M.; Chow, D.-C.; Palzkill, T.; Xu, J.; O'Malley, B.W.; et al. Development of potent small-molecule inhibitors to drug the undruggable steroid receptor coactivator-3. *Proc. Natl. Acad. Sci. USA* **2016**, *113*, 4970–4975. [CrossRef]
78. Wang, Y.; Lonard, D.M.; Yu, Y.; Chow, D.-C.; Palzkill, T.G.; Wang, J.; Qi, R.; Matzuk, A.J.; Song, X.; Madoux, F.; et al. Bufalin is a potent small-molecule inhibitor of the steroid receptor coactivators SRC-3 and SRC-1. *Cancer Res.* **2014**, *74*, 1506–1517. [CrossRef]

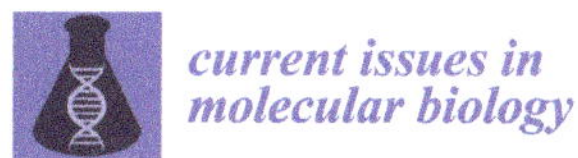

current issues in molecular biology

MDPI

Review

miRNA Expression Profiling in Human Breast Cancer Diagnostics and Therapy

Iga Dziechciowska [1], Małgorzata Dąbrowska [1], Anna Mizielska [1], Natalia Pyra [1], Natalia Lisiak [1], Przemysław Kopczyński [2], Magdalena Jankowska-Wajda [3] and Błażej Rubiś [1,*]

[1] Department of Clinical Chemistry and Molecular Diagnostics, Poznan University of Medical Sciences, Rokietnicka 3, 60-806 Poznan, Poland; iga_dz@op.pl (I.D.); gosia.guciol@gmail.com (M.D.); amizielska@ump.edu.pl (A.M.)
[2] Centre for Orthodontic Mini-Implants, Department and Clinic of Maxillofacial Orthopedics and Orthodontics, Poznan University of Medical Sciences, Bukowska 70 Str., 60-812 Poznan, Poland
[3] Faculty of Chemistry, Adam Mickiewicz University, Uniwersytetu Poznanskiego 8 Str., 61-614 Poznan, Poland; magdajw@amu.edu.pl
* Correspondence: blazejr@ump.edu.pl; Tel.: +48-616418304

Abstract: Breast cancer is one of the most commonly diagnosed cancer types worldwide. Regarding molecular characteristics and classification, it is a heterogeneous disease, which makes it more challenging to diagnose. As is commonly known, early detection plays a pivotal role in decreasing mortality and providing a better prognosis for all patients. Different treatment strategies can be adjusted based on tumor progression and molecular characteristics, including personalized therapies. However, dealing with resistance to drugs and recurrence is a challenge. The therapeutic options are limited and can still lead to poor clinical outcomes. This review aims to shed light on the current perspective on the role of miRNAs in breast cancer diagnostics, characteristics, and prognosis. We discuss the potential role of selected non-coding RNAs most commonly associated with breast cancer. These include miR-21, miR-106a, miR-155, miR-141, let-7c, miR-335, miR-126, miR-199a, miR-101, and miR-9, which are perceived as potential biomarkers in breast cancer prognosis, diagnostics, and treatment response monitoring. As miRNAs differ in expression levels in different types of cancer, they may provide novel cancer therapy strategies. However, some limitations regarding dynamic alterations, tissue-specific profiles, and detection methods must also be raised.

Keywords: breast cancer; miRNA profiling; cancer diagnostics; cancer therapy

Citation: Dziechciowska, I.; Dąbrowska, M.; Mizielska, A.; Pyra, N.; Lisiak, N.; Kopczyński, P.; Jankowska-Wajda, M.; Rubiś, B. miRNA Expression Profiling in Human Breast Cancer Diagnostics and Therapy. *Curr. Issues Mol. Biol.* **2023**, *45*, 9500–9525. https://doi.org/10.3390/cimb45120595

Academic Editor: Dumitru A. Iacobas

Received: 29 September 2023
Revised: 16 November 2023
Accepted: 21 November 2023
Published: 25 November 2023

1. Introduction

Breast cancer (BC) is the most commonly diagnosed cancer worldwide [1,2]. This disease constitutes approximately 11.7% of total cancer cases. It has recently overtaken lung cancer, which is assessed at 11.4%. Consequently, breast carcinoma is the leading cause of death among females and the fifth among both sexes (although rare in males) [3–5]. Non-invasive breast cancer ductal and lobular carcinomas in situ represent 15% of total breast malignancies. The first type develops in milk ducts, whereas the latter originates in breast lobules. However, in both cases, cells can transform and become invasive [6].

Without a doubt, breast malignancy is heterogeneous, which makes it more challenging to diagnose. The entities of neoplasms differ regarding how they were triggered, how they respond to therapy, and the outcome. According to the World Health Organization (WHO), there are at least 18 histological types of breast carcinoma. There is a significant difference between histological and molecular classification. Luminal, HER2-enriched (human epidermal growth factor receptor 2), basal-like, and normal breast-like are identified as four molecular subtypes. The luminal is further divided into subgroups: luminal A and luminal B [7]. The molecular taxonomy involves profiling of gene expression, evaluated at the mRNA levels, which provides a high sensitivity of detection. Early detection plays a crucial role in decreasing mortality and providing a better prognosis for patients.

Breast malignancy has numerous risk factors such as sex, age, having children, age at first birth, family history, genetic background, taking birth control pills, alcohol consumption, smoking, etc. [4,8]. The prognosis and treatment of breast carcinoma depends, among other things, on tumor-node-metastasis staging. The other crucial factors are lymphovascular spread, age and menopausal status of the patient, histological grade, hormone receptor status, and overexpression of ERBB2/HER2 (erythroblastic oncogene B2/HER2) [9]. Altogether, dealing with unequivocal diagnostics, metastasis, resistance to drugs, and recurrence are the burdens of breast cancer treatment that result in severe limitations in therapy efficacy. That is why developing novel diagnostics and treatment strategies is so valid. It seems that an epigenetics assessment covering micro-ribonucleic acid (miRNA) profiling might significantly contribute to better patient outcomes as it refers to many pathways involving oncogenes or tumor suppressor genes [10–12].

2. Breast Cancer Treatment Strategies Based on Molecular Characteristics

Early cancer diagnosis increases treatment options and patients' survival. Available diagnostic strategies are based on medical imaging and biomarker analysis. In 20 to 30% of invasive breast cancer cases, the overexpression or amplification of HER2 is observed [13–15]. Dimerization of the receptor initiates various signaling pathways, leading to cell proliferation and tumorigenesis. This is why HER2-positive cells are more aggressive than HER2-negative breast cancer cells [16]. There are some known upregulated miRNAs in HER2+ breast cancer patients, e.g., miR-4728 or miR-146a-5p, and also there are miRNAs which are HER2 cell signaling inhibitors, like miR-33b, miR-491-5p, miR-634, and miR-637 [17–20]. Furthermore, miR-342-5p and miR-744 are significantly downregulated in HER2-positive breast tumors as compared to HER2-negative tumors, and higher expression of miR-342-5p is associated with better survival in both HER2-positive and HER2-negative breast cancer patients [17].

Targeted therapy for HER2-positive breast cancer uses monoclonal antibodies such as trastuzumab (known as Herceptin), pertuzumab, and margetuximab [16]. They bind HER2 protein, attenuate proliferation signaling, and decrease cancer growth. This protein is a kinase that can also be blocked with drugs such as lapatinib or neratinib, which are kinase inhibitors [16]. Alternatively, another strategy can be involved that is based on an antibody–drug conjugate (ADC) [16]. One of the examples is ado-trastuzumab emtansine (Kadcyla), a combination of Herceptin and the chemotherapeutic drug emtansine [16]. It is used to treat early-stage breast cancer after surgery or in advanced stage after chemotherapy. The FDA also approved Enhertu (a HER2-directed antibody and topoisomerase inhibitor conjugate) that works for patients with an inoperable or metastasized tumor [16]. Another strategy is based on targeting hormone receptors (in estrogen-positive cancers). Cancer can also be targeted with drugs such as CDK4/6 inhibitors (palbociclib, ribociclib, and abemaciclib) that enable the slowing down of cancer development. The drugs block cyclin-dependent kinases (CDKs) and stop cells from dividing [18]. Similarly, some well-known mTOR inhibitors (e.g., sirolimus, everolimus, and temsirolimus) attenuate the malignancy potential of cancer cells [19]. Another signaling pathway effectively blocked in cancer is the PI3K pathway (found to control the proliferation and survival of breast cancer that results in tumor growth inhibition) [16].

However, one of the most critical pathways in cancer development is associated with BRCA genes (BRCA 1 and BRCA 2), which are related to DNA repair and cell cycle control [20]. The human BRCA1 mRNA 3′UTR region is predicted to bind 20–100 miRNAs, whereas some of these, e.g., miR-146a, miR-146b-5p, miR-182, miR-16, miR-17, miR-15a, and miR-638, were shown to regulate BRCA1 expression [21]. BRCA1 epigenetically represses miR-155, and overexpression of miR-155 accelerates tumor cell line growth in vitro [22]. Moreover, there are also miRNAs such as miR-155, miR-148, miR-152, miR-205, miR-99b, and miR-146a, which are targeted by BRCA1 [21]. BRCA1 was shown to be associated with the expressions of both precursor and mature forms of let-7a-1, miR-16-1, miR-145, and miR-34a [23].

The risk of developing breast (or ovarian) cancer in carriers of mutations in these tumor suppressor genes is significantly higher than in non-carriers. Precisely, in the general population, about 13% of women will eventually develop breast cancer [24], while 55–72% of women who inherit a harmful BRCA1 variant and 45–69% of women who inherit a harmful BRCA2 variant will develop breast cancer by 70–80 years of age [25–27]. Noteworthy, the risk depends on several factors, some of which have not been fully characterized. Another protein associated with the DNA repair pathway is PARP (poly (ADP-ribose) polymerase), which can be targeted with olaparib and talazoparib [16,20]. However, these strategies are not efficient enough in diagnostics (mutation detection) or therapy. Thus, providing novel diagnostic and therapeutic targets is still highly important, and miRNA could be a promising area.

3. miRNAs—The Mechanism of Action

miRNAs are defined as endogenous, 21–25 nucleotide single-stranded RNAs (ssRNAs) that are produced from hairpin-shaped precursors [25]. These molecules are involved in processes crucial for development and general metabolism. The role of these non-coding RNAs covers cell proliferation, differentiation, apoptosis, and tumorigenesis [26]. As shown in *Drosophila melanogaster*, miRNAs controlled cell death, proliferation, and Notch signaling [27–29]. In mice, they were shown to contribute to T-cell development and innate immunity [30–32]. In humans, miRNAs were shown to participate in the regulation of granulocyte maturation [33], development and function of the immune system [34], adipocyte differentiation [35], antiviral defense [36], gene downregulation in colon adenocarcinoma [37] and upregulation in B-cell lymphoma [38,39], and many other functions. Recent studies have shown the pivotal role of specific miRNAs in the development, progression, and cancer response to treatment [40–42]. It was also suggested that miRNAs could function as breast cancer biomarkers due to their aberrant expression [7].

The miRNA-related mechanism of gene expression modulation is related to the post-transcriptional effect, possibly due to the base pair complementarity with mRNA molecules. The gene silencing process can be conducted with mRNA cleavage or inhibition of transcript translation [43]. Predominantly, miRNAs bind to the sequence at the 3′ UTR of their target mRNAs, but binding to other mRNA regions such as the 5′ UTR and the coding sequence was also revealed [43]. Significantly, miRNA binding to 3′ UTR and coding regions contributes to gene expression silencing, while binding the promoter region results in transcription induction [43]. miRNA does not usually show complete complementarity to the 3′ UTR, which enables the targeting of many of genes, some of which may be involved in carcinogenesis [44]. The miRNA genes are situated directly at or near mutation-prone sites of chromosomes. Thus, DNA damage influences the expression of tumor suppressors and miRNAs. The variety of miRNA expression that regulates cancer-related genes make miRNAs a new class of oncogenes and tumor suppressor genes [26].

The miRNA mechanism of action and transcription starts in the nucleus, from miRNA genes. Formation of pri-miRNA is operated by RNA polymerase II (Pol II), Drosha and Pasha cofactor, into 60- to 110-nucleotide pre-miRNA hairpins. It is exported to the cytosol, where it is cleaved by the RNase activity of Dicer into a transient, 22-nucleotide miRNA/miRNA duplex intermediate. The duplex loads onto components in the RNA-induced silencing complex (RISC) and separates. Further, the miRNA-RISC complex leads to double-stranded helix formation by complementing the antisense strand with the mRNA sequence target. If mRNA is bound with complete complementarity, it encounters endonucleolytic cleavage (Figure 1). Partial complementarity leads to translational repression, probably by forming a bulge sequence in the middle of the helix [45].

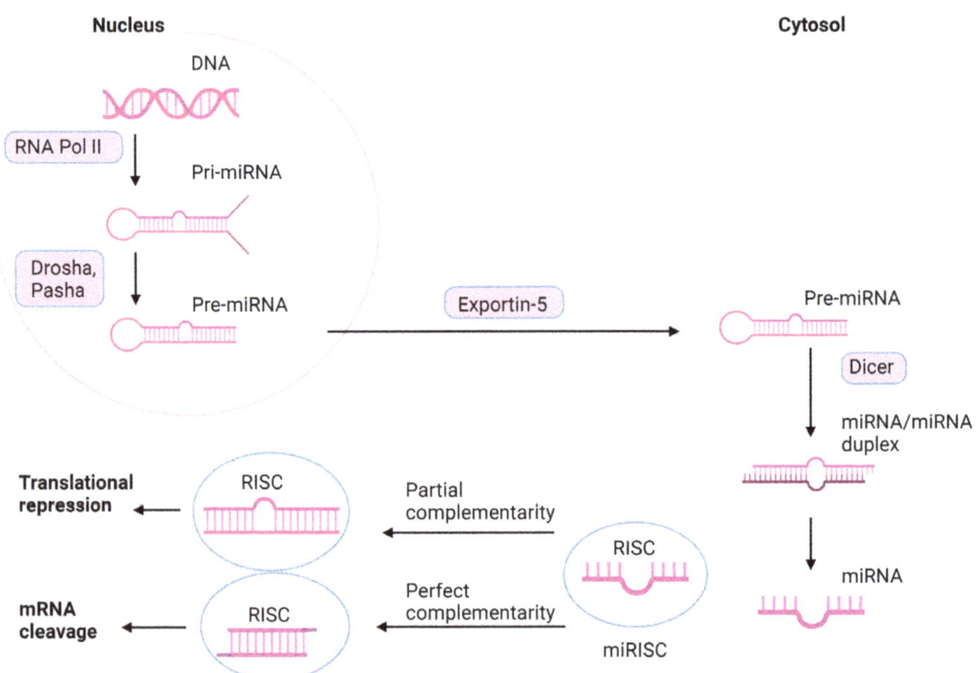

Figure 1. The transcription and mechanism of miRNA action (according to [45], created in BioRender.com). The microRNA gene is operated by RNA polymerase II (Pol II) into primary microRNA (pri-miRNA), which is prepared into pre-miRNA hairpins by Drosha and Pasha (RNase III enzyme and its cofactor). Then, the pri-miRNA is exported by Exportin-5 from the nucleus to the cytoplasm, where Dicer cleaves it by the ribonuclease (RNase) activity into a transient miRNA/miRNA duplex intermediate. The duplex loads onto components in the RNA-induced silencing complex (RISC), separate, and leads to double-stranded helix formation by complementing the antisense strand with the mRNA sequence target. mRNA could bind with complete complementarity, which leads to endonucleolytic cleavage, or just with partial complementarity, which drives translational repression, probably by a bulge sequence formation.

4. The Potential of miRNAs in Current Cancer Diagnostics and Therapy

By targeting numerous transcripts, miRNAs affect pathways, leading to different phenotypic status [46]. Over the years, abnormal levels of various miRNAs have been reported in many cancers, including breast cancer. There is a strong belief that miRNA expression profiles could become predictive and prognostic biomarkers, similar to protein-coding gene expression assessment. The levels of individual miRNAs differ between breast cancer stages, which could also be used in future diagnostics. Depending on the type of cells, the same miRNAs can show oncogenic or tumor suppressor properties [47–49].

As reported, miRNAs interact with cell signaling pathways and affect breast cancer metastasis and progression. They can regulate TGF-β (miR-106b, miR-200 family, miR-106b; the effects are mainly mediated by the downregulation of SMAD), Wnt/β-catenin (miR-4469 is a main inducer of the pathway in breast cancer cells while miR-34a is the main repressor), NF-κB (directly targeted by miR-29a, which controls proliferation, cell cycle, and apoptosis, but also controlled by miR-1246, miR-449a, or miR-200b), PI3K/Akt/mTOR (miR-21 is the main inducer of the pathway), Notch (miR-34a plays a critical function in sustaining breast cancer stem cells), and ERK/MAPK (controlled by miR-543, miR-200c, or miR-148a/152) pathways [50]. Significantly, the latter can contribute to many pathways related to the differentiation and migration of breast cancer cells. Various studies showed a correlation between individual miRNAs and ERK/MAPK and the ability of miRNAs to downregulate or upregulate this pathway [50]. For example, miR-543 was shown to impair tumor growth and proliferation in breast cancer cell lines (MCF-7, MDA-MB-231, MDA-MB-453, HCC-1937) by inhibiting ERK2 activity [51]. Noteworthy, abnormal activation of receptor tyrosine kinases (RTKs, one of the mediators in the ERK/MAPK pathway) results in the progression of various cancers. This pathway (also known as the Ras-Raf-MEK-ERK pathway) conveys extracellular information to the DNA in the nucleus and takes part in cell proliferation and differentiation control [51,52]. The pathway can be initiated with growth factors, cytokines, or hormones. They bind to the two subunits of a RTK, followed by the dimer formation. RTK binds to adaptor proteins, which attract guanine–nucleotide exchange factors (GEFs). These factors displace GDP from RAF proteins and allow GTP to bind, which causes RAS activation. Then, further protein kinases, RAF, and MEK are activated. The final enzyme MAPK (ERK) is translocated to the nucleus and activates transcription factors [51–53]. RTK are transmembrane proteins that mediate cell-to-cell communication. The aberrant activation of RTKs is found as a cancer progression factor [52]. Thus, they are used as therapeutic targets. Unfortunately, tyrosine kinase inhibitors have multiple side effects, including diarrhea, nausea, vomiting, oral ulceration, headache, and dizziness [54]. In addition, the drug-resistance cases of tyrosine kinase inhibitors (TKI) are already known [54]. Therefore, the use of RTK inhibitors has become limited. Thus, novel, more specific strategies, e.g., miRNA-based strategies, are expected.

5. microRNA Profiling

miRNAs were shown to modulate the chemosensitivity of cancer cells to therapeutic agents, but this relationship is still unclear. Due to the diagnostic potential in breast cancer, miRNA profiling has become of interest in many studies. However, first, high-efficacy isolation must be provided. miRNA isolation can be performed from various biological samples such as cells in culture, tissue, blood plasma, serum, and other body fluids. miRNAs are more stable than mRNAs in blood plasma and serum, contributing to their potential use in gene regulation research. There are some challenges in selecting a method for miRNA profiling. microRNAs represent a small part of the total RNA fraction and can differ by a single nucleotide, which makes them more challenging to identify and distinguish. Thus, it is crucial to select the appropriate microRNA profiling method. Each method has its advantages, disadvantages, and limitations. They differ in the required amount of RNA, sensitivity, specificity, and costs.

One established method is real-time quantitative reverse transcription PCR (qRT-PCR), based on the reverse transcription of miRNA to cDNA and polymerase chain reaction (SybrGreen- or probe-labeled systems). Another method, microarray, relies on the hybridization of the labeled miRNA using capture probes, but this method cannot be used to determine absolute quantification. Another hybridization-based method is Nanostring nCounter, but one of the most modern methods is based on next-generation sequencing that allows the distinguishing of different miRNAs with very high accuracy [55–58]. Eventually, based on the collected literature data, several upregulated or downregulated miRNAs can be listed in breast cancer (Table 1). We consider them as potential cancer biomarkers that can become useful in medical diagnostics.

Table 1. The list of critical miRNAs associated with breast cancer (according to [48], modified) reported as good diagnostic or therapeutic candidates. The candidates were selected based on the latest reports indicating the role of miRNA in breast cancer. Consequently, a broad analysis of selected miRNA targets suggested some good candidate markers based on the global data at TargetScanHuman 8.0 [59].

miRNA	Regulation in Breast Cancer Cells	Source and Detection Method	Target	Target Effects/Action	Metabolic Consequences
miR-21	Upregulated	Serum, qRT-PCR [60]	PTEN [44,58]	Downregulation of PTEN expression [44,58,61]	Drug resistance to doxorubicin in HER2- BC cells [61]
				* miR-21 inhibition induces PTEN expression [62]	* Restored trastuzumab sensitivity in the resistant BC xenografts in vivo [62]
			PTEN/Akt [63]	Downregulation of PTEN expression and Akt activation [63]	Induction of EMT and gemcitabine resistance [63]
			PI3K/Akt, MEK/ERK [58]	Activation PI3K/Akt and MEK/ERK signaling pathways [58]	Development of MDR [58]
			TPM1, TGF-β [64]	Repression of expression TPM1 [65,66]	Increased BC cells proliferation, migration, invasion, survival, and EMT [64]
			Mesenchymal cell markers (N-cadherin, Vimentin, α-SMA) [67]	Activation of mesenchymal cell markers [67]	Re-expression of miR-21 is responsible for migration and invasion by activating the EMT process in MCF7 cells [67]
			Epithelial cell marker (E-cadherin) [67]	Inhibition of epithelial cell marker [67]	
miR-106a	Upregulated	Serum, qRT-PCR [68]	Bcl-2,ABCG2, BAX, P53, RUNX3 [69]	Upregulation of Bcl-2 protein and multidrug transporter ABCG2. Downregulation of BAX protein and genes products: P53, RUNX3 [69]	Promotes BC cells proliferation and invasion [69]
				* Inhibition of miR-106a downregulates the expression of Bcl-2, ABCG2 and upregulates the BAX, P53, RUNX3 expression [69]	

Table 1. *Cont.*

miRNA	Regulation in Breast Cancer Cells	Source and Detection Method	Target	Target Effects/Action	Metabolic Consequences
miR-106a	Upregulated	Serum, qRT-PCR [68]	RAF-1 [68]	Decreases RAF-1 levels and RAF-1 is a part of MAPK/ERK signaling pathway [68]	Possibly induces proliferation and decreases apoptosis in BC cells through regulation of the MAPK/ERK signaling pathway, which controls gene expression [68]
			ZBTB4 [70–72]	Negative regulation of ZBTB4 gene, which functions as a tumor suppressor gene [70–72]	* Restoration of ZBTB4 suppress Sp1, Sp3, Sp4 expression resulting in inhibition of BC cells proliferation, invasion [70–72]
miR-155	Upregulated	Serum, qRT-PCR [73]	TERF1 [74]	Reduction in the shelterin component TRF1 expression. TRF1 regulates telomere length and suppresses DNA breakage [74]	Antagonization of telomere integrity in BC cells and increased genomic instability [74]
			SOCS1 [75]	Repression of SOCS1 (negative feedback regulator of JAK/STAT signaling) [75]	Constitutive activation of STAT3 in BC cells, promotion of cell proliferation and colony formation [75]
			C/EBPβ [76,77]	Loss of CCAAT-enhancer binding protein beta (C/EBPβ) [76,77]	Modification of TGF-β response; from growth inhibition to EMT, invasion, and metastasis in BC. Promotion of BC progression [76,77]
mir-141	Downregulated	Tissue, qRT-PCR, Microarray [78]	ANP32E [78]	Regulation of ANP32E (positive regulator of tumor growth and metastasis) [78,79]	ANP32E induces tumorigenesis of BC by upregulating E2F1 and promoting the G1/S transition [79]
				** Overexpression of miR-141 downregulated ANP32E expression [78]	** Inhibition of BC cells proliferation, migration, and invasion [78]

Table 1. *Cont.*

miRNA	Regulation in Breast Cancer Cells	Source and Detection Method	Target	Target Effects/Action	Metabolic Consequences
mir-141	Downregulated	Tissue, qRT-PCR, Microarray [78]	SIP1 [80]	Regulation of EMT [80]	EMT plays a crucial role in early tumor metastasis and SIP1 is a promoter of cancer progression [80]
let-7c	Downregulated	Serum, qRT-PCR [81]	ERCC6 [82]	Upregulation of ERCC6 [82]	Intensified cancer growth ability and lower rate of apoptosis; DNA damage accumulation [82]
			BCL2, BAX [83]	** Overexpression of let-7c decreases level of Bcl-2 and increases the level of BAX, TP53, PTEN [83]	** Promotion of apoptotic cell death, suppression of cancer progression [83]
			ERα and Wnt signaling [84]	** Overexpression of let-7c inhibits estrogen induction in ERα and Wnt signaling [84]	** Inhibition of BCSCs self-renew and suppresses tumor formation [84]
miR-335	Downregulated	Serum, qRT-PCR [85]	BRCA1 [86,87]	Downregulation of BRCA1 [86]	Accelerated tumor growth, genomic instability, BC progression [86]
				** Overexpression of miR-335 upregulates the level of BRCA1 [86,87]	** Decreased cell viability and increased apoptosis [86,87]
miR-126	Downregulated	Tissue, qRT-PCR [88]	VEGFA [88], PIK3R2 [89]	Inactivation of the PIK3R2/PI3K/Akt/mTOR signaling pathway [89]	Vasculogenesis, angiogenesis resulting in tumor growth [88] Resistance to trastuzumab [89] in SKBR3 and BT747 cell lines
			ADAM9 [90]	** Up-regulation of miR-126 is silencing ADAM9 gene [90]	** Inhibition of BC cells invasion and metastasis [90]
miR-199a	Downregulated	Tissue, qRT-PCR [91]	PAK4/MEK/ERK signaling pathway [92]	Regulation of PAK4/MEK/ERK signaling pathway [92]	PAK4 activates the ERK pathway, and MEK/ERK pathway plays a part in PAK4-induced cell growth regulation [92]
				** MiR-199a/b-3p downregulates PAK4 expression and PAK4/MEK/ERK signaling pathway [92]	** Suppression of BC cells migration and invasion [92]

Table 1. *Cont.*

miRNA	Regulation in Breast Cancer Cells	Source and Detection Method	Target	Target Effects/Action	Metabolic Consequences
miR-101	Downregulated	Tissue, qRT-PCR [93]	COX-2/MMP1 signaling pathway [94]	Upregulation of COX-2/MMP1 signaling pathway [94] ** Restoring miR-101-3p in BC cells reduces COX-2/MMP1 expression [94]	Promotes transmigration of metastatic BC cells through the brain endothelium [94] ** Reduction in transmigratory ability [94]
miR-9	Upregulated	Cell culture, qRT-PCR [95]	FOXO1 [96]	Downregulation of FOXO1 expression [96]	Promotion of proliferation, migration, and invasion of BC cells [96]
			STARD13 [97]	Repression of STARD13 [97]	Upon stimulation of PDGFRβ signaling, miR-9 could promote the formation of vascular-like structures of TNBC [97]
			E-cadherin [96,98]	E-cadherin downregulation [98,99]	Increased tumor angiogenesis [99] Primes BC cells to EMT and invasion [98]

* refers to report showing effects of miR inhibition. ** refers to report showing effects of miR activation.

5.1. miR-21

In 2019, a study investigating miR-21 levels in the plasma of breast cancer patients and breast cancer cell lines was published. The study included 127 healthy patients (controls), 82 patients with benign breast tumors, and 252 with breast cancer. The levels of miR-21 were found to be different between these groups—the lowest miR-21 levels were found in healthy controls, while an increase in miR-21 levels was observed in patients with breast cancer. miR-21 levels were also compared between patients with different stages of the disease. Plasma miR-21 levels of breast cancer patients were correlated with the tumor, node, and metastasis (TNM) stage. In particular, an increase in miR-21 level was observed in the T3 stage, meaning the tumor is bigger than 5 cm. Then, the breast cancer cell lines Hs578T and MDA-MB-231 were transfected with a miR-21 inhibitor. After 14 days, it was found that colony formation ability was reduced in transfected cells compared with the controls. Transwell and wound healing tests were performed using the same cell lines to assess the ability of the cells to migrate. The tests confirmed that the miR-21 inhibitor reduced cell migration capacity. These results showed that inhibition of miR-21 could reduce metastasis and breast cancer proliferation. This means that therapies with miR-21 inhibitors might constitute a promising strategy for breast cancer patients [100,101]. miR-21 was also shown to play a crucial role in regulating drug resistance in breast cancer, and its overexpression was correlated with the development of multidrug resistance (MDR) [58]. Specifically, the research investigating the association of miR-21 expression with drug resistance in breast cancer indicated that miR-21 modulated the resistance of breast cancer cells to doxorubicin [58]. The study used a breast cancer cell line (MCF-7) and a doxorubicin-resistant breast cancer cell line (MCF-7/ADR). As reported, miR-21 expression was increased in MCF-7/ADR cells relative to MCF-7 cells. Importantly, one of the targets for miR-21 is the tumor suppressor gene PTEN, and in this study, PTEN expression was downregulated in MCF-7/ADR cells. This study suggested that miR-21 overexpression was associated with doxorubicin resistance to breast cancer and mediated by targeting phosphatase and tensin homolog (PTEN) [58,61]. Similarly, other research showed that miR-21 targeted insulin-like growth factor binding protein 3 (IGFBP3), which can be associated with brain metastases of BC cells. miR-21 was shown to cause an increase in cancer cell proliferation, migration, and the epithelial-to-mesenchymal transition (EMT) mediated by targeting TPM1, PCD4, and TGF-beta1 [64]. Knockdown of this particular miRNA was reported to induce cell apoptosis and inhibit proliferation and invasion of EMT [64].

5.2. miR-106a

It was reported that miR-106a was overexpressed in breast cancer tissue compared with normal tissue and was correlated with enhanced breast cancer cell proliferation. It was also associated with the downregulation of P53, BAX, and RUNX3 and the upregulation of Bcl-2 and ABCG2, which promote breast cancer cell proliferation. In addition, it was reported that upregulation of miR-106a decreased cell sensitivity to cisplatin [69]. The study in the mouse model also showed that miR-106a overexpression affected drug chemosensitivity. MDA-MB-231 and MCF-7 cell lines were treated with miR-106a inhibitor and miR-106a mimic. The mice were used to make a transplanted tumor model, and then the cisplatin treatment was added. The inhibition of tumor growth was observed when the inhibitor was applied, suggesting an association between miR106a and tumor sensitivity to cisplatin [102]. Altogether, it was reported that miR-106a could contribute to enhanced cell proliferation due to lowered sensitivity to chemotherapeutic agents [69,102]. Another study revealed a possible correlation between miR-106a levels and breast cancer cell proliferation mediated by RAF-1, activating the MAPK/ERK signaling pathway [68].

5.3. miR-155

miR-155 is an oncogenic miRNA involved in breast cancer growth regulation and is upregulated in breast cancer specimens. It was shown to contribute to telomere destabilization due to targeting TRF1 (shelterin component) [74] in MCF-7, SK-BR-3, and MDA-MB-468 cells. miR-155 overexpression was reported to reduce the expression of TRF1, leading to increased chromosome instability. Interestingly, reducing miR-155 levels showed an opposite effect [73,74,103,104].

5.4. miR-141

It was reported that the levels of miR-141 were decreased in breast cancer cells relative to surrounding tissues (qPCR) [78]. Additionally, the miR-141 level was correlated with the tumor stage. As demonstrated, overexpression of this miRNA was associated with decreased cell proliferation and enhanced apoptosis. Moreover, wound healing, assays, showed that miR-141 overexpression was accompanied by the inhibition of MDA-MB-231 cell migration. More detailed analysis revealed that one of the miR-141 targets, acidic nuclear phosphoprotein 32 family member E (ANP32E), was manifested by a significantly decreased level of ANP32E both at mRNA and protein levels in the miR-141 mimics transfected group [78]. Consequently, experiments with specific vshRNAs revealed that ANP32E knockdown inhibited MDA-MB-231 cell proliferation [78]. The relationship between ANP32E and triple-negative breast cancer was studied and it was demonstrated that ANP32E promotes tumor proliferation and the G1/S transition [79].

Another study demonstrated that miR-141-3p overexpression was correlated with aggressive breast carcinoma cases. The miRNA expression was compared between different breast tissues (malignant and benign), and significantly high miR-141-3p expression was demonstrated in grade III breast cancer compared to grade II [105]. These results suggested that miR-141-3p could discriminate malignant from benign breast tissues and, even more, could distinguish TNBC (triple-negative breast cancer) from other molecular subtypes of breast cancer. Altogether, miR-141-3p expression was correlated with shorter overall patient survival [105]. Additionally, assessment of the combination of miR-141-3p, miR-181b1-5p, and miR-23b-3p was suggested as a useful approach in cancer molecular subtypes identification.

5.5. Let-7c miRNA

The let-7 family of microRNAs are known to act as tumor suppressors [81]. Specifically, the let-7c level in the breast cancer patients' serum and tissues was lower than in the controls. The association between let-7c expression levels and ER/PR status was investigated, but no significant difference was detected [81]. Interestingly, the upregulation of let-7c in premenopausal patients compared with postmenopausal patients was shown [81]. Another study suggested that ERCC6 (this gene encodes a DNA-binding protein that is important in transcription-coupled excision repair [106]) was also a target for let-7c-5p that led to the downregulation of the encoded protein in MCF-7 cells [82].

Interestingly, the ERCC6 mRNA was unaltered, suggesting transcription degradation instead of mRNA degradation. Another study showed the downregulation of let-7c-5p in breast cancer tissues. Furthermore, it was found that let-7c-5p overexpression could inhibit breast cancer cell proliferation [82].

5.6. miR-335

miR-335 coding sequence is located on the chromosome 7q32.2 locus and controlled by DNA methylation and was reported to act as an oncogene showing both tumor promoter and suppressor effects depending on the tumor stage and type. BC surpasses tumor invasion and metastasis by downregulating several signal pathways. It also affects the tumor environment and drug sensitivity [85]. Studies show that the overexpression of miR-335 affects cell proliferation, viability, and apoptosis by being a crucial factor in the BRCA1 regulatory network [86]. Interestingly, BRCA1/2 increases the transcription levels of miR-335, which leads to increased cell plasticity and growth [107]. In addition, BRCA1 and EGFR/HER2 can inhibit mRNA maturation, enhancing cell survival and invasiveness [107].

New evidence reports that the downregulation miR-335 in BC suppresses cell metastasis and migration by targeting transcription factor SOX4 and extracellular matrix component tenascin C [108].

The hepatocyte growth factor (HGF)/c-Met pathway contributes to tumor invasion and metastasis and is an essential factor in the progression and prognosis of BC patients [108]. C-Met, being an oncogene, can bind HGF, which induces autophosphorylation of tyrosine residues in c-Met [108]. Studies conducted by Gao et al. in 2014 showed that the forced overexpression of miR-335 revokes HGF-stimulated c-Met phosphorylation and, in consequence, cell migration due to reducing c-Met expression [108]. The same studies indicated that 5-AZA-CdR treatment (DNA methyltransferase inhibitor) significantly increased miR-335 expression, which later influences the HGF/c-Met pathway, and, simultaneously, the level of miR-335 that can play a significant role in breast cancer diagnosis and prognosis and novel strategies for BC therapy [107–109]. Research carried out on MDA-MB-231 cells showed that the overexpression of miR-335 could increase the sensitivity of triple-negative breast cancer (BC with negative immunohistochemical results of estrogen receptor, progesterone receptor, and proto-oncogene HER-2) to cisplatin and doxorubicin, which improved the efficacy of chemotherapy [87]. The mechanism involved in increased cell sensitivity still needs to be investigated. It may play an essential role in breast cancer treatment.

5.7. miR-126

miR-126, located in the EGFL7 region (a natural negative regulator of vascular elastogenesis), is exclusively expressed in endothelial cells and regulates angiogenic signaling and vascular integrity [88]. Furthermore, it reduces the proliferation and metastasis of tumors by targeting vascular endothelial growth factor (VEGF), which positively regulates vasculogenesis and angiogenesis [88]. It was reported that the expression of miR-126 is downregulated in breast cancer, whereas the VEGF signaling pathway is activated in these cells, which leads to the acceleration of the growth of the tumor [88].

Studies conducted on breast cancer cells MCF7 treated with miR-126 lipofectamine showed evident downregulation of VEGF-A, which is consistent with other studies and shows a negative correlation between upregulation of the VEGF-A expression level and downregulation of the miR-126 expression level. It leads to the conclusion that miR-126 acts as a tumor-suppressive gene and that VEGF-A may be a promising target in breast cancer therapy [110]. One of the drugs used to treat BC is trastuzumab, a monoclonal antibody targeting HER2 receptors, leading to reduced BC cell division, migration, and differentiation [89]. In a study performed by Fu et al., 2020, trastuzumab-resistant SK-BR-3 (SKBR3/TR) cells transfected with miR-126 mimic showed attenuated resistance to trastuzumab while the parental line SK-BR-3 transfected with miR-126 inhibitor showed increased trastuzumab resistance [89]. The same study found that miR-126 directly targets PIK3R2 and is partially involved in the inactivation of the PIK3R2/PI3K/Akt/mTOR signaling pathway responsible for mediated trastuzumab resistance in BC [89]. In conclusion, the overexpression of miR-126 in cells resistant to trastuzumab with inhibition of PIK3R2 and the downstream PIK3R2/PI3K/Akt/mTOR signaling pathway causes a decreased drug resistance [89].

Research conducted by The Affiliated Tumor Hospital of Zhengzhou University showed a correlation between the expression of miR-126 and the regulation of critical metastatic molecule ADAM9 (ADAM metallopeptidase domain 9, a component of cell–cell junctions). Overexpression of this miRNA inhibited breast cancer cell invasion by silencing ADAM9 [90].

Clinical evidence shows that due to increased or decreased expression of specific genes in breast cancer tissue, miR-126 can be used as a biomarker to predict and diagnose breast cancer and therapy response [111].

5.8. miR-199a

Recent studies show that overexpression of miR-199a-3p suppresses proliferation, multidrug resistance, migration, and invasion, and it might suppress metastasis progression in breast cancer cells [92]. It also leads to inhibition of PAK4 expression, which has been connected to tumorigenesis and increased cell survival, which is believed to interfere with an aggressive breast cancer phenotype. Targeting the PAK4/MEK/ERK pathway can repress breast cancer progression by inducing G1 phase arrest [92].

Triple-negative breast cancer, accounting for 10–15% of BC, is plagued by significant drug resistance [112]. Studies indicate that in this type of BC, the level of miR-199a-3p is downregulated. It was found that this particular miRNA targets mTOR, which regulates cell proliferation, autophagy, and apoptosis and plays an essential role in cancer cell metabolism [19,112]. Overexpression of miR-199a-3p targets c-Met and mTOR, affecting increased sensitivity to doxorubicin and also leading to G1 phase arrest, resulting in reduced invasion and increased doxorubicin-induced apoptosis in BC cells [112]. The study on MDA-MB-231 cells indicated that miR-199a-3p could downregulate mitochondrial transcription factor A (TFAM) by promoting the sensitivity of BC cells to chemotherapy resistance [91]. In turn, inhibition of TFAM expression could attenuate cisplatin resistance in breast cancer cells and induce apoptotic and proliferative effects [113]. A study regarding the cardiotoxicity of doxorubicin showed that upon doxorubicin exposure, the level of miR-199a expression was upregulated [112]. Considering these findings, miR-199a-3p might be an excellent prognostic and predictive biomarker in breast cancer [114].

5.9. miR-101

miR-101 is acknowledged to be a tumor suppressor, and its expression is downregulated in BC [115]. It affects cancer-related processes: proliferation, apoptosis, angiogenesis, drug resistance, invasion, and metastasis. It targets proteasome maturation protein (POMP), stathmin (Stmn1), and DNA (cytosine-5)-methyltransferase 3A (DNMT3A), which suppress the proliferation of BC cells by decreasing the expression levels of Jak2, EYA1, and SOX2 and by reducing levels of VHL, which negatively regulates hypoxia-inducible factor 1-alpha (HIF1alpha), leading to the apoptosis of cancer cells [115]. It was also reported that its high levels in TNBC increase chemotherapeutic sensitivity to paclitaxel by decreasing the level of MCL-1 expression [115].

Brain metastasis is a late event in breast cancer patients. It is a cascade in which metastatic cells detach from the tumor and travel through the bloodstream or lymphatics to arrest into the capillary bed and attach to the brain endothelium, passing through the blood–brain barrier and colonizing the brain [94]. Studies show that overexpression of miR-101-3p reduces the migration of BC cells through the brain endothelium by restraining the COX-2/MMP1 signaling pathway [94].

The experiments conducted in SK-BR-3 and MCF-7 cells showed significant upregulation of an oncogene EZH2, which promotes carcinogenesis and is related to poor prognosis and aggressiveness of breast cancer. Studies have shown that simultaneous induction of miR-101 and treatment with Syn-cal14.1a, a synthetic peptide acquired from Californiconus californicus, suppresses EZH2-induced breast cancer cell migration, invasion, and proliferation and promotes apoptosis of BC cells [116]. Additionally, studies reported that miR-101 played a critical role in the pathological grade in TNM classification in BC cells, making it a promising biomarker [115]. When the miR-101-5p-associated pathways in breast cancer were assessed using RNA-seq, a particular group of genes, HMGB3, ESRP1, GINS1, TPD52, SRPK1, VANGL1, and MAGOHB, were suggested to be associated with a poor prognosis of BC [93].

5.10. miR-9

Recent studies have shown the promoting role of miR-9 in breast cancer development [98]. Its upregulation is associated with high malignancy invasive epithelial-to-mesenchymal transition, which enables cells to gain the ability of self-renewal and have the characteristics of stem cells, promoting the production of cancer stem cells (CSCs) which generate an invasive phenotype leading to poor outcome, high tumor stage and histologic grade, poor overall survival, and distant metastasis-free survival [98]. Low miR-9 expression was associated with improved overall survival, smaller tumors, earlier stage, and ER-positive cancers due to the enrichment of estrogen response genes [117]. Furthermore, miR-9 is highly expressed in HER2+ and triple-negative breast cancer and tumors displaying CD44+/CD24- phenotype and E-cadherin loss [98]. Because of the significant engagement of miR-9 in CSCs metabolism, which is considered the origin of tumorigenesis, drug resistance, and development, this miRNA seems a good predictor marker of cancer metastasis and chemoresistance [98]. Studies show that the upregulated expression of miR-9 is induced by MYC and MYCN, which leads to angiogenesis through activation of beta-catenin signaling and elevating the expression of VEGF. It also leads to increased EMT invasiveness and motility by targeting FOXO1 and STRD13, which are also associated with vascular sprouting and promoting tumor metastasis [98]. Another research conducted by Wang et al. indicated that lncTUG1 (taurine-upregulated gene 1) could modulate the susceptibility of BC cells to doxorubicin by regulating the expression of eIF5A2 (eukaryotic translation initiation factor 5A-2) via interacting with miR-9, indicating a novel potential pathway that could be targeted to overcome doxorubicin resistance in BC [95]. Interestingly, NGS results show that miR-9 directly targets HMGA2, EGR1, and IGFBP3, which are closely related to the invasion and metastasis of breast cancer [64].

6. miRNA as a Therapy Target

Most miRNAs are found inside the cell but also migrate in body fluids such as blood, urine, saliva, or breast milk. Thus, these short RNA particles are considered diagnostic and therapeutic markers, especially in cancer, neurology, or cardiology [118]. It is noteworthy that miRNA dysregulation is common in many cancer cases as they can act as both tumor suppressors or oncogenes.

miRNA as a therapy target is gaining extensive attention due to its various effects on cancer development. For example, supplementation of miRNA mimics (miR-15a) in prostate cancer cell lines induced apoptosis and blocked cell proliferation [119]. Another study showed that miR-99a reduced breast cancer cell proliferation, invasion, and migration in vitro and in vivo [120]. Numerous studies showed that targeting miRNA with its antagonists might lead to tumor suppression and efficient, personalized cancer therapy [121,122]. Significantly, miRNA-targeted therapy may influence a single gene and whole cellular pathways, which can be particularly beneficial [123]. Specifically, the latest approach in miRNA therapeutics is mainly based on two strategies, i.e., the inhibition of oncogenic miRNAs and, hence, the restoration of the expression of tumor-suppressing genes that they target, or restoring the expression of tumor-suppressing miRNAs and consequently inhibiting the oncogenes that they target. Downregulation of tumor miRNA suppressors leads to the overexpression of their target oncogenes. To restore the expression of tumor-suppressing miRNAs, promising areas are the mimic miRNAs. They are small, chemically modified (2'-O'methoxy) double-stranded RNA molecules that mimic the endogenous mature miRNA molecules [121,124].

Because oncogenic miRNAs are usually upregulated in tumors, their suppression enables tumor suppressors to be active and inhibit tumorigenesis or its progression [121]. For that reason, a few therapeutic strategies based on oncomiR inhibition were created, and one of them is AMOs (anti-mRNA oligonucleotides). AMOs are single-stranded oligonucleotides (17–22 nt long) that prevent mature miRNA interaction with the target gene by complementary binding. As a result, the AMO-miRNA duplex will be cleaved by RNAse-H [121]. By complementary binding with the target mRNA, they exert transcriptional downregulation. Another therapeutic strategy based on oncomiR inhibition is miRNA sponges, which are competitive inhibitors with multiple binding sites for an endogenous miRNA and prevent the interaction between the miRNA and its target mRNA. There is also a strategy based on inhibiting miRNA biogenesis or target interactions via small molecules, like azobenzene [121].

Due to miRNA's inability to passively diffuse through cell membranes, there is a barrier to miRNA clinical implementation and a need for effective and safe delivery systems development. Nowadays, miRNA delivery systems may be divided into two main categories: non-viral and viral vectors. Non-viral miRNA vectors are based on organic, inorganic, and polymer materials, while viral vectors usually use lentiviruses, retroviruses, or adenoviruses [121]. Other challenges of miRNA therapeutics are associated with its degradation by nucleases, endosomal entrapment, poor target tissue delivery, innate immune reaction activation, and poor binding affinity for complementary sequences [122]. Despite these difficulties, miRNA clinical implications are highly promising [121,122].

6.1. The Role of miRNA in Breast Cancer Chemoresistance

Numerous factors, including late diagnosis or resistance to therapeutic agents, may cause therapy failure in cancer therapy. The two basic types of drug resistance, i.e., innate or acquired, constitute a severe challenge in oncology. Recently, both these mechanisms were reported to be associated with miRNAs that modulate drug-resistance-related genes or affect genes related to cell proliferation, cell cycle, DNA damage repair, and apoptosis [125]. Hence, the miRNA-based therapeutic approach seems to provide an interesting and efficient perspective in cancer therapy. Specifically, in breast cancer, several miRNAs were suggested to play a critical role in therapy response, showing a tumor-type-dependent effect. miR-200c, miR-155, and miR-218 were shown to mediate the therapeutic effect of selected drugs, i.e., (i) trastuzumab, (ii) aclitaxel, VP16, doxorubicin, and (iii) cisplatin, respectively [126]. Another study demonstrated 123 miRNAs that were dysregulated in vinorelbine (NVB)-resistant breast cancer cell lines (MDA-MB-231/NVB). A total of 31 of these miRNAs were downregulated, and 92 were upregulated in those cells, suggesting complex regulation [127]. It was also demonstrated that 17 specific miRNAs were involved in oncogenic pathways, including TGFβ, mTOR, Wnt, and MAPK. It is noteworthy that elevated TGFβ signaling and downregulation of miR-200c were also demonstrated in trastuzumab-resistant breast cancer cells while increased miR-200c or the blockade of TNFβ signaling increased trastuzumab sensitivity and inhibited invasiveness of breast cancer cells [128].

Similarly, miR-494 and miR-141 were shown to suppress the progression of breast cancer by repressing β-catenin expression [129,130]. Recently, Yu et al. reported that the miR-17/20 cluster increased tamoxifen sensitivity and attenuated doxorubicin resistance in MCF-7 cells via Akt1 [131]. Another study showed that miR-218, which targets BRCA1, was downregulated in cisplatin-resistant breast cancer cell lines and, interestingly, the restoration of miR-218-sensitized MCF-7 breast cancer cells to this drug [132]. Numerous studies show other miRNAs that are capable of modifying the response of breast cancer cells to different therapeutic agents, including 5-fluorouracil, trastuzumab, lapatinib, cisplatin, fulvestrant, tamoxifen, paclitaxel, doxorubicin, and palbociclib. The most commonly reported BC-related miRNAs (and probably the most critical ones) are presented in Table 1. Recent data suggest that the function of some miRNAs may be involved in the epithelial–mesenchymal transition process that mediates multidrug resistance (MDR) phenotype promotion. A thoroughly revised contribution of miRNAs to individual ABC family transporters was shown elsewhere [133]. Thus, further screening and miRNA profiling in cancer tissues is highly required as it may provide in-depth information regarding critical genes expression regulation. It may be, however, that similarly to wide-genome sequencing that aims to evaluate the role of individual SNPs in genomic DNA, miRNA profiling will not be sufficient to evaluate the risk or monitor disease progression and therapy efficacy. The only possible way seems to be the further assessment of clinical samples that show real mechanistic networks in vivo. Importantly, some clinical trials are being carried out—more than 50 refer to miRNA application in breast cancer [134].

Some translational potential shows the studies that involve a combination of miRNA modulators with anti-cancer chemotherapeutics (specifically, a combination of antagomiRs with therapeutic agents). Alternatively, mimics could be applied that reinforce the function and expression of miRNAs. By affecting the expression of endogenous microRNAs in tumor cells and consequently leading to the modulation of target pathways, they may affect chemotherapy efficacy. However, there are still many difficulties to overcome before we should be able to use miRNAs in the clinical setting, including effective delivery systems, stability, etc.

6.2. The Role of miRNA in Breast Cancer Stem Cells

Some recent studies revealed that both cancer stem-like properties and drug resistance were associated with EMT. As mentioned above, miRNAs play a pivotal role in regulating EMT phenotype. As a result, some miRNAs impact cancer stemness and drug resistance [135], which might show some benefits to clinical treatment. Breast cancer stem cells (BCSCs) show self-renewal and differentiation capacities that contribute to the aggressiveness of metastatic lesions, and all these mechanisms can be controlled by regulatory miRNAs [136]. As demonstrated, the expression of microRNAs can be deregulated in BCSCs [137]. Specifically, mir-21, mir-22, mir-29a, and mir-221/222 were shown to increase tumorigenesis, while miR-34a, miR-628, miRNA-140-5p, and miRNA-4319 were reported to decrease metastasis in BCSCs [46,76,138]. The specific pathways targeted by miRNAs are mediated by the key players in cancer development and proliferation, including HIF-1 alpha, PI3K/Akt, and STAT3 signaling, which play critical roles in the prognosis and survival of BCSCs [136].

6.3. The Role of miRNA in Cancer Cell Cycle Control

Cell cycle dysregulation is a recognized hallmark of cancer, and its aberrant activation has been related to poor prognosis and drug resistance. Different miRNAs have been described to target genes involved in cell cycle regulation, leading to drug resistance or sensitivity. They were reported not only to target specific pathways but also were shown to be cell cycle step-specific [133].

Several miRNAs have been shown to induce cell cycle arrest due to targeting cyclins. One of them is miR-34a, which was demonstrated to increase resistance to docetaxel (DTX) in luminal BC cells, probably through the inhibition of cyclin D1 (CCND1) and B-cell lymphoma 2 (Bcl-2), inducing G1 arrest and blocking DTX effectiveness as a consequence [139]. miR-93 has also been linked to cell cycle arrest in the G1/S phase. Moreover, some other miRNAs have been shown to modulate drug resistance through targeting CDKs. One of them is miR-29c (targeting directly CDK6), which was downregulated in BC compared to normal tissues [140]. miR-29c overexpression decreased CDK6 level, inducing cell cycle arrest and PTX sensitivity.

Additionally, Citron et al. [141] showed that miR-223 expression levels could predict the effect of CDK4/6 inhibitors and palbociclib (PAB), as well as patients' prognosis for invasive ductal carcinoma. It was demonstrated that miR-223 was downregulated in luminal and HER2+ BC subtypes. Its low expression was correlated with cell cycle deregulation, poor prognosis, PAB resistance, and low survival in BC patients. Significantly, miRNAs were also shown to affect one of the essential response pathways that are triggered by cancer drugs, i.e., DNA repair pathways, including ATM [142].

6.4. miRNAs and Cell Death

Sooner or later, applying specific miRNAs in cancer therapy is supposed to provoke cancer cell death. As demonstrated, it can be caused in a particular manner, also due to miRNA involvement. This makes it again a promising strategy to consider, especially since the miRNA-target gene interactions show numerous effects that directly involve cell death modulators. Some examples are miR-125b, which confers resistance to PTX by suppressing the expression of BAK1 [143], miR-149-5p, which was found to be downregulated in PTX-resistant cells and its overexpression demonstrated to increase BAX expression [144], or miR-663b that confers TAM resistance by indirectly upregulating BAX [145]. Additional miRNAs modulate drug response by regulating the expression of Bcl-2 family members [146]. Moreover, miR-203a-3p and miR-203b-3p have been reported to decrease the antiapoptotic protein Bcl-XL and to be correlated to PTX sensitivity in BC positively regulated by MYC in cell line models of PTX-responsive BC [147].

Interestingly, miR-100 was found to be downregulated in BC cell lines with acquired resistance to CIS. In turn, overexpression of miR-100 showed increased sensitivity to CIS due to modulation of the HCLS1-associated protein X-1(HAX-1), an inhibitor of mitochondrial apoptosis that maintains mitochondrial membrane potential in cancer cells [148]. miR-944 inhibitors facilitated CIS-induced loss of mitochondrial membrane potential in resistant models, resulting in intrinsic apoptosis via targeting Bcl-2 interacting protein 3 (BNIP3) [148].

Similarly, miRNAs control critical mediators of apoptosis [149] and autophagy [150] at different levels, including PI3K/Akt/mTOR, ATGs, and LC3 [150]. Primary reports showed some specific miRNAs that affected STAT3 and ATG12 targets [151], while further studies demonstrated broader roles of autophagy-related microRNAs in cancer cells [152], showing numerous miRNAs acting at the levels of induction, nucleation, expansion, fusion, degradation, and recycling. With so many miRNA particles and the dynamics of autophagy, it is difficult to show a specific pattern that would apply to any specific cancer type. However, as miRNAs target specific genes, monitoring their expression during promoting (e.g., rapamycin, everolimus) or inhibiting autophagy (e.g., chloroquine, hydroxychloroquine) may reflect metabolic alterations that accompany different stages of therapy. Thus, we can evaluate the therapy efficacy and indicate molecular targets for more efficient therapeutic strategies. However, the pool of the genes that effectively affect pathways associated with autophagy, i.e., energy, growth, starvation response, etc., can be modulated by over 250 miRNA-target gene interactions in different cellular stress response mechanisms [151], which may make the whole idea more complex.

7. Tools in miRNA-Based Therapy Adjustment

Modulation of gene expression seems to be one of the best ways to control cell metabolism against all odds, including mutations or epigenetic factors. Overcoming these obstacles enables controlling of the phenotype, i.e., metabolism, structure, enzyme activity, substrate affinity, and protein stability. Altogether, it provides quantity and quality of cell metabolism that eventually affects the quality and the length of human life. However, using miRNA or targeting this non-coding RNA requires first identification of specific interactions as well as tissue-type and personalized profiling. This can be achieved by RNA-seq or spatial transcriptomics that deliver information on the whole transcriptome. Another step is to find a pattern—an assessment of association analysis that enables distinguishing health and disease. This approach can be obtained using different data systems, e.g., TargetScanHuman 8.0 [59], that can predict biological targets of miRNAs by searching for the presence of conserved 8mer, 7mer, and 6mer sites that match the seed region of each miRNA. The results demonstrate predictions with adjustable high and low conserved sites ranked based on the predicted efficacy of targeting.

Another option is to use Xena Browser [153] or Gepia2 (http://gepia2.cancer-pku.cn/#index, accessed on 9 February 2023) [154] to identify any alterations in the levels of selected miRNAs in different cancer tissues. Similarly, another system, On-coLnc (http://www.oncolnc.org/, accessed on 27 September 2023) [155], can link TCGA survival data to mRNA, miRNA, or lncRNA expression levels. Altogether, we have some sophisticated and advanced tools that enable prediction and assessment of the miRNA profiles. The main goal would be to find a characteristic and unique profile of the oligonucleotides that show significant association with clinical characteristics and patient outcomes.

8. Conclusions: Challenges in miRNA Modulation Approach

Since miRNAs control the expression of numerous target genes, it is unsurprising that they play critical roles in regulating cell metabolism. Thus, they have recently become the primary candidates for markers in cell homeostasis imbalance detection, disease diagnostics, and prognostics. We still study the miRNA-target gene interactions' role, mechanism, and specificity. Importantly, it was demonstrated that these short oligonucleotides showed significant stability in the extracellular space and were reported to mediate functional communication between cells. It is mainly associated with their ability to transfer between cells via extracellular vesicles (EVs) or other cell-free miRNA carriers [156,157]. This, in turn, raises the question about the tissue specificity of their expression/localization. Another critical challenge is that miRNAs target multiple genes with different efficacies that may not show specific effects after target miRNA modulation. The sequence complementarity of endogenous miRNAs ranges between 20 and 90% [158]. Surprisingly, in specific conditions (e.g., starvation), some miRNAs can upregulate the expression of target genes or lead to induction of the immune system and provoke severe adverse effects (e.g., miR-34a mimic and targeting miR-122 evaluation was discontinued after phase I and phase II, respectively) [158].

A single miRNA can target many mRNAs, and a single mRNA can be targeted by many miRNAs (many in this case means at least hundreds), which makes identification of precise interactions or using a specific miRNA as a target extremely difficult. Theoretically, using in silico algorithms, we can predict the miRNA–mRNA binding strength. However, the biological effect will depend on multiple factors, such as the level/stability of selected miRNA, the level of other miRNAs that target the same mRNA, the level of mRNA/target gene expression, and the availability of AGO2. Additionally, the complexity level significantly increases due to the earlier-mentioned ability of miRNAs to be transferred between different cells.

However, more questions refer not only to the specificity aspect but also to safety and side effects issues, formulation and bioavailability problems, and efficacy challenges. These aspects are also important when miRNAs are combined with certain drugs, which may lead to some metabolic interactions [159]. However, the issues appear also at the delivery step. It results from the fact that most miRNA modulators are negatively charged, which leads to nonspecific binding to blood proteins and decreases urinary clearance [160]. On the other hand, oligonucleotides that lack a charge weakly bind to plasma proteins and exhibit a rapid clearance either due to metabolism in the blood or excretion via urine, leading to a lower tissue uptake [160,161]. Although numerous clinical trials using miRNAs are being carried out, they have yet to show efficient solutions for the above-mentioned reasons. From the diagnostic perspective, miRNAs also show some limitations mainly associated with the overlapping activities and effects of selected miRNAs, which show limited specificity in diagnosing a specific cancer type [162]. Thus, we should instead focus on profiling miRNA levels and creating some diagnostic panels that could be used to improve the classification system and therapy planning.

Significantly, novel, personalized, and precise medicine is based on the identification of specific biomarkers but also on robust and versatile analytical technologies that improve patient outcomes [162]. The group of methods capable of identifying miRNAs and meeting the high sensitivity criteria includes quantitative reverse transcriptase PCR, digital PCR, microarray, or next-generation sequencing modified to miRNA-seq. All these methods have some limitations (including technical, standardization, reference controls, etc.) that were thoroughly discussed elsewhere [163]. Altogether, miRNA biosynthesis control and extracellular trafficking pathways constitute a challenging aspect of miRNA-based therapeutic or diagnostic strategies, significantly since they can be affected by environmental and uncontrollable factors (such as smoking, diet, circadian cycles, etc.) [164].

Author Contributions: Conceptualization, B.R.; methodology, I.D. and A.M.; resources, P.K. and M.J.-W.; writing—original draft preparation, M.D. and N.P.; writing—review and editing, N.L.; supervision, B.R. All authors have read and agreed to the published version of the manuscript.

Funding: The project is financed by the Minister of Science and Higher Education SKN/SP/496721/2021 "Student science clubs create innovations".

Conflicts of Interest: The authors declare no conflict of interest.

Abbreviations

ABCG2	adenosine triphosphate binding cassette subfamily G member 2
ADAM9	disintegrin and metalloproteinase domain-containing protein 9
ADC	antibody-drug conjugate
Akt	protein kinase B
AMOs	anti-messenger ribonucleic acid oligonucleotides
ANP32E	acidic nuclear phosphoprotein 32 family member E
ATG12	autophagy-related gene 12
ATM	ataxia telangiectasia mutated kinase
BAK1	Bcl-2 homologous antagonist killer 1
BAX	Bcl-2 associated X protein, apoptosis regulator
BC	breast cancer
Bcl-2	B-cell lymphoma 2
BCSCs	breast cancer stem cells
BNIP3	BCL2 interacting protein 3
BRCA1/2	breast cancer gene 1/2
CCND1	cyclin D1
CDKs	cyclin-dependent kinases
cDNA	complementary deoxyribonucleic acid
CIS	cisplatin
COX-2	cyclooxygenase 2
CSCs	cancer stem cells
Dicer	endoribonuclease Dicer
DNA	deoxyribonucleic acid
DNMT3A	deoxyribonucleic acid methyltransferase 3 alpha
DOX	doxorubicin
DTX	docetaxel
c-Met	mesenchymal-epithelial transition factor
EGFL7	epidermal growth factor-like protein 7
EMT	epithelial-to-mesenchymal transition
ER	estrogen receptor
ERBB2	erythroblastic oncogene B2
ERCC6	deoxyribonucleic acid excision repair protein
ERKs	extracellular signal-regulated kinases
FDA	Food and Drug Administration
FOXO1	forkhead box protein O1
GDP	guanosine diphosphate
GEFs	guanine–nucleotide exchange factors
GTP	guanosine triphosphate
HAX-1	HCLS1-associated protein X-1
HER2	human epidermal growth factor receptor 2
HGF	hepatocyte growth factor
IGFBP3	insulin-like growth factor binding protein 3
JAK/STAT	janus kinase/signal transducer and activator of transcription
MAPKs	mitogen-activated protein kinases
MDR	multidrug resistance

MMP1	matrix metallopeptidase 1
mRNA	messenger ribonucleic acid
miRNA	micro-ribonucleic acid
mTOR	mammalian target of rapamycin
NF-κB	nuclear factor kappa-light-chain-enhancer of activated B cells
NVB	vinorelbine
P53	tumor protein p53
PAB	palbociclib
PARP	poly (ADP-ribose) polymerase
PAK4	serine/threonine-protein kinase
PDCD4	programmed cell death protein 4
PI3K	phosphoinositide 3-kinase
PIK3R2	phosphoinositide 3-kinase regulatory subunit 2
Pol II	ribonucleic acid polymerase II
POMP	proteasome maturation protein
PR	progesterone receptor
pri-miRNA	primary micro-ribonucleic acid
pre-miRNA	precursor micro-ribonucleic acid
PTEN	phosphatase and tensin homolog
PTX	paclitaxel
RAF	rapidly accelerated fibrosarcoma
RISC	ribonucleic acid-induced silencing complex
RNA	ribonucleic acid
RNase	ribonuclease
RTKs	receptor tyrosine kinases
RUNX3	runt-related transcription factor 3
ssRNAs	single-stranded ribonucleic acids
STAT3	signal transducer and activator of transcription 3
TFAM	transcription factor A, mitochondrial
TGF-β	transforming growth factor-beta
TKI	tyrosine kinase inhibitors
TNM	tumor: node, metastasis
TPM1	tropomyosin 1
VEGF	vascular endothelial growth factor
WHO	World Health Organization
RT-qPCR	quantitative reverse transcription polymerase chain reaction
3′UTR	three prime untranslated region
5′UTR	five prime untranslated region

References

1. Akram, M.; Iqbal, M.; Daniyal, M.; Khan, A.U. Awareness and Current Knowledge of Breast Cancer. *Biol. Res.* **2017**, *50*, 33. [CrossRef] [PubMed]
2. Macias, H.; Hinck, L. Mammary Gland Development. *Wiley Interdiscip. Rev. Dev. Biol.* **2012**, *1*, 533–557. [CrossRef] [PubMed]
3. Yao, N.; Shi, W.; Liu, T.; Siyin, S.T.; Wang, W.; Duan, N.; Xu, G.; Qu, J. Clinicopathologic Characteristics and Prognosis for Male Breast Cancer Compared to Female Breast Cancer. *Sci. Rep.* **2022**, *12*, 220. [CrossRef] [PubMed]
4. Sung, H.; Ferlay, J.; Siegel, R.L.; Laversanne, M.; Soerjomataram, I.; Jemal, A.; Bray, F. Global Cancer Statistics 2020: GLOBOCAN Estimates of Incidence and Mortality Worldwide for 36 Cancers in 185 Countries. *Cancer J. Clin.* **2021**, *71*, 209–249. [CrossRef] [PubMed]
5. Wilkinson, L.; Gathani, T. Understanding Breast Cancer as a Global Health Concern. *Br. J. Radiol.* **2022**, *95*, 20211033. [CrossRef]
6. Klimaszewska, K.; Krajewska-Kułak, E. *Rola Zespołu Interdyscyplinarnego w Opiece nad Pacjentami Onkologicznymi. T. 1*; Uniwersytet Medyczny w Białymstoku: Białystok, Poland, 2021.
7. Łukasiewicz, S.; Czeczelewski, M.; Forma, A.; Baj, J.; Sitarz, R.; Stanisławek, A. Breast Cancer-Epidemiology, Risk Factors, Classification, Prognostic Markers, and Current Treatment Strategies-An Updated Review. *Cancers* **2021**, *13*, 4287. [CrossRef] [PubMed]
8. Komuravelli, M.; Gaddala, A.; Nama, S.; Matli, P.; Anchuri, K. Awareness and Attitude Regarding the Risk Factors, and Self Examination for Breast Cancer in the Students of an Engineering College of South India. *Eur. J. Mol. Clin. Med.* **2022**, *9*.

9.   Maughan, K.L.; Lutterbie, M.A.; Ham, P.S. Treatment of Breast Cancer. *Am. Fam. Physician* **2010**, *81*, 1339–1346.
10.   Burguin, A.; Diorio, C.; Durocher, F. Breast Cancer Treatments: Updates and New Challenges. *J. Pers. Med.* **2021**, *11*, 808. [CrossRef]
11.   Advances in Breast Cancer Research—NCI. Available online: https://www.cancer.gov/types/breast/research (accessed on 26 September 2023).
12.   Sher, G.; Salman, N.A.; Khan, A.Q.; Prabhu, K.S.; Raza, A.; Kulinski, M.; Dermime, S.; Haris, M.; Junejo, K.; Uddin, S. Epigenetic and Breast Cancer Therapy: Promising Diagnostic and Therapeutic Applications. *Semin. Cancer Biol.* **2022**, *83*, 152–165. [CrossRef]
13.   Hagemann, I.S. Molecular Testing in Breast Cancer: A Guide to Current Practices. *Arch. Pathol. Lab. Med.* **2016**, *140*, 815–824. [CrossRef]
14.   Breast Cancer Hormone Receptor Status | Estrogen Receptor. Available online: https://www.cancer.org/cancer/types/breast-cancer/understanding-a-breast-cancer-diagnosis/breast-cancer-hormone-receptor-status.html (accessed on 26 September 2023).
15.   Iqbal, N.; Iqbal, N. Human Epidermal Growth Factor Receptor 2 (HER2) in Cancers: Overexpression and Therapeutic Implications. *Mol. Biol. Int.* **2014**, *2014*, 852748. [CrossRef] [PubMed]
16.   Targeted Drug Therapy | Breast Cancer Treatment. Available online: https://www.cancer.org/cancer/types/breast-cancer/treatment/targeted-therapy-for-breast-cancer.html (accessed on 26 September 2023).
17.   Leivonen, S.-K.; Sahlberg, K.K.; Mäkelä, R.; Due, E.U.; Kallioniemi, O.; Børresen-Dale, A.-L.; Perälä, M. High-Throughput Screens Identify microRNAs Essential for HER2 Positive Breast Cancer Cell Growth. *Mol. Oncol.* **2014**, *8*, 93–104. [CrossRef] [PubMed]
18.   Uras, I.Z.; Sexl, V.; Kollmann, K. CDK6 Inhibition: A Novel Approach in AML Management. *Int. J. Mol. Sci.* **2020**, *21*, 2528. [CrossRef]
19.   Zou, Z.; Tao, T.; Li, H.; Zhu, X. mTOR Signaling Pathway and mTOR Inhibitors in Cancer: Progress and Challenges. *Cell Biosci.* **2020**, *10*, 31. [CrossRef]
20.   Varol, U.; Kucukzeybek, Y.; Alacacioglu, A.; Somali, I.; Altun, Z.; Aktas, S.; Oktay Tarhan, M. BRCA Genes: BRCA 1 and BRCA 2. *J. BUON Off. J. Balk. Union Oncol.* **2018**, *23*, 862–866.
21.   Chang, S.; Sharan, S.K. BRCA1 and MicroRNAs: Emerging Networks and Potential Therapeutic Targets. *Mol. Cells* **2012**, *34*, 425–432. [CrossRef]
22.   Chang, S.; Wang, R.-H.; Akagi, K.; Kim, K.-A.; Martin, B.K.; Cavallone, L.; Haines, D.C.; Basik, M.; Mai, P.; Poggi, E.; et al. Tumor Suppressor BRCA1 Epigenetically Controls Oncogenic microRNA-155. *Nat. Med.* **2011**, *17*, 1275–1282. [CrossRef]
23.   Kawai, S.; Amano, A. BRCA1 Regulates microRNA Biogenesis via the DROSHA Microprocessor Complex. *J. Cell Biol.* **2012**, *197*, 201–208. [CrossRef]
24.   SEER. SEER Cancer Statistics Review, 1975–2017. Available online: https://seer.cancer.gov/csr/1975_2017/index.html (accessed on 26 September 2023).
25.   Ambros, V.; Bartel, B.; Bartel, D.P.; Burge, C.B.; Carrington, J.C.; Chen, X.; Dreyfuss, G.; Eddy, S.R.; Griffiths-Jones, S.; Marshall, M.; et al. A Uniform System for microRNA Annotation. *RNA* **2003**, *9*, 277–279. [CrossRef]
26.   Budzyński, M.; Grenda, A.; Filip, A.A. Cząsteczki mikroRNA Jako Istotny Składnik Mechanizmów Regulacji Ekspresji Genów Związanych z Nowotworami. *Nowotw. J. Oncol.* **2014**, *64*, 48. [CrossRef]
27.   Brennecke, J.; Hipfner, D.R.; Stark, A.; Russell, R.B.; Cohen, S.M. Bantam Encodes a Developmentally Regulated microRNA That Controls Cell Proliferation and Regulates the Proapoptotic Gene Hid in Drosophila. *Cell* **2003**, *113*, 25–36. [CrossRef] [PubMed]
28.   PLoS Biology. Identification of Drosophila MicroRNA Targets. Available online: https://journals.plos.org/plosbiology/article?id=10.1371/journal.pbio.0000060 (accessed on 26 September 2023).
29.   Lai, E.C.; Tam, B.; Rubin, G.M. Pervasive Regulation of Drosophila Notch Target Genes by GY-Box-, Brd-Box-, and K-Box-Class microRNAs. *Genes Dev.* **2005**, *19*, 1067–1080. [CrossRef] [PubMed]
30.   Thai, T.-H.; Calado, D.P.; Casola, S.; Ansel, K.M.; Xiao, C.; Xue, Y.; Murphy, A.; Frendewey, D.; Valenzuela, D.; Kutok, J.L.; et al. Regulation of the Germinal Center Response by microRNA-155. *Science* **2007**, *316*, 604–608. [CrossRef] [PubMed]
31.   Vigorito, E.; Perks, K.L.; Abreu-Goodger, C.; Bunting, S.; Xiang, Z.; Kohlhaas, S.; Das, P.P.; Miska, E.A.; Rodriguez, A.; Bradley, A.; et al. microRNA-155 Regulates the Generation of Immunoglobulin Class-Switched Plasma Cells. *Immunity* **2007**, *27*, 847–859. [CrossRef] [PubMed]
32.   Rodriguez, A.; Vigorito, E.; Clare, S.; Warren, M.V.; Couttet, P.; Soond, D.R.; van Dongen, S.; Grocock, R.J.; Das, P.P.; Miska, E.A.; et al. Requirement of Bic/microRNA-155 for Normal Immune Function. *Science* **2007**, *316*, 608–611. [CrossRef]
33.   Fazi, F.; Rosa, A.; Fatica, A.; Gelmetti, V.; De Marchis, M.L.; Nervi, C.; Bozzoni, I. A Minicircuitry Comprised of microRNA-223 and Transcription Factors NFI-A and C/EBPalpha Regulates Human Granulopoiesis. *Cell* **2005**, *123*, 819–831. [CrossRef]
34.   Chang, T.-C.; Yu, D.; Lee, Y.-S.; Wentzel, E.A.; Arking, D.E.; West, K.M.; Dang, C.V.; Thomas-Tikhonenko, A.; Mendell, J.T. Widespread microRNA Repression by Myc Contributes to Tumorigenesis. *Nat. Genet.* **2008**, *40*, 43–50. [CrossRef]
35.   Esau, C.; Kang, X.; Peralta, E.; Hanson, E.; Marcusson, E.G.; Ravichandran, L.V.; Sun, Y.; Koo, S.; Perera, R.J.; Jain, R.; et al. MicroRNA-143 Regulates Adipocyte Differentiation. *J. Biol. Chem.* **2004**, *279*, 52361–52365. [CrossRef]
36.   Lecellier, C.-H.; Dunoyer, P.; Arar, K.; Lehmann-Che, J.; Eyquem, S.; Himber, C.; Saïb, A.; Voinnet, O. A Cellular microRNA Mediates Antiviral Defense in Human Cells. *Science* **2005**, *308*, 557–560. [CrossRef]
37.   Michael, M.Z.; O'Connor, S.M.; van Holst Pellekaan, N.G.; Young, G.P.; James, R.J. Reduced Accumulation of Specific microRNAs in Colorectal Neoplasia. *Mol. Cancer Res. MCR* **2003**, *1*, 882–891. [PubMed]

38. He, L.; Thomson, J.M.; Hemann, M.T.; Hernando-Monge, E.; Mu, D.; Goodson, S.; Powers, S.; Cordon-Cardo, C.; Lowe, S.W.; Hannon, G.J.; et al. A microRNA Polycistron as a Potential Human Oncogene. *Nature* **2005**, *435*, 828–833. [CrossRef] [PubMed]

39. O'Donnell, K.A.; Wentzel, E.A.; Zeller, K.I.; Dang, C.V.; Mendell, J.T. C-Myc-Regulated microRNAs Modulate E2F1 Expression. *Nature* **2005**, *435*, 839–843. [CrossRef] [PubMed]

40. Mizielska, A.; Dziechciowska, I.; Szczepański, R.; Cisek, M.; Dąbrowska, M.; Ślężak, J.; Kosmalska, I.; Rymarczyk, M.; Wilkowska, K.; Jacczak, B.; et al. Doxorubicin and Cisplatin Modulate miR-21, miR-106, miR-126, miR-155 and miR-199 Levels in MCF7, MDA-MB-231 and SK-BR-3 Cells That Makes Them Potential Elements of the DNA-Damaging Drug Treatment Response Monitoring in Breast Cancer Cells-A Preliminary Study. *Genes* **2023**, *14*, 702. [CrossRef]

41. Ye, Q.; Raese, R.A.; Luo, D.; Feng, J.; Xin, W.; Dong, C.; Qian, Y.; Guo, N.L. MicroRNA-Based Discovery of Biomarkers, Therapeutic Targets, and Repositioning Drugs for Breast Cancer. *Cells* **2023**, *12*, 1917. [CrossRef]

42. Ahmadi, S.M.; Amirkhanloo, S.; Yazdian-Robati, R.; Ebrahimi, H.; Pirhayati, F.H.; Almalki, W.H.; Ebrahimnejad, P.; Kesharwani, P. Recent Advances in Novel miRNA Mediated Approaches for Targeting Breast Cancer. *J. Drug Target.* **2023**, *31*, 777–793. [CrossRef] [PubMed]

43. O'Brien, J.; Hayder, H.; Zayed, Y.; Peng, C. Overview of MicroRNA Biogenesis, Mechanisms of Actions, and Circulation. *Front. Endocrinol.* **2018**, *9*, 402. [CrossRef]

44. Fang, H.; Xie, J.; Zhang, M.; Zhao, Z.; Wan, Y.; Yao, Y. miRNA-21 Promotes Proliferation and Invasion of Triple-Negative Breast Cancer Cells through Targeting PTEN. *Am. J. Transl. Res.* **2017**, *9*, 953–961.

45. Magri, F.; Vanoli, F.; Corti, S. miRNA in Spinal Muscular Atrophy Pathogenesis and Therapy. *J. Cell. Mol. Med.* **2018**, *22*, 755–767. [CrossRef]

46. Singh, R.; Mo, Y.-Y. Role of microRNAs in Breast Cancer. *Cancer Biol. Ther.* **2013**, *14*, 201–212. [CrossRef]

47. Shah, M.Y.; Ferrajoli, A.; Sood, A.K.; Lopez-Berestein, G.; Calin, G.A. microRNA Therapeutics in Cancer—An Emerging Concept. *EBioMedicine* **2016**, *12*, 34–42. [CrossRef] [PubMed]

48. Hamam, R.; Hamam, D.; Alsaleh, K.A.; Kassem, M.; Zaher, W.; Alfayez, M.; Aldahmash, A.; Alajez, N.M. Circulating microRNAs in Breast Cancer: Novel Diagnostic and Prognostic Biomarkers. *Cell Death Dis.* **2017**, *8*, e3045. [CrossRef] [PubMed]

49. Bautista-Sánchez, D.; Arriaga-Canon, C.; Pedroza-Torres, A.; De La Rosa-Velázquez, I.A.; González-Barrios, R.; Contreras-Espinosa, L.; Montiel-Manríquez, R.; Castro-Hernández, C.; Fragoso-Ontiveros, V.; Álvarez-Gómez, R.M.; et al. The Promising Role of miR-21 as a Cancer Biomarker and Its Importance in RNA-Based Therapeutics. *Mol. Ther. Nucleic Acids* **2020**, *20*, 409–420. [CrossRef] [PubMed]

50. Abolghasemi, M.; Tehrani, S.S.; Yousefi, T.; Karimian, A.; Mahmoodpoor, A.; Ghamari, A.; Jadidi-Niaragh, F.; Yousefi, M.; Kafil, H.S.; Bastami, M.; et al. MicroRNAs in Breast Cancer: Roles, Functions, and Mechanism of Actions. *J. Cell. Physiol.* **2020**, *235*, 5008–5029. [CrossRef] [PubMed]

51. Safa, A.; Abak, A.; Shoorei, H.; Taheri, M.; Ghafouri-Fard, S. MicroRNAs as Regulators of ERK/MAPK Pathway: A Comprehensive Review. *Biomed. Pharmacother.* **2020**, *132*, 110853. [CrossRef] [PubMed]

52. Sudhesh Dev, S.; Zainal Abidin, S.A.; Farghadani, R.; Othman, I.; Naidu, R. Receptor Tyrosine Kinases and Their Signaling Pathways as Therapeutic Targets of Curcumin in Cancer. *Front. Pharmacol.* **2021**, *12*, 772510. [CrossRef] [PubMed]

53. McCain, J. The MAPK (ERK) Pathway: Investigational Combinations for the Treatment of BRAF-Mutated Metastatic Melanoma. *P T Peer-Rev. J. Formul. Manag.* **2013**, *38*, 96–108.

54. Xuhong, J.-C.; Qi, X.-W.; Zhang, Y.; Jiang, J. Mechanism, Safety and Efficacy of Three Tyrosine Kinase Inhibitors Lapatinib, Neratinib and Pyrotinib in HER2-Positive Breast Cancer. *Am. J. Cancer Res.* **2019**, *9*, 2103–2119.

55. Pritchard, C.C.; Cheng, H.H.; Tewari, M. MicroRNA Profiling: Approaches and Considerations. *Nat. Rev. Genet.* **2012**, *13*, 358–369. [CrossRef]

56. Real-Time qRT-PCR. Available online: https://www.ncbi.nlm.nih.gov/probe/docs/techqpcr/ (accessed on 26 September 2023).

57. Beneš, V.; Castoldi, M. Expression Profiling of microRNA Using Real-Time Quantitative PCR, How to Use It and What Is Available. *Methods* **2010**, *50*, 244–249. [CrossRef]

58. Najjary, S.; Mohammadzadeh, R.; Mokhtarzadeh, A.; Mohammadi, A.; Kojabad, A.B.; Baradaran, B. Role of miR-21 as an Authentic Oncogene in Mediating Drug Resistance in Breast Cancer. *Gene* **2020**, *738*, 144453. [CrossRef] [PubMed]

59. TargetScanHuman 8.0. Available online: https://www.targetscan.org/vert_80/ (accessed on 26 September 2023).

60. Si, H.; Sun, X.; Chen, Y.; Cao, Y.; Chen, S.; Wang, H.; Hu, C. Circulating microRNA-92a and microRNA-21 as Novel Minimally Invasive Biomarkers for Primary Breast Cancer. *J. Cancer Res. Clin. Oncol.* **2013**, *139*, 223–229. [CrossRef] [PubMed]

61. Wang, Z.-X.; Lu, B.-B.; Wang, H.; Cheng, Z.-X.; Yin, Y.-M. MicroRNA-21 Modulates Chemosensitivity of Breast Cancer Cells to Doxorubicin by Targeting PTEN. *Arch. Med. Res.* **2011**, *42*, 281–290. [CrossRef]

62. Gong, C.; Yao, Y.; Wang, Y.; Liu, B.; Wu, W.; Chen, J.; Su, F.; Yao, H.; Song, E. Up-Regulation of miR-21 Mediates Resistance to Trastuzumab Therapy for Breast Cancer. *J. Biol. Chem.* **2011**, *286*, 19127–19137. [CrossRef]

63. Wu, Z.-H.; Tao, Z.-H.; Zhang, J.; Li, T.; Ni, C.; Xie, J.; Zhang, J.-F.; Hu, X.-C. MiRNA-21 Induces Epithelial to Mesenchymal Transition and Gemcitabine Resistance via the PTEN/AKT Pathway in Breast Cancer. *Tumor Biol.* **2016**, *37*, 7245–7254. [CrossRef]

64. Shi, Y.; Ye, P.; Long, X. Differential Expression Profiles of the Transcriptome in Breast Cancer Cell Lines Revealed by Next Generation Sequencing. *Cell. Physiol. Biochem. Int. J. Exp. Cell. Physiol. Biochem. Pharmacol.* **2017**, *44*, 804–816. [CrossRef] [PubMed]

65. Zhu, S.; Si, M.-L.; Wu, H.; Mo, Y.-Y. MicroRNA-21 Targets the Tumor Suppressor Gene Tropomyosin 1 (TPM1). *J. Biol. Chem.* **2007**, *282*, 14328–14336. [CrossRef]

66. Zhu, S.; Wu, H.; Wu, F.; Nie, D.; Sheng, S.; Mo, Y.-Y. MicroRNA-21 Targets Tumor Suppressor Genes in Invasion and Metastasis. *Cell Res.* **2008**, *18*, 350–359. [CrossRef]

67. Han, M.; Liu, M.; Wang, Y.; Mo, Z.; Bi, X.; Liu, Z.; Fan, Y.; Chen, X.; Wu, C. Re-Expression of miR-21 Contributes to Migration and Invasion by Inducing Epithelial-Mesenchymal Transition Consistent with Cancer Stem Cell Characteristics in MCF-7 Cells. *Mol. Cell Biochem.* **2012**, *363*, 427–436. [CrossRef]

68. Ahmed Mohmmed, E.; Shousha, W.G.; El-Saiid, A.S.; Ramadan, S.S. A Clinical Evaluation of Circulating MiR-106a and Raf-1 as Breast Cancer Diagnostic and Prognostic Markers. *Asian Pac. J. Cancer Prev. APJCP* **2021**, *22*, 3513–3520. [CrossRef]

69. You, F.; Luan, H.; Sun, D.; Cui, T.; Ding, P.; Tang, H.; Sun, D. miRNA-106a Promotes Breast Cancer Cell Proliferation, Clonogenicity, Migration, and Invasion Through Inhibiting Apoptosis and Chemosensitivity. *DNA Cell Biol.* **2019**, *38*, 198–207. [CrossRef] [PubMed]

70. Kawaguchi, T.; Yan, L.; Qi, Q.; Peng, X.; Edge, S.B.; Young, J.; Yao, S.; Liu, S.; Otsuji, E.; Takabe, K. Novel MicroRNA-Based Risk Score Identified by Integrated Analyses to Predict Metastasis and Poor Prognosis in Breast Cancer. *Ann. Surg. Oncol.* **2018**, *25*, 4037–4046. [CrossRef] [PubMed]

71. Kim, K.; Chadalapaka, G.; Lee, S.-O.; Yamada, D.; Sastre-Garau, X.; Defossez, P.-A.; Park, Y.-Y.; Lee, J.-S.; Safe, S. Identification of Oncogenic microRNA-17-92/ZBTB4/Specificity Protein Axis in Breast Cancer. *Oncogene* **2012**, *31*, 1034–1044. [CrossRef] [PubMed]

72. Yang, W.S.; Chadalapaka, G.; Cho, S.-G.; Lee, S.; Jin, U.-H.; Jutooru, I.; Choi, K.; Leung, Y.-K.; Ho, S.-M.; Safe, S.; et al. The Transcriptional Repressor ZBTB4 Regulates EZH2 through a MicroRNA-ZBTB4-Specificity Protein Signaling Axis. *Neoplasia* **2014**, *16*, 1059–1069. [CrossRef] [PubMed]

73. Roth, C.; Rack, B.; Müller, V.; Janni, W.; Pantel, K.; Schwarzenbach, H. Circulating microRNAs as Blood-Based Markers for Patients with Primary and Metastatic Breast Cancer. *Breast Cancer Res. BCR* **2010**, *12*, R90. [CrossRef] [PubMed]

74. Dinami, R.; Ercolani, C.; Petti, E.; Piazza, S.; Ciani, Y.; Sestito, R.; Sacconi, A.; Biagioni, F.; le Sage, C.; Agami, R.; et al. miR-155 Drives Telomere Fragility in Human Breast Cancer by Targeting TRF1. *Cancer Res.* **2014**, *74*, 4145–4156. [CrossRef] [PubMed]

75. Jiang, S.; Zhang, H.-W.; Lu, M.-H.; He, X.-H.; Li, Y.; Gu, H.; Liu, M.-F.; Wang, E.-D. MicroRNA-155 Functions as an OncomiR in Breast Cancer by Targeting the *Suppressor of Cytokine Signaling 1* Gene. *Cancer Res.* **2010**, *70*, 3119–3127. [CrossRef]

76. Ding, L.; Gu, H.; Xiong, X.; Ao, H.; Cao, J.; Lin, W.; Yu, M.; Lin, J.; Cui, Q. MicroRNAs Involved in Carcinogenesis, Prognosis, Therapeutic Resistance, and Applications in Human Triple-Negative Breast Cancer. *Cells* **2019**, *8*, 1492. [CrossRef]

77. Johansson, J.; Berg, T.; Kurzejamska, E.; Pang, M.-F.; Tabor, V.; Jansson, M.; Roswall, P.; Pietras, K.; Sund, M.; Religa, P.; et al. MiR-155-Mediated Loss of C/EBPβ Shifts the TGF-β Response from Growth Inhibition to Epithelial-Mesenchymal Transition, Invasion and Metastasis in Breast Cancer. *Oncogene* **2013**, *32*, 5614–5624. [CrossRef]

78. Li, P.; Xu, T.; Zhou, X.; Liao, L.; Pang, G.; Luo, W.; Han, L.; Zhang, J.; Luo, X.; Xie, X.; et al. Downregulation of miRNA-141 in Breast Cancer Cells Is Associated with Cell Migration and Invasion: Involvement of ANP32E Targeting. *Cancer Med.* **2017**, *6*, 662–672. [CrossRef]

79. Xiong, Z.; Ye, L.; Zhenyu, H.; Li, F.; Xiong, Y.; Lin, C.; Wu, X.; Deng, G.; Shi, W.; Song, L.; et al. ANP32E Induces Tumorigenesis of Triple-negative Breast Cancer Cells by Upregulating E2F1. *Mol. Oncol.* **2018**, *12*, 896–912. [CrossRef]

80. Gregory, P.A.; Bert, A.G.; Paterson, E.L.; Barry, S.C.; Tsykin, A.; Farshid, G.; Vadas, M.A.; Khew-Goodall, Y.; Goodall, G.J. The miR-200 Family and miR-205 Regulate Epithelial to Mesenchymal Transition by Targeting ZEB1 and SIP1. *Nat. Cell Biol.* **2008**, *10*, 593–601. [CrossRef] [PubMed]

81. Li, X.-X.; Gao, S.-Y.; Wang, P.-Y.; Zhou, X.; Li, Y.-J.; Yu, Y.; Yan, Y.-F.; Zhang, H.-H.; Lv, C.-J.; Zhou, H.-H.; et al. Reduced Expression Levels of Let-7c in Human Breast Cancer Patients. *Oncol. Lett.* **2015**, *9*, 1207–1212. [CrossRef]

82. Fu, X.; Mao, X.; Wang, Y.; Ding, X.; Li, Y. Let-7c-5p Inhibits Cell Proliferation and Induces Cell Apoptosis by Targeting ERCC6 in Breast Cancer. *Oncol. Rep.* **2017**, *38*, 1851–1856. [CrossRef]

83. Bozgeyik, E. Bioinformatic Analysis and In Vitro Validation of Let-7b and Let-7c in Breast Cancer. *Comput. Biol. Chem.* **2020**, *84*, 107191. [CrossRef]

84. Sun, X.; Xu, C.; Tang, S.-C.; Wang, J.; Wang, H.; Wang, P.; Du, N.; Qin, S.; Li, G.; Xu, S.; et al. Let-7c Blocks Estrogen-Activated Wnt Signaling in Induction of Self-Renewal of Breast Cancer Stem Cells. *Cancer Gene Ther.* **2016**, *23*, 83–89. [CrossRef] [PubMed]

85. Swellam, M.; Mahmoud, M.S.; Hashim, M.; Hassan, N.M.; Sobeih, M.E.; Nageeb, A.M. Clinical Aspects of Circulating miRNA-335 in Breast Cancer Patients: A Prospective Study. *J. Cell. Biochem.* **2019**, *120*, 8975–8982. [CrossRef]

86. Heyn, H.; Engelmann, M.; Schreek, S.; Ahrens, P.; Lehmann, U.; Kreipe, H.; Schlegelberger, B.; Beger, C. MicroRNA miR-335 Is Crucial for the BRCA1 Regulatory Cascade in Breast Cancer Development. *Int. J. Cancer* **2011**, *129*, 2797–2806. [CrossRef]

87. Hao, J.; Lai, M.; Liu, C. Expression of miR-335 in Triple-Negative Breast Cancer and Its Effect on Chemosensitivity. *J. BUON Off. J. Balk. Union Oncol.* **2019**, *24*, 1526–1531.

88. Zhu, N.; Zhang, D.; Xie, H.; Zhou, Z.; Chen, H.; Hu, T.; Bai, Y.; Shen, Y.; Yuan, W.; Jing, Q.; et al. Endothelial-Specific Intron-Derived miR-126 Is down-Regulated in Human Breast Cancer and Targets Both VEGFA and PIK3R2. *Mol. Cell. Biochem.* **2011**, *351*, 157–164. [CrossRef]

89. Fu, R.; Tong, J.-S. miR-126 Reduces Trastuzumab Resistance by Targeting PIK3R2 and Regulating AKT/mTOR Pathway in Breast Cancer Cells. *J. Cell. Mol. Med.* **2020**, *24*, 7600–7608. [CrossRef]

90. Wang, C.-Z.; Yuan, P.; Li, Y. MiR-126 Regulated Breast Cancer Cell Invasion by Targeting ADAM9. *Int. J. Clin. Exp. Pathol.* **2015**, *8*, 6547–6553.
91. Zuo, Y.; Qu, C.; Tian, Y.; Wen, Y.; Xia, S.; Ma, M. The HIF-1/SNHG1/miR-199a-3p/TFAM Axis Explains Tumor Angiogenesis and Metastasis under Hypoxic Conditions in Breast Cancer. *BioFactors Oxf. Engl.* **2021**, *47*, 444–460. [CrossRef]
92. Li, S.-Q.; Wang, Z.-H.; Mi, X.-G.; Liu, L.; Tan, Y. MiR-199a/b-3p Suppresses Migration and Invasion of Breast Cancer Cells by Downregulating PAK4/MEK/ERK Signaling Pathway. *IUBMB Life* **2015**, *67*, 768–777. [CrossRef]
93. Toda, H.; Seki, N.; Kurozumi, S.; Shinden, Y.; Yamada, Y.; Nohata, N.; Moriya, S.; Idichi, T.; Maemura, K.; Fujii, T.; et al. RNA-Sequence-Based microRNA Expression Signature in Breast Cancer: Tumor-Suppressive miR-101-5p Regulates Molecular Pathogenesis. *Mol. Oncol.* **2020**, *14*, 426–446. [CrossRef]
94. Harati, R.; Mohammad, M.G.; Tlili, A.; El-Awady, R.A.; Hamoudi, R. Loss of miR-101-3p Promotes Transmigration of Metastatic Breast Cancer Cells through the Brain Endothelium by Inducing COX-2/MMP1 Signaling. *Pharmaceuticals* **2020**, *13*, 144. [CrossRef]
95. Wang, S.; Cheng, M.; Zheng, X.; Zheng, L.; Liu, H.; Lu, J.; Liu, Y.; Chen, W. Interactions Between lncRNA TUG1 and miR-9-5p Modulate the Resistance of Breast Cancer Cells to Doxorubicin by Regulating eIF5A2. *OncoTargets Ther.* **2020**, *13*, 13159–13170. [CrossRef]
96. Liu, D.-Z.; Chang, B.; Li, X.-D.; Zhang, Q.-H.; Zou, Y.-H. MicroRNA-9 Promotes the Proliferation, Migration, and Invasion of Breast Cancer Cells via down-Regulating FOXO1. *Clin. Transl. Oncol.* **2017**, *19*, 1133–1140. [CrossRef]
97. D'Ippolito, E.; Plantamura, I.; Bongiovanni, L.; Casalini, P.; Baroni, S.; Piovan, C.; Orlandi, R.; Gualeni, A.V.; Gloghini, A.; Rossini, A.; et al. miR-9 and miR-200 Regulate PDGFRβ-Mediated Endothelial Differentiation of Tumor Cells in Triple-Negative Breast Cancer. *Cancer Res.* **2016**, *76*, 5562–5572. [CrossRef]
98. Li, X.; Zeng, Z.; Wang, J.; Wu, Y.; Chen, W.; Zheng, L.; Xi, T.; Wang, A.; Lu, Y. MicroRNA-9 and Breast Cancer. *Biomed. Pharmacother.* **2020**, *122*, 109687. [CrossRef]
99. Ma, L.; Young, J.; Prabhala, H.; Pan, E.; Mestdagh, P.; Muth, D.; Teruya-Feldstein, J.; Reinhardt, F.; Onder, T.T.; Valastyan, S.; et al. miR-9, a MYC/MYCN-Activated microRNA, Regulates E-Cadherin and Cancer Metastasis. *Nat. Cell Biol.* **2010**, *12*, 247–256. [CrossRef]
100. Wang, H.; Tan, Z.; Hu, H.; Liu, H.; Wu, T.; Zheng, C.; Wang, X.; Luo, Z.; Wang, J.; Liu, S.; et al. microRNA-21 Promotes Breast Cancer Proliferation and Metastasis by Targeting LZTFL1. *BMC Cancer* **2019**, *19*, 738. [CrossRef]
101. TNM Staging for Breast Cancer. Available online: https://www.cancerresearchuk.org/about-cancer/breast-cancer/stages-grades/tnm-staging (accessed on 26 September 2023).
102. You, F.; Li, J.; Zhang, P.; Zhang, H.; Cao, X. miR106a Promotes the Growth of Transplanted Breast Cancer and Decreases the Sensitivity of Transplanted Tumors to Cisplatin. *Cancer Manag. Res.* **2020**, *12*, 233–246. [CrossRef]
103. Loh, H.-Y.; Norman, B.P.; Lai, K.-S.; Rahman, N.M.A.N.A.; Alitheen, N.B.M.; Osman, M.A. The Regulatory Role of MicroRNAs in Breast Cancer. *Int. J. Mol. Sci.* **2019**, *20*, 4940. [CrossRef] [PubMed]
104. TERF1 Telomeric Repeat Binding Factor 1 [Homo Sapiens (Human)]—Gene—NCBI. Available online: https://www.ncbi.nlm.nih.gov/gene/7013 (accessed on 26 September 2023).
105. Taha, M.; Mitwally, N.; Soliman, A.S.; Yousef, E. Potential Diagnostic and Prognostic Utility of miR-141, miR-181b1, and miR-23b in Breast Cancer. *Int. J. Mol. Sci.* **2020**, *21*, 8589. [CrossRef]
106. GeneCards—Human Genes | Gene Database | Gene Search. Available online: https://www.genecards.org/ (accessed on 26 September 2023).
107. Bertoli, G.; Cava, C.; Castiglioni, I. MicroRNAs: New Biomarkers for Diagnosis, Prognosis, Therapy Prediction and Therapeutic Tools for Breast Cancer. *Theranostics* **2015**, *5*, 1122–1143. [CrossRef]
108. Gao, Y.; Zeng, F.; Wu, J.-Y.; Li, H.-Y.; Fan, J.-J.; Mai, L.; Zhang, J.; Ma, D.-M.; Li, Y.; Song, F. MiR-335 Inhibits Migration of Breast Cancer Cells through Targeting Oncoprotein c-Met. *Tumour Biol. J. Int. Soc. Oncodevelopmental. Biol. Med.* **2015**, *36*, 2875–2883. [CrossRef]
109. Ye, L.; Wang, F.; Wu, H.; Yang, H.; Yang, Y.; Ma, Y.; Xue, A.; Zhu, J.; Chen, M.; Wang, J.; et al. Functions and Targets of miR-335 in Cancer. *OncoTargets Ther.* **2021**, *14*, 3335–3349. [CrossRef]
110. Alhasan, L. MiR-126 Modulates Angiogenesis in Breast Cancer by Targeting VEGF-A -mRNA. *Asian Pac. J. Cancer Prev. APJCP* **2019**, *20*, 193–197. [CrossRef]
111. Soofiyani, S.R.; Hosseini, K.; Ebrahimi, T.; Forouhandeh, H.; Sadeghi, M.; Beirami, S.M.; Ghasemnejad, T.; Tarhriz, V.; Montazersaheb, S. Prognostic Value and Biological Role of miR-126 in Breast Cancer. *MicroRNA Shariqah United Arab Emir.* **2022**, *11*, 95–103. [CrossRef]
112. Qattan, A.; Al-Tweigeri, T.; Alkhayal, W.; Suleman, K.; Tulbah, A.; Amer, S. Clinical Identification of Dysregulated Circulating microRNAs and Their Implication in Drug Response in Triple Negative Breast Cancer (TNBC) by Target Gene Network and Meta-Analysis. *Genes* **2021**, *12*, 549. [CrossRef]
113. Fan, X.; Zhou, S.; Zheng, M.; Deng, X.; Yi, Y.; Huang, T. MiR-199a-3p Enhances Breast Cancer Cell Sensitivity to Cisplatin by Downregulating TFAM (TFAM). *Biomed. Pharmacother.* **2017**, *88*, 507–514. [CrossRef] [PubMed]
114. Fuso, P.; Di Salvatore, M.; Santonocito, C.; Guarino, D.; Autilio, C.; Mulè, A.; Arciuolo, D.; Rinninella, A.; Mignone, F.; Ramundo, M.; et al. Let-7a-5p, miR-100-5p, miR-101-3p, and miR-199a-3p Hyperexpression as Potential Predictive Biomarkers in Early Breast Cancer Patients. *J. Pers. Med.* **2021**, *11*, 816. [CrossRef] [PubMed]

115. Wang, C.-Z.; Deng, F.; Li, H.; Wang, D.-D.; Zhang, W.; Ding, L.; Tang, J.-H. MiR-101: A Potential Therapeutic Target of Cancers. *Am. J. Transl. Res.* **2018**, *10*, 3310–3321. [PubMed]
116. Jiang, H.; Li, L.; Zhang, J.; Wan, Z.; Wang, Y.; Hou, J.; Yu, Y. MiR-101-3p and Syn-Cal14.1a Synergy in Suppressing EZH2-Induced Progression of Breast Cancer. *OncoTargets Ther.* **2020**, *13*, 9599–9609. [CrossRef]
117. Sporn, J.C.; Katsuta, E.; Yan, L.; Takabe, K. Expression of MicroRNA-9 Is Associated with Overall Survival in Breast Cancer Patients. *J. Surg. Res.* **2019**, *233*, 426–435. [CrossRef]
118. Condrat, C.E.; Thompson, D.C.; Barbu, M.G.; Bugnar, O.L.; Boboc, A.; Cretoiu, D.; Suciu, N.; Cretoiu, S.M.; Voinea, S.C. miRNAs as Biomarkers in Disease: Latest Findings Regarding Their Role in Diagnosis and Prognosis. *Cells* **2020**, *9*, 276. [CrossRef] [PubMed]
119. Bonci, D.; Coppola, V.; Musumeci, M.; Addario, A.; Giuffrida, R.; Memeo, L.; D'Urso, L.; Pagliuca, A.; Biffoni, M.; Labbaye, C.; et al. The miR-15a-miR-16-1 Cluster Controls Prostate Cancer by Targeting Multiple Oncogenic Activities. *Nat. Med.* **2008**, *14*, 1271–1277. [CrossRef]
120. Long, X.; Shi, Y.; Ye, P.; Guo, J.; Zhou, Q.; Tang, Y. MicroRNA-99a Suppresses Breast Cancer Progression by Targeting FGFR3. *Front. Oncol.* **2020**, *9*, 1473. [CrossRef]
121. Menon, A.; Abd-Aziz, N.; Khalid, K.; Poh, C.L.; Naidu, R. miRNA: A Promising Therapeutic Target in Cancer. *Int. J. Mol. Sci.* **2022**, *23*, 11502. [CrossRef]
122. Shah, V.; Shah, J. Recent Trends in Targeting miRNAs for Cancer Therapy. *J. Pharm. Pharmacol.* **2020**, *72*, 1732–1749. [CrossRef]
123. Baumann, V.; Winkler, J. miRNA-Based Therapies: Strategies and Delivery Platforms for Oligonucleotide and Non-Oligonucleotide Agents. *Future Med. Chem.* **2014**, *6*, 1967–1984. [CrossRef]
124. Rupaimoole, R.; Slack, F.J. MicroRNA Therapeutics: Towards a New Era for the Management of Cancer and Other Diseases. *Nat. Rev. Drug Discov.* **2017**, *16*, 203–222. [CrossRef]
125. Si, W.; Shen, J.; Zheng, H.; Fan, W. The Role and Mechanisms of Action of MicroRNAs in Cancer Drug Resistance. *Clin. Epigenetics* **2019**, *11*, 25. [CrossRef] [PubMed]
126. Magee, P.; Shi, L.; Garofalo, M. Role of MicroRNAs in Chemoresistance. *Ann. Transl. Med.* **2015**, *3*, 332. [CrossRef] [PubMed]
127. Zhong, S.; Ma, T.; Zhang, X.; Lv, M.; Chen, L.; Tang, J.; Zhao, J. MicroRNA Expression Profiling and Bioinformatics Analysis of Dysregulated MicroRNAs in Vinorelbine-Resistant Breast Cancer Cells. *Gene* **2015**, *556*, 113–118. [CrossRef] [PubMed]
128. Bai, W.-D.; Ye, X.-M.; Zhang, M.-Y.; Zhu, H.-Y.; Xi, W.-J.; Huang, X.; Zhao, J.; Gu, B.; Zheng, G.-X.; Yang, A.-G.; et al. MiR-200c Suppresses TGF-β Signaling and Counteracts Trastuzumab Resistance and Metastasis by Targeting ZNF217 and ZEB1 in Breast Cancer: MiR-200c/ZNF217/TGF-β Regulates Trastuzumab Resistance. *Int. J. Cancer* **2014**, *135*, 1356–1368. [CrossRef]
129. Song, L.; Liu, D.; Wang, B.; He, J.; Zhang, S.; Dai, Z.; Ma, X.; Wang, X. MiR-494 Suppresses the Progression of Breast Cancer In Vitro by Targeting CXCR4 through the Wnt/β-Catenin Signaling Pathway. *Oncol. Rep.* **2015**, *34*, 525–531. [CrossRef]
130. Abedi, N.; Mohammadi-Yeganeh, S.; Koochaki, A.; Karami, F.; Paryan, M. MiR-141 as Potential Suppressor of β-Catenin in Breast Cancer. *Tumour Biol. J. Int. Soc. Oncodevelopmental. Biol. Med.* **2015**, *36*, 9895–9901. [CrossRef]
131. Yu, Z.; Xu, Z.; Disante, G.; Wright, J.; Wang, M.; Li, Y.; Zhao, Q.; Ren, T.; Ju, X.; Gutman, E.; et al. MiR-17/20 Sensitization of Breast Cancer Cells to Chemotherapy-Induced Apoptosis Requires Akt1. *Oncotarget* **2014**, *5*, 1083–1090. [CrossRef]
132. He, X.; Xiao, X.; Dong, L.; Wan, N.; Zhou, Z.; Deng, H.; Zhang, X. MiR-218 Regulates Cisplatin Chemosensitivity in Breast Cancer by Targeting BRCA1. *Tumour Biol. J. Int. Soc. Oncodevelopmental. Biol. Med.* **2015**, *36*, 2065–2075. [CrossRef] [PubMed]
133. Garrido-Cano, I.; Pattanayak, B.; Adam-Artigues, A.; Lameirinhas, A.; Torres-Ruiz, S.; Tormo, E.; Cervera, R.; Eroles, P. MicroRNAs as a Clue to Overcome Breast Cancer Treatment Resistance. *Cancer Metastasis Rev.* **2022**, *41*, 77–105. [CrossRef] [PubMed]
134. ClinicalTrials. Available online: https://clinicaltrials.gov/ (accessed on 27 September 2023).
135. Pan, G.; Liu, Y.; Shang, L.; Zhou, F.; Yang, S. EMT-Associated MicroRNAs and Their Roles in Cancer Stemness and Drug Resistance. *Cancer Commun. Lond. Engl.* **2021**, *41*, 199–217. [CrossRef] [PubMed]
136. Song, K.; Farzaneh, M. Signaling Pathways Governing Breast Cancer Stem Cells Behavior. *Stem Cell Res. Ther.* **2021**, *12*, 245. [CrossRef] [PubMed]
137. Ali Syeda, Z.; Langden, S.S.S.; Munkhzul, C.; Lee, M.; Song, S.J. Regulatory Mechanism of MicroRNA Expression in Cancer. *Int. J. Mol. Sci.* **2020**, *21*, 1723. [CrossRef]
138. Yousefnia, S.; Seyed Forootan, F.; Seyed Forootan, S.; Nasr Esfahani, M.H.; Gure, A.O.; Ghaedi, K. Mechanistic Pathways of Malignancy in Breast Cancer Stem Cells. *Front. Oncol.* **2020**, *10*, 452. [CrossRef]
139. Kastl, L.; Brown, I.; Schofield, A.C. MiRNA-34a Is Associated with Docetaxel Resistance in Human Breast Cancer Cells. *Breast Cancer Res. Treat.* **2012**, *131*, 445–454. [CrossRef]
140. Zhao, X.; Li, J.; Huang, S.; Wan, X.; Luo, H.; Wu, D. MiRNA-29c regulates cell growth and invasion by targeting CDK6 in bladder cancer. *Am. J. Transl. Res.* **2015**, *7*, 1382–1389.
141. Citron, F.; Segatto, I.; Vinciguerra, G.L.R.; Musco, L.; Russo, F.; Mungo, G.; D'Andrea, S.; Mattevi, M.C.; Perin, T.; Schiappacassi, M.; et al. Downregulation of miR-223 Expression Is an Early Event during Mammary Transformation and Confers Resistance to CDK4/6 Inhibitors in Luminal Breast Cancer. *Cancer Res.* **2020**, *80*, 1064–1077. [CrossRef]
142. Bisso, A.; Faleschini, M.; Zampa, F.; Capaci, V.; De Santa, J.; Santarpia, L.; Piazza, S.; Cappelletti, V.; Daidone, M.; Agami, R.; et al. Oncogenic MiR-181a/b Affect the DNA Damage Response in Aggressive Breast Cancer. *Cell Cycle Georget. Tex* **2013**, *12*, 1679–1687. [CrossRef]

143. Feng, D.; Zhang, H.; Zhang, P.; Zheng, Y.; Zhang, X.; Han, B.; Luo, X.; Xu, L.; Zhou, H.; Qu, L.; et al. Down-regulated MiR-331–5p and MiR-27a Are Associated with Chemotherapy Resistance and Relapse in Leukaemia. *J. Cell. Mol. Med.* **2011**, *15*, 2164–2175. [CrossRef]

144. Zou, Z.; Zou, R.; Zong, D.; Shi, Y.; Chen, J.; Huang, J.; Zhu, J.; Chen, L.; Bao, X.; Liu, Y.; et al. MiR-495 Sensitizes MDR Cancer Cells to the Combination of Doxorubicin and Taxol by Inhibiting MDR1 Expression. *J. Cell. Mol. Med.* **2017**, *21*, 1929–1943. [CrossRef] [PubMed]

145. Lu, X.; Liu, R.; Wang, M.; Kumar, A.K.; Pan, F.; He, L.; Hu, Z.; Guo, Z. MicroRNA-140 Impedes DNA Repair by Targeting FEN1 and Enhances Chemotherapeutic Response in Breast Cancer. *Oncogene* **2020**, *39*, 234–247. [CrossRef]

146. Zhao, W.; Li, H.; Yang, S.; Guo, D.; Chen, J.; Miao, S.; Xin, Y.; Liang, M. MicroRNA-152 Suppresses Cisplatin Resistance in A549 Cells. *Oncol. Lett.* **2019**, *18*, 4613–4620. [CrossRef] [PubMed]

147. Xie, Q.; Wang, S.; Zhao, Y.; Zhang, Z.; Qin, C.; Yang, X. MiR-519d Impedes Cisplatin-Resistance in Breast Cancer Stem Cells by down-Regulating the Expression of MCL-1. *Oncotarget* **2017**, *8*, 22003–22013. [CrossRef] [PubMed]

148. He, H.; Tian, W.; Chen, H.; Jiang, K. MiR-944 Functions as a Novel Oncogene and Regulates the Chemoresistance in Breast Cancer. *Tumor Biol.* **2016**, *37*, 1599–1607. [CrossRef] [PubMed]

149. Bao, X.; Ren, T.; Huang, Y.; Sun, K.; Wang, S.; Liu, K.; Zheng, B.; Guo, W. Knockdown of Long Non-Coding RNA HOTAIR Increases MiR-454-3p by Targeting Stat3 and Atg12 to Inhibit Chondrosarcoma Growth. *Cell Death Dis.* **2017**, *8*, e2605. [CrossRef]

150. Chong, Z.X.; Yeap, S.K.; Ho, W.Y. Regulation of autophagy by microRNAs in human breast cancer. *J. Biomed. Sci.* **2021**, *28*, 21. [CrossRef]

151. Akkoc, Y.; Gozuacik, D. MicroRNAs as Major Regulators of the Autophagy Pathway. *Biochim. Biophys. Acta BBA Mol. Cell Res.* **2020**, *1867*, 118662. [CrossRef]

152. Shan, C.; Chen, X.; Cai, H.; Hao, X.; Li, J.; Zhang, Y.; Gao, J.; Zhou, Z.; Li, X.; Liu, C.; et al. The Emerging Roles of Autophagy-Related MicroRNAs in Cancer. *Int. J. Biol. Sci.* **2021**, *17*, 134–150. [CrossRef]

153. UCSC Xena. Available online: https://xenabrowser.net/ (accessed on 27 September 2023).

154. GEPIA 2. Available online: http://gepia2.cancer-pku.cn/#index (accessed on 9 February 2023).

155. OncoLnc. Available online: http://www.oncolnc.org/ (accessed on 27 September 2023).

156. Boon, R.A.; Vickers, K.C. Intercellular Transport of microRNAs. *Arterioscler. Thromb. Vasc. Biol.* **2013**, *33*, 186–192. [CrossRef] [PubMed]

157. Makarova, J.; Turchinovich, A.; Shkurnikov, M.; Tonevitsky, A. Extracellular miRNAs and Cell–Cell Communication: Problems and Prospects. *Trends Biochem. Sci.* **2021**, *46*, 640–651. [CrossRef] [PubMed]

158. Zhang, S.; Cheng, Z.; Wang, Y.; Han, T. The Risks of miRNA Therapeutics: In a Drug Target Perspective. *Drug Des. Devel. Ther.* **2021**, *15*, 721–733. [CrossRef] [PubMed]

159. Momin, M.Y.; Gaddam, R.R.; Kravitz, M.; Gupta, A.; Vikram, A. The Challenges and Opportunities in the Development of MicroRNA Therapeutics: A Multidisciplinary Viewpoint. *Cells* **2021**, *10*, 3097. [CrossRef] [PubMed]

160. Geary, R.S.; Norris, D.; Yu, R.; Bennett, C.F. Pharmacokinetics, Biodistribution and Cell Uptake of Antisense Oligonucleotides. *Adv. Drug Deliv. Rev.* **2015**, *87*, 46–51. [CrossRef]

161. McMahon, B.M.; Mays, D.; Lipsky, J.; Stewart, J.A.; Fauq, A.; Richelson, E. Pharmacokinetics and Tissue Distribution of a Peptide Nucleic Acid after Intravenous Administration. *Antisense Nucleic Acid Drug Dev.* **2002**, *12*, 65–70. [CrossRef]

162. Cacheux, J.; Bancaud, A.; Leichlé, T.; Cordelier, P. Technological Challenges and Future Issues for the Detection of Circulating MicroRNAs in Patients with Cancer. *Front. Chem.* **2019**, *7*, 815. [CrossRef] [PubMed]

163. Ho, P.T.B.; Clark, I.M.; Le, L.T.T. MicroRNA-Based Diagnosis and Therapy. *Int. J. Mol. Sci.* **2022**, *23*, 7167. [CrossRef]

164. Nagaraj, S. Pros and Cons of miRNAs as Non-Invasive Circulatory Biomarkers. *Res. Rev. Res. J. Biol.* **2017**, *5*, 1–7.

current issues in molecular biology

MDPI

Communication

DDIT4 Downregulation by siRNA Approach Increases the Activity of Proteins Regulating Fatty Acid Metabolism upon Aspirin Treatment in Human Breast Cancer Cells

Aistė Savukaitytė [1,*], Agnė Bartnykaitė [1], Justina Bekampytė [1], Rasa Ugenskienė [1,2] and Elona Juozaitytė [3]

[1] Oncology Research Laboratory, Institute of Oncology, Lithuanian University of Health Sciences, LT-50161 Kaunas, Lithuania; agne.bartnykaite@lsmuni.lt (A.B.); justina.bekampyte@lsmuni.lt (J.B.); rasa.ugenskiene@lsmuni.lt (R.U.)

[2] Department of Genetics and Molecular Medicine, Lithuanian University of Health Sciences, LT-50161 Kaunas, Lithuania

[3] Institute of Oncology, Lithuanian University of Health Sciences, LT-50161 Kaunas, Lithuania; elona.juozaityte@lsmuni.lt

* Correspondence: aiste.savukaityte@lsmuni.lt

Abstract: Repositioning of aspirin for a more effective breast cancer (BC) treatment requires identification of predictive biomarkers. However, the molecular mechanism underlying the anticancer activity of aspirin remains fully undefined. Cancer cells enhance de novo fatty acid (FA) synthesis and FA oxidation to maintain a malignant phenotype, and the mechanistic target of rapamycin (mTORC1) is required for lipogenesis. We, therefore, aimed to test if the expression of mTORC1 suppressor DNA damage-inducible transcript (DDIT4) affects the activity of main enzymes in FA metabolism after aspirin treatment. MCF-7 and MDA-MB-468 human BC cell lines were transfected with siRNA to downregulate DDIT4. The expression of carnitine palmitoyltransferase 1 A (CPT1A) and serine 79-phosphorylated acetyl-CoA carboxylase 1 (ACC1) were analyzed by Western Blotting. Aspirin enhanced ACC1 phosphorylation by two-fold in MCF-7 cells and had no effect in MDA-MB-468 cells. Aspirin did not change the expression of CPT1A in either cell line. We have recently reported DDIT4 itself to be upregulated by aspirin. DDIT4 knockdown resulted in 1.5-fold decreased ACC1 phosphorylation (dephosphorylation activates the enzyme), 2-fold increased CPT1A expression in MCF-7 cells, and 2.8-fold reduced phosphorylation of ACC1 following aspirin exposure in MDA-MB-468 cells. Thus, DDIT4 downregulation raised the activity of main lipid metabolism enzymes upon aspirin exposure which is an undesired effect as FA synthesis and oxidation are linked to malignant phenotype. This finding may be clinically relevant as DDIT4 expression has been shown to vary in breast tumors. Our findings justify further, more extensive investigation of the role of DDIT4 in aspirin's effect on fatty acid metabolism in BC cells.

Keywords: aspirin; breast cancer; DDIT4; ACC1; CPT1A

Citation: Savukaitytė, A.; Bartnykaitė, A.; Bekampytė, J.; Ugenskienė, R.; Juozaitytė, E. DDIT4 Downregulation by siRNA Approach Increases the Activity of Proteins Regulating Fatty Acid Metabolism upon Aspirin Treatment in Human Breast Cancer Cells. *Curr. Issues Mol. Biol.* **2023**, *45*, 4665–4674. https://doi.org/10.3390/cimb45060296

Academic Editor: Dumitru A. Iacobas

Received: 25 April 2023
Revised: 22 May 2023
Accepted: 26 May 2023
Published: 28 May 2023

1. Introduction

Aspirin, or acetylsalicylic acid, has been used to treat pain, fever, and inflammation and prevent heart attacks, strokes, and pathological clot formation [1]. A considerable body of data demonstrates that it also exerts antitumor action in various cancer types, including breast cancer (BC) [2–16], via a not fully clear mechanism [17]. BC is the most frequent cancer worldwide [18] and the leading cause of cancer death [19]. Thus, management of patients with BC requires new solutions to improve patients' outcomes [20]. Repurposing aspirin for a more successful cancer treatment needs identification of predictive biomarkers to select patients who are most likely to benefit from the therapy [21].

Metabolic reprogramming is a well-established hallmark of cancer [22]. In addition to the well-known aerobic glycolysis (Warburg effect), cancer cells rewire their metabolism via additional strategies to sustain growth and survival [22]. It is increasingly recognized that cancer cells reprogram their lipid metabolism [23]. Tumor cells markedly elevate de novo fatty acid (FA) synthesis despite the availability of exogenous lipid sources to satisfy the demand for energy and macromolecules [24]. Increased expression of enzymes which regulate de novo FA synthesis have been associated with cancer risk and prognosis [25]. Elevated expression of sterol regulatory element-binding transcription factor 1 (SREBP1), the transcriptional activator of lipid biogenesis genes, has been linked to poor prognosis in BC [26]. Cancer cells are also known to dysregulate FA oxidation [27]. Mitochondrial FA oxidation produces adenosine triphosphate (ATP) to fuel cancer cells [28]. Genes involved in FA oxidation encourage cell proliferation and survival [25]. In vitro experiments have shown that the inhibition of FA oxidation suppresses the metastatic properties of BC cells [29]. Thus, targeting lipid metabolism may be an attractive therapeutic option, especially in BC, where it plays a central role in tumor biology [30].

Acetyl-CoA carboxylases (ACC) catalyze the ATP-dependent conversion of acetyl-CoA into malonyl-CoA [31]. Cytosolic ACC1 isoform is thought to be the critical enzyme in FA synthesis, since its enzymatic product malonyl-CoA serves as substrate for the synthesis of FAs [28]. The activity of ACC1 is regulated transcriptionally and post-transcriptionally [32]. Phosphorylation at a major regulatory site Ser79 by AMP-activated protein kinase (AMPK) results in the inhibition of ACC1 activity through the blockage of homodimer formation [32]. Breast cancer gene 1 (BRCA1) can also increase ACC1 phosphorylation at Ser79 by associating with and thus preventing ACC1 dephosphorylation [33]. It has been demonstrated that ACC1 silencing induces apoptosis in BC cells [34]. ACC1 mRNA expression is decreased in triple-negative BC compared to receptor-positive BC in both tissue samples and cell lines [35].

The key enzyme in mitochondrial FA oxidation is carnitine palmitoyltransferase 1 (CPT1) [23]. CPT1A is the most prevailing isoform which accomplishes the rate-limiting step in FA oxidation [36]. While the inner mitochondrial membrane is impermeable to fatty acyl-CoA thioesters, CPT1 catalyzes the formation of long-chain acylcarnitines which are able to cross the inner mitochondrial membrane [37]. An integrated genomics approach based on the use of gene expression signatures of oncogenic pathway activity identified CPT1A as a driver of proliferation in luminal BCs [38]. Long-term clinical follow-up data suggested high CPT1A mRNA expression in tumors to promote distant BC metastasis [39]. CPT1A has been recently suggested as a biomarker for BC disease-monitoring [37,40].

Activation of transcriptional lipogenesis regulator SREBP requires mechanistic target of rapamycin (mTORC1) [41]. mTORC1 was also shown to be necessary for de novo lipogenesis activated by serine/threonine kinase (AKT) [41]. We have recently shown that DNA damage-inducible transcript 4 (DDIT4) expression determines the phosphorylation of mTORC1 target eukaryotic translation initiation factor 4E-binding protein 1 (4E-BP1) upon aspirin treatment [42]. Considering that de novo lipogenesis requires mTORC1 and mTORC1 activity (assessed by 4E-BP1 phosphorylation) is affected by intracellular DDIT4 level after aspirin treatment, we aimed to investigate whether DDIT4 knockdown impacts aspirin effect on the activity of main enzymes in lipid metabolism. We found that DDIT4 downregulation increased the activity of ACC1 and CPT1A in MCF-7 cells and activated ACC1 in the MDA-MB-468 BC cell line upon aspirin treatment.

2. Materials and Methods

2.1. Cell Lines, Culture Conditions, and Treatment

Human BC cell lines MCF-7 (luminal subtype) and MDA-MB-468 (triple-negative subtype) were purchased from CLS Cell Line Service (Eppelheim, Germany) and grown as monolayers in *Dulbecco's Modified Eagle's medium* (DMEM; Sigma-Aldrich, St. Louis, MO, USA) supplemented with 10% fetal bovine serum (Gibco, Gaithersburg, MD, USA) and 100 U/mL penicillin with 100 µg/mL streptomycin (Gibco, Gaithersburg, MD, USA) and 2 mM L-glutamine (Gibco, Gaithersburg, MD, USA) at 37 °C in a humidified environment with 5% CO_2. Exponentially growing cells were plated into 25 cm^2 flasks for 24 h and then treated with 2 mM of aspirin for 24 h. For the transfection experiments, the cells were reverse-transfected at the time of plating into flasks. We used 2 mM of aspirin since it is an achievable plasma salicylate concentration obtained from hydrolysis of aspirin [43].

2.2. Chemicals and Antibodies

Aspirin was purchased from Sigma-Aldrich (St. Louis, MO, USA) and 0.5 M stock solution was prepared in water (pH 7) and frozen at $-20\ °C$ in small quantities to prevent freeze–thaw cycles. Working solutions were prepared before each experiment.

Antibodies against DDIT4 (#ab191871) and CPT1A (#ab128568) were obtained from Abcam (Cambridge, UK). Anti-phospho-ACC1 Ser79 (#PA5-17725) antibody and anti-β-actin (#AM4302) were purchased from Invitrogen (Waltham, MA, USA). Horseradish peroxidase-conjugated anti-rabbit (#65-6120), anti-mouse (#61-6520) and alkaline phosphatase-conjugated anti-mouse (#WP20006) secondary antibodies were bought from Invitrogen (Waltham, MA, USA).

2.3. Western Blot Analysis

Cells were lysed in radioimmunoprecipitation assay (RIPA) buffer (Abcam, Cambridge, UK), supplemented with protease (Sigma-Aldrich, St. Louis, MO, USA) and phosphatase inhibitor cocktails (Sigma-Aldrich, St. Louis, MO, USA) for 20 min on ice. Cell lysate was centrifuged at $8000 \times g$ for 20 min, and the supernatants were further assayed or stored at $-20\ °C$ until use. Protein concentration was determined using Pierce™ BCA Protein Assay Kit (Thermo Fisher Scientific, Waltham, MA, USA), according to the manufacturer's protocol. Forty micrograms of proteins were loaded in each lane for electrophoresis in 12% or 4–12% gradient SDS-PAGE. The resolved proteins were electrophoretically transferred onto a PVDF (polyvinylidene fluoride) membrane using a semi-wet transfer unit Mini Blot module (Invitrogen, Waltham, MA, USA). Membranes were blocked in optimal blocking buffer for each antibody for 45 min and incubated overnight at 4 °C with primary antibody against DDIT4 (1:1000), phospho-ACC1 Ser79 (1:1000), CPT1A (1:1000), or β-actin (1:2000). After washing three times using TBST (Tris buffered saline with tween 20), the blots were further incubated with HRP (for probing DDIT4 and phospho-ACC1) or AP-conjugated secondary antibodies (for β-actin and CPT1A). The blots were washed again with TBST three times, and the proteins were detected using SuperSignal™ West Atto Ultimate Sensitivity Substrate(Thermo Fisher Scientific, Waltham, MA, USA; for DDIT4 and phospho-ACC1) or CDP-Star® chemiluminescent substrate (for β-actin and CPT1A) from WesternBreeze Chemiluminescent Kit (Invitrogen, Waltham, MA, USA). The chemiluminescent reaction was captured using the iBright™ CL750 Imaging System (Invitrogen, Waltham, MA, USA). Densitometry analysis was performed with the iBright™ Analysis Software (Invitrogen, Waltham, MA, USA) in the Thermo Fisher Connect Platform. B-actin was used as a loading control.

2.4. Cell Transfection

Cells were reverse-transfected using Lipofectamine™ RNAiMAX transfection reagent (Invitrogen, Waltham, MA, USA) and 8 nM Silencer Select siRNA as per manufacturer's instructions. The following siRNAs were employed: siRNA targeting DDIT4 (#s29166; Ambion, Austin, TX, USA), non-targeting siRNA (#4390843; Invitrogen, Waltham, MA, USA), and positive control siRNA against GAPDH (#4390849; Invitrogen, Waltham, MA, USA). Target gene knockdown efficiency was first measured in time course experiments (24–72 h) to select the time point with efficient silencing for the drug treatment. Transfection conditions were optimized to induce at least 70% target mRNA knockdown while minimizing cytotoxicity.

Gene knockdown efficiency was evaluated by quantitative reverse transcription-polymerase chain reaction (qRT-PCR) on a Quanstudio 3 Real-Time PCR System (Applied Biosystems, Foster City, CA, USA). RNA was extracted with the PureLink™ RNA Mini Kit (Invitrogen, Waltham, MA, USA) and quantified using a Qubit® 3.0 fluorometer (Life Technologies, Carlsbad, CA, USA) and the Qubit^{TM} RNA HS Assay Kit (Invitrogen, Waltham, MA, USA) following the manufacturer's recommendations. cDNA was synthesized using the High-Capacity RNA-to-cDNA kit (Applied Biosystems, Foster City, CA, USA), as per manufacturer's instructions. PCR reaction was performed with Taqman^{TM} Gene Expression Assays (Hs99999905_m1 for GAPDH and Hs01111686_g1 for DDIT4 quantification) and Taqman^{TM} Universal Master Mix II with UNG from Applied Biosystems (Foster City, CA, USA), following the manufacturer's instructions. Relative quantification of target gene expression was determined by the comparative Ct method with β-actin as internal control. The transfection efficiency was verified by Western Blotting analysis.

2.5. Comparison of Protein Fold Change in Transfected Cells

Fold change due to aspirin (ASA) treatment was compared in DDIT4 siRNA-transfected cells (siDDIT4) to control siRNA transfected cells (siCT) as follows:

$$\left(\frac{expression\ post\ ASA\ treatment\ in\ siDDIT4\ cells}{expression\ post\ control\ treatment\ in\ siDDIT4\ cells} \right) versus \left(\frac{expression\ post\ ASA\ treatment\ in\ siCT\ cells}{expression\ post\ control\ treatment\ in\ siCT\ cells} \right)$$

2.6. Statistical Analysis

Results have been expressed as mean $\perp$ standard deviation (SD) from independent experiments carried out in triplicate. Statistical significance was tested via Student's *t*-test using SPSS 27.0.1 software (SPSS Inc., Chicago, IL, USA). Results were considered statistically significant at $p < 0.05$.

3. Results

3.1. Aspirin Inhibits ACC1 Activity in MCF-7 Cells

We tested if aspirin affects the activity of ACC1 and expression of CPT1A in MCF-7 and MDA-MB-468 human BC cell lines. We observed that 24 h aspirin exposure increases ACC1 phosphorylation (at serine 79) by two-fold in the MCF-7 cell line and has no effect in MDA-MB-468 cells (Figure 1). Aspirin treatment for 24 h did not change the expression of CPT1A in either cell line (Figure 1).

3.2. DDIT4 Knockdown Enhances ACC1 Activity following Aspirin Treatment in BC Cell Lines

We further asked whether DDIT4 expression determines ACC1 activity post-aspirin treatment in these BC cell lines. To manipulate DDIT4 expression, we adopted the RNA interference strategy. Compared to the control siRNA, DDIT4-targeting siRNA decreased ACC1 phosphorylation 1.5-fold in the MCF-7 cell line (Figure 2). In MDA-MB-468 cells, downregulation of DDIT4 caused a 2.8-fold reduction of the phosphorylated ACC1 amount following aspirin exposure (Figure 2). Thus, a lower DDIT4 level within these cell lines leads to a higher ACC1 activity after aspirin treatment as phosphorylation deactivates ACC1.

3.3. DDIT4 Knockdown Increases CPT1 Expression after Aspirin Treatment in MCF-7 Cells

Having established that DDIT4 downregulation using siRNA approach influences the effect of aspirin on phosphorylation of FA biosynthesis enzyme ACC1, we then examined whether it decides the impact on the rate-limiting enzyme in FA oxidation. We found that DDIT4 knockdown increased CPT1A expression by two-fold after treatment with aspirin in MCF-7 cells (Figure 3). DDIT4-targeting siRNA had no effect on CPT1A expression post aspirin treatment in MDA-MB-468 cells compared to control siRNA (Figure 3).

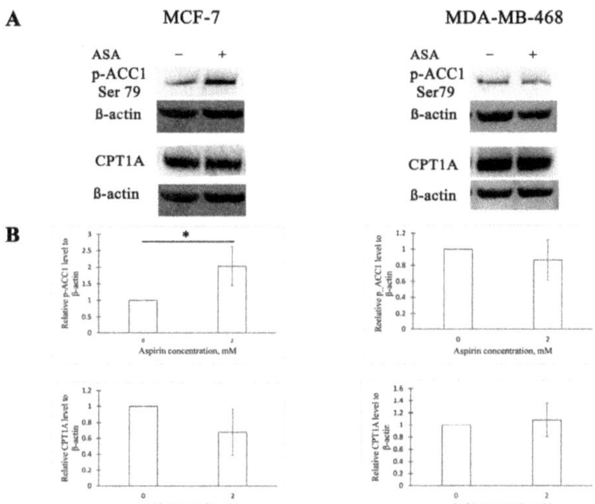

Figure 1. Aspirin increases acetyl-CoA carboxylase 1 (ACC1) phosphorylation in the MCF-7 cell line. (**A**) Representative Western blots of ACC1 phosphorylation and carnitine palmitoyltransferase 1A (CPT1A) expression in MCF-7 and MDA-MB-468 cells after 24 h aspirin (ASA) treatment. (**B**) Densitometric quantifications of p-ACC1 and CPT1A levels normalized to β-actin. The results are expressed as the mean ± SD (n = 3 in all graphs except CPT1A quantification in MCF-7 cells, where n = 4), * p value < 0.05.

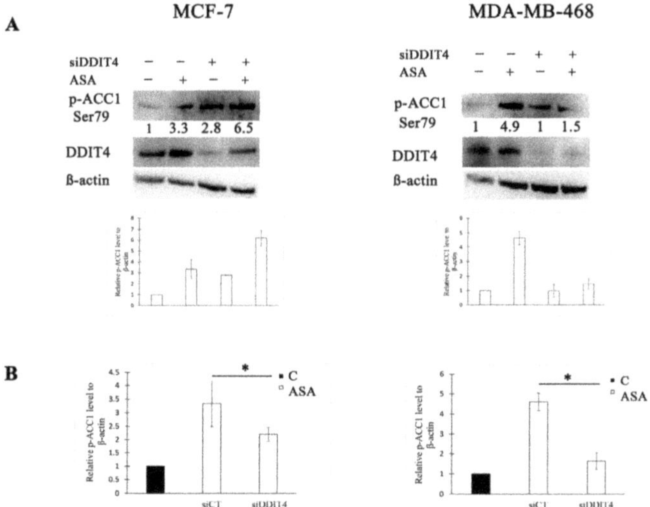

Figure 2. DNA damage-inducible transcript 4 (DDIT4) knockdown reduces ACC1 phosphorylation after aspirin treatment in breast cancer (BC) cell lines. (**A**) Representative Western blots and densitometric quantifications of p-ACC1 levels normalized to β-actin. MCF-7 and MDA-MB-468 cells were transfected with non-targeting (siCT) or DDIT4 siRNA (siDDIT4) for 24 h and treated with vehicle control (C) or 2 mM of aspirin (ASA) for the next 24 h before cell lysis. DDIT4 was probed on a different blot than p-ACC1. The numbers under the bands indicate densitometric quantifications of relative p-ACC1 level to β-actin. The results in the graphs are expressed as the mean ± SD (n = 3). (**B**) Comparison of p-ACC1 fold change after aspirin treatment between cells transfected with non-targeting siRNA and DDIT4 siRNA. The results are expressed as the mean ± SD (n = 3), * p value < 0.05.

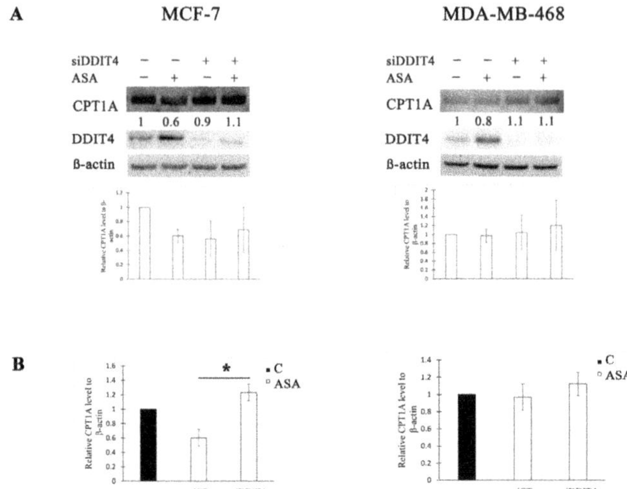

Figure 3. DDIT4 knockdown enhances CPT1A expression post aspirin treatment in MCF-7 cells. (**A**) Representative Western blots and densitometric quantifications of CPT1A levels normalized to β-actin. MCF-7 and MDA-MB-468 cells were transfected with non-targeting (siCT) or DDIT4 siRNA (siDDIT4) for 24 h and treated with vehicle control (C) or 2 mM of aspirin (ASA) for the next 24 h before cell lysis. DDIT4 was probed on a different blot than CPT1A in MCF-7 cell lysates. The numbers under the bands indicate densitometric quantifications of relative CPT1A level to β-actin. The results in the graphs are expressed as the mean ± SD (*n* = 3). (**B**) Comparison of CPT1A fold change after aspirin treatment between cells transfected with non-targeting siRNA and DDIT4 siRNA. The results are expressed as the mean ± SD (*n* = 3). * *p* value < 0.05.

4. Discussion

Dysregulated lipid metabolism impacts multiple intracellular processes, such as membrane synthesis, energy metabolism, and cell signaling to support tumorigenesis and cancer progression [44]. Therefore, targeting abnormal lipid metabolism in cancer treatment is of substantial clinical interest [44].

The de novo pathway of FA synthesis plays an important role in mammary tumorigenesis, and ACC1 is the rate-limiting enzyme in this process [35]. Inhibition of lipogenic enzymes, including ACC1, has been recently reported as one of anticancer therapy strategies [45]. Aspirin has been demonstrated to reduce ACC activity through activation of its repressor AMPK in human hepatoma HepG2 cells, and the effect was ascribed to salicylic acid [46]. Aspirin-mediated inhibitory phosphorylation at serine 79 of ACC has also been shown in SUM159-PT human BC cell lines [47]. Downregulation of other critical enzymes in FA synthesis including SREBP1, stearoyl-CoA desaturase 1 (SCD1), fatty acid synthase (FASN), and ATP citrate lyase (ACLY) has been reported for aspirin in AU-565 and SKBR-3 human BC cells [48]. Quantification of fatty acid production through liquid chromatography–tandem mass spectrometry analysis also showed that aspirin attenuated the levels of lipid formation in BC cells [48]. Accordingly, suppression of FA synthesis via blockage of active state ACC1 and other lipogenesis enzymes may at least partly explain the antitumor properties of aspirin.

CPT1A is the rate-limiting enzyme in FA oxidation which is also implicated in malignant phenotype [35,38,39]. On the contrary to aspirin's inhibitory effect on ACC1 and other FA synthesis enzymes, aspirin has been reported to promote the expression of FA oxidation genes CPT1 and MCAD at RNA and protein levels in human hepatoma HepG2 cell line [46]. The effect of aspirin on the increase in fatty acid oxidation has been also documented in mouse embryonic fibroblasts through analysis of ^{3}H-labeled palmitate oxidation [49]. The authors suggested that the metabolic outcome of aspirin in vivo may depend on the expression of thioesterases that hydrolyze acetylsalicylic acid (aspirin) to salicylic acid since they found an opposite effect of salicylic acid on 3H-palmitate oxidation. Aspirin-mediated activation of fatty acid oxidation was further confirmed through the analysis of human urine samples collected after 7 days of aspirin treatment [50]. Promotion of FA oxidation upon aspirin treatment is consistent with the reported AMPK activation and resulting deactivation of ACC [46,47], because the catalytic product of ACC2 isoform malonyl-CoA is the natural inhibitor of CPT1A [28,51]. However, since FA oxidation plays a role in maintenance of the malignant phenotype, one may speculate that the induction of lipid oxidation may limit the effectiveness of aspirin in cancer treatment.

In this study, we first examined if aspirin influences the activity of ACC1 and CPT1A in MCF-7 and MDA-MB-468 human BC cell lines. We found an aspirin-mediated increase in ACC1 phosphorylation at serine 79 in the MCF-7 cell line, which is in line with previous reports [46,47]. However, aspirin exposure had no effect on ACC1 phosphorylation in MDA-MB-468 cells. Varying activities of the tested drug may be explained by different genetic backgrounds within these cells.

We did not detect alterations in CPT1A expression due to aspirin treatment in either of the tested cell lines. This is in contrast with a previous report where aspirin was noted to promote the expression of CPT1 in the human hepatoma cell line [46]. The different cancer types analyzed may be the cause of observed contrasting effects on CPT1. However, we cannot rule out the possibility that aspirin enhances FA oxidation in the tested BC cells via other mechanisms.

Next, we asked whether downregulation of DDIT4 expression alters the effect of aspirin on these enzymes. We chose to investigate the effect of DDIT4 expression based on a report by Porstmann et al. [41] which announced the necessity of mTORC1 for de novo lipid synthesis and on our recently published results uncovering the DDIT4 effect on mTORC1 activity upon aspirin treatment [42]. Our present data have shown for the first time that DDIT4 knockdown promotes the expression of CPT1A at the protein level and reduces phosphorylation of ACC1 (phosphorylation deactivates the enzyme) following aspirin treatment in MCF-7 BC cells and lowers the phosphorylation of ACC1 in MDA-MB-468 cells. Thus, DDIT4 downregulation leads to an undesired effect of aspirin on the activity of these lipid metabolism enzymes, as FA synthesis and oxidation are linked to malignant phenotype. These findings may be clinically relevant as DDIT4 expression has been shown to vary in breast tumors [52,53]. In this manner, the present findings are the basis for further, more extensive studies on the role of DDIT4 in aspirin's effect on fatty acid metabolism.

Author Contributions: Conceptualization, A.S., R.U. and E.J.; methodology, A.S., A.B. and J.B.; validation, A.S., A.B. and J.B.; formal analysis, A.S.; investigation, A.S., A.B. and J.B.; resources, A.S. and A.B.; data curation, A.S.; writing—original draft preparation, A.S.; writing—review and editing, A.S.; visualization, A.S.; supervision, R.U. and E.J.; project administration, A.S. and E.J.; funding acquisition, E.J. All authors have read and agreed to the published version of the manuscript.

Funding: This research was funded by a grant (No. S-MIP-17-56) from the Research Council of Lithuania.

Institutional Review Board Statement: Not applicable.

Informed Consent Statement: Not applicable.

Data Availability Statement: Data are contained within the article.

Conflicts of Interest: The authors declare no conflict of interest. The funder had no role in the design of the study; in the collection, analyses, or interpretation of data; in the writing of the manuscript; or in the decision to publish the results.

References

1. Tatham, M.H.; Cole, C.; Scullion, P.; Wilkie, R.; Westwood, N.J.; Stark, L.A.; Hay, R.T. A proteomic approach to analyze the aspirin-mediated lysine acetylome. *Mol. Cell. Proteom.* **2017**, *16*, 310–326. [CrossRef] [PubMed]
2. Bosetti, C.; Rosato, V.; Gallus, S.; Cuzick, J.; La Vecchia, C. Aspirin and cancer risk: A quantitative review to 2011. *Ann. Oncol.* **2012**, *23*, 1403–1415. [CrossRef]
3. Zhong, S.; Zhang, X.; Chen, L.; Ma, T.; Tang, J.; Zhao, J. Association between aspirin use and mortality in breast cancer patients: A meta-analysis of observational studies. *Breast Cancer Res. Treat.* **2015**, *150*, 199–207. [CrossRef] [PubMed]
4. Liu, J.F.; Jamieson, G.G.; Wu, T.C.; Zhu, G.J.; Drew, P.A. A preliminary study on the postoperative survival of patients given aspirin after resection for squamous cell carcinoma of the esophagus or adenocarcinoma of the cardia. *Ann. Surg. Oncol.* **2009**, *16*, 1397–1402. [CrossRef]
5. Van Staalduinen, J.; Frouws, M.; Reimers, M.; Bastiaannet, E.; van Herk-Sukel, M.P.; Lemmens, V.; de Steur, W.O.; Hartgrink, H.H.; van de Velde, C.J.; Liefers, G.J. The effect of aspirin and nonsteroidal anti-inflammatory drug use after diagnosis on survival of oesophageal cancer patients. *Br. J. Cancer.* **2016**, *114*, 1053–1059. [CrossRef]
6. Zaorsky, N.G.; Buyyounouski, M.K.; Li, T.; Horwitz, E.M. Aspirin and statin nonuse associated with early biochemical failure after prostate radiation therapy. *Int. J. Radiat. Oncol. Biol. Phys.* **2012**, *84*, e13–e17. [CrossRef]
7. Choe, K.S.; Cowan, J.E.; Chan, J.M.; Carroll, P.R.; D'Amico, A.V.; Liauw, S.L. Aspirin use and the risk of prostate cancer mortality in men treated with prostatectomy or radiotherapy. *J. Clin. Oncol.* **2012**, *30*, 3540–3544. [CrossRef]
8. Jacobs, E.J.; Newton, C.C.; Stevens, V.L.; Campbell, P.T.; Freedland, S.J.; Gapstur, S.M. Daily aspirin use and prostate cancer–specific mortality in a large cohort of men with nonmetastatic prostate cancer. *J. Clin. Oncol.* **2014**, *32*, 3716–3722. [CrossRef]
9. Holmes, M.D.; Chen, W.Y.; Li, L.; Hertzmark, E.; Spiegelman, D.; Hankinson, S.E. Aspirin intake and survival after breast cancer. *J. Clin. Oncol.* **2010**, *28*, 1467. [CrossRef] [PubMed]
10. Fraser, D.M.; Sullivan, F.M.; Thompson, A.M.; McCowan, C. Aspirin use and survival after the diagnosis of breast cancer: A population-based cohort study. *Br. J. Cancer* **2014**, *111*, 623–627. [CrossRef]
11. Chan, A.T.; Ogino, S.; Fuchs, C.S. Aspirin use and survival after diagnosis of colorectal cancer. *JAMA* **2009**, *302*, 649–658. [CrossRef]
12. Bastiaannet, E.; Sampieri, K.; Dekkers, O.M.; De Craen, A.J.; van Herk-Sukel, M.P.; Lemmens, V.; Van Den Broek, C.B.; Coebergh, J.W.; Herings, R.M.; Van De Velde, C.J.; et al. Use of aspirin postdiagnosis improves survival for colon cancer patients. *Br. J. Cancer* **2012**, *106*, 1564–1570. [CrossRef] [PubMed]
13. McCowan, C.; Munro, A.J.; Donnan, P.T.; Steele, R.J. Use of aspirin post-diagnosis in a cohort of patients with colorectal cancer and its association with all-cause and colorectal cancer specific mortality. *Eur. J. Cancer* **2013**, *49*, 1049–1057. [CrossRef] [PubMed]
14. Walker, A.J.; Grainge, M.J.; Card, T.R. Aspirin and other non-steroidal anti-inflammatory drug use and colorectal cancer survival: A cohort study. *Br. J. Cancer* **2012**, *107*, 1602–1607. [CrossRef] [PubMed]
15. Bains, S.; Mahic, M.; Cvancarova, M.; Yaqub, S.; Dørum, L.M.; Bjørnbeth, B.A.; Møller, B.; Brudvik, K.W.; Tasken, K. Impact of aspirin as secondary prevention in an unselected cohort of 25,644 patients with colorectal cancer: A population-based study. *J. Clin. Oncol.* **2015**, *33*, 3504. [CrossRef]
16. Ng, K.; Meyerhardt, J.A.; Chan, A.T.; Sato, K.; Chan, J.A.; Niedzwiecki, D.; Saltz, L.B.; Mayer, R.J.; Benson, A.B., III; Schaefer, P.L.; et al. Aspirin and COX-2 inhibitor use in patients with stage III colon cancer. *J. Natl. Cancer Inst.* **2015**, *107*, dju345. [CrossRef]
17. Bashir, A.U.; Kankipati, C.S.; Jones, S.; Newman, R.M.; Safrany, S.T.; Perry, C.J.; Nicholl, I.D. A novel mechanism for the anticancer activity of aspirin and salicylates. *Int. J. Oncol.* **2019**, *54*, 1256–1270. [CrossRef]
18. Heer, E.; Harper, A.; Escandor, N.; Sung, H.; McCormack, V.; Fidler-Benaoudia, M.M. Global burden and trends in premenopausal and postmenopausal breast cancer: A population-based study. *Lancet Glob. Health* **2020**, *8*, e1027–e1037. [CrossRef]
19. Cardoso, F.; Kyriakides, S.; Ohno, S.; Penault-Llorca, F.; Poortmans, P.; Rubio, I.T.; Zackrisson, S.; Senkus, E. Early breast cancer: ESMO Clinical Practice Guidelines for diagnosis, treatment and follow-up. *Ann. Oncol.* **2019**, *30*, 1194–1220. [CrossRef]
20. Foulon, A.; Theret, P.; Rodat-Despoix, L.; Kischel, P. Beyond chemotherapies: Recent strategies in breast cancer treatment. *Cancers* **2020**, *12*, 2634. [CrossRef]
21. Grancher, A.; Michel, P.; Di Fiore, F.; Sefrioui, D. Colorectal cancer chemoprevention: Is aspirin still in the game? *Cancer Biol. Ther.* **2022**, *23*, 446–461. [CrossRef]
22. Ward, A.V.; Anderson, S.M.; Sartorius, C.A. Advances in analyzing the breast cancer lipidome and its relevance to disease progression and treatment. *J. Mammary Gland Biol. Neoplasia* **2021**, *26*, 399–417. [CrossRef]
23. Schlaepfer, I.R.; Joshi, M. CPT1A-mediated fat oxidation, mechanisms, and therapeutic potential. *Endocrinology* **2020**, *161*, bqz046. [CrossRef] [PubMed]
24. Bian, X.; Liu, R.; Meng, Y.; Xing, D.; Xu, D.; Lu, Z. Lipid metabolism and cancer. *J. Exp. Med.* **2021**, *218*, e20201606. [CrossRef] [PubMed]
25. Gómez-Cebrián, N.; Domingo-Ortí, I.; Poveda, J.L.; Vicent, M.J.; Puchades-Carrasco, L.; Pineda-Lucena, A. Multi-omic approaches to breast cancer metabolic phenotyping: Applications in diagnosis, prognosis, and the development of novel treatments. *Cancers* **2021**, *13*, 4544. [CrossRef]
26. Bao, J.; Zhu, L.; Zhu, Q.; Su, J.; Liu, M.; Huang, W. SREBP-1 is an independent prognostic marker and promotes invasion and migration in breast cancer. *Oncol. Lett.* **2016**, *12*, 2409–2416. [CrossRef] [PubMed]

27. Sebestyén, A.; Dankó, T.; Sztankovics, D.; Moldvai, D.; Raffay, R.; Cervi, C.; Krencz, I.; Zsiros, V.; Jeney, A.; Petővári, G. The role of metabolic ecosystem in cancer progression—Metabolic plasticity and mTOR hyperactivity in tumor tissues. *Cancer Metastasis Rev.* **2021**, *40*, 989–1033. [CrossRef] [PubMed]
28. Melone, M.A.; Valentino, A.; Margarucci, S.; Galderisi, U.; Giordano, A.; Peluso, G. The carnitine system and cancer metabolic plasticity. *Cell Death Dis.* **2018**, *9*, 228. [CrossRef]
29. Park, J.H.; Vithayathil, S.; Kumar, S.; Sung, P.L.; Dobrolecki, L.E.; Putluri, V.; Bhat, V.B.; Bhowmik, S.K.; Gupta, V.; Arora, K.; et al. Fatty acid oxidation-driven Src links mitochondrial energy reprogramming and oncogenic properties in triple-negative breast cancer. *Cell Rep.* **2016**, *14*, 2154–2165. [CrossRef]
30. Hao, P.; Yu, J.; Ward, R.; Liu, Y.; Hao, Q.; An, S.; Xu, T. Eukaryotic translation initiation factors as promising targets in cancer therapy. *Cell Commun. Signal.* **2020**, *18*, 175. [CrossRef]
31. Zhu, Y.; Lin, X.; Zhou, X.; Prochownik, E.V.; Wang, F.; Li, Y. Posttranslational control of lipogenesis in the tumor microenvironment. *J. Hematol. Oncol.* **2022**, *15*, 120. [CrossRef]
32. Wang, Y.; Yu, W.; Li, S.; Guo, D.; He, J.; Wang, Y. Acetyl-CoA carboxylases and diseases. *Front. Oncol.* **2022**, *12*, 836058. [CrossRef] [PubMed]
33. Koobotse, M.; Holly, J.; Perks, C. Elucidating the novel BRCA1 function as a non-genomic metabolic restraint in ER-positive breast cancer cell lines. *Oncotarget* **2018**, *9*, 33562. [CrossRef]
34. Chajes, V.; Cambot, M.; Moreau, K.; Lenoir, G.M.; Joulin, V. Acetyl-CoA carboxylase α is essential to breast cancer cell survival. *Cancer Res.* **2006**, *66*, 5287–5294. [CrossRef] [PubMed]
35. Monaco, M.E. Fatty acid metabolism in breast cancer subtypes. *Oncotarget* **2017**, *8*, 29487. [CrossRef]
36. Qu, Q.; Zeng, F.; Liu, X.; Wang, Q.J.; Deng, F. Fatty acid oxidation and carnitine palmitoyltransferase I: Emerging therapeutic targets in cancer. *Cell Death Dis.* **2016**, *7*, e2226. [CrossRef]
37. Das, M.; Giannoudis, A.; Sharma, V. The role of CPT1A as a biomarker of breast cancer progression: A bioinformatic approach. *Sci. Rep.* **2022**, *12*, 16441. [CrossRef]
38. Gatza, M.L.; Silva, G.O.; Parker, J.S.; Fan, C.; Perou, C.M. An integrated genomics approach identifies drivers of proliferation in luminal-subtype human breast cancer. *Nat. Genet.* **2014**, *46*, 1051–1059. [CrossRef] [PubMed]
39. Kessler, J.D.; Kahle, K.T.; Sun, T.; Meerbrey, K.L.; Schlabach, M.R.; Schmitt, E.M.; Skinner, S.O.; Xu, Q.; Li, M.Z.; Hartman, Z.C.; et al. A SUMOylation-dependent transcriptional subprogram is required for Myc-driven tumorigenesis. *Science* **2012**, *335*, 348–353. [CrossRef]
40. Tan, Z.; Zou, Y.; Zhu, M.; Luo, Z.; Wu, T.; Zheng, C.; Xie, A.; Wang, H.; Fang, S.; Liu, S.; et al. Carnitine palmitoyl transferase 1A is a novel diagnostic and predictive biomarker for breast cancer. *BMC Cancer* **2021**, *21*, 409. [CrossRef]
41. Porstmann, T.; Santos, C.R.; Griffiths, B.; Cully, M.; Wu, M.; Leevers, S.; Griffiths, J.R.; Chung, Y.L.; Schulze, A. SREBP activity is regulated by mTORC1 and contributes to Akt-dependent cell growth. *Cell Metab.* **2008**, *8*, 224–236. [CrossRef]
42. Savukaitytė, A.; Gudoitytė, G.; Bartnykaitė, A.; Ugenskienė, R.; Juozaitytė, E. siRNA knockdown of REDD1 facilitates aspirin-mediated dephosphorylation of mTORC1 target 4E-BP1 in MDA-MB-468 human breast cancer cell line. *Cancer Manag. Res.* **2021**, *13*, 1123–1133. [CrossRef]
43. Dovizio, M.; Tacconelli, S.; Sostres, C.; Ricciotti, E.; Patrignani, P. Mechanistic and pharmacological issues of aspirin as an anticancer agent. *Pharmaceuticals* **2012**, *5*, 1346–1371. [CrossRef] [PubMed]
44. Koundouros, N.; Poulogiannis, G. Reprogramming of fatty acid metabolism in cancer. *Br. J. Cancer* **2020**, *122*, 4–22. [CrossRef] [PubMed]
45. Svensson, R.U.; Parker, S.J.; Eichner, L.J.; Kolar, M.J.; Wallace, M.; Brun, S.N.; Lombardo, P.S.; Van Nostrand, J.L.; Hutchins, A.; Vera, L.; et al. Inhibition of acetyl-CoA carboxylase suppresses fatty acid synthesis and tumor growth of non-small-cell lung cancer in preclinical models. *Nat. Med.* **2016**, *22*, 1108–1119. [CrossRef] [PubMed]
46. He, Z.; Peng, Y.; Duan, W.; Tian, Y.; Zhang, J.; Hu, T.; Cai, Y.; Feng, Y.; Li, G. Aspirin regulates hepatocellular lipid metabolism by activating AMPK signaling pathway. *J. Toxicol. Sci.* **2015**, *40*, 127–136. [CrossRef]
47. Henry, W.S.; Laszewski, T.; Tsang, T.; Beca, F.; Beck, A.H.; McAllister, S.S.; Toker, A. Aspirin Suppresses Growth in PI3K-Mutant Breast Cancer by Activating AMPK and Inhibiting mTORC1 SignalingAspirin and PI3K in Breast Cancer. *Cancer Res.* **2017**, *77*, 790–801. [CrossRef]
48. Wu, Y.; Yan, B.; Xu, W.; Guo, L.; Wang, Z.; Li, G.; Hou, N.; Zhang, J.; Ling, R. Compound C enhances the anticancer effect of aspirin in HER-2-positive breast cancer by regulating lipid metabolism in an AMPK-independent pathway. *Int. J. Biol. Sci.* **2020**, *16*, 583. [CrossRef]
49. Uppala, R.; Dudiak, B.; Beck, M.E.; Bharathi, S.S.; Zhang, Y.; Stolz, D.B.; Goetzman, E.S. Aspirin increases mitochondrial fatty acid oxidation. *Biochem. Biophys. Res. Commun.* **2017**, *482*, 346–351. [CrossRef]
50. Di Minno, A.; Porro, B.; Turnu, L.; Manega, C.M.; Eligini, S.; Barbieri, S.; Chiesa, M.; Poggio, P.; Squellerio, I.; Anesi, A.; et al. Untargeted metabolomics to go beyond the canonical effect of acetylsalicylic acid. *J. Clin. Med.* **2019**, *9*, 51. [CrossRef]
51. Steinberg, G.R.; Carling, D. AMP-activated protein kinase: The current landscape for drug development. *Nat. Rev. Drug Discov.* **2019**, *18*, 527–551. [CrossRef] [PubMed]

52. Pinto, J.A.; Rolfo, C.; Raez, L.E.; Prado, A.; Araujo, J.M.; Bravo, L.; Fajardo, W.; Morante, Z.D.; Aguilar, A.; Neciosup, S.P.; et al. In silico evaluation of DNA Damage Inducible Transcript 4 gene (DDIT4) as prognostic biomarker in several malignancies. *Sci. Rep.* **2017**, *7*, 1526. [CrossRef] [PubMed]

53. DeYoung, M.P.; Horak, P.; Sofer, A.; Sgroi, D.; Ellisen, L.W. Hypoxia regulates TSC1/2–mTOR signaling and tumor suppression through REDD1-mediated 14–3–3 shuttling. *Genes Dev.* **2008**, *22*, 239–251. [CrossRef] [PubMed]

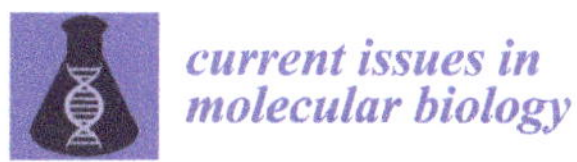
current issues in molecular biology

MDPI

Review

Molecular Profiling of Circulating Tumour Cells and Circulating Tumour DNA: Complementary Insights from a Single Blood Sample Utilising the Parsortix® System

Gabrielle Wishart *, Amy Templeman , Francesca Hendry, Karen Miller and Anne-Sophie Pailhes-Jimenez

ANGLE plc, Guildford GU2 7QB, UK; a.templeman@angleplc.com (A.T.); f.hendry@angleplc.com (F.H.); k.miller@angleplc.com (K.M.); a.pailhes-jimenez@angleplc.com (A.-S.P.-J.)
* Correspondence: g.wishart@angleplc.com

Abstract: The study of molecular drivers of cancer is an area of rapid growth and has led to the development of targeted treatments, significantly improving patient outcomes in many cancer types. The identification of actionable mutations informing targeted treatment strategies are now considered essential to the management of cancer. Traditionally, this information has been obtained through biomarker assessment of a tissue biopsy which is costly and can be associated with clinical complications and adverse events. In the last decade, blood-based liquid biopsy has emerged as a minimally invasive, fast, and cost-effective alternative, which is better suited to the requirement for longitudinal monitoring. Liquid biopsies allow for the concurrent study of multiple analytes, such as circulating tumour cells (CTCs) and circulating tumour DNA (ctDNA), from a single blood sample. Although ctDNA assays are commercially more advanced, there is an increasing awareness of the clinical significance of the transcriptome and proteome which can be analysed using CTCs. Herein, we review the literature in which the microfluidic, label-free Parsortix® system is utilised for CTC capture, harvest and analysis, alongside the analysis of ctDNA from a single blood sample. This detailed summary of the literature demonstrates how these two analytes can provide complementary disease information.

Keywords: blood; cancer; liquid biopsy; circulating tumor cells; circulating tumor DNA; cell-free DNA; microfluidic devices; neoplastic cells

check for updates

Citation: Wishart, G.; Templeman, A.; Hendry, F.; Miller, K.; Pailhes-Jimenez, A.-S. Molecular Profiling of Circulating Tumour Cells and Circulating Tumour DNA: Complementary Insights from a Single Blood Sample Utilising the Parsortix® System. *Curr. Issues Mol. Biol.* **2024**, *46*, 773–787. https://doi.org/10.3390/cimb46010050

Academic Editor: Dumitru A. Iacobas

Received: 5 December 2023
Revised: 12 January 2024
Accepted: 15 January 2024
Published: 17 January 2024

1. Background

1.1. Tumour Burden and Heterogeneity

The global burden of cancer challenges human health and the economy and was responsible for nearly 10 million deaths in 2020 [1] (accessed on 18 September 2023). Rising prevalence and incidence rates call for effective diagnostics and treatment selection strategies. Furthermore, the dynamic landscape of cancer demands continuous up-to-date and accurate monitoring methods for effective patient care [2]. This intertumoral and intratumoral heterogeneity of cancer as a basis for tumour evolution, treatment resistance and subsequent treatment failure, is an area of growing understanding [3]. Recent advances in high-throughput, relatively low-cost sequencing techniques (for example, next generation sequencing (NGS)) have shed light on molecular drivers of cancer, actionable mutations and the continuous process of clonal evolution from selective pressure of cancer therapies [4]. As such, it is widely accepted that personalised or precision medicine will optimise response to cancer therapy and improve quality of life for the patient.

The standard of care for evaluating patient specific biomarkers, mutations, and genetic signatures for the appropriate selection of targeted cancer treatments is to conduct a tissue biopsy. Despite being the most widely used technique, tissue biopsies have numerous disadvantages such as being invasive, costly, failing to capture tumour heterogeneity, and being harmful to the patient [5]. Moreover, tissue biopsies are rarely suitable for longitudinal monitoring where the patient is too unwell and where the patient's primary tumour has been excised, or metastasised to two or more sites [6]. Liquid biopsy techniques are advancing to provide a less invasive, safer, less costly alternative that provides results faster than tissue biopsies [7]. Furthermore, this technique is better suited for longitudinal disease monitoring and captures heterogeneity and the clonal evolution responsible for treatment failure and drug resistance [7].

1.2. Liquid Biopsy

A liquid biopsy is a minimally invasive test performed using bodily fluids, such as blood, and it has received growing clinical interest for its applications in personalised medicine [8]. Blood-based liquid biopsies allow for the analysis of circulating tumour cells (CTCs), cell-free DNA (cfDNA), circulating tumour DNA (ctDNA), or other plasma components, such as cell-free RNA, proteins, and exosomes, to provide clinically relevant and actionable information. More specifically, liquid biopsies have shown significant relevance across the cancer care pathway, informing cancer diagnosis, prognosis, treatment selection, the monitoring of disease evolution, and disease relapse [7,8] (Figure 1). The advancement of sequencing technology is fuelling a revolution in liquid biopsy analysis, providing genomic and transcriptomic characterisation for personalised therapy selection [9]. Liquid biopsies are also emerging as valuable tools for drug discovery and development having applications as prognostic and pharmacodynamic biomarkers, with several consortiums founded to analyse, implement, and develop standards for liquid biopsy in clinical trials and drug development. These include Friends of Cancer research ctMONiTR, the International Liquid Biopsy Standardization Alliance (ILSA), the Blood profiling Atlas in Cancer (BloodPAC) Consortium and Cancer ID.

1.3. Circulating Tumour DNA

Cell-free DNA (cfDNA) consists of DNA fragments found in the body fluids of healthy and non-healthy patients and is thought to be derived from cellular breakdown mechanisms [10]. cfDNA circulates in fragments typically ranging in size from 120 to 140 base pairs [10]. In cancer patients, circulating tumour DNA (ctDNA) accounts for a very low percentage (0.01–1%) of total cfDNA and is specifically derived from the tumour [11]. The origin of ctDNA is not fully understood, but is believed to be from apoptotic cells, necrotic cells or to enter the bloodstream via active secretion [8,12,13]. The profiling of ctDNA has received attention for early diagnosis, treatment selection, the identification of resistance mechanisms and detection of post-surgical minimal residual disease in numerous cancer types [4,12,14]. The analysis of ctDNA provides a simple method of obtaining genetic information; however, this is limited to point mutations, structural rearrangements, copy number variants (CNVs) and changes in DNA methylation [4]. It has been reported in the literature that genetic signatures in ctDNA can be derived from the major clone in a tumour and thus, subclonal signatures may be missed when studying this analyte [5]. However, ctDNA and cfDNA are the most established liquid biopsy analytes in the oncology market with five FDA-approved companion diagnostics for targeted treatment selection and residual disease monitoring [15] (accessed on 1 December 2023). For example, the cobas epidermal growth factor receptor (EGFR) mutation test V2 (Roche) to detect EGFR mutations (Exon 19 deletion or exon 21 L858R substitution mutation) in non-small-cell lung cancer (NSCLC) patients for treatment with Tagrisso (osimertinib) and Tarceva (erlotinib) [16,17] (accessed on 9 January 2024).

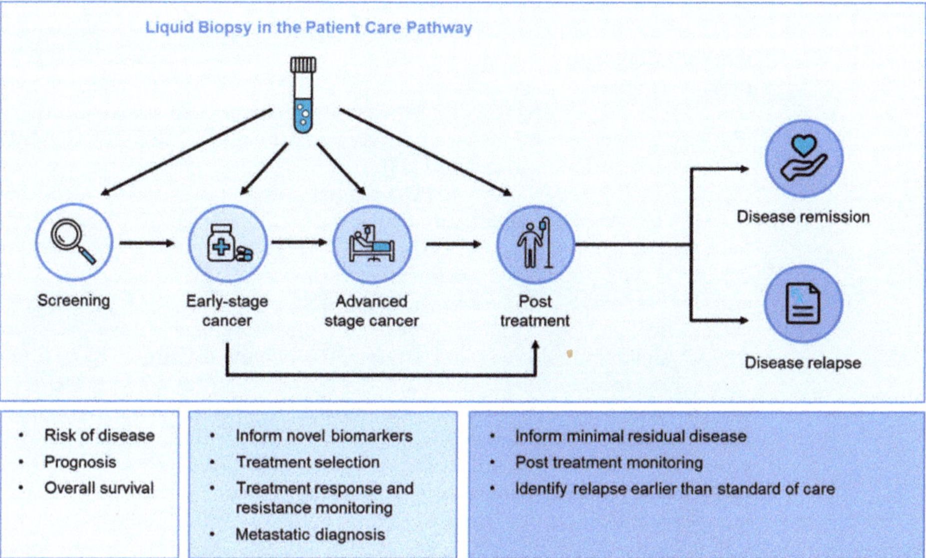

Figure 1. Clinical utility of liquid biopsies across the patient care pathway. Liquid biopsies are minimally invasive tools used (1) in patient screening to predict risk of disease, prognosis, and overall survival; (2) in early-stage cancer to inform targeted therapies for first-line treatment, identify novel biomarkers, and to monitor treatment response and to provide an early predictor of treatment resistance; (3) at disease progression in advanced stage cancer to confirm metastatic diagnosis, inform targeted treatment selection, monitor treatment response and treatment resistance, and identify new drug targets as the tumour evolves (clonal evolution); (4) post treatment to identify minimal residual disease, monitor the patient during remission, and identify risk of relapse. Liquid biopsies allow the analysis of different blood-based analytes including circulating tumour cells (CTCs) and cell-free DNA (cfDNA). The latter provides genomic information from fragmented DNA, whereas CTCs are whole cells providing not only genomic, but transcriptomic and proteomic information for a more inclusive view of the current state of tumour mutations and biomarkers towards personalised therapy.

1.4. Circulating Tumour Cells

CTCs are whole cells released by a tumour into the bloodstream and are responsible for metastatic seeding [18,19]. CTC enumeration provides robust prognostic information; increased CTC presence correlates to metastatic burden, with a strong association with overall survival in numerous cancer types [4]. Beyond CTC enumeration, CTCs can provide functional genomic, transcriptomic and proteomic information, providing accurate tumour phenotypic information at the time of sampling [2,3,8]. This presents a unique real-time window into clinically relevant information towards personalised treatment. This analyte has been reported to reflect high levels of tumour heterogeneity [3] and represent clonal evolution that may be responsible for treatment failure and drug resistance [2,8,19]. As such, CTCs are suitable for treatment selection [2], real-time longitudinal disease monitoring, treatment monitoring, and relapse monitoring [5]. Furthermore, harvesting CTCs from blood facilitates research into the complex landscape of cancer including CTC clustering, cellular invasion, and metastasis [4] and is suitable for in vitro/in vivo culture research [20]. CTCs are an area of growing interest across multiple cancer types [2,8] and are emerging as a tool to address challenges of the complex landscape of heterogeneity in the clinic. As such, there is a demand for enrichment technologies that are able to successfully isolate rare CTCs from whole blood. Numerous CTC enrichment technologies are emerging based on a cell's physical properties, biological properties and a combination of the two [19,21]. These include membrane microfilters, microfluidic technologies, non-microfluidic technologies, positive selection by tumour marker technologies, and negative selection by non-tumour marker technologies [19,21]. Technologies face challenges in isolating CTCs given their rarity, their phenotype, size heterogeneity and the need for downstream analysis [22,23]. Often, there is a reported trade-off between CTC recovery and sample purity [22]. Microfluidic CTC isolation technologies have received attention for high throughput, sensitivity, low sample consumption and cost [23].

Currently there are only two FDA-cleared medical devices for the enrichment of CTCs. These include the CellSearch® Circulating Tumor Cell (CTC) Test (Menarini-Silicon Biosystems, Huntingdon Valley, PA, USA): *for the enumeration of CTCs of epithelial origin for the monitoring of prognostic information of patients with metastatic breast, colorectal, or prostate cancer,* and the Parsortix® PC1 System (ANGLE plc, Guildford, UK): *for the capture and harvest of CTCs from the blood of metastatic breast cancer (MBC) patients for subsequent, user-validated analysis.* CellSearch® isolates and detects CTCs of epithelial origin via an immunoaffinity-based enrichment method. However, CTCs can exist in three subtypes including epithelial, mesenchymal, and epithelial/mesenchymal CTCs, thus cells undergoing or having undergone epithelial to mesenchymal transition (EMT) (a process that increases metastatic properties of cancer cells, enhancing cellular migration and invasion [24]) may be missed by such enrichment technologies. The Parsortix® system overcomes this issue with epitope-independent CTC capture, isolating epithelial, mesenchymal, and epithelial/mesenchymal CTCs. Furthermore, the subsequent downstream analysis of CTCs provides a wealth of information as compared to CTC enumeration.

1.5. The Parsortix® System

The Parsortix® system is a liquid biopsy platform that uses a patented microfluidic technology enabling label-free (epitope-independent) capture of all CTC phenotypes based on cell size and deformability, allowing for CTC enumeration and subsequent downstream analysis [25]. More specifically, the Parsortix® system can facilitate the capture [25] of CTCs, as well as the harvest of CTCs for subsequent downstream analysis methods, [26] including individual gene expression analysis (messenger RNA [mRNA]) and protein evaluation (e.g., cytological/immunofluorescent [IF] staining) [27–29], the evaluation of DNA aberrations [30], and whole genomic [31] and transcriptomic sequencing [32], amongst others. These subsequent downstream methods have been utilised in the literature as tools for studying CTCs and the tumour microenvironment [33,34], identifying clinically actionable targets [30] towards therapeutic screening [31]/patient cohort selection and personalised treatment, resistance profiling [35], and drug discovery [33] and development [14].

These applications and techniques used in tandem with the Parsortix® system are explored in 92 peer reviewed publications from 38 independent study centres across 18 cancer types [36] (accessed on 28 September 2023). The Parsortix® PC1 Clinical System's analytical performance [37] and multi-centre clinical performance [26] has been demonstrated to capture and harvest CTCs, and provide specific, user-validated downstream analysis in MBC. Moreover, the Parsortix® system is currently under evaluation in clinical trials to investigate therapeutic influence on CTC clusters [38], the role of sleep in the spread of CTCs in lung cancer patients [39], and to investigate the intestinal polyp secretion of tumour cells and circulating factors [40].

2. Molecular Advances: The Omics Revolution

In the last 20 years, there have been exponential advances in the understanding and application of molecular analysis and computational tools as genomic sequencing has become well-established and affordable. More recently, it is understood that studying the genome provides a basis of information that is just the beginning of a complex biological landscape and that we are able to look beyond the genome [41]. Genomic information can be supplemented with transcriptomic and proteomic data for closer evaluation of tumour phenotype towards more accurate, real-time information for personalised treatment (the study of multi-omics). Genomic analysis provides information on past mutations acquired during the evolutionary history of the tumour, whereas transcriptomic analysis provides a window into epigenetic influence on gene expression and thus the current state of the tumour [42]. This interplay between the genome and the transcriptome is relevant for identifying up-to-date and accurate treatment options [42]. The importance of studying the transcriptome has been demonstrated in real-world clinical data, in which tissue-derived RNA sequencing discovered more clinically actionable targets than DNA sequencing alone, increasing the number of patients eligible for matched therapies by 24% [43]. Similarly, other research has shown that utilising transcriptomics can increase the number of patients administered for matched therapy [44]. Moreover, it is predicted that by harnessing NGS tools and the nature of transcriptomics, it is possible to head towards a new era of personalised medicine, something which is recognised by the National Institute of Health [45]. As such, we are entering an omics revolution that aims to progress personalised medicine [41]. Furthermore, this evolution of molecular technology has necessitated the concurrent development and application of artificial intelligence and machine learning for the integration of big data into the clinic [46].

These molecular advances are fuelling liquid biopsy analysis [9]. Advances in digital polymerase chain reaction (dPCR) and sequencing technologies are facilitating low-cost, rapid analysis, with limited starting material to provide clinically relevant multi-omic information [47]. Notably, the application of NGS technology is enabling the identification of druggable targets, clonal selection, and metastatic information from liquid biopsy analytes as real-time tools [30]. There are 20 peer reviewed publications that utilise the Parsortix® system and NGS technology to study the genome or transcriptome. More specifically, eight of these publications perform bulk harvest NGS analysis, and 12 study the use of single-cell analysis. As such, the ability of CTCs to provide both genomic and transcriptomic information in addition to genomic information from ctDNA as dual analytes from the same patient sample is an exciting prospect. Currently, there is no single device or companion diagnostic approved for the combined analysis of CTCs and ctDNA or multi-analyte analysis from a single blood sample, but dual analysis is emerging in the literature.

3. Complementary Insights

In the literature, the enumeration of CTCs and analysis of ctDNA in tandem have previously informed prognosis across various cancer types [48,49]. Rapid advances in CTC isolation technologies and the omics revolution have enabled the molecular analysis of both CTCs and ctDNA as a minimally invasive approach to define tumour heterogeneity and clonal evolution to study metastasis [4]. As this information is imperative for treatment success, CTCs and ctDNA have been described as cornerstones of liquid biopsy diagnosis, paving the way for new diagnostic opportunities [8]. Until recently, the analysis of CTCs and ctDNA have been referred to in the literature as competing sources of information [50]; however, there has been a shift in understanding that the two analytes can provide complementary insights [3,8], expanding the amount of clinically actionable information to inform the patient care pathway.

Aoki et al. (2020) describe the dual analysis of these analytes to increase genomic mutation profiling sensitivity without decreasing specificity [51] and alludes to the unique ability of CTCs to provide novel genomic, transcriptomic, proteomic, metabolomic, and secretomic information in the future. Onidani et al. (2019) conducted NGS research into the genomic profiles of CTCs and ctDNA via the targeted sequencing of 37 head and neck or gastrointestinal cancer patients [52]. They reported that in both cancer types, patients identified with both concordant and discordant clinically actionable information within CTCs and ctDNA (Figure 2). For example, in some head and neck cancer patients, mutations in *ALK* and *KIT* were present in both analytes, whereas mutations in *TP53* and *SMAD4* were exclusive to CTCs and mutations in *MET* were exclusive to ctDNA. Similarly, in colorectal cancer patients, mutations in *TP53* and *SMAD4* were present in both analytes, whereas *EGFR* mutations were exclusive to CTCs, and *BRAF*, *KRAS* and *PIK3CA* mutations were exclusive to ctDNA. The authors state that CTCs and ctDNA exhibited genetic heterogeneity and that dual analysis is more informative than using one analyte alone, outlining the relevance of this tool for real-time monitoring of disease progression, treatment selection and personalised care [52]. Similarly, Manier et al., (2018), performed research into 28 multiple myeloma patients to report that whole exome sequencing (WES) revealed mutations exclusive to either CTCs or cfDNA. These analytes presented different genetic profiles for the cross-evaluation of mutations, and the research infers that this complementary information provided a comprehensive profile of clonal heterogeneity in multiple myeloma [53]. This research also reports that in specific cases, the actionable biomarker *TP53* was mutated in both CTCs and ctDNA but not in primary tissue samples, highlighting the benefit of liquid biopsy [53].

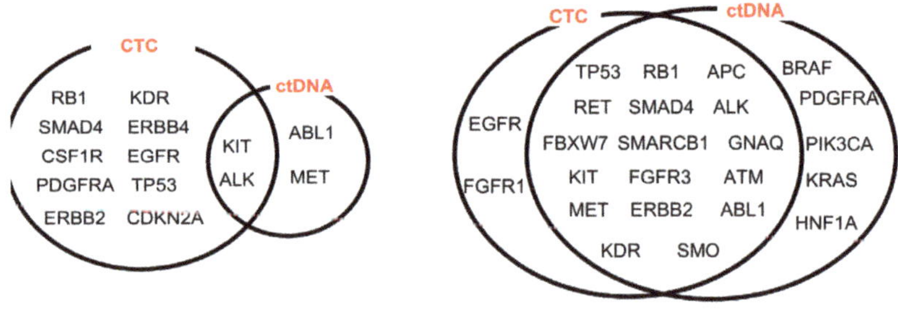

Figure 2. CTC and ctDNA analysis of genomic alterations in head and neck and colorectal cancer patients. Figure reproduced from Onidani et al. (2019) [52].

Kong et al. (2020) performed CTC and ctDNA mutation profiling via qPCR and Sanger sequencing in 16 lung adenocarcinoma and 21 breast ductal carcinoma patients. This research reported that higher degrees of genomic heterogeneity were present in CTCs as compared to ctDNA. More specifically, in some breast cancer patients, clinically actionable mutations such as *JAK3*, *BRAF* or *MTOR* amplifications were present at specific timepoints in CTC analysis but absent in matched ctDNA. The authors hypothesise that the difference may stem from the origin of the analytes; CTCs may have evolved and survived treatment whereas the ctDNA may be presenting genetic information of apoptotic tumour cells. Furthermore, when analysed together, CTCs and ctDNA displayed higher degrees of concordance with the metastatic tumour as compared to the primary tumour, representing clonal evolution. In detail, this evidence indicates that dual analysis detected evolving signatures during the progression of disease and throughout treatment, highlighting the potential for use as treatment guides in personalised therapy [5]. Other research articles support the findings that dual analysis of these analytes provides complementary profiling information [54,55]. Some authors state that single-cell profiling of CTCs allows tumour heterogeneity insights beyond that of ctDNA alone, and that the addition of CTCs to the study of cfDNA is clinically relevant for monitoring clonal evolution and relapse [3].

Keup et al., report on a project named ELIMA ('evaluation of multiple liquid biopsy analytes in metastatic breast cancer patients all from one blood sample') in which they published a series of investigations assessing the mutation profiles in three or more blood-based analytes. Keup et al. (2021) evaluated CTC mRNA, extracellular vesicle (EV) mRNA and cfDNA profiles in 27 hormone receptor positive, HER2 negative MBC patients, reporting that the largest and most diverse number of overexpression signals occurred within CTCs [56]. The authors state that this diversity mirrors spatial tumour heterogeneity, a leading cause of treatment failure. Moreover, EV signals fluctuated greatly showing that temporal heterogeneity and cfDNA provided a source for actionable variants. Thus, all three analytes were complementary and together provided longitudinal, multiparametric information to capture heterogeneity and tumour evolution [56]. In a similar ELIMA study, Keup et al. (2021) evaluated CTC mRNA, CTC gDNA, EV mRNA and cfDNA from 26 hormone receptor positive, HER2 negative MBC patients via qPCR, finding that a combination of two analytes resulted in 81–92% of patients presenting with actionable signals, a combination of three resulted in 92–96%, and all four resulted in 96% of patients presenting with an actionable mutation signal [57]. Thus, these analytes are complementary as opposed to competitive, and enable genomic and transcriptomic disease characterisation towards more effective personalised medicine.

In the literature, the number of articles published including both CTCs and cfDNA/ctDNA blood analytes is low in comparison to the analytes studied alone. The rapid evolution of this research field may influence this in the future. Currently, clinical trials undertaking the dual assessment of CTCs and ctDNA are underway in a number of cancer types to assess patterns in diagnosis [58], to monitor biomarker response to treatment [59,60], and to test if dual analysis is more sensitive than standard parameters and imaging for disease monitoring [61]. It is suggested that the current limited access to both CTC enrichment platforms and ctDNA sequencing platforms in the same laboratory is responsible for the rarity of dual analysis research articles [5]. Moreover, it is reported that in some studies, the dual analysis of CTCs and ctDNA has taken place, but only epithelial CTCs have been isolated, thus mesenchymal or EMT phenotypes were missing. Furthermore, some studies have only focused on a single mutation and therefore lack comprehensive profiling, and others study CTCs and ctDNA from different blood samples, failing to account for inter-sample heterogeneity [5]. The Parsortix® system can address these issues as a label-free tool for the isolation and harvest of CTCs, facilitating the analysis of CTCs in conjunction with ctDNA from the same blood sample. Herein, we review the literature in which the Parsortix® system has been utilised for this dual analysis.

4. The Parsortix® System and Dual Analysis

The Parsortix® system has been used in studies investigating complementary information from CTCs and ctDNA in NSCLC [62–66], triple negative breast cancer (TNBC) [67], head and neck cancer, colorectal cancer, and melanoma [68]. These studies include dual analysis towards the evaluation of prognosis [62,68], biomarker treatment selection [62,67] and to inform treatment resistance [64,66] and relapse [62,63] faster than the standard of care [69]. This showcases the clinical utility of liquid biopsy dual analysis throughout the patient care pathway. A selection of these peer reviewed publications is discussed below and listed in Table 1.

Markou et al. (2023) investigated Parsortix®-enriched CTCs and also cfDNA for hotspot mutations in four therapeutically relevant genes (*BRAF, KRAS, EGFR,* and *PIK3CA: E545K* and *H1045R*) from 49 early-stage NSCLC patients via droplet digital PCR (ddPCR) to find complementary genomic information from the same blood sample. The prevalence of the mutations tested was higher in CTCs as compared to cfDNA (38.8% and 24.5%, respectively), and high heterogeneity was present both within and between the analytes. The combined analyses of CTCs and cfDNA increased the percentage of patients identified with actionable mutations to 53%, highlighting the benefit of dual analysis (Figure 3). Moreover, this research showed that the incidence of progression and relapse was higher when at least one mutation was detected in either sample, as compared to no detectable mutation, revealing important stratification factors for early-stage NSCLC. As such, these samples provided diverse genomic information regarding the prognosis and treatment of NSCLC [62].

Table 1. Publications using the Parsortix® system: CTC and ctDNA analysis.

Study	Cancer	Patients	Analysis	Key Message	Reference
Markou et al. (2023)	Early-stage NSCLC	49	ddPCR of hotspot mutations *BRAF, KRAS, EGFR,* and *PIK3CA*	Dual analysis provided complementary molecular information and greater diversity in genomic information for cancer prognosis and treatment.	[62]
Markou et al. (2022)	Early-stage NSCLC	42	MSP of *APC, RASSFIA1, FOXA1, SLFN11, SHOX2*	Methylation profiles varied between CTCs, ctDNA, and primary tissue, suggesting that dual analysis allowed real-time monitoring of tumour evolution. A higher incidence of relapse was reported when at least one gene promoter is methylated in CTCs or cfDNA, highlighting the prognostic value of dual analysis.	[63]
Ntzifa et al. (2021)	NSCLC	42	DNA methylation patterns of *RASSF1A, RASSF10, APC, WIF-1, BRMS1, SLFN11, RARβ, SHISA3,* and *FOXA1*	CTCs and cfDNA provided complementary information and showed that methylation was associated with disease progression and identified as a potential resistance mechanism.	[64]
Ntzifa et al. (2021)	NSCLC	48	Crystal dPCR genotyping of EGFR mutations including *T790M*	Differences between ctDNA and CTCs show heterogeneity and could be predictive of resistance mechanisms useful for evolution tracking and treatment monitoring.	[66]
Ortolan et al. (2021)	TNBC	42	ddPCR and NGS, personalised panels including *TP53, PIK3CA, FGFR3* and more.	ctDNA and CTCs represent both spatial and temporal heterogeneity and allow dynamic monitoring of cancer progression.	[67]
Mondelo-Macia et al. (2020)	Lung, colon, prostate, melanoma, breast, and gastric	30	ddPCR for MET amplification in cfDNA and IF for MET overexpression in CTCs	CTC and cfDNA MET status analysis is a tool for monitoring resistance to anti-EGFR therapy and can inform overall survival.	[65]
Gorges et al. (2019)	Melanoma	84	Analysis of 61 clinically relevant variants across 13 genes including *BRAF, NRAS* and *MAP2K1*	Combined CTC and ctDNA analyses can reveal synergistic information, as well as predict relapse earlier than imaging and the standard of care in some cases.	[69]
Aya-Bonilla et al. (2020)	Melanoma	37	Immunocytochemistry of CTCs and ddPCR of *MLANA, TYR, MAGEA3, ABCB5* and *PAX3*	CTCs are a complementary feature to cfDNA monitoring and can be associated with shorter overall and progression-free survival.	[70]

The same research group investigated the prognostic value of DNA methylation detection in five gene promoters (*APC, RASSFIA1, FOXA1, SLFN11, SHOX2*) in early-stage NSCLC patients via real-time methylation-specific PCR (MSP). Beyond DNA mutations, epigenetic changes in methylation patterns can influence tumour suppressor gene expression and can be identified as an early event in tumorigenesis. This study reports differences in DNA methylation patterns in CTCs, cfDNA and the primary tumour, as well as a higher incidence of relapse when at least one methylated gene promoter was detected in CTCs or cfDNA, highlighting the complementary nature and prognostic benefit to dual analyte analysis. The authors state that the dual analysis of CTCs and cfDNA allow for real-time monitoring of tumour evolution [63]. Similarly, Ntzifa et al. (2021) investigated the DNA methylation patterns of nine genes (*RASSF1A, RASSF10, APC, WIF-1, BRMS1, SLFN11, RARβ, SHISA3,* and *FOXA1*) in NSCLC patients during osimertinib treatment to find complementary information in CTCs and cfDNA. This research reported that the presence of at least one gene methylation pattern was associated with progressive disease and identified methylation as a potential resistance mechanism [64].

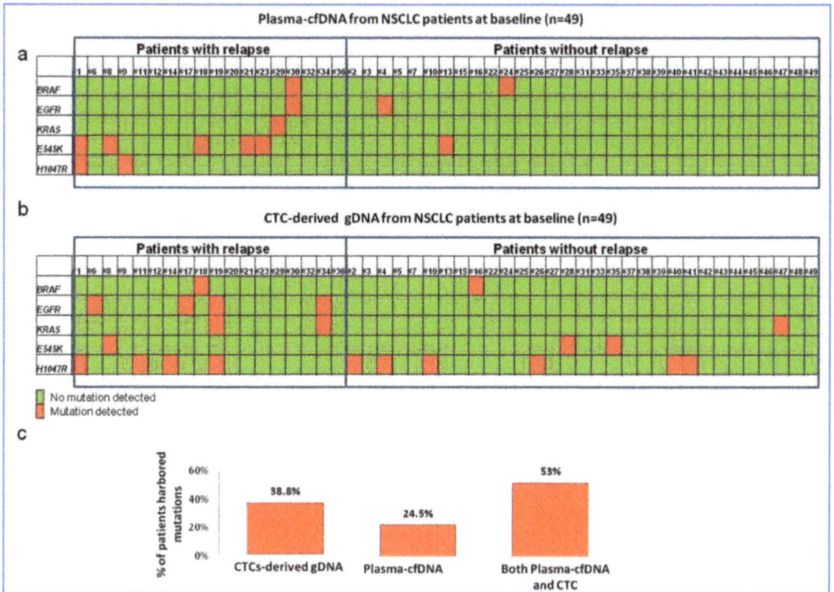

Figure 3. Mutation status in four therapeutically relevant genes (*BRAF, KRAS, EGFR,* and *PIK3CA: E545K* and *H1045R*) from 49 early-stage NSCLC patients in (**a**) CTC-derived DNA and (**b**) plasma ctDNA and (**c**) the percentage of patient mutations from CTC-derived DNA alone, plasma ctDNA alone or analysed in combination. Red represents mutation. Green represents wildtype. Figure reproduced from Markou, A.N. et al. (2023) [62].

Mondelo-Macía et al. (2020) reported the successful detection of MET (hepatocyte growth factor receptor) expression in CTCs (via immunofluorescence) and amplification in cfDNA (via ddPCR) in a variety of cancer types (lung, colon, prostate, melanoma, breast, and gastric cancer patients) towards the characterisation of tumours and for the detection of treatment resistance [68]. More specifically, a correlation between cfDNA concentration and *MET* copy number was determined. Furthermore, an association between CTCs that were MET positive and poor survival in head and neck cancer patients was reported, an association not observed for MET amplification determined by cfDNA analysis [68]. This research highlights the potential for both CTC and cfDNA analysis as useful tools for characterising tumours and guiding personalised treatment upon detection of treatment resistance, through longitudinal monitoring.

In a study by Ntzifa et al. (2021), the presence of EGFR mutations in tissue, cfDNA and CTCs in NSCLC patients undergoing osimertinib therapy was determined using Crystal Digital PCR^{TM} and subsequently compared (Figure 4). Of note, two patients (#11 and #38) with a T790M mutation (a mutation associated with resistance to EGFR inhibitors) detected in CTCs at the baseline but not in cfDNA or tissue had significantly lower progression-free survival. Moreover, the presence of the T790M mutation was detected in CTCs from three patients (#12, #17 and #18) at disease progression, which was absent at this time in cfDNA.

The authors reported that this may be indicative of tumour heterogeneity and could also be predictive of resistance mechanisms occurring under selective treatment pressure. The authors conclude that analysis of EGFR mutations in both CTCs and cfDNA could be more informative for treatment monitoring in these patients [66].

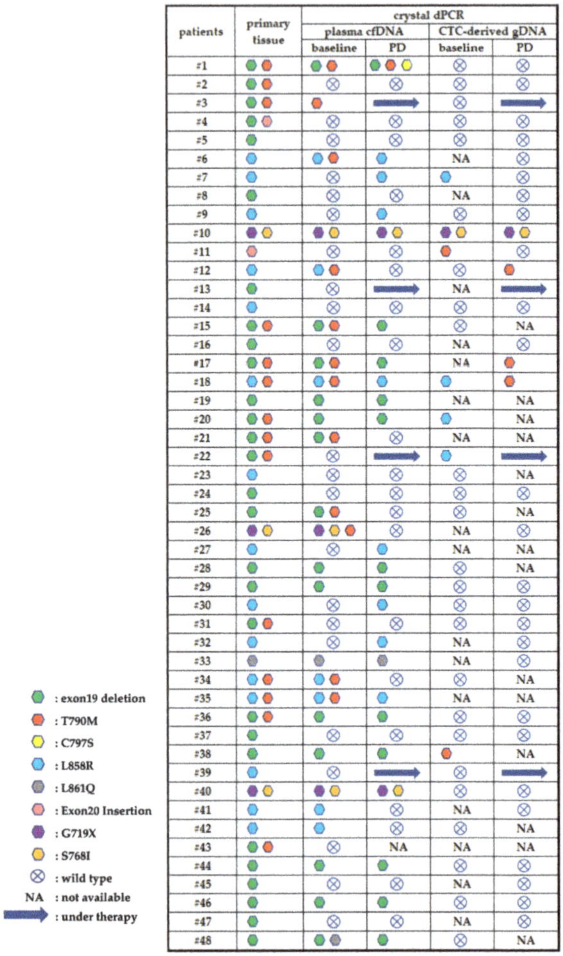

patients	primary tissue	crystal dPCR			
		plasma cfDNA		CTC-derived gDNA	
		baseline	PD	baseline	PD
#1	E19del, T790M	E19del, T790M	E19del, T790M, C797S	⊗	⊗
#2	E19del, T790M	⊗	⊗	⊗	⊗
#3	E19del, T790M	T790M	→	⊗	→
#4	E19del, Ex20ins	⊗	⊗	⊗	⊗
#5	E19del	⊗	⊗	⊗	⊗
#6	L858R	L858R, T790M	L858R	NA	⊗
#7	L858R	⊗	L858R	L858R	⊗
#8	E19del	⊗	⊗	NA	⊗
#9	L858R	⊗	L858R	⊗	⊗
#10	G719X, C797S	G719X, C797S	G719X, C797S	G719X, C797S	G719X, C797S
#11	Ex20ins	⊗	⊗	T790M	⊗
#12	L858R	L858R, T790M	⊗	⊗	T790M
#13	E19del	⊗	→	NA	→
#14	L858R	⊗	⊗	⊗	⊗
#15	E19del, T790M	E19del, T790M	E19del	⊗	NA
#16	E19del	⊗	⊗	NA	⊗
#17	E19del, T790M	E19del, T790M	E19del	NA	T790M
#18	L858R, T790M	L858R, T790M	L858R	L858R	T790M
#19	E19del	E19del	E19del	NA	NA
#20	E19del, T790M	E19del	E19del	L858R	NA
#21	E19del, T790M	E19del, T790M	⊗	NA	NA
#22	E19del, T790M	⊗	→	L858R	→
#23	L858R	⊗	⊗	⊗	NA
#24	E19del	⊗	⊗	⊗	⊗
#25	E19del	E19del, T790M	⊗	⊗	NA
#26	G719X, C797S	G719X, C797S, T790M	⊗	NA	⊗
#27	L858R	⊗	L858R	NA	NA
#28	E19del	E19del	E19del	⊗	NA
#29	E19del	E19del	E19del	⊗	⊗
#30	L858R	⊗	L858R	⊗	⊗
#31	E19del, T790M	⊗	⊗	⊗	⊗
#32	L858R	⊗	L858R	NA	⊗
#33	L861Q	L861Q	L861Q	NA	⊗
#34	L858R, T790M	L858R, T790M	⊗	⊗	NA
#35	L858R, T790M	L858R, T790M	L858R	NA	NA
#36	E19del, T790M	E19del	E19del	⊗	⊗
#37	E19del	⊗	⊗	⊗	⊗
#38	E19del	E19del	E19del	T790M	NA
#39	L858R	⊗	→	⊗	→
#40	G719X, C797S	G719X, C797S	G719X, C797S	⊗	⊗
#41	L858R	L858R	⊗	NA	⊗
#42	L858R	L858R	⊗	⊗	NA
#43	E19del, T790M	⊗	NA	NA	NA
#44	E19del	E19del	E19del	⊗	⊗
#45	E19del	⊗	⊗	NA	⊗
#46	E19del	E19del	E19del	⊗	⊗
#47	E19del	⊗	⊗	NA	⊗
#48	E19del	E19del, L861Q	E19del	⊗	NA

Legend:
- exon19 deletion
- T790M
- C797S
- L858R
- L861Q
- Exon20 Insertion
- G719X
- S768I
- ⊗ : wild type
- NA : not available
- → : under therapy

Figure 4. EGFR mutation comparison in primary tissue, plasma cfDNA and CTC-derived gDNA. Figure reproduced from Ntzifa, A. et al. (2021) [66].

Ortolan et al. (2021) evaluated CTCs and ctDNA in 42 patients with early-stage TNBC, via ddPCR and NGS. The authors state that ctDNA presence was indicative of relapse events and may help stratify patients suitable for intensification or alterative treatment post therapy to prevent metastasis development. Furthermore, CTCs analysed at disease progression revealed unique genetic abnormalities such as gain/loss of chromosome 10 and 21q. Network analysis of these altered regions identified actionable pathways including PI3K/Akt, erbB, Raf, platinum-resistance signalling, and regulation of immune response. This research states that CTCs were non-epithelial in most cases, as such they would not have been detected by epithelial dependent CTC enrichment technologies. Overall, the research team endorsed blood-based genomic analyses to utilise ctDNA as a tool for treatment response monitoring and CTCs as a tool to explore druggable targets in disease progression in TNBC patients [67].

In a study by Gorges et al. (2019), CTCs and ctDNA samples from 84 melanoma patients underwent the analysis of 61 clinically relevant variants across 13 genes including *BRAF, NRAS* and *MAP2K1*. The study reported that ctDNA and CTCs provided complementary information, indicated relapse prior to standard of care imaging, and were more accurate than the current melanoma staging system and biomarkers in some patients. More specifically, in one case, CTCs presented with BRAFV600E and EGFRI491M mutations at patient relapse, guiding targeted therapy; however, neither ctDNA, LDH (lactate dehydrogenase), or S100 (a melanoma marker gene) levels were elevated at this time. This research concludes that CTCs and ctDNA together provide real-time, complementary information on the mutational status of RNA and protein expression, with clinical significance for melanoma patients [69]. Aya-Bonilla, et al. (2020) also studied CTCs and cfDNA from melanoma patients (37 patients). The researchers reported that although immunocytochemistry showed a vast heterogeneity of CTC morphology and phenotype, gene expression analysis via ddPCR of five melanoma-associated genes revealed a comparable trend in CTC and cfDNA scores. However, in some cases CTC analysis revealed changes in molecular signatures at the baseline and in post treatment that were complementary to ctDNA monitoring. Furthermore, this research describes the Parsortix® yield as a suitable platform for potential downstream transcriptomic analysis due to its low white blood cell background yield as compared to other technologies [70].

5. Future Directions

As the omics revolution continues, we expect the further uptake of transcriptomics and proteomics to shape the future of liquid biopsies and personalised medicine for a comprehensive picture of tumour biology and clinical insights. The integration of multi-omics from laboratory bench to patient bedside faces challenges in translating vast and complex datasets into clinical benefit. Liquid biopsy-based multi-omics analysis is in its infancy, and standardisation and clinical feasibility are key to the successful integration of this tool into the clinic. This includes a need for improved access to microfluidic CTC isolation devices and sequencing platforms. However, the future of liquid biopsies is bright, with promising data emerging to support the use of whole blood as a source for multiple analytes providing information on disease prognosis, treatment selection, the monitoring of tumour evolution, and disease relapse. Furthermore, the use of dual analytes to discover complementary information will continue to emerge in the future literature, uncovering exclusive actionable insights to better inform personalised medicine. The future of the Parsortix® system involves the development and commercialisation of a breadth of downstream assays to expand CTC analysis, via immunofluorescent and molecular solutions, to provide clinically actionable insight, as well as continued investigation into the dual analysis of CTCs and ctDNA.

6. Conclusions

Liquid biopsies are emerging as a less invasive, less costly, and safer tool that provide faster results and are more suited for longitudinal disease monitoring for cancer care. CTCs and ctDNA are described as cornerstones of liquid biopsy analysis, providing minimally invasive, real-time clinical information throughout the patient care pathway. Rapid advances in technology and the affordability of NGS continue to excel, paving the way for a new era of liquid biopsy. The omics revolution is driving the dual analysis of CTCs and ctDNA as complementary sources of genomic and transcriptomic information, as RNA emerges as a tool for more accurate phenotypical sampling. The Parsortix® system is a versatile microfluidic device that facilitates epitope-independent capture and the analysis of CTCs in conjunction with the analysis of ctDNA from a single blood sample towards real-time personalised medicine, overcoming the shortfalls of immunoaffinity-based enrichment technologies that rely on epithelial surface markers known to understate CTC capture.

In particular, the Parsortix® system has been utilised in studies investigating the complementary information from CTCs and ctDNA for the evaluation of prognosis, to inform treatment selection and assess resistance and relapse, in some cases faster than the standard of care. This system addresses issues of sample heterogeneity and epitope-dependent CTC capture with label-free microfluidic isolation.

Author Contributions: Conceptualization, G.W. and F.H.; writing—original draft preparation, G.W; writing—review and editing, G.W., A.T., F.H., K.M. and A.-S.P.-J. supervision, A.T. and F.H.; project administration, F.H.; funding acquisition, F.H. All authors have read and agreed to the published version of the manuscript.

Funding: This research was funded by ANGLE plc.

Conflicts of Interest: All contributors are paid employees of ANGLE plc.

References

1. The World Health Organization. Cancer. Available online: https://www.who.int/news-room/fact-sheets/detail/cancer (accessed on 18 September 2023).
2. Rupp, B.; Ball, H.; Wuchu, F.; Nagrath, D.; Nagrath, S. Circulating Tumor Cells in Precision Medicine: Challenges and Opportunities. *Trends Pharmacol. Sci.* **2022**, *43*, 378–391. [CrossRef]
3. Keller, L.; Pantel, K. Unravelling Tumour Heterogeneity by Single-Cell Profiling of Circulating Tumour Cells. *Nat. Rev. Cancer* **2019**, *19*, 553–567. [CrossRef] [PubMed]
4. Bersani, F.; Morena, D.; Picca, F.; Morotti, A.; Tabbò, F.; Bironzo, P.; Righi, L.; Taulli, R. Future Perspectives from Lung Cancer Pre-Clinical Models: New Treatments Are Coming? *Transl. Lung Cancer Res.* **2020**, *9*, 6. [CrossRef]
5. Kong, S.L.; Liu, X.; Tan, S.J.; Tai, J.A.; Phua, L.Y.; Poh, H.M.; Yeo, T.; Chua, Y.W.; Haw, Y.X.; Ling, W.H.; et al. Complementary Sequential Circulating Tumor Cell (CTC) and Cell-Free Tumor DNA (CtDNA) Profiling Reveals Metastatic Heterogeneity and Genomic Changes in Lung Cancer and Breast Cancer. *Front. Oncol.* **2021**, *11*, 698551. [CrossRef]
6. Tay, T.K.Y.; Tan, P.H. Liquid Biopsy in Breast Cancer: A Focused Review. *Arch. Pathol. Lab. Med.* **2020**, *145*, 678–686. [CrossRef] [PubMed]
7. Narayan, P.; Ghosh, S.; Philip, R.; Barrett, J.C.; McCormack, R.T.; Odegaard, J.I.; Oxnard, G.R.; Pracht, L.J.; Williams, P.M.; Kelloff, G.J.; et al. State of the Science and Future Directions for Liquid Biopsies in Drug Development. *Oncologist* **2020**, *25*, 730–732. [CrossRef] [PubMed]
8. Alix-Panabières, C.; Pantel, K. Clinical Applications of Circulating Tumor Cells and Circulating Tumor DNA as Liquid Biopsy. *Cancer Discov.* **2016**, *6*, 479–491. [CrossRef]
9. Castro-Giner, F.; Aceto, N. Tracking Cancer Progression: From Circulating Tumor Cells to Metastasis. *Genome Med.* **2020**, *12*, 31. [CrossRef]
10. Alcaide, M.; Cheung, M.; Hillman, J.; Rassekh, S.R.; Deyell, R.J.; Batist, G.; Karsan, A.; Wyatt, A.W.; Johnson, N.; Scott, D.W.; et al. Evaluating the Quantity, Quality and Size Distribution of Cell-Free DNA by Multiplex Droplet Digital PCR. *Sci. Rep.* **2020**, *10*, 12564. [CrossRef]
11. Sant, M.; Bernat-Peguera, A.; Felip, E.; Margelí, M. Role of CtDNA in Breast Cancer. *Cancers* **2022**, *14*, 310. [CrossRef]
12. Keller, L.; Belloum, Y.; Wikman, H.; Pantel, K. Clinical Relevance of Blood-Based CtDNA Analysis: Mutation Detection and Beyond. *Br. J. Cancer* **2021**, *124*, 345–358. [CrossRef]
13. Hu, Z.; Chen, H.; Long, Y.; Li, P.; Gu, Y. The Main Sources of Circulating Cell-Free DNA: Apoptosis, Necrosis and Active Secretion. *Crit. Rev. Oncol. Hematol.* **2021**, *157*, 103166. [CrossRef] [PubMed]
14. Cohen, S.A.; Liu, M.C.; Aleshin, A. Practical Recommendations for Using CtDNA in Clinical Decision Making. *Nature* **2023**, *619*, 259–268. [CrossRef]
15. FDA. In *List of Cleared or Approved Companion Diagnostic Devices (In Vitro and Imaging Tools)*; FDA: Silver Spring, MD, USA, 2022.
16. Premarket Approval (PMA) Cobas EGFR Mutation Test. Available online: https://www.accessdata.fda.gov/scripts/cdrh/cfdocs/cfpma/pma.cfm?id=P120019s018 (accessed on 9 January 2024).
17. Premarket Approval (PMA) Cobas Mutation Test 2. Available online: https://www.accessdata.fda.gov/scripts/cdrh/cfdocs/cfpma/pma.cfm?id=P150047 (accessed on 9 January 2024).
18. Kim, M.-Y.; Oskarsson, T.; Acharyya, S.; Nguyen, D.X.; Zhang, X.H.-F.; Norton, L.; Massagué, J. Tumor Self-Seeding by Circulating Cancer Cells. *Cell* **2009**, *139*, 1315–1326. [CrossRef]
19. Tretyakova, M.S.; Menyailo, M.E.; Schegoleva, A.A.; Bokova, U.A.; Larionova, I.V.; Denisov, E.V. Technologies for Viable Circulating Tumor Cell Isolation. *Int. J. Mol. Sci.* **2022**, *23*, 15979. [CrossRef] [PubMed]
20. Maheswaran, S.; Haber, D.A. Ex Vivo Culture of CTCs: An Emerging Resource to Guide Cancer Therapy. *Cancer Res.* **2015**, *75*, 2411–2415. [CrossRef]
21. Deng, Z.; Wu, S.; Wang, Y.; Shi, D. Circulating Tumor Cell Isolation for Cancer Diagnosis and Prognosis. *eBioMedicine* **2022**, *83*, 104237. [CrossRef] [PubMed]

22. Descamps, L.; Le Roy, D.; Deman, A.-L. Microfluidic-Based Technologies for CTC Isolation: A Review of 10 Years of Intense Efforts towards Liquid Biopsy. *Int. J. Mol. Sci.* **2022**, *23*, 1981. [CrossRef]
23. Bhat, M.P.; Thendral, V.; Uthappa, U.T.; Lee, K.-H.; Kigga, M.; Altalhi, T.; Kurkuri, M.D.; Kant, K. Recent Advances in Microfluidic Platform for Physical and Immunological Detection and Capture of Circulating Tumor Cells. *Biosensors* **2022**, *12*, 220. [CrossRef]
24. Ribatti, D.; Tamma, R.; Annese, T. Epithelial-Mesenchymal Transition in Cancer: A Historical Overview. *Transl. Oncol.* **2020**, *13*, 100773. [CrossRef]
25. Miller, M.C.; Robinson, P.S.; Wagner, C.; O'Shannessy, D.J. The Parsortix™ Cell Separation System—A Versatile Liquid Biopsy Platform. *Cytometry A* **2018**, *93*, 1234–1239. [CrossRef]
26. Cohen, E.N.; Jayachandran, G.; Moore, R.G.; Cristofanilli, M.; Lang, J.E.; Khoury, J.D.; Press, M.F.; Kim, K.K.; Khazan, N.; Zhang, Q.; et al. A Multi-Center Clinical Study to Harvest and Characterize Circulating Tumor Cells from Patients with Metastatic Breast Cancer Using the Parsortix® PC1 System. *Cancers* **2022**, *14*, 5238. [CrossRef]
27. Grašič Kuhar, C.; Silvester, J.; Mencinger, M.; Ovčariček, T.; Čemažar, M.; Miceska, S.; Modic, Ž.; Kuhar, A.; Jesenko, T.; Kloboves Prevodnik, V. Association of Circulating Tumor Cells, Megakaryocytes and a High Immune-Inflammatory Environment in Metastatic Breast Cancer. *Cancers* **2023**, *15*, 3397. [CrossRef]
28. Janning, M.; Kobus, F.; Babayan, A.; Wikman, H.; Velthaus, J.-L.; Bergmann, S.; Schatz, S.; Falk, M.; Berger, L.-A.; Böttcher, L.-M.; et al. Determination of PD-L1 Expression in Circulating Tumor Cells of NSCLC Patients and Correlation with Response to PD-1/PD-L1 Inhibitors. *Cancers* **2019**, *11*, 835. [CrossRef] [PubMed]
29. Gorges, T.M.; Kuske, A.; Röck, K.; Mauermann, O.; Müller, V.; Peine, S.; Verpoort, K.; Novosadova, V.; Kubista, M.; Riethdorf, S.; et al. Accession of Tumor Heterogeneity by Multiplex Transcriptome Profiling of Single Circulating Tumor Cells. *Clin. Chem.* **2016**, *62*, 1504–1515. [CrossRef] [PubMed]
30. Silvestri, M.; Dugo, M.; Vismara, M.; De Cecco, L.; Lanzoni, D.; Vingiani, A.; Folli, S.; De Santis, M.C.; de Braud, F.; Pruneri, G.; et al. Copy Number Alterations Analysis of Primary Tumor Tissue and Circulating Tumor Cells from Patients with Early-Stage Triple Negative Breast Cancer. *Sci. Rep.* **2022**, *12*, 1470. [CrossRef] [PubMed]
31. Alves, J.M.; Estévez-Gómez, N.; Valecha, M.; Prado-López, S.; Tomás, L.; Alvariño, P.; Piñeiro, R.; Muinelo-Romay, L.; Mondelo-Macía, P.; Salgado, M.; et al. Comparative Analysis of Capture Methods for Genomic Profiling of Circulating Tumor Cells in Colorectal Cancer. *Genomics* **2022**, *114*, 110500. [CrossRef]
32. Ring, A.; Campo, D.; Porras, T.B.; Kaur, P.; Forte, V.A.; Tripathy, D.; Lu, J.; Kang, I.; Press, M.F.; Jeong, Y.J.; et al. Circulating Tumor Cell Transcriptomics as Biopsy Surrogates in Metastatic Breast Cancer. *Ann. Surg. Oncol.* **2022**, *29*, 2882–2894. [CrossRef] [PubMed]
33. Donato, C.; Kunz, L.; Castro-Giner, F.; Paasinen-Sohns, A.; Strittmatter, K.; Szczerba, B.M.; Scherrer, R.; Di Maggio, N.; Heusermann, W.; Biehlmaier, O.; et al. Hypoxia Triggers the Intravasation of Clustered Circulating Tumor Cells. *Cell Rep.* **2020**, *32*, 108105. [CrossRef]
34. Thompson, K.N.; Ju, J.A.; Ory, E.C.; Pratt, S.J.P.; Lee, R.M.; Mathias, T.J.; Chang, K.T.; Lee, C.J.; Goloubeva, O.G.; Bailey, P.C.; et al. Microtubule Disruption Reduces Metastasis More Effectively than Primary Tumor Growth. *Breast Cancer Res.* **2022**, *24*, 13. [CrossRef]
35. Ntzifa, A.; Strati, A.; Kallergi, G.; Kotsakis, A.; Georgoulias, V.; Lianidou, E. Gene Expression in Circulating Tumor Cells Reveals a Dynamic Role of EMT and PD-L1 during Osimertinib Treatment in NSCLC Patients. *Sci. Rep.* **2021**, *11*, 2313. [CrossRef] [PubMed]
36. ANGLE plc. Publications. Available online: https://angleplc.com/publications/ (accessed on 28 September 2023).
37. Templeman, A.; Miller, M.C.; Cooke, M.J.; O'Shannessy, D.J.; Gurung, Y.; Pereira, T.; Peters, S.G.; Piano, M.D.; Teo, M.; Khazan, N.; et al. Analytical Performance of the FDA-Cleared Parsortix® PC1 System. *J. Circ. Biomark.* **2023**, *12*, 26–33. [CrossRef] [PubMed]
38. University Hospital, Basel, Switzerland. Effect of Digoxin on Clusters of Circulating Tumor Cells (CTCs) in Breast Cancer Patients; Clinical Trial Registration NCT03928210. clinicaltrials.gov; 2023. Available online: https://clinicaltrials.gov/study/NCT03928210 (accessed on 1 January 2023).
39. Institut Claudius Regaud. Impact of Circadian Rhythm on the Spread of Circulating Tumor Cells in Lung Cancer Patients; Clinical Trial Registration NCT05988970. clinicaltrials.gov; 2023. Available online: https://clinicaltrials.gov/study/NCT05988970 (accessed on 1 January 2023).
40. Centre Hospitalier Universitaire de Nîmes. Proof-of-Concept Study of Blood Markers of Tumor Dissemination in Patients with versus without Intestinal Polyps. Clinical Trial Registration NCT05648240; clinicaltrials.gov; 2023. Available online: https://clinicaltrials.gov/study/NCT05648240 (accessed on 1 January 2023).
41. Mathé, E.; Hays, J.L.; Stover, D.G.; Chen, J.L. The Omics Revolution Continues: The Maturation of High-Throughput Biological Data Sources. *Yearb. Med. Inform.* **2018**, *27*, 211–222. [CrossRef]
42. Martínez-Ruiz, C.; Black, J.R.M.; Puttick, C.; Hill, M.S.; Demeulemeester, J.; Larose Cadieux, E.; Thol, K.; Jones, T.P.; Veeriah, S.; Naceur-Lombardelli, C.; et al. Genomic–Transcriptomic Evolution in Lung Cancer and Metastasis. *Nature* **2023**, *616*, 543–552. [CrossRef] [PubMed]
43. Michuda, J.; Park, B.H.; Cummings, A.L.; Devarakonda, S.; O'Neil, B.; Islam, S.; Parsons, J.; Ben-Shachar, R.; Breschi, A.; Blackwell, K.L.; et al. Use of Clinical RNA-Sequencing in the Detection of Actionable Fusions Compared to DNA-Sequencing Alone. *J. Clin. Oncol.* **2022**, *40* (Suppl. 16), 3077. [CrossRef]

44. Rodon, J.; Soria, J.-C.; Berger, R.; Miller, W.H.; Rubin, E.; Kugel, A.; Tsimberidou, A.; Saintigny, P.; Ackerstein, A.; Braña, I.; et al. Genomic and Transcriptomic Profiling Expands Precision Cancer Medicine: The WINTHER Trial. *Nat. Med.* **2019**, *25*, 751–758. [CrossRef]

45. Analyzing Tumor RNA to Improve Cancer Precision Medicine—NCI. Available online: https://www.cancer.gov/news-events/cancer-currents-blog/2020/tumor-rna-cancer-precision-medicine (accessed on 29 September 2023).

46. Perrier, A.; Hainaut, P.; Guenoun, A.; Nguyen, D.-P.; Lamy, P.-J.; Guerber, F.; Troalen, F.; Denis, J.A.; Boissan, M. En marche vers une oncologie personnalisée: L'apport des techniques génomiques et de l'intelligence artificielle dans l'usage des biomarqueurs tumoraux circulants. *Bull. Cancer* **2022**, *109*, 170–184. [CrossRef]

47. Russano, M.; Napolitano, A.; Ribelli, G.; Iuliani, M.; Simonetti, S.; Citarella, F.; Pantano, F.; Dell'Aquila, E.; Anesi, C.; Silvestris, N.; et al. Liquid Biopsy and Tumor Heterogeneity in Metastatic Solid Tumors: The Potentiality of Blood Samples. *J. Exp. Clin. Cancer Res.* **2020**, *39*, 95. [CrossRef]

48. Rossi, G.; Mu, Z.; Rademaker, A.W.; Austin, L.K.; Strickland, K.S.; Costa, R.L.B.; Nagy, R.J.; Zagonel, V.; Taxter, T.J.; Behdad, A.; et al. Cell-Free DNA and Circulating Tumor Cells: Comprehensive Liquid Biopsy Analysis in Advanced Breast Cancer. *Clin. Cancer Res.* **2018**, *24*, 560–568. [CrossRef]

49. Alama, A.; Coco, S.; Genova, C.; Rossi, G.; Fontana, V.; Tagliamento, M.; Dal Bello, M.G.; Rosa, A.; Boccardo, S.; Rijavec, E.; et al. Prognostic Relevance of Circulating Tumor Cells and Circulating Cell-Free DNA Association in Metastatic Non-Small Cell Lung Cancer Treated with Nivolumab. *J. Clin. Med.* **2019**, *8*, 1011. [CrossRef]

50. Calabuig-Fariñas, S.; Jantus-Lewintre, E.; Herreros-Pomares, A.; Camps, C. Circulating Tumor Cells versus Circulating Tumor DNA in Lung Cancer-Which One Will Win? *Transl. Lung Cancer Res.* **2016**, *5*, 466–482. [CrossRef] [PubMed]

51. Aoki, M.; Shoji, H.; Kashiro, A.; Takeuchi, K.; Shimizu, Y.; Honda, K. Prospects for Comprehensive Analyses of Circulating Tumor Cells in Tumor Biology. *Cancers* **2020**, *12*, 1135. [CrossRef] [PubMed]

52. Onidani, K.; Shoji, H.; Kakizaki, T.; Yoshimoto, S.; Okaya, S.; Miura, N.; Sekikawa, S.; Furuta, K.; Lim, C.T.; Shibahara, T.; et al. Monitoring of Cancer Patients via Next-Generation Sequencing of Patient-Derived Circulating Tumor Cells and Tumor DNA. *Cancer Sci.* **2019**, *110*, 2590–2599. [CrossRef]

53. Manier, S.; Park, J.; Capelletti, M.; Bustoros, M.; Freeman, S.S.; Ha, G.; Rhoades, J.; Liu, C.J.; Huynh, D.; Reed, S.C.; et al. Whole-Exome Sequencing of Cell-Free DNA and Circulating Tumor Cells in Multiple Myeloma. *Nat. Commun.* **2018**, *9*, 1691. [CrossRef] [PubMed]

54. Kidess-Sigal, E.; Liu, H.E.; Triboulet, M.M.; Che, J.; Ramani, V.C.; Visser, B.C.; Poultsides, G.A.; Longacre, T.A.; Marziali, A.; Vysotskaia, V.; et al. Enumeration and Targeted Analysis of KRAS, BRAF and PIK3CA Mutations in CTCs Captured by a Label-Free Platform: Comparison to CtDNA and Tissue in Metastatic Colorectal Cancer. *Oncotarget* **2016**, *7*, 85349–85364. [CrossRef]

55. Tzanikou, E.; Markou, A.; Politaki, E.; Koutsopoulos, A.; Psyrri, A.; Mavroudis, D.; Georgoulias, V.; Lianidou, E. PIK3CA Hotspot Mutations in Circulating Tumor Cells and Paired Circulating Tumor DNA in Breast Cancer: A Direct Comparison Study. *Mol. Oncol.* **2019**, *13*, 2515–2530. [CrossRef]

56. Keup, C.; Suryaprakash, V.; Storbeck, M.; Hoffmann, O.; Kimmig, R.; Kasimir-Bauer, S. Longitudinal Multi-Parametric Liquid Biopsy Approach Identifies Unique Features of Circulating Tumor Cell, Extracellular Vesicle, and Cell-Free DNA Characterization for Disease Monitoring in Metastatic Breast Cancer Patients. *Cells* **2021**, *10*, 212. [CrossRef]

57. Keup, C.; Suryaprakash, V.; Hauch, S.; Storbeck, M.; Hahn, P.; Sprenger-Haussels, M.; Kolberg, H.-C.; Tewes, M.; Hoffmann, O.; Kimmig, R.; et al. Integrative Statistical Analyses of Multiple Liquid Biopsy Analytes in Metastatic Breast Cancer. *Genome Med.* **2021**, *13*, 85. [CrossRef]

58. Li, K. Application of the Detection of Circulating Tumor Cell and Circulating Tumor DNA in the Diagnosis of Metastasis in Gastric Cancer. Clinical Trial Registration NCT05208372; clinicaltrials.gov; 2022. Available online: https://clinicaltrials.gov/study/NCT05208372 (accessed on 9 August 2023).

59. Jonsson Comprehensive Cancer Center. A Single-Arm, Open-Label, Phase II Study of Systemic and Tumor Directed Therapy for Recurrent Oligometastatic M1 Prostate Cancer. Clinical Trial Registration NCT03902951; clinicaltrials.gov; 2023. Available online: https://clinicaltrials.gov/study/NCT03902951 (accessed on 9 August 2023).

60. Zhao, J. Liquid Biopsy in Monitoring the Neoadjuvant Chemotherapy and Operation in Patients with Resectable or Locally Advanced Gastric or Gastro-Oesophageal Junction Cancer. Clinical Trial Registration NCT03957564; clinicaltrials.gov; 2020. Available online: https://clinicaltrials.gov/study/NCT03957564 (accessed on 9 August 2023).

61. Lygre, K.B. Open D3 Right Colectomy Compared to Laparoscopic CME Right Colectomy for Right Sided Colon Cancer; an Open Randomized Controlled Study. Clinical Trial Registration NCT03776591; clinicaltrials.gov; 2021. Available online: https://clinicaltrials.gov/study/NCT03776591 (accessed on 9 August 2023).

62. Markou, A.N.; Londra, D.; Stergiopoulou, D.; Vamvakaris, I.; Potaris, K.; Pateras, I.S.; Kotsakis, A.; Georgoulias, V.; Lianidou, E. Preoperative Mutational Analysis of Circulating Tumor Cells (CTCs) and Plasma-CfDNA Provides Complementary Information for Early Prediction of Relapse: A Pilot Study in Early-Stage Non-Small Cell Lung Cancer. *Cancers* **2023**, *15*, 1877. [CrossRef]

63. Markou, A.; Londra, D.; Tserpeli, V.; Kollias, I.; Tsaroucha, E.; Vamvakaris, I.; Potaris, K.; Pateras, I.; Kotsakis, A.; Georgoulias, V.; et al. DNA Methylation Analysis of Tumor Suppressor Genes in Liquid Biopsy Components of Early Stage NSCLC: A Promising Tool for Early Detection. *Clin. Epigenetics* **2022**, *14*, 61. [CrossRef]

64. Ntzifa, A.; Londra, D.; Rampias, T.; Kotsakis, A.; Georgoulias, V.; Lianidou, E. DNA Methylation Analysis in Plasma Cell-Free DNA and Paired CTCs of NSCLC Patients before and after Osimertinib Treatment. *Cancers* **2021**, *13*, 5974. [CrossRef]
65. Mondelo-Macía, P.; García-González, J.; León-Mateos, L.; Anido, U.; Aguín, S.; Abdulkader, I.; Sánchez-Ares, M.; Abalo, A.; Rodríguez-Casanova, A.; Díaz-Lagares, Á.; et al. Clinical Potential of Circulating Free DNA and Circulating Tumour Cells in Patients with Metastatic Non-Small-Cell Lung Cancer Treated with Pembrolizumab. *Mol. Oncol.* **2021**, *15*, 2923–2940. [CrossRef] [PubMed]
66. Ntzifa, A.; Kotsakis, A.; Georgoulias, V.; Lianidou, E. Detection of EGFR Mutations in Plasma CfDNA and Paired CTCs of NSCLC Patients before and after Osimertinib Therapy Using Crystal Digital PCR. *Cancers* **2021**, *13*, 2736. [CrossRef] [PubMed]
67. Ortolan, E.; Appierto, V.; Silvestri, M.; Miceli, R.; Veneroni, S.; Folli, S.; Pruneri, G.; Vingiani, A.; Belfiore, A.; Cappelletti, V.; et al. Blood-Based Genomics of Triple-Negative Breast Cancer Progression in Patients Treated with Neoadjuvant Chemotherapy. *ESMO Open* **2021**, *6*, 100086. [CrossRef] [PubMed]
68. Mondelo-Macía, P.; Rodríguez-López, C.; Valiña, L.; Aguín, S.; León-Mateos, L.; García-González, J.; Abalo, A.; Rapado-González, O.; Suárez-Cunqueiro, M.; Díaz-Lagares, A.; et al. Detection of MET Alterations Using Cell Free DNA and Circulating Tumor Cells from Cancer Patients. *Cells* **2020**, *9*, 522. [CrossRef]
69. Gorges, K.; Wiltfang, L.; Gorges, T.M.; Sartori, A.; Hildebrandt, L.; Keller, L.; Volkmer, B.; Peine, S.; Babayan, A.; Moll, I.; et al. Intra-Patient Heterogeneity of Circulating Tumor Cells and Circulating Tumor DNA in Blood of Melanoma Patients. *Cancers* **2019**, *11*, 1685. [CrossRef]
70. Aya-Bonilla, C.A.; Morici, M.; Hong, X.; McEvoy, A.C.; Sullivan, R.J.; Freeman, J.; Calapre, L.; Khattak, M.A.; Meniawy, T.; Millward, M.; et al. Detection and Prognostic Role of Heterogeneous Populations of Melanoma Circulating Tumour Cells. *Br. J. Cancer* **2020**, *122*, 1059–1067. [CrossRef]

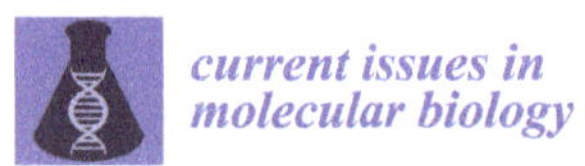

Communication

HLA-G 14 bp Ins/Del (rs66554220) Variant Is Not Associated with Breast Cancer in Women from Western Mexico

Denisse Stephania Becerra-Loaiza [1,2,†], Luisa Fernanda Roldan Flores [1,†], Luis Antonio Ochoa-Ramírez [3], Bricia M. Gutiérrez-Zepeda [1,2], Alicia Del Toro-Arreola [1], Ramón Antonio Franco-Topete [4], Andrés Morán-Mendoza [5], Antonio Oceguera-Villanueva [6], Antonio Topete [1], David Javalera [7], Antonio Quintero-Ramos [1,8,*] and Adrián Daneri-Navarro [1,*]

1   Centro Universitario de Ciencias de la Salud, Laboratorio de Inmunología, Departamento de Fisiología, Universidad de Guadalajara, Sierra Mojada #950, Guadalajara 44340, Mexico
2   Doctorado en Genética Humana, Centro Universitario de Ciencias de la Salud, Departamento de Biología Molecular y Genómica, Universidad de Guadalajara, Sierra Mojada #950, Guadalajara 44340, Mexico
3   Laboratorio de Medicina Genómica, Hospital General de Culiacán, Culiacán 80230, Mexico
4   Centro Universitario de Ciencias de la Salud, Laboratorio de Patología, Departamento de Microbiología y Patología, Universidad de Guadalajara, Sierra Mojada #950, Guadalajara 44340, Mexico
5   Centro Médico Nacional de Occidente, Hospital de Gineco Obstetricia, Instituto Mexicano del Seguro Social, Av. Belisario Domínguez #1000, Guadalajara 44340, Mexico
6   Instituto Jalisciense de Cancerología, Secretaría de Salud, Coronel Calderón #715, Guadalajara 44280, Mexico
7   Departamento de Aparatos y Sistemas II, Universidad Autónoma de Guadalajara, Av. Patria #1201, Zapopan 45129, Mexico
8   Centro Médico Nacional de Occidente, Unidad de Investigación Biomédica 02, Hospital de Especialidades, Instituto Mexicano del Seguro Social, Av. Belisario Domínguez #999, Guadalajara 44340, Mexico
*   Correspondence: antonio.qramos@academicos.udg.mx (A.Q.-R.); adrian.daneri@academicos.udg.mx (A.D.-N.); Tel.: +52-33-1058-5200 (ext. 34045) (A.Q.-R.); +52-33-1058-5200 (ext. 33668) (A.D.-N.)
†   These authors contributed equally to this study.

Citation: Becerra-Loaiza, D.S.; Roldan Flores, L.F.; Ochoa-Ramírez, L.A.; Gutiérrez-Zepeda, B.M.; Del Toro-Arreola, A.; Franco-Topete, R.A.; Morán-Mendoza, A.; Oceguera-Villanueva, A.; Topete, A.; Javalera, D.; et al. *HLA-G* 14 bp Ins/Del (rs66554220) Variant Is Not Associated with Breast Cancer in Women from Western Mexico. *Curr. Issues Mol. Biol.* **2023**, *45*, 6842–6850. https://doi.org/10.3390/cimb45080432

Academic Editor: Dumitru A. Iacobas

Received: 4 July 2023
Revised: 9 August 2023
Accepted: 14 August 2023
Published: 16 August 2023

Abstract: HLA-G is a physiology and pathologic immunomodulator detrimentally related to cancer. Its gene is heavily transcriptionally and post-transcriptionally regulated by variants located in regulator regions like 3′UTR, being the most studied Ins/Del of 14-bp (rs66554220), which is known to influence the effects of endogen cell factors; nevertheless, the reports are discrepant and controversial. Herein, the relationship of the 14-bp Ins/Del variant (rs66554220) with breast cancer (BC) and its clinical characteristics were analyzed in 182 women with non-familial BC and 221 disease-free women as a reference group. Both groups from western Mexico and sex–age-matched (sm-RG). The rs66554220 variant was amplified by SSP-PCR and the fragments were visualized in polyacrylamide gel electrophoresis. The variant rs66554220 was not associated with BC in our population. However, we suggest the Ins allele as a possible risk factor for developing BC at clinical stage IV (OR = 3.05, 95% CI = 1.16–7.96, p = 0.01); nevertheless, given the small stratified sample size (n = 11, statistical power = 41%), this is inconclusive. In conclusion, the 14-bp Ins/Del (rs66554220) variant of *HLA-G* is not associated with BC in the Mexican population, but might be related to advanced breast tumors. Further studies are required.

Keywords: breast cancer; HLA-G; Mexican population; rs66554220; 14-bp Ins/Del; western Mexico; stage IV; clinical

1. Introduction

Breast cancer (BC) is the most common and fatal type of cancer among the female population worldwide [1]. The same is true for Mexico where, in the last three decades, BC incidence and mortality have increased importantly [1]. The risk of developing BC depends on ethnicity, family history and environmental factors, with genetics representing a central element to identify possible biomarkers [2].

Phenotypic heterogeneity in BC shows a considerable challenge for tumor therapy [3]. Recently, Anna et al. evidenced a component from non-classical MHC-I, named HLA-G, as an excellent and effective target for CAR-T immunotherapy [4]. HLA-G constitutive expression is usually restricted to a few human tissues, but its ectopic expression has been demonstrated in different kinds of tumors [5,6]. Moreover, it has been related to disease stage and outcomes, metastatic status and response to different therapies [3]. Nonetheless, it is still difficult to determine a clear correlation between the HLA-G isoforms and the disease features [7].

The *HLA-G* gene is within the *MHC* cluster in 6p21.3, encoding for a non-classical HLA-I molecule with (1) immunosuppressive properties, (2) the capacity for ectopic expression in pathological conditions and (3) low rate of variants even when it is in linkage disequilibrium (LD) with *HLA-A* [8]. Emphasizing the third point, *HLA-G* genetic variation affects its expression at the transcriptional and post-transcriptional level [9]; for that, it is important to explore the association between its genetic determinants and cancer susceptibility and progression [10].

The 3′ untranslated region (UTR) of the *HLA-G* gene contains signals that regulate the spatial and temporal expression of its mRNA [11]. It is notable that this region is polymorphic, which may impact the response to endogen cellular factors according to cellular type [12]. One of the most studied genetic variants in the 3′UTR of the *HLA-G* gene is rs66554220 [8,13], which produces a 14-bp insertion (Ins)/deletion (Del) of the sequence 5′-ATT TGT TCA TGC CT-3′ between the +2960 and +2961 position in exon 8 [14]. The Ins (wild-type) allele is associated with low *HLA-G* gene expression and low levels of free soluble HLA-G (sHLA-G) given the strong LD with other single-nucleotide variants (SNVs) [10]. On the other hand, the Del allele is related to an increased *HLA-G* gene expression and higher levels of sHLA-G [12,14].

A previously published meta-analysis suggested that the rs66554220 variant may not influence cancer susceptibility in an overall context [15,16]. Nevertheless, the role of *HLA-G* SNVs in BC has already been suggested based on their biological interactions, although their precise mechanisms of action remain unclear [17]. In this respect, the *HLA-G* gene rs66554220 variant has been studied in different populations, with discordant results regarding its association with BC [15,18,19]. In Mexican people, however, there are no published studies on its association with cancer.

Considering the aforementioned context and the lack of reports on other admixture populations such as Latin-Americans, we aimed to study the potential role of variant rs66554220 in the susceptibility and clinical outcome of BC in a Mexican population.

2. Materials and Methods

2.1. Subjects

We conducted a case-control study at Universidad de Guadalajara, in Jalisco, Mexico. The patient group included 182 Mexican women (due to the low incidence of breast cancer in men, only women were included) aged $\geq$18, diagnosed with de novo non-familial BC, clinically and histologically confirmed by medical oncologists and pathologists and recruited as a part of the earlier "ELLA Binational Breast Cancer Study" [20]. Their clinical features, including menopausal status; body mass index; molecular phenotype according to ER, PR and HER2/neu; clinical and pathological stage; and metastasis were obtained through medical records.

Also, we included a disease-free sex-age ($\pm$5 years)-matched reference group (sm-RG) which was composed of 221 healthy women, who, upon questioning, did not report having breast cancer; were randomly selected; without any history or laboratory evidence of infectious, heart-related, inflammatory and renal diseases; and without background of surgery or blood transfusions for at least one year at the time of sampling, with a mean age of 50.50 $\pm$ 11.43 years old. All participants were born in the state of Jalisco with ethnic ancestry of three generations from western Mexico and provided signed informed consent.

The study was approved by the ethical and investigation committee from Universidad de Guadalajara (CI-9708) and conducted according to the Declaration of Helsinki, 1964.

2.2. Genotyping

Genomic DNA was obtained from peripheral blood using the salting-out method [21]. The rs66554220 variant was amplified by PCR using primer sequences modified from García-González et al., 2014 [22], to which two nucleotides (GT) were added at the beginning of the forward primer and one nucleotide (A) was added at the end of the reverse primer in order to adjust the two primers to the same alignment temperature, as follows: F: 5′-GTG ATG GGC TGT TTA AAG TGT CAC C-3′ and R: 5′-GGA AGG AAT GCA GTT CAG CAT GA-3′. The PCR reactions were performed using 20 ng of genomic DNA in a total volume of 10 μL, containing 1X PCR buffer, 1.5 mM $MgCl_2$, 100 mM of each dNTP, 0.3 mM of each primer and 0.025 U of recombinant Taq DNA polymerase recombinant, all reagents from Invitrogen (Life Technologies Corporation, Carlsbad, CA, USA). Later, the reaction was carried out in a thermal cycler Aeris (Esco® Lifesciences group, Changi, Singapore) with the following conditions: initial denaturation at 94 °C for 4 min; followed by 30 cycles of 26 s, each one at 94 °C, 65 °C and 72 °C; and final extension at 72 °C for 7 min. Fragments of 210 pb (Deletion) or 224 pb (Insertion) were obtained. These fragments were visualized in polyacrylamide gel electrophoresis (Golden Bell reagents, Jalisco, Mexico) at 6% in an OWL P9DS camera (Thermo Fisher Scientific, Waltham, MA, USA) and stained with silver nitrate (Golden Bell reagents, Jalisco, Mexico). As a quality control, 10% of all samples were selected, reanalyzed and all results were confirmed by an independent blinded observer.

2.3. Statistical Analysis

The allele and genotype frequencies were calculated via direct counting in both study groups. Hardy–Weinberg Equilibrium (HWE), χ^2 and logistic regression were performed in the online SNPstats software: https://www.snpstats.net/ (accessed on 6 June 2023). Also, comparisons between allelic and genotypic frequencies vs. clinical characteristics in the BC group were made in IBM SPSS Statistics (v27.0). Values of $p < 0.05$ were considered significant. "Finally, the statistic power $(1-b)$ was calculated in Post-hoc power calculator online (https://clincalc.com/Stats/Power.aspx, accessed on 6 June 2023) according to the sample size of the study, respectively".

3. Results

3.1. Description of Clinical Variables

The clinical data of the BC patients are shown in Table 1. At the time of BC diagnosis, most of the patients were postmenopausal and 54.53 ± 12.53 years old. Moreover, according to their body mass index, most BC patients presented overweight (37.9%) or obesity (28.6%), and 33% of those were ≥60 years old. Regarding the molecular phenotype of cancer, Luminal A was the most predominant (29.1%). Additionally, the clinical and the pathological stage II were the most prevalent, with around 16% of metastasis reported.

3.2. Genetic Association

Allele and genotype frequencies for the rs66554220 variant were in agreement with HWE in BC patients ($p = 0.61$) and sm-RG ($p = 0.79$). Electrophoretic patterns of *HLA-G* 14-bp Ins/Del (rs66554220) variant genotypes are demonstrated in Figure 1. As shown in Table 2, according to χ^2 and logistic regression with different inheritance models, no statistically significant differences were observed between groups. In addition to this, we stratified BC patients according to clinical features and we found the Ins allele as a possible risk factor for clinical stage IV in the BC group vs. the sm-RG (Table 3). However, the statistical power (41%) was insufficient to be a conclusive association. The remaining clinical features (menopausal status, body mass index, molecular phenotype, pathological stage and metastatic status showed no association).

Table 1. Clinical features of breast cancer patients.

Modifiable Risk Factors	n (%)
Menopausal status	
Pre-menopause	62 (34.1)
Post-menopause	118 (64.8)
Unknown	2 (1.1)
Body mass index	
Underweight	1 (0.5)
Normal weight	48 (26.4)
Overweight	69 (37.9)
Obesity class I	40 (22)
Obesity class II	6 (3.3)
Obesity class III	6 (3.3)
Unknown	12 (6.6)
Mechanisms/Pathophysiology	**n (%)**
Molecular phenotype	
Luminal A	53 (29.1)
Luminal B	22 (12.1)
HER2/neu	27 (14.9)
TNBC	23 (12.6)
Unknown	57 (31.3)
Clinical stage	
I	19 (10.4)
II	77 (42.3)
III	58 (31.9)
IV	11 (6.0)
Unknown	17 (9.4)
Pathological stage	
I	21 (11.5)
II	68 (37.4)
III	65 (35.7)
IV	9 (5)
Unknown	19 (10.4)
Metastatic status	
Presence	29 (15.9)
Absent	126 (69.2)
Unknown	27 (14.9)

TNBC, triple-negative breast cancer.

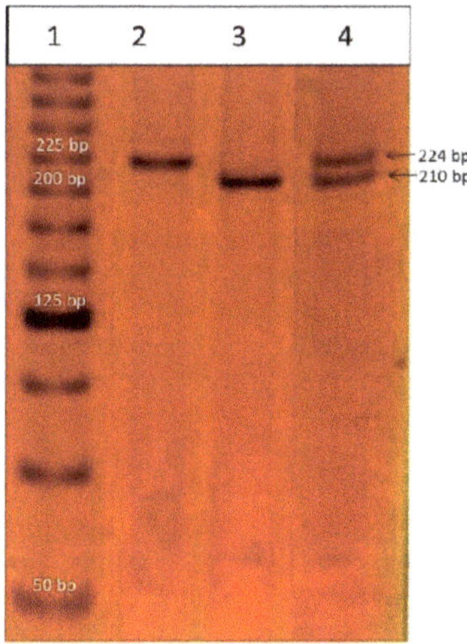

Figure 1. Visualization of *HLA-G* 14-bp Ins/Del (rs66554220) variant genotypes via 6% polyacrylamide gel electrophoresis. Lane 1, 25 base-pair molecular-weight markers indicating 125 bp as the most intense band; lane 2, homozygous Ins/Ins genotype; lane 3, homozygous Del/Del genotype; and lane 4, heterozygous Del/Ins genotype.

Table 2. Allelic and genotypic frequencies and the results of the association test of the *HLA-G* 14-bp Del/Ins (rs66554220) variant.

Inheritance Model	Sex-Matched Reference Group n = 221 (%)	Breast Cancer n = 182 (%)	*p*-Value *
Allele			
Del	(54)	(53)	Reference
Ins	(46)	(47)	n.s.
Co-dominant			
Del/Del	65 (29.4)	50 (27.5)	Reference
Del/Ins	108 (48.9)	94 (51.6)	n.s.
Ins/Ins	48 (21.7)	38 (20.9)	n.s.
Dominant			
Del/Del	65 (29.4)	50 (27.5)	Reference
Del/Ins + Ins/Ins	156 (70.6)	132 (72.5)	n.s.
Recessive			
Del/Del + Del/Ins	173 (78.3)	144 (79.1)	Reference
Ins/Ins	48 (21.7)	38 (20.9)	n.s.

p-value: significance defined by the χ^2 test; n.s.: non-significant. * Due to the lack of statistical significance, the OR and 95% CI were omitted.

Table 3. Allelic frequencies and association test of the *HLA-G* 14-bp Del/Ins (rs66554220) variant in clinical stage IV of breast cancer clinical features.

Allele	Sex-Matched Reference Group n = 221 (%)	Clinical Stage IV of Breast Cancer n = 11 (%)	*p*-Value, OR (95% CI) *
Del	(54)	(27)	Reference
Ins	(46)	(73)	0.01, 3.05 (1.16–7.96)

p-value: significance defined by the χ^2 test; OR: odds ratio; 95% CI: 95% confidence interval. * The statistic power $(1-\beta)$ derived from this comparison was 41%.

4. Discussion

The *HLA-G* gene 3'UTR 14-bp (rs66554220) variant is involved in HLA-G production via modulating the mRNA stability of mechanisms that have not been yet fully elucidated [23,24], albeit the insertion of 14-bp (Ins) has been associated with low *HLA-G* mRNA production [23]. On the other hand, with the deletion of 14-bp (Del), the transcripts could be further processed by the removal of the first 92 bases of exon 8, producing a smaller and stable transcript as compared with the complete mRNA forms [23,24]. Nevertheless, it should be emphasized that the presence of the 14-bp insertion is always related to the presence of another two variants, rs1063320 and rs9382142 (+3142G and +3187A, respectively), in strong LD and is associated with a low quantity of *HLA-G* mRNA [25].

In the present study, genetic frequencies of the rs66554220 variant were not associated with BC, which is like other studies [26]. It is important to mention that the frequency of rs6655420 allele/genotypes was in accordance with earlier reports from the West of Mexico [22,27,28]. Furthermore, according to Farias-Rodrigues et al., the insertion allele has a similar distribution around the world, indicating a possible action of balancing selection [29]. This fact is important because HLA-G's pivotal function in the immune system and its putative beneficial role in maintaining genetic variability in different stages of immunological processes [29]. In this way, it allows for a better understanding of the role of genetic variability in complex diseases such as BC.

Studies of different pathological conditions have indicated that the *HLA-G* gene might serve as a clinical marker for the diagnosis or prediction of the clinical outcomes of breast cancer [30–34]. In the present study, we suggest that the Ins allele could granted up to three times the risk of developing clinical stage IV BC, where the microenvironment changes from anti-tumor to tumor-promoting [2]. At this point, *HLA-G* gene expression can act as a checkpoint and as a critical marker of immune tolerance in cancer-cell immune evasion, disease progression and prognosis, given that the heterogeneity of their expression in immune-suppressive microenvironments and the isoform profiles vary among tumor type and patients [3] (Figure 2).

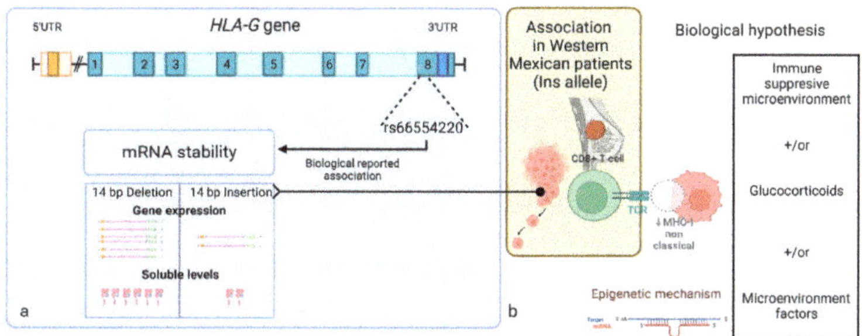

Figure 2. Biological hypothesis about risk association of Ins allele of rs66554220 in *HLA-G* gene with clinical stage IV breast cancer. (**a**) As a background, *HLA-G* gene contains 8 exons; their last exon, the 3'UTR, contains the rs66554220 variant related to mRNA stability given the Ins/Del of 14-bp associated with decreased/augmented gene expression and soluble levels of the HLA-G protein, respectively. (**b**) According to this research, we proposed the biological hypothesis derived from the clinical suggestion of association between Ins allele and clinical stage IV in patients from western Mexico with breast cancer as a risk factor, because their biological function is related to epigenetic mechanisms, an immune-suppressive microenvironment and other microenvironmental factors that allow the immune surveillance and evasion of the system MHC-I non classical. Created with BioRender.com.

Our results are similar to the findings in South Indian women, in which the Ins allele is proposed as an important factor in the pathogenesis of BC. Nevertheless, they lacked an analysis of clinical variables [18]. Contrary to our findings, in populations from Tunisia and Iran, the Del allele is proposed as a risk factor for developing BC [19,32–34]. Also, in Brazilians, the Del/Del genotype is associated with higher levels of soluble HLA-G in invasive breast ductal carcinoma, poor prognosis of life and metastasis [35]. Moreover, in a recent meta-analysis in Caucasic and Asiatic populations, Tizaoui et al. suggested the rs66554220 variant as a risk factor and sHLA-G level as a biomarker for BC [6]. These similarities and differences remark the importance of studying different populations to ascertain the validity of a gene as a possible risk factor for a pathology, particularly in populations with high genetic admixture such as the Mexican population [36].

In the context of cancer, the immunoediting process includes the gain of expression of immune-inhibitory molecules such as HLA-G [37]. Furthermore, *HLA-G* gene expression can be induced by glucocorticoids or microenvironmental factors such as low oxygen tension or tryptophan starvation, both characteristic of cancer, along with it being regulated by epigenetic mechanisms [38]. Owing to the latter, the presence of the Ins allele could be an important regulator of the union of different microRNAs that allow HLA-G to function as a bipartite immune checkpoint, contributing in complex diseases such as BC.

The evaluation of the clinical characteristics such as menopause was exclusive for patients, and they were not comparable with the sm-RG; in addition, the stratification of patients based on their clinical characteristics reduced the statistic power to half of the statistical significance.

It is important to emphasize that this first approach is focused on the role of the rs66554220 variant in our population, and future case–control studies will be performed with other characteristics of women, with the inclusion of a larger number of patients to increase the robustness of what is proposed here. Also, future investigations with new methodologies will be needed. Finally, the prevalence of the clinicopathological stage and metastasis in our patients could be related to an early or late diagnosis more than a genetic factor in its entirety, taking into consideration that, in the Mexican population, the diagnostic rate of BC is above <50 years compared with the United States and Europe and it is also mostly in advanced stages (III, IV, N.C) in two out of three patients [39,40].

5. Conclusions

We concluded that the 14-bp Ins/Del (rs66554220) variant of *HLA-G* is not associated with BC in the Mexican population, but might be related to advanced breast tumors; further studies are required. We suggest the Ins allele as a possible risk factor for developing BC at clinical stage IV; nevertheless, a bigger stratified sample size might verify this. We propose the integration of clinical features in the association studies due to the possibility of identifying possible genetic factors involved in the etiopathogenesis of complex diseases like cancer and, as in other publications, we also suggest that the 3'UTR of the *HLA-G* gene segment should be analyzed in a wider approach because of its strong linkage disequilibrium with other variants. This could be useful in future clinical practice settings or in generating new strategies for the diagnosis–prognosis of cancer.

Author Contributions: Conceptualization, D.S.B.-L., L.F.R.F. and A.Q.-R.; Data curation, R.A.F.-T., A.M.-M. and A.O.-V.; Formal analysis, L.A.O.-R., A.T. and D.J.; Funding acquisition, A.D.-N. and A.Q.-R.; Methodology, D.S.B.-L., L.F.R.F. and B.M.G.-Z.; Visualization, R.A.F.-T., A.M.-M. and A.O.-V.; Writing—original draft, D.S.B.-L., L.F.R.F. and A.D.T.-A.; Writing—review and editing, A.D.-N., R.A.F.-T. and A.Q.-R. All authors have read and agreed to the published version of the manuscript.

Funding: This work was supported by grants awarded from PRO-SNI 2020/2021 funding of the Universidad de Guadalajara.

Institutional Review Board Statement: This study was conducted in accordance with the Declaration of Helsinki and approved by the Ethical and Investigation Committee of Universidad de Guadalajara (code CI-9708 and date of approval was 8 November 2008) for studies involving humans.

Informed Consent Statement: Informed consent was obtained from all subjects involved in the study.

Data Availability Statement: Research data are stored in our laboratory storage unit.

Acknowledgments: The authors would like to express their gratitude to Ella Binational Breast Cancer Study, Avon Mexico: Promesa Cáncer 2019, Doctorado en Genética Humana, Centro Universitario de Ciencias de la Salud, Universidad de Guadalajara and to the PRO-SNI funding of the Universidad de Guadalajara.

Conflicts of Interest: The authors declare no conflict of interest.

References

1. Ferlay, J.; Ervik, M.; Lam, F.; Colombet, M.; Mery, L.; Piñeros, M. *Global Cancer Observatory*; International Agency for Research on Cancer: Lyon, France, 2020. Available online: https://gco.iarc.fr/today (accessed on 7 December 2022).
2. Harbeck, N.; Penault-Llorca, F.; Cortes, J.; Gnant, M.; Houssami, N.; Poortmans, P.; Ruddy, K.; Tsang, J.; Cardoso, F. Breast cancer. *Nat. Rev. Dis. Primers* **2019**, *5*, 66. [CrossRef] [PubMed]
3. Lin, A.; Yan, W.H. Heterogeneity of HLA-G Expression in Cancers: Facing the Challenges. *Front. Immunol.* **2018**, *9*, 2164. [CrossRef] [PubMed]
4. Anna, F.; Bole-Richard, E.; LeMaoult, J.; Escande, M.; Lecomte, M.; Certoux, J.-M.; Souque, P.; Garnache, F.; Adotevi, O.; Langlade-Demoyen, P.; et al. First immunotherapeutic CAR-T cells against the immune checkpoint protein HLA-G. *J. Immunother. Cancer* **2021**, *9*, e001998. [CrossRef] [PubMed]
5. Carosella, E.D.; Rouas-Freiss, N.; Tronik-Le Roux, D.; Moreau, P.; LeMaoult, J. HLA-G: An Immune Checkpoint Molecule. *Adv. Immunol.* **2015**, *127*, 33–144. [PubMed]
6. Tizaoui, K.; Jalouli, M.; Ouzari, H.I.; Harrath, A.H.; Rizzo, R.; Boujelbene, N.; Zidi, I. 3'UTR-HLA-G polymorphisms and circulating sHLA-G are associated with breast cancer: Evidence from a meta-analysis. *Immunol. Lett.* **2022**, *248*, 78–89. [CrossRef] [PubMed]
7. Loustau, M.; Anna, F.; Dréan, R.; Lecomte, M.; Langlade-Demoyen, P.; Caumartin, J. HLA-G Neo-Expression on Tumors. *Front. Immunol.* **2020**, *11*, 1685. [CrossRef]

8. Castelli, E.C.; de Almeida, B.S.; Muniz, Y.C.; Silva, N.S.; Passos, M.R.; Souza, A.S.; Page, A.E.; Dyble, M.; Smith, D.; Aguileta, G.; et al. HLA-G genetic diversity and evolutive aspects in worldwide populations. *Sci. Rep.* **2021**, *11*, 23070. [CrossRef]

9. Krijgsman, D.; Roelands, J.; Hendrickx, W.; Bedognetti, D.; Kuppen, P.J. HLA-G: A New Immune Checkpoint in Cancer? *Int. J. Mol. Sci.* **2020**, *21*, 4528. [CrossRef]

10. Chen, X.Y.; Yan, W.H.; Lin, A.; Xu, H.H.; Zhang, J.G.; Wang, X.X. The 14 bp deletion polymorphisms in HLA-G gene play an important role in the expression of soluble HLA-G in plasma. *Tissue Antigens* **2008**, *72*, 335–341. [CrossRef]

11. Donadi, E.A.; Castelli, E.C.; Arnaiz-Villena, A.; Roger, M.; Rey, D.; Moreau, P. Implications of the polymorphism of HLA-G on its function, regulation, evolution and disease association. *Cell Mol. Life Sci.* **2011**, *68*, 369–395. [CrossRef]

12. Poras, I.; Yaghi, L.; Martelli-Palomino, G.; Mendes-Junior, C.T.; Muniz, Y.C.N.; Cagnin, N.F.; de Almeida, B.S.; Castelli, E.C.; Carosella, E.D.; Donadi, E.A.; et al. Haplotypes of the HLA-G 3′ Untranslated Region Respond to Endogenous Factors of HLA-G+ and HLA-G- Cell Lines Differentially. *PLoS ONE* **2017**, *12*, e0169032. [CrossRef] [PubMed]

13. de Almeida, B.S.; Muniz, Y.C.; Prompt, A.H.; Castelli, E.C.; Mendes-Junior, C.T.; Donadi, E.A. Genetic association between HLA-G 14-bp polymorphism and diseases: A systematic review and meta-analysis. *Hum. Immunol.* **2018**, *79*, 724–735. [CrossRef]

14. Castelli, E.C.; Veiga-Castelli, L.C.; Yaghi, L.; Moreau, P.; Donadi, E.A. Transcriptional and posttranscriptional regulations of the HLA-G gene. *J. Immunol. Res.* **2014**, *2014*, 734068. [CrossRef] [PubMed]

15. Li, T.; Huang, H.; Liao, D.; Ling, H.; Su, B.; Cai, M. Genetic polymorphism in HLA-G 3′UTR 14-bp ins/del and risk of cancer: A meta-analysis of case-control study. *Mol. Genet. Genom.* **2015**, *290*, 1235–1245. [CrossRef]

16. Ge, Y.Z.; Ge, Q.; Li, M.H.; Shi, G.M.; Xu, X.; Xu, L.W.; Xu, Z.; Lu, T.Z.; Wu, R.; Zhou, L.H.; et al. Association between human leukocyte antigen-G 14-bp insertion/deletion polymorphism and cancer risk: A meta-analysis and systematic review. *Hum. Immunol.* **2014**, *75*, 827–832. [CrossRef]

17. Zheng, G.; Jia, L.; Yang, A.G. Roles of HLA-G/KIR2DL4 in Breast Cancer Immune Microenvironment. *Front. Immunol.* **2022**, *3*, 791975. [CrossRef] [PubMed]

18. Kadiam, S.; Ramasamy, T.; Ramakrishnan, R.; Mariakuttikan, J. Association of HLA-G 3′UTR 14-bp Ins/Del polymorphism with breast cancer among South Indian women. *J. Clin. Pathol.* **2020**, *73*, 456–462. [CrossRef]

19. Ouni, N.; Chaaben, A.B.; Kablouti, G.; Ayari, F.; Douik, H.; Abaza, H.; Gara, S.; Elgaaied-Benammar, A.; Guemira, F.; Tamouza, R. The Impact of HLA-G 3′UTR Polymorphisms in Breast Cancer in a Tunisian Population. *Immunol. Investig.* **2019**, *48*, 521–532. [CrossRef]

20. Martínez, M.E.; Gutiérrez-Millan, L.E.; Bondy, M.; Daneri-Navarro, A.; Meza-Montenegro, M.M.; Anduro-Corona, I.; Aramburo-Rubio, M.I.; Balderas-Peña, L.M.A.; Barragan-Ruiz, J.A.; Brewster, A.; et al. Comparative study of breast cancer in Mexican and Mexican-American women. *Health* **2010**, *2*, 1040–1048. [CrossRef]

21. Miller, S.A.; Dykes, D.D.; Polesky, H.F. A simple salting out procedure for extracting DNA from human nucleated cells. *Nucleic Acids Res.* **1988**, *16*, 1215. [CrossRef]

22. García-González, I.J.; Valle, Y.; Rivas, F.; Figuera-Villanueva, L.E.; Muñoz-Valle, J.F.; Flores-Salinas, H.E.; Gutiérrez-Amavizca, B.E.; Dávalos-Rodríguez, N.O.; Padilla-Gutiérrez, J.R. The 14 bp Del/Ins HLA-G polymorphism is related with high blood pressure in acute coronary syndrome and type 2 diabetes mellitus. *Biomed Res. Int.* **2014**, *2014*, 898159. [CrossRef] [PubMed]

23. Rousseau, P.; Le Discorde, M.; Mouillot, G.; Marcou, C.; Carosella, E.D.; Moreau, P. The 14 bp deletion-insertion polymorphism in the 3′ UT region of the HLA-G gene influences HLA-G mRNA stability. *Hum. Immunol.* **2003**, *64*, 1005–1010. [CrossRef] [PubMed]

24. Hviid, T.V.F.; Hylenius, S.; Rørbye, C.; Nielsen, L.G. HLA-G allelic variants are associated with differences in the HLA-G mRNA isoform profile and HLA-G mRNA levels. *Immunogenetics* **2003**, *55*, 63–79. [CrossRef] [PubMed]

25. Arnaiz-Villena, A.; Juarez, I.; Suarez-Trujillo, F.; López-Nares, A.; Vaquero, C.; Palacio-Gruber, J.; Martin-Villa, J.M. HLA-G: Function, polymorphisms and pathology. *Int. J. Immunogenet.* **2021**, *48*, 172–192. [CrossRef] [PubMed]

26. Jeong, S.; Park, S.; Park, B.W.; Park, Y.; Kwon, O.J.; Kim, H.S. Human leukocyte antigen-G (HLA-G) polymorphism and expression in breast cancer patients. *PLoS ONE* **2014**, *9*, e98284. [CrossRef]

27. Becerra-Loaiza, D.S.; Ochoa-Ramírez, L.A.; Velarde-Félix, J.S.; Sánchez-Zazueta, J.G.; Quintero-Ramos, A. Association of the HLA-G rs66554220 variant with non-segmental vitiligo and its clinical features in Northwestern Mexico population. *Postepy Dermatol. Alergol.* **2023**, *40*, 246–252. [CrossRef]

28. Porras, D.A.; Benita Lazcano, C.A.; Donizete Da Silva, J.T.; Juárez, C.I.; Juárez, J.A.; Perea, F.J; García, J.E. Análisis de asociación de los polimorfismos -725C>G (rs1233334), -201G>A (rs1233333) y 14 bp deleción/inserción (14-pb del/ins) (rs66554220) del gen HLA-G en mujeres mexicanas con pérdida gestacional recurrente. *Rev. Chil. Obstet. Ginecol.* **2014**, *79*, 86–91. Available online: http://www.scielo.cl/scielo.php?script=sci_arttext&pid=S0717-75262014000200004&lng=es (accessed on 20 September 2022). [CrossRef]

29. Farias-Rodrigues, J.K.; Crovella, S.; da Silva, R.C. The HLA-G 14 bp allele frequency in different populations: A global meta-analysis. *Meta Gene* **2020**, *23*, 100624. [CrossRef]

30. Yang, Y.C.; Chang, T.Y.; Chen, T.C.; Lin, W.S.; Chang, S.C.; Lee, Y.J. Human leucocyte antigen-G polymorphisms are associated with cervical squamous cell carcinoma risk in Taiwanese women. *Eur. J. Cancer* **2014**, *50*, 469–474. [CrossRef]

31. Xu, H.H.; Wang, H.L.; Xing, T.J.; Wang, X.Q. A Novel Prognostic Risk Model for Cervical Cancer Based on Immune Checkpoint HLA-G-Driven Differentially Expressed Genes. *Front. Immunol.* **2022**, *18*, 851622. [CrossRef]

32. Haghi, M.; Feizi, M.A.H.; Sadeghizadeh, M.; Lotfi, A.S. 14-bp Insertion/Deletion Polymorphism of the HLA-G gene in Breast Cancer among Women from North Western Iran. *Asian Pac. J. Cancer Prev.* **2015**, *16*, 6155–6158. [CrossRef] [PubMed]

33. Eskandari-Nasab, E.; Hashemi, M.; Hasani, S.-S.; Omrani, M.; Taheri, M.; Mashhadi, M.-A. Association between HLA-G 3'UTR 14-bp ins/del polymorphism and susceptibility to breast cancer. *Cancer Biomark.* **2013**, *13*, 253–259. [CrossRef] [PubMed]
34. Haghi, M.; Ranjbar, M.; Karari, K.; Samadi-Miandoab, S.; Eftekhari, A.; Hosseinpour-Feizi, M.A. Certain haplotypes of the 3'-UTR region of the HLA-G gene are linked to breast cancer. *Br. J. Biomed. Sci.* **2021**, *78*, 87–91. [CrossRef] [PubMed]
35. Ramos, C.S.; Gonçalves, A.S.; Marinho, L.C.; Avelino, M.A.G.; Saddi, V.A.; Lopes, A.C.; Simões, R.T.; Wastowski, I.J. Analysis of HLA-G gene polymorphism and protein expression in invasive breast ductal carcinoma. *Hum. Immunol.* **2014**, *75*, 667–672. [CrossRef] [PubMed]
36. Martínez-Cortés, G.; Salazar-Flores, J.; Haro-Guerrero, J.; Rubi-Castellanos, R.; Velarde-Félix, J.S.; Muñoz-Valle, J.F.; López-Casamichana, M.; Carrillo-Tapia, E.; Canseco-Avila, L.M.; Bravi, C.M.; et al. Maternal admixture and population structure in Mexican-Mestizos based on mtDNA haplogroups. *Am. J. Phys. Anthropol.* **2013**, *151*, 526–537. [CrossRef]
37. Steven, A.; Seliger, B. The Role of Immune Escape and Immune Cell Infiltration in Breast Cancer. *Breast Care* **2018**, *13*, 16–21. [CrossRef]
38. González, A.; Rebmann, V.; LeMaoult, J.; Horn, P.A.; Carosella, E.D.; Alegre, E. The immunosuppressive molecule HLA-G and its clinical implications. *Crit. Rev. Clin. Lab. Sci.* **2012**, *49*, 63–84. [CrossRef]
39. Rodríguez-Cuevas, S.; Macías Martínez, C.G.; Labastida Almendaro, S. Cáncer de mama en México¿ Enfermedad de mujeres jóvenes? *Ginecol. Obstet. Mex.* **2000**, *68*, 185–190.
40. Rodríguez-Cuevas, S.; Macías, C.G.; Franceschi, D.; Labastida, S. Breast carcinoma presents a decade earlier in Mexican women than in women in the United States or European countries. *Cancer* **2001**, *91*, 863–868. [CrossRef]

current issues in molecular biology

MDPI

Review

The Role of Amino Acids in the Diagnosis, Risk Assessment, and Treatment of Breast Cancer: A Review

Lyudmila V. Bel'skaya [1,*], Ivan A. Gundyrev [1] and Denis V. Solomatin [2]

1   Biochemistry Research Laboratory, Omsk State Pedagogical University, 644099 Omsk, Russia; ivangundyrev@yandex.ru
2   Department of Mathematics and Mathematics Teaching Methods, Omsk State Pedagogical University, 644043 Omsk, Russia; denis_2001j@bk.ru
*   Correspondence: belskaya@omgpu.ru

Abstract: This review summarizes the role of amino acids in the diagnosis, risk assessment, imaging, and treatment of breast cancer. It was shown that the content of individual amino acids changes in breast cancer by an average of 10–15% compared with healthy controls. For some amino acids (Thr, Arg, Met, and Ser), an increase in concentration is more often observed in breast cancer, and for others, a decrease is observed (Asp, Pro, Trp, and His). The accuracy of diagnostics using individual amino acids is low and increases when a number of amino acids are combined with each other or with other metabolites. Gln/Glu, Asp, Arg, Leu/Ile, Lys, and Orn have the greatest significance in assessing the risk of breast cancer. The variability in the amino acid composition of biological fluids was shown to depend on the breast cancer phenotype, as well as the age, race, and menopausal status of patients. In general, the analysis of changes in the amino acid metabolism in breast cancer is a promising strategy not only for diagnosis, but also for developing new therapeutic agents, monitoring the treatment process, correcting complications after treatment, and evaluating survival rates.

Keywords: amino acids; breast cancer; diagnosis; imaging; risk; prognosis; treatment

check for updates

Citation: Bel'skaya, L.V.; Gundyrev, I.A.; Solomatin, D.V. The Role of Amino Acids in the Diagnosis, Risk Assessment, and Treatment of Breast Cancer: A Review. *Curr. Issues Mol. Biol.* **2023**, *45*, 7513–7537. https://doi.org/10.3390/cimb45090474

Academic Editor: Stergios Boussios

Received: 29 August 2023
Revised: 5 September 2023
Accepted: 12 September 2023
Published: 13 September 2023

1. Introduction

Breast cancer (BC) remains the leading malignant neoplasm in women and the second leading cause of death from cancer among women worldwide [1,2]. Treatment for breast cancer has improved in recent years, but its high mortality remains a concern. Existing treatments for breast cancer often lead to intractable drug resistance. Therefore, the need for methods of early diagnosis and the identification of risk factors and groups are still relevant tasks. To solve these problems, scientists began to study the metabolic pathway in breast cancer. The connection between breast cancer and metabolic pathways may also lead to the discovery of new therapeutic possibilities and targets [3,4].

Recently, much attention has been paid to metabolomics as an effective tool for differentiating samples from patients with breast cancer from normal samples [5–11]. For example, a statistically significant difference in 24 metabolites in breast cancer compared with the control group was found by Dougan et al. [12], and a statistically significant difference in 78 metabolites was confirmed by Shen et al. [8]. Interestingly, the levels of metabolites are unique for different types of cancer, which could be a new way to classify tumors [13]. Jobard et al. identified nine metabolites to differentiate between metastatic and early breast cancer, although a separation of breast cancer phenotypes based on metabolomics is still controversial [14,15]. Several metabolomic studies of plasma/serum breast cancer have been performed, mainly aimed at distinguishing between the subtypes of breast cancer [16], metastatic breast cancer [8,14,17–20], recurrence [21,22], response to neoadjuvant chemotherapy [23,24], etc. An important part of the metabolomic profile, which makes it possible to discriminate between BC patients and healthy controls, is represented by amino acids. Amino acids in breast cancer are determined in the tissue [25–27], serum [7,28–30], plasma [6,31–34], saliva [35–38], and urine [39–41]. Amino acids not only play a vital role in the synthesis of biological molecules such as proteins in malignant cancer cells, but they are also important metabolites for immune cell activation and antitumor activity in the tumor microenvironment. Abnormal changes in the amino acid metabolism are closely associated with the onset and development of tumors and immunity [42]. An amino acid can supply sources of nitrogen and carbon for biosynthesis or satisfy the energy requirement for the rapid growth of tumor cells [43].

In this review, we focus on the role of amino acids in the diagnosis, risk assessment, imaging, and treatment of breast cancer.

2. Diagnostic Value of Determination of Amino Acids in Biological Fluids

2.1. Amino Acid Composition of Serum and Blood Plasma in Breast Cancer

The determination of the content of amino acids in serum and blood plasma was carried out in a large number of studies (Table 1).

Table 1. Basic studies of the amino acid composition of plasma/serum in breast cancer.

No.	Author, Year	BC/Control	BC Stages	Method	Up-Regulated AAs	Down-Regulated AAs
1	Kubota A., 1992 [44]	22/11	St I + II—22	AA analyzer	Ala, Arg, Thr	Cys, Gln
2	Cascino A., 1995 [45]	33/28	-	AA analyzer	Glu, Orn, Trp	-
3	Proenza A.M.A., 2003 [46]	16/18	St I—2, St II—5, St III—3, St IV—5, Unknown—1	HPLC	Asn, Gln, Pro (Hydroxyproline)	Asp
4	Minet-Quinard R., 2004 [47]	19/18	T0—16%, T1—42%, T2—42%	AA analyzer	Ser, Glu, Orn	-
5	Vissers Y.L.J., 2005 [48]	22/22	St I—6, St II—7, St III—8	HPLC	-	Arg
6	Okamoto N., 2009 [49]	61/51	St 0—8, St I—30, St II—18, St III—5	HPLC–ESI–MS	Thr, Ser, Glu, Orn	Met, Ile, Phe, Arg
7	Miyagi Y., 2011 [34]	196/976	St II—95, St III—19, St IV—15, Unknown—5	HPLC–ESI–MS	Thr, Pro, Ser, Gly, Ala, Orn	Gln, Trp, His, Phe, Tyr
8	Shen J., 2013 [8]	60/60	ER+/PR+—30, triple negative—30 African Americans and Caucasian Americans data sets	UPLC-MS/MS or GC-MS	-	Ala, His, Asp, Lys, Tyr, Trp, Met, Arg, Pro
9	Poschke I., 2013 [50]	41/9	-	HPLC	Glu, Ser, Gln, Ala, Val, Phe, Ile, Leu	-
10	Barnes T., 2014 [51]	8/8	St I + II—8	HPLC	Ala, Asp, Gln, Lys, Met, Tyr	Arg, Gly, Pro, Ser, Ile, Val, Orn
11	Gu Y., 2015 [52]	28/137	St I—7, St II—18, St III—3	AA analyzer	Thr, Arg	Asp, Gln, Gly, His
12	Jové M., 2017 [5]	91/20	St I—2, St IIA—40, St IIB—30, St IIIA—13, St IIIB—7	ESI-Q-TOF MS/MS	-	Gln, Arg, Lys, Val

Table 1. *Cont.*

No.	Author, Year	BC/Control	BC Stages	Method	Up-Regulated AAs	Down-Regulated AAs
13	Wang X., 2018 [7]	44/34	-	UPLC-MS	Ala, Asp, Cys, Gly, Glu, Gln, His, Ile, Met, Pro, Phe, Ser, Tyr, Val	Arg
14	Jasbi P., 2018 [6]	102/99	St I—24, St II—42, St III—36	UPLC-MS	Asp	Pro
15	Eniu D.T., 2018 [28]	30/26	St I—3, St II—17, St III—10	HPLC-MS	-	Tyr, Arg, Ala, Ile, Trp, Leu
16	Cala M.P., 2018 [33]	29/29	St I—3, St II—15, St III—11 Colombian Hispanic data set	GC-MS	Val, Ala, Ile, Ser, Glu, 4-Hydroxyproline	-
17	Park J., 2019 [53]	40/30	St I—15, St II—15, St III—10	HPLC-MS	5-oxoproline, Phe, Ile + Leu	-
18	Li L., 2020 [54]	105/35	St I—65, St II—40	NMR spectroscopy	Leu, Phe	Arg, Glu, Lys, Tyr, His
19	Politi C., 2021 [55]	38/10	-	GC-MS	Glu, Ile, Leu, Phe, Pro, Ser	Cys
20	An R., 2022 [56]	75/20	St I—31, St II—33, St III—11	UPLC-MS	Gln, Arg	Glu, Asp, Cys
21	Baranovicova E., 2022 [57]	50/46	St I + II—50	NMR spectroscopy	-	Leu, Ile, Val, Ala, His
22	Han X., 2022 [58]	30/30	THPMЖ	MALDI-TOF-MS	-	Ala, Ser, Pro, Val, Thr, His, Phe, Arg, Tyr, Trp
23	Santaliz-Casiano A., 2023 [59]	103/150	ER + breast cancer African American (AA) and non-Hispanic White (NHW) data sets	GC-MS	-	Arg (AA), His (AA), Met (AA)
24	Panigoro S.S., 2023 [60]	29/28	St I—13, St II—13, St III—3	HPLC	Cys	Glu, His, Orn, Thr, Tyr, Val

Note: HPLC—high performance liquid chromatography; GC-MS—gas chromatography coupled to a mass spectrometer; MALDI-TOF-MS—matrix-assisted laser desorption/ionization time-of-flight mass spectrometry; HPLC-ESI-MS—high performance liquid chromatography–electrospray ionization-mass spectrometry; UPLC-MS—ultra-performance liquid chromatography coupled to a mass spectrometer; UPLC-QTOF-MS—ultra-performance liquid chromatography coupled with quadrupole/time-of-flight mass spectrometry; NMR—nuclear magnetic resonance spectroscopy.

For example, Kubota et al. showed that in breast cancer, the total content of amino acids (TAA) in plasma increases (3035 ± 118 vs. 2529 ± 115 nmol/mL), as do the contents of all major groups of amino acids: essential amino acids (EAAs), branched-chain amino acids (BCAAs), aromatic amino acids (AAAs), and gluconeogenic amino acids (GAAs) [44]. Similarly, a number of studies have also noted an increase in the amino acid content compared to healthy controls [50,61]. According to other data, in patients with breast cancer, a decrease in the plasma levels of branched-chain amino acids (BCAAs) was observed [57]. Lai et al. [62] showed that a decrease in the content of seven amino acids (Ala, His, Thr, Arg, Pro, Glu, and Gly) occurs more than six times more often than their increase. In this regard, it is noteworthy that these seven amino acids play more important roles than others during changes in the protein metabolism in cancer patients. In a review by Yang et al., an analysis of the most common biomarkers of breast cancer was carried out [63]. Based on clinical metabolic studies, Tyr and Ala shared the highest frequency, indicating that they may be sensitive metabolites in the diagnosis of breast cancer. Other amino acids are mentioned less frequently, and the frequency of occurrence decreases in the following order: Tyr (6/6), Ala (6/5), Glu (6/4), Val (4/6), Phe (5/4), Gln (4/5), Lys (3/6), Ile (4/4), His (3/4), Gly (4/2), Arg (3/3), Asn (4/2), Pro (3/3), Ser (5/1), Leu (2/4), Trp (1/5), Thr (2/3), Asp (3/1), Orn (2/2), and Cys (2/2). It should be noted that in different studies, the level of amino acids changes in different directions: the first digit corresponds to up-regulation, and the second digit corresponds to down-regulation [63]. Thus, the existing data from the literature provide conflicting information about the nature of changes in the content of amino acids in breast cancer.

In different studies, both the contents of individual amino acids and the complete amino acid profile of plasma and/or serum were evaluated. It has been shown, for example, that the arginine levels were significantly lower and the ornithine levels were significantly higher in breast cancer patients than in control patients [64]. Differences between breast cancer cases and controls were observed for the levels of Trp, Thr, Ala, Gly, Pro, Ile, Leu, and Val [34]. Miyagi et al. observed significant changes in the amino acid profiles between cancer patients and controls. They found that the concentrations of Thr, Pro, Ser, Gly, Ala, and Orn were significantly higher, but the concentrations of Gln, Trp, His, Phe, and Tyr were significantly lower compared to the control. Compared with the data from Lai et al., they found similarities in decreasing His and Gln levels and increasing Pro and Ala levels in breast cancer patients. Budczies et al. demonstrated that changes in the amino acid metabolism were associated with at least a 1.9-fold increase in 16 amino acids in breast cancer compared with healthy breast tissue [50,65].

The summarized results on the amino acid composition of serum/plasma in breast cancer compared with healthy controls are shown in Table 1.

To compare the change in the contents of individual amino acids in breast cancer according to different authors, we compared the relative contents of amino acids in relation to healthy controls (Table 2). It can be seen that a significant change in the concentration is rarely observed, for example, an increase in the contents of Arg [52,56], Thr [52], and Ser [33] by 2 times, a decrease in the content of Asp by 1.5 times [46,56], etc. In most cases, the change in concentration is up to 10–15% compared to the control. We calculated the average change in concentration and the interval of variation for each amino acid and plotted them on a diagram (Figure 1). Thus, it is possible to assess the nature of changes in the concentrations of individual amino acids in breast cancer.

Table 2. Fold change BC/control according to different authors.

Amino acid	Kubota, 1992 [44]	Proenza, 2003 [46]	Minet-Quinard, 2004 [47]	Vissers, 2005 [48]	Okamoto, 2009 [49]	Miyagi, 2011 [34]	Shen, 2013 [8]	Barnes, 2014 [51]	Gu, 2015 [52]	Wang, 2018 [7]	Eniu, 2018 [28]	Cala, 2018 [33]	Park, 2019 [53]	Baranovicova, 2022 [57]	An, 2022 [56]	Panigoro, 2023 [60]
Ala	1.96	0.93	1.11	0.94	1.06	1.08	0.82/0.80 *	1.16	0.93	1.11	0.59	1.25		0.80	0.92	0.88
Arg	1.33		0.86		0.84	0.98	0.90/0.86	0.91	1.78	0.81	0.32				2.35	0.99
Asn		1.54	0.95	0.91	0.97	0.99	0.84/0.82	1.16		1.23	0.65				0.91	
Asp		0.45	0.80					1.33	0.67		0.71				0.43	
Cys	0.58						0.93/1.01		1.04	1.31	0.97				0.66	1.35
Gln	0.73	1.16	1.05	0.92	1.00	0.97	0.94/1.00	1.08		1.53	0.77				1.21	
Glu		1.02	1.42	1.24	1.40		1.05/1.01		0.46	1.81	0.42	1.27			0.58	0.86
Gly		0.94		1.01	1.08	1.12	0.90/0.91	0.91	0.90	1.16	0.68				0.94	
His	1.08	1.04	1.17	1.04	0.95	0.97	0.88/0.91		0.87	1.20	0.61		0.99	0.63	0.93	0.75

Table 2. *Cont.*

Amino acid	Kubota, 1992 [44]	Proenza, 2003 [46]	Minet-Quinard, 2004 [47]	Vissers, 2005 [48]	Okamoto, 2009 [49]	Miyagi, 2011 [34]	Shen, 2013 [8]	Barnes, 2014 [51]	Gu, 2015 [52]	Wang, 2018 [7]	Eniu, 2018 [28]	Cala, 2018 [33]	Park, 2019 [53]	Baranovicova, 2022 [57]	An, 2022 [56]	Panigoro, 2023 [60]
Ile		0.94	1.00	0.93	0.85	1.02	0.93/0.94	0.72	0.93	1.11	0.55	1.21	1.32	0.82	1.12	0.96
Leu		1.00		0.81	0.94	1.00	0.94/0.93		0.95	1.15	0.64			0.85	0.95	0.88
Lys		1.04	1.07	1.02	0.99	1.03	0.91/0.86	1.19	0.92	0.98	0.62		0.94		0.84	
Met		1.01	1.04	0.87	0.93	0.99	0.88/0.87	1.08	1.05	2.02	0.67				1.36	
Phe		1.09	1.08	0.83	0.89	0.98	0.92/0.90	1.00	1.00	1.21	0.78		1.30		1.04	
Pro		1.07	0.85		0.97	1.12	0.85/0.87	0.95	0.51	1.19	0.59				1.17	
Ser		1.01	1.13	1.11	1.10	1.04	0.92/0.89	0.86	0.98	1.14	0.79	2.10			0.87	0.93
Thr	1.48	1.01	0.94	0.92	1.07	1.08	0.89/0.94		2.46	1.00	0.63		1.08		0.93	0.82
Trp				0.82	0.93	0.94	0.94/0.88			1.02	0.61		1.15		1.02	
Tyr		0.98	1.09	0.89	0.92	0.96	0.92/0.80	1.55	1.07	1.20	0.56				1.24	0.78
Val		0.93	1.02	0.79	1.01	1.01	0.95/0.94	0.90	0.93	1.15	0.68	1.36		0.85	0.96	0.79
Cit			1.23	0.97	0.90	1.01					0.47				1.04	
Orn	1.28	1.07	1.47	0.65	1.25	1.12		0.98			0.53					0.68

Note: *—ER + PR+/TNBC.

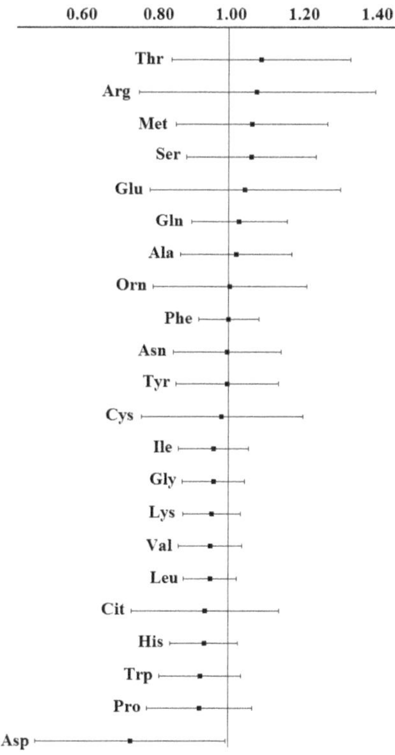

Figure 1. Fold change BC/control (data were averaged over [7,8,28,33,34,44,46–49,51–53,56,57,60]).

It can be seen that for some amino acids (Thr, Arg, Met, and Ser), an increase in concentration is more often observed in breast cancer compared with a healthy control, and for other amino acids, a decrease is observed (Asp, Pro, Trp, and His) (Figure 1). However, in each case, to assess the change in the amino acid profile, it is necessary to take into account the characteristics of the sample due to the high heterogeneity of breast cancer.

The construction of the ROC curve makes it possible to assess the classification accuracy for separating BC patients from healthy controls. According to Park et al. [53], for Lys, His, Thr, and Trp, the area under the curve varies in the range of 0.527–0.583; for Leu/Ile, the area is 0.728; for Phe, the area is 0.838; and for 5-hydroxyproline, the area is 0.968 on the training set. However, in the test set for 5-hydroxyproline, the area under the curve decreases to 0.744. With the simultaneous evaluation of many metabolites, which include amino acids, the accuracy of diagnosis increases. Thus, An et al. [56] showed that with the simultaneous evaluation of 47 metabolites using the random forest method, AUC values of 0.998 were obtained. Jasby et al. [6], using 30 metabolites, showed that AUC = 0.89 (95% CI: 0.85–0.93), sensitivity = 0.80, and specificity = 0.75. Baranovicova et al. [57] showed that AUC > 0.91 for 10 metabolites, while only for the BCAAs, AUC = 0.733. Only for the combination of 10 amino acids, Han et al. [58] showed that AUC = 0.91. Thus, individual amino acids have a low diagnostic value for the detection of breast cancer, while combinations of a number of amino acids with each other or with other metabolites provide a higher accuracy. However, in each specific case, it is necessary to check the data obtained during the construction of the model on a test sample in order to obtain an objective idea of the diagnostic capabilities of the algorithm.

2.2. Features of the Amino Acid Composition of Serum/Plasma in Different Molecular Biological Subtypes of Breast Cancer

In a number of studies, special attention was paid to the influence of the molecular biological characteristics of breast cancer on the amino acid profile of blood serum/plasma. Thus, Ala had a significant difference between ER-positive and ER-negative breast cancer [16,66]. Fan Y. identified a panel of eight potential low-molecular-weight biomarkers for diagnosing breast cancer subtypes, which included three amino acids: Pro, Ala, and Val [16]. Elevated Pro levels (FC = 1.217, p = 0.007) may indicate the suppression of proline oxidase in the human growth factor receptor 2 (HER2) positive group [67]. Ala was the most significantly reduced metabolite (FC = 0.544, p < 0.001) in the ER-positive participants compared to the ER-negative group [66]. Val was significantly increased in the HER2-positive groups compared to the HER2-negative groups, but was markedly decreased in the ER-positive groups compared to the ER-negative groups. Val anomalies indicated a violation of energy supply in the HER2-positive (fold change = 1.187, p = 0.002) and ER-positive (fold change = 0.682, p < 0.001) patients. The authors emphasized the clinical predictive potential of the identified eight biomarkers for breast cancer subtypes. An average prediction accuracy of 88.5% (95% CI 83.3–93.7%) was obtained for the training set and an average prediction accuracy of 85.6% (95% CI 80.9–90.1%) was obtained for the test set.

Han et al. [58] studied TNBC (tumor control) as well as the non-malignant pathologies of mammary glands. Four amino acids (Phe, Tyr, Ser, and Arg) were found to be more common in the triple-negative breast cancer (TNBC) group, while six amino acids (Pro, Trp, Val, Thr, Ala, and His) were more common in the benign group. The sensitivity, specificity, and accuracy of the classification model for identifying healthy controls and patients with TNBC were 91%, 100%, and 95%, respectively. Shen et al. observed eight amino acids whose levels were significantly lower in patients with triple-negative breast cancer than in healthy controls (Asp, Ala, His, Tyr, Trp, Met, Arg, and Pro), and two amino acids whose levels were significantly lower in patients with ER+/PR+ breast cancer than healthy individuals (Ala and His) [8].

Baranovicova et al. [57] showed that the levels of circulating metabolites are not associated with breast cancer molecular subtypes (luminal A/luminal B), histological findings, or stage. According to the authors, in the early stage of breast cancer, patients have common metabolic fingerprints in blood plasma, regardless of the degree, stage, or molecular subtype of breast cancer. However, statistically significant correlations were found between the level of tumor proliferation marker Ki-67 and circulating metabolites: Ala, Tyr, Gln, His, and Pro [57].

A significant decrease in the Trp concentration was observed in the plasma of luminal A, triple-negative, and HER2-positive breast cancer compared with healthy controls [68].

2.3. Racial Characteristics of the Serum/Plasma Amino Acid Profile in Breast Cancer

Breast cancer is associated with marked metabolic changes. However, these metabolic shifts in tumors may differ between stages, subtypes, and race [8,69]. Racial differences may not only influence metabolite levels, but also modify associations between metabolites and breast cancer. Racial differences in the incidence of breast cancer are well documented [70–72]. Compared to Caucasian American (CA) women, African American (AA) women tend to develop more aggressive tumors that are characterized by an earlier age at diagnosis, a higher mitotic index, and a lower prevalence of ER/PR expression, and subsequently have a lower survival rate. It has been suggested that genetic predisposition plays a role in racial/ethnic disparity in breast cancer [73,74]. For example, Gieger et al. found that up to 12% of the observed dispersion of the metabolic homeostasis of the human body could be explained by genetic variants [75].

In a study by Shen et al., when stratified by race, the difference in the plasma amino acid composition between healthy and sick patients was more evident in the AA participants than in the CA participants [8]. However, due to the small sample size, no significant difference remained in either the AA or CA participants after adjusting for multiple comparisons [8]. A study by Cala et al. described the metabolic features in breast cancer in Latin woman of Hispanic origin [33].

Santaliz Casiano et al. found that amino acid levels are significantly lower in AA breast cancer patients than in healthy individuals [59]. The pathways associated with energy metabolism are glycolysis, amino acid metabolism, and the TCA cycle, which dominate in AA women with ER+ tumors, potentially indicating the aggressiveness of their tumors [76]. In AA individuals, metabolites associated with aminoacyl-tRNA biosynthesis, Arg metabolism, branched amino acid metabolism, and His metabolism were differently distributed in the plasma in individuals with breast cancer. It is known that tumor cells need amino acids both as alternative fuel and for DNA synthesis, building new blood vessels and supporting their rapid growth and proliferation. Amino acids provide metabolic intermediates for epigenetic regulation [77]. In particular, Met levels were found to be lower in plasma samples from AA women who had breast cancer. Met is a methyl group donor for methylation and is a major contributor to epigenetic regulation [78,79]. Systemic or cellular metabolic changes affect the epigenetic landscape, which is important for ERα activity and the response to clinical drugs [80,81]. In AA women, poverty rates are correlated with the hypermethylation of cancer-associated pathways, including the glucocorticoid receptor, p53, estrogen-dependent breast cancer signaling, and cell proliferation [82]. Therefore, hypermethylation is a possible biological mechanism that may explain the poorer outcomes in AA women with live-caused breast cancer in areas of low socioeconomic status.2.4. Serum/Plasma Amino Acids in Breast Cancer Risk Assessment.

Amino acids are most often positively associated with breast cancer risk, and among them are the branched-chain amino acids Val and Leu, as well as Lys, Arg, Phe, and Gln [83]. They are also positively associated with the risk of His as a necessary precursor of histamine, the release of which is an early event in inflammatory responses and is a regulator of cell proliferation [84].

The branched-chain amino acids (BCAAs) Leu, Val, and Ile are dietary essential amino acids and are important metabolites that are involved in cell signaling pathways and muscle protein synthesis [85]. Elevated plasma BCAA concentrations are strongly positively correlated with body mass index and insulin resistance and are markers of a dysfunctional metabolism [86]. In premenopause, elevated levels of circulating BCAAs have been associated with a lower risk of breast cancer (Table 3) [86]. In contrast, among postmenopausal women, elevated levels of circulating BCAAs have been associated with an increased risk of breast cancer [87].

Table 3. Metabolites associated with breast cancer risk.

No.	Author, Year	BC/Control	Subgroup	AAs	HR (95% CI)
1	Nagata C. et al., 2014 [88]	350	premenopausal	Arg, Leu, Tyr, Asp	-
2	Lécuyer L. et al., 2018 [89]	206/396	-	Val ↑ Gln ↑ Lys + Creatine + Creatinine ↑ Lys + Arg ↑	1.83 (1.15–2.92) 1.61 (1.02–2.55) 1.84 (1.19–2.85) 1.62 (1.05–2.48)
3	His M. et al., 2019 [90]	1624/1624	-	Arg ↑ Asp ↑ Gln ↑ Gly ↑ His ↑ Lys ↑ Thr ↑	0.89 (0.80–0.99) 0.87 (0.80–0.95) 0.91 (0.84–0.99) 0.90 (0.83–0.97) 0.91 (0.84–0.99) 0.90 (0.83–0.98) 0.92 (0.85–0.99)
4	Zhang J. et al., 2020 [91]	735/735	-	Orn ↑	0.70 (0.53, 0.94)
5	Zeleznik O.A. et al., 2021 [87]	1997/1997	premenopausal postmenopausal	Ile ↑ Ile ↑	0.86 (0.65–1.13) 1.63 (1.12–2.39)
6	Jobard E. et al., 2021 [92]	791/791	premenopausal	His ↑ Orn ↑ Leu ↑ Gln ↑ Glu ↑	1.70 (1.19–2.41) 1.43 (1.06–1.95) 1.37 (1.01–1.86) 1.33 (1.00–1.78) 1.34 (1.00–1.79)
7	Stevens V.L. et al., 2023 [93]	1687/1983	-	Ser ↓ Asp ↓ Gln ↓	0.89 (0.83–0.96) 0.91 (0.84–0.97) 0.91 (0.85–0.98)

Note: ↑—amino acid concentration in breast cancer increases; ↓—concentration decreases.

Several studies have evaluated the effects of BCAAs on breast cancer risk with conflicting results, and only one study has assessed menopausal status [90,94,95]. A survival analysis showed that the expression of the catabolic BCAA gene is closely associated with long-term oncological outcomes [96]. High BCAA levels suppressed both tumor growth and breast cancer metastasis, demonstrating the potential benefits of increasing dietary BCAA intake during breast cancer therapy [97,98]. After stratification based on menopausal status, there was a significant inverse relationship between BCAA intake and the likelihood of postmenopausal breast cancer (RR = 0.22; 95% CI 0.13–0.39), although this significant association was not found in premenopausal breast cancer (RR = 2.57, 95% CI 0.51–12.73) [99].

Jobard et al. showed that in the premenopausal subgroup, breast cancer can be predicted by several risk-related metabolites, including His with moderate accuracy (AUC = 0.61, 95% CI: 0.49–0.73) (Table 3) [92]. Predictive power or significant metabolites have not been found in general or in postmenopausal women. Lecuyer et al. found high levels of Gln, Arg, Lys, and Val to be closely associated with a higher risk of BC [89]. Another study reported that higher levels of Gln/isoglutamine, Val/norvaline, Trp, and Phe were related to an increased risk of BC [95]. A study by Nagata et al. showed that the plasma levels of certain specific amino acids, such as Arg, Leu, Tyr, and Asp, were associated with endogenous sex hormone levels, sex-hormone-binding globulin (SHBG), or insulin-like growth factor (IGF-1), as determined by biomarkers of breast cancer risk [88].

Stevens et al. [93] found that Gln was associated with a reduced risk of breast cancer, as in studies involving pre- and postmenopausal women in EPIC [93], but other studies produced conflicting results. Gln was associated with an increased risk in the SU.VI.MAX [95] and E3N [92] cohorts, where the association was limited to premenopausal women. Mrowiec et al. [100] showed that the serum Asn concentration was significantly higher in late-detected breast cancer compared with early-detected breast cancer only in a subgroup of older women. The authors found that several metabolic pathways, including BCAA degradation and glutathione metabolism, differed between younger and older women with a cut-off point of 45 years. The overrepresentation of pathways associated with metabolites that differentiate early- and late-diagnosed cancer (His and Ala metabolism) was observed only in a subgroup of older women, which, once again, confirmed the age-related nature of the metabolic features associated with the risk of developing breast cancer. The results of His et al. [90] show that the concentration of amino acids in plasma is inversely related to the risk of breast cancer; however, these results contradict those of other studies, where, on the contrary, an increase in the concentration of amino acids contributed to an increase in the risk of breast cancer (Table 3).

Risi et al. conducted a study on the risk of breast cancer recurrence [101]. Higher levels of Phe and lower serum levels of Leu, Ile, Val, and His were shown to be associated with the presence of advanced breast cancer. Despite the fact that no statistically significant correlation was found between the levels of individual metabolites and the risk of recurrence in the entire cohort of breast cancer patients, Phe was significantly associated with advanced breast cancer that increased with breast cancer recurrence. However, the authors proposed a metabolomics model that demonstrated strong predictive power. Patients with a "high risk" had a significantly increased likelihood of disease recurrence compared to patients with a low-risk metabolomics fingerprint (HR = 3.42, 95% CI 1.58–7.37, $p < 0.001$). These results are consistent with those of other studies [102].

The AminoIndex Cancer Screening (AICS) technology has been described and used as a new cancer risk calculation method for early cancer diagnosis [34,49,103–105]. AICS (breast) detects breast cancer by detecting abnormal plasma concentrations of Thr, Ala, Orn, His, and Trp [106]. In breast cancer, the Thr, Ala, and Orn concentrations are elevated, while the His and Trp concentrations are reduced, and a multivariate analysis shows an overall accuracy of 88.2% [107].

Thus, it was shown that the risk of breast cancer is associated with the amino acid composition of blood plasma/serum; however, the patterns identified by the authors are closely related to the characteristics of the sample, menopausal status, and age. Nevertheless, according to most authors, Gln/Glu, Asp, Arg, Leu/Ile, Lys, and Orn have the greatest significance in assessing the risk of breast cancer.

3. Amino Acid Metabolism in Breast Cancer

3.1. Metabolic Features of Breast Cancer

The amino acid metabolism plays a critical role in the proliferation of breast cancer cells [108,109]. The intake and use of amino acids helps to support the growth of cancer cells [110–112]. A number of studies have shown that a decrease in the content of a number of amino acids may be the result of their excessive consumption or preferential use to support the uncontrolled growth of breast cancer cells [5,26,37,111–114].

The Glu/Gln, Ala, Asp, and Arg metabolisms were the most important biosynthetic pathways in breast cancer, suggesting extensive metabolic disturbances during breast cancer progression [115,116]. Huang S. et al. showed that the alanine, aspartate, and glutamate pathways are critical biological pathways for the early diagnosis of breast cancer [30]. Most of the metabolites in these three metabolic pathways were reduced in breast cancer patients compared to healthy controls. Thus, Ala promotes the proliferation of breast cancer cells, which indicates the potential role of Ala as a marker for cancer diagnosis [52]. The down-regulation of Gln indicated that Glu can accumulate in the body, which contributes to the development of breast cancer due to the increased proliferation of mammary epithelial cells [117] due to ATP production and nucleotide biosynthesis [118]. Moreover, Glu activation via glutaminolysis can maintain the citric acid cycle [119]. The metabolism of Gln is closely related to the process of providing energy to the cancer cell. Gln is transported into cancer cells by means of multiple transporters [116,120]. It was previously shown that the expressions of the Gln transporters ASCT2, SNAT1, SNAT2, and SNAT5 are increased in tumor tissue [121]. On the other hand, the inhibition of ASCT2 reduces the growth of TNBC [122]. It is known that the reductive metabolism of Gln supports tumor growth under conditions of hypoxia, mitochondrial dysfunction [123], and in an environment with a low nutrient content [124]. It was noted that Gln/Glu reversibility decreased in MCF-7 cells, indicating that breast cancer cells may be partially associated with irreversible glutaminase [125]. The Gln/Glu ratio can be used as a biomarker for the diagnosis of breast cancer [126]. The enzyme glutaminase I (GLS-I), which converts Gln to Glu via glutaminolysis, can be considered as a target for breast cancer therapy [127].

The change in the Gln level can be reflected in fluctuating levels of Ala and Asp due to the abnormal transport of ammonia. Higher histidine decarboxylase activity may lead to a decrease in the His levels [128], and low concentrations of His may be associated with increases in Asp and Glu, which can be converted to oxaloacetic and α-ketoglutaric acids, which are intermediates of the tricarboxylic acid cycle. Asp has been shown to be more sensitive to breast cancer [129]. Therefore, an increase in the use of Asp by BC cells can lead to a decrease in the levels of Asp and oxaloacetate in the blood. It is noteworthy that, as a transamination product of aspartic acid, Asp has an important effect on breast cancer metastasis [130].

The dysregulation of branched-chain amino acid (BCAA) metabolism, including Leu, Ile, and Val, has been reported to be associated with specific cancer phenotypes. BCAAs can inhibit tumor growth and metastasis [97], so changes in the BCAA levels can often reflect systemic changes in cancer patients compared to healthy controls [131]. Plasma Arg, Pro, and Trp metabolites decreased in BC patients [6]. Huang et al. [30] revealed a decrease in the Ser and Thr levels in the serum of patients with breast cancer. Harvie et al. showed that a Tyr deficiency can lead to the stunting of breast cancer cells [132], and the inhibition of tumor growth was confirmed with diets that were low in Phe and Tyr in an animal study [133]. It is known that sharp metabolic shifts in the levels of choline and Pro are characteristic of metastatic breast cancer [134]. Breast cancer has been shown to be highly dependent on Arg [6]. It has been shown to enhance the immune response, both innate and adaptive, with the administration of Arg supplements [135]. Conversely, a decrease in the dietary Asn intake or the suppression of asparagine synthetase reduced breast cancer metastasis [130].

Many authors agree that the concentration of Trp in the plasma and serum of breast cancer patients is reduced [26,28,113]. Previously, it was shown that Trp indirectly promotes the degradation of the extracellular matrix and the invasion of cancer cells [136]. Two major enzymes catalyze Trp into metabolites of the kynurenine (Kyn) pathway: indolamine 2,3 dioxygenase (IDO1) and tryptophan 2,3 dioxygenase (TDO2) [137]. Kyn activates the aryl hydrocarbon receptor, which promotes the evasion of the cancer immune response by increasing IL-10 and suppressing immune activation cells [138]. Thus, with IDO1/TDO2 overexpression, increased Trp catabolism can lead to a decrease in its serum concentration and an accumulation of Kyn metabolites [139].

Ser is transported into cells by means of ASCT1, which is highly expressed in breast cancer [140]. An increase in the rate of tumor cell proliferation depends on the presence of extracellular Ser. In an experiment with mice, it was shown that decreases in the Ser and Gly levels suppresses tumor growth and increases the lifespan [141]. It is interesting to note that, depending on the type of cancer, either Ser or Gly can contribute to the rapid proliferation of cancer cells [142].

Budhu A. et al. showed that a decrease in plasma Cys was inversely associated with an increase in Cys in breast cancer tissues [13], suggesting that breast cancer cells use more Cys. Cys is involved in the redox reaction of glutathione. With an increase in the concentration of Cys, oxidative damage and the production of free radicals also increase, which leads to gene mutation [143]. On the other hand, Cys can be considered as a substrate for the production of hydrogen sulfide, which stimulates cellular bioenergetics [144]. It has been established that the Cys-associated metabolic pathway of cysteinyl leukotrienes (CysLT) is also closely associated with cancer in enhancing the ability to survive and proliferate cancer cells [145].

3.2. Amino Acid Metabolism as a Target for Breast Cancer Imaging

Elevated levels of Met, Gln, Cys, Trp, Tyr, and other amino acids in tissues have been noted in many malignancies, including breast cancer [146]. Cancer cells with the up-regulation of amino acid metabolism stimulate the increased transport of amino acids into the cell [147]. The increased consumption of amino acids and the overexpression of amino acid transporters in malignant tumors make radiolabeled amino acids attractive imaging agents [114,148].

Multiple amino acid transporters have been demonstrated to be up-regulated in breast cancer families, including L-type amino acid transporter (LAT1), ASC transporter 2 (ASCT2), $ATB^{0,+}$, SNAT1, and xc- [149,150]. LAT1 is required for the transport of large neutral amino acids and is overexpressed in many types of malignancies, including breast cancer [151]. The expression of ASCT2 also has prognostic associations in breast cancer [120]. The xc-system transporter, which mediates cysteine uptake, is up-regulated in some breast cancer tumors [152]. The two amino acid transporters SLC7A5 and SLC7A11 are considered essential for the growth of breast cancer cells in a cell-dependent manner [153].

Met is a naturally occurring large neutral amino acid that is readily labeled with the 11C radioactive isotope. ^{11}C-methionine serves as a metabolic marker for Met uptake via L-type amino acid transporters. PET with ^{11}C-methionine makes it possible to visualize primary and metastatic lesions, as well as to predict the response to breast cancer treatment [154]. Limitations of ^{11}C-methionine include the ability to detect metastases only in the liver and bone marrow and its relatively short half-life (20 min). On the basis of Met, an MR contrast agent based on Met-MSN-Gd^{3+} was developed that targets methionine receptors, which are overexpressed in tumor cells (MSN—Mesoporous Silica Nanoparticles) [155,156].

Trans-1-amino-3-^{18}F-fluorocyclobutanecarboxylic acid (anti-^{18}F-FACBC, also known as ^{18}F-fluciclovine) is a synthetic analogue of Leu, which is transported into the cell by the ASCT2 transporter with the additional involvement of LAT1 [157]. Uptake by cells is most similar to uptake of the natural amino acid Gln [158]. ^{18}F-fluciclovine can visualize breast lesions [159,160], axillary lymph node metastases [161], and previously undetected extra-axillary nodular metastases [161,162]. Preclinical studies have demonstrated significant success with 18F-fluciclovine in detecting bone metastases.

(4S)-4-(3-[^{18}F]fluoropropyl)-l-glutamate (BAY 94-9392, also known as [^{18}F]FSPG), is a synthetic amino acid analogue of SLC7A11 [152,163]. Histological or molecular subtypes of breast cancer may affect ^{18}F-FSPG uptake. ^{18}F-5-fluoroaminosuberic acid, a synthetic amino acid substrate of SLC7A11, exhibited tumor uptake in three breast cancer cell lines (MDA-MB-231, MCF-7, and ZR-75-1), with the highest uptake observed in MDA-MB- 231, the TNBC cell line [164].

Several radioactively labeled Tyr analogues have been developed for tumor imaging, including L-[1–^{11}C]tyrosine [165]. Technetium-labeled tyrosine analogs have also been synthesized in high yield and can distinguish malignant breast neoplasms from benign breast tissues [166]. O-(2-^{18}F-fluoroethyl)-l-tyrosine (^{18}F-FET) is a synthetic amino acid transported by SLC7A5 [167]. Animal experiments using rats and mice have shown that ^{18}F-FET can distinguish between inflammation and malignancy [168,169] as well as in 75% of breast cancer patients.

In preclinical studies, several propanoic acid derivatives have demonstrated good tumor uptake in human breast cancer cells as well as mouse tumor xenografts [170]. 2-Amino-5-(4-[^{18}F]fluorophenyl)pent-4-ynoic acid ([^{18}F]FPhPA) is a synthetic amino acid that targets SLC1A5 and SLC7A5. A high uptake of the radiopharmaceutical [^{18}F]FPhPA was detected in the mouse breast cancer cell line EMT6 by PET [171]. Trp analogs, primarily using L-type amino acid transport, have been developed by several research groups and have also been shown to be taken up by breast cancer cells in small animal studies [172]. An analogue of Leu, 5-[^{18}F]fluoroleucine, was synthesized with primary transport via LAT1 [173]. Unlike ^{18}F-fluciclovine, the absorption of 5-[^{18}F]fluoroleucine gradually increases over time. The glutaminolysis pathway is very active in many malignancies, including triple-negative breast cancer. Preclinical work with the Gln analog [^{18}F](2S,4R)4-fluoroglutamine demonstrated the ability of this indicator to track changes in the size of the cellular Gln pool and the glutaminolysis pathway after glutaminase inhibition [174].

Amino acid transporters can also be used to image tumors using single-photon emission computed tomography (SPECT). 3-[^{123}I]-α-methyl-l-tyrosine (IMT) is an artificial amino acid that is transported through SLC7A5 [175] and is also a suitable metabolic indicator for SPECT in extracranial tumors, including breast cancer [176]. Using IMT SPECT, primary and metastatic breast cancer, as well as tumor regression after radiotherapy, were detected and were consistent with the results of the clinical assessment [175]. ^{99m}Tc-labeled diethylenetriaminepentaaceticacid (DTPA-bis)-methionine scintymammography has shown 96% sensitivity and a 96% positive predictive value for the detection of breast cancer [177], and therefore may be an alternative to conventional SPECT using non-specific mitochondrial uptake.

4. Amino Acids in Potential Strategies for the Treatment of Breast Cancer

One of the therapeutic approaches to cancer treatment is aimed at changing the metabolism of the tumor [178,179].

Some essential amino acids (EAAs)—Trp, His, Met, and branched-chain amino acids—are directly associated with tumor growth and treatment resistance [180–183]. EAAs provide important raw materials for protein synthesis, influence the biological behavior of cells, and induce therapeutic resistance in breast cancer cells. Strekalova et al. [184] found that Met plays a critical role in maintaining the self-renewal and survival of cancer stem cells. Met restriction reduced the population of cancer stem cells in TNBC. Saito et al. [185] showed that Leu not only promotes cell proliferation in ER-positive breast cancer, but also participates in the mechanism of resistance to tamoxifen. In addition, essential amino acids, as important nutrients, mediate the development of an immunosuppressive tumor microenvironment in breast cancer. For example, kynurenine, a product of Trp metabolism, can inhibit T cell proliferation and differentiation, leading to immune evasion and tumor progression in breast cancer [186]. Interestingly, both Trp and kynurenine were lower in plasma in breast cancer patients compared to controls, especially in women with estrogen-receptor-negative and advanced breast cancers [187]. These results show an intrinsic relationship between EAA metabolism and the immune microenvironment in breast cancer. Zhao et al. [188] found that the ratio of SLC7A5 to SLC7A8 (SSR) is significantly correlated with the level of EAA and the metabolic activity of EAA in breast cancer, and therefore, the SSR index can be used as a biomarker to assess the degree of metabolism of EAA in breast cancer. In addition, breast cancer patients with a high EAA metabolism had a shorter overall survival time, a higher PD-L1 expression, and higher T-regulatory cell infiltration, indicating that a high EAA metabolism was associated with a poor prognosis and immunosuppression in breast cancer.

Zhang, L. showed that the level of BCAA in the plasma and cancer tissues of BC patients was increased, which was accompanied by an increase in the expression of BCAA catabolism enzymes [189]. The stimulation of BCAA catabolism by modulating BCAT1 enhanced the growth and formation of colonies of breast cancer cells. BCAT1 promoted mitochondrial biogenesis and enhanced mitochondrial function by promoting ATP production and protection against oxidative stress by activating mTOR signaling, but not AMPK or SIRT1. The inhibition of mTOR by rapamycin neutralizes the role of BCAT1 in mitochondrial function and cancer cell growth.

For some cancer subtypes, such as TNBC, there is no specific therapy, resulting in a poor prognosis that is associated with invasion and metastasis.

Under physiological conditions, Gln is transported into cells by many transporters such as Ala, Ser, Cys-preferential transporter 2 (ASCT2, also known as SLC1A5), and L-type amino acid transporter 1 (LAT1, also known as SLC7A5). In TNBC, both ASCT2 and LAT1 are overexpressed [122,190]. Compared to other subtypes of breast cancer, TNBC is more Gln dependent and sensitive to glutaminolysis-targeted therapy due to glutaminase overexpression (GLS) [191,192], which is associated with high-grade metastatic breast cancer. Several small-molecule GLS inhibitors, such as CB-839, have been developed to combat the dysregulation of glutaminolysis [193]. Other combination therapies, such as the combination of GLS inhibition and bevacizumab (an anti-angiogenesis monoclonal antibody targeting VEGF), also show antitumor effects on TNBC [194]. Ginsenoside was described to effectively inhibit TNBC by suppressing Gln uptake and Glu production by down-regulating glutaminase 1 (GLS1) expression [195]. Ginsenoside treatment further reduced cellular ATP production, decreased amino acid utilization associated with Gln metabolism, and induced glutathione depletion and reactive oxygen species accumulation, which consequently triggered apoptosis in TNBC. Morotti et al. showed that the knockdown of the Gln transporter SLC38A2, which was identified as a highly expressed amino acid transporter in six breast cancer cell lines [196], reduced Gln uptake, inhibited cell growth, induced autophagy, and resulted in the production of reactive oxygen species in a subgroup of Gln-sensitive cell lines. A high expression of the SLC38A2 protein was associated with poor breast cancer survival in a large group of patients ($p = 0.004$), especially in TNBC ($p = 0.02$).

Increased Glu production by GLS may support the uptake of exogenous cystine via the cystine/glutamate antiporter xCT to maintain redox balance. As a clinically approved anti-inflammatory drug, sulfasalazine (SASP) was found to inhibit xCT activity and retard the growth of TNBC [197]. By immunizing mice with a DNA-based vaccine expressing the xCT protein, the cell surface immunotargeting of the xCT antigen effectively attenuated tumor growth and lung metastasis, and increased chemosensitivity to doxorubicin [198]. Additionally, virus-like particle immunotherapy was developed, which elicits a stronger humoral response against xCT [199].

In addition to Gln and Cys, TNBC cells are also somewhat dependent on the availability of several other amino acids such as Met, Asp, and Arg, suggesting that restricting these amino acids may have a therapeutic effect [200]. The depletion of either Met or Gln can increase the cell surface expression of the pro-apoptotic TNF-related apoptosis-inducing ligand receptor-2 (TRAIL-R2) and increase the sensitivity of TNBC cells to TRAIL-induced apoptosis [130,201]. Met deprivation in the diet increases cellular susceptibility to lexatumumab, an agonistic monoclonal antibody targeting TRAIL-R2, and reduces the rate of lung metastasis [202]. In addition, many tumor and stem cells depend on the biosynthesis of the universal methyl donor S-adenosylmethionine from the exogenous Met via methionine adenosyltransferase 2α (MAT2A) to maintain their epigenome [203,204]. A restriction on Met is enough to undermine the ability of TNBC to initiate a tumor, which is in part due to the impaired formation of S-adenosylmethionine. The combination of methionine restriction and the MAT2A inhibitor cycloleucine has a synergistic antitumor effect [184]. Under normal physiological conditions, the serum levels of Asp are lower than in the mammary gland, making Asp bioavailability a key regulator of circulating tumor cells and the metastatic potential of breast cancer. The restriction of Asp intake by the suppression of asparagine synthetase, treatment with L-asparaginase, or the restriction of Asp intake in the diet inhibits breast cancer metastasis [201]. The depletion of Arg by recombinant human arginase (rhArg) leads to the apoptosis of TNBC cells via reactive oxygen species and induces adaptive autophagy, while blocking the flow of autophagy via autophagy-targeting drugs enhances rhArg cytotoxicity [205]. The deprivation of Arg by L-arginase impairs tumor growth, leading to cell death [206]. Much attention has been paid to epigenetic modifications caused by enzymes of protein arginine methyltransferases, in which methylate Arg make a great contribution to the process of breast carcinogenesis and tumor suppression [207] and are targets for many types of cancer [208].

Leu uptake is predominantly mediated by the L-type amino acid transporter (LAT) family, a group of four Na+-independent transporters (LAT1, SLC7A5; LAT2, SLC7A8; LAT3, SLC43A1; and LAT4, SLC43A2) with an affinity for branched and neutral amino acid transporters [130]. Glutamine transport is largely mediated by Ala, Ser, and Cys-preferential transporter 2 (ASCT2; SLC1A5). The authors proposed using the ASCT2 inhibitor, benzylserine (BenSer), to doubly inhibit Gln [209] and Leu [210] uptake. It has been shown that a double inhibition with the pharmacological inhibitor BenSer can reduce the growth of breast cancer cells and limit the progression of the cell cycle.

Amino acid deprivation (AADT) is becoming a promising strategy for the development of new therapeutic agents against cancer [211]. The rapid growth of tumors leads to a decrease in the expression of certain enzymes, which leads to the auxotrophy of some specific amino acids. Amino acid depletion selectively inhibits tumor growth because normal cells can synthesize amino acids through their normal machinery. The enzymes used in AADT are primarily obtained from microbes due to their easy availability. Thus, the deprivation of Gln leads to a decrease in cell proliferation and cell death in breast cancer cell lines [212].

Endocrine therapy is the standard treatment for estrogen-receptor-positive (ER+) breast cancer, but 40% of women experience a recurrence of the disease during therapy. A general analysis of transcription in cells revealed a suppression of the neutral and basic amino acid transporter SLC6A14, which is regulated by the increased expression of miR-23b-3p, which leads to impaired amino acid metabolism [213]. This altered cellular amino acid metabolism is supported by autophagy activation and the increased import of acidic amino acids (Asp and Glu) mediated by the SLC1A2 transporter. Targeting these amino acid metabolic dependencies increases the sensitivity of cells to endocrine therapy.

Anticancer agents delivered to cancer cells often exhibit multidrug resistance due to a displacement of the agents. One way to solve this problem is to increase the accumulation of anticancer agents in cells with the help of amino acid transporters. Val-lapatinib and Tyr-lapatinib were newly synthesized by adding Val and Tyr fragments, respectively, to the parent anticancer agent lapatinib. Val-lapatinib and Tyr-lapatinib demonstrated enhanced anticancer activity compared to parental lapatinib in various cancer cell lines (MDA-MB-231 and MCF7) [214]. Both Val-lapatinib and Tyr-lapatinib, but not the parent lapatinib, inhibit glutamine transport in MDA-MB-231 and MCF7 cells, suggesting the involvement of amino acid transporters. Thus, amino acid transporters can be effective drug delivery targets to increase the uptake of anticancer agents, leading to one method of overcoming multidrug resistance in cancer cells.

Mello-Andrade et al. studied the effect of ruthenium(II) complexes associated with the amino acids methionine (RuMet) and tryptophan (RuTrp) on the induction of cell death, clonogenic survival, the inhibition of angiogenesis, and the migration of MDA-MB-231 cells [215]. The study also showed that RuMet and RuTrp complexes induce cell cycle arrest and the apoptosis of MDA-MB-231 cells, as evidenced by an increase in the number of annexin V-positive cells, p53 phosphorylation, caspase 3 activation, and poly(ADP-ribose) polymerase cleavage. RuMet and RuTrp complexes act directly on breast tumor cells, leading to cell death and suppressing their metastatic potential; this reveals the potential therapeutic effect of these drugs.

The use of various drugs based on platinum nanoparticles leads to a disruption of the amino acid metabolism, a disruption of tRNA aminoacylation, and protein synthesis [216]. Mitrevska et al. [217] analyzed the effect on amino acid metabolism in MDA-MB-231 cells upon treatment with cisplatin, PtNP-10, and PtNP-40, and revealed a marked contrast between the effects of cisplatin and PtNP. The results indicate a higher sensitivity of MDA-MB-231 cells to PtNP compared to cisplatin, since the increase in the number of amino acids was associated with the degree of insensitivity to various chemotherapeutic agents [217].

In general, combinations of anticancer drugs and amino acids can improve the intra-tumoral distribution of the active substance and increase its bioavailability. In particular, amino acid-based poly(ether urea ester) (AA-PEUU), as a nanocarrier for the systemic delivery of gamboginic acid, demonstrates the effectiveness of engineered AA-PEUU nanocarriers with custom structures and universal customization for the systemic delivery of therapeutics in the treatment of TNBC [218].

Plasma amino acid analysis can be used to monitor treatment progress. So, Minet-Quinard et al. [47] showed that the plasma levels of Ser and Glu returned to normal six months after the surgical removal of the tumor. Dunstan et al. [219] considered the changes in the amino acid homeostasis during radiation therapy. The urinary histidine and alanine levels were shown to be elevated prior to radiotherapy, while the Thr, Met, Ala, Ser, Asp, and Gln levels were higher after 5 weeks of radiotherapy. Many complications such as cachexia, anorexia, and fatigue occur in the treatment of problems associated with breast cancer, and many studies have considered the addition of amino acids with BCAAs [220] or a single amino acid or its derivative [221] to reduce the effects of treatment. Li et al. provided preliminary data to support the correction of the Trp metabolism for the treatment of neuropsychiatric symptoms [222].

Changes in the concentration of metabolites may also be useful in predicting the overall survival of patients with breast cancer. Thus, two metabolites differ significantly depending on the previous therapy: Met and Ser [223]. The blood Met levels were higher in the patients treated with anti-Her2 therapy, while the Ser levels were lower in the patients treated with endocrine therapy alone. Patients with TNBC were previously shown to have higher Ser levels, while patients with luminal cancer A, on the contrary, had low blood Ser concentrations [224], which is consistent with the therapy and Ser accumulation with anti-Her2 therapy [225]. In addition, Possemato et al. found an increased flux of Ser synthesis in patients with estrogen-negative breast cancer, which is also associated with a poor 5-year survival [226].

5. Conclusions

A fairly large number of studies have been devoted to the study of amino acid metabolism in breast cancer. At the same time, both the contents of individual amino acids and the combinations of amino acids with each other or with other metabolites determined in the course of obtaining the metabolomic profiles of biological fluids were determined. We have shown that for some amino acids (Thr, Arg, Met, and Ser) an increase in concentration is more often observed in breast cancer compared with a healthy control, and for other amino acids, there is a decrease (Asp, Pro, Trp, and His). However, the amino acid profile must be analyzed while taking into account the high heterogeneity of breast cancer, as well as age and race. The accuracy of the diagnosis using amino acids depends on the number of metabolites in the algorithm and varies from 52 to 98%. The contents of amino acids in biological fluids are used to assess the risk of breast cancer; however, the identified patterns are closely related to the characteristics of the sample, menopausal status, and age. According to most authors, Gln/Glu, Asp, Arg, Leu/Ile, Lys, and Orn have the greatest significance in assessing the risk of breast cancer. An analysis of the changes in the amino acid metabolism in breast cancer is a promising strategy for developing new therapeutic agents, monitoring the treatment process, correcting complications after treatment, and evaluating the survival rates. This, once again, emphasizes the high importance of research in this area.

Author Contributions: Conceptualization, L.V.B. and D.V.S.; methodology, L.V.B. and D.V.S.; software, I.A.G.; validation, L.V.B., I.A.G. and D.V.S.; formal analysis, I.A.G.; resources, L.V.B.; data curation, I.A.G. and D.V.S.; writing—original draft preparation, I.A.G. and D.V.S.; writing—review and editing, L.V.B.; visualization, I.A.G.; supervision, D.V.S.; project administration, L.V.B.; funding acquisition, L.V.B. All authors have read and agreed to the published version of the manuscript.

Funding: This research was funded by the Russian Science Foundation, grant number 23-15-00188; https://rscf.ru/project/23-15-00188/, accessed on 13 September 2023.

Institutional Review Board Statement: Not applicable.

Informed Consent Statement: Not applicable.

Data Availability Statement: Not applicable.

Conflicts of Interest: The authors declare no conflict of interest.

References

1. Siegel, R.L.; Miller, K.D.; Wagle, N.S.; Jemal, A. Cancer statistics, 2023. *CA Cancer J. Clin.* **2023**, *73*, 17–48. [CrossRef] [PubMed]
2. Sung, H.; Ferlay, J.; Siegel, R.L.; Laversanne, M.; Soerjomataram, I.; Jemal, A.; Bray, F. Global cancer statistics 2020: GLOBOCAN estimates of incidence and mortality worldwide for 36 cancers in 185 countries. *CA Cancer J. Clin.* **2021**, *71*, 209–249. [CrossRef] [PubMed]
3. Li, S.; Zeng, H.; Fan, J.; Wang, F.; Xu, C.; Li, Y.; Tu, J.; Nephew, K.P.; Long, X. Glutamine metabolism in breast cancer and possible therapeutic targets. *Biochem. Pharmacol.* **2023**, *210*, 115464. [CrossRef]
4. Jiao, Z.; Pan, Y.; Chen, F. The Metabolic Landscape of Breast Cancer and Its Therapeutic Implications. *Mol. Diagn. Ther.* **2023**, *27*, 349–369. [CrossRef] [PubMed]
5. Jové, M.; Collado, R.; Quiles, J.L.; Ramírez-Tortosa, M.C.; Sol, J.; Ruiz-Sanjuan, M.; Fernandez, M.; de la Torre Cabrera, C.; Ramírez-Tortosa, C.; Granados-Principal, S.; et al. A plasma metabolomic signature discloses human breast cancer. *Oncotarget* **2017**, *8*, 19522–19533. [CrossRef]
6. Jasbi, P.; Wang, D.; Cheng, S.L.; Fei, Q.; Cui, J.Y.; Liu, L.; Wei, Y.; Raftery, D.; Gu, H. Breast cancer detection using targeted plasma metabolomics. *J. Chromatogr. B* **2019**, *1105*, 26–37. [CrossRef]
7. Wang, X.; Zhao, X.; Chou, J.; Yu, J.; Yang, T.; Liu, L.; Zhang, F. Taurine, glutamic acid and ethylmalonic acid as important metabolites for detecting human breast cancer based on the targeted metabolomics. *Cancer Biomarkers* **2018**, *23*, 255–268. [CrossRef]
8. Shen, J.; Yan, L.; Liu, S.; Ambrosone, C.B.; Zhao, H. Plasma metabolomic profiles in breast cancer patients and healthy controls: By race and tumor receptor subtypes. *Transl. Oncol.* **2013**, *6*, 757–765. [CrossRef]
9. Da Cunha, P.A.; Nitusca, D.; Canto, L.M.D.; Varghese, R.S.; Ressom, H.W.; Willey, S.; Marian, C.; Haddad, B.R. Metabolomic Analysis of Plasma from Breast Cancer Patients Using Ultra-High Performance Liquid Chromatography Coupled with Mass Spectrometry: An Untargeted Study. *Metabolites* **2022**, *12*, 447. [CrossRef]
10. Neagu, A.-N.; Whitham, D.; Bruno, P.; Morrissiey, H.; Darie, C.A.; Darie, C.C. Omics-Based Investigations of Breast Cancer. *Molecules* **2023**, *28*, 4768. [CrossRef]
11. Wei, Y.; Jasbi, P.; Shi, X.; Turner, C.; Hrovat, J.; Liu, L.; Rabena, Y.; Porter, P.; Gu, H. Early Breast Cancer Detection Using Untargeted and Targeted Metabolomics. *J. Proteome Res.* **2021**, *20*, 3124–3133. [CrossRef]
12. Dougan, M.M.; Li, Y.; Chu, L.W.; Haile, R.W.; Whittemore, A.S.; Han, S.S.; Moore, S.C.; Sampson, J.N.; Andrulis, I.L.; John, E.M.; et al. Metabolomic profiles in breast cancer: A pilot case-control study in the breast cancer family registry. *BMC Cancer* **2018**, *18*, 532. [CrossRef]
13. Budhu, A.; Terunuma, A.; Zhang, G.; Hussain, S.P.; Ambs, S.; Wang, X.W. Metabolic profiles are principally different between cancers of the liver, pancreas and breast. *Int. J. Biol. Sci.* **2014**, *10*, 966–972. [CrossRef]
14. Jobard, E.; Pontoizeau, C.; Blaise, B.J.; Bachelot, T.; Elena-Herrmann, B.; Trédan, O. A serum nuclear magnetic resonance-based metabolomic signature of advanced metastatic human breast cancer. *Cancer Lett.* **2014**, *343*, 33–41. [CrossRef] [PubMed]
15. Mishra, P.; Ambs, S. Metabolic Signatures of Human Breast Cancer. *Mol. Cell. Oncol.* **2015**, *2*, e992217. [CrossRef] [PubMed]
16. Fan, Y.; Zhou, X.; Xia, T.S.; Chen, Z.; Li, J.; Liu, Q.; Alolga, R.N.; Chen, Y.; Lai, M.D.; Li, P.; et al. Human plasma metabolomics for identifying differential metabolites and predicting molecular subtypes of breast cancer. *Oncotarget* **2016**, *7*, 9925–9938. [CrossRef] [PubMed]
17. Cui, M.; Wang, Q.; Chen, G. Serum metabolomics analysis reveals changes in signaling lipids in breast cancer patients. *Biomed. Chromatogr.* **2016**, *30*, 42–47. [CrossRef]
18. Gu, H.; Pan, Z.; Xi, B.; Asiago, V.; Musselman, B.; Raftery, D. Principal component directed partial least squares analysis for combining nuclear magnetic resonance and mass spectrometry data in metabolomics: Application to the detection of breast cancer. *Anal. Chim. Acta* **2011**, *686*, 57–63. [CrossRef]
19. Oakman, C.; Tenori, L.; Claudino, W.M.; Cappadona, S.; Nepi, S.; Battaglia, A.; Bernini, P.; Zafarana, E.; Saccenti, E.; Fornier, M.; et al. Identification of a serum-detectable metabolomic fingerprint potentially correlated with the presence of micrometastatic disease in early breast cancer patients at varying risks of disease relapse by traditional prognostic methods. *Ann. Oncol.* **2011**, *22*, 1295–1301. [CrossRef]
20. Tenori, L.; Oakman, C.; Claudino, W.M.; Bernini, P.; Cappadona, S.; Nepi, S.; Biganzoli, L.; Arbushites, M.C.; Luchinat, C.; Bertini, I.; et al. Exploration of serum metabolomic profiles and outcomes in women with metastatic breast cancer: A pilot study. *Mol. Oncol.* **2012**, *6*, 437–444. [CrossRef]
21. Asiago, V.M.; Alvarado, L.Z.; Shanaiah, N.; Gowda, G.A.N.; Owusu-Sarfo, K.; Ballas, R.A.; Raftery, D. Early detection of recurrent breast cancer using metabolite profiling. *Cancer Res.* **2010**, *70*, 8309–8318. [CrossRef]
22. Tenori, L.; Oakman, C.; Morris, P.G.; Gralka, E.; Turner, N.; Cappadona, S.; Fornier, M.; Hudis, C.; Norton, L.; Luchinat, C.; et al. Serum metabolomic profiles evaluated after surgery may identify patients with oestrogen receptor negative early breast cancer at increased risk of disease recurrence. Results from a retrospective study. *Mol. Oncol.* **2015**, *9*, 128–139. [CrossRef]

23. Wei, S.; Liu, L.; Zhang, J.; Bowers, J.; Gowda, G.A.N.; Seeger, H.; Fehm, T.; Neubauer, H.J.; Vogel, U.; Clare, S.E.; et al. Metabolomics approach for predicting response to neoadjuvant chemotherapy for breast cancer. *Mol. Oncol.* **2013**, *7*, 297–307. [CrossRef] [PubMed]

24. Zapater-Moros, A.; Díaz-Beltrán, L.; Gámez-Pozo, A.; Trilla-Fuertes, L.; Lumbreras-Herrera, M.I.; López-Camacho, E.; González-Olmedo, C.; Espinosa, E.; Zamora, P.; Sánchez-Rovira, P.; et al. Metabolomics unravels subtype-specific characteristics related to neoadjuvant therapy response in breast cancer patients. *Metabolomics* **2023**, *19*, 60. [CrossRef] [PubMed]

25. Torata, N.; Kubo, M.; Miura, D.; Ohuchida, K.; Mizuuchi, Y.; Fujimura, Y.; Hayakawa, E.; Kai, M.; Oda, Y.; Mizumoto, K.; et al. Visualizing energy charge in breast carcinoma tissues by MALDI mass-spectrometry imaging profiles of low-molecular-weight metabolites. *Anticancer Res.* **2018**, *38*, 4267–4272. [CrossRef] [PubMed]

26. More, T.H.; RoyChoudhury, S.; Christie, J.; Taunk, K.; Mane, A.; Santra, M.K.; Chaudhury, K.; Rapole, S. Metabolomic alterations in invasive ductal carcinoma of breast: A comprehensive metabolomic study using tissue and serum samples. *Oncotarget* **2018**, *9*, 2678–2696. [CrossRef]

27. Bathen, T.F.; Geurts, B.; Sitter, B.; Fjosne, H.E.; Lundgren, S.; Buydens, L.M.; Gribbestad, I.S.; Postma, G.; Giskeødegård, G.F. Feasibility of MR metabolomics for immediate analysis of resection margins during breast cancer surgery. *PLoS ONE* **2013**, *8*, e61578. [CrossRef]

28. Eniu, D.T.; Romanciuc, F.; Moraru, C.; Goidescu, I.; Eniu, D.; Staicu, A.; Rachieriu, C.; Buiga, R.; Socaciu, C. The decrease of some serum free amino acids can predict breast cancer diagnosis and progression. *Scand. J. Clin. Lab. Investig.* **2019**, *79*, 17–24. [CrossRef]

29. Zhou, J.; Wang, Y.; Zhang, X. Metabolomics studies on serum and urine of patients with breast cancer using 1H-NMR spectroscopy. *Oncotarget* **2017**, *5*. [CrossRef]

30. Huang, S.; Chong, N.; Lewis, N.E.; Jia, W.; Xie, G.; Garmire, L.X. Novel personalized pathway-based metabolomics models reveal key metabolic pathways for breast cancer diagnosis. *Genome Med.* **2016**, *8*, 34. [CrossRef]

31. Yuan, B.W.; Schafferer, S.; Tang, Q.Q.; Scheffler, M.; Nees, J.; Heil, J.; Schott, S.; Golatta, M.; Wallwiener, M.; Sohn, C.; et al. A plasma metabolite panel as biomarkers for early primary breast cancer detection. *Int. J. Cancer* **2019**, *144*, 2833–2842. [CrossRef]

32. Suman, S.; Sharma, R.K.; Kumar, V.; Sinha, N.; Shukla, Y. Metabolic fingerprinting in breast cancer stages through (1)H NMR spectroscopy-based metabolomic analysis of plasma. *J. Pharm. Biomed. Anal.* **2018**, *160*, 38–45. [CrossRef] [PubMed]

33. Cala, M.P.; Aldana, J.; Medina, J.; Sanchez, J.; Guio, J.; Wist, J.; Meesters, R.J.W. Multiplatform plasma metabolic and lipid fingerprinting of breast cancer: A pilot control-case study in Colombian Hispanic women. *PLoS ONE* **2018**, *13*, e0190958. [CrossRef] [PubMed]

34. Miyagi, Y.; Higashiyama, M.; Gochi, A.; Akaike, M.; Ishikawa, T.; Takeshi, M.; Saruki, N.; Bando, E.; Kimura, H.; Imamura, F.; et al. Plasma free amino acid profiling of five types of cancer patients and its application for early detection. *PLoS ONE* **2011**, *6*, e24143. [CrossRef] [PubMed]

35. Murata, T.; Yanagisawa, T.; Kurihara, T.; Kaneko, M.; Ota, S.; Enomoto, A.; Tomita, M.; Sugimoto, M.; Sunamura, M.; Hayashida, T.; et al. Salivary metabolomics with alternative decision tree-based machine learning methods for breast cancer discrimination. *Breast Cancer Res. Treat.* **2019**, *177*, 591–601. [CrossRef]

36. Zhong, L.; Cheng, F.; Lu, X.; Duan, Y.; Wang, X. Untargeted saliva metabonomics study of breast cancer based on ultra-performance liquid chromatography coupled to mass spectrometry with HILIC and RPLC separations. *Talanta* **2016**, *158*, 351–360. [CrossRef]

37. Sugimoto, M.; Wong, D.T.; Hirayama, A.; Soga, T.; Tomita, M. Capillary electrophoresis mass spectrometry-based saliva metabolomics identified oral, breast and pancreatic cancer-specific profiles. *Metabolomics* **2010**, *6*, 78–95. [CrossRef]

38. Bel'skaya, L.V.; Sarf, E.A.; Loginova, A.I. Diagnostic Value of Salivary Amino Acid Levels in Cancer. *Metabolites* **2023**, *13*, 950. [CrossRef]

39. Cala, M.P.; Aldana, J.; Sanchez, J.; Guio, J.; Meesters, R.J.W. Urinary metabolite and lipid alterations in Colombian Hispanic women with breast cancer: A pilot study. *J. Pharm. Biomed. Anal.* **2018**, *152*, 234–241. [CrossRef]

40. Slupsky, C.M.; Steed, H.; Wells, T.H.; Dabbs, K.; Schepansky, A.; Capstick, V.; Faught, W.; Sawyer, M.B. Urine metabolite analysis offers potential early diagnosis of ovarian and breast cancers. *Clin. Cancer Res.* **2010**, *16*, 5835–5841. [CrossRef]

41. Chen, Y.; Zhang, R.; Song, Y.; He, J.; Sun, J.; Bai, J.; An, Z.; Dong, L.; Zhan, Q.; Abliz, Z. RRLC-MS/MS-based metabonomics combined with in-depth analysis of metabolic correlation network: Finding potential biomarkers for breast cancer. *Analyst* **2009**, *134*, 2003–2011. [CrossRef]

42. Hussain, A.; Xie, L.; Deng, G.; Kang, X. Common alterations in plasma free amino acid profiles and gut microbiota-derived tryptophan metabolites of five types of cancer patients. *Amino Acids* **2023**. [CrossRef]

43. Liu, N.; Shi, F.; Yang, L.; Liao, W.; Cao, Y. Oncogenic viral infection and amino acid metabolism in cancer progression: Molecular insights and clinical implications. *Biochim. Biophys. Acta (BBA)-Rev. Cancer* **2022**, *1877*, 188724. [CrossRef]

44. Kubota, A.; Meguid, M.M.; Hitch, D.C. Amino acid profiles correlate diagnostically with organ site in three kinds of malignant tumors. *Cancer* **1992**, *69*, 2343–2348. [CrossRef]

45. Cascino, A.; Muscaritoli, M.; Cangiano, C.; Conversano, L.; Laviano, A.; Ariemma, S.; Meguid, M.M.; Rossi Fanelli, F. Plasma amino acid imbalance in patients with lung and breast cancer. *Anticancer Res.* **1995**, *15*, 507–510. [PubMed]

46. Proenza, A.M.A.; Oliver, J.J.; Palou, A.A.; Roca, P.P. Breast and lung cancer are associated with a decrease in blood cell amino acid content. *J. Nutr. Biochem.* **2003**, *14*, 133–138. [CrossRef]

47. Minet-Quinard, R.; Van Praagh, I.; Kwiatkowski, F.; Beaujon, G.; Feillel, V.; Beaufrère, B.; Bargnoux, P.J.; Cynober, L.; Vasson, M.P. Pre- and postoperative aminoacidemia in breast cancer: A study vs. matched healthy subjects. *Cancer Investig.* **2004**, *22*, 203–210. [CrossRef] [PubMed]
48. Vissers, Y.L.J.; Dejong, C.H.C.; Luiking, Y.C.; Fearon, K.C.H.; von Meyenfeldt, M.F.; Deutz, N.E.P. Plasma arginine concentrations are reduced in cancer patients: Evidence for arginine deficiency? *Am. J. Clin. Nutr.* **2005**, *81*, 1142–1146. [CrossRef]
49. Okamoto, N.; Miyagi, Y.; Chiba, A.; Akaike, M.; Shiozawa, M.; Imaizumi, A.; Yamamoto, H.; Ando, T.; Yamakado, M.; Tochikubo, O. Diagnostic modeling with differences in plasma amino acid profiles between non-cachectic colorectal/breast cancer patients and healthy individuals. *Int. J. Med. Med. Sci.* **2009**, *1*, 1–8.
50. Poschke, I.; Mao, Y.; Kiessling, R.; Boniface, J. Tumor-dependent increase of serum amino acid levels in breast cancer patients has diagnostic potential and correlates with molecular tumor subtypes. *J. Transl. Med.* **2013**, *11*, 290. [CrossRef]
51. Barnes, T.; Bell, K.; DiSebastiano, K.M.; Vance, V.; Hanning, R.; Russell, C.; Dubin, J.A.; Bahl, M.; Califaretti, N.; Campbell, C.; et al. Plasma amino acid profiles of breast cancer patients early in the trajectory of the disease differ from healthy comparison groups. *Appl. Physiol. Nutr. Metab.* **2014**, *39*, 740–744. [CrossRef] [PubMed]
52. Gu, Y.; Chen, T.; Fu, S.; Sun, X.; Wang, L.; Lu, Y.; Ding, S.; Ruan, G.; Teng, L.; Wang, M. Perioperative dynamics and significance of amino acid profiles in patients with cancer. *J. Transl. Med.* **2015**, *13*, 35. [CrossRef]
53. Park, J.; Shin, Y.; Kim, T.H.; Kim, D.-H.; Lee, A. Plasma metabolites as possible biomarkers for diagnosis of breast cancer. *PLoS ONE* **2019**, *14*, e0225129. [CrossRef] [PubMed]
54. Li, L.; Zhang, M.; Men, Y.; Wang, W.; Zhang, W. Heavy metals interfere with plasma metabolites, including lipids and amino acids, in patients with breast cancer. *Oncol. Lett.* **2020**, *19*, 2925–2933. [CrossRef]
55. Politi, C.; Fattuoni, C.; Serra, A.; Noto, A.; Loi, S.; Casanova, A.; Faa, G.; Ravarino, A.; Saba, L. Metabolomic analysis of plasma from breast tumour patients. A pilot study. *J. Public Health Res.* **2021**, *10*, 2304. [CrossRef]
56. An, R.; Yu, H.; Wang, Y.; Lu, J.; Gao, Y.; Xie, X.; Zhang, J. Integrative analysis of plasma metabolomics and proteomics reveals the metabolic landscape of breast cancer. *Cancer Metab.* **2022**, *10*, 13. [CrossRef]
57. Baranovicova, E.; Racay, P.; Zubor, P.; Smolar, M.; Kudelova, E.; Halasova, E.; Dvorska, D.; Dankova, Z. Circulating metabolites in the early stage of breast cancer were not related to cancer stage or subtypes but associated with ki67 level. Promising statistical discrimination from controls. *Mol. Cell. Probes* **2022**, *66*, 101862. [CrossRef]
58. Han, X.; Li, D.; Wang, S.; Lin, Y.; Liu, Y.; Lin, L.; Qiao, L. Serum amino acids quantification by plasmonic colloidosome-coupled MALDI-TOF MS for triple-negative breast cancer diagnosis. *Mater. Today Bio* **2022**, *17*, 100486. [CrossRef] [PubMed]
59. Santaliz-Casiano, A.; Mehta, D.; Danciu, O.C.; Patel, H.; Banks, L.; Zaidi, A.; Buckley, J.; Rauscher, G.H.; Schulte, L.; Weller, L.R.; et al. Identification of metabolic pathways contributing to ER+ breast cancer disparities using a machine-learning pipeline. *Sci. Rep.* **2023**, *13*, 12136. [CrossRef] [PubMed]
60. Panigoro, S.S.; Kurniawan, A.; Ramadhan, R.; Sukartini, N.; Herqutanto, H.; Paramita, R.I.; Sandra, F. Amino Acid Profile of Luminal A and B Subtypes Breast Cancer. *Indones. Biomed. J.* **2023**, *15*, 194–295. [CrossRef]
61. Morad, H.M.; Abou-Elzahab, M.M.; Aref, S.; EL-Sokkary, A.M.A. Diagnostic Value of 1H NMR-Based Metabolomics in Acute Lymphoblastic Leukemia, Acute Myeloid Leukemia, and Breast Cancer. *ACS Omega* **2022**, *7*, 8128–8140. [CrossRef] [PubMed]
62. Lai, H.S.; Lee, J.C.; Lee, P.H.; Wang, S.T.; Chen, W.J. Plasma free amino acid profile in cancer patients. *Semin. Cancer Biol.* **2005**, *15*, 267–276. [CrossRef] [PubMed]
63. Yang, L.; Wang, Y.; Cai, H.; Wang, S.; Shen, Y.; Ke, C. Application of metabolomics in the diagnosis of breast cancer: A systematic review. *J. Cancer* **2020**, *11*, 2540–2551. [CrossRef] [PubMed]
64. Hu, L.; Gao, Y.; Cao, Y.; Zhang, Y.; Xu, M.; Wang, Y.; Jing, Y.; Guo, S.; Jing, F.; Hu, X.; et al. Identification of arginine and its "Downstream" molecules as potential markers of breast cancer. *IUBMB Life* **2016**, *68*, 817–822. [CrossRef]
65. Budczies, J.; Denkert, C.; Muller, B.M.; Brockmoller, S.F.; Klauschen, F.; Gyorffy, B.; Dietel, M.; Richter-Ehrenstein, C.; Marten, U.; Salek, R.M.; et al. Remodeling of central metabolism in invasive breast cancer compared to normal breast tissue—A GC-TOFMS based metabolomics study. *BMC Genom.* **2012**, *13*, 334. [CrossRef]
66. Budczies, J.; Brockmöller, S.F.; Müller, B.M.; Barupal, D.K.; Richter-Ehrenstein, C.; Kleine-Tebbe, A.; Griffin, J.L.; Orešič, M.; Dietel, M.; Denkert, C.; et al. Comparative metabolomics of estrogen receptor positive and estrogen receptor negative breast cancer: Alterations in glutamine and beta-alanine metabolism. *J. Proteom.* **2013**, *94*, 279–288. [CrossRef]
67. Togashi, Y.; Arao, T.; Kato, H.; Matsumoto, K.; Terashima, M.; Hayashi, H.; de Velasco, M.A.; Fujita, Y.; Kimura, H.; Yasuda, T. Frequent amplification of ORAOV1 gene in esophageal squamous cell cancer promotes an aggressive phenotype via proline metabolism and ROS production. *Oncotarget* **2014**, *5*, 2962. [CrossRef] [PubMed]
68. Díaz-Beltrán, L.; González-Olmedo, C.; Luque-Caro, N.; Díaz, C.; Martín-Blázquez, A.; Fernández-Navarro, M.; Ortega-Granados, A.L.; Gálvez-Montosa, F.; Vicente, F.; Pérez del Palacio, J.; et al. Human Plasma Metabolomics for Biomarker Discovery: Targeting the Molecular Subtypes in Breast Cancer. *Cancers* **2021**, *13*, 147. [CrossRef] [PubMed]
69. Tayyari, F.; Gowda, G.A.N.; Olopade, O.F.; Berg, R.; Yang, H.H.; Lee, M.P.; Ngwa, W.F.; Mittal, S.K.; Raftery, D.; Mohammed, S.I. Metabolic profiles of triple-negative and luminal A breast cancer subtypes in African-American identify key metabolic differences. *Oncotarget* **2018**, *9*, 11677–11690. [CrossRef]
70. Chlebowski, R.T.; Chen, Z.; Anderson, G.L.; Rohan, T.; Aragaki, A.; Lane, D.; Dolan, N.C.; Paskett, E.D.; McTiernan, A.; Hubbell, F.A.; et al. Ethnicity and breast cancer: Factors influencing differences in incidence and outcome. *J. Natl. Cancer Inst.* **2005**, *97*, 439–448. [CrossRef]

71. Kwan, M.L.; Kushi, L.H.; Weltzien, E.; Maring, B.; Kutner, S.E.; Fulton, R.S.; Lee, M.M.; Ambrosone, C.B.; Caan, B.J. Epidemiology of breast cancer subtypes in two prospective cohort studies of breast cancer survivors. *Breast Cancer Res.* **2009**, *11*, R31. [CrossRef] [PubMed]
72. Amirikia, K.C.; Mills, P.; Bush, J.; Newman, L.A. Higher population-based incidence rates of triple-negative breast cancer among young African-American women: Implications for breast cancer screening recommendations. *Cancer* **2011**, *117*, 2747–2753. [CrossRef] [PubMed]
73. Mavaddat, N.; Antoniou, A.C.; Easton, D.F.; Garcia-Closas, M. Genetic susceptibility to breast cancer. *Mol. Oncol.* **2010**, *4*, 174–191. [CrossRef]
74. Pharoah, P.D.P.; Antoniou, A.C.; Easton, D.F.; Ponder, B.A.J. Polygenes, risk prediction, and targeted prevention of breast cancer. *N. Engl. J. Med.* **2008**, *358*, 2796–2803. [CrossRef] [PubMed]
75. Gieger, C.; Geistlinger, L.; Altmaier, E.; Hrabé de Angelis, M.; Kronenberg, F.; Meitinger, T.; Mewes, H.W.; Wichmann, H.E.; Weinberger, K.M.; Adamski, J.; et al. Genetics meets metabolomics: A genome-wide association study of metabolite profiles in human serum. *PLoS Genet.* **2008**, *4*, e1000282. [CrossRef] [PubMed]
76. Zhao, H.; Shen, J.; Moore, S.C.; Ye, Y.; Wu, X.; Esteva, F.J.; Tripathy, D.; Chow, W.H. Breast cancer risk in relation to plasma metabolites among Hispanic and African American women. *Breast Cancer Res. Treat.* **2019**, *176*, 687–696. [CrossRef] [PubMed]
77. Huo, M.; Zhang, J.; Huang, W.; Wang, Y. Interplay among metabolism, epigenetic modifcations, and gene expression in cancer. *Front. Cell Dev. Biol.* **2021**, *9*, 793428. [CrossRef]
78. Jones, P.A. Functions of DNA methylation: Islands, start sites, gene bodies and beyond. *Nat. Rev. Genet.* **2012**, *13*, 484–492. [CrossRef]
79. Herman, J.G.; Baylin, S.B. Gene silencing in cancer in association with promoter hypermethylation. *N. Engl. J. Med.* **2003**, *349*, 2042–2054. [CrossRef]
80. Garcia-Martinez, L.; Zhang, Y.; Nakata, Y.; Chan, H.L.; Morey, L. Epigenetic mechanisms in breast cancer therapy and resistance. *Nat. Commun.* **2021**, *12*, 1786. [CrossRef]
81. Mogol, A.N.; Zuo, Q.; Yoo, J.Y.; Kaminsky, A.Z.; Imir, O.B.; Landesman, Y.; Walker, C.J.; Erdogan, Z.M. NAD+ metabolism generates a metabolic vulnerability in endocrine-resistant metastatic breast tumors in females. *Endocrinology* **2023**, *164*, bqad073. [CrossRef] [PubMed]
82. Hsu, P.C.; Kadlubar, S.; Su, L.J.; Acheampong, D.; Rogers, L.J.; Runnells, G.; McElfish, P.A.; Schootman, M. County poverty levels infuence genome-wide DNA methylation profiles in African American and European American women. *Transl. Cancer Res.* **2019**, *8*, 683–692. [CrossRef]
83. Moore, S.C.; Playdon, M.C.; Sampson, J.N.; Hoover, R.N.; Trabert, B.; Matthews, C.E.; Ziegler, R.G. A metabolomics analysis of body mass index and postmenopausal breast cancer risk. *J. Natl. Cancer Inst.* **2018**, *110*, 588–597. [CrossRef]
84. Medina, V.; Cricco, G.; Nunez, M.; Martin, G.; Mohamad, N.; Correa-Fiz, F.; Sanchez-Jimenez, F.; Bergoc, R.; Rivera, E.S. Histamine-mediated signaling processes in human malignant mammary cells. *Cancer Biol. Ther.* **2006**, *11*, 1462–1471. [CrossRef]
85. Wolfe, R.R. Branched-chain amino acids and muscle protein synthesis in humans: Myth or reality? *J. Int. Soc. Sports Nutr.* **2017**, *14*, 30. [CrossRef]
86. Newgard, C.B.; An, J.; Bain, J.R.; Muehlbauer, M.J.; Stevens, R.D.; Lien, L.F.; Haqq, A.M.; Shah, S.H.; Arlotto, M.; Slentz, C.A.; et al. A branched-chain amino acid-related metabolic signature that differentiates obese and lean humans and contributes to insulin resistance. *Cell Metab.* **2009**, *9*, 311–326. [CrossRef] [PubMed]
87. Zeleznik, O.A.; Balasubramanian, R.; Ren, Y.; Tobias, D.K.; Rosner, B.A.; Peng, C.; Bever, A.M.; Frueh, L.; Jeanfavre, S.; Avila-Pacheco, J.; et al. Branched-Chain Amino Acids and Risk of Breast Cancer. *JNCI Cancer Spectr.* **2021**, *5*, pkab059. [CrossRef] [PubMed]
88. Nagata, C.; Wada, K.; Tsuji, M.; Hayashi, M.; Takeda, N.; Yasuda, K. Plasma amino acid profiles are associated with biomarkers of breast cancer risk in premenopausal Japanese women. *Cancer Causes Control* **2014**, *25*, 143–149. [CrossRef]
89. Lécuyer, L.; Victor Bala, A.; Deschasaux, M.; Bouchemal, N.; Nawfal Triba, M.; Vasson, M.P.; Rossary, A.; Demidem, A.; Galan, P.; Hercberg, S.; et al. NMR metabolomic signatures reveal predictive plasma metabolites associated with long-term risk of developing breast cancer. *Int. J. Epidemiol.* **2018**, *47*, 484–494. [CrossRef]
90. His, M.; Viallon, V.; Dossus, L.; Gicquiau, A.; Achaintre, D.; Scalbert, A.; Ferrari, P.; Romieu, I.; Onland-Moret, N.C.; Weiderpass, E.; et al. Prospective analysis of circulating metabolites and breast cancer in EPIC. *BMC Med.* **2019**, *17*, 178. [CrossRef]
91. Zhang, J.; Tao, B.; Chong, Y.; Ma, S.; Wu, G.; Zhu, H.; Zhao, Y.; Zhao, S.; Niu, M.; Zhang, S.; et al. Ornithine and breast cancer: A matched case-control study. *Sci. Rep.* **2020**, *10*, 15502. [CrossRef] [PubMed]
92. Jobard, E.; Dossus, L.; Baglietto, L.; Fornili, M.; Lécuyer, L.; Mancini, F.R.; Gunter, M.J.; Trédan, O.; Boutron-Ruault, M.C.; Elena-Herrmann, B.; et al. Investigation of circulating metabolites associated with breast cancer risk by untargeted metabolomics: A case-control study nested within the French E3N cohort. *Br. J. Cancer* **2021**, *124*, 1734–1743. [CrossRef] [PubMed]
93. Stevens, V.L.; Carter, B.D.; Jacobs, E.J.; McCullough, M.L.; Teras, L.R.; Wang, Y. A prospective case–cohort analysis of plasma metabolites and breast cancer risk. *Breast Cancer Res.* **2023**, *25*, 5. [CrossRef] [PubMed]
94. Kühn, T.; Floegel, A.; Sookthai, D.; Johnson, T.; Rolle-Kampczyk, U.; Otto, W.; von Bergen, M.; Boeing, H.; Kaaks, R. Higher plasma levels of lysophosphatidylcholine 18:0 are related to a lower risk of common cancers in a prospective metabolomics study. *BMC Med.* **2016**, *14*, 13. [CrossRef] [PubMed]

95. Lécuyer, L.; Dalle, C.; Lyan, B.; Demidem, A.; Rossary, A.; Vasson, M.P.; Petera, M.; Lagree, M.; Ferreira, T.; Centeno, D.; et al. Plasma metabolomic signatures associated with long-term breast cancer risk in the SU. VI. MAX prospective cohort. *Cancer Epidemiol. Biomark. Prev.* **2019**, *28*, 1300–1307. [CrossRef]

96. Jung, M.K.; Okekunle, A.P.; Lee, J.E.; Sung, M.K.; Lim, Y.J. Role of Branched-chain Amino Acid Metabolism in Tumor Development and Progression. *J. Cancer Prev.* **2021**, *26*, 237–243. [CrossRef]

97. Chi, R.; Yao, C.; Chen, S.; Liu, Y.; He, Y.; Zhang, J.; Ellies, L.G.; Wu, X.; Zhao, Q.; Zhou, C.; et al. Elevated BCAA suppresses the development and metastasis of breast cancer. *Front. Oncol.* **2022**, *12*, 887257. [CrossRef]

98. Tobias, D.K.; Chai, B.; Tamimi, R.M.; Manson, J.E.; Hu, F.B.; Willett, W.C.; Eliassen, A.H. Dietary Intake of Branched Chain Amino Acids and Breast Cancer Risk in the NHS and NHS II Prospective Cohorts. *JNCI Cancer Spectr.* **2021**, *5*, pkab032. [CrossRef]

99. Nouri-Majd, S.; Salari-Moghaddam, A.; Benisi-Kohansal, S.; Azadbakht, L.; Esmaillzadeh, A. Dietary intake of branched-chain amino acids in relation to the risk of breast cancer. *Breast Cancer* **2022**, *29*, 993–1000. [CrossRef]

100. Mrowiec, K.; Kurczyk, A.; Jelonek, K.; Debik, J.; Giskeødegård, G.F.; Bathen, T.F.; Widłak, P. Association of serum metabolome profile with the risk of breast cancer in participants of the HUNT2 study. *Front. Oncol.* **2023**, *13*, 1116806. [CrossRef]

101. Risi, E.; Lisanti, C.; Vignoli, A.; Biagioni, C.; Paderi, A.; Cappadona, S.; Monte, F.D.; Moretti, E.; Sanna, G.; Livraghi, L.; et al. Risk assessment of disease recurrence in early breast cancer: A serum metabolomic study focused on elderly patients. *Transl. Oncol.* **2023**, *27*, 101585. [CrossRef]

102. Hart, C.D.; Vignoli, A.; Tenori, L.; Uy, G.L.; Van To, T.; Adebamowo, C.; Hossain, S.M.; Biganzoli, L.; Risi, E.; Love, R.R.; et al. Serum Metabolomic Profiles Identify ER-Positive Early Breast Cancer Patients at Increased Risk of Disease Recurrence in a Multicenter Population. *Clin. Cancer Res.* **2017**, *23*, 1422–1431. [CrossRef]

103. Yatabe, J.; Yatabe, M.S.; Ishibashi, K.; Nozawa, Y.; Sanada, H. Early detection of colon cancer by amino acid profiling using AminoIndex Technology: A case report. *Diagn. Pathol.* **2013**, *8*, 203. [CrossRef]

104. Maeda, J.; Higashiyama, M.; Imaizumi, A.; Nakayama, T.; Yamamoto, H.; Daimon, T.; Yamakado, M.; Imamura, F.; Kodama, K. Possibility of multivariate function composed of plasma amino acid profiles as a novel screening index for non-small cell lung cancer: A case control study. *BMC Cancer* **2010**, *10*, 690. [CrossRef]

105. Okamoto, N. Use of 'AminoIndex Technology' for cancer screening. *Ningen Dock* **2012**, *26*, 911–922.

106. Jikuzono, T.; Ishibashi, O.; Kure, S.; Ohmae, Y.; Ohmae, T. Associations of AminoIndex Cancer Screening (Breast) Grade with Clinical and Laboratory Variables. *J. Nippon Med. Sch.* **2022**, *89*, 377–383. [CrossRef] [PubMed]

107. Okamoto, N.; Miyagi, Y.; Chiba, A.; Shiozawa, M.; Akaike, M.; Imaizumi, A.; Ando, T.; Tochikubo, O. Multivariate discrimination functions composed with amino acid profiles (Amino Index®) as a novel diagnostic marker for breast and colon cancer. *Cancer Prev.* **2008**, *6*, 47–48. [CrossRef]

108. Wei, Z.; Liu, X.; Cheng, C.; Yu, W.; Yi, P. Metabolism of Amino Acids in Cancer. *Front. Cell Dev. Biol.* **2020**, *8*, 603837. [CrossRef] [PubMed]

109. Geck, R.C.; Toker, A. Nonessential amino acid metabolism in breast cancer. *Adv. Biol. Regul.* **2016**, *62*, 11–17. [CrossRef]

110. Subramani, R.; Poudel, S.; Smith, K.D.; Estrada, A.; Lakshmanaswamy, R. Metabolomics of Breast Cancer: A Review. *Metabolites* **2022**, *12*, 643. [CrossRef]

111. Willmann, L.; Schlimpert, M.; Halbach, S.; Erbes, T.; Stickeler, E.; Kammerer, B. Metabolic profiling of breast cancer: Differences in central metabolism between subtypes of breast cancer cell lines. *J. Chromatogr B Anal. Technol. Biomed. Life Sci.* **2015**, *1000*, 95–104. [CrossRef] [PubMed]

112. Du, S.; Wang, Y.; Alatrash, N.; Weatherly, C.A.; Roy, D.; MacDonnell, F.M.; Armstrong, D.W. Altered profiles and metabolism of land d-amino acids in cultured human breast cancer cells vs. non-tumorigenic human breast epithelial cells. *J. Pharm. Biomed. Anal.* **2019**, *164*, 421–429. [CrossRef] [PubMed]

113. Wang, Q.; Sun, T.; Cao, Y.; Gao, P.; Dong, J.; Fang, Y.; Fang, Z.; Sun, X.; Zhu, Z. A dried blood spot mass spectrometry metabolomic approach for rapid breast cancer detection. *Onco Targets Ther.* **2016**, *9*, 1389–1398. [PubMed]

114. Jain, M.; Nilsson, R.; Sharma, S.; Madhusudhan, N.; Kitami, T.; Souza, A.L.; Kafri, R.; Kirschner, M.W.; Clish, C.B.; Mootha, V.K. Metabolite profiling identifies a key role for glycine in rapid cancer cell proliferation. *Science* **2012**, *336*, 1040–1044. [CrossRef]

115. Craze, M.L.; Cheung, H.; Jewa, N.; Coimbra, N.D.M.; Soria, D.; El-Ansari, R.; Aleskandarany, M.A.; Wai Cheng, K.; Diez-Rodriguez, M.; Nolan, C.C.; et al. MYC regulation of glutamine-proline regulatory axis is key in luminal B breast cancer. *Br. J. Cancer* **2018**, *118*, 258–265. [CrossRef]

116. Cha, Y.J.; Kim, E.S.; Koo, J.S. Amino Acid Transporters and Glutamine Metabolism in Breast Cancer. *Int. J. Mol. Sci.* **2018**, *19*, 907. [CrossRef]

117. Coloff, J.L.; Murphy, J.P.; Braun, C.R.; Harris, I.S.; Shelton, L.M.; Kami, K.; Gygi, S.P.; Selfors, L.M.; Brugge, J.S. Differential glutamate metabolism in proliferating and quiescent mammary epithelial cells. *Cell Metab.* **2016**, *23*, 867–880. [CrossRef]

118. El Ansari, R.; McIntyre, A.; Craze, M.L.; Ellis, I.O.; Rakha, E.A.; Green, A.R. Altered glutamine metabolism in breast cancer; subtype dependencies and alternative adaptations. *Histopathology* **2018**, *72*, 183–190. [CrossRef]

119. Dowling, P.; Henry, M.; Meleady, P.; Clarke, C.; Gately, K.; O'Byrne, K.; Connolly, E.; Lynch, V.; Ballot, J.; Gullo, G.; et al. Metabolomic and proteomic analysis of breast cancer patient samples suggests that glutamate and 12-HETE in combination with CA15-3 may be useful biomarkers reflecting tumour burden. *Metabolomics* **2015**, *11*, 620–635. [CrossRef]

120. Kandasamy, P.; Gyimesi, G.; Kanai, Y.; Hediger, M.A. Amino acid transporters revisited: New views in health and disease. *Trends Biochem. Sci.* **2018**, *43*, 752–789. [CrossRef]

121. Bhutia, Y.D.; Babu, E.; Ramachandran, S.; Ganapathy, V. Amino Acid transporters in cancer and their relevance to "glutamine addiction": Novel targets for the design of a new class of anticancer drugs. *Cancer Res.* **2015**, *75*, 1782–1788. [CrossRef] [PubMed]
122. van Geldermalsen, M.; Wang, Q.; Nagarajah, R.; Marshall, A.D.; Thoeng, A.; Gao, D.; Ritchie, W.; Feng, Y.; Bailey, C.G.; Deng, N.; et al. ASCT2/SLC1A5 controls glutamine uptake and tumour growth in triple-negative basal-like breast cancer. *Oncogene* **2016**, *35*, 3201–3208. [CrossRef] [PubMed]
123. Metallo, C.M.; Gameiro, P.A.; Bell, E.L.; Mattaini, K.R.; Yang, J.; Hiller, K.; Jewell, C.M.; Johnson, Z.R.; Irvine, D.J.; Guarente, L.; et al. Reductive glutamine metabolism by IDH1 mediates lipogenesis under hypoxia. *Nature* **2011**, *481*, 380–384. [CrossRef] [PubMed]
124. Le, A.; Lane, A.N.; Hamaker, M.; Bose, S.; Gouw, A.; Barbi, J.; Tsukamoto, T.; Rojas, C.J.; Slusher, B.S.; Zhang, H.; et al. Glucoseindependent glutamine metabolism via TCA cycling for proliferation and survival in B cells. *Cell Metab.* **2012**, *15*, 110–121. [CrossRef] [PubMed]
125. Meadows, A.L.; Kong, B.; Berdichevsky, M.; Roy, S.; Rosiva, R.; Blanch, H.W.; Clark, D.S. Metabolic and morphological differences between rapidly proliferating cancerous and normal breast epithelial cells. *Biotechnol. Prog.* **2008**, *24*, 334–341. [CrossRef]
126. Budczies, J.; Pfitzner, B.M.; Györffy, B.; Winzer, K.-J.; Radke, C.; Dietel, M.; Fiehn, O.; Denkert, C. Glutamate enrichment as new diagnostic opportunity in breast cancer. *Int. J. Cancer* **2015**, *136*, 1619–1628. [CrossRef]
127. Zhou, W.-X.; Chen, C.; Liu, X.-Q.; Li, Y.; Lin, Y.-L.; Wu, X.-T.; Kong, L.-Y.; Luo, J.-G. Discovery and optimization of withangulatin A derivatives as novel glutaminase 1 inhibitors for the treatment of triple-negative breast cancer. *Eur. J. Med. Chem.* **2021**, *210*, 112980. [CrossRef]
128. Garcia-Caballero, M.; Neugebauer, E.; Campos, R.; Nunez de Castro, I.; Vara-Thorbeck, C. Increased histidine decarboxylase (HDC) activity in human colorectal cancer: Results of a study on ten patients. *Agents Actions* **1988**, *23*, 357–360. [CrossRef]
129. Xie, G.; Zhou, B.; Zhao, A.; Qiu, Y.; Zhao, X.; Garmire, L.; Shvetsov, Y.B.; Yu, H.; Yen, Y.; Jia, W. Lowered circulating aspartate is a metabolic feature of human breast cancer. *Oncotarget* **2015**, *6*, 33369–33381. [CrossRef]
130. Knott, S.R.V.; Wagenblast, E.; Khan, S.; Kim, S.Y.; Soto, M.; Wagner, M.; Turgeon, M.-O.; Fish, L.; Erard, N.; Gable, A.L.; et al. Asparagine bioavailability governs metastasis in a model of breast cancer. *Nature* **2018**, *554*, 378–381. [CrossRef]
131. Sivanand, S.; Vander Heiden, M.G. Emerging roles for branched-chain amino acid metabolism in cancer. *Cancer Cell* **2020**, *37*, 147–156. [CrossRef] [PubMed]
132. Harvie, M.N.; Campbell, I.T.; Howell, A.; Thatcher, N. Acceptability and tolerance of a low tyrosine and phenylalanine diet in patients with advanced cancer—A pilot study. *J. Hum. Nutr. Diet.* **2002**, *15*, 193–202. [CrossRef] [PubMed]
133. Atoum, M.; Abdel-Fattah, M.; Nimer, N.; Abdel-Rahman, S.; Abdeldayem, S.A. Association of alanine-valine manganese superoxide dismutase gene polymorphism and microheterogeneity manganese superoxide dismutase activity in breast cancer and benign breast tissue. *J. Breast Cancer* **2012**, *15*, 157–161. [CrossRef]
134. Nittoli, A.C.; Costantini, S.; Sorice, A.; Capone, F.; Ciarcia, R.; Marzocco, S.; Budillon, A.; Severino, L. Effects of α-zearalenol on the metabolome of two breast cancer cell lines by 1H-NMR approach. *Metabolomics* **2018**, *14*, 33. [CrossRef] [PubMed]
135. Cao, Y.; Feng, Y.; Zhang, Y.; Zhu, X.; Jin, F. L-Arginine supplementation inhibits the growth of breast cancer by enhancing innate and adaptive immune responses mediated by suppression of MDSCs in vivo. *BMC Cancer* **2016**, *16*, 343. [CrossRef] [PubMed]
136. Opitz, C.A.; Litzenburger, U.M.; Sahm, F.; Ott, M.; Tritschler, I.; Trump, S.; Schumacher, T.; Jestaedt, L.; Schrenk, D.; Weller, M.; et al. An endogenous tumour-promoting ligand of the human aryl hydrocarbon receptor. *Nature* **2011**, *478*, 197–203. [CrossRef] [PubMed]
137. Ye, Z.; Yue, L.; Shi, J.; Shao, M.; Wu, T. Role of IDO and TDO in Cancers and Related Diseases and the Therapeutic Implications. *J. Cancer* **2019**, *10*, 2771–2782. [CrossRef]
138. Cheong, J.E.; Sun, L. Targeting the IDO1/TDO2–KYN–AhR Pathway for Cancer Immunotherapy—Challenges and Opportunities. *Trends Pharmacol. Sci.* **2018**, *39*, 307–325. [CrossRef]
139. Platten, M.; Wick, W.; Van de Eynde, B.J. Tryptophan Catabolism in Cancer: Beyond IDO and Tryptophan Depletion. *Cancer Res.* **2012**, *72*, 5435–5440. [CrossRef]
140. Pollari, S.; Käkönen, S.M.; Edgren, H.; Wolf, M.; Kohonen, P.; Sara, H.; Guise, T.; Nees, M.; Kallioniemi, O. Enhanced serine production by bone metastatic breast cancer cells stimulates osteoclastogenesis. *Breast Cancer Res. Treat.* **2011**, *125*, 421–430. [CrossRef]
141. Maddocks, O.D.K.; Athineos, D.; Cheung, E.C.; Lee, P.; Zhang, T.; van den Broek, N.J.F.; Mackay, G.M.; Labuschagne, C.F.; Gay, D.; Kruiswijk, F.; et al. Modulating the therapeutic response of tumours to dietary serine and glycine starvation. *Nature* **2017**, *544*, 372–376. [CrossRef] [PubMed]
142. Vettore, L.; Westbrook, R.L.; Tennant, D.A. New aspects of amino acid metabolism in cancer. *Br. J. Cancer* **2020**, *122*, 150–156. [CrossRef] [PubMed]
143. Lin, J.; Lee, I.M.; Song, Y.; Cook, N.R.; Selhub, J.; Manson, J.E.; Buring, J.E.; Zhang, S.M. Plasma homocysteine and cysteine and risk of breast cancer in women. *Cancer Res.* **2010**, *70*, 2397–2405. [CrossRef]
144. Bonifácio, V.D.B.; Pereira, S.A.; Serpa, J.; Vicente, J.B. Cysteine metabolic circuitries: Druggable targets in cancer. *Br. J. Cancer* **2021**, *124*, 862–879. [CrossRef] [PubMed]
145. Tsai, M.J.; Chang, W.A.; Chuang, C.H.; Wu, K.L.; Cheng, C.H.; Sheu, C.C.; Hsu, Y.L.; Hung, J.Y. Cysteinyl leukotriene pathway and cancer. *Int. J. Mol. Sci.* **2021**, *23*, 120. [CrossRef] [PubMed]

146. Haukaas, T.H.; Euceda, L.R.; Giskeødegård, G.F.; Bathen, T.F. Metabolic Portraits of Breast Cancer by HR MAS MR Spectroscopy of Intact Tissue Samples. *Metabolites* **2017**, *7*, 18. [CrossRef] [PubMed]
147. Liang, Z.; Cho, H.T.; Williams, L.; Zhu, A.; Liang, K.; Huang, K.; Wu, H.; Jiang, C.; Hong, S.; Crowe, R.; et al. Potential biomarker of L-type amino acid transporter 1 in breast cancer progression. *Nucl. Med. Mol. Imaging* **2011**, *45*, 93–102. [CrossRef]
148. Ulaner, G.A.; Schuster, D.M. Amino Acid Metabolism as a Target for Breast Cancer Imaging. *PET Clin.* **2018**, *13*, 437–444. [CrossRef]
149. Hayashi, K.; Anzai, N. L-type amino acid transporter 1 as a target for inflammatory disease and cancer immunotherapy. *J. Pharmacol. Sci.* **2022**, *148*, 31–40. [CrossRef]
150. Bhat, H.K.; Vadgama, J.V. Role of estrogen receptor in the regulation of estrogen induced amino acid transport of System A in breast cancer and other receptor positive tumor cells. *Int. J. Mol. Med.* **2002**, *9*, 271–279. [CrossRef]
151. Shennan, D.B.; Thomson, J. Inhibition of system L (LAT1/CD98hc) reduces the growth of cultured human breast cancer cells. *Oncol. Rep.* **2008**, *20*, 885–889. [CrossRef] [PubMed]
152. Baek, S.; Choi, C.M.; Ahn, S.H.; Lee, J.W.; Gong, G.; Ryu, J.S.; Oh, S.J.; Bacher-Stier, C.; Fels, L.; Koglin, N.; et al. Exploratory clinical trial of (4s)-4-(3-[18f]fluoropropyl)-l-glutamate for imaging xc-transporter using positron emission tomography in patients with non-small cell lung or breast cancer. *Clin. Cancer Res.* **2012**, *18*, 5427–5437. [CrossRef]
153. Saito, Y.; Soga, T. Amino acid transporters as emerging therapeutic targets in cancer. *Cancer Sci.* **2021**, *112*, 2958–2965. [CrossRef] [PubMed]
154. Lindholm, P.; Lapela, M.; Någren, K.; Lehikoinen, P.; Minn, H.; Jyrkkiö, S. Preliminary study of carbon-11 methionine PET in the evaluation of early response to therapy in advanced breast cancer. *Nucl. Med. Commun.* **2009**, *30*, 30–36. [CrossRef]
155. Mehravi, B.; Ardestani, M.S.; Damercheli, M.; Soltanghoraee, H.; Ghanaldarlaki, N.; Alizadeh, A.M.; Oghabian, M.A.; Shirazi, M.S.; Mahernia, S.; Amanlou, M. Breast Cancer Cells Imaging by Targeting Methionine Transporters with Gadolinium-Based Nanoprobe. *Mol. Imaging Biol.* **2014**, *16*, 519–528. [CrossRef]
156. Taylor, K.M.; Kim, J.S.; Rieter, W.J.; An, H.; Lin, W.; Lin, W. Mesoporous silica nanospheres as highly efficient MRI contrast agents. *J. Am. Chem. Soc.* **2008**, *130*, 2154–2155. [CrossRef]
157. Okudaira, H.; Nakanishi, T.; Oka, S.; Kobayashi, M.; Tamagami, H.; Schuster, D.M.; Goodman, M.M.; Shirakami, Y.; Tamai, I.; Kawai, K. Kinetic analyses of trans-1-amino-3-[18F]fluorocyclobutanecarboxylic acid transport in Xenopus laevis oocytes expressing human ASCT2 and SNAT2. *Nucl. Med. Biol.* **2013**, *40*, 670–675. [CrossRef] [PubMed]
158. Oka, S.; Okudaira, H.; Ono, M.; Schuster, D.M.; Goodman, M.M.; Kawai, K.; Shirakami, Y. Differences in transport mechanisms of trans-1-amino-3-[18F] fluorocyclobutanecarboxylic acid in inflammation, prostate cancer, and glioma cells: Comparison with L-[methyl-11C]methionine and 2-deoxy-2-[18F]fluoro-D-glucose. *Mol. Imaging Biol.* **2014**, *16*, 322–329. [CrossRef] [PubMed]
159. Tade, F.I.; Cohen, M.A.; Styblo, T.M.; Odewole, O.A.; Holbrook, A.I.; Newell, M.S.; Savir-Baruch, B.; Li, X.; Goodman, M.M.; Nye, J.A.; et al. Anti-3-18FFACBC (18F-Fluciclovine) PET/CT of breast cancer: An exploratory study. *J. Nucl. Med.* **2016**, *57*, 1357–1363. [CrossRef]
160. Ulaner, G.A.; Goldman, D.A.; Corben, A.; Lyashchenko, S.K.; Gönen, M.; Lewis, J.S.; Dickler, M. Prospective clinical trial of 18F-fluciclovine PET/CT for determining the response to neoadjuvant therapy in invasive ductal and invasive lobular breast cancers. *J. Nucl. Med.* **2017**, *58*, 1037–1042. [CrossRef]
161. Ulaner, G.A.; Goldman, D.A.; Gonen, M.; Pham, H.; Castillo, R.; Lyashchenko, S.K.; Lewis, J.S.; Dang, C. Initial results of a prospective clinical trial of 18f-fluciclovine pet/ct in newly diagnosed invasive ductal and invasive lobular breast cancers. *J. Nucl. Med.* **2016**, *57*, 1350–1356. [CrossRef] [PubMed]
162. Oka, S.; Kanagawa, M.; Doi, Y.; Schuster, D.M.; Goodman, M.M.; Yoshimura, H. Fasting enhances the contrast of bone metastatic lesions in 18F-fluciclovine-PET: Preclinical study using a rat model of mixed osteolytic/osteoblastic bone metastases. *Int. J. Mol. Sci.* **2017**, *18*, 934. [CrossRef] [PubMed]
163. Koglin, N.; Mueller, A.; Berndt, M.; Schmitt-Willich, H.; Toschi, L.; Stephens, A.W.; Gekeler, V.; Friebe, M.; Dinkelborg, L.M. Specific PET imaging of xC-transporter activity using a (1)(8)Flabeled glutamate derivative reveals a dominant pathway in tumor metabolism. *Clin. Cancer Res.* **2011**, *17*, 6000–6011. [CrossRef]
164. Yang, H.; Jenni, S.; Colovic, M.; Merkens, H.; Poleschuk, C.; Rodrigo, I.; Miao, Q.; Johnson, B.F.; Rishel, M.J.; Sossi, V.; et al. (18)f-5-fluoroaminosuberic acid as a potential tracer to gauge oxidative stress in breast cancer models. *J. Nucl. Med.* **2017**, *58*, 367–373. [CrossRef]
165. Kole, A.C.; Nieweg, O.E.; Pruim, J.; Paans, A.M.; Plukker, J.T.; Hoekstra, H.J.; Schraffordt Koops, H.; Vaalburg, W. Standardized uptake value and quantification of metabolism for breast cancer imaging with FDG and L-[1-11C]tyrosine PET. *J. Nucl. Med.* **1997**, *38*, 692–696. [PubMed]
166. Kong, F.L.; Ali, M.S.; Rollo, A.; Smith, D.L.; Zhang, Y.; Yu, D.F.; Yang, D.J. Development of tyrosine-based radiotracer 99mTc-N4-Tyrosine for breast cancer imaging. *J. Biomed. Biotechnol.* **2012**, *2012*, 671708. [CrossRef]
167. Kaim, A.H.; Weber, B.; Kurrer, M.O.; Westera, G.; Schweitzer, A.; Gottschalk, J.; von Schulthess, G.K.; Buck, A. (18)f-fdg and (18)f-fet uptake in experimental soft tissue infection. *Eur. J. Nucl. Med. Mol. Imaging* **2002**, *29*, 648–654. [CrossRef] [PubMed]
168. Rau, F.C.; Weber, W.A.; Wester, H.J.; Herz, M.; Becker, I.; Kruger, A.; Schwaiger, M.; Senekowitsch-Schmidtke, R. O-(2-[(18)f]fluoroethyl)-l-tyrosine (fet): A tracer for differentiation of tumour from inflammation in murine lymph nodes. *Eur. J. Nucl. Med. Mol. Imaging* **2002**, *29*, 1039–1046. [CrossRef]

169. Ren, W.; Liu, G.; Yin, J.; Tan, B.; Wu, G.; Bazer, F.W.; Peng, Y.; Yin, Y. Amino-acid transporters in t-cell activation and differentiation. *Cell Death Dis.* **2017**, *8*, e2655. [CrossRef]

170. Yu, W.; McConathy, J.; Olson, J.J.; Goodman, M.M. System a amino acid transport-targeted brain and systemic tumor PET imaging agents 2-amino-3-[(18)F]fluoro-2-methylpropanoic acid and 3-[(18)F]fluoro-2-methyl2-(methylamino)propanoic acid. *Nucl. Med. Biol.* **2015**, *42*, 8–18. [CrossRef]

171. Way, J.D.; Wang, M.; Hamann, I.; Wuest, M.; Wuest, F. Synthesis and evaluation of 2-amino-5-(4-[(18)f]fluorophenyl)pent-4-ynoic acid ([(18)f]fphpa): A novel (18)f-labeled amino acid for oncologic pet imaging. *Nucl. Med. Biol.* **2014**, *41*, 660–669. [CrossRef] [PubMed]

172. Michelhaugh, S.K.; Muzik, O.; Guastella, A.R.; Klinger, N.V.; Polin, L.A.; Cai, H.; Xin, Y.; Mangner, T.J.; Zhang, S.; Juhász, C.; et al. Assessment of tryptophan uptake and kinetics using 1-(2-18F-fluoroethyl)-l-tryptophan and alpha-11Cmethyl-l-tryptophan PET imaging in mice implanted with patient-derived brain tumor xenografts. *J. Nucl. Med.* **2017**, *58*, 208–213. [CrossRef] [PubMed]

173. Chin, B.B.; McDougald, D.; Weitzel, D.H.; Hawk, T.; Reiman, R.E.; Zalutsky, M.R.; Vaidyanathan, G. Synthesis and preliminary evaluation of 5-[18F]fluoroleucine. *Curr. Radiopharm.* **2017**, *10*, 41–50. [CrossRef] [PubMed]

174. Zhou, R.; Pantel, A.R.; Li, S.; Lieberman, B.P.; Ploessl, K.; Choi, H.; Blankemeyer, E.; Lee, H.; Kung, H.F.; Mach, R.H.; et al. [18F](2S,4R)4-fluoroglutamine PET detects glutamine pool size changes in triple-negative breast cancer in response to glutaminase inhibition. *Cancer Res.* **2017**, *77*, 1476–1484. [CrossRef] [PubMed]

175. Kopka, K.; Riemann, B.; Friedrich, M.; Winters, S.; Halfter, H.; Weckesser, M.; Stögbauer, F.; Bernd Ringelstein, E.; Schober, O. Characterization of 3-[123i]iodo-l-α-methyl tyrosine transport in astrocytes of neonatal rats. *J. Neurochem.* **2001**, *76*, 97–104. [CrossRef] [PubMed]

176. Jager, P.L.; Franssen, E.J.; Kool, W.; Szabo, B.G.; Hoekstra, H.J.; Groen, H.J.; de Vries, E.G.; van Imhoff, G.W.; Vaalburg, W.; Piers, D.A. Feasibility of tumor imaging using l-3-[iodine-123]-iodo-alpha-methyl-tyrosine in extracranial tumors. *J. Nucl. Med.* **1998**, *39*, 1736–1743.

177. Sharma, S.; Singh, B.; Mishra, A.K.; Rathod, D.; Hazari, P.P.; Chuttani, K.; Chopra, S.; Singh, P.M.; Abrar, M.L.; Mittal, B.R.; et al. Lat-1 based primary breast cancer detection by [99m]tc-labeled dtpa-bis-methionine scintimammography: First results using indigenously developed single vial kit preparation. *Cancer Biotherapy Radiopharm.* **2014**, *29*, 283–288. [CrossRef]

178. Brandon, F.; Ashley, S. Deberardinis RJ. Metabolic reprogramming and cancer progression. *Science* **2020**, *368*, eaaw5473.

179. Chowdhry, S.; Zanca, C.; Rajkumar, U.; Koga, T.; Diao, Y.; Raviram, R.; Liu, F.; Turner, K.; Yang, H.; Brunk, E.; et al. NAD metabolic dependency in cancer is shaped by gene amplification and enhancer remodelling. *Nature* **2019**, *569*, 570–575. [CrossRef]

180. Bian, Y.; Li, W.; Kremer, D.M.; Sajjakulnukit, P.; Li, S.; Crespo, J.; Nwosu, Z.C.; Zhang, L.; Czerwonka, A.; Pawłowska, A.; et al. Cancer SLC43A2 alters T cell methionine metabolism and histone methylation. *Nature* **2020**, *585*, 277–282. [CrossRef]

181. Gao, X.; Sanderson, S.M.; Dai, Z.; Reid, M.A.; Cooper, D.E.; Lu, M.; Richie, J.P., Jr.; Ciccarella, A.; Calcagnotto, A.; Mikhael, P.G.; et al. Dietary methionine influences therapy in mouse cancer models and alters human metabolism. *Nature* **2019**, *572*, 397–401. [CrossRef] [PubMed]

182. Kanarek, N.; Keys, H.R.; Cantor, J.R.; Lewis, C.A.; Chan, S.H.; Kunchok, T.; Abu-Remaileh, M.; Freinkman, E.; Schweitzer, L.D.; Sabatini, D.M. Histidine catabolism is a major determinant of methotrexate sensitivity. *Nature* **2018**, *559*, 632–636. [CrossRef] [PubMed]

183. Sadik, A.; Somarribas Patterson, L.F.; Öztürk, S.; Mohapatra, S.R.; Panitz, V.; Secker, P.F.; Pfänder, P.; Loth, S.; Salem, H.; Prentzell, M.T.; et al. IL4I1 is a metabolic immune checkpoint that activates the AHR and promotes tumor progression. *Cell* **2020**, *182*, 1252–1270. [CrossRef] [PubMed]

184. Strekalova, E.; Malin, D.; Weisenhorn, E.M.M.; Russell, J.D.; Hoelper, D.; Jain, A.; Coon, J.J.; Lewis, P.W.; Cryns, V.L. S-adenosylmethionine biosynthesis is a targetable metabolic vulnerability of cancer stem cells. *Breast Cancer Res. Treat.* **2019**, *175*, 39–50. [CrossRef] [PubMed]

185. Saito, Y.; Li, L.; Coyaud, E.; Luna, A.; Sander, C.; Raught, B.; Asara, J.M.; Brown, M.; Muthuswamy, S.K. LLGL2 rescues nutrient stress by promoting leucine uptake in ER+ breast cancer. *Nature* **2019**, *569*, 275–279. [CrossRef] [PubMed]

186. Chen, J.Y.; Li, C.F.; Kuo, C.C.; Tsai, K.K.; Hou, M.F.; Hung, W.C. Cancer/stroma interplay via cyclooxygenase-2 and indoleamine 2,3-dioxygenase promotes breast cancer progression. *Breast Cancer Res.* **2014**, *16*, 410. [CrossRef]

187. Greene, L.I.; Bruno, T.C.; Christenson, J.L.; D'Alessandro, A.; Culp-Hill, R.; Torkko, K.; Borges, V.F.; Slansky, J.E.; Richer, J.K. A Role for Tryptophan-2,3-dioxygenase in CD8 T-cell Suppression and Evidence of Tryptophan Catabolism in Breast Cancer Patient Plasma. *Mol. Cancer Res.* **2019**, *17*, 131–139. [CrossRef]

188. Zhao, Y.; Pu, C.; Liu, Z. Essential amino acids deprivation is a potential strategy for breast cancer treatment. *Breast* **2022**, *62*, 152–161. [CrossRef]

189. Zhang, L.; Han, J. Branched-chain amino acid transaminase 1 (BCAT1) promotes the growth of breast cancer cells through improving mTOR-mediated mitochondrial biogenesis and function. *Biochem. Biophys. Res. Commun.* **2017**, *486*, 224–231. [CrossRef]

190. El Ansari, R.; Craze, M.L.; Miligy, I.; Diez-Rodriguez, M.; Nolan, C.C.; Ellis, I.O.; Rakha, E.A.; Green, A.R. The amino acid transporter SLC7A5 confers a poor prognosis in the highly proliferative breast cancer subtypes and is a key therapeutic target in luminal B tumours. *Breast Cancer Res.* **2018**, *20*, 21. [CrossRef]

191. Cao, M.D.; Lamichhane, S.; Lundgren, S.; Bofin, A.; Fjøsne, H.; Giskeødegård, G.F.; Bathen, T.F. Metabolic characterization of triple negative breast cancer. *BMC Cancer* **2014**, *14*, 941. [CrossRef]

192. Lampa, M.; Arlt, H.; He, T.; Ospina, B.; Reeves, J.; Zhang, B.; Murtie, J.; Deng, G.; Barberis, C.; Hoffmann, D.; et al. Glutaminase is essential for the growth of triple-negative breast cancer cells with a deregulated glutamine metabolism pathway and its suppression synergizes with mTOR inhibition. *PLoS ONE* **2017**, *12*, e0185092. [CrossRef] [PubMed]

193. Mates, J.M.; Campos-Sandoval, J.A.; Marquez, J. Glutaminase isoenzymes in the metabolic therapy of cancer. *Biochim. Biophys. Acta Rev. Cancer* **2018**, *1870*, 158–164. [CrossRef] [PubMed]

194. Huang, Q.; Stalnecker, C.; Zhang, C.; McDermott, L.A.; Iyer, P.; O'Neill, J.; Reimer, S.; Cerione, R.A.; Katt, W.P. Characterization of the interactions of potent allosteric inhibitors with glutaminase C, a key enzyme in cancer cell glutamine metabolism. *J. Biol. Chem.* **2018**, *293*, 3535–3545. [CrossRef] [PubMed]

195. Zhang, B.; Fu, R.; Duan, Z.; Shen, S.; Zhu, C.; Fan, D. Ginsenoside CK induces apoptosis in triple-negative breast cancer cells by targeting glutamine metabolism. *Biochem. Pharmacol.* **2022**, *202*, 115101. [CrossRef]

196. Morotti, M.; Zois, C.E.; El-Ansari, R.; Craze, M.L.; Rakha, E.A.; Fan, S.J.; Valli, A.; Haider, S.; Goberdhan, D.C.I.; Green, A.R.; et al. Increased expression of glutamine transporter SNAT2/SLC38A2 promotes glutamine dependence and oxidative stress resistance, and is associated with worse prognosis in triple-negative breast cancer. *Br. J. Cancer* **2021**, *124*, 494–505. [CrossRef]

197. Timmerman, L.A.; Holton, T.; Yuneva, M.; Louie, R.J.; Padró, M.; Daemen, A.; Hu, M.; Chan, D.A.; Ethier, S.P.; van 't Veer, L.J.; et al. Glutamine sensitivity analysis identifies the xCT antiporter as a common triple-negative breast tumor therapeutic target. *Cancer Cell* **2013**, *24*, 450–465. [CrossRef]

198. Lanzardo, S.; Conti, L.; Rooke, R.; Ruiu, R.; Accart, N.; Bolli, E.; Arigoni, M.; Macagno, M.; Barrera, G.; Pizzimenti, S.; et al. Immunotargeting of Antigen xCT Attenuates Stem-like Cell Behavior and Metastatic Progression in Breast Cancer. *Cancer Res.* **2016**, *76*, 62–72. [CrossRef]

199. Bolli, E.; O'Rourke, J.P.; Conti, L.; Lanzardo, S.; Rolih, V.; Christen, J.M.; Barutello, G.; Forni, M.; Pericle, F.; Cavallo, F. A Virus-Like-Particle immunotherapy targeting Epitope-Specific anti-xCT expressed on cancer stem cell inhibits the progression of metastatic cancer in vivo. *Oncoimmunology* **2017**, *7*, e1408746. [CrossRef]

200. Wang, Z.; Jiang, Q.; Dong, C. Metabolic reprogramming in triple-negative breast cancer. *Cancer Biol. Med.* **2020**, *17*, 44–59. [CrossRef]

201. Strekalova, E.; Malin, D.; Good, D.M.; Cryns, V.L. Methionine deprivation induces a targetable vulnerability in triple-negative breast cancer cells by enhancing TRAIL receptor-2 expression. *Clin. Cancer Res.* **2015**, *21*, 2780–2791. [CrossRef] [PubMed]

202. Jeon, H.; Kim, J.H.; Lee, E.; Jang, Y.J.; Son, J.E.; Kwon, J.Y.; Lim, T.G.; Kim, S.; Park, J.H.; Kim, J.E.; et al. Methionine deprivation suppresses triple-negative breast cancer metastasis in vitro and in vivo. *Oncotarget* **2016**, *7*, 67223–67234. [CrossRef] [PubMed]

203. Borrego, S.L.; Fahrmann, J.; Datta, R.; Stringari, C.; Grapov, D.; Zeller, M.; Chen, Y.; Wang, P.; Baldi, P.; Gratton, E.; et al. Metabolic changes associated with methionine stress sensitivity in MDA-MB-468 breast cancer cells. *Cancer Metab.* **2016**, *4*, 9. [CrossRef]

204. Wang, Z.; Yip, L.Y.; Lee, J.H.J.; Wu, Z.; Chew, H.Y.; Chong, P.K.W.; Teo, C.C.; Ang, H.Y.; Peh, K.L.E.; Yuan, J.; et al. Methionine is a metabolic dependency of tumor-initiating cells. *Nat. Med.* **2019**, *25*, 825–837. [CrossRef] [PubMed]

205. Wang, Z.; Shi, X.; Li, Y.; Fan, J.; Zeng, X.; Xian, Z.; Wang, Z.; Sun, Y.; Wang, S.; Song, P.; et al. Blocking autophagy enhanced cytotoxicity induced by recombinant human arginase in triple-negative breast cancer cells. *Cell Death Dis.* **2014**, *5*, e1563. [CrossRef] [PubMed]

206. Hassabo, A.A.; Abdelraof, M.; Allam, R.M. L-arginase from Streptomyces diastaticus MAM5 as a potential therapeutic agent in breast cancer: Purification, characterization, G1 phase arrest and autophagy induction. *Int. J. Biol. Macromol.* **2023**, *224*, 634–645. [CrossRef]

207. Wang, S.M.; Dowhan, D.H.; Muscat, G.E.O. Epigenetic arginine methylation in breast cancer: Emerging therapeutic strategies. *J. Mol. Endocrinol.* **2019**, *62*, R223–R237. [CrossRef]

208. Oh, T.G.; Wang, S.M.; Muscat, G.E. Therapeutic implications of epigenetic signaling in breast cancer. *Endocrinology* **2017**, *158*, 431–447. [CrossRef]

209. Jeon, Y.J.; Khelifa, S.; Ratnikov, B.; Scott, D.A.; Feng, Y.; Parisi, F.; Ruller, C.; Lau, E.; Kim, H.; Brill, L.M.; et al. Regulation of glutamine carrier proteins by RNF5 determines breast Cancer response to ER stress-inducing chemotherapies. *Cancer Cell* **2015**, *27*, 354–369. [CrossRef]

210. Broer, A.; Rahimi, F.; Broer, S. Deletion of amino acid transporter ASCT2 (SLC1A5) reveals an essential role for transporters SNAT1 (SLC38A1) and SNAT2 (SLC38A2) to sustain Glutaminolysis in Cancer cells. *J. Biol. Chem.* **2016**, *291*, 13194–13205. [CrossRef]

211. Dhankhar, R.; Gupta, V.; Kumar, S.; Kapoor, R.K.; Gulati, P. Microbial enzymes for deprivation of amino acid metabolism in malignant cells: Biological strategy for cancer treatment. *Appl. Microbiol. Biotechnol.* **2020**, *104*, 2857–2869. [CrossRef] [PubMed]

212. Gwangwa, M.V.; Joubert, A.M.; Visagie, M.H. Effects of glutamine deprivation on oxidative stress and cell survival in breast cell lines. *Biol. Res.* **2019**, *52*, 15. [CrossRef] [PubMed]

213. Bacci, M.; Lorito, N.; Ippolito, L.; Ramazzotti, M.; Luti, S.; Romagnoli, S.; Parri, M.; Bianchini, F.; Cappellesso, F.; Virga, F.; et al. Reprogramming of Amino Acid Transporters to Support Aspartate and Glutamate Dependency Sustains Endocrine Resistance in Breast Cancer. *Cell Rep.* **2019**, *28*, 104–118. [CrossRef] [PubMed]

214. Maeng, H.J.; Kim, E.S.; Chough, C.; Joung, M.; Lim, J.W.; Shim, C.K.; Shim, W.S. Addition of amino acid moieties to lapatinib increases the anti-cancer effect via amino acid transporters. *Biopharm. Drug Dispos.* **2014**, *35*, 60–69. [CrossRef] [PubMed]

215. Mello-Andrade, F.; Guedes, A.P.M.; Pires, W.C.; Velozo-Sá, V.S.; Delmond, K.A.; Mendes, D.; Molina, M.S.; Matuda, L.; de Sousa, M.A.M.; Melo-Reis, P.; et al. Ru(II)/amino acid complexes inhibit the progression of breast cancer cells through multiple mechanism-induced apoptosis. *J. Inorg. Biochem.* **2022**, *226*, 111625. [CrossRef]

216. Yoo, H.C.; Han, J.M. Amino acid metabolism in cancer drug resistance. *Cells* **2022**, *11*, 140. [CrossRef]
217. Mitrevska, K.; Rodrigo, M.A.M.; Cernei, N.; Michalkova, H.; Splichal, Z.; Hynek, D.; Zitka, O.; Heger, Z.; Kopel, P.; Adam, V.; et al. Chick chorioallantoic membrane (CAM) assay for the evaluation of the antitumor and antimetastatic activity of platinum-based drugs in association with the impact on the amino acid metabolism. *Mater. Today Bio* **2023**, *19*, 100570. [CrossRef]
218. Ji, Y.; Li, J.; Xiao, S.; Kwan, H.Y.; Bian, Z.; Chu, C.C. Optimization of amino acid-based poly(ester urea urethane) nanoparticles for the systemic delivery of gambogic acid for treating triple negative breast cancer. *Biomater. Sci.* **2023**, *11*, 4370–4384. [CrossRef]
219. Dunstan, R.H.; Sparkes, D.L.; Macdonald, M.M.; Roberts, T.K.; Wratten, C.; Kumar, M.B.; Baines, S.; Denham, J.W.; Gallagher, S.A.; Rothkirch, T. Altered amino acid homeostasis and the development of fatigue by breast cancer radiotherapy patients: A pilot study. *Clin. Biochem.* **2011**, *44*, 208–215. [CrossRef]
220. Poon, R.T.P.; Yu, W.C.; Fan, S.T.; Wong, J. Long-term oral branched chain amino acids in patients undergoing chemoembolization for hepatocellular carcinoma: A randomized trial. *Aliment. Pharmacol. Ther.* **2004**, *19*, 779–788. [CrossRef]
221. Gramignano, G.; Lusso, M.R.; Madeddu, C.; Massa, E.; Serpe, R.; Deiana, L.; Lamonica, G.; Dessì, M.; Spiga, C.; Astara, G.; et al. Efficacy of L-carnitine administration on fatigue, nutritional status, oxidative stress, and related quality of life in 12 advanced cancer patients undergoing anticancer therapy. *Nutrition* **2006**, *22*, 136–145. [CrossRef] [PubMed]
222. Li, H.; Lockwood, M.B.; Schlaeger, J.M.; Liu, T.; Danciu, O.C.; Doorenbos, A.Z. Tryptophan and Kynurenine Pathway Metabolites and Psychoneurological Symptoms Among Breast Cancer Survivors. *Pain Manag. Nurs.* **2023**, *24*, 52–59. [CrossRef] [PubMed]
223. Nees, J.; Schafferer, S.; Yuan, B.; Tang, Q.; Scheffler, M.; Hartkopf, A.; Golatta, M.; Schneeweiß, A.; Burwinkel, B.; Wallwiener, M. How previous treatment changes the metabolomic profile in patients with metastatic breast cancer. *Arch. Gynecol. Obstet.* **2022**, *306*, 2115–2122. [CrossRef] [PubMed]
224. Kim, S.K.; Jung, W.H.; Koo, J.S. Diferential expression of enzymes associated with serine/glycine metabolism in diferent breast cancer subtypes. *PLoS ONE* **2014**, *9*, e101004.
225. Prokhorova, E.; Zobel, F.; Smith, R.; Zentout, S.; Gibbs-Seymour, I.; Schützenhofer, K.; Peters, A.; Groslambert, J.; Zorzini, V.; Agnew, T.; et al. Serine-linked PARP1 auto-modification controls PARP inhibitor response. *Nat. Commun.* **2021**, *12*, 4055. [CrossRef] [PubMed]
226. Possemato, R.; Marks, K.M.; Shaul, Y.D.; Pacold, M.E.; Kim, D.; Birsoy, K.; Sethumadhavan, S.; Woo, H.K.; Jang, H.G.; Jha, A.K.; et al. Functional genomics reveal that the serine synthesis pathway is essential in breast cancer. *Nature* **2011**, *476*, 346–350. [CrossRef]

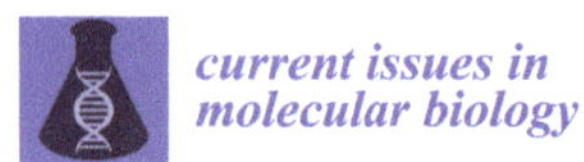

current issues in molecular biology

Article

Analysis of Somatic Mutations in the TCGA-LIHC Whole Exome Sequence to Identify the Neoantigen for Immunotherapy in Hepatocellular Carcinoma

Swetha Pulakuntla [1], Khajamohiddin Syed [2,*] and Vaddi Damodara Reddy [1,2,*]

[1] School of Applied Sciences, REVA University, Bangalore 560064, Karnataka, India; swethabiochem@gmail.com

[2] Department of Biochemistry and Microbiology, Faculty of Science, Agriculture and Engineering, University of Zululand, KwaDlangezwa 3886, South Africa

* Correspondence: khajamohiddinsyed@gmail.com (K.S.); damodara.reddyv@reva.edu.in (V.D.R.); Tel.: +27-035-902-6857 (K.S.); +91-950-263-9348 (V.D.R.)

Abstract: There are numerous clinically proven methods for treating cancer worldwide. Immunotherapy has been used to treat cancer with significant success in the current studies. The purpose of this work is to identify somatically altered target gene neoantigens and investigate liver cancer-related immune cell interaction and functional changes for potential immunotherapy in future clinical trials. Clinical patient data from the Cancer Genome Atlas (TCGA) database were used in this investigation. The R maf utility package was used to perform somatic analysis. The 17-mer peptide neoantigens were extracted using an in-house Python software called Peptide.py. Additionally, the epitope analysis was conducted using NetMHCpan4.1 program. Neopeptide immunogenicity was assessed using DeepCNN-Ineo, and tumor immune interaction, association with immune cells, correlation, and survival analysis were assessed using the TIMER web server. Based on somatic mutation analysis, we have identified the top 10 driver genes (TP53, TNN, CTNNB1, MUC16, ALB, PCLO, MUC4, ABCA13, APOB, and RYR2). From the superfamily of 20 HLA (Human leukocyte antigens) allele epitopes, we discovered 5653 neopeptides. Based on T cell receptor face hydrophobic analysis, these neopeptides were subjected to immunogenicity investigation. A mutation linked to tumor growth may have an impact on immune cells. According to this study's correlation and survival analysis, all driver genes may function as immune targets for liver cancer. These genes are recognized to be immune targets. In the future, immune checkpoint inhibitors may be developed to prolong patient survival times and prevent hepatocellular carcinoma (HCC) through immunotherapy.

Keywords: liver hepatocellular carcinoma; somatic mutations; neoantigens; immune checkpoint inhibitors; immunotherapy

check for **updates**

Citation: Pulakuntla, S.; Syed, K.; Reddy, V.D. Analysis of Somatic Mutations in the TCGA-LIHC Whole Exome Sequence to Identify the Neoantigen for Immunotherapy in Hepatocellular Carcinoma. *Curr. Issues Mol. Biol.* **2024**, *46*, 106–120. https://doi.org/10.3390/cimb46010009

Received: 25 October 2023
Revised: 8 December 2023
Accepted: 19 December 2023
Published: 22 December 2023

1. Introduction

Approximately 1.80 billion cases and 830,000 fatalities from liver cancer were predicted for 2020; by 2025, more than 1 million people may be affected [1–3]. Liver cancer is currently the third most common cause of cancer-related deaths globally. Out of the total liver malignancies, 85–90% of them are hepatocellular carcinomas (HCC). Chronic hepatitis B and C virus infections include a number of significant risk factors, the most dangerous of which is cirrhosis [4,5]. As the pathogenic co-factors in HCC, alcohol and tobacco are additional related risk factors [6].

Depending on the etiologies and gene alterations, different pathogenic molecular studies have been conducted [7]. Molecular parthenogenesis could be used to identify the disease's underlying mutations, but the therapeutic use of this newfound understanding is still far in the future [8]. Finding driver genes with oncogenic and suppressive properties in HCC may be improved by the high throughput gene sequence [9]. Telomerase activation mutations, viral insertions, chromosomal changes, and gene duplications are characteristics that set HCC apart [10]. Novel proteins (neoantigens) or tumor-specific proteins that are attached to major histocompatibility complexes on the cell surface and that can be recognized by T cell receptors (TCRs) for additional cell response may be generated by mutations and viral oncogenes [11].

The immune classification of liver cancer has been established by numerous investigations utilizing biological, immunological, genomics, and epigenomics techniques [12,13]. The phrases immune-active, immune-exhausted, and immune-classification were utilized in our study, which was immune-classification oriented. Tumor microenvironment activation of immune-exhausted tumors is a major factor in HCC. It has an increased concentration of helper T(CD4+) cells, and it may cause cytotoxic T (CD8+) cells to react negatively to immune checkpoint inhibitors (ICIs) [14]. The immune system's current reaction to a tumor attack is known as the ICI, which activates T cells and has demonstrated increased effectiveness in treating a number of solid tumors [12]. For the treatment of liver cancer, the Food and Drug Administration (FDA) has approved ICIs such as ipilimumab, nivolumab, pembrolizumab, and atezolimumab [15,16]. These immune checkpoint inhibitors (ICIs) target T cell immunoglobulin and mucin domain –3 (TIM3), lymphocyte activation gene 3 (LAG-3), programmed cell death protein-1 (PD-1) and ligand (PDL-1), and cytotoxic T-lymphocyte protein-4 (CTLA4). ICIs can shrink tumors and increase survival rates by reactivating repressed T cells that attack cancer [17,18]. Checkpoint blockade therapy is successful, although only a small percentage of patients benefit from it. There are currently no immunological targets that can be used to predict a patient's response. The identification of unique and uncommon cancer antigens as well as co-inhibitory signaling molecules that coordinate T cell immunotherapy thus constitutes the novelty of this work.

2. Materials and Methods

2.1. Data Collection

The TCGA database (accessed on 15 July 2022, https://portal.gdc.cancer.gov/) provided the whole exome sequencing (WES) open-source data for LIHC. The mutation annotation format (MAF) file containing all of the patient's clinical data was obtained. With the Illumina HiSeq 2000 Whole Exome Sequencing Platform, 358 LIHC patients' sample sequencing was completed. All patients with hepatocellular carcinoma had the liver and intrahepatic bile ducts as their primary sites of cancer. The MAF files were analyzed using the R maftools and TCGA bio links packages.

2.2. Identification of Neoantigens

Protein sequences for the top 10 driver genes (P53, TNN, CTNNB1, MUC16, ALB, PCLO, MUC4, ABCA13, APOB, and RYR2) were obtained from the Uniport database (https://www.uniprot.org/, accessed on 15 July 2022). The 17-mer peptide length, where the mutated type (MT) amino acid was in the middle of the other eight amino acids from upstream and downstream, and wildtype (WT) amino acid sequences for the top 10 genes (S2) were extracted using our proprietary Python script (Peptide.py) (Table S1) with pVAC-seq (v.4.08) as the reference [19]. Artificial neural networks are used by NetMHCpan v4.1 software, Lyngby, Denmark to train an epitope analysis algorithm. We selected a super-family of HLA class-I 20 alleles (HLA-A*01:01, HLA-A*02:01, HLA-A*02:03, HLA-A*02:07, HLA-A*03:01, HLA-A*11:01, HLA-A*24:02, HLA-A*29:02, HLA-A*31:01, HLA-A*32:01, HLA-A*68:02, HLA-A*07:02, HLA-B*15:01, HLA-B*35:01, HLA-B*40:01, HLA-B*44:02, HLA-B* 44:03, HLA-B*51:01, HLA-B*54:01, and HLA-B*57:01) from earlier research [20]. For epitope analysis, the top 10 driver genes were tested against 20 alleles. The key selection and filtering mechanism for all of the peptide binding affinity of MHC (major histocompatibility) molecules is the 9-mer amino acid chain. For this investigation, we selected 9-mer peptides determined by inhibitory concentration (IC50). These IC50 values are assumed to have 500 nM for weak binding and 50 nM for strong binding [21–23].

2.3. Potential Neoantigen Analysis

The class-I HLA neoantigens' immunogenicity were predicted by the DeepCNN-Ineo (accessed on 15 July 2022, http://119.3.70.71/dbPepNeo2/deepcnn-ineo.html) based on the score. This application is based on a convolutional neural network-based deep learning model that was generated utilizing curated MHC-I epitope data from the Immune Epitope Database (accessed on 15 July 2022, IEDB, https://www.iedb.org/). The recommended score for high immunogenicity is 0.8, with 0.5–0.8 for low immunogenicity and less than 0.5 for non-immunogenicity.

2.4. Immune Profile Studies with Timer Web Server

The tumor immune estimation resource (TIMER) is a web resource for systematic evaluation of the clinical impact of different immune cells in diverse cancer types that (accessed on 15 July 2022, https://cistrome.shinyapps.io/timer) may be used to analyze the relationship and survival analysis between immune gene markers and liver cancer, as well as to determine the infiltrating status for six immune cell types: B cells, CD8+ T cells, CD4+ T cells, macrophages, neutrophils, and dendritic cells [24].

2.5. Statistical Analysis

All patient somatic mutation analysis was conducted using statistical significance analysis, and presentations were performed using R v4.0. Fisher's exact test was used to compare categorical variables. We used Spearman's correlation and statistical significance to evaluate the correlation of gene expression. The Cox proportional hazards regression model was used to assess the risk factors in the overall survival analysis.

3. Results

3.1. LIHC Data and Clinical Information Selection

We obtained LIHC data for 358 patient samples including normal and tumor samples from the TCGA using clinical information. Whole exome sequencing (WES) data were used for all of these samples. As seen in Figure 1, this clinical data included the patient ID, gender, age, and survival status. The number of male patients (241) was more than the number of female patients (117) in this instance (Table S2). Using Fisher's exact test, the death rate in the late stage was considerably high (p = 0.0436). The overall survival (OS) of males and females differed slightly.

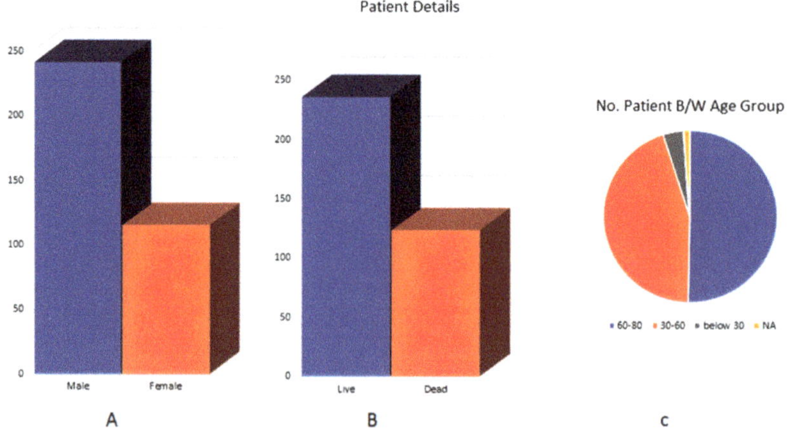

Figure 1. Clinical information on the included LIHC-TCGA samples (**A**) of gender differences between male and female LIHC patients, (**B**) survival status, whether life or death, and (**C**) no. of patients in each age group.

3.2. Analyzing the MAF File for the Somatic Mutation Analysis

High-level platforms include the full exome sequencing, TCGA-LIHC MAF file data-specific variant calling with MuSE or MuTech. We utilized the R package maftools to statistically summarize and visualize the mutation study. The TCGA sample barcode in the MAF file may be used to identify somatic mutations and determine the frequency of mutations for each patient's suggested clinical data. Plot maf was utilized to show the variation categorization and kind in a boxplot, with the number of variants in each sample in a stacked barplot. The summary of multiple hits, annotated variants, and mutated genes is shown in Figure 2A. The number of variants in each gene divided by the total number of patients (358) with at least one mutation identified provides the frequency of the genes. Of the 358 total samples in Figure 2B, the top 10 genes altered 268 samples, or 74.86% of the total. The functional plots in boxplot Figure 2C indicate the number of variations in allele mean frequency 50 for each sample, which may help identify the most important driver genes (P53, TNN, CTNNB1, MUC16, ALB, PCLO, MUC4, ABCA13, APOB, and RYR2).

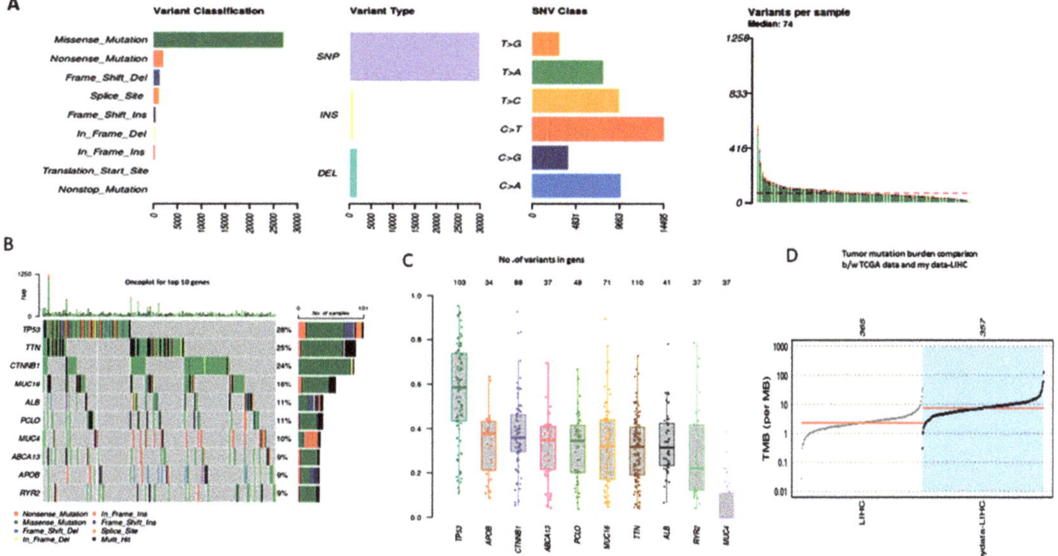

Figure 2. Somatic variant analysis of TCGA-LIHC data. (**A**) Variants per sample in stacked barplot, with variant types as a box plot; (**B**) onco plot for the TOP 10 driver genes; (**C**) driver genes variant allele frequency in the box plot; (**D**) comparison of the mutation load against the TCGA cohorts.

3.3. TMB (Tumor Mutational Burden)

The liver cancer mutation landscape has been revealed by NextGen Sequencing (NGS) technology. The TMB measures assess how many non-synonymous somatic mutations there are in each patient's sample per million base pairs. Previous studies have demonstrated that TMB reacts to solid tumors and may be used to target liver cancer biomarkers with immune treatment. The TMB determines how many mutations there are in each megabase (log10per) of the genomic sequences. It is believed that the TMB is a major factor in the production of immunogenic neopeptides. One sample had no mutations, making the total 358 samples' mutation burden one. Figure 2D shows a plot of the data, which are the 357 LIHC samples compared with 363 TCGA cohorts.

3.4. Mutational Signatures

Most hepatocytes constantly accumulate several DNA mutations and epigenetic modifications along with other risk factors when liver disorders arise. Six DNA substitutions

(C > T, C > A, T > C, T > A, C > G, T > G) were found by the somatic mutation analysis of LIHC. Additional classifications of the single nucleotide polymorphisms (SNPs) into transitions (Ti) and transversions (Tv) are displayed in the stacked bar graph in Figure 3. The C > T transition has the highest number of base mutations, but transversions in the C > A and T > C transitions also have the highest number of base mutations. Notably, there are significantly fewer transversions in the base pairs T > A, C > G, and T > G. Transversions, on the other hand, encompass far more than just transitions.

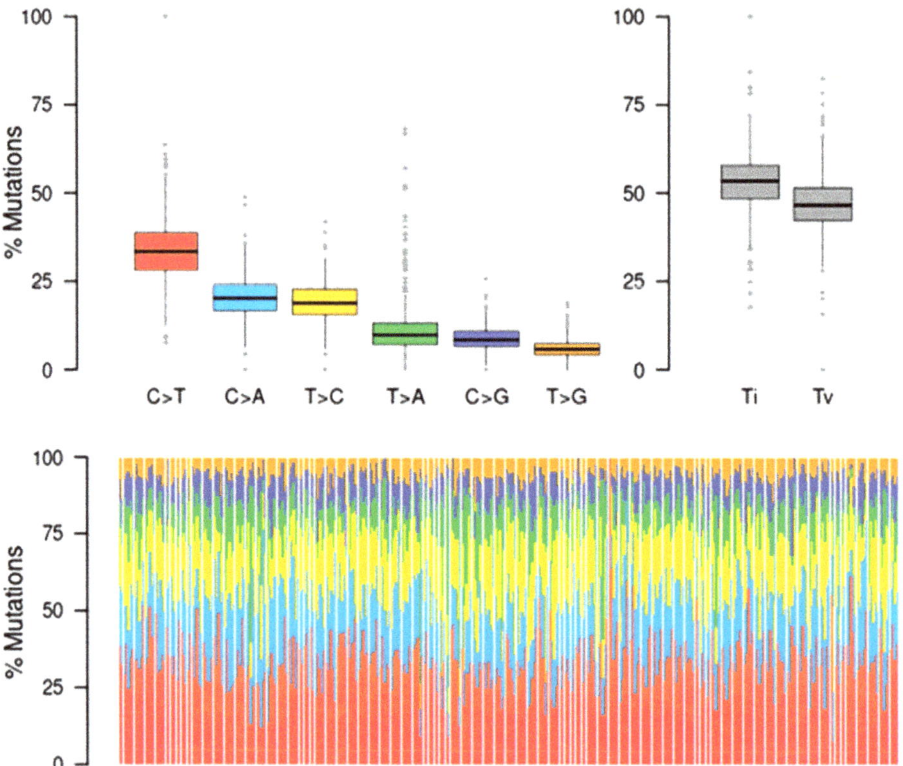

Figure 3. Mutational signatures of liver cancer patients. The functional classification of SNPs is summarized; statistical data were visualized as box plots, with the overall distribution of six different conversions of transitions and transversions.

3.5. Pathways of Oncogenic Signaling

In the present study, ten well-known signaling pathways such as the RTK-RAS, WNT, NOTCH, Hippo, PIK3, Cell Cycle, MYC, TGF-, TP53, and NRF2 were examined. In addition, we examined the processes behind these somatic alterations. Here, using 358 TCGA-LIHC samples with a fraction of mutations in clusters on the X-axis, as displayed in Figure 4, we worked on a framework to design routes evenly. Numerous studies have shown that TCGA-LIHC commonly alters a wide range of significant pathways. More changed pathways, such as 80% RTK-RAS or cell-cycle pathways in numerous tumor types, are also present in tumors with the highest tumor mutation burden. In total, 30% to 50% of the Wnt signaling pathways are caused by the CTNNB1 gene mutation. In 70% of other driver genes, P53 genomic alterations changed gene-centricity as well as intra- and inter-pathway interactions. This pathway particularly responds to immune checkpoint inhibitors and may be enhanced by CD4+ and CD8+ cell infiltrations and immunological categorization of HCC tumors. It might help in the treatment of cancer.

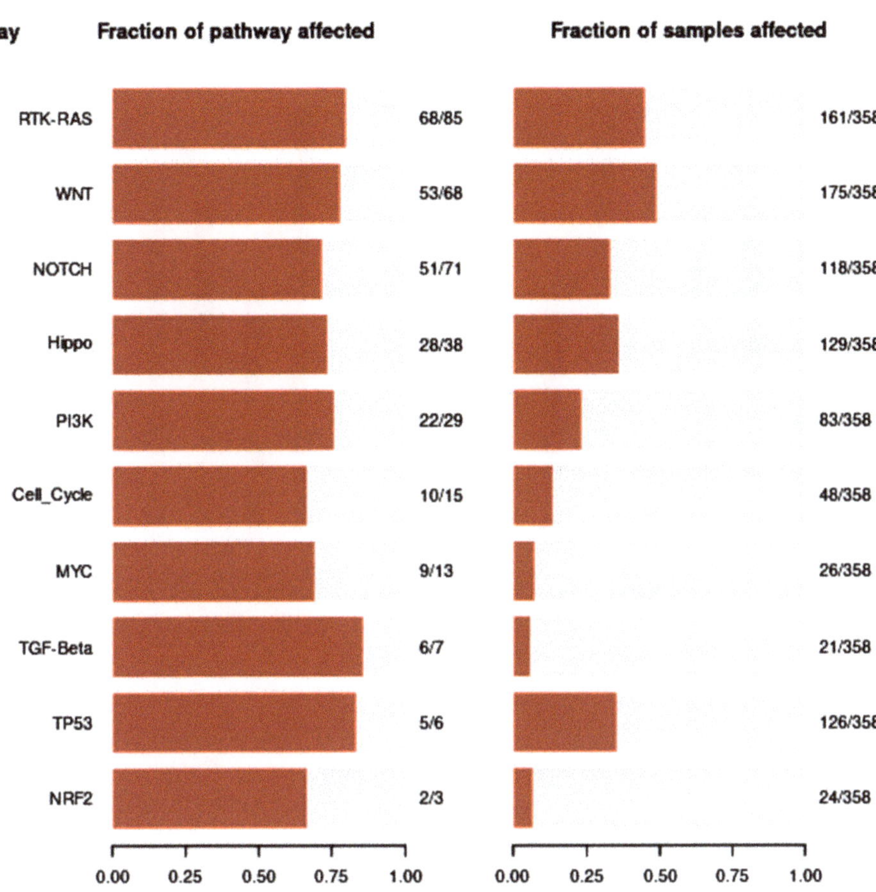

Figure 4. Enrichment of known oncogenic signaling pathways in TCGA-LIHC cohorts.

3.6. Identification of Neoantigens (Peptide Selection and Epitope Analysis)

To choose the 17-mer peptide length MT and WT types in FASTA format, we created a bespoke Python tool (Table S3). For the epitope analysis, we selected the top 10 somatic mutation driver genes from the TCGA-LIHC. The Uniport database was the source of the driver gene protein sequence. We chose super family HLA-Class-I (20) alleles, enabling us to provide results that may account for 95% of the world's human population. Additionally, using the elution ligand method (EL) and the academically licensed NetMHCpan-4.1 software, epitope analysis was carried out using the HLA alleles versus peptides as inputs. A large amount of allele training data was used to accurately quantify the peptide's prediction of the HLA binding affinity. On the basis of the IC50 threshold (500 nM) values, we were able to forecast all potential mutant epitopes by strong and weak binding affinity. Since 9-mer peptides may comprise 90% of neoantigens, we exclusively took them into consideration. Human T cell responses may demonstrate this.

Using the default NetMhcpan4.1 IC50 values, we filtered 5653 neopeptides of driver gene missense mutations (Table S4). The altered peptides are recognized by cytotoxic CD8+ T cells when they bind to class-I MHC molecules and nucleotides. Following the expected extension and characterization of the investigation, there appeared to be a correlation between the observed number of neopeptides and the mutation burden. Figure 5 shows the findings of the largest amount of neopeptides' predicted frequency driving genes.

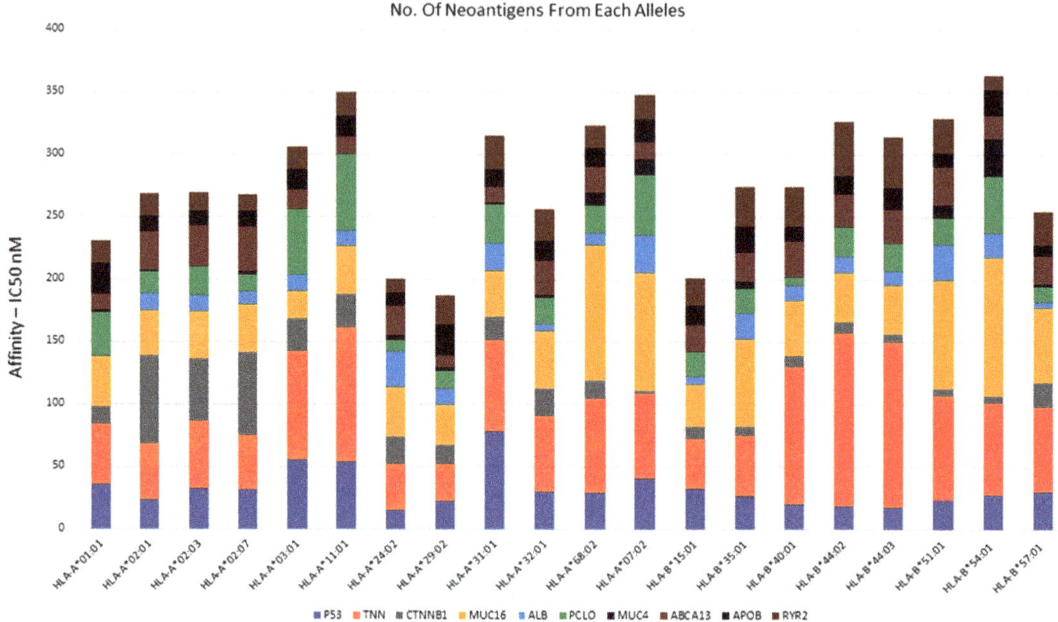

Figure 5. High throughput prediction of MHC class-I alleles: the neoantigen-specific prediction based on IC50 binding affinity (IC50 < 500 nM) and the top mutated genes are indicated with different color bar graphs.

3.7. Identification of Potential Neoantigens

Ultimately, neoantigens and epitopes from the NetMHCpan study were confirmed by HLA-class-I 8–9 amino acid length. Because they are more sensitive, a few amino acids choose to add N- and C-terminal anchor sites. TCR binds to hydrophobic amino acids most of the time. Epitopes with this frequency are very immunogenic. The frequency of the neopeptide immunogenicity features was validated in the IEDB based on the charge. To predict the TCR epitopes for both immunogenic and non-immunogenic neopeptides, we employed the website DeepCNN-Ineo program. To identify neoantigens using the top 10 driver genes, we employed nine superfamilies (20) of HLA class-I alleles (Table S5). The suggested peptide length is eight or twelve amino acids, with position two at the N-terminus and position C-terminal being used as the terminal anchor sites by default. The prediction scores were less than 0.5 when contemplating non-immunogenicity (negative high), higher than 0.5 when considering high positive immunogenicity, and between 0.5 and 0.8 when indicating low positive immunogenicity. These values show that the immunogenicity score is indicated on the Y-axis of all neoantigen immunogenicity values that were plotted using the Python script results. Every driver gene with 20 HLA alleles on the X-axis is a neoantigen (Figure 6).

3.8. Immune Profile Data Analysis

Genetic changes are linked to the growth and spread of tumors, and these changes may have an impact on the immune cells that infiltrate tumors (TIIC). All six immune cell types—B cells, CD8+ T cells, CD4+ T cells, macrophages, neutrophils, and dendritic cells—in their WT and mutant states could be compared with the driver gene somatic mutation. P53 (28%), TNN (25%), CTNNB1 (24%), MUC16 (16%), PCLO (11%), ALB (11%), MUC4 (10), ABCA13 (9%), APOB (9%), and RYR2 (9%) are the 10 most frequently mutated genes. With the use of the dynamic web interface tool TIMER, the mutation module was used to examine the immunological infiltration. The box plots produced for each immune group are shown in Figure 7, where the immune infiltration distribution level of each gene mutation is compared using statistically significant values and a 95% confidence interval, determined using the two-sided Wilcoxon rank sum test.

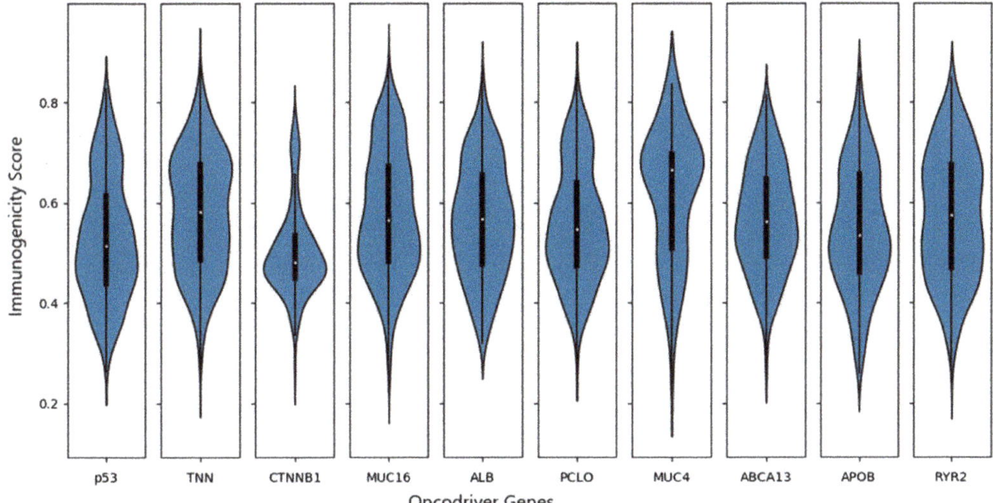

Figure 6. Immunogenicity of recurrent oncogenic genes non-synonymous mutations' distribution shown in Violin plots. The points show the median values, where 0.8 indicates high immunogenicity, 0.5–0.8 means low immunogenicity, and less than 0.5 is non-immunogenic.

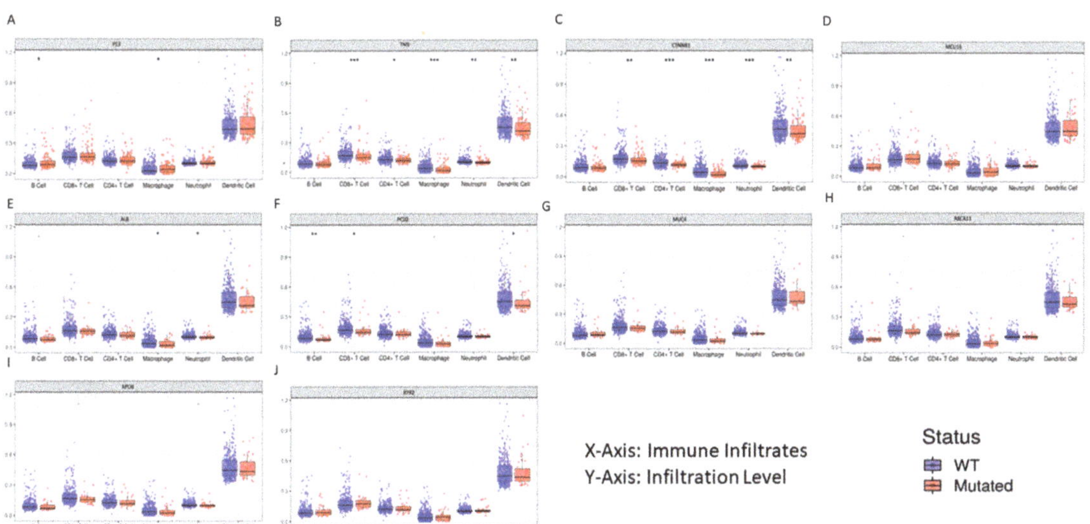

Figure 7. The box plots from the mutation module display the difference in the tumor immune estimation resource (TIMER)-estimated in six immune cell (B Cell, CD8+T Cell, CD4+ T Cell, macrophage, neutrophil, and dendritic cell) infiltrates level estimated between mutant and wildtype driver genes (**A**). TP53, (**B**). TNN, (**C**). CTNNB1, (**D**). MUC16, (**E**). ALB, (**F**). PCLO, (**G**). MUC4, (**H**). ABCA13, (**I**). APOB, and (**J**). RYR2, respectively, in liver cancer.

3.9. Investigate Immune Checkpoint Inhibitors

Immunotherapy using immune checkpoint inhibitors is currently the most successful therapeutic treatment for metastatic liver cancer. This is an important development in cancer biomarker prediction. In this work, we examined the relationship between immune cells and tumor immune infiltration, and we found that immune inhibitory receptors such as PDCD1, CTLA4, TIM3, and LAG3 may be important for T cell activation in tumor cells. Dostarlimab (TSR-042), an antibody that blocks the PDCD1 receptor, was recently approved by the FDA to treat endometrial cancer. Advanced solid tumors have been used in this clinical trial (NCT02715284) to test the antibody mismatch repair and DNA repair functions [25]. There are ongoing phase-II clinical trials (NCT03680508) using TSR-042 PDCD1 (PD-1) and TSR-022 anti-HAVCR2 (TIM3) antibodies for primary liver cancer. We examined the computational relationship between the widespread ICB receptors PDCD1/HAVCR2 and LAG3 and the TCGA-LIHC driver gene. According to earlier research, CD8+ T cells are associated with inhibitor receptors, which trigger T cell activation for therapeutic liver cancer treatment. Not on a normal liver tissue sample but on CD8+ T cell malignancies, elevated expression of TIM-3 and LAG-3 may facilitate immune evasion and poor prognosis. The TIMER analysis revealed that PDCD1, CTLA4, HAVCR2, and LAG-3 correlated with all ten of the TCGA-LIHC's top genes, with correlation values ranging from 0.19 to 1. The significant p values for this statistical connection, which were determined using Spearman's rho value, are displayed in Figure 8A,B.

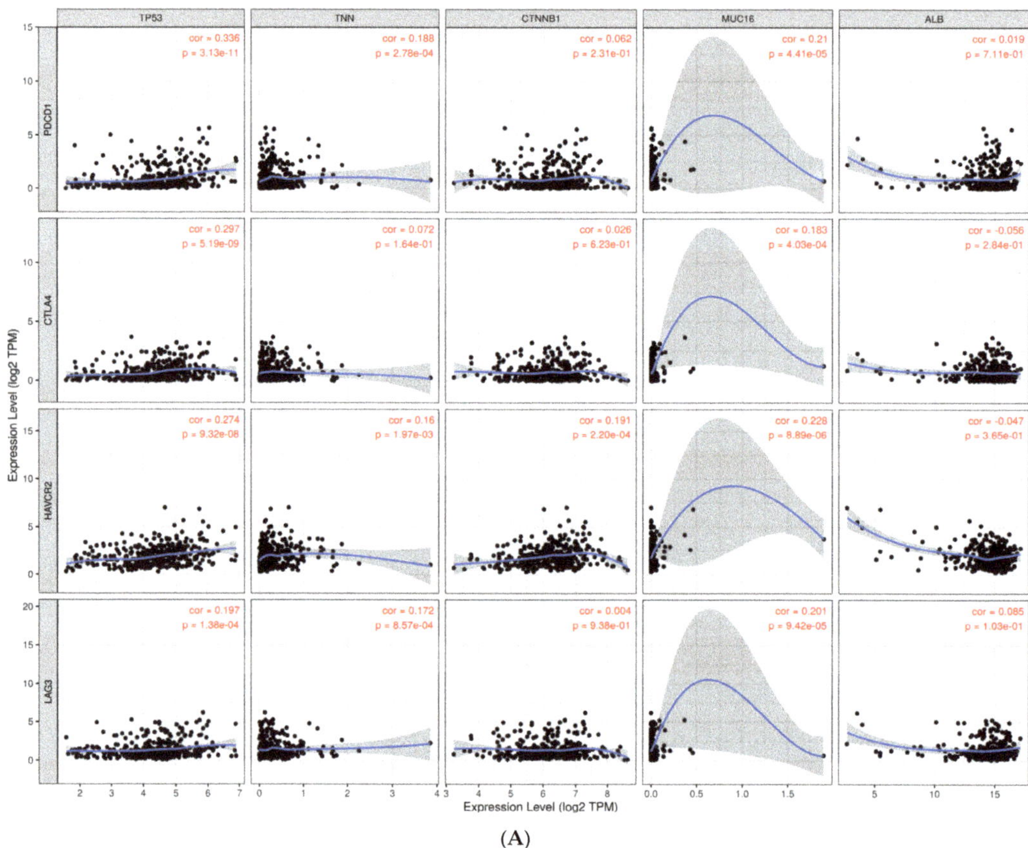

(A)

Figure 8. *Cont.*

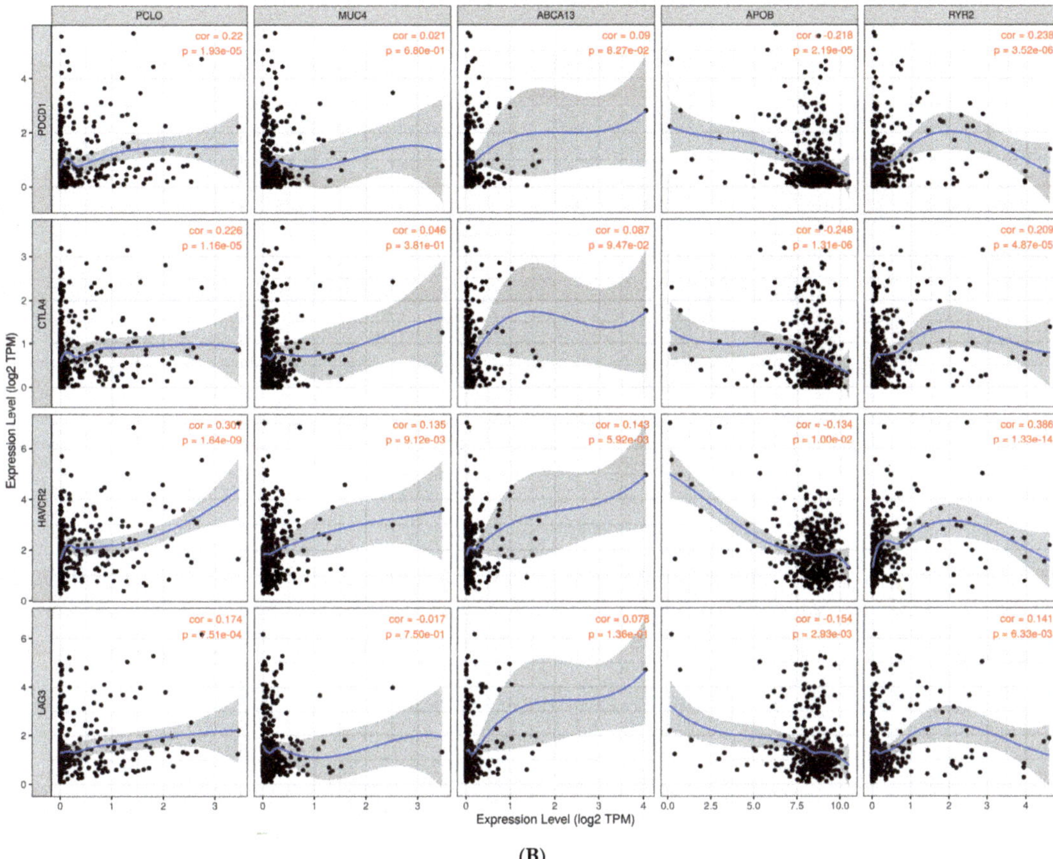

Figure 8. (**A**). Scatter plots derived from the relationship between the expression of driver genes (TP53, TNN, CTNNB1, MUC16, and ALB) on the X-axis and known immune targets (PDCD1, CTLA4, HAVCR2, and LAG3) on the Y-axis. (**B**). Scatter plots derived from the association between driver genes (PCLO, MUC4, ABCA13, APOB, and RYR2) and known immunological targets (PDCD1, CTLA4, HAVCR2, and LAG3) on the X- and Y-axes.

The median group computed the hazard ratio (HR) on the Cox PH model with a 95% confidence interval (CI) by analyzing the TCGA-LIHC of each driver gene's overall survival. Red denotes a high group risk, and blue denotes a low group risk in the group survival plot displayed in Figure 9. An examination of the tumor mutation burden survival shows a correlation between these driver genes. Future immunotherapy research, including ICI clinical studies, will be necessary.

Overall Survival

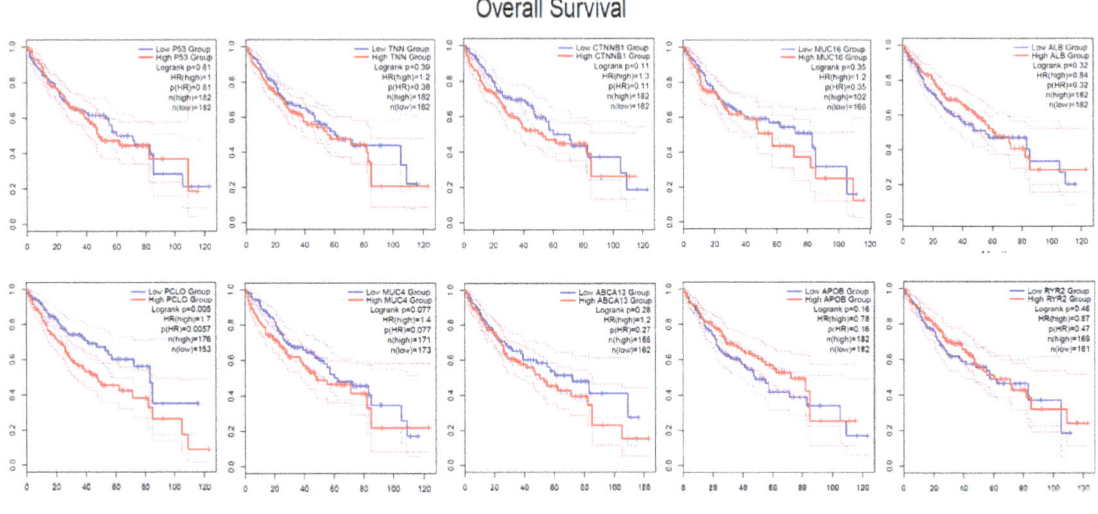

Figure 9. The overall survival analysis plot shows the median values on the Y-axis and months on the X-axis, indicating all driver genes TP53, TNN, CTNNB1, MUC16, ALB, PCLO, MUC4, ABCA13, APOB, and RYR2, respectively.

4. Discussion

The study group is gradually developing next-generation sequencing technology to combine massive amounts of data for immune therapy-related mutation studies in various malignancies. The entire exome sequences of 358 patients' sample maf files, which were obtained from the TCGA database, were used in the current investigation. Using R maf tools, we conducted a mutation study and discovered a large number of mutant genes. The number of neoantigens with somatic mutations per megabase in cancer correlates with the tumor mutation burden [26]. Using the netMHCpan4.1 program, an epitope analysis was conducted on a modified peptide sequence against 20 superfamily HLA-I alleles. Based on the IC50 values of 500 nM strong interaction with T cells, MHC molecules were predicted for the 8–9 mer epitopes/neoantigens. We identified possible neoantigens based on the immunogenicity score of the neopeptides using the DeepCNN-Ineo website. Data from the IEDB repository were used to train this program. The CD8+ T cells were classified as immunogenic or non-immunogenic based on the relative hydrophobicity of amino acids in this study. The T cell receptors could identify the antigenic character of the epitopes. With low G+C genomic codons potentially having an impact on the amino acid use case, the majority of the data included in this study came from intracellular diseases such as viruses [27]. These more hydrophobic potential pathogens may be employed to identify T cell receptors. Based on their close proximity to antigens, hydrophobic areas have been found to greatly boost the rate of proteasomal breakdown and the immunogenic T cell epitopes [28].

Current research on immunotherapy is leading to a greater understanding of how the immune system infiltrates cancer. It is difficult to analyze and visualize the vast amounts of clinical and genetic data. Developing a unique computational technique to deconvolve complex data is essential to investigate tumor–immune interactions [29]. Using six immune cells—B cells, CD8+ T cells, CD4+ T cells, macrophages, neutrophils, and dendritic cells—we comprehensively analyzed the immunological and genomic characteristics of tumors in our study using the TIMER online server-based technology [30]. Mutations in the top driver genes identified by TCGA-LIHC (TP53, TNN, CTNNB1, MUC16, ALB, PCLO, MUC4, ABCA13, APOB, and RYR2) were examined. Box plots were constructed for each immune cell subgroup, and the expression displayed the statistically significant values that were assessed using the two-sided Wilcoxon rank sum test (Figure 7). We found that CD8+ immune infiltration enhanced the tumors with high mutant gene counts in all TCGA-LIHC patients. We postulated that the immunogenic mutations in these predictions may reflect the expression of CD8+ [31].

The relationship between liver cancer driver genes and immune cell receptors was analyzed using TIMER in another module, and the results indicated a modest correlation with Spearman's statistical significance (0.9–1). The most significant immune biomarkers for hepatocellular carcinoma were employed in immunotherapy, according to earlier research: PDCD1, CTLA4, HAVCR2/TIM3, and LAG3 [32]. Liver cancer patients now have access to immune checkpoint inhibitor therapy. Both the tumor cells and the surrounding environment should be affected by the immunotherapy [33]. The immune response of CD8+ immune cells and the PDCD1 tumor association may be the targets of ICI molecules, which can also indicate the prognosis of cancer. TIM3 and LAG3 receptors are present on CD4+ and CD8+ immune cells [34]. Hepatocytes manufacture a unique functional fibrinogen-like protein-1 ligand that is exclusive to LAG3. These research preclinical findings back up the exploration of TIM3, LAG-3, and PDCD1 inhibition in liver cancer cases [35].

The best mean survival was 22.8 months based on the initial study of immunological biomarkers (CTLA4, PDCD1) and ICI combinations of ipilimumab and nivolumab [36]. The FDA approved the combinational ICI and additional combinations of durvalumab and tremelimumab for the PDCD1 in light of these positive results [37]. Atemalizumab and bevacizumab were licensed by the FDA for use in immunotherapy for the first HCC combination on a worldwide scale [38]. These findings have led to the present clinical trials of a number of ICI combinations on patients receiving systemic immune treatment. Adoptive cell therapy for liver cancer also involves lymphocyte sensitization therapy, which is carried out in vivo prior to being reinfused into the patient [39]. The cytokine killer cells are triggered by the lymphokine. Another extremely effective and promising treatment for hematological malignancies with solid tumors that is still in the research phase is the CAR T cell [40]. The antigen domain of this treatment is made up of a monoclonal antibody signal fragment that is tailored to a particular tumor cell target. The target cell's toxicity is one drawback of this CAR T cell treatment. The target cells express themselves in normal cells as a result. GPC3 CAR T cell intravenous injunction is undergoing safe and effective clinical trials (NCT04121273) [41].

The use of vaccines to prevent cancer is another important factor. The potency may rise due to the tumor-specific response. This demonstrates how well T lymphocytes are prepared to combat antigens produced by cancer cells [42]. Historically, vaccinations have been administered as a standalone treatment; however, it is now known that immuno-suppression in the tumor microenvironment increases T cell activity [43]. Additional immunological vaccinations may detect the antigen specific to a tumor. HLA typing is the best method to use for immune-related peptide identification in this case. Furthermore, specific immunological signatures associated with cytotoxic T-lymphocytes (CTLs) are expressed along with this HLA peptidome. Based on these immunological signature results on board, these vaccination clinical studies for liver cancer are still ongoing [44].

5. Conclusions

It is necessary to expand the gene mutation analysis from the sequence source in this investigation of immunotherapy for liver cancer. Novel neoantigens unique to tumor antigens recognized by T lymphocytes may be found as a result of somatic mutations. Further research on immune cell profiles and immune infiltration expression CD8+ predictions are indicated by the altered genes. In order to identify novel immune biomarkers from the TCGA-LIHC and examine associated expression with established immune biomarkers for T cell activation with an immune checkpoint inhibitor in liver cancer, this study examined PDCD1, CTLA4, HAVCR2, and LAG3. These investigations on immune receptors and altered genes may prove useful in the future for developing vaccines and ICIs for the immunotherapy of liver cancer.

Supplementary Materials: The following supporting information can be downloaded at: https://www.mdpi.com/article/10.3390/cimb46010009/s1, Table S1: In-house c program for the 18-mer neoantigen peptide filter from the wildtype and normal peptide sequences; Table S2: XLS file of 358 patients' clinical information; Table S3: XLS file of the top 10 genes' 18-mer peptide sequence (wildtype and mutant); Table S4: XLS file of the 18-mer protein sequence with superfamily HLA class-1 alleles using NetMhcpan software extensive data; Table S5: XLS file for the potential neoantigen analysis data.

Author Contributions: Conceptualization, S.P., K.S. and V.D.R.; methodology, S.P., K.S. and V.D.R.; software, S.P., K.S. and V.D.R.; validation, S.P., K.S. and V.D.R.; formal analysis, S.P., K.S. and V.D.R.; investigation, S.P., K.S. and V.D.R.; resources, S.P., K.S. and V.D.R.; data curation, S.P., K.S. and V.D.R.; writing—original draft preparation, S.P., K.S. and V.D.R.; writing—review and editing, S.P., K.S. and V.D.R.; visualization, S.P., K.S. and V.D.R.; supervision, K.S. and V.D.R.; project administration, K.S. and V.D.R.; funding acquisition, K.S. and V.D.R. All authors have read and agreed to the published version of the manuscript.

Funding: V. Damodara Reddy thanks SERB, New Delhi, India, for providing financial support under Ramanujan Fellowship schemes (SB/S2/RJN-043/2014 dated 17 December 2015), and Swetha Pulakuntla thanks ICMR, India for a Senior Research Fellowship (ISRM/11 (32)/2022). Khajamohiddin Syed sincerely thanks the University of Zululand for financial aid (Grant number P419).

Institutional Review Board Statement: Not applicable.

Informed Consent Statement: Not applicable.

Data Availability Statement: All data are available in the main manuscript and supplementary information.

Acknowledgments: Vaddi Damodara Reddy thanks the University of Zululand, South Africa, for appointing him as a visiting fellow.

Conflicts of Interest: The authors declare no conflicts of interest, and the funders had no role in the design of the study, in the collection, analyses, or interpretation of data, in the writing of the manuscript, or in the decision to publish the results.

References

1. Ferlay, J.; Ervik, M.; Lam, F.; Colombet, M.; Mery, L.; Piñeros, M.; Znaor, A.; Soerjomataram, I.; Bray, F. *Global Cancer Observatory: Cancer Today*; International Agency for Research on Cancer: Lyon, France, 2020. Available online: https://gco.iarc.fr/today/home (accessed on 18 December 2023).
2. Villanueva, A. Hepatocellular Carcinoma. *N. Engl. J. Med.* **2019**, *380*, 1450–1462. [CrossRef] [PubMed]
3. Renne, S.L.; Sarcognato, S.; Sacchi, D.; Guido, M.; Roncalli, M.; Terracciano, L.; Di Tommaso, L. Hepatocellular carcinoma: A clinical and pathological overview. *Pathologica* **2021**, *113*, 203. [CrossRef] [PubMed]
4. Llovet, J.; Kelley, R.; Villanueva, A.; Singal, A.; Pikarsky, E.; Roayaie, S.; Lencioni, R.; Koike, K.; Zucman-Rossi, J.; Finn, R. Hepatocellular carcinoma. *Nat. Rev. Dis. Primers* **2021**, *7*, 6. [CrossRef] [PubMed]
5. Gilles, H.; Garbutt, T.; Landrum, J. Hepatocellular Carcinoma. *Crit. Care Nurs. Clin. N. Am.* **2022**, *34*, 289–301. [CrossRef]
6. Rich, N.E.; Yopp, A.C.; Singal, A.G.; Murphy, C.C. Hepatocellular carcinoma incidence is decreasing among younger adults in the United States. *Clin. Gastroenterol. Hepatol.* **2020**, *18*, 242–248.e5. [CrossRef] [PubMed]
7. Gao, X.; Zhao, C.; Zhang, N.; Cui, X.; Ren, Y.; Su, C.; Wu, S.; Yao, Z.; Yang, J. Genetic expression and mutational profile analysis in different pathologic stages of hepatocellular carcinoma patients. *BMC Cancer* **2021**, *21*, 786. [CrossRef] [PubMed]
8. Liver, E.A.F.T.S.O.T. EASL clinical practice guidelines: Management of hepatocellular carcinoma. *J. Hepatol.* **2018**, *69*, 182–236.
9. Zeng, H.; Hui, Y.; Qin, W.; Chen, P.; Huang, L.; Zhong, W.; Lin, L.; Lv, H.; Qin, X. High-throughput sequencing-based analysis of gene expression of hepatitis B virus infection-associated human hepatocellular carcinoma. *Oncol. Lett.* **2020**, *20*, 18. [CrossRef]
10. in der Stroth, L.; Tharehalli, U.; Günes, C.; Lechel, A. Telomeres and telomerase in the development of liver cancer. *Cancers* **2020**, *12*, 2048. [CrossRef]
11. Xie, N.; Shen, G.; Gao, W.; Huang, Z.; Huang, C.; Fu, L. Neoantigens: Promising targets for cancer therapy. *Signal Transduct. Target. Ther.* **2023**, *8*, 9. [CrossRef]
12. Sia, D.; Jiao, Y.; Martinez-Quetglas, I.; Kuchuk, O.; Villacorta-Martin, C.; de Moura, M.C.; Putra, J.; Camprecios, G.; Bassaganyas, L.; Akers, N. Identification of an immune-specific class of hepatocellular carcinoma, based on molecular features. *Gastroenterology* **2017**, *153*, 812–826. [CrossRef] [PubMed]

13. Rebouissou, S.; Nault, J.-C. Advances in molecular classification and precision oncology in hepatocellular carcinoma. *J. Hepatol.* **2020**, *72*, 215–229. [CrossRef] [PubMed]

14. Ding, L.; Sun, L.; Bu, M.T.; Zhang, Y.; Scott, L.N.; Prins, R.M.; Su, M.A.; Lechner, M.G.; Hugo, W. Antigen presentation by clonally diverse CXCR5+ B cells to CD4 and CD8 T cells is associated with durable response to immune checkpoint inhibitors. *Front. Immunol.* **2023**, *14*, 1176994. [CrossRef] [PubMed]

15. Lachenmayer, A.; Alsinet, C.; Savic, R.; Cabellos, L.; Toffanin, S.; Hoshida, Y.; Villanueva, A.; Minguez, B.; Newell, P.; Tsai, H.-W. Wnt-pathway activation in two molecular classes of hepatocellular carcinoma and experimental modulation by sorafenib. *Clin. Cancer Res.* **2012**, *18*, 4997–5007. [CrossRef] [PubMed]

16. Brunet, J.-P.; Tamayo, P.; Golub, T.R.; Mesirov, J.P. Metagenes and molecular pattern discovery using matrix factorization. *Proc. Natl. Acad. Sci. USA* **2004**, *101*, 4164–4169. [CrossRef] [PubMed]

17. Wu, M.; Huang, Q.; Xie, Y.; Wu, X.; Ma, H.; Zhang, Y.; Xia, Y. Improvement of the anticancer efficacy of PD-1/PD-L1 blockade via combination therapy and PD-L1 regulation. *J. Hematol. Oncol.* **2022**, *15*, 24. [CrossRef] [PubMed]

18. Wang, B.; Zhao, Q.; Zhang, Y.; Liu, Z.; Zheng, Z.; Liu, S.; Meng, L.; Xin, Y.; Jiang, X. Targeting hypoxia in the tumor microenvironment: A potential strategy to improve cancer immunotherapy. *J. Exp. Clin. Cancer Res.* **2021**, *40*, 24. [CrossRef]

19. Hundal, J.; Carreno, B.M.; Petti, A.A.; Linette, G.P.; Griffith, O.L.; Mardis, E.R.; Griffith, M. pVAC-Seq: A genome-guided in silico approach to identifying tumor neoantigens. *Genome Med.* **2016**, *8*, 11. [CrossRef]

20. Bulik-Sullivan, B.; Busby, J.; Palmer, C.D.; Davis, M.J.; Murphy, T.; Clark, A.; Busby, M.; Duke, F.; Yang, A.; Young, L. Deep learning using tumor HLA peptide mass spectrometry datasets improves neoantigen identification. *Nat. Biotechnol.* **2019**, *37*, 55–63. [CrossRef]

21. Abelin, J.G.; Keskin, D.B.; Sarkizova, S.; Hartigan, C.R.; Zhang, W.; Sidney, J.; Stevens, J.; Lane, W.; Zhang, G.L.; Eisenhaure, T.M. Mass spectrometry profiling of HLA-associated peptidomes in mono-allelic cells enables more accurate epitope prediction. *Immunity* **2017**, *46*, 315–326. [CrossRef]

22. Bassani-Sternberg, M.; Chong, C.; Guillaume, P.; Solleder, M.; Pak, H.; Gannon, P.O.; Kandalaft, L.E.; Coukos, G.; Gfeller, D. Deciphering HLA-I motifs across HLA peptidomes improves neo-antigen predictions and identifies allostery regulating HLA specificity. *PLoS Comput. Biol.* **2017**, *13*, e1005725. [CrossRef] [PubMed]

23. Lund, O.; Nielsen, M.; Kesmir, C.; Petersen, A.G.; Lundegaard, C.; Worning, P.; Sylvester-Hvid, C.; Lamberth, K.; Røder, G.; Justesen, S. Definition of supertypes for HLA molecules using clustering of specificity matrices. *Immunogenetics* **2004**, *55*, 797–810. [CrossRef] [PubMed]

24. Li, T.; Fan, J.; Wang, B.; Traugh, N.; Chen, Q.; Liu, J.S.; Li, B.; Liu, X.S. TIMER: A web server for comprehensive analysis of tumor-infiltrating immune cells. *Cancer Res.* **2017**, *77*, e108–e110. [CrossRef] [PubMed]

25. Oaknin, A.; Tinker, A.V.; Gilbert, L.; Samouëlian, V.; Mathews, C.; Brown, J.; Barretina-Ginesta, M.-P.; Moreno, V.; Gravina, A.; Abdeddaim, C. Clinical activity and safety of the anti–programmed death 1 monoclonal antibody dostarlimab for patients with recurrent or advanced mismatch repair–deficient endometrial cancer: A nonrandomized phase 1 clinical trial. *JAMA Oncol.* **2020**, *6*, 1766–1772. [CrossRef] [PubMed]

26. Chan, T.A.; Yarchoan, M.; Jaffee, E.; Swanton, C.; Quezada, S.A.; Stenzinger, A.; Peters, S. Development of tumor mutation burden as an immunotherapy biomarker: Utility for the oncology clinic. *Ann. Oncol.* **2019**, *30*, 44–56. [CrossRef] [PubMed]

27. Calis, J.J.; Sanchez-Perez, G.F.; Keşmir, C. MHC class I molecules exploit the low G+ C content of pathogen genomes for enhanced presentation. *Eur. J. Immunol.* **2010**, *40*, 2699–2709. [CrossRef] [PubMed]

28. Kim, Y.; Yewdell, J.W.; Sette, A.; Peters, B. Positional bias of MHC class I restricted T-cell epitopes in viral antigens is likely due to a bias in conservation. *PLoS Comput. Biol.* **2013**, *9*, e1002884. [CrossRef]

29. Li, B.; Li, T.; Pignon, J.-C.; Wang, B.; Wang, J.; Shukla, S.A.; Dou, R.; Chen, Q.; Hodi, F.S.; Choueiri, T.K. Landscape of tumor-infiltrating T cell repertoire of human cancers. *Nat. Genet.* **2016**, *48*, 725–732. [CrossRef]

30. Li, B.; Severson, E.; Pignon, J.-C.; Zhao, H.; Li, T.; Novak, J.; Jiang, P.; Shen, H.; Aster, J.C.; Rodig, S. Comprehensive analyses of tumor immunity: Implications for cancer immunotherapy. *Genome Biol.* **2016**, *17*, 174. [CrossRef]

31. Flecken, T.; Schmidt, N.; Hild, S.; Gostick, E.; Drognitz, O.; Zeiser, R.; Schemmer, P.; Bruns, H.; Eiermann, T.; Price, D.A. Immunodominance and functional alterations of tumor-associated antigen-specific CD8+ T-cell responses in hepatocellular carcinoma. *Hepatology* **2014**, *59*, 1415–1426. [CrossRef]

32. Sangro, B.; Sarobe, P.; Hervás-Stubbs, S.; Melero, I. Advances in immunotherapy for hepatocellular carcinoma. *Nat. Rev. Gastroenterol. Hepatol.* **2021**, *18*, 525–543. [CrossRef] [PubMed]

33. Onuma, A.E.; Zhang, H.; Huang, H.; Williams, T.M.; Noonan, A.; Tsung, A. Immune checkpoint inhibitors in hepatocellular cancer: Current understanding on mechanisms of resistance and biomarkers of response to treatment. *Gene Expr.* **2020**, *20*, 53. [CrossRef] [PubMed]

34. Zhou, G.; Sprengers, D.; Boor, P.P.; Doukas, M.; Schutz, H.; Mancham, S.; Pedroza-Gonzalez, A.; Polak, W.G.; De Jonge, J.; Gaspersz, M. Antibodies against immune checkpoint molecules restore functions of tumor-infiltrating T cells in hepatocellular carcinomas. *Gastroenterology* **2017**, *153*, 1107–1119.e10. [CrossRef] [PubMed]

35. Wang, J.; Sanmamed, M.F.; Datar, I.; Su, T.T.; Ji, L.; Sun, J.; Chen, L.; Chen, Y.; Zhu, G.; Yin, W. Fibrinogen-like protein 1 is a major immune inhibitory ligand of LAG-3. *Cell* **2019**, *176*, 334–347.e12. [CrossRef] [PubMed]

36. Yau, T.; Kang, Y.-K.; Kim, T.-Y.; El-Khoueiry, A.B.; Santoro, A.; Sangro, B.; Melero, I.; Kudo, M.; Hou, M.-M.; Matilla, A. Nivolumab (NIVO)+ ipilimumab (IPI) combination therapy in patients (pts) with advanced hepatocellular carcinoma (aHCC): Results from CheckMate 040. *J. Clin. Oncol.* **2019**, *39*, 4012. [CrossRef]

37. Kelley, R.K.; Sangro, B.; Harris, W.P.; Ikeda, M.; Okusaka, T.; Kang, Y.-K.; Qin, S.; Tai, W.M.D.; Lim, H.Y.; Yau, T. Efficacy, tolerability, and biologic activity of a novel regimen of tremelimumab (T) in combination with durvalumab (D) for patients (pts) with advanced hepatocellular carcinoma (aHCC). *J. Clin. Oncol.* **2020**, *38*, 4508. [CrossRef]

38. Sinner, F.; Pinter, M.; Scheiner, B.; Ettrich, T.J.; Sturm, N.; Gonzalez-Carmona, M.A.; Waidmann, O.; Finkelmeier, F.; Himmelsbach, V.; De Toni, E.N. Atezolizumab plus bevacizumab in patients with advanced and progressing hepatocellular carcinoma: Retrospective multicenter experience. *Cancers* **2022**, *14*, 5966. [CrossRef] [PubMed]

39. Rosenberg, S.A.; Restifo, N.P. Adoptive cell transfer as personalized immunotherapy for human cancer. *Science* **2015**, *348*, 62–68. [CrossRef]

40. Jiang, Z.; Jiang, X.; Chen, S.; Lai, Y.; Wei, X.; Li, B.; Lin, S.; Wang, S.; Wu, Q.; Liang, Q. Anti-GPC3-CAR T cells suppress the growth of tumor cells in patient-derived xenografts of hepatocellular carcinoma. *Front. Immunol.* **2017**, *7*, 690. [CrossRef]

41. June, C.H.; Sadelain, M. Chimeric antigen receptor therapy. *N. Engl. J. Med.* **2018**, *379*, 64–73. [CrossRef]

42. Hu, Z.; Ott, P.A.; Wu, C.J. Towards personalized, tumour-specific, therapeutic vaccines for cancer. *Nat. Rev. Immunol.* **2018**, *18*, 168–182. [CrossRef] [PubMed]

43. Tagliamonte, M.; Petrizzo, A.; Mauriello, A.; Tornesello, M.L.; Buonaguro, F.M.; Buonaguro, L. Potentiating cancer vaccine efficacy in liver cancer. *Oncoimmunology* **2018**, *7*, e1488564. [CrossRef] [PubMed]

44. Vercher, E.; Tamayo, I.; Mancheño, U.; Elizalde, E.; Conde, E.; Reparaz, D.; Lasarte, J.J.; Villar, V.; Iñarrairaegui, M.; Sangro, B.; et al. AS051—Identification of neoantigen-reactive T cells in hepatocellular carcinoma: Implication in adoptive T cell therapy. *J. Hepatol.* **2020**, *73*, S39–S40. [CrossRef]

current issues in molecular biology

MDPI

Review

Clinical Significance of SOX10 Expression in Human Pathology

Hisham F. Bahmad [1,*] , Aran Thiravialingam [2], Karthik Sriganeshan [2] , Jeffrey Gonzalez [2] , Veronica Alvarez [2], Stephanie Ocejo [2], Alvaro R. Abreu [2], Rima Avellan [2], Alejandro H. Arzola [2], Sana Hachem [3] and Robert Poppiti [1,4,†]

1. The Arkadi M. Rywlin M.D. Department of Pathology and Laboratory Medicine, Mount Sinai Medical Center, Miami Beach, FL 33140, USA; robert.poppiti@msmc.com
2. Herbert Wertheim College of Medicine, Florida International University, Miami, FL 33199, USA; athir006@med.fiu.edu (A.T.); ksrig002@med.fiu.edu (K.S.); jgonz1074@med.fiu.edu (J.G.); socej001@med.fiu.edu (S.O.); aabre070@med.fiu.edu (A.R.A.); ravel001@med.fiu.edu (R.A.); aarzo006@med.fiu.edu (A.H.A.)
3. Department of Anatomy, Cell Biology, and Physiological Sciences, Faculty of Medicine, American University of Beirut, Beirut 1107, Lebanon; sih12@mail.aub.edu
4. Department of Pathology, Herbert Wertheim College of Medicine, Florida International University, Miami, FL 33199, USA
* Correspondence: hisham.bahmad@msmc.com; Tel.: +1-305-674-2277
† Senior author.

Abstract: The embryonic development of neural crest cells and subsequent tissue differentiation are intricately regulated by specific transcription factors. Among these, *SOX10*, a member of the *SOX* gene family, stands out. Located on chromosome 22q13, the *SOX10* gene encodes a transcription factor crucial for the differentiation, migration, and maintenance of tissues derived from neural crest cells. It plays a pivotal role in developing various tissues, including the central and peripheral nervous systems, melanocytes, chondrocytes, and odontoblasts. Mutations in *SOX10* have been associated with congenital disorders such as Waardenburg–Shah Syndrome, PCWH syndrome, and Kallman syndrome, underscoring its clinical significance. Furthermore, SOX10 is implicated in neural and neuroectodermal tumors, such as melanoma, malignant peripheral nerve sheath tumors (MPNSTs), and schwannomas, influencing processes like proliferation, migration, and differentiation. In mesenchymal tumors, SOX10 expression serves as a valuable marker for distinguishing between different tumor types. Additionally, SOX10 has been identified in various epithelial neoplasms, including breast, ovarian, salivary gland, nasopharyngeal, and bladder cancers, presenting itself as a potential diagnostic and prognostic marker. However, despite these associations, further research is imperative to elucidate its precise role in these malignancies.

Keywords: SOX10; neural crest cells; melanoma; neuroectodermal tumors; mesenchymal tumors

Citation: Bahmad, H.F.; Thiravialingam, A.; Sriganeshan, K.; Gonzalez, J.; Alvarez, V.; Ocejo, S.; Abreu, A.R.; Avellan, R.; Arzola, A.H.; Hachem, S.; et al. Clinical Significance of SOX10 Expression in Human Pathology. *Curr. Issues Mol. Biol.* **2023**, *45*, 10131–10158. https://doi.org/10.3390/cimb45120633

Received: 20 November 2023
Revised: 10 December 2023
Accepted: 12 December 2023
Published: 15 December 2023

1. Introduction

During embryonic stages, the formation of the primitive neural crest gives rise to diverse neural structures. These neural crest cells undergo differentiation into various tissues, a process regulated by specific transcription factors with varying expression levels. The SRY-related HMG box, also known as the *SOX* gene, plays a multifaceted role in differentiating embryological and biological processes among neural crest cells. The *SOX* gene family comprises eight subfamilies.

Within the SoxE subfamily, SOX10 emerges as a distinctive transcription factor that significantly contributes to the enhancement of differentiation, migration, and maintenance of tissues derived from neural crest cells. Initially expressed in the dorsal neural tube, SOX10 guides the differentiation of tissues within the peripheral nervous system [1,2]. This gene's pivotal role in embryonic development facilitates neural crest cell differentiation, giving rise to several sublineages, including the arachnoid and pia mater, melanocytes, odontoblasts, tracheal cartilage, laryngeal cartilage, and Schwann cells (Figure 1).

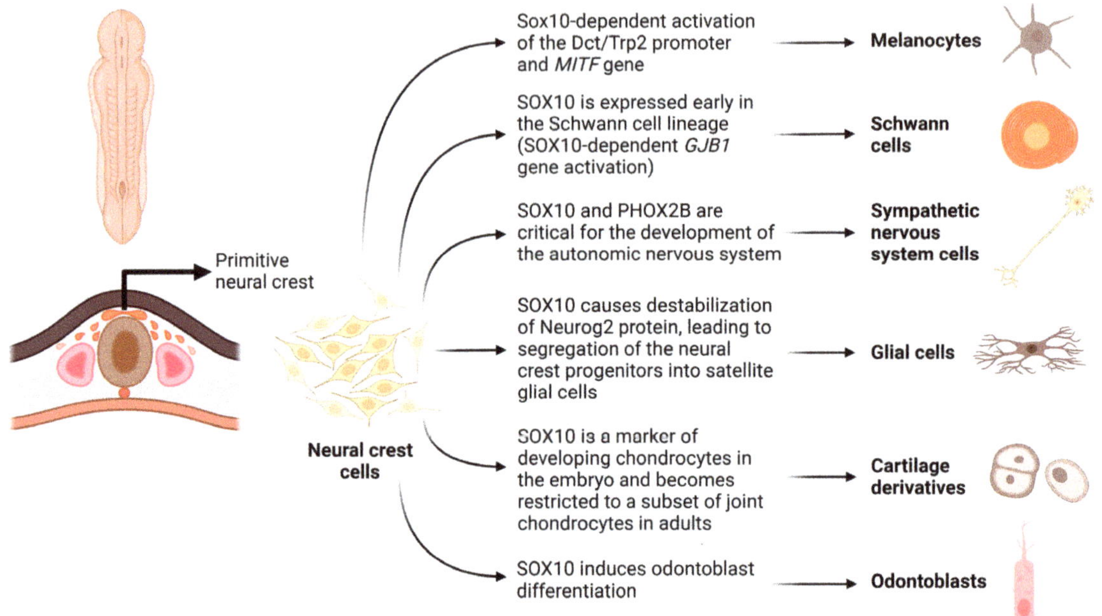

Figure 1. Schematic demonstrating the role of SOX10 in embryonic development, facilitating neural crest cell differentiation, and giving rise to several sublineages, including melanocytes, Schwann cells, sympathetic nervous system cells, glial cells, odontoblasts, and cartilage derivatives (chondrocytes). Created with BioRender.com (2023).

The *SOX10* gene is located on chromosome 22q13, encoding the SOX10 protein with an open reading frame consisting of 466 amino acids and a weight of 51 kDa [3]. The protein possesses a highly conserved dimerization domain at its N-terminus within the SoxE subfamily. Comprising 40 amino acids, this region facilitates the protein's dimerization ability for binding target genes. Adjacent to the N-terminus is the high mobility group (HMG) domain, spanning 79 amino acids and maintaining a consistent structure across all SOX family members. This domain, characterized by three alpha helices forming an L-shape, is designed for binding DNA sequences within the minor groove, specifically containing the nucleotide sequence of C[A/T]TTG[A/T][A/T]. This binding modulates DNA molecules, creating a compatible structure for active transcriptional complexes [3].

Within the domain, an intron and K2 domain are present, along with a nuclear localization and export signal [3]. The K2 domain functions as a promoter-specific transactivation domain, TAM (transactivation domain in the middle of the protein), crucial for SOX10 expression in the peripheral nervous system [4]. On the opposite end of the protein, in the C-terminal region, 66 amino acids are located, marked by a high expression of serine, prolines, and glutamine sequences [5]. This C-terminus is essential for SOX10's interaction with specific binding targets during tissue differentiation, facilitated by a transactivation domain (TA or TAC) [3].

The distinctive composition of SOX10 enables it to exist as a monomer or dimer, exerting influence on various DNA binding targets with differing affinities. Beyond this dual functionality, SOX10 also serves as a nucleocytoplasmic shuttle protein for transcriptional activation, potentially binding to cis elements on target genes to regulate their expression [6,7]. These specific functions are intricately regulated through the modification and expression of SOX10, involving various signal transduction pathways such as Wnt, BMPs, and FGFs pathways [2,3,8].

Wnt signaling, in particular, plays a crucial role in neural crest formation. Decreased levels of Wnt signaling inhibit neural crest formation, underscoring its necessity in this developmental process. A study demonstrated that blocking Wnt using a second messenger resulted in the suppression of SOX10 expression [2]. Moreover, over a dozen transcription factors bind to the N-terminus of the SOX10 HMG domain, regulating its transcriptional activity [9]. *SOX9* and *Slug* are implicated in the regulation of SOX10, showing their necessity in neural crest cell development. Manipulating Slug and Sox9 expression, whether wild type or mutant, resulted in high or absent SOX10 expression, suggesting a mutual relationship between Slug and SOX10 [2].

Various modifications, including phosphorylation, acetylation, SUMOylation, and methylation, have been identified in different amino acid residues of SOX10. SUMOylation at three lysine residues (K55, K246, and K357) represses the transcriptional activation of target genes crucial for cell development and maintenance, such as *MITF* in melanocytes and *GJB1* in Schwann cells [10]. Additionally, phosphorylation of Ser24 and Thr240, two highly conserved sites within the SoxE family, has been associated with melanoma [11].

SUMOylation of SoxE proteins is integral to the development of the inner ear. A yeast two-hybrid screen identified UCB9 and SUMO-1 in SoxE proteins, including SOX10 and SOX9, crucial for inner ear regulation. Both SOX10 and SOX9 feature two conserved SUMOylation sites—one at the N-terminal of an E1 domain and the other at the C-terminal of the activation domain. Specifically, SOX10 undergoes SUMOylation at K44 and K333, at the N-terminus and activation domain, respectively, in addition to other conserved sites [12]. SUMOylation may also occur at K55 and K357 sites within the SOX10 due to their involvement in the interaction of UBC9 and SOX10 [13]. Consequently, the absence of a SUMOylated site may indicate the non-expression of a lysine residue in a SOX10 variant.

The expression of SOX10 varies in response to SUMOylation or the absence of necessary residues in SOX9 [12]. This evolved ability of SOX10 to undergo SUMOylation plays a pivotal role in regulating the protein, enabling it to modulate distant proteins, up- and downregulate various cellular functions, and modify protein complex interactions.

Given the highly conserved expression of SOX10 within neural crest cells and their derivatives, the presence of mutated variants can result in a spectrum of severe to lethal diseases. Over half of the variations within the SOX10 family result from truncations. The remaining variants include non-truncating, missense, in-frame insertions or deletions, and partial copy number variants. Missense mutations typically cluster in the HMG domain [3]. These mutations can lead to conditions such as deafness, dysregulation of the peripheral and central nervous systems, embryonic lethality, colonic issues, and various neoplasms.

In cases of sensorineural hearing loss, various *SOX10* mutations may lead to the agenesis or hypoplasia of semicircular canals and enlarged vestibules. Imaging modalities, including computed tomography (CT) and magnetic resonance imaging (MRI), have revealed a connection between *SOX10* mutations and the absence or hypoplasia of these structures [3]. These malformations associated with *SOX10* mutations have also been linked to dysregulation of *WNT1* (regulating cell fate), *KCNQ4* (potassium voltage-gated channel), *STRC* (stereocilin, associated with the hair bundle of the ear), and *PAX6* (paired box 6) [3].

Considering the crucial role of *SOX10* in myelin-containing glial cells, various mutations have been identified. Two frameshift mutations within the carboxy-terminal, resulting in truncations (*SOX10Dom* and *SOX10-59*), have been associated with dominant megacolon and Waardenburg–Hirschsprung disease [14]. A group of disorders collectively labeled as PCWH (peripheral demyelinating neuropathy, central demyelinating leukodystrophy, Waardenburg syndrome, and Hirschsprung disease) result from variants within the nervous system. Clinical presentations may include delayed motor and cognitive development, cerebral palsy, ataxia, spasticity, congenital nystagmus, hyporeflexia, distal sensory impairments, and distal muscle wasting [3]. Signs of Kallmann syndrome (KS) have also been observed in Waardenburg syndrome, suggesting that KS may result from *SOX10* mutations. KS manifests with hypogonadotropic hypogonadism and anosmia. Many patients with KS may also present with hearing deficits and harbor *SOX10* mutations [4]. The physiological basis of this disorder in relation to *SOX10* is believed to involve the dysregulation of GnRH (gonadotropin-releasing hormone) as it travels through the neurons of the peripheral olfactory nerve, up to and through the olfactory bulb [3].

SOX10 plays a crucial role in the embryogenesis of neural crest cells, and deviations from its normal function can give rise to various congenital disorders. However, the impact of *SOX10* variants extends beyond developmental disorders, contributing to the initiation and progression of different cancers due to its involvement in numerous tissues.

SOX10 expression has been identified in various cancer types, including breast tumors, glioma, glioblastoma, salivary adenoid cystic tumors, melanoma, and hepatocellular carcinoma. Intriguingly, *SOX10* exhibits dual roles in these tumors, acting as a tumor suppressor and promoter. For instance, it functions as an oncogene in hepatocellular carcinoma and nasopharyngeal carcinoma while exerting tumor-suppressive effects in gastrointestinal neoplasms. Urothelial carcinoma shows an overexpression of *SOX10*, indicating its role as a tumor promoter [15]. The significance of SOX10 expression becomes evident when comparing its levels in different bladder cancers to normal bladder tissue [16].

These varied expressions underscore the need to study SOX10's role and levels in both normal and pathological tissues. This comprehensive understanding is crucial for unraveling its precise role in cell biology and appreciating its clinical significance (Figure 2).

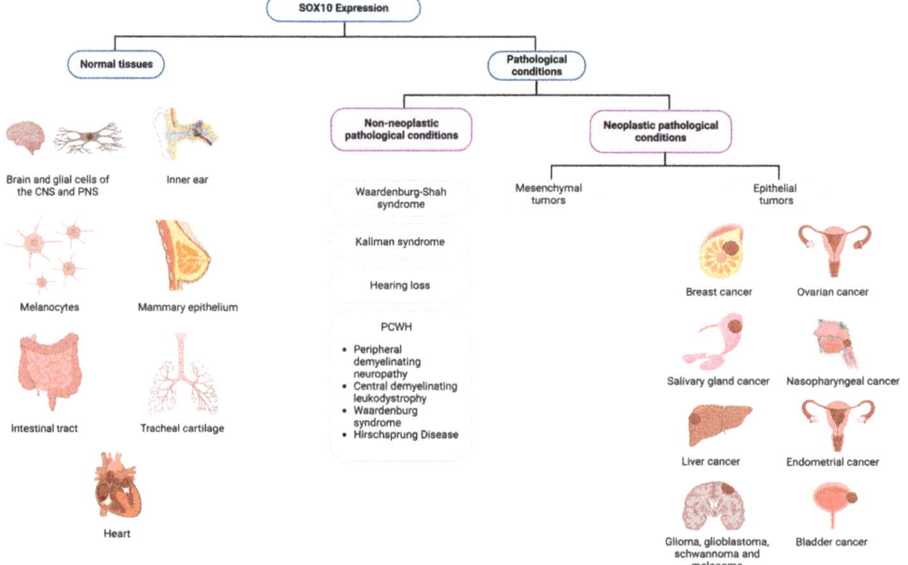

Figure 2. SOX10 expression in normal tissues and across different pathological conditions. Created with BioRender.com (2023).

2. SOX10 Expression in Normal Tissues

The differentiation of various tissues from neural progenitor crest cells involves distinct processes. SOX10 expression remains elevated in tissues such as the brain, inner ear, intestinal tract, tracheal cartilage, and heart. In the early embryonic development of the inner ear, SOX10 shows high expression, gradually declining as hair cells mature. At this stage, SOX10 becomes specific to supporting cells, and an inability to express either SOX10 or SOX9 may result in the development of an enlarged or cystic otocyst [12,17,18].

Conversely, lower levels of SOX10 expression are observed in the prostate, testis, bladder, pancreas, and stomach [19]. During peripheral nervous system development, some neurons lose SOX10 expression while all mature glial cells maintain its expression. In the central nervous system, oligodendrocytes exhibit a high level of SOX10 expression. Similarly, melanocytes heavily rely on SOX10 for their specialization, maturation, and maintenance [2].

2.1. SOX10 Expression within the Peripheral Nervous System

Within the peripheral nervous system, *SOX10* plays a pivotal role in facilitating the differentiation of both Schwann cells and glial cells, employing distinct biochemical processes in each cell type. In Schwann cell development, SOX10 directly targets the protein zero (*P0*) gene coding region, a myelin gene exclusively expressed in Schwann cells, tightly regulated by *SOX10* [14].

Analysis of mouse embryos with mutated binding sequences on *P0* for *SOX10*, compared to those with normal binding sequences, revealed robust SOX10 expression in mature Schwann cells with high P0 expression. This expression was further intensified when a *SOX10* induction signal was introduced into these sequences, resulting in a ten-fold increase in P0 expression [14].

Neurogulin-1 has been identified as a key player in controlling the differentiation of neural crest cells into glia via the activation of ErbB receptors [20]. The absence of interaction between Neurogulins binding to the EGF receptor tyrosine kinase, ErbB3, has been associated with developmental defects in neural crest cells and their derivatives. The relationship between *SOX10* and ErbB3 was investigated using the tet-on system, inducing SOX10 expression, leading to a significant increase in ErbB3 expression. However, whether this effect was direct or indirect remained unclear. Supporting this relationship, *SOX10* mutant variations were found in ErbB3 mutant mice [20].

It is important to note that in certain cells, there was a high expression of SOX10 coupled with a low expression of P0, particularly in non-myelinating cells. This suggests that *SOX10* typically does not function independently but, instead, interacts with different protein complexes. In unmyelinated Schwann cells, the downregulation of myelination may be attributed to *SOX10's* regulation of various transcription factors, including *SOX5, SOX6, NOTCH1, HMGA2, HES1, MYCN, ID4,* and *ID2*. These regulators were found to oppose the process of myelination within Schwann cells [21]. Furthermore, in an experimental study on SOX10 expression within mammary glands, mouse embryos were manipulated to be homozygous dominant knockout for *SOX10*. In these specific mice, death was quickly encountered, in addition to the complete absence of Schwann cell production [22].

2.2. SOX10 Expression within the Inner Ear

Moving beyond the peripheral nervous system to the inner ear, there is meticulous regulation of *SOX10* and *SOX9*, which is crucial for normal development. During gastrulation and neural crest development, SOX10 is expressed in the otic vesicle, reaching its peak around stage 25 [2]. Both SoxE proteins, SOX10 and SOX9, undergo SUMOylation at different lysine residues and two conserved sequences [12]. The regulation of these modifications may have subsequent consequences, leading to progressive hearing loss.

SOX10 exhibits high expression in the otic vesicle from E9.5 onward until it becomes exclusively expressed in the supporting cells later in development. This sustained expression of SOX10 facilitates the maintenance of cochlear progenitors during the development of the organ of Corti and the otocyst [3].

2.3. SOX10 Expression in Melanocytes

The precise expression of SOX10 in melanocytes is indispensable for gene regulation within these cells. Before melanocyte development, SOX10 is highly expressed in the neural crest region, initially across all axial areas and later progressing to the expression only in the truncal region. Overexpression of SOX10 at this stage results in high expression in the Slug domain, both playing a role in the development of pigmented melanocytes [3].

Through a complex network, *SOX10* collaborates with *PAX3* to activate *MITF*, enhancing its expression. Increased *MITF*, in turn, works with *SOX10* to promote *DCT/TRP2* expression. Dominant *SOX10* mutant mice exhibit a decrease in melanocyte markers Dct/Trp2, underscoring the pivotal role of *SOX10* in pigment production [10]. Consistent with these cell markers, it has been shown that mutant *SOX10* or low expressions lead to a proportional decrease in markers Trp2, c-kit, and Mitf [3]. The varying levels of these markers depend on the stage of melanocyte development within the embryo, starting with nonpigmented melanoblasts and eventually transitioning to melanocytes. It has been demonstrated that SOX10 could produce pigment at injected sites, while Slug alone could not [3].

2.4. SOX10 Expression in the Mammary Epithelium

An exception to the typical expression of SOX10 in neural crest cell derivatives is observed in the mammary epithelium, which originates from the ectoderm. The mammary gland houses epithelium that bifurcates into the ductal epithelial tree during puberty. The mammary epithelium undergoes dynamic changes in growth due to hormonal stimulation during puberty, pregnancy, lactation, and menopause. SOX10 expression in these cells initiates prenatally during the development of stem cells.

Within these stem cells, *SOX10* responds to *FGF* signaling, facilitating their progression to mesenchymal tissue. A study manipulating mice embryos analyzed the effects of homozygous, heterozygous *SOX10* knockouts, and wild type. Both heterozygous and homozygous knockout mice exhibited decreased mammary branching growth development. Furthermore, postnatal mammary development revealed that these adult mice were unable to lactate after pregnancy.

Continuing through the female reproductive process, mice were further analyzed during involution. Compared to wild-type mice, knockout mice started with substantially fewer epithelial cells in the mammary glands. However, during involution, the epithelial cell count decreased significantly more in wild-type mice. These findings suggest the involvement of *SOX10* throughout the entire process, including the involution of expanded mammary epithelia. Although *SOX10* may play a crucial role in this process, the presence of mammary growth indicates that *SOX9* and *SOX10* may work synergistically, with *SOX9* contributing to the absence of *SOX10* [22].

3. SOX10 Expression in Non-Neoplastic Pathological Conditions

3.1. SOX10 in Waardenburg–Shah Syndrome

SOX10 mutations have been implicated in disrupting neural crest development, leading to a diverse range of clinical phenotypes. The association of the *SOX10* gene with congenital disorders was initially recognized in the context of Waardenburg–Shah syndrome, a subtype of Waardenburg syndrome (WS), also known as Waardenburg–Hirschsprung syndrome and WS type 4 [8]. WS4 is characterized by sensorineural hearing loss, depigmentation of hair, skin, and eyes, and Hirschsprung's disease. The *SOX10* gene was first identified as the mutant gene responsible for megacolon and depigmentation in Dom mutant mice (SOX10^{Dom}) [1,23]. Specifically, a frameshift mutation in *SOX10* causing haploinsufficiency was found to be the cause, with a homozygous mutation in mice proving lethal [24]. Based on this discovery, *SOX10* mutations were screened for in human patients with Waardenburg–Hirschsprung disease, in whom a causative mutation had not yet been identified. Several cases were found to have a *SOX10* mutation, confirming its involvement in the Waardenburg–Hirschsprung disease [8].

Waardenburg syndrome has been classified into four main presentations. Type I (WS1) presents with pigmentary abnormalities of the hair, heterochromia irides, sensorineural hearing loss, and the characteristic dystopia canthorum. Type 2 (WS2) has similar features with the absence of dystopia canthorum. Type 3 (WS3) is distinguished by abnormalities of the upper limb. While Waardenburg syndrome was initially classified by phenotypic presentation, detected mutations in patients with WS have been integrated into further subclassifications. For instance, WS4 has been split into WS4A, WS4B, and WS4C, with mutations in *EDNRB*, *EDN3*, and *SOX10*, respectively [25]. Another subtype, WS2E, is also caused by a *SOX10* mutation [26].

3.2. PCWH

PWCH (Peripheral demyelinating neuropathy, central demyelinating leukodystrophy, Waardenburg syndrome, and Hirschsprung disease) represents a neurological variant of the previously discussed WS4, where a *SOX10* mutation is also implicated. Patients with PWCH exhibit a similar presentation, including heterochromia irides, sensorineural hearing loss, and Hirschsprung's disease, as observed in Waardenburg–Shah syndrome. Additionally, they experience neurological symptoms such as peripheral neuropathy, ataxia, and intellectual disability [27,28]. The syndrome was first described shortly after the discovery of *SOX10* mutations in WS4. Due to the shared features, mutations in the *SOX10* gene were investigated in patients with what is now termed PCWH. A de novo deletion mutation was identified in the coding region of *SOX10*, leading to an extension of the peptide and a toxic gain of function [29]. The discovery of a *SOX10* mutation as a perpetrator (and the exclusion of other known mutations such as *PMP22*) in PCWH, a demyelinating disease, further supported the role of *SOX10* in Schwann cell differentiation.

3.3. Kallman Syndrome

Due to the presence of hypogonadism and anosmia in subtypes of Waardenburg syndrome (e.g., WS2E), *SOX10* was investigated as a potential candidate gene for Kallman syndrome (KS), which falls under the umbrella of congenital hypogonadotropic hypogonadism (HH). KS is characterized by anosmia, distinguishing it from idiopathic hypogonadotropic hypogonadism, which lacks anosmia. Both are considered manifestations of the same syndrome, and instances of each may coexist within the same family [30]. Although nine genes have been implicated in HH, demonstrating extensive genetic heterogeneity, they only account for 30% of KS cases. Therefore, *SOX10* appeared to be a likely candidate mutation to explain the presence of anosmia within the disease spectrum.

In a mouse model study, *SOX10* deficient mice exhibited an almost complete absence of olfactory ensheathing cells (OECs), misrouting of nerve fibers, impaired migration of GnRH cells, and disorganization of the olfactory nerve layer in the olfactory bulbs [31]. In the same study, a cohort of KS patients without known mutations were screened for *SOX10* mutations. Six patients had novel *SOX10* mutations, and five out of the six also had deafness.

The diversity and overlap of clinical features in patients with Waardenburg–Shah syndrome, PCWH, and Kallman syndrome underscore the role of *SOX10* as a common factor for pathogenesis. However, the phenotypic variability among patients with the same mutations or in the same families emphasizes the need for further study of intermediate and downstream factors [3].

3.4. Hearing Loss

SOX10 mutations in the inner ear explain abnormalities in hearing, such as hypoplasia of semicircular canals, enlarged vestibular canals, vestibulocochlear nerve agenesis, and cochlear deformities [32,33]. Although the presence of sensorineural hearing loss among patients with Waardenburg–Shah syndrome varied among genotypes, Song et al. found that the prevalence of hearing loss in patients with a *SOX10* mutation was 100% [33]. In mouse models, the expression of SOX10 in vestibulocochlear development has been studied, revealing an increase in SOX10 expression in the maturing cochleovestibular ganglion. In *SOX10*-deficient mice, there was a lack of glial cell development in this area [34]. Hearing loss is such a penetrant phenotype in patients with *SOX10* mutations that it can manifest without any other features of WS or KS, resembling isolated hearing loss [35].

As of yet, there is no clear role for *SOX10* in genetic screening or counseling for the discussed conditions. More effort is necessary to consolidate the range of phenotypes into one disease spectrum rather than individual syndromes.

4. SOX10 Expression in Neural and Neuroectodermal Tumors

SOX10's role is of interest in the development of certain malignancies and as a potential differentiating marker with diagnostic use.

4.1. SOX10 Expression in Melanoma

The expression of SOX10 in melanoma has been conducted due to its significance in both diagnostic and therapeutic applications (Table 1). Bakos et al. investigated the expression of SOX10 through immunohistochemistry (IHC) in primary and metastatic melanoma cells and its association with neistin coexpression [36]. Nestin is an intermediate filament present in neural progenitor cells, melanomas, and melanocytic nevi. This study disclosed a significant co-expression of SOX10, SOX9, and neistin in early primary melanoma. However, no statistically significant co-expression was observed in the metastatic melanoma [36]. These results align with their in vitro findings, suggesting that SOX10 plays a crucial role in neistin activation during early melanoma development but is not associated with its expression in the more advanced stages of the disease [36]. These findings suggest that SOX10 may serve as a potential marker for determining melanoma stage.

In a separate study by Zhongyuan et al., the role of *SOX10* in melanoma development was similarly investigated. According to that study, *SOX10* plays an important role in regulating various factors involved in melanocyte proliferation and survival, including melanocyte inhibitory activity (*MIA*), *MITF*, *p21/WAF1*, and *E2F1* [37]. A reduction in SOX10 expression resulted in reduced melanoma formation, and the knockout of the *SOX10* gene led to the elimination of new tumor formation [37]. These findings provide additional evidence supporting the role of *SOX10* in melanocyte proliferation. That study also aimed to establish the downstream pathway through which *SOX10* affects melanocyte proliferation by observing its effects on the expression of the minichromosomal maintenance complex component (MCM5). The results demonstrated that the overexpression of *MCM5* in *SOX10*-negative cells partially rescued the proliferation defect observed when *SOX10* was absent [37]. Overall, these findings indicate that *SOX10* is involved in multiple melanocyte proliferation pathways, with the *SOX10-MCM5* axis playing a critical, though not exclusive, role in the proliferation [37].

Further evidence on the role of *SOX10* on melanoma cell proliferation was reported in a study by Cronin et al., which revealed that the loss of SOX10 in melanoma cells resulted in cell arrest in the G1 phase [38]. Molecular studies of melanoma cells with absent *SOX10* showed reduced expression of MITF, elevated expression of p21/WAF1 and p27KIP2, hypophosphorylated RB, and reduced levels of E2F1 [38]. These results suggest that the removal of *SOX10* leads to cell arrest in the G1 phase [38]. Another study by Rosenbaum et al. examined the role of *SOX10* in the regulation of the melanoma cell cycle, finding that knocking out *SOX10* in immune-competent models led to a reduced expression of immune checkpoint proteins HVEM and CEACAM1 [39]. The loss of these immune checkpoint proteins promotes the proliferation of malignant melanoma cells by preventing cellular senesce and apoptosis [39].

Studies on SOX10 have extended beyond its role in proliferation with investigations into its involvement in the migration of melanoma cells. Seong et al. explored this aspect by studying the migration of B16F10 melanoma cell lines following the introduction of siRNA specific for *SOX10*. This was compared to a control group of the same cell line. That study demonstrated a significant reduction in migration in the experimental cell line with downregulated *SOX10*, as confirmed through a TUNEL assay. Additionally, microarray screening revealed a three-fold decrease in *SOX10* and one of its downstream targets, *MITF* [40]. These findings highlight the significant role of *SOX10* expression and its effect on MITF in B16F10 melanoma cells, suggesting a crucial role in cell migration and, consequently, metastasis [40]. Attempts to replicate these results using different melanoma cell lines (Cloudman S9 and Melan-A melanoma cell lines) yielded no statistically significant effects on cell migration, emphasizing the variability of *SOX10* effects depending on the specific cell line being studied [40].

In light of the diverse yet persistent role of *SOX10* in melanoma cells, its potential as a diagnostic histopathological marker has been explored. Clevenger et al. conducted a comparative study using a pan-melanoma cocktail, a SOX10 stain, and an MITF stain to identify melanoma cells of epithelioid origin, those with a predominantly spindle appearance. That study revealed a 100% SOX10 positive staining pattern in both epithelioid and spindle-shaped cells, demonstrating nuclear staining with a strong and diffused pattern. In contrast, the pan melanoma cocktail and MITF stain showed positive staining in 86% and 93% of cases, respectively, for epithelioid cells, and 86% for spindle-shaped melanoma cells [41]. The high rate of detection using SOX10 staining suggests its utility in detecting metastasis in locations where a small number of cells would be expected, such as the cerebrospinal fluid (CSF). However, caution is advised, and a more sensitive stain for melanoma should be considered due to the non-exclusive expression of SOX10 [41].

MITF, downstream of the *SOX10* gene, plays a crucial role in the transcription control of melanocytes and retinal pigment cells and is strongly associated with malignancies [42]. Studies have shown that the absence or reduced activity of *SOX10* consistently leads to cell death in melanocytic descent, particularly at the G1 stage of the cell cycle. The impact on lineage is associated with the type of knockout, whether it involves a complete knockout or an interruption in the product's structure, or a reduction in the half-life [39,43].

Notably, the knockout of *SOX10*, when simultaneously treating advanced melanoma, can confer resistive mechanisms against chemotherapeutic medications. Using Vemurafenib to treat advanced melanoma with an observed *BRAFV600E* mutation, cells acquiring a somatic *SOX10* mutation that hinders proper gene product formation allow the tumor to grow unchallenged by therapeutic treatments that would otherwise be effective [42,44]. This underscores the intricacies and complex integration of *SOX10*, which primarily directs proliferation and steers cells toward differentiated paths. Acting as an oversight system for downstream transcription factors, such as *MITF* [45], the gene gains unregulated function to promote transcriptive and translative efforts within the cell, allowing malignancies to establish their proliferative roots [39,44]. However, the knockout of *SOX10* in existing cancers can lead to acquired resistance against chemotherapeutic efforts. In other iterations of malignancies, knocking out the gene has been found to suspend cell proliferation, restrain cell growth, and reduce overall tumor size [39,42,43]. As melanomas approach their proliferative limits or the threshold for potential invasion, *SOX10* has been observed to become downregulated within the tumor cells. This change induces a phenotypic shift from melanocytic cell lineages to undifferentiated mesenchymal cell lines, characterized by their invasive nature and ability to resist targeted therapeutic regiments against malignancy [42].

Not all mutations of the *SOX10* locus are somatic. In studies focusing on childhood melanoma, almost all congenital melanomas were found to be *SOX10* positive. The significance of this positivity, whether it represents an unhindered function or a gain-of-function mutation, is yet to be determined. Regardless, its presence signifies its key integration in the early stages of skin neoplasms [46]. Studies supporting *SOX10* as a more sensitive marker for melanoma, compared to *MITF*, the previous standard marker for neoplastic testing within this sector, further highlight its diagnostic potential [3,45].

In research by Shakova et al. concerning the significance of SOX10 in melanoma and congenital giant melanocytic nevus, a pre-cancerous lesion heavily associated with melanoma formation, it was confirmed in mouse subjects and later human cell lines that the knockout of this transcription factor showed effective results in blocking tumorigenesis. Furthermore, the knockout or inactivation of the *SOX10* gene established its role as a prerequisite for the formation and maintenance of pre-melanoma lesions [46]. In observed human cell lines, the absence of *SOX10* activity resulted in an estimated nine-fold increase in apoptotic cells due to the disrupted regulation of apoptotic control factors. Examples of this dysregulation were noted from the increases in these control factors, such as caspases and proteins related to the tumor necrosis factor (TNF) pathway [46].

A study by Capparelli et al. demonstrated that *SOX10* plays a crucial role in mediating phenotypic switching in cutaneous melanoma. The loss of *SOX10* led to the development of an invasive, slow-cycling state in melanoma cells, promoting tolerance to *BRAF* and/or *MEK* inhibitors, which are commonly used in melanoma treatment. That study also identified a vulnerability in *SOX10*-deficient melanoma cells, specifically an up-regulation of cellular inhibitors of apoptosis-2 (*cIAP2*). The use of *cIAP1/2* inhibitors selectively induced cell death in *SOX10*-deficient cells, providing a potential therapeutic strategy to target and eliminate these cells. Additionally, combining *cIAP1/2* inhibitors with *BRAF/MEK* inhibitors delayed the onset of acquired resistance in melanomas in vivo [47].

4.2. SOX10 Expression in Malignant Peripheral Nerve Sheath Tumor and Schwannomas

While much of the existing data on the role of *SOX10* in neoplasms primarily focuses on melanoma, this gene is also implicated in other neural and neuroectodermal tumors. Malignant peripheral nerve sheath tumor (MPNST) is one such malignancy where the role of *SOX10* has been investigated. A study by Kang et al. aimed to assess SOX10 as a marker for distinguishing MPNST from synovial sarcoma, given the histopathological similarities that can make differentiation challenging [48]. SOX10 staining revealed a 67% positivity rate in MPNST cells compared to only 7% in synovial sarcomas. The overall results demonstrated a 67% sensitivity rate and a high specificity rate of 93% for SOX10 staining in MPNST, with a positive predictive value of 82% and a negative predictive value of 89% [48]. These findings suggest that SOX10 staining is moderately sensitive but highly specific, serving as a valuable marker for differentiating MPNST from synovial sarcomas in cases where there is a diagnostic discrepancy [48].

Another study by Pekmezci et al. investigated the use of SOX10 as a differentiating marker between MPNST and schwannomas, revealing a positive diffuse SOX10 expression pattern seen only in cellular schwannomas [49]. The results imply that SOX10 expression is significantly more prevalent in cellular schwannomas, and its loss of expression is indicative of MPNST when compared to cellular schwannomas [49]. Doddrell et al. explored SOX10 expression in merlin-null schwannomas, finding reduced expression of SOX10 and two proteins crucial for the myelinating function of Schwann cells: KROX20 and OCT6 [50]. Reintroducing the *SOX10* gene in schwannoma cells showed a small increase in KROX20 expression, which significantly increased with the introduction of cAMP [50]. Overall, the results suggest that the loss of *SOX10* in Schwann cells leads to cellular abnormalities resembling schwannomas [50]. Collectively, these studies indicate that SOX10 expression is a relatively effective marker for differentiating between specific malignancies that may pose diagnostic challenges. Moreover, a recurring pattern in the reported results suggests that SOX10 expression tends to decrease as cells undergo a transition from normal to malignant states in tumors.

Table 1. Studies demonstrating SOX10 expression in neural and neuroectodermal tumors.

Tumor	References	Findings
Melanoma	[51]	Role of SOX10 in Melanoma: • *SOX10* serves as a crucial regulator of melanoma invasion and survival by influencing the expression of key factors such as MIA (melanocyte inhibitory activity), MITF, p21/WAF1, and E2F1; • The absence of SOX10 expression has been linked to a reduction in melanoma formation, and silencing *SOX10* leads to the elimination of tumor formation in vivo; • *SOX10* plays a pivotal role in regulating melanocyte proliferation through its interaction with the minichromosomal maintenance complex component 5 (MCM5); • Inducing overexpression of MCM5 in *SOX10* knockout cells partially rescues the cell's impaired proliferation capacity.

Table 1. *Cont.*

Tumor	References	Findings
	[45]	• Loss of *SOX10* in melanoma cells results in cell cycle arrest in the G1 phase, accompanied by molecular changes such as reduced MITF expression, elevated p21/WAF1 and p27KIP2 expression, hypo-phosphorylated RB, and reduced levels of E2F1; • *SOX10* is essential for melanogenesis.
	[39]	• *SOX10* knockout in immune-competent models leads to reduced expression of immune checkpoint proteins HVEM and CEACAM1, facilitating tumor growth.
	[40]	SOX10 in Melanoma Cell Migration and Metastasis: • siRNA specific for *SOX10* demonstrates that downregulation of *SOX10* in B16F10 melanoma cells significantly reduces cell migration compared to control cells in a Transwell migration assay; • TUNEL assay results indicate that the lower migration in the experimental cell line is not due to apoptosis or senescence; • Microarray screening reveals a three-fold decrease in *SOX10* and in *MITF*, a known target of *SOX10*; • The gene expression cascade initiated by *SOX10* and mediated by *MITF* plays a significant role in melanoma cell migration and metastasis; • These effects are not reproducible in Cloudman S9 and Melan-A cells, suggesting that the SOX10/MITF effects on migration and metastasis vary depending on the melanoma cell line.
	[41]	SOX10 as a Diagnostic Marker for Melanoma: • SOX10 staining is highly effective in identifying melanoma cells, with a 100% positivity rate in both epithelioid melanoma and melanoma with a predominantly spindle cell appearance; • Given its high detection rate and strong staining intensity, SOX10 is a valuable marker for detecting melanoma cell metastasis in locations like the cerebrospinal fluid (CSF), where a large number of cells are not expected. However, a more sensitive melanoma stain should be used for confirmation, considering SOX10 is not exclusively specific for melanoma
	[36]	SOX10 and Nestin in Melanoma Development: • Nestin, an intermediate filament found in neural progenitor cells, melanomas, and melanocytic nevi, shows statistically significant co-expression with SOX10 in primary melanomas; • SOX10 plays a key role in Nestin activation in primary melanoma cells, suggesting that SOX10 is a major mediator of early melanoma development.
Malignant peripheral nerve sheath tumor	[48,49]	SOX10 in Differentiating MPNSTs and Synovial Sarcomas: • SOX10 staining demonstrates a 67% positivity rate for MPNST, compared to only 7% in synovial sarcomas; • SOX10 staining exhibits a sensitivity of 67% and specificity of 93%, making it a moderately sensitive but highly specific marker for distinguishing MPNST from synovial sarcoma.
Merlin-null schwannoma	[50]	SOX10 in Schwannomas and Normal Schwann Cell Function: • Loss of SOX10 expression strongly supports the diagnosis of MPNST; • Schwannoma cells show reduced SOX10 expression, as well as diminished expression of KROX20 and OCT6, crucial proteins in the myelination process; • Reintroduction of the *SOX10* gene in schwannoma cells increases KROX20 expression, particularly in the absence of cAMP, with a significant boost upon cAMP introduction; • Removal of *SOX10* from normal Schwann cells in a mouse model results in minimal expression of KROX20 and OCT6, irrespective of cAMP levels; • SOX10 expression is necessary for normal Schwann cell function, and its loss leads to abnormalities resembling those seen in schwannoma cells.

5. SOX10 Expression in Mesenchymal Tumors

In a study conducted by Miettinen et al., the expression of SOX10 was analyzed in 1645 non-neurogenic mesenchymal tumors. Among non-nerve sheath tumors, positive SOX10 tumor cells were identified only in alveolar rhabdomyosarcoma (2/27) and ossifying fibromyxoid tumors (2/47). Thirty-three other types of mesenchymal tissues analyzed

(1571 samples), including fibroblastic-myofibroblastic tumors, benign fibrous histiocytoma and subtypes, solitary fibrous tumor/hemangiopericytoma of the peripheral soft tissues and intracranial space, and undifferentiated pleomorphic sarcomas, were negative for SOX10. Synovial sarcomas, desmoid fibromatosis, and glomus tumors showed fewer than 5% of SOX10-positive nuclei, possibly representing entangled neural components [52].

Research by Karamchandani et al. aimed to validate the use of SOX10 and S100 protein as reliable markers in soft tissue neoplasms of both neural crest and non-neural crest origin. SOX10 and S100 mRNA levels were evaluated in 122 cases of peripheral nerve sheath tumors and synovial sarcomas, and IHC was used for SOX10 and S100 protein expression in 1012 tissue specimens [53]. Synovial sarcomas expressed significantly higher levels of S100 than SOX10, and no significant SOX10 mRNA expression was identified in synovial sarcoma [53]. The majority of schwannomas and neurofibromas showed increased expression of both SOX10 and S100 mRNA [53]. MPNSTs revealed highly correlated, variable levels of SOX10 and S100 mRNA expression. Of the non-neural, nonmelanocytic sarcomas, only one rhabdomyosarcoma sample was positive for SOX10. In summary, SOX10 was positive in only 5 of 668 cases with a 99% specificity for non-schwannian, nonmelanocytic tumors [53].

Kang et al. evaluated the diagnostic utility of SOX10 IHC in differentiating between synovial sarcoma and MPNST due to similar histomorphology and immunophenotype [48]. Forty-eight cases of MPNST and 97 cases of synovial sarcoma, including four intraneural synovial sarcomas, were stained for SOX10. Sixty-seven percent of MPNST (32/48) and only 7% (7/97) of synovial sarcomas were positive for SOX10. Nevertheless, there is uncertainty as to whether SOX10-positive cells in intraneural synovial sarcoma represent entangled Schwann cells, synovial sarcoma cells, or both [48].

In an attempt to demonstrate the clinical and morphological heterogeneity between gastrointestinal mesenchymal tumors with neurotrophic tyrosine receptor kinase (*NTRK*) gene rearrangements and gastrointestinal stromal tumors, Atiq et al. reported consistently absent SOX10 expression in eight mesenchymal tumors in the gastrointestinal tract with *NTRK1* or *NTRK3* rearrangements [54].

Research by Chiang et al. focused on classifying a newly discovered category of high-grade uterine sarcomas. Four *NTRK* fusion-positive uterine sarcomas were identified and distinguished from both undifferentiated uterine sarcomas and more commonly aggressive leiomyosarcomas. All four mesenchymal tumors lacked SOX10 expression [55].

6. SOX10 Expression in Epithelial Neoplasms

When examining the impact of SOX10 on epithelial neoplasms, its influence is extensive and continues to unfold with further investigations. This transcription factor plays a crucial role in regulating the proliferation and specialization processes of melanocyte and Schwannian lineages, exhibiting high expression levels in melanoma malignancies and those affecting the central nervous system [3]. While many of the studied mutations indicate somatic changes, there are instances of inherited cases [46]. In the observed cases, SOX10 expression is more prevalent in malignancies during proliferative stages compared to those found in invasive or metastatic stages [44] (Table 2).

Beyond tumors involving melanocytic lineages, research has provided substantial evidence of SOX10 expression in the salivary gland, breast, and ovarian neoplasms affecting epithelial cells. Although this evidence has accumulated in recent years, sensitivity for diagnostic differentiation, particularly in salivary gland tumors, remains less reliable [56]. Conversely, concerning ovarian cancers, distinguishing between SOX10 expression within the nucleus and cytoplasm has shown promise in estimating grade and prognosis.

In the context of salivary gland neoplasms, SOX10 expression has been identified in tumors arising from acinar and intercalated ductal cells [57–59]. Notably, tumors lacking SOX10 have been associated with the appearance of excretory ducts or striated ducts [59]. SOX10-expressing neoplasms in the salivary glands include acinic cell carcinoma, epithelial-myoepithelial carcinoma, adenoid cystic carcinoma, and polymorphous adenocarcinoma [59,60]. Adenoid cystic carcinoma and polymorphous adenocarcinoma have consistently demonstrated SOX10 expression in virtually all cases studied [57,58]. Distinctively, acinic cell carcinoma can be differentiated from metastatic renal cell carcinoma in the parotid gland, as the latter does not express SOX10 on staining [58].

However, certain salivary gland neoplasms either show no SOX10 representation or exhibit focal expression in staining. These include salivary duct carcinoma, mucoepidermoid carcinoma (MEC), squamous cell carcinoma (SCC), and oncocytic carcinoma, which originate from excretory and serous ductal cells within the salivary glands. While MEC tumors were initially considered SOX10-negative, further investigation revealed a subgroup of SOX10-positive MEC with distinct morphology and colloid-like secretion [57,59]. Additionally, SOX10 has been found to be positive in other tumors, such as basal cell carcinomas (BCCs) and low-grade salivary duct carcinomas [57]. In the case of SCC secondary to HPV infection, SOX10 is not a reliable diagnostic marker due to similar staining distributions with HPV-related multiphenotypic sinonasal carcinoma [61]. While SOX10 staining can aid in categorizing tumors based on cell origins, negative staining does not necessarily imply the absence of SOX10 mutation, as inactivating or truncating mutations can result in reduced or absent SOX10 expression [59,61,62]. Despite this, SOX10 is considered a valuable protein expression marker for the diagnostic identification of salivary gland neoplasms, contributing to increased diagnostic accuracy [58].

SOX10 protein expression has also been observed in breast carcinomas, particularly in approximately 66–74% of triple-negative breast carcinomas [63]. Triple-negative breast carcinoma has shown SOX10 expression in a substantial number of cases, ranging from 38 to 67% in the literature [62]. SOX10 has been associated with CD117 and vimentin expression in triple-negative breast carcinomas, though its prognostic value remains inconclusive and is mainly considered a marker for aiding in differential diagnoses [62]. While evidence suggests a potential prognostic value due to associations with malignant characteristics of triple-negative breast carcinomas, further research is needed to establish its definitive prognostic significance [64]. In cases where homozygous deletions and point mutations eliminate SOX10 staining presence, GATA3, a common marker in breast carcinoma, has been used in conjunction with SOX10 to address this limitation. Approximately 60% of triple-negative breast carcinomas have been identified using this dual-staining method, making SOX10 a useful marker in identifying epithelial neoplasms of the breast [62,65].

In the context of ovarian epithelial tumors, SOX10 has shown value in differentiating cell origin and estimating prognosis. Contrary to previous claims that suggested no application for SOX10 in the study and diagnosis of ovarian epithelial tumors, Kwon et al. demonstrated its utility. Ovarian epithelial neoplasms, including serous, mucinous, and endometrioid subtypes, can be differentiated based on the localization of staining. Serous neoplasms show nuclear localization, while mucinous and endometrioid neoplasms exhibit cytoplasmic localization [66]. Staining in both regions is possible, but the diffuse characteristic of SOX10 staining helps distinguish between subtypes. The intensity of the stain within the nucleus correlates with the prognosis of the patient, emphasizing its potential as a prognostic marker [66]. While SOX10's involvement in ovarian carcinomas was assessed, other common expression markers studied for ovarian cancer include SOX8 and, notably, SOX9, which has been implicated in various signaling pathways in ovarian cancer development [67–71].

In nasopharyngeal carcinomas, SOX10 is markedly overexpressed, and this overexpression is associated with a poorer prognosis, particularly in T classification and lymph node metastasis. The correlation with poor prognosis is linked to SOX10's involvement in tumor development and metastatic-seeding ability in breast cancer cells. The overexpression of SOX10 in nasopharyngeal tumors highlights its potential importance as a diagnostic and prognostic marker for patients with nasopharyngeal carcinoma [72].

SOX10 has also been associated with metaplastic bladder cancers, where it exhibits elevated expression in bladder cancer tissues compared to healthy tissue. Knockdown experiments targeting *SOX10* confirmed its prognostic value by significantly impacting the growth and metastatic ability of bladder cancer. The suspected mechanism involves SOX10 influencing the expression of other components such as B-catenin and Met. Targeting SOX10 as a marker for diagnosis, prognosis, and treatment may prove useful in the context of bladder cancer [16].

In summary, while further research is needed to fully understand the extent of SOX10 expression in various epithelial neoplasms, it remains a promising marker for the diagnosis and prognosis development of several carcinomas. Additionally, it shows potential as a treatment target in certain cancers.

Table 2. SOX10 expression in epithelial neoplasms.

Epithelial Neoplasm	SOX10 Expression	Implications
Ovarian serous, mucinous, and endometrioid carcinoma	Overexpressed	• SOX10 exhibits stem cell-supporting properties in both normal and cancerous cells [66]; • SOX10 presence is associated with chemoresistance, possibly contributing to poorer prognoses in certain cancers [66]; • SOX10 helps differentiate cell origin and estimate prognosis in ovarian epithelial tumors; • Subtypes such as serous, mucinous, and endometrioid can be differentiated based on the localization and intensity of SOX10 staining [66].
Triple-negative breast cancer	Overexpressed	• Clinical significance of SOX10 in breast carcinomas is not fully understood, but it is considered a useful marker [62]; • Approximately 66–74% of triple-negative breast carcinomas express SOX10 [63]; • Prognostic value is unclear, but there is evidence suggesting a possible contribution to malignant characteristics [62]; • High sensitivity in identifying triple-negative breast carcinomas [62–64].
Nasopharyngeal carcinomas	Overexpressed	• Marked overexpression of SOX10 is observed in nasopharyngeal carcinomas [72]; • Higher expression is associated with a poorer prognosis, and SOX10 is believed to be involved in tumor growth and metastasis [72]; • Potential importance as a diagnostic and prognostic marker [72].
Bladder carcinomas	Overexpressed	• SOX10 is significantly elevated in bladder carcinomas compared to surrounding healthy tissues [16]; • Knockdown of *SOX10* impacts cancer growth and spread, making it a potential treatment target [16]; • SOX10 inhibition may affect cancer progression by influencing other components in development pathways such as B-catenin and Met [16]; • Potential usefulness as a diagnostic marker for bladder cancers [16].
Salivary gland neoplasms	Overexpressed	• SOX10 helps distinguish between various types of salivary gland neoplasms [57]; • SOX10 rules out mimic lesions, differentiates between high- and low-grade adenocarcinomas, and is a reliable marker against certain similar-appearing tumors [57]; • Tumors lacking SOX10 are associated with specific histological features, such as the appearance of excretory or striated ducts; • SOX10 expression varies among different subtypes of salivary gland neoplasms [57–59].
Gastrointestinal Mesenchymal Tumors	Lost	• SOX10 expression is absent or minimal in gastrointestinal mesenchymal tumors with *NTRK* gene rearrangements, distinguishing them from gastrointestinal stromal tumors [54].
Uterine Sarcomas	Lost	• SOX10 lacks expression in a category of uterine sarcomas with NTRK fusions, distinguishing them from undifferentiated uterine sarcomas and aggressive leiomyosarcomas [55].

7. Expression of Other Members of the "SRY-Related HMG Box" in Cancers

7.1. The HMG Box Family

The HMG box is a versatile protein domain consisting of about 75 amino acids that plays a crucial role in DNA binding and various transcription and translation processes. The name "High Mobility" originates from the initial discovery of these proteins in the acid extracts of mammalian chromatin, where they exhibited significant electrophoretic mobility [73].

HMG box domains can be broadly categorized into two types based on their DNA binding specificity: non-sequence specific; and sequence specific [74]. Both types of HMG box domains exhibit a high affinity for non-B-type DNA structures, which include bent, kinked, and unwound DNA. Additionally, these domains are involved in diverse protein-protein interactions, such as DNA bending, looping, and unwinding [74,75].

7.1.1. Non-Sequence Specific HMG Box Domains

- Proteins in this category, such as HMGB1-4, typically possess two HMG boxes or four to six HMG boxes in the presence of transcription factor UBF [75];
- Mammals have four HMGB proteins (HMGB1-4), and they function as DNA chaperones, contributing to processes like transcription and DNA repair. However, each of these proteins has distinct characteristics [75].

7.1.2. Sequence Specific HMG Box Domains

- Proteins classified as sequence-specific usually have a single HMG box and lack acidic C-tails, which are common in non-sequence-specific HMG box proteins [74];
- Examples of proteins in this category include TCF, SRY, and SOX [75];
- Despite recognizing specific DNA sequences, these proteins form few base-specific hydrogen bonds, resulting in less sequence specificity [75].

7.2. SRY-Related HMG Box

The *SOX* genes, a subset of HMG box-type proteins, are encoded by 20 different genes in both humans and mice. These genes, located within the *SRY* gene on the Y chromosome, play pivotal roles in various cellular processes, including stemness maintenance, cell lineage determination, differentiation, proliferation, and even cell death. Unlike typical DNA modification mechanisms, *SOX* genes achieve their functions by binding specifically to the minor groove of pre-existing DNA, thereby influencing its shape and facilitating higher affinity binding of DNA to various transcription factors [76]. Key features of the *SOX* genes include the below.

7.2.1. Genetic Organization

- *SOX* genes are organized into eight groups (A–H), with group B further divided into subgroups B1 and B2 [76];
- Within the same group, SOX proteins share a high degree of structural and identity similarity, ranging from 70% to 95%, both in the HMG box domain and in external characteristics;
- Groups outside the same group have partial similarities in identities (>46%) in the HMG box domain and none in the external domains [76];

7.2.2. Functions and Mechanisms

- *SOX* genes play crucial roles in DNA replication and mutations, contributing to diverse cellular processes [76];

7.2.3. Individual SOX Genes

- The specific locus and schematic of the different *SOX* genes are detailed in Table 3

The following sections will provide insights into the implications of individual *SOX* gene groups in the genesis and progression of common cancers. The diversity within the *SOX* gene family allows for a wide range of functions and regulatory roles in cellular processes, making them essential players in normal development as well as potential contributors to cancer development [76].

Table 3. Specific locus and schematic of the different *SOX* genes. The blue oval represents the HMG box domain. The text within the blue oval indicates which SOX gene the schematic correlates to. The hexagon with "TA" indicates a transactivation domain. The hexagon with "TR" indicates a trans-repression domain. The gray diamond with "D" indicates a dimerization domain. Schematics were created with BioRender.com (2023).

Group	Gene	Locus	Schematic
A	*SRY*	YC3	
B1	*SOX1*	8 A1–A2	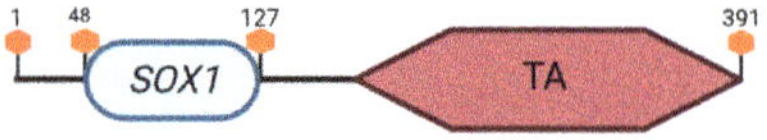
	SOX2	3 A2–B	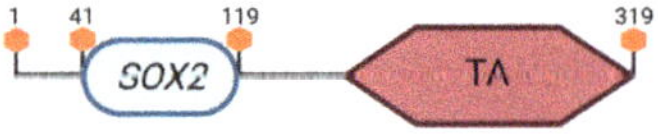
	SOX3	X A7.3–B	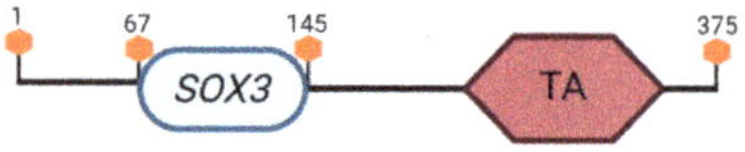
B2	*SOX14*	9 E3.3	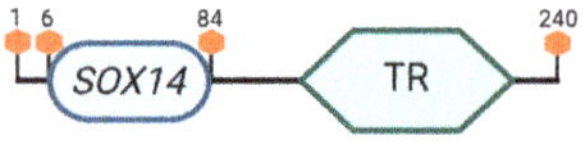
	SOX21	14 E4	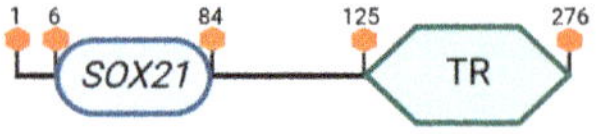
C	*SOX4*	13 A3–A5	
	SOX11	12 A3	
	SOX12	2 G3	
D	*SOX5*	6 G3	
	L-SOX5	6 G3	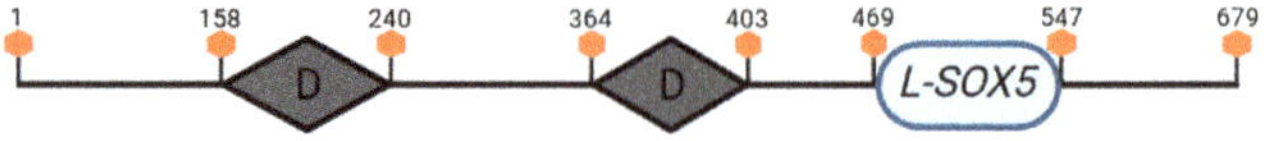
	SOX6	7 F1	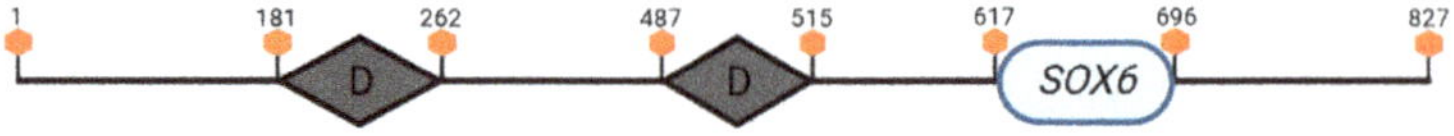
	SOX13	1 E4	

Table 3. *Cont.*

Group	Gene	Locus	Schematic
	SOX8	17 A3	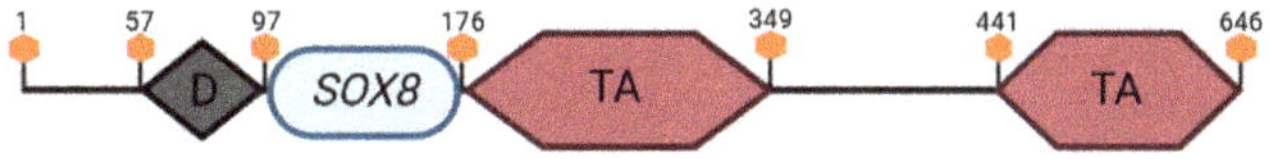 1 · 57 · 97 · D · *SOX8* · 176 · TA · 349 · 441 · TA · 646
E	*SOX9*	11 E2	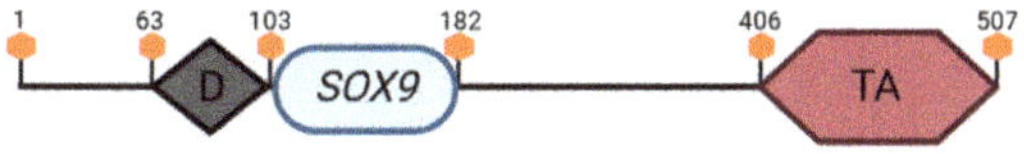 1 · 63 · 103 · D · *SOX9* · 182 · 406 · TA · 507
	SOX10	15 E1	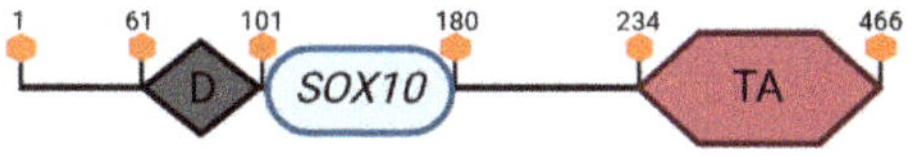 1 · 61 · 101 · D · *SOX10* · 180 · 234 · TA · 466
	SOX7	14 C3	 1 · 43 · *SOX7* · 122 · TA · 380
F	*SOX17*	1 A1	 1 · 68 · *SOX17* · 147 · 343 · TA · 419
	SOX18	2 H4	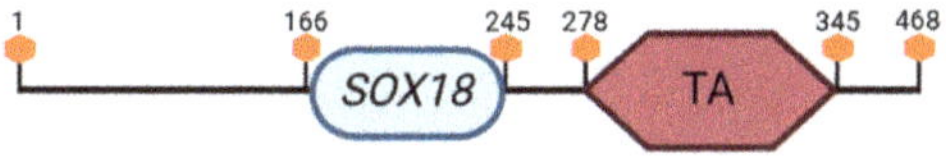 1 · 166 · *SOX18* · 245 · 278 · TA · 345 · 468
G	*SOX15*	11 B3	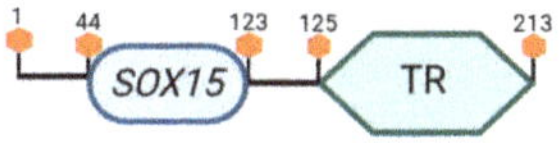 1 · 44 · *SOX15* · 123 · 125 · TR · 213
H	*SOX30*	11 B1.1	 1 · 365 · *SOX30* · 435 · TA · 782

Group A

Group A of the *SOX* gene family consists of a single member, *SRY* (Sex-determining Region Y), and its primary function is to determine sex in mammals [77,78]. In the context of cancer, particularly in prostate cancer, the role of *SRY* is not well understood, and it is unclear whether *SRY* acts as a tumor suppressor or has other implications in cancer development [79]. Downregulation of *SRY* has been observed in prostate cancer, but it often occurs concurrently with the downregulation of other Y chromosome-specific genes [79]. Therefore, it would be premature to attribute the development and proliferation of prostate cancer solely to the downregulation of *SRY*. Further research is needed to elucidate the specific role of *SRY* and its potential contributions to prostate cancer and other cancers.

Group B (B1 + B2)

Group B of the *SOX* gene family consists of *SOX1*, *SOX2*, *SOX3*, *SOX14*, and *SOX21*. *SOX1*, *SOX2*, and *SOX3* belong to subgroup B1, while *SOX14* and *SOX21* fall into subgroup B2 [77,80–82].

SOX1

- Function: *SOX1* plays a crucial role in maintaining stem cell lineage, particularly in embryogenesis, differentiation, and mammalian brain development. It is essential for the survival and function of dopaminergic neurons [80];
- Oncogenic properties: *SOX1* has been implicated in the development of small cell lung, central nervous system, breast, and ovarian cancers. In small-cell lung cancer, *SOX1* collaborates with *NKX2.1* to maintain its identity and function. In central nervous

system tumors like glioblastomas, *SOX1* extends the survivability of cancer cells [83]. In breast and ovarian cancer, *SOX1* acts as a tumor suppressor by inhibiting the Wnt/B-Catenin and *STAT3* signaling pathways [84,85];

SOX2

- Function: *SOX2* is a transcription factor that prolongs stemness in both embryonic and adult stem cells [86];
- Oncogenic properties: Dysregulation of *SOX2* expression is associated with increased proliferation and metastasis in the central nervous system and lung carcinomas [86];

SOX3

- Function: *SOX3* is upregulated in esophageal SCC, ovarian carcinoma, and osteosarcoma, promoting proliferation and migration [87]. It induces apoptosis in human breast cancer cell lines [87];

SOX14

- Function: *SOX14* is involved in the development of cervical cancer, inducing *P53* activation, which leads to apoptosis in cervical carcinoma cell lines [88]. It also promotes proliferation and invasion through the Wnt/B-catenin pathway [88,89];

SOX21

- Function: *SOX21* has a tumor suppressor-like function in central nervous system cancers [90], inhibiting the carcinogenic properties of *SOX2* [91]. Forced expression of *SOX21* induces cellular apoptosis in glioma cells and enables differentiation, preventing glioma formation [90].

These *SOX* genes in Group B exhibit diverse functions and play critical roles in various cancers, either promoting or inhibiting oncogenic processes. Their involvement underscores the complexity of *SOX* gene functions in different cellular contexts and cancer types;

Group C

The *SOX* genes that have been classified into group C include *SOX4, SOX11,* and *SOX12;*

SOX4

- Function: *SOX4* is implicated in embryogenesis and tissue development [92–97];
- Cancer associations: Elevated *SOX4* expression is observed in leukemia, colorectal, lung, breast, and hepatocellular cancers [92–97]. In hepatocellular carcinoma, increased *SOX4* expression inhibits *P53*-directed apoptosis by restricting *BAX* expression [96];

SOX11

- Function: *SOX11* serves as both a causative and protective agent in various tumors;
- Cancer associations: Upregulation of *SOX11* is seen in medulloblastoma, mantle cell lymphoma, endometrial and breast cancer, Burkitt's lymphoma, colorectal cancer, lung adenocarcinoma, lung SCC, and ovarian cancer [98–104]. *SOX11* expression is a unique feature in certain cancers and helps distinguish them from other malignancies [105,106];
- Prognostic factors: High *SOX11* expression in gastric and ovarian cancers is linked to higher survival rates, while in breast cancers, the opposite is observed [102,107];

SOX12

- Function: Hepatocellular carcinomas positive for *SOX12* exhibit increased proliferation, malignant potential, and higher resistance to cisplatin, a common chemotherapy agent [108];
- Cancer associations: *SOX12* is involved in gastric, lung, hepatocellular, colorectal, renal carcinomas, and thyroid cancers (elevated levels) [51,108–110]. Increased *SOX12* expression in thyroid cancer cells is associated with promoting carcinogenic properties [51].

The *SOX* genes in Group C play diverse roles in embryonic development and tissue maintenance and are implicated in various cancers. They contribute to the complexity of cancer biology by either promoting or inhibiting tumorigenesis depending on the specific context and cancer type;

Group D

The *SOX* genes that have been classified into group D include *SOX5*, *SOX6*, and *SOX13*;

SOX5

- Function: *SOX5* plays a role in the development and differentiation of embryonic germ cell lines [111];
- Cancer associations: Similar to other *SRY*-related HMG box genes, *SOX5* elevates the ability of cancer to grow, metastasize, and invade through angiogenesis. It is implicated in hepatocellular, breast, and gastric cancer [112–114];
- Unique properties: *SOX5* can mediate the epithelial-to-mesenchymal transition (EMT), a fundamental process in metastasis, by regulating the expression of E-cadherin and vimentin [112,115,116];

SOX6

- Function: *SOX6* exhibits both tumor suppressor and oncogenic properties depending on the cancer type;
- Cancer associations: *SOX6* is downregulated in osteosarcoma [117], esophageal SCC [118], hepatocellular carcinoma [119], and pancreatic β-cell cancers [120]. It shows oncogenic properties in gliomas [121] and endometrial cancers [122];
- Unique properties: *SOX6* induces autophagy in cervical cancer cell lines, leading to increased resistance to cisplatin chemotherapy and enhanced survivability [123];

SOX13

- Cancer associations: *SOX13* is highly expressed in oligodendrogliomas, gliomas, gastric carcinomas, and hepatocellular carcinomas [123–126]. *SOX13* overexpression in hepatocellular carcinoma activates *TWIST1*, a major transcription factor in embryonic development, promoting cancer metastasis [126]. *SOX13* supports stem-like properties in hepatocellular carcinoma, contributing to increased self-renewal, resistance to chemotherapy, and tumorgenicity [127].

These *SOX* genes, namely *SOX5*, *SOX6*, and *SOX13*, demonstrate diverse roles in embryonic development and cancer biology. Their involvement in processes like EMT, autophagy induction, and support for stem-like properties highlights their significance in the complex landscape of cancer progression and metastasis;

Group E

The *SOX* genes that have been classified into group E include *SOX8*, *SOX9*, and *SOX10*;

SOX8

- Function: *SOX8* has some minor effects on the specification and differentiation of glial cells;
- Cancer associations: *SOX8* expression is greatest during central nervous system development in immature cells. Elevated levels of *SOX8* indicate an undifferentiated state in the gliomas [124];

SOX9

- Function: *SOX9* is involved in multiple cancers in a variety of ways;
- Cancer associations: In some breast cancer subtypes, *SOX9* is involved in a positive feedback loop through Wnt/β-catenin activation [128]. Prostate cancer tends to be correlated with elevated levels of SOX9 [129]. *SOX9* contributes to cell proliferation and invasion in renal cell carcinoma. MiRNA-138-induced *SOX9* suppression prevents renal cell carcinoma progression [130]. Through the WNT/β-catenin pathway, *SOX9* is involved in cancer cell proliferation and invasion in papillary thyroid cancer [131]. *SOX9* increases *LGR5* expression, imparting the ability of glioblastoma cells to undergo tumorigenesis [77]. Elevated levels of *SOX9* expression in colorectal cancers are

associated with lower 5-year survival rates [132]. *SOX9* levels are increased in non-small lung cancer [77] due to tumor-associated macrophages, which release TGF-β [133]. In skin cancers, *SOX9* levels are elevated too [77]. Increased *SOX9* levels cause melanoma cells to metastasize [134]. *SOX9*-involved keratinocyte proliferation also occurs in cutaneous BCC and cutaneous SCC [135].

SOX8, *SOX9*, and *SOX10* play diverse roles in different cancers, influencing processes such as differentiation, proliferation, invasion, and tumorigenesis. Understanding their specific functions in various cancer types is crucial for developing targeted therapeutic approaches;

Group F

The *SOX* genes that have been classified into group F include *SOX7*, *SOX17*, and *SOX18*;

SOX7

- Cancer associations: *SOX7* is implicated in several cancers. In breast cancer, *SOX7* functions as a tumor suppressor [136]. Hypermethylation-mediated silencing of the *SOX7* promoter is associated with greater carcinogenesis in breast cancer [136]. *SOX7* can be used as a marker for prognosis in prostate cancer. Its downregulation may be involved in the castration-resistant progression of prostate cancer [129]. *SOX7* also exhibits tumor-suppressive effects in gastric cancer through potential involvement in abnormalities with the *SOX7*-associated WNT/β-catenin pathway [137]. *SOX7's* tumor suppressor effects have also been delineated in non-small-cell lung cancer, targeted by microRNA-9 [138];

SOX17

- Cancer associations: *SOX17* is associated with several cancers. Hypermethylation-dependent silencing of the *SOX17* promoter may induce inappropriate activation of the Wnt pathway, giving rise to breast cancer, thyroid cancer, gliomas, and gastrointestinal tumors [139–142]. Melanoma pathogenesis is also associated with decreased SOX17 expression; however, the mechanism is unclear [143];

SOX18

- Function: *SOX18* takes part in the development of blood and lymphatic vessels, as well as hair follicles [144]. Wound healing also involves SOX18 [145].
- Cancer associations: *SOX18* is associated with breast, lung, and skin cancers. In breast cancer, there is a positive correlation between *SOX18* and vascular endothelial growth factor D (*VEGF-D*), suggesting that *SOX18* positively influences angiogenesis [144]. In non-small-cell lung cancer, *SOX18* expression is noted in cells and vessels, and its expression may be used as a prognostic marker [145]. In skin cancers, elevated *SOX18* expression is involved in the formation of BCC and SCC [146].

Understanding the roles of *SOX7*, *SOX17*, and *SOX18* in various cancers provides insights into their potential as diagnostic markers and therapeutic targets in cancer treatment;

Group G

The sole member of this group is *SOX15*. Compared to the other members of the *SOX* family, it has been relatively understudied. Overexpression of *SOX15* is linked to lower proliferation of testicular embryonic cancer cell lines [147]. *SOX15* serves as a potential tumor suppressor gene and is negatively associated with the development of pancreatic ductal adenocarcinoma through the Wnt/B-catenin pathway [148]. Additionally, *SOX15* is repeatedly underexpressed among cancer cell lines, including colon, prostate, stomach, and uterine cancers, and overexpressed in some subsets of lung carcinomas [149].

SOX15, despite being relatively understudied compared to other *SOX* family members, demonstrates potential significance in regulating proliferation and acting as a tumor suppressor in specific cancer types, such as testicular embryonic cancer and pancreatic ductal adenocarcinoma. Its differential expression across various cancers suggests a context-dependent role, and further research may unveil its precise mechanisms and therapeutic implications;

Group H

SOX30 is the sole member of Group H [77]. It acts as a tumor suppressor by activating *P53* transcription, leading to apoptosis. *SOX30* inhibits T-cell factor (TCF) either by binding to β-catenin or inhibiting β-catenin transcription [77,150,151]. Regarding lung adenocarcinoma specifically, the latter can be associated with hypermethylation of the *SOX30* gene. *SOX30's* inhibition of TCF can contribute to the development of lung cancer. It also functions as a tumor suppressor by activating desmosomal genes, impeding cancer growth and spread [77,152].

8. Conclusions and Future Directions

In conclusion, *SOX10* emerges as a pivotal transcription factor with a multifaceted role extending from embryonic development to the pathogenesis of diverse pathological conditions. Its critical significance is exemplified by its association with congenital disorders such as Waardenburg–Shah Syndrome, PCWH syndrome, and Kallman syndrome, where mutations disrupt neural crest development. Within neural and neuroectodermal tumors, *SOX10* serves as a key player influencing proliferation and differentiation, making it a promising diagnostic and therapeutic marker.

The spotlight on *SOX10* intensifies in melanoma, where its impact on crucial factors like MITF and cell migration shapes tumor progression and treatment responses. In mesenchymal tumors, *SOX10* expression becomes a valuable tool for distinguishing between different tumor types, thereby facilitating accurate diagnoses and informed treatment decisions.

Epithelial neoplasms further underscore *SOX10's* clinical relevance. Its expression or absence provides crucial insights into tumor cell origins, prognosis, and treatment responses. Particularly in ovarian cancer, *SOX10's* involvement in chemoresistance highlights its significance in clinical settings.

The multifunctionality of *SOX10* positions it as a promising candidate for extensive research and clinical applications across various pathological conditions. As we delve deeper into its intricacies, there is potential for improved diagnostic accuracy and the development of more effective therapeutic strategies. *SOX10* stands at the intersection of basic research and clinical utility, holding promise for advancements that could reshape our approach to a spectrum of diseases.

Author Contributions: Conceptualization, H.F.B.; methodology, H.F.B., A.T., K.S., J.G., V.A., S.O., A.R.A., R.A., A.H.A. and S.H.; investigation, H.F.B., A.T.; K.S., J.G., V.A., S.O., A.R.A., R.A. and A.H.A.; resources, H.F.B., A.T., K.S., J.G., V.A., S.O., A.R.A., R.A., A.H.A. and S.H.; data curation, H.F.B., A.T., K.S., J.G., V.A., S.O., A.R.A., R.A. and A.H.A.; writing—original draft preparation, H.F.B., A.T., K.S., J.G., V.A., S.O., A.R.A., R.A., A.H.A. and S.H.; writing—review and editing, R.P.; visualization, R.P.; supervision, R.P.; project administration, H.F.B.; funding acquisition, H.F.B. All authors have read and agreed to the published version of the manuscript.

Funding: This research received no external funding.

Institutional Review Board Statement: Not applicable.

Informed Consent Statement: Not applicable.

Data Availability Statement: No new data were created or analyzed in this study. Data sharing is not applicable to this article.

Acknowledgments: We would like to thank all members of the Department of Pathology and Laboratory Medicine, Mount Sinai Medical Center of Florida (Miami Beach, FL, USA) for their help with this work. Figures were created with BioRender.com (accessed on 7 October 2023). All rights and ownership of BioRender content are reserved by BioRender (Agreement numbers *JK25Y0GP09, JF25Y0I2WN, CJ25Y0KPLG,* and *YD25Y0KPPH*). BioRender content included in the completed graphic is not licensed for any commercial uses beyond publication in a journal.

Conflicts of Interest: The authors declare no conflict of interest.

References

1. Herbarth, B.; Pingault, V.; Bondurand, N.; Kuhlbrodt, K.; Hermans-Borgmeyer, I.; Puliti, A.; Lemort, N.; Goossens, M.; Wegner, M. Mutation of the Sry-related *Sox10* gene in Dominant megacolon, a mouse model for human Hirschsprung disease. *Proc. Natl. Acad. Sci. USA* **1998**, *95*, 5161–5165. [CrossRef]

2. Aoki, Y.; Saint-Germain, N.; Gyda, M.; Magner-Fink, E.; Lee, Y.H.; Credidio, C.; Saint-Jeannet, J.P. Sox10 regulates the development of neural crest-derived melanocytes in Xenopus. *Dev. Biol.* **2003**, *259*, 19–33. [CrossRef]

3. Pingault, V.; Zerad, L.; Bertani-Torres, W.; Bondurand, N. SOX10: 20 years of phenotypic plurality and current understanding of its developmental function. *J. Med. Genet.* **2022**, *59*, 105–114. [CrossRef]

4. Schreiner, S.; Cossais, F.; Fischer, K.; Scholz, S.; Bösl, M.R.; Holtmann, B.; Sendtner, M.; Wegner, M. Hypomorphic Sox10 alleles reveal novel protein functions and unravel developmental differences in glial lineages. *Development* **2007**, *134*, 3271–3281. [CrossRef] [PubMed]

5. Pusch, C.; Hustert, E.; Pfeifer, D.; Südbeck, P.; Kist, R.; Roe, B.; Wang, Z.; Balling, R.; Blin, N.; Scherer, G. The *SOX10/Sox10* gene from human and mouse: Sequence, expression, and transactivation by the encoded HMG domain transcription factor. *Hum. Genet.* **1998**, *103*, 115–123. [CrossRef]

6. Sommer, L. Generation of melanocytes from neural crest cells. *Pigment. Cell Melanoma Res.* **2011**, *24*, 411–421. [CrossRef] [PubMed]

7. Yalan, L.; Hua, Z.; Yong, F. Progress in the study of syndromic hearing loss resulted from neural crest abnormalities. *Yi Chuan* **2014**, *36*, 1131–1144. [PubMed]

8. Pingault, V.; Bondurand, N.; Kuhlbrodt, K.; Goerich, D.E.; Préhu, M.O.; Puliti, A.; Herbarth, B.; Hermans-Borgmeyer, I.; Legius, E.; Matthijs, G.; et al. SOX10 mutations in patients with Waardenburg-Hirschsprung disease. *Nat. Genet.* **1998**, *18*, 171–173. [CrossRef]

9. Wissmüller, S.; Kosian, T.; Wolf, M.; Finzsch, M.; Wegner, M. The high-mobility-group domain of Sox proteins interacts with DNA-binding domains of many transcription factors. *Nucleic Acids Res.* **2006**, *34*, 1735–1744. [CrossRef] [PubMed]

10. Schock, E.N.; LaBonne, C. Sorting Sox: Diverse Roles for Sox Transcription Factors During Neural Crest and Craniofacial Development. *Front. Physiol.* **2020**, *11*, 606889. [CrossRef] [PubMed]

11. Williams, C.A.C.; Soufi, A.; Pollard, S.M. Post-translational modification of SOX family proteins: Key biochemical targets in cancer? *Semin. Cancer Biol.* **2020**, *67*, 30–38. [CrossRef]

12. Taylor, K.M.; Labonne, C. SoxE factors function equivalently during neural crest and inner ear development and their activity is regulated by SUMOylation. *Dev. Cell* **2005**, *9*, 593–603. [CrossRef] [PubMed]

13. Girard, M.; Goossens, M. Sumoylation of the SOX10 transcription factor regulates its transcriptional activity. *FEBS Lett.* **2006**, *580*, 1635–1641. [CrossRef] [PubMed]

14. Peirano, R.I.; Goerich, D.E.; Riethmacher, D.; Wegner, M. Protein zero gene expression is regulated by the glial transcription factor Sox10. *Mol. Cell Biol.* **2000**, *20*, 3198–3209. [CrossRef] [PubMed]

15. Amer, S.; Ibrahim, H.; Elkordy, M. The Immunohistochemical Expression of SOX-10 in Urothelial Carcinoma and the Non Neoplastic Urothelium; and a Correlation with the Tumor Features. *Asian Pac. J. Cancer Prev.* **2022**, *23*, 1425–1432. [CrossRef] [PubMed]

16. Yin, H.; Qin, C.; Zhao, Y.; Du, Y.; Sheng, Z.; Wang, Q.; Song, Q.; Chen, L.; Liu, C.; Xu, T. SOX10 is over-expressed in bladder cancer and contributes to the malignant bladder cancer cell behaviors. *Clin. Transl. Oncol.* **2017**, *19*, 1035–1044. [CrossRef] [PubMed]

17. Watanabe, K.; Takeda, K.; Katori, Y.; Ikeda, K.; Oshima, T.; Yasumoto, K.; Saito, H.; Takasaka, T.; Shibahara, S. Expression of the Sox10 gene during mouse inner ear development. *Brain Res. Mol. Brain Res.* **2000**, *84*, 141–145. [CrossRef]

18. Locher, H.; Frijns, J.H.; van Iperen, L.; de Groot, J.C.; Huisman, M.A.; Chuva de Sousa Lopes, S.M. Neurosensory development and cell fate determination in the human cochlea. *Neural Dev.* **2013**, *8*, 20. [CrossRef]

19. Qi, J.; Ma, L.; Guo, W. Recent advances in the regulation mechanism of SOX10. *J. Otol.* **2022**, *17*, 247–252. [CrossRef]

20. Britsch, S.; Goerich, D.E.; Riethmacher, D.; Peirano, R.I.; Rossner, M.; Nave, K.A.; Birchmeier, C.; Wegner, M. The transcription factor Sox10 is a key regulator of peripheral glial development. *Genes. Dev.* **2001**, *15*, 66–78. [CrossRef] [PubMed]

21. Gopinath, C.; Law, W.D.; Rodríguez-Molina, J.F.; Prasad, A.B.; Song, L.; Crawford, G.E.; Mullikin, J.C.; Svaren, J.; Antonellis, A. Stringent comparative sequence analysis reveals SOX10 as a putative inhibitor of glial cell differentiation. *BMC Genom.* **2016**, *17*, 887. [CrossRef]

22. Mertelmeyer, S.; Weider, M.; Baroti, T.; Reiprich, S.; Fröb, F.; Stolt, C.C.; Wagner, K.U.; Wegner, M. The transcription factor Sox10 is an essential determinant of branching morphogenesis and involution in the mouse mammary gland. *Sci. Rep.* **2020**, *10*, 17807. [CrossRef]

23. Southard-Smith, E.M.; Kos, L.; Pavan, W.J. Sox10 mutation disrupts neural crest development in Dom Hirschsprung mouse model. *Nat. Genet.* **1998**, *18*, 60–64. [CrossRef] [PubMed]

24. Lane, P.W.; Liu, H.M. Association of megacolon with a new dominant spotting gene (Dom) in the mouse. *J. Hered.* **1984**, *75*, 435–439. [CrossRef] [PubMed]

25. Veronique, P.; Dorothée, E.; Florence Dastot-Le, M.; Michel, G.; Sandrine, M.; Nadege, B. Review and update of mutations causing Waardenburg syndrome. *Hum. Mutat.* **2010**, *31*, 391–406. [CrossRef]

26. Bondurand, N.; Dastot-Le Moal, F.; Stanchina, L.; Collot, N.; Baral, V.; Marlin, S.; Attie-Bitach, T.; Giurgea, I.; Skopinski, L.; Reardon, W.; et al. Deletions at the SOX10 gene locus cause Waardenburg syndrome types 2 and 4. *Am. J. Hum. Genet.* **2007**, *81*, 1169–1185. [CrossRef] [PubMed]
27. Inoue, K.; Shilo, K.; Boerkoel, C.F.; Crowe, C.; Sawady, J.; Lupski, J.R.; Agamanolis, D.P. Congenital hypomyelinating neuropathy, central dysmyelination, and Waardenburg-Hirschsprung disease: Phenotypes linked by SOX10 mutation. *Ann. Neurol.* **2002**, *52*, 836–842. [CrossRef] [PubMed]
28. Touraine, R.L.; Attié-Bitach, T.; Manceau, E.; Korsch, E.; Sarda, P.; Pingault, V.; Encha-Razavi, F.; Pelet, A.; Augé, J.; Nivelon-Chevallier, A.; et al. Neurological phenotype in Waardenburg syndrome type 4 correlates with novel SOX10 truncating mutations and expression in developing brain. *Am. J. Hum. Genet.* **2000**, *66*, 1496–1503. [CrossRef] [PubMed]
29. Inoue, K.; Tanabe, Y.; Lupski, J.R. Myelin deficiencies in both the central and the peripheral nervous systems associated with a SOX10 mutation. *Ann. Neurol.* **1999**, *46*, 313–318. [CrossRef]
30. Lieblich, J.M.; Rogol, A.D.; White, B.J.; Rosen, S.W. Syndrome of anosmia with hypogonadotropic hypogonadism (*Kallmann syndrome*): Clinical and laboratory studies in 23 cases. *Am. J. Med.* **1982**, *73*, 506–519. [CrossRef]
31. Veronique, P.; Virginie, B.; Viviane, B.; Séverine, M.; Yuli, W.; Asma, C.; Corinne, F.; Chrystel, L.; Verier-Mine, O.; Christine, F.; et al. Loss-of-Function Mutations in SOX10 Cause Kallmann Syndrome with Deafness. *Am. J. Hum. Genet.* **2013**, *92*, 707–724. [CrossRef]
32. Elmaleh-Bergès, M.; Baumann, C.; Noël-Pétroff, N.; Sekkal, A.; Couloigner, V.; Devriendt, K.; Wilson, M.; Marlin, S.; Sebag, G.; Pingault, V. Spectrum of Temporal Bone Abnormalities in Patients with Waardenburg Syndrome and SOX10 Mutations. *Am. J. Neuroradiol.* **2013**, *34*, 1257–1263. [CrossRef] [PubMed]
33. Jian, S.; Yong, F.; Frederic, R.A.; Frederic, A.; Paul, C.; Kris, V.; Ingeborg, D. Hearing loss in Waardenburg syndrome: A systematic review. *Clin. Genet.* **2016**, *89*, 416–425. [CrossRef]
34. Breuskin, I.; Bodson, M.; Thelen, N.; Thiry, M.; Borgs, L.; Nguyen, L.; Stolt, C.; Wegner, M.; Lefebvre, P.P.; Malgrange, B. Glial but not neuronal development in the cochleo-vestibular ganglion requires Sox10. *J. Neurochem.* **2010**, *114*, 1827–1839. [CrossRef] [PubMed]
35. Veronique, P.; Emmanuelle, F.; Viviane, B.; Souad, G.; Natalie, L.; Vincent, C.; Françoise, D.; Noël-Pétroff, N.; Pointe, H.D.L.; Monique, E.; et al. SOX10 mutations mimic isolated hearing loss. *Clin. Genet.* **2015**, *88*, 352–359. [CrossRef]
36. Bakos, R.M.; Maier, T.; Besch, R.; Mestel, D.S.; Ruzicka, T.; Sturm, R.A.; Berking, C. Nestin and SOX9 and SOX10 transcription factors are coexpressed in melanoma. *Exp. Dermatol.* **2010**, *19*, e89–e94. [CrossRef] [PubMed]
37. Su, Z.; Zheng, X.; Zhang, X.; Wang, Y.; Zhu, S.; Lu, F.; Qu, J.; Hou, L. Sox10 regulates skin melanocyte proliferation by activating the DNA replication licensing factor MCM5. *J. Dermatol. Sci.* **2017**, *85*, 216–225. [CrossRef]
38. Cronin, J.C.; Watkins-Chow, D.E.; Incao, A.; Hasskamp, J.H.; Schönewolf, N.; Aoude, L.G.; Hayward, N.K.; Bastian, B.C.; Dummer, R.; Loftus, S.K.; et al. SOX10 Ablation Arrests Cell Cycle, Induces Senescence, and Suppresses Melanomagenesis. *Cancer Res.* **2013**, *73*, 5709–5718. [CrossRef]
39. Rosenbaum, S.R.; Tiago, M.; Caksa, S.; Capparelli, C.; Purwin, T.J.; Kumar, G.; Glasheen, M.; Pomante, D.; Kotas, D.; Chervoneva, I.; et al. SOX10 requirement for melanoma tumor growth is due, in part, to immune-mediated effects. *Cell Rep.* **2021**, *37*, 110085. [CrossRef]
40. Seong, I.; Min, H.J.; Lee, J.H.; Yeo, C.Y.; Kang, D.M.; Oh, E.S.; Hwang, E.S.; Kim, J. Sox10 controls migration of B16F10 melanoma cells through multiple regulatory target genes. *PLoS ONE* **2012**, *7*, e31477. [CrossRef]
41. Clevenger, J.; Joseph, C.; Dawlett, M.; Guo, M.; Gong, Y. Reliability of immunostaining using pan-melanoma cocktail, SOX10, and microphthalmia transcription factor in confirming a diagnosis of melanoma on fine-needle aspiration smears. *Cancer Cytopathol.* **2014**, *122*, 779–785. [CrossRef]
42. Cronin, J.C.; Loftus, S.K.; Baxter, L.L.; Swatkoski, S.; Gucek, M.; Pavan, W.J. Identification and functional analysis of SOX10 phosphorylation sites in melanoma. *PLoS ONE* **2018**, *13*, e0190834. [CrossRef]
43. Wouters, J.; Kalender-Atak, Z.; Minnoye, L.; Spanier, K.I.; De Waegeneer, M.; Bravo González-Blas, C.; Mauduit, D.; Davie, K.; Hulselmans, G.; Najem, A.; et al. Robust gene expression programs underlie recurrent cell states and phenotype switching in melanoma. *Nat. Cell Biol.* **2020**, *22*, 986–998. [CrossRef]
44. Verfaillie, A.; Imrichova, H.; Atak, Z.K.; Dewaele, M.; Rambow, F.; Hulselmans, G.; Christiaens, V.; Svetlichnyy, D.; Luciani, F.; Van den Mooter, L.; et al. Decoding the regulatory landscape of melanoma reveals TEADS as regulators of the invasive cell state. *Nat. Commun.* **2015**, *6*, 6683. [CrossRef] [PubMed]
45. Cronin, J.C.; Wunderlich, J.; Loftus, S.K.; Prickett, T.D.; Wei, X.; Ridd, K.; Vemula, S.; Burrell, A.S.; Agrawal, N.S.; Lin, J.C.; et al. Frequent mutations in the MITF pathway in melanoma. *Pigment. Cell Melanoma Res.* **2009**, *22*, 435–444. [CrossRef] [PubMed]
46. Shakhova, O.; Zingg, D.; Schaefer, S.M.; Hari, L.; Civenni, G.; Blunschi, J.; Claudinot, S.; Okoniewski, M.; Beermann, F.; Mihic-Probst, D.; et al. Sox10 promotes the formation and maintenance of giant congenital naevi and melanoma. *Nat. Cell Biol.* **2012**, *14*, 882–890. [CrossRef] [PubMed]
47. Capparelli, C.; Purwin, T.J.; Glasheen, M.; Caksa, S.; Tiago, M.; Wilski, N.; Pomante, D.; Rosenbaum, S.; Nguyen, M.Q.; Cai, W.; et al. Targeting SOX10-deficient cells to reduce the dormant-invasive phenotype state in melanoma. *Nat. Commun.* **2022**, *13*, 1381. [CrossRef] [PubMed]
48. Kang, Y.; Pekmezci, M.; Folpe, A.L.; Ersen, A.; Horvai, A.E. Diagnostic utility of SOX10 to distinguish malignant peripheral nerve sheath tumor from synovial sarcoma, including intraneural synovial sarcoma. *Mod. Pathol.* **2014**, *27*, 55–61. [CrossRef] [PubMed]

49. Pekmezci, M.; Reuss, D.E.; Hirbe, A.C.; Dahiya, S.; Gutmann, D.H.; von Deimling, A.; Horvai, A.E.; Perry, A. Morphologic and immunohistochemical features of malignant peripheral nerve sheath tumors and cellular schwannomas. *Mod. Pathol.* **2015**, *28*, 187–200. [CrossRef] [PubMed]

50. Doddrell, R.D.; Dun, X.P.; Shivane, A.; Feltri, M.L.; Wrabetz, L.; Wegner, M.; Sock, E.; Hanemann, C.O.; Parkinson, D.B. Loss of SOX10 function contributes to the phenotype of human Merlin-null schwannoma cells. *Brain* **2013**, *136*, 549–563. [CrossRef]

51. Su, Z.; Bao, W.; Yang, G.; Liu, J.; Zhao, B. SOX12 promotes thyroid cancer cell proliferation and invasion by regulating the expression of POU2F1 and POU3F1. *Yonsei Med. J.* **2022**, *63*, 591. [CrossRef] [PubMed]

52. Miettinen, M.; McCue, P.A.; Sarlomo-Rikala, M.; Biernat, W.; Czapiewski, P.; Kopczynski, J.; Thompson, L.D.; Lasota, J.; Wang, Z.; Fetsch, J.F. Sox10—A marker for not only schwannian and melanocytic neoplasms but also myoepithelial cell tumors of soft tissue: A systematic analysis of 5134 tumors. *Am. J. Surg. Pathol.* **2015**, *39*, 826–835. [CrossRef] [PubMed]

53. Karamchandani, J.R.; Nielsen, T.O.; van de Rijn, M.; West, R.B. Sox10 and S100 in the diagnosis of soft-tissue neoplasms. *Appl. Immunohistochem. Mol. Morphol.* **2012**, *20*, 445–450. [CrossRef]

54. Atiq, M.A.; Davis, J.L.; Hornick, J.L.; Dickson, B.C.; Fletcher, C.D.M.; Fletcher, J.A.; Folpe, A.L.; Mariño-Enríquez, A. Mesenchymal tumors of the gastrointestinal tract with NTRK rearrangements: A clinicopathological, immunophenotypic, and molecular study of eight cases, emphasizing their distinction from gastrointestinal stromal tumor (GIST). *Mod. Pathol.* **2021**, *34*, 95–103. [CrossRef] [PubMed]

55. Chiang, S.; Cotzia, P.; Hyman, D.M.; Drilon, A.; Tap, W.D.; Zhang, L.; Hechtman, J.F.; Frosina, D.; Jungbluth, A.A.; Murali, R.; et al. NTRK Fusions Define a Novel Uterine Sarcoma Subtype With Features of Fibrosarcoma. *Am. J. Surg. Pathol.* **2018**, *42*, 791–798. [CrossRef] [PubMed]

56. Lee, J.H.; Kang, H.J.; Yoo, C.W.; Park, W.S.; Ryu, J.S.; Jung, Y.S.; Choi, S.W.; Park, J.Y.; Han, N. PLAG1, SOX10, and Myb Expression in Benign and Malignant Salivary Gland Neoplasms. *J. Pathol. Transl. Med.* **2019**, *53*, 23–30. [CrossRef]

57. Hsieh, M.S.; Lee, Y.H.; Chang, Y.L. SOX10-positive salivary gland tumors: A growing list, including mammary analogue secretory carcinoma of the salivary gland, sialoblastoma, low-grade salivary duct carcinoma, basal cell adenoma/adenocarcinoma, and a subgroup of mucoepidermoid carcinoma. *Hum. Pathol.* **2016**, *56*, 134–142. [CrossRef]

58. Schmitt, A.C.; Cohen, C.; Siddiqui, M.T. Expression of SOX10 in Salivary Gland Oncocytic Neoplasms: A Review and a Comparative Analysis with Other Immunohistochemical Markers. *Acta Cytol.* **2015**, *59*, 384–390. [CrossRef]

59. Zhu, S.; Schuerch, C.; Hunt, J. Review and updates of immunohistochemistry in selected salivary gland and head and neck tumors. *Arch. Pathol. Lab. Med.* **2015**, *139*, 55–66. [CrossRef]

60. Ohtomo, R.; Mori, T.; Shibata, S.; Tsuta, K.; Maeshima, A.M.; Akazawa, C.; Watabe, Y.; Honda, K.; Yamada, T.; Yoshimoto, S.; et al. SOX10 is a novel marker of acinus and intercalated duct differentiation in salivary gland tumors: A clue to the histogenesis for tumor diagnosis. *Mod. Pathol.* **2013**, *26*, 1041–1050. [CrossRef]

61. Rooper, L.M.; McCuiston, A.M.; Westra, W.H.; Bishop, J.A. SOX10 Immunoexpression in Basaloid Squamous Cell Carcinomas: A Diagnostic Pitfall for Ruling out Salivary Differentiation. *Head. Neck Pathol.* **2019**, *13*, 543–547. [CrossRef]

62. Kriegsmann, K.; Flechtenmacher, C.; Heil, J.; Kriegsmann, J.; Mechtersheimer, G.; Aulmann, S.; Weichert, W.; Sinn, H.P.; Kriegsmann, M. Immunohistological Expression of SOX-10 in Triple-Negative Breast Cancer: A Descriptive Analysis of 113 Samples. *Int. J. Mol. Sci.* **2020**, *21*, 6407. [CrossRef]

63. Cimino-Mathews, A. Novel uses of immunohistochemistry in breast pathology: Interpretation and pitfalls. *Mod. Pathol.* **2021**, *34*, 62–77. [CrossRef]

64. Liu, J.L.; Chen, D.S.; Cheng, Z.Q.; Hu, J.T. Expression of SOX10 and GATA3 in breast cancer and their significance. *Zhonghua Bing Li Xue Za Zhi* **2022**, *51*, 536–541. [PubMed]

65. Adkins, B.D.; Geromes, A.; Zhang, L.Y.; Chernock, R.; Kimmelshue, K.; Lewis, J., Jr.; Ely, K. SOX10 and GATA3 in Adenoid Cystic Carcinoma and Polymorphous Adenocarcinoma. *Head. Neck Pathol.* **2020**, *14*, 406–411. [CrossRef] [PubMed]

66. Kwon, A.Y.; Heo, I.; Lee, H.J.; Kim, G.; Kang, H.; Heo, J.H.; Kim, T.H.; An, H.J. Sox10 expression in ovarian epithelial tumors is associated with poor overall survival. *Virchows Arch.* **2016**, *468*, 597–605. [CrossRef] [PubMed]

67. Raspaglio, G.; Petrillo, M.; Martinelli, E.; Li Puma, D.D.; Mariani, M.; De Donato, M.; Filippetti, F.; Mozzetti, S.; Prislei, S.; Zannoni, G.F.; et al. Sox9 and Hif-2α regulate TUBB3 gene expression and affect ovarian cancer aggressiveness. *Gene* **2014**, *542*, 173–181. [CrossRef]

68. Siu, M.K.Y.; Jiang, Y.X.; Wang, J.J.; Leung, T.H.Y.; Han, C.Y.; Tsang, B.K.; Cheung, A.N.Y.; Ngan, H.Y.S.; Chan, K.K.L. Hexokinase 2 Regulates Ovarian Cancer Cell Migration, Invasion and Stemness via FAK/ERK1/2/MMP9/NANOG/SOX9 Signaling Cascades. *Cancers* **2019**, *11*, 813. [CrossRef]

69. Lu, R.; Tang, P.; Zhang, D.; Lin, S.; Li, H.; Feng, X.; Sun, M.; Zhang, H. SOX9/NFIA promotes human ovarian cancer metastasis through the Wnt/β-catenin signaling pathway. *Pathol. Res. Pract.* **2023**, *248*, 154602. [CrossRef]

70. Hou, R.; Jiang, L. LINC00115 promotes stemness and inhibits apoptosis of ovarian cancer stem cells by upregulating SOX9 and inhibiting the Wnt/β-catenin pathway through competitively binding to microRNA-30a. *Cancer Cell Int.* **2021**, *21*, 360. [CrossRef]

71. Malki, S.; Bibeau, F.; Notarnicola, C.; Roques, S.; Berta, P.; Poulat, F.; Boizet-Bonhoure, B. Expression and biological role of the prostaglandin D synthase/SOX9 pathway in human ovarian cancer cells. *Cancer Lett.* **2007**, *255*, 182–193. [CrossRef]

72. Zhao, Y.; Liu, Z.G.; Tang, J.; Zou, R.F.; Chen, X.Y.; Jiang, G.M.; Qiu, Y.F.; Wang, H. High expression of Sox10 correlates with tumor aggressiveness and poor prognosis in human nasopharyngeal carcinoma. *Onco Targets Ther.* **2016**, *9*, 1671–1677. [CrossRef]

73. Goodwin, G.H.; Sanders, C.; Johns, E.W. A new group of chromatin-associated proteins with a high content of acidic and basic amino acids. *Eur. J. Biochem.* **1973**, *38*, 14–19. [CrossRef]
74. Štros, M.; Launholt, D.; Grasser, K.D. The HMG-box: A versatile protein domain occurring in a wide variety of DNA-binding proteins. *Cell. Mol. Life Sci.* **2007**, *64*, 2590–2606. [CrossRef]
75. Štros, M. HMGB proteins: Interactions with DNA and chromatin. *Biochim. Biophys. Acta (BBA)-Gene Regul. Mech.* **2010**, *1799*, 101–113. [CrossRef]
76. Lefebvre, V.; Dumitriu, B.; Penzo-Méndez, A.; Han, Y.; Pallavi, B. Control of cell fate and differentiation by Sry-related high-mobility-group box (Sox) transcription factors. *Int. J. Biochem. Cell Biol.* **2007**, *39*, 2195–2214. [CrossRef] [PubMed]
77. Grimm, D.; Bauer, J.; Wise, P.; Krüger, M.; Simonsen, U.; Wehland, M.; Infanger, M.; Corydon, T.J. The role of SOX family members in solid tumours and metastasis. *Semin. Cancer Biol.* **2020**, *67*, 122–153. [CrossRef] [PubMed]
78. Kashimada, K.; Koopman, P. Sry: The master switch in mammalian sex determination. *Development* **2010**, *137*, 3921–3930. [CrossRef] [PubMed]
79. Lau, Y.F.C.; Zhang, J. Expression analysis of thirty one Y chromosome genes in human prostate cancer. *Mol. Carcinog. Publ. Coop. Univ. Tex. MD Anderson Cancer Cent.* **2000**, *27*, 308–321. [CrossRef]
80. Kanwore, K.; Guo, X.-X.; Abdulrahman, A.A.; Kambey, P.A.; Nadeem, I.; Gao, D. SOX1 is a backup gene for brain neurons and glioma stem cell protection and proliferation. *Mol. Neurobiol.* **2021**, *58*, 2634–2642. [CrossRef] [PubMed]
81. Li, M.; Zou, Y.; Lu, Q.; Tang, N.; Heng, A.; Islam, I.; Tong, H.J.; Dawe, G.S.; Cao, T. Efficient derivation of dopaminergic neurons from SOX1—Floor plate cells under defined culture conditions. *J. Biomed. Sci.* **2016**, *23*, 34. [CrossRef] [PubMed]
82. Nitta, K.R.; Takahashi, S.; Haramoto, Y.; Fukuda, M.; Onuma, Y.; Asashima, M. Expression of Sox1 during Xenopus early embryogenesis. *Biochem. Biophys. Res. Commun.* **2006**, *351*, 287–293. [CrossRef] [PubMed]
83. Garcia, I.; Aldaregia, J.; Marjanovic Vicentic, J.; Aldaz, P.; Moreno-Cugnon, L.; Torres-Bayona, S.; Carrasco-Garcia, E.; Garros-Regulez, L.; Egaña, L.; Rubio, A. Oncogenic activity of SOX1 in glioblastoma. *Sci. Rep.* **2017**, *7*, 46575. [CrossRef] [PubMed]
84. Yin, L.; Liu, T.; Li, C.; Yan, G.; Li, C.; Zhang, J.; Wang, L. The MRTF-A/miR-155/SOX1 pathway mediates gastric cancer migration and invasion. *Cancer Cell Int.* **2020**, *20*, 1–13. [CrossRef]
85. Guan, Z.; Zhang, J.; Wang, J.; Wang, H.; Zheng, F.; Peng, J.; Xu, Y.; Yan, M.; Liu, B.; Cui, B. SOX1 down-regulates β-catenin and reverses malignant phenotype in nasopharyngeal carcinoma. *Mol. Cancer* **2014**, *13*, 1–12. [CrossRef] [PubMed]
86. Novak, D.; Hüser, L.; Elton, J.J.; Umansky, V.; Altevogt, P.; Utikal, J. SOX2 in development and cancer biology. *Semin. Cancer Biol.* **2020**, *67*, 74–82. [CrossRef] [PubMed]
87. Silva, F.H.d.S.; Underwood, A.; Almeida, C.P.; Ribeiro, T.S.; Souza-Fagundes, E.M.; Martins, A.S.; Eliezeck, M.; Guatimosim, S.; Andrade, L.O.; Rezende, L. Transcription factor SOX3 upregulated pro-apoptotic genes expression in human breast cancer. *Med. Oncol.* **2022**, *39*, 212. [CrossRef]
88. Zhao, J.; Cao, H.; Zhang, W.; Fan, Y.; Shi, S.; Wang, R. SOX14 hypermethylation as a tumour biomarker in cervical cancer. *BMC Cancer* **2021**, *21*, 675. [CrossRef]
89. Li, F.; Wang, T.; Tang, S. SOX14 promotes proliferation and invasion of cervical cancer cells through Wnt/β-catenin pathway. *Int. J. Clin. Exp. Pathol.* **2015**, *8*, 1698.
90. Caglayan, D.; Lundin, E.; Kastemar, M.; Westermark, B.; Ferletta, M. SOX21 inhibits glioma progression in vivo by forming complexes with SOX2 and stimulating aberrant differentiation. *Int. J. Cancer* **2013**, *133*, 1345–1356. [CrossRef]
91. Ferletta, M.; Caglayan, D.; Mokvist, L.; Jiang, Y.; Kastemar, M.; Uhrbom, L.; Westermark, B. Forced expression of SOX21 inhibits SOX2 and induces apoptosis in human glioma cells. *Int. J. Cancer* **2011**, *129*, 45–60. [CrossRef] [PubMed]
92. Vervoort, S.J.; van Boxtel, R.; Coffer, P.J. The role of SRY-related HMG box transcription factor 4 (SOX4) in tumorigenesis and metastasis: Friend or foe? *Oncogene* **2013**, *32*, 3397–3409. [CrossRef] [PubMed]
93. Zhang, J.; Liang, Q.; Lei, Y.; Yao, M.; Li, L.; Gao, X.; Feng, J.; Zhang, Y.; Gao, H.; Liu, D.-X. SOX4 induces epithelial–mesenchymal transition and contributes to breast cancer progression. *Cancer Res.* **2012**, *72*, 4597–4608. [CrossRef]
94. Hanieh, H.; Ahmed, E.A.; Vishnubalaji, R.; Alajez, N.M. SOX4: Epigenetic regulation and role in tumorigenesis. *Semin. Cancer Biol.* **2020**, *67*, 91–104. [CrossRef]
95. Medina, P.P.; Castillo, S.D.; Blanco, S.; Sanz-Garcia, M.; Largo, C.; Alvarez, S.; Yokota, J.; Gonzalez-Neira, A.; Benitez, J.; Clevers, H.C. The SRY-HMG box gene, SOX4, is a target of gene amplification at chromosome 6p in lung cancer. *Hum. Mol. Genet.* **2009**, *18*, 1343–1352. [CrossRef] [PubMed]
96. Hur, W.; Rhim, H.; Jung, C.K.; Kim, J.D.; Bae, S.H.; Jang, J.W.; Yang, J.M.; Oh, S.-T.; Kim, D.G.; Wang, H.J. SOX4 overexpression regulates the p53-mediated apoptosis in hepatocellular carcinoma: Clinical implication and functional analysis in vitro. *Carcinogenesis* **2010**, *31*, 1298–1307. [CrossRef]
97. Jafarnejad, S.M.; Ardekani, G.S.; Ghaffari, M.; Li, G. Pleiotropic function of SRY-related HMG box transcription factor 4 in regulation of tumorigenesis. *Cell. Mol. Life Sci.* **2013**, *70*, 2677–2696. [CrossRef]
98. Weigle, B.; Ebner, R.; Temme, A.; Schwind, S.; Schmitz, M.; Kiessling, A.; Rieger, M.A.; Schackert, G.; Schackert, H.K.; Rieber, E.P. Highly specific overexpression of the transcription factor SOX11 in human malignant gliomas. *Oncol. Rep.* **2005**, *13*, 139–144. [CrossRef]

99. de Bont, J.M.; Kros, J.M.; Passier, M.M.; Reddingius, R.E.; Smitt, P.A.S.; Luider, T.M.; Boer, M.L.d.; Pieters, R. Differential expression and prognostic significance of *SOX* genes in pediatric medulloblastoma and ependymoma identified by microarray analysis. *Neuro-Oncol.* **2008**, *10*, 648–660. [CrossRef]

100. Ek, S.; Dictor, M.; Jerkeman, M.; Jirström, K.; Borrebaeck, C.A. Nuclear expression of the non-B-cell lineage SOX11 transcription factor identifies mantle cell lymphoma. *Blood J. Am. Soc. Hematol.* **2008**, *111*, 800–805. [CrossRef]

101. Dictor, M.; Ek, S.; Sundberg, M.; Warenholt, J.; György, C.; Sernbo, S.; Gustavsson, E.; Abu-Alsoud, W.; Wadström, T.; Borrebaeck, C. Strong lymphoid nuclear expression of SOX11 transcription factor defines lymphoblastic neoplasms, mantle cell lymphoma and Burkitt's lymphoma. *Haematologica* **2009**, *94*, 1563. [CrossRef] [PubMed]

102. Brennan, D.J.; Ek, S.; Doyle, E.; Drew, T.; Foley, M.; Flannelly, G.; O'Connor, D.P.; Gallagher, W.M.; Kilpinen, S.; Kallioniemi, O.-P. The transcription factor Sox11 is a prognostic factor for improved recurrence-free survival in epithelial ovarian cancer. *Eur. J. Cancer* **2009**, *45*, 1510–1517. [CrossRef] [PubMed]

103. Zvelebil, M.; Oliemuller, E.; Gao, Q.; Wansbury, O.; Mackay, A.; Kendrick, H.; Smalley, M.J.; Reis-Filho, J.S.; Howard, B.A. Embryonic mammary signature subsets are activated in Brca1-/-and basal-like breast cancers. *Breast Cancer Res.* **2013**, *15*, R25. [CrossRef] [PubMed]

104. Chang, L.; Yuan, Z.; Shi, H.; Bian, Y.; Guo, R. miR-145 targets the SOX11 3′UTR to suppress endometrial cancer growth. *Am. J. Cancer Res.* **2017**, *7*, 2305. [PubMed]

105. Prat, A.; Adamo, B.; Fan, C.; Peg, V.; Vidal, M.; Galván, P.; Vivancos, A.; Nuciforo, P.; Palmer, H.G.; Dawood, S. Genomic analyses across six cancer types identify basal-like breast cancer as a unique molecular entity. *Sci. Rep.* **2013**, *3*, 3544. [CrossRef] [PubMed]

106. Tsang, S.M.; Oliemuller, E.; Howard, B.A. Regulatory roles for SOX11 in development, stem cells and cancer. *Semin. Cancer Biol.* **2020**, *67*, 3–11. [CrossRef] [PubMed]

107. Qu, Y.; Zhou, C.; Zhang, J.; Cai, Q.; Li, J.; Du, T.; Zhu, Z.; Cui, X.; Liu, B. The metastasis suppressor SOX11 is an independent prognostic factor for improved survival in gastric cancer. *Int. J. Oncol.* **2014**, *44*, 1512–1520. [CrossRef]

108. Zou, S.; Wang, C.; Liu, J.; Wang, Q.; Zhang, D.; Zhu, S.; Xu, S.; Kang, M.; He, S. Sox12 is a cancer stem-like cell marker in hepatocellular carcinoma. *Mol. Cells* **2017**, *40*, 847.

109. Qu, M.; Zhu, Y.; Jin, M. MicroRNA-138 inhibits SOX12 expression and the proliferation, invasion and migration of ovarian cancer cells. *Exp. Ther. Med.* **2018**, *16*, 1629–1638. [CrossRef]

110. Du, F.; Feng, W.; Chen, S.; Wu, S.; Cao, T.; Yuan, T.; Tian, D.; Nie, Y.; Wu, K.; Fan, D. Sex determining region Y-box 12 (SOX12) promotes gastric cancer metastasis by upregulating MMP7 and IGF1. *Cancer Lett.* **2019**, *452*, 103–118. [CrossRef]

111. Yuan, W.-M.; Fan, Y.-G.; Cui, M.; Luo, T.; Wang, Y.-E.; Shu, Z.-J.; Zhao, J.; Zheng, J.; Zeng, Y. SOX5 regulates cell proliferation, apoptosis, migration and invasion in KSHV-infected cells. *Virol. Sin.* **2021**, *36*, 449–457. [CrossRef] [PubMed]

112. Wang, D.; Han, S.; Wang, X.; Peng, R.; Li, X. SOX5 promotes epithelial–mesenchymal transition and cell invasion via regulation of Twist1 in hepatocellular carcinoma. *Med. Oncol.* **2015**, *32*, 461. [CrossRef] [PubMed]

113. Sun, C.; Ban, Y.; Wang, K.; Sun, Y.; Zhao, Z. SOX5 promotes breast cancer proliferation and invasion by transactivation of EZH2. *Oncol. Lett.* **2019**, *17*, 2754–2762. [CrossRef]

114. You, J.; Zhao, Q.; Fan, X.; Wang, J. SOX5 promotes cell invasion and metastasis via activation of twist-mediated epithelial–mesenchymal transition in gastric cancer. *OncoTargets Ther.* **2019**, *12*, 2465. [CrossRef]

115. Lamouille, S.; Xu, J.; Derynck, R. Molecular mechanisms of epithelial–mesenchymal transition. *Nat. Rev. Mol. Cell Biol.* **2014**, *15*, 178–196. [CrossRef]

116. Niknami, Z.; Muhammadnejad, A.; Ebrahimi, A.; Harsani, Z.; Shirkoohi, R. Significance of E-cadherin and Vimentin as epithelial-mesenchymal transition markers in colorectal carcinoma prognosis. *EXCLI J.* **2020**, *19*, 917. [PubMed]

117. Wang, Z.; Li, J.; Li, K.; Xu, J. SOX6 is downregulated in osteosarcoma and suppresses the migration, invasion and epithelial-mesenchymal transition via TWIST1 regulation. *Mol. Med. Rep.* **2018**, *17*, 6803–6811. [CrossRef] [PubMed]

118. Qin, Y.-R.; Tang, H.; Xie, F.; Liu, H.; Zhu, Y.; Ai, J.; Chen, L.; Li, Y.; Kwong, D.L.; Fu, L. Characterization of tumor-suppressive function of SOX6 in human esophageal squamous cell carcinoma. *Clin. Cancer Res.* **2011**, *17*, 46–55. [CrossRef]

119. Guo, X.; Yang, M.; Gu, H.; Zhao, J.; Zou, L. Decreased expression of SOX6 confers a poor prognosis in hepatocellular carcinoma. *Cancer Epidemiol.* **2013**, *37*, 732–736. [CrossRef]

120. Iguchi, H.; Urashima, Y.; Inagaki, Y.; Ikeda, Y.; Okamura, M.; Tanaka, T.; Uchida, A.; Yamamoto, T.T.; Kodama, T.; Sakai, J. SOX6 suppresses cyclin D1 promoter activity by interacting with β-catenin and histone deacetylase 1, and its down-regulation induces pancreatic β-cell proliferation. *J. Biol. Chem.* **2007**, *282*, 19052–19061. [CrossRef]

121. Ueda, R.; Ohkusu-Tsukada, K.; Fusaki, N.; Soeda, A.; Kawase, T.; Kawakami, Y.; Toda, M. Identification of HLA-A2-and A24-restricted T-cell epitopes derived from SOX6 expressed in glioma stem cells for immunotherapy. *Int. J. Cancer* **2010**, *126*, 919–929. [CrossRef]

122. Lin, M.; Lei, T.; Zheng, J.; Chen, S.; Du, L.; Xie, H. UBE2S mediates tumor progression via SOX6/β-Catenin signaling in endometrial cancer. *Int. J. Biochem. Cell Biol.* **2019**, *109*, 17–22. [CrossRef]

123. Huang, H.; Han, Q.; Zheng, H.; Liu, M.; Shi, S.; Zhang, T.; Yang, X.; Li, Z.; Xu, Q.; Guo, H. MAP4K4 mediates the SOX6-induced autophagy and reduces the chemosensitivity of cervical cancer. *Cell Death Dis.* **2021**, *13*, 13. [CrossRef]

124. Schlierf, B.; Friedrich, R.; Roerig, P.; Felsberg, J.; Reifenberger, G.; Wegner, M. Expression of *SOXE* and *SOXD* genes in human gliomas. *Neuropathol. Appl. Neurobiol.* **2007**, *33*, 621–630. [CrossRef]

125. Bie, L.-Y.; Li, D.; Wei, Y.; Li, N.; Chen, X.-B.; Luo, S.-X. SOX13 dependent PAX8 expression promotes the proliferation of gastric carcinoma cells. *Artif. Cells Nanomed. Biotechnol.* **2019**, *47*, 3180–3187. [CrossRef] [PubMed]
126. Feng, M.; Fang, F.; Fang, T.; Jiao, H.; You, S.; Wang, X.; Zhao, W. SOX13 promotes hepatocellular carcinoma metastasis by transcriptionally activating Twist1. *Lab. Investig.* **2020**, *100*, 1400–1410. [CrossRef] [PubMed]
127. Jiao, H.; Fang, F.; Fang, T.; You, Y.; Feng, M.; Wang, X.; Yin, Z.; Zhao, W. SOX13 regulates cancer stem-like properties and tumorigenicity in hepatocellular carcinoma cells. *Am. J. Cancer Res.* **2021**, *11*, 760. [PubMed]
128. Wang, H.; He, L.; Ma, F.; Regan, M.M.; Balk, S.P.; Richardson, A.L.; Yuan, X. SOX9 regulates low density lipoprotein receptor-related protein 6 (LRP6) and T-cell factor 4 (TCF4) expression and Wnt/β-catenin activation in breast cancer. *J. Biol. Chem.* **2013**, *288*, 6478–6487. [CrossRef]
129. Zhong, W.-d.; Qin, G.-q.; Dai, Q.-s.; Han, Z.-d.; Chen, S.-m.; Ling, X.-h.; Fu, X.; Cai, C.; Chen, J.-h.; Chen, X.-b. SOXs in human prostate cancer: Implication as progression and prognosis factors. *BMC Cancer* **2012**, *12*, 248. [CrossRef]
130. Hu, B.; Wang, J.; Jin, X. MicroRNA-138 suppresses cell proliferation and invasion of renal cell carcinoma by directly targeting SOX9. *Oncol. Lett.* **2017**, *14*, 7583–7588. [CrossRef]
131. Huang, J.; Guo, L. Knockdown of SOX9 inhibits the proliferation, invasion, and EMT in thyroid cancer cells. *Oncol. Res. Featur. Preclin. Clin. Cancer Ther.* **2017**, *25*, 167–176. [CrossRef]
132. Lü, B.; Fang, Y.; Xu, J.; Wang, L.; Xu, F.; Xu, E.; Huang, Q.; Lai, M. Analysis of SOX9 expression in colorectal cancer. *Am. J. Clin. Pathol.* **2008**, *130*, 897–904. [CrossRef] [PubMed]
133. Zhang, S.; Che, D.; Yang, F.; Chi, C.; Meng, H.; Shen, J.; Qi, L.; Liu, F.; Lv, L.; Li, Y. Tumor-associated macrophages promote tumor metastasis via the TGF-β/SOX9 axis in non-small cell lung cancer. *Oncotarget* **2017**, *8*, 99801. [CrossRef] [PubMed]
134. Yang, X.; Liang, R.; Liu, C.; Liu, J.A.; Cheung, M.P.L.; Liu, X.; Man, O.Y.; Guan, X.-Y.; Lung, H.L.; Cheung, M. SOX9 is a dose-dependent metastatic fate determinant in melanoma. *J. Exp. Clin. Cancer Res.* **2019**, *38*, 17. [CrossRef] [PubMed]
135. Shi, G.; Sohn, K.-C.; Li, Z.; Choi, D.-K.; Park, Y.M.; Kim, J.-H.; Fan, Y.-M.; Nam, Y.H.; Kim, S.; Im, M. Expression and functional role of SOX9 in human epidermal keratinocytes. *PLoS ONE* **2013**, *8*, e54355. [CrossRef] [PubMed]
136. Stovall, D.B.; Wan, M.; Miller, L.D.; Cao, P.; Maglic, D.; Zhang, Q.; Stampfer, M.R.; Liu, W.; Xu, J.; Sui, G. The regulation of SOX7 and its tumor suppressive role in breast cancer. *Am. J. Pathol.* **2013**, *183*, 1645–1653. [CrossRef] [PubMed]
137. Cui, J.; Xi, H.; Cai, A.; Bian, S.; Wei, B.; Chen, L. Decreased expression of SOX7 correlates with the upregulation of the Wnt/β-catenin signaling pathway and the poor survival of gastric cancer patients. *Int. J. Mol. Med.* **2014**, *34*, 197–204. [CrossRef] [PubMed]
138. Han, L.; Wang, W.; Ding, W.; Zhang, L. MiR-9 is involved in TGF-β1-induced lung cancer cell invasion and adhesion by targeting SOX7. *J. Cell. Mol. Med.* **2017**, *21*, 2000–2008. [CrossRef]
139. Fu, D.-Y.; Wang, Z.-M.; Wang, B.-L.; Shen, Z.-Z.; Huang, W.; Shao, Z.-M. SOX17, the canonical Wnt antagonist, is epigenetically inactivated by promoter methylation in human breast cancer. *Breast Cancer Res. Treat.* **2010**, *119*, 601–612. [CrossRef]
140. Li, J.-Y.; Han, C.; Zheng, L.-L.; Guo, M.-Z. Epigenetic regulation of Wnt signaling pathway gene SRY-related HMG-box 17 in papillary thyroid carcinoma. *Chin. Med. J.* **2012**, *125*, 3526–3531.
141. Majchrzak-Celińska, A.; Słocińska, M.; Barciszewska, A.-M.; Nowak, S.; Baer-Dubowska, W. Wnt pathway antagonists, SFRP1, SFRP2, SOX17, and PPP2R2B, are methylated in gliomas and SFRP1 methylation predicts shorter survival. *J. Appl. Genet.* **2016**, *57*, 189–197. [CrossRef]
142. Du, Y.C.; Oshima, H.; Oguma, K.; Kitamura, T.; Itadani, H.; Fujimura, T.; Piao, Y.S.; Yoshimoto, T.; Minamoto, T.; Kotani, H. Induction and down-regulation of SOX17 and its possible roles during the course of gastrointestinal tumorigenesis. *Gastroenterology* **2009**, *137*, 1346–1357. [CrossRef]
143. Lu, J.; Zhang, G.; Cheng, Y.; Tang, Y.; Dong, Z.; McElwee, K.J.; Li, G. Reduced expression of SRY-box containing gene 17 correlates with an unfavorable melanoma patient survival. *Oncol. Rep.* **2014**, *32*, 2571–2579. [CrossRef] [PubMed]
144. Pula, B.; Olbromski, M.; Wojnar, A.; Gomulkiewicz, A.; Witkiewicz, W.; Ugorski, M.; Dziegiel, P.; Podhorska-Okolow, M. Impact of SOX18 expression in cancer cells and vessels on the outcome of invasive ductal breast carcinoma. *Cell. Oncol.* **2013**, *36*, 469–483. [CrossRef] [PubMed]
145. Jethon, A.; Pula, B.; Olbromski, M.; Werynska, B.; Muszczynska-Bernhard, B.; Witkiewicz, W.; Dziegiel, P.; Podhorska-Okolow, M. Prognostic significance of SOX18 expression in non-small cell lung cancer. *Int. J. Oncol.* **2015**, *46*, 123–132. [CrossRef] [PubMed]
146. Ornat, M.; Kobierzycki, C.; Grzegrzolka, J.; Pula, B.; Zamirska, A.; Bieniek, A.; Szepietowski, J.C.; Dziegiel, P.; Okolow, M.P. SOX18 expression in non-melanoma skin cancer. *Anticancer. Res.* **2016**, *36*, 2379–2383. [PubMed]
147. Yan, H.-T.; Shinka, T.; Sato, Y.; Yang, X.-J.; Chen, G.; Sakamoto, K.; Kinoshita, K.; Aburatani, H.; Nakahori, Y. Overexpression of SOX15 inhibits proliferation of NT2/D1 cells derived from a testicular embryonal cell carcinoma. *Mol. Cells* **2007**, *24*, 323–328. [PubMed]
148. Thu, K.; Radulovich, N.; Becker-Santos, D.; Pikor, L.; Pusic, A.; Lockwood, W.; Lam, W.; Tsao, M. SOX15 is a candidate tumor suppressor in pancreatic cancer with a potential role in Wnt/β-catenin signaling. *Oncogene* **2014**, *33*, 279–288. [CrossRef]
149. Thu, K.L.; Becker-Santos, D.D.; Radulovich, N.; Pikor, L.A.; Lam, W.L.; Tsao, M.-S. SOX15 and other SOX family members are important mediators of tumorigenesis in multiple cancer types. *Oncoscience* **2014**, *1*, 326. [CrossRef]
150. Han, F.; Liu, W.; Jiang, X.; Shi, X.; Yin, L.; Ao, L.; Cui, Z.; Li, Y.; Huang, C.; Cao, J. SOX30, a novel epigenetic silenced tumor suppressor, promotes tumor cell apoptosis by transcriptional activating p53 in lung cancer. *Oncogene* **2015**, *34*, 4391–4402. [CrossRef]

151. Han, F.; Liu, W.-B.; Shi, X.-Y.; Yang, J.-T.; Zhang, X.; Li, Z.-M.; Jiang, X.; Yin, L.; Li, J.-J.; Huang, C.-S. SOX30 inhibits tumor metastasis through attenuating Wnt-signaling via transcriptional and posttranslational regulation of β-catenin in lung cancer. *EBioMedicine* **2018**, *31*, 253–266. [CrossRef] [PubMed]
152. Hao, X.; Han, F.; Ma, B.; Zhang, N.; Chen, H.; Jiang, X.; Yin, L.; Liu, W.; Ao, L.; Cao, J. SOX30 is a key regulator of desmosomal gene suppressing tumor growth and metastasis in lung adenocarcinoma. *J. Exp. Clin. Cancer Res.* **2018**, *37*, 111. [CrossRef] [PubMed]

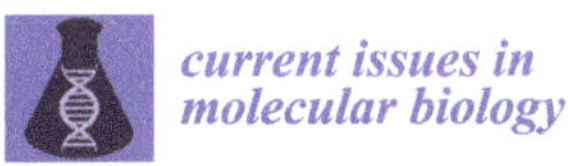
current issues in molecular biology

Article

Activation of Adrenoceptor Alpha-2 (ADRA2A) Promotes Chemosensitization to Carboplatin in Ovarian Cancer Cell Lines

Haya Albanna [†], Alesia Gjoni [†], Danielle Robinette, Gerardo Rodriguez, Lora Djambov, Margaret E. Olson * and Peter C. Hart *

College of Science, Health and Pharmacy, Roosevelt University, 1400 N Roosevelt Blvd, Schaumburg, IL 60173, USA; halbanna@mail.roosevelt.edu (H.A.); agjoni@mail.roosevelt.edu (A.G.); drobinette@mail.roosevelt.edu (D.R.); grodriguez24@mail.roosevelt.edu (G.R.); ldjambov@mail.roosevelt.edu (L.D.)
* Correspondence: mkurbanov01@roosevelt.edu (M.E.O.); phart02@roosevelt.edu (P.C.H.); Tel.: +1-(847)-330-4542 (M.E.O.); +1-(847)-330-4505 (P.C.H.)
† These authors contributed equally to this work.

Abstract: Recurrence of ovarian cancer (OvCa) following surgery and standard carboplatin/paclitaxel first-line therapy signifies poor median progression-free survival (<24 months) in the majority of patients with OvCa. The current study utilized unbiased high-throughput screening (HTS) to evaluate an FDA-approved compound library for drugs that could be repurposed to improve OvCa sensitivity to carboplatin. The initial screen revealed six compounds with agonistic activity for the adrenoceptor alpha-2a (ADRA2A). These findings were validated in multiple OvCa cell lines (TYKnu, CAOV3, OVCAR8) using three ADRA2A agonists (xylazine, dexmedetomidine, and clonidine) and two independent viability assays. In all the experiments, these compounds enhanced the cytotoxicity of carboplatin treatment. Genetic overexpression of ADRA2A was also sufficient to reduce cell viability and increase carboplatin sensitivity. Taken together, these data indicate that ADRA2A activation may promote chemosensitivity in OvCa, which could be targeted by widely used medications currently indicated for other disease states.

Keywords: ovarian cancer; OvCa; carboplatin resistance; ADRA2A; adrenoceptor alpha-2a

Citation: Albanna, H.; Gjoni, A.; Robinette, D.; Rodriguez, G.; Djambov, L.; Olson, M.E.; Hart, P.C. Activation of Adrenoceptor Alpha-2 (ADRA2A) Promotes Chemosensitization to Carboplatin in Ovarian Cancer Cell Lines. *Curr. Issues Mol. Biol.* **2023**, *45*, 9566–9578. https://doi.org/10.3390/cimb45120598

Academic Editor: Dumitru A. Iacobas

Received: 30 October 2023
Revised: 23 November 2023
Accepted: 26 November 2023
Published: 28 November 2023

1. Introduction

Ovarian cancer (OvCa) commonly presents at advanced stages and has a poor prognosis, with an overall 5-year survival rate of 50% across all subtypes and stages at diagnosis [1]. The standard of care for high-grade serous ovarian carcinoma (HGSOC), the most prevalent subtype of OvCa, includes cytoreductive debulking surgery and chemotherapy consisting of a platinum agent such as carboplatin (CP) and a taxane such as paclitaxel [2]. Following active transport, CP promotes DNA lesions that include direct platinum–DNA adducts as well as inter- and intra-strand crosslinking that ultimately result in single- and double-strand breaks, inducing apoptosis [3]. In a clinical context, the use of CP/paclitaxel (CarboTaxol) allows for the ablation of unresectable or visually undetectable tumor cells to prevent recurrence of the disease.

While the majority of patients initially respond to CP therapy, the majority will relapse and inevitably succumb to their disease [2,4,5]. Multiple mechanisms have been attributed to CP treatment failure, including dysfunctional transport, compensatory DNA repair, and the upregulation of pro-survival and anti-apoptotic pathways [6]. The use of paclitaxel in combination to impede progression through the M phase of cells that evade apoptosis despite DNA adduct accumulation by carboplatin may be central to its ability to enhance tumor cell ablation following surgery and to improve the initial therapeutic response. Conceivably, drugs targeting other processes that evade apoptosis or promote survival may complement the current therapeutic approaches [7]. In line with this, paclitaxel has been used in conjunction with carboplatin as first-line therapy to improve survival in patients with advanced HGSOC due to its inhibition of microtubule disassembly and subsequent cell cycle arrest to further enhance the apoptosis caused by CP-dependent DNA damage [8]. Emerging molecular targets have been identified that may delay recurrence following first-line treatment in OvCa, such as VEGF (e.g., bevacizumab) and PARP (e.g., olaparib), which have provided new therapeutic strategies to improve progression-free survival [9]. In the case of PARP inhibitors, suppression of nucleotide excision repair may further enhance the efficacy of CP by preventing the removal of intrastrand crosslinking to promote apoptosis [6,10]; however, PARylation is critical in regulating numerous processes involved in DNA repair and replication that may be involved in preventing the proliferation of CP-treated cells [11]. While promising, in both cases of VEGF and PARP inhibitors, it has been observed that treatment failure may still occur due to resistance that develops in response to therapy [12,13]. Thus, it is critical to continue expanding the repertoire of nonredundant treatment options available to clinicians that may facilitate personalized treatment plans to overcome any acquired drug resistance and to prolong patient survival.

To expedite the implementation of novel targetable mechanisms for drug therapies, a drug repurposing paradigm is often used in the drug discovery process for many disease states. This approach allows researchers to proceed more rapidly to clinical trials to determine efficacy and has facilitated the common use of multiple drugs as mainstays in current cancer therapy, such as tamoxifen in breast cancer [14,15]. Many of the drugs repositioned for off-label use in these contexts have been evaluated as a result of retrospective cohort analyses that have identified associations between disease onset or progression and medications taken for other indications [14]. Although these retrospective associations have, in many cases, facilitated preclinical experimentation and the pursuit of clinical trials to determine efficacy, their application in new clinical contexts is not always successful [14]. An alternative strategy for drug repurposing has been through the prospective preclinical assessment of drugs already on the market. In particular, in vitro high-throughput screening (HTS) using curated compound libraries to evaluate thousands of compounds rapidly [16] allows for lead-compound identification and in vivo testing prior to clinical application. Indeed, a recent study using a synthetic lethality HTS approach identified FDA-approved drugs that may improve the response to ATP8B-dependent cisplatin resistances in OvCa [17], indicating the potential use of these compounds (e.g., tranilast) in ATP8B-overexpressing ovarian tumors. Using a similar unbiased strategy, the current study aimed to enhance the cytotoxicity of carboplatin to prevent treatment resistance by utilizing an FDA-approved compound library.

2. Materials and Methods

2.1. Reagents

High-throughput screening (HTS) was performed using the DiscoveryProbe L1021 FDA-approved drug library (APExBIO L1021). Carboplatin (#c2538), xylazine (#x1126), dexmedetomidine (#sml0956), and clonidine (#c7897) were purchased from Sigma-Aldrich.

2.2. Cell Lines

The OvCa cell lines TYKnu, CAOV3, and OVCAR8 were a gift from Dr. Ernst Lengyel and Dr. Iris Romero of the University of Chicago Section of Gynecologic Oncology (Chicago, IL, USA). All cell lines were passaged at least 3 times prior to their use in experiments. Cell lines were banked in liquid nitrogen and were confirmed to be mycoplasma-free by the donating investigators as described previously [18]. Viability was confirmed to be 98% or higher (using trypan exclusions with Countess II, ThermoFisher, Waltham MA, USA) prior to counting and seeding for the HTS as well as the primary and secondary assays. All three cell lines were maintained in RPMI1640 with L-glutamine (VWR) supplemented with 10% FB-Essence (VWR), 1% penicillin/streptomycin (VWR), 1% MEM vitamins (VWR), and 1% nonessential amino acids (VWR).

2.3. High-Throughput Screening (HTS)

The initial high-throughput screening (HTS) was performed using a 96-well format with the DiscoveryProbe L1021 FDA-approved drug library (ApexBio, Houston, TX, USA). The compound library was reconstituted to 2 mM (in DMSO) so that the addition of 5 μL of each drug would yield 50 μM per drug per well. A total of 88 compounds were tested per plate, with 1584 compounds tested. For each plate, internal controls included 4 untreated samples and 4 samples treated only with the IC_{50} of carboplatin. Due to their relatively high resistance to carboplatin, as indicated by the highest IC_{50} of the cell lines tested, OVCAR8 cells were used in the pilot HTS assay. Cells were seeded at 3.0×10^4/well and allowed to grow for 24 h prior to treatment with carboplatin (at IC_{50}) $\pm$ the L1021 drug library for 48 h. After 48 h of treatment, the MTT assay was performed (as described below) to determine the effect on cell viability. Absorbance measurements were then processed in which the absorbance of each drug with carboplatin was compared to the positive carboplatin control (i.e., the IC_{50} of carboplatin alone). Any well which was observed to increase the cytotoxicity of carboplatin by 80% or higher was considered a potential lead (marked as red; Figure 1E and Supplementary Table S1). These statistics were then cross-referenced per well per plate to identify compounds of interest for subsequent primary (MTT viability) and secondary (colony formation) confirmatory assays.

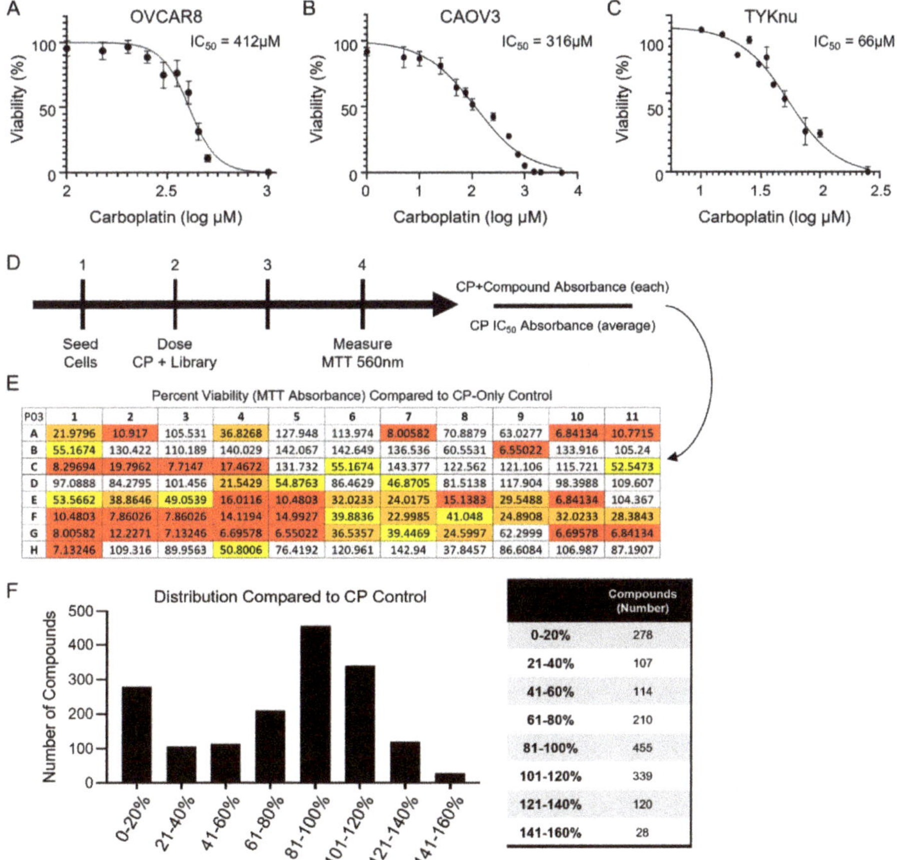

P03	1	2	3	4	5	6	7	8	9	10	11
A	21.9796	10.917	105.531	36.8268	127.948	113.974	8.00582	70.8879	63.0277	6.84134	10.7715
B	55.1674	130.422	110.189	140.029	142.067	142.649	136.536	60.5531	6.55022	133.916	105.24
C	8.29694	19.7962	7.7147	17.4672	131.732	55.1674	143.377	122.562	121.106	115.721	52.5473
D	97.0888	84.2795	101.456	21.5429	54.8763	86.4629	46.8705	81.5138	117.904	98.3988	109.607
E	53.5662	38.8646	49.0539	16.0116	10.4803	32.0233	24.0175	15.1383	29.5488	6.84134	104.367
F	10.4803	7.86026	7.86026	14.1194	14.9927	39.8836	22.9985	41.048	24.8908	32.0233	28.3843
G	8.00582	12.2271	7.13246	6.69578	6.55022	36.5357	39.4469	24.5997	62.2999	6.69578	6.84134
H	7.13246	109.316	89.9563	50.8006	76.4192	120.961	142.94	37.8457	86.6084	106.987	87.1907

	Compounds (Number)
0–20%	278
21–40%	107
41–60%	114
61–80%	210
81–100%	455
101–120%	339
121–140%	120
141–160%	28

Figure 1. Determination of carboplatin resistance in OvCa cell lines. Cells were plated and allowed to adhere for 24 h prior to treatments. Cells were treated with 12 escalating doses of carboplatin (0–5000 μM)

for 48 h ((**A**) OVCAR8, (**B**) CAOV3, and (**C**) TYKnu cell lines). N = 4 for each data point. (**D**) Experimental timeline for treating cells with carboplatin at its determined IC_{50} along with the DiscoveryProbe L1021 FDA-approved drug library (ApexBio) (88 agents per plate at 50 µM final concentration). (**E**) MTT absorbance was normalized to the carboplatin-only group to determine the impact on cell death. Numbers presented refer to the percentage (%) of carboplatin-only control. Wells marked in red had an absorbance of <20% when divided by the average absorbance of the carboplatin-only controls for each plate. (**F**) The frequency distribution of compounds + carboplatin binned in 20% increments, with 100% representing having an absorbance equal to the carboplatin-only controls. The number of compounds identified in each category is shown on the right.

2.4. Primary Assay—MTT Viability Assay

The thiazolyl blue tetrazolium bromide (MTT) colorimetric metabolism assay was used to determine cell viability (GoldBio, St. Louis, MO, USA; #T-030). Briefly, cells were seeded at 3.0×10^4/well by dispensing 100 µL of a 3.0×10^5/mL suspension of cells into a 96-well plate (Pipette.com, San Diego, CA, USA) and allowed to adhere for 24 h prior to treatment. Media was aspirated and fresh growth media was added so that the addition of carboplatin ± secondary compounds would result in a final volume of 200 µL. Carboplatin was diluted fresh from 10 mM stocks (never previously thawed) to working stocks to allow for 5 µL of drug to result in the desired concentrations (ranging from 10–5000 µM). In the context of the HTS, only the IC_{50} calculated for OVCAR8 was used in combination with each drug. In the context of the MTT confirmatory experiments, a dose range was used to reestablish the IC_{50} per plate (carboplatin alone) and to determine IC_{50} values when used in combination with each alpha-2 agonist (set to 50 µM and 100 µM per agonist). Multichannel pipettes were used to limit variability between wells. After 48 h of treatment, 12 mM MTT was added to a final concentration of 300 µM and allowed to incubate for 1.5 h (OVCAR8) or 2 h (CAOV3 or TYKnu). After incubation, the plates were decanted and patted dry, and MTT-reduced formazan crystals were suspended in 200 µL DMSO. The absorbance was then measured at 560 nm using SpectraMax ABS (MolecularDevices, San Jose, CA, USA). The absorbance values were exported via Microsoft Excel and analyzed using GraphPad Prism (described below).

2.5. Secondary Assay—Colony-Formation Clonogenicity Assay

As a secondary assay to confirm the impact of each drug combination on cell chemoresistance, a colony formation (clonogenicity) assay was performed. Cells were seeded at 500 (0.5×10^3; OVCAR8) or 1000 (1.0×10^3; CAOV3, TYKnu) cells per well of a 6-well plate (VWR, Radnor, PA, USA) and allowed to recover for 24 h. After 24 h, drug treatments were initiated. At days 5 and 7, the drug treatments were refreshed, with a full media change performed on day 5. On day 10, the cells were washed with 1× PBS, fixed using 4% paraformaldehyde for 15 min, washed again with 1× PBS, then stained using crystal violet (Sigma-Aldrich, St. Louis, MO, USA, #61135) for 30 min. The crystal violet was then moved to a waste container, and the cells were rinsed with water until the stain was washed away. Plates were then imaged using ChemiDoc XR+ (Bio-Rad, Hercules, CA, USA) while colonies were counted manually using ImageJ/FIJI (NIH, Bethesda, MD, USA), and statistical analyses were performed using GraphPad Prism.

2.6. Plasmid Transfection

To overexpress adrenoceptor alpha-a2, transfection was performed using pmEGFP-ADRA2A on the pcDNA3.1 backbone (Addgene, Watertown, MA, USA; plasmid #190753). pmEGFP-ADRA2A was a gift from Dave Piston of Washington University. OVCAR8 cells were seeded at 4.0×10^6 per 10 cm plate and allowed to proliferate for 24 h prior to transfection. Cells were transfected using Lipofectamine 2000 (Invitrogen, Waltham, MA, USA) according to the manufacturer's instructions. At 24 h post transfection, cells were seeded into either 96-well or 6-well plates for the MTT or colony formation assays, respectively. After 48 h post transfection, treatments were initiated as described for the primary and secondary assays. For quantitative real-time PCR, the reaction was scaled down to a 6-well format with $1\times$ reaction per well.

2.7. Quantitative Real-Time Polymerase Chain Reactions

The expression of ADRA2A following transfection was confirmed by quantitative real-time PCR (qRT-PCR). RNA isolation was performed using Trizol (Invitrogen) with chloroform extraction. Complementary DNA (cDNA) synthesis was performed using the High-Capacity cDNA Reverse Transcription Kit (Applied Biosystems, Waltham, MA, USA). qRT-PCR was performed using the POWRUP SYBR Master Mix (Life Technologies, Carlsbad, CA, USA) according to the manufacturer's instructions. Primers were designed using PrimerBank (NP_000672, #194353969c2), and 25 nmole quantities of DNA oligonucleotides were purchased from IDT (Newark, NJ, USA; primers were prepared as LabReady, normalized to 100 µM in IDTE pH 8.0). Forward primer: AGA AGT GGT ACG TCA TCT CGT; Reverse primer: CGC TTG GCG ATC TGG TAG A.

2.8. TCGA Assessment of ADRA2A Expression in Ovarian Cancer

Expression of ADRA2A in ovarian cancer was determined using the Gene Expression Profiling Interactive Analysis (GEPIA) database (GEPIA.cancer-pku.cn). The "OV" (ovarian cancer) dataset from The Cancer Genome Atlas (TCGA) was selected with a $\log_2$FC cutoff of 1 and a *p*-value cutoff of 0.01. The data are presented on a log scale.

2.9. Statistical Analysis

Data was analyzed using Graphpad Prism 10 (Graphpad, La Jolla, CA, USA) by 1-way ANOVA with the post hoc Tukey t-test or Student's unpaired t-test as appropriate. For the IC_{50} calculations, XY data tables and graphs were used. Absorbance measurements from the MTT assays were imported into Prism. The data were then transformed to log(x) and normalized as a percentage within each group. The transformed/normalized data were then modeled using nonlinear regression (dose response: log(inhibitor) with four parameters, with constraints for "top" (100%) and "bottom" (0%)). Representative curve fit and calculated IC_{50} values are presented for each experiment. All experiments were repeated at least once.

3. Results

3.1. High-Throughput Screening of an FDA-Approved Drug Library Identifies Adrenoceptor Alpha-2 (ADRA2A) as a Potential Target for Enhancing Carboplatin Treatment

To identify novel compounds that could enhance the cytotoxicity of carboplatin and, thus, possibly reduce resistance to the current standard of care, three cell lines representative of HGSOC were assessed for their potential resistance to carboplatin (CP). Treatment of TYKnu, CAOV3, and OVCAR8 cells with increasing doses of carboplatin for 48 h revealed significant resistance of OVCAR8 cells compared with either CAOV3 or TYKnu, with an IC_{50} value of 412 µM (Figure 1A) compared with either 316 µM (Figure 1B) or 66 µM (Figure 1C), respectively, as determined by the MTT cell viability assay. OV-CAR8 cells were therefore selected for high-throughput screening (HTS) utilizing an FDA-approved drug library (APExBIO L1021). Cells were treated with the relative IC_{50} of CP (400 µM) $\pm$ compound library (50 µM) for 48 h prior to the MTT assay (Figure 1D; example plate analysis shown in Figure 1E). Of the 1651 compounds tested, 278 demonstrated a marked enhancement in cell death that increased the cytotoxicity of CP by more than 80% compared with the carboplatin-only controls (Figure 1F and Supplementary Table S1).

Of the 278 compounds observed to have potential synergy with CP, 4 were identified to be agonists of the ADRA2A receptor, including xylazine, tizanidine, medetomidine, and dexmedetomidine (Figure 2A). Two agents with mixed agonism of ADRA2A and significant I_1-imidazoline receptor activity were also identified: moxonidine and rilmenidine (Figure 2A). This trend indicated that ADRA2A could be an important molecular target that may regulate chemosensitivity in OvCa tumor cells. Evaluation of ADRA2A expression in patient tumors from the TCGA database revealed that ADRA2A is significantly repressed in OvCa tumors (Figure 2B), further supporting the hypothesis that inhibition of its signaling may be important in tumor development or progression.

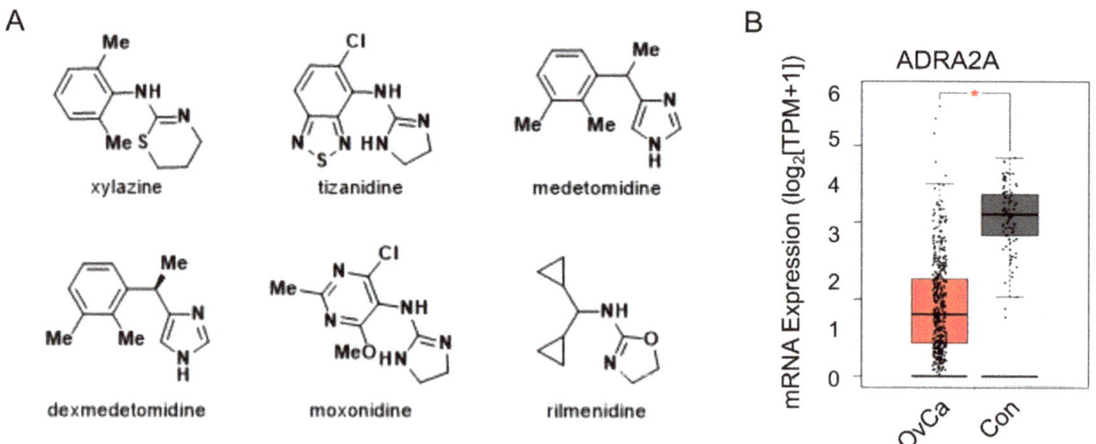

Figure 2. Identification of ADRA2A as a potential therapeutic target in OvCa. (**A**) High-throughput screening (HTS) of carboplatin + the FDA library identified several alpha-2 adrenoceptor (ADRA2A) agonists that increased cytotoxicity by >80% in combination with carboplatin; these included xylazine, tizanidine, medetomidine, and dexmedetomidine. Drugs with partial ADRA2A agonism, namely moxonidine and rilmenidine, were also identified in the HTS. (**B**) Gene (mRNA) expression of ADRA2A in TCGA clinical data (GEPIA database) comparing control tissue (gray) to ovarian tumor tissue (pink). * $p < 0.01$.

3.2. Activation of ADRA2A Using Pharmacologic and Genetic Manipulation Enhances Carboplatin Toxicity in Viability and Clonogenicity Assays

To validate the findings from the HTS in OVCAR8 cells, TYKnu, CAOV3, and OVCAR8 cells were treated with increasing doses of carboplatin ± the ADRA2A agonists xylazine, dexmedetomidine, or clonidine. While not observed in the HTS, clonidine was chosen due to its widespread clinical use for hypertension, relatively high specificity, and low toxicologic profile. Primary confirmatory MTT assays in these cell lines demonstrated consistent reductions in the IC_{50} of carboplatin compared with the carboplatin-only groups when 50 or 100 µM of any of the ADRA2A agonists was co-administered (Figure 3A–C). This effect was largely dose-dependent and was consistent across the three cell lines tested (Figure 3D). To determine whether this effect was assay-specific, a secondary colony-formation clonogenicity assay was performed using the approximate IC_{50} value of CP (1 µM) in combination with each compound. Extended treatment of colonies for 10d with a lower dose (25 µM) of each compound suppressed colony formation, an effect that was particularly significant with xylazine and dexmedetomidine (Figure 4A–C). Clonidine appeared to have weak activity in this assay at the test dose and reached significance in CAOV3 cells (Figure 4B) but did not meet statistical significance in TYKnu (Figure 4A) or OVCAR8 (Figure 4C) cells, despite showing a similar trend [$p < 0.09$].

To further evaluate our findings using the pharmacologic activation of ADRA2A, genetic upregulation of ADRA2A in the most resistant cell line (OVCAR8) was performed (Figure 5A). Following the ectopic upregulation of ADRA2A, cells were subjected to either MTT or colony formation assays. ADRA2A overexpression in OVCAR8 cells strikingly reduced the IC_{50} of carboplatin needed in these cells (Figure 5B). Consistently, ADRA2A expression induced an overall inhibition of colony formation in OVCAR8 cells and promoted a further reduction in colonies compared with the carboplatin-only group (Figure 5C). In line with the enhancement in carboplatin sensitivity observed using xylazine, dexmedetomidine, and clonidine, these findings suggest that ADRA2A activation and the consequent inhibition of cyclic AMP (cAMP) production may downregulate processes necessary for tumor cells to escape carboplatin-induced cytotoxicity. Thus, taken together, these data further support the idea that the molecular activity of ADRA2A may suppress the resistance of OvCa tumor cells to carboplatin.

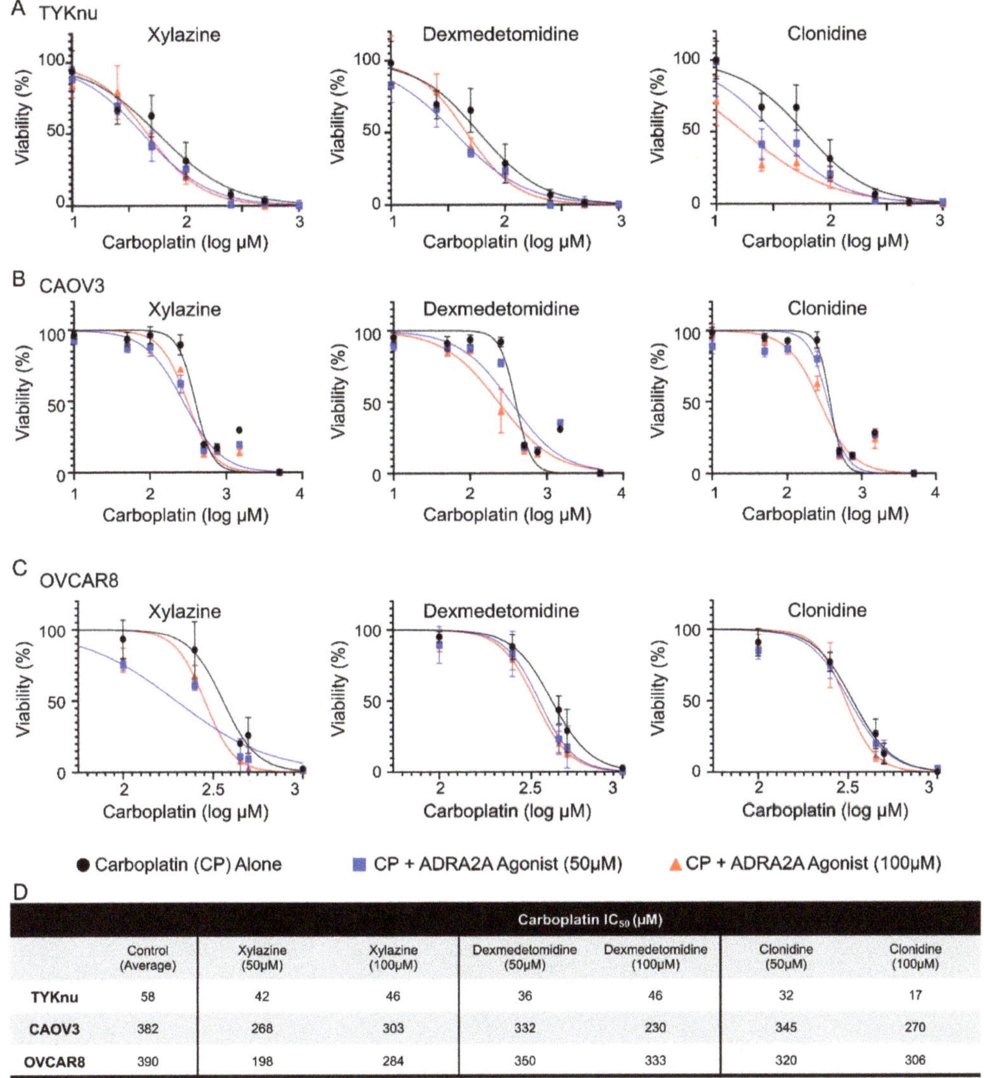

	Carboplatin IC$_{50}$ (µM)						
	Control (Average)	Xylazine (50µM)	Xylazine (100µM)	Dexmedetomidine (50µM)	Dexmedetomidine (100µM)	Clonidine (50µM)	Clonidine (100µM)
TYKnu	58	42	46	36	46	32	17
CAOV3	382	268	303	332	230	345	270
OVCAR8	390	198	284	350	333	320	306

Figure 3. Validation of ADRA2A agonists in enhancing carboplatin-dependent cytotoxicity. Cells were plated and allowed to adhere for 24 h prior to treatments. Cells were treated with escalating doses of carboplatin (0–5000 µM) ± 50 or 100 µM of the indicated ADRA2A agonist for 48 h ((**A**) TYKnu, (**B**) CAOV3, (**C**) OVCAR8). N = 4 for each data point. (**D**) Carboplatin IC$_{50}$ values for carboplatin ± ADRA2A agonists (xylazine, dexmedetomidine, clonidine) for each cell line.

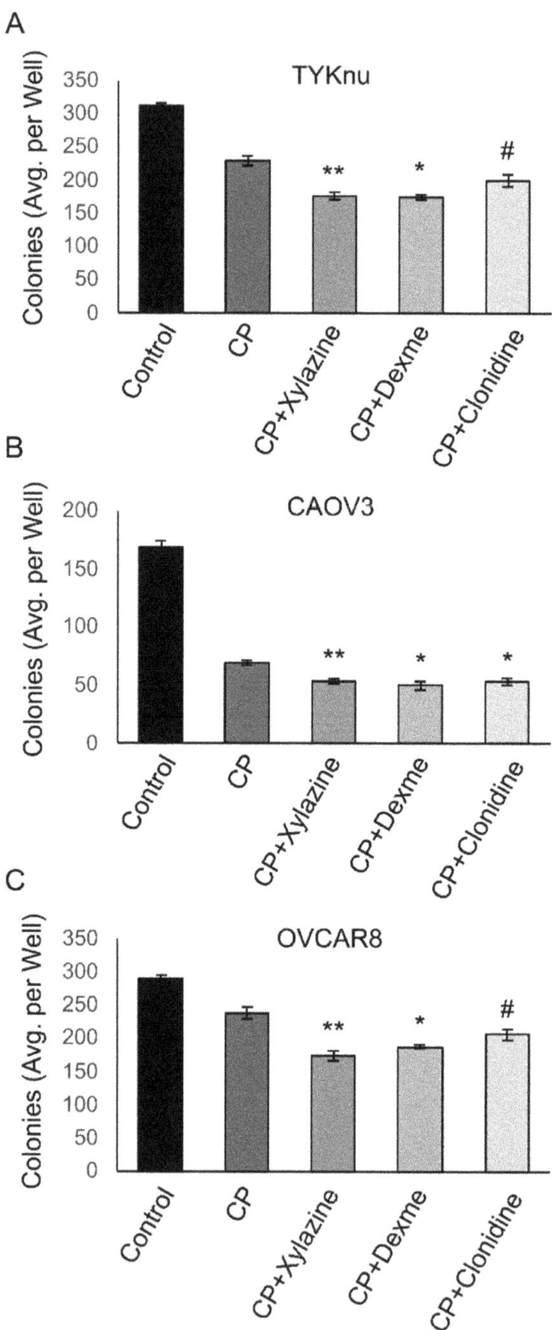

Figure 4. ADRA2A activation further enhances the inhibition of clonogenicity by carboplatin. Cells were allowed to adhere for 24 h prior to treatments. Cells were treated with 1 μM carboplatin ± 25 μM of the indicated ADRA2A agonist for 10 days. ((**A**) TYKnu, (**B**) CAOV3, (**C**) OVCAR8). N = 3 for each condition. * $p < 0.05$; ** $p < 0.01$; # $p < 0.07$. Dexme = dexmedetomidine; CP = carboplatin.

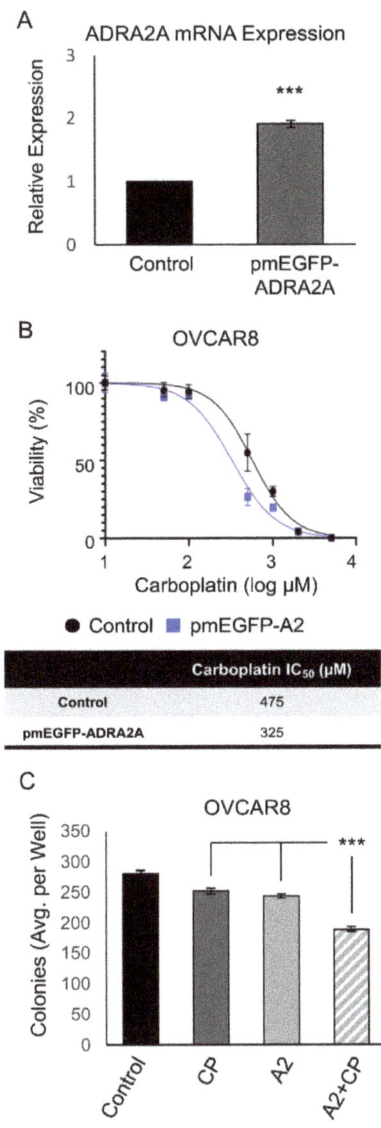

	Carboplatin IC$_{50}$ (µM)
Control	475
pmEGFP-ADRA2A	325

Figure 5. Overexpression of ADRA2A promotes chemosensitivity to carboplatin. (**A**) Expression of ADRA2A mRNA at 48 h following the transfection of OVCAR8 cells with pmEGFP-ADRA2A (N = 3). (**B**) MTT viability assay of OVCAR8 cells ± pmEGFP-ADRA2A treated with an escalating dose curve of carboplatin (0–5000 µM) for 48 h (N = 6). (**C**) Colony formation assay of OVCAR8 cells ± pmEGFP-ADRA2A [A2] treated with or without carboplatin (1 µM) for 10 days (N = 6). *** $p < 0.005$.

4. Discussion

In summary, an unbiased HTS cytotoxicity assay identified that several modulators of ADRA2A activity could enhance the sensitivity of OvCa cell lines to carboplatin. The initial HTS data was confirmed using the compounds xylazine, dexmedetomidine, and clonidine sourced from a different manufacturer for validation in both primary (MTT viability) and secondary (colony formation) assays. These compounds demonstrated a reduction in the IC_{50} of CP in all three cell lines tested by the MTT viability assay. When combined with CP, xylazine and dexmedetomidine also showed a marked reduction in the number of colonies formed by all three cell lines. While the use of the three compounds in these assays was largely consistent, clonidine did not reach the same statistical significance compared with xylazine and dexmedetomidine in the colony formation assay in two of the cell lines. This could be in part due to differences in its potency, a difference in specificity for the three $\alpha2$ subunits (A–C), or its mixed activity on the I_1 imidazoline receptor; however, in the case of the latter, dexmedetomidine has also been shown to be pharmacologically active in a nonadrenergic I_1 receptor manner [19]. In addition to assessing the impact of $\alpha2$ B and C subunits on OvCa chemoresistance, further evaluation of the I_1 receptor may indicate its usefulness as a target in OvCa and may indicate whether increased specificity for ADRA2A may alter its potential anticancer effects; this may be best accomplished using moxoni-dine and rilmenidine due to their higher selectivity for I_1 compared with clonidine [19]. Regardless, the data presented herein indicate the potential for ADRA2A-modulating anti-hypertensives (clonidine) or sedative analgesics (dexmedetomidine), which are relatively well tolerated and have well defined pharmacokinetics, to serve as an adjunct to improve the efficacy of carboplatin. This notion was further supported by the genetic upregulation of ADRA2A, which demonstrated a striking sensitization of the highly chemoresistant OVCAR8 cell line to carboplatin.

While the present study did not delineate the exact mechanisms underlying the potential CP sensitization, multiple downstream effectors of ADRA2A have been well described in multiple cancers. ADRA2A belongs to the adrenoceptor alpha-2 family of GPCRs coupled to the inhibitory G-alpha subunit (Gi) that suppresses adenylyl cyclase activity, thereby suppressing the cyclization of AMP to cAMP [20]. cAMP is a key second messenger that impacts multiple intracellular pathways, canonically regulating protein kinase A (PKA) and cAMP-response-element-binding protein (CREB) activity, but also calcium-dependent signaling and mitogenic pathways such as MAPK [21]. While the functional role of cAMP varies by tissue, in OvCa, cAMP signaling has been associated with enhanced proliferation and the suppression of pro-apoptotic signals, frequently associated with the activity of PKA or CREB1 (reviewed in [22]). Notably, PKA activation has been observed to promote platinum resistance in ovarian and other cancers [22–24], and CREB1 inhibition potently sensitized OvCa tumor cells to cisplatin [25]. In a recent study, ADRA2A overexpression was observed to suppress proliferation and to promote apoptosis through the repression of PI3K/AKT/mTOR signaling in cervical cancer [26], another signaling axis associated with platinum resistance in several contexts [6]; however, to date, no work has demonstrated the impact of ADRA2A activity in OvCa development or treatment specifically. Further evaluation is needed given the many potential downstream pathways that may be attributed to the effects of ADRA2A activation observed in the current work.

To date, sparse data exist that show the clinical impact of the role of alpha adrenoceptor activity in tumor development and progression. Although antihypertensives, including beta blockers, have not been associated with OvCa patient survival [27], the effects of alpha adrenoceptor activation have yet to be evaluated in this clinical context. As clonidine has been used to mitigate symptoms in OvCa patients following bilateral oophorectomy [28], it may be possible to retrospectively analyze whether the inclusion of this drug may delay the time to recurrence in these patients. To increase the clinical relevancy of our findings, an evaluation of whether ADRA2A activity is relevant in a more physiologic model system that includes components of the OvCa tumor microenvironment (TME) (as performed in a recent study [18]) is required to improve the translatability of these findings, considering that cancer-associated fibroblasts in the TME are thought to significantly contribute to platinum resistance in multiple contexts [29,30]. This may be especially important given recent findings observing that dexmedetomidine induced IL-6 secretion from stromal stellate cells and pro-tumorigenic signaling in hepatocellular cancer cells [31], although whether this is relevant to that distinct TME is to be determined. Moreover, the beta-adrenergic antagonist propranolol has been shown to suppress tumor growth by inhibiting tumor angiogenesis and promoting T-cell recruitment in sarcoma [32], and the nonselective blockade of beta-adrenoceptors in prostate and pancreatic cancer in vivo suppressed tumor growth [33]. Additionally, beta-adrenergic activation has been associated with remodeling of the extracellular matrix to promote invasive phenotypes in breast cancer [34], which may be, in part, due to the potential cAMP-mediated epithelial–mesenchymal transition (EMT) [35,36]; however, whether ADRA2A inhibition of this signaling pathway results in suppression of the EMT is still to be determined. Those results and ours here suggest a complex relationship among adrenergic receptor subtypes and the processes involved in tumor development in a context-dependent manner. Further evaluation of the effects of specific adrenoceptor isoforms in tumor and stromal cells within the TME may provide a clearer understanding of the possible therapeutic value of targeting noradrenergic signaling. Lastly, in vivo data utilizing an orthotopic approach combined with a CarboTaxol regimen will ultimately determine whether the agents identified here would be suitable for inclusion in the therapy for advanced HGSOC.

5. Conclusions

Our HTS findings indicate that a number of currently approved compounds may enhance the chemosensitivity of OvCa cells to CP. Of these, ADRA2A-targeted compounds were highly represented in the HTS. Validation of several of these compounds supports the hypothesis that the activation of ADRA2A may reduce the chemoresistance of OvCa tumor cells and improve the response to CP. These pharmacologic data were in agreement with the effect of genetically upregulating ADRA2A, which resulted in the sensitization of chemoresistant OvCa cells to CP.

Supplementary Materials: The following supporting information can be downloaded at: https://www.mdpi.com/article/10.3390/cimb45120598/s1, Table S1: Viability of OvCa cells in response to carboplatin and the FDA -approved compound library. Percentage of viable cells treated with carboplatin + the FDA library relative to the carboplatin-only control.

Author Contributions: The manuscript was written through contributions from all authors. M.E.O. and P.C.H. led the conceptual and experimental design as well as the development of the methodology. H.A., A.G., D.R., G.R., L.D. and P.C.H. performed and analyzed the experiments, and contributed to the manuscript writing and editing. M.E.O. contributed to the manuscript writing, editing, and revision. M.E.O. and P.C.H. contributed equally to the supervision of the study. All authors have read and agreed to the published version of the manuscript.

Funding: This work was supported, in part, by the American Association of Colleges of Pharmacy (New Investigator Award to PCH) and the Howard Hughes Medical Institute (Inclusive Excellence Grant to Roosevelt University).

Institutional Review Board Statement: Not applicable.

Informed Consent Statement: Not applicable.

Data Availability Statement: Original data is available from the corresponding authors upon request.

Acknowledgments: The authors thank Melissa Hogan and Lawrence A. Potempa for their support in acquisition of funds for the reagents and equipment used to procure data in this manuscript.

Conflicts of Interest: There are no conflict of interest to report.

References

1. Matulonis, U.A.; Sood, A.K.; Fallowfield, L.; Howitt, B.E.; Sehouli, J.; Karlan, B.Y. Ovarian cancer. *Nat. Rev. Dis. Primers* **2016**, *2*, 16061. [CrossRef] [PubMed]
2. Kim, A.; Ueda, Y.; Naka, T.; Enomoto, T. Therapeutic strategies in epithelial ovarian cancer. *J. Exp. Clin. Cancer Res.* **2012**, *31*, 14. [CrossRef] [PubMed]
3. Rabik, C.A.; Dolan, M.E. Molecular mechanisms of resistance and toxicity associated with platinating agents. *Cancer Treat. Rev.* **2007**, *33*, 9–23. [CrossRef] [PubMed]

4.   Pignata, S.; Cecere, S.C.; Du Bois, A.; Harter, P.; Heitz, F. Treatment of recurrent ovarian cancer. *Ann. Oncol.* **2017**, *28* (Suppl. 8), viii51–viii56. [CrossRef] [PubMed]
5.   du Bois, A.; Reuss, A.; Pujade-Lauraine, E.; Harter, P.; Ray-Coquard, I.; Pfisterer, J. Role of surgical outcome as prognostic factor in advanced epithelial ovarian cancer: A combined exploratory analysis of 3 prospectively randomized phase 3 multicenter trials: By the Arbeitsgemeinschaft Gynaekologische Onkologie Studiengruppe Ovarialkarzinom (AGO-OVAR) and the Groupe d'Investigateurs Nationaux Pour les Etudes des Cancers de l'Ovaire (GINECO). *Cancer* **2009**, *115*, 1234–1244. [PubMed]
6.   Zhou, J.; Kang, Y.; Chen, L.; Wang, H.; Liu, J.; Zeng, S.; Yu, L. The Drug-Resistance Mechanisms of Five Platinum-Based Antitumor Agents. *Front. Pharmacol.* **2020**, *11*, 343. [CrossRef] [PubMed]
7.   Bayat Mokhtari, R.; Homayouni, T.S.; Baluch, N.; Morgatskaya, E.; Kumar, S.; Das, B.; Yeger, H. Combination therapy in combating cancer. *Oncotarget* **2017**, *8*, 38022–38043. [CrossRef]
8.   Alqahtani, F.Y.; Aleanizy, F.S.; El Tahir, E.; Alkahtani, H.M.; AlQuadeib, B.T. Paclitaxel. *Profiles Drug Subst. Excip. Relat. Methodol.* **2019**, *44*, 205–238.
9.   Wang, Q.; Peng, H.; Qi, X.; Wu, M.; Zhao, X. Targeted therapies in gynecological cancers: A comprehensive review of clinical evidence. *Signal Transduct. Target. Ther.* **2020**, *5*, 137. [CrossRef]
10.   Pines, A.; Vrouwe, M.G.; Marteijn, J.A.; Typas, D.; Luijsterburg, M.S.; Cansoy, M.; Hensbergen, P.; Deelder, A.; de Groot, A.; Matsumoto, S.; et al. PARP1 promotes nucleotide excision repair through DDB2 stabilization and recruitment of ALC1. *J. Cell Biol.* **2012**, *199*, 235–249. [CrossRef]
11.   Ray Chaudhuri, A.; Nussenzweig, A. The multifaceted roles of PARP1 in DNA repair and chromatin remodelling. *Nat. Rev. Mol. Cell Biol.* **2017**, *18*, 610–621. [CrossRef]
12.   Montemagno, C.; Pages, G. Resistance to Anti-angiogenic Therapies: A Mechanism Depending on the Time of Exposure to the Drugs. *Front. Cell Dev. Biol.* **2020**, *8*, 584. [CrossRef]
13.   Klotz, D.M.; Wimberger, P. Overcoming PARP inhibitor resistance in ovarian cancer: What are the most promising strategies? *Arch. Gynecol. Obstet.* **2020**, *302*, 1087–1102. [CrossRef] [PubMed]
14.   Cavalla, D. Using human experience to identify drug repurposing opportunities: Theory and practice. *Br. J. Clin. Pharmacol.* **2019**, *85*, 680–689. [CrossRef] [PubMed]
15.   Jordan, V.C. 50th anniversary of the first clinical trial with ICI 46,474 (tamoxifen): Then what happened? *Endocr. Relat. Cancer* **2021**, *28*, R11–R30. [CrossRef] [PubMed]
16.   Ediriweera, M.K.; Tennekoon, K.H.; Samarakoon, S.R. In vitro assays and techniques utilized in anticancer drug discovery. *J. Appl. Toxicol.* **2019**, *39*, 38–71. [CrossRef] [PubMed]
17.   Mariniello, M.; Petruzzelli, R.; Wanderlingh, L.G.; La Montagna, R.; Carissimo, A.; Pane, F.; Amoresano, A.; Ilyechova, E.Y.; Galagudza, M.M.; Catalano, F.; et al. Synthetic Lethality Screening Identifies FDA-Approved Drugs that Overcome ATP7B-Mediated Tolerance of Tumor Cells to Cisplatin. *Cancers* **2020**, *12*, 608. [CrossRef] [PubMed]
18.   Kenny, H.A.; Hart, P.C.; Kordylewicz, K.; Lal, M.; Shen, M.; Kara, B.; Chen, Y.J.; Grassl, N.; Alharbi, Y.; Pattnaik, B.R.; et al. The Natural Product beta-Escin Targets Cancer and Stromal Cells of the Tumor Microenvironment to Inhibit Ovarian Cancer Metastasis. *Cancers* **2021**, *13*, 3931. [CrossRef]
19.   Bousquet, P.; Hudson, A.; García-Sevilla, J.A.; Li, J.-X. Imidazoline Receptor System: The Past, the Present, and the Future. *Pharmacol. Rev.* **2020**, *72*, 50–79. [CrossRef]
20.   Pettinger, W.A.; Jackson, E.K. α_2-Adrenoceptors: Challenges and Opportunities-Enlightenment from the Kidney. *Cardiovasc. Ther.* **2020**, *2020*, 2478781. [CrossRef]
21.   Ahmed, M.B.; Alghamdi, A.A.A.; Islam, S.U.; Lee, J.-S.; Lee, Y.-S. cAMP Signaling in Cancer: A PKA-CREB and EPAC-Centric Approach. *Cells* **2022**, *11*, 2020. [CrossRef]
22.   Kilanowska, A.; Ziółkowska, A.; Stasiak, P.; Gibas-Dorna, M. cAMP-Dependent Signaling and Ovarian Cancer. *Cells* **2022**, *11*, 3835. [CrossRef] [PubMed]
23.   Stewart, D.J. Mechanisms of resistance to cisplatin and carboplatin. *Crit. Rev. Oncol. Hematol.* **2007**, *63*, 12–31.
24.   Dempke, W.; Voigt, W.; Grothey, A.; Hill, B.T.; Schmoll, H.-J. Cisplatin resistance and oncogenes—A review. *Anticancer. Drugs* **2000**, *11*, 225–236. [CrossRef] [PubMed]
25.   Dimitrova, N.; Nagaraj, A.B.; Razi, A.; Singh, S.; Kamalakaran, S.; Banerjee, N.; Joseph, P.; Mankovich, A.; Mittal, P.; DiFeo, A.; et al. InFlo: A novel systems biology framework identifies cAMP-CREB1 axis as a key modulator of platinum resistance in ovarian cancer. *Oncogene* **2017**, *36*, 2472–2482. [CrossRef] [PubMed]
26.   Wang, W.; Guo, X.; Dan, H. alpha2A-Adrenergic Receptor Inhibits the Progression of Cervical Cancer through Blocking PI3K/AKT/mTOR Pathway. *Onco Targets Ther.* **2020**, *13*, 10535–10546. [CrossRef] [PubMed]
27.   Huang, T.; Townsend, M.K.; Dood, R.L.; Sood, A.K.; Tworoger, S.S. Antihypertensive medication use and ovarian cancer survival. *Gynecol. Oncol.* **2021**, *163*, 342–347. [CrossRef]
28.   Van Le, L.; McCormack, M. Enhancing Care of the Survivor of Gynecologic Cancer: Managing the Menopause and Radiation Toxicity. *Am. Soc. Clin. Oncol. Educ. Book* **2016**, *35*, e270–e275. [CrossRef]
29.   Long, X.; Xiong, W.; Zeng, X.; Qi, L.; Cai, Y.; Mo, M.; Jiang, H.; Zhu, B.; Chen, Z.; Li, Y. Cancer-associated fibroblasts promote cisplatin resistance in bladder cancer cells by increasing IGF-1/ERbeta/Bcl-2 signalling. *Cell Death Dis.* **2019**, *10*, 375. [CrossRef]
30.   Wang, L.; Li, X.; Ren, Y.; Geng, H.; Zhang, Q.; Cao, L.; Meng, Z.; Wu, X.; Xu, M.; Xu, K. Cancer-associated fibroblasts contribute to cisplatin resistance by modulating ANXA3 in lung cancer cells. *Cancer Sci.* **2019**, *110*, 1609–1620. [CrossRef]

31. Chen, P.; Luo, X.; Dai, G.; Jiang, Y.; Luo, Y.; Peng, S.; Wang, H.; Xie, P.; Qu, C.; Lin, W.; et al. Dexmedetomidine promotes the progression of hepatocellular carcinoma through hepatic stellate cell activation. *Exp. Mol. Med.* **2020**, *52*, 1062–1074. [CrossRef] [PubMed]
32. Fjaestad, K.Y.; Rømer, A.M.; Goitea, V.; Johansen, A.Z.; Thorseth, M.L.; Carretta, M.; Engelholm, L.H.; Grøntved, L.; Junker, N.; Madsen, D.H. Blockade of beta-adrenergic receptors reduces cancer growth and enhances the response to anti-CTLA4 therapy by modulating the tumor microenvironment. *Oncogene* **2022**, *41*, 1364–1375. [CrossRef] [PubMed]
33. Zhou, S.; Li, J.; Yu, J.; Wang, Y.; Wang, Z.; He, Z.; Ouyang, D.; Liu, H.; Wang, Y. Tumor microenvironment adrenergic nerves blockade liposomes for cancer therapy. *J. Control. Release* **2022**, *351*, 656–666. [CrossRef] [PubMed]
34. Conceicao, F.; Sousa, D.M.; Paredes, J.; Lamghari, M. Sympathetic activity in breast cancer and metastasis: Partners in crime. *Bone Res.* **2021**, *9*, 9. [CrossRef]
35. Shaikh, D.; Zhou, Q.; Chen, T.; Ibe, J.C.; Raj, J.U.; Zhou, G. cAMP-dependent protein kinase is essential for hypoxia-mediated epithelial-mesenchymal transition, migration, and invasion in lung cancer cells. *Cell Signal* **2012**, *24*, 2396–2406. [CrossRef]
36. Wang, X.; Cui, H.; Lou, Z.; Huang, S.; Ren, Y.; Wang, P.; Weng, G. Cyclic AMP responsive element-binding protein induces metastatic renal cell carcinoma by mediating the expression of matrix metallopeptidase-2/9 and proteins associated with epithelial-mesenchymal transition. *Mol. Med. Rep.* **2017**, *15*, 4191–4198. [CrossRef]

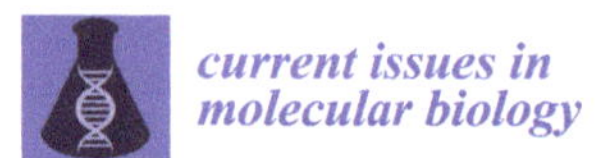
current issues in molecular biology

Review

Clinical Characteristics of Molecularly Defined Renal Cell Carcinomas

Xinfeng Hu, Congzhu Tan and Guodong Zhu *

Department of Urology, The First Affiliated Hospital of Xi'an Jiaotong University, Xi'an 710061, China
* Correspondence: gdzhu@xjtufh.edu.cn; Tel.: +86-029-85323940

Abstract: Kidney tumors comprise a broad spectrum of different histopathological entities, with more than 0.4 million newly diagnosed cases each year, mostly in middle-aged and older men. Based on the description of the 2022 World Health Organization (WHO) classification of renal cell carcinoma (RCC), some new categories of tumor types have been added according to their specific molecular typing. However, studies on these types of RCC are still superficial, many types of these RCC currently lack accurate diagnostic standards in the clinic, and treatment protocols are largely consistent with the treatment guidelines for clear cell RCC (ccRCC), which might result in worse treatment outcomes for patients with these types of molecularly defined RCC. In this article, we conduct a narrative review of the literature published in the last 15 years on molecularly defined RCC. The purpose of this review is to summarize the clinical features and the current status of research on the detection and treatment of molecularly defined RCC.

Keywords: renal cell carcinoma; molecular; pathology; treatment; clinical

check for updates

Citation: Hu, X.; Tan, C.; Zhu, G. Clinical Characteristics of Molecularly Defined Renal Cell Carcinomas. *Curr. Issues Mol. Biol.* **2023**, *45*, 4763–4777. https://doi.org/10.3390/cimb45060303

Academic Editor: Dumitru A. Iacobas

Received: 5 April 2023
Revised: 23 May 2023
Accepted: 26 May 2023
Published: 31 May 2023

1. Introduction

Kidney cancer is the 14th most common cancer worldwide, and its incidence has continued to increase in recent years [1]. To date, more than 0.4 million new cases of kidney cancer are diagnosed each year [2,3]. Among them, more than 85% of patients present with renal cell carcinoma (RCC) [4]. Based on the traditional histopathological classification, RCC can be divided into three main categories: clear cell carcinoma (ccRCC, 75%), papillary renal cell carcinoma (PRCC, 15–20%), and chromophobe cell renal carcinoma (chRCC, 5%) [5]. Studies in recent years have found that RCC mostly occurs in older men [6] and most cases are localized tumors, with only 17% of RCC patients having distant metastases at the time of diagnosis, which are mainly found in lung, bone, liver, lymph nodes, and adrenal gland [1,4]. In 2020, a statistic by Padala SA showed that the 5-year survival rate for kidney cancer patients with metastatic disease was only 12% [7]. Currently, there are more and more treatment modalities for patients with metastatic RCC (mRCC), with targeted therapies and immune checkpoint inhibitor-based immunotherapy gradually proving to be effective in the treatment of patients with mRCC and the survival rate of those patients greatly improving recently [8,9].

Epigenetic alterations are considered to be a hallmark of cancer [10]. However, recent studies found that RCC has multiple molecular alterations, such as DNA methylation and micro-RNA alterations in ccRCC, which could greatly affect the biological progression of these tumors [11,12]. The 2022 World Health Organization (WHO) classification of pathological kidney tumors added new histopathological subtypes, including molecularly defined RCC [5,8]. It includes transcription factor binding to *IGHM* enhancer 3 (*TFE3*)-rearranged renal cell carcinomas, transcription factor *EB* (*TFEB*)-altered renal cell carcinomas, elongin C (*ELOC*)-mutated renal cell carcinoma, fumarate hydratase (*FH*)-deficient renal cell carcinoma, succinate dehydrogenase (*SDH*)-deficient renal cell carcinoma, anaplastic lymphoma kinase (*ALK*)-rearranged renal cell carcinomas, and *SWI/SNF*-related, matrix-associated, actin-dependent regulator of chromatin subfamily B member 1 (*SMARCB1*)-deficient renal medullary carcinoma (see Table 1) [5]. These different molecularly defined histopathological subtypes of RCC are easily confused and may lead to suboptimal treatment outcomes as a result of misdiagnoses [12]. In this article, we summarize the pathological and clinical characteristics of each molecularly defined RCC subtypes and present their molecular features and the current treatment strategy status. We hope that this will be helpful for physicians to develop accurate diagnostic and therapeutic options for those RCC patients in clinical practice.

Table 1. Genes of molecularly defined renal cell carcinoma and associated clinical syndromes.

Molecularly Defined Renal Cell Carcinoma Types	*TFE3*-Rearranged Renal Cell Carcinomas	*TFEB*-Altered Renal Cell Carcinomas	Elongin C (*ELOC*, Formerly *TCEB1*)-Mutated Renal Cell Carcinoma	Fumarate Hydratase-Deficient Renal Cell Carcinoma	Succinate Dehydrogenase-Deficient Renal Cell Carcinoma	*ALK*-Rearranged Renal Cell Carcinomas	*SMARCB1*-Deficient Renal Medullary Carcinoma
Mutated genes	Transcription factor binding to *IGHM* enhancer 3 (*TFE3*)	Transcription factor EB (*TFEB*)	Elongin C (*ELOC*)	Fumarate hydratase (*FH*) gene	Succinate dehydrogenase (*SDH*)	Anaplastic lymphoma kinase (*ALK*)	Subfamily B member 1 (*SMARCB1*)
Location of genes	Xp11.23	6p21	8q21.11	1q43	SDHA: 5p15 SDHB: 1p35-p36.1 SDHC: 1q21 SDHD: 11q23	2p23	22q11.2
Prevalence age	Childhood	Childhood	Middle and old age	Adult	All ages	Childhood	Teenage
Clinical Syndromes	None	None	None	Hereditary leiomyomatosis and renal cell carcinoma (HLRCC)	*SDH*-deficient tumor syndrome	None	Rhabdoid tumor predisposition syndrome; familial schwannomatosis syndrome
Chaperone genes	*ASPL, PRCC, SFPQ, CLTC, PARP14, RBM10, NONO, MED15*	*MALAT1, CLTC, KHDRBS2, CADM2*	None	None	None	*VCL, TPM3, EML4, STRN, HOOK1*	None
Mode of inheritance	Dominant inheritance	Dominant inheritance	Dominant inheritance	Dominant inheritance	Dominant inheritance	Dominant inheritance	Dominant inheritance
Morphological characteristics	Transparent eosinophils; papillary architecture and psammoma bodies under the microscope	*TFEB*-translocated RCC: the biphasic growth pattern consisting of large and small tumor cells; smaller cells around the basement membrane-like structures; extensive hyalinization; papillary architecture; clear cell morphology. *TFEB*-amplified RCC: above pattern was less common	A clear cellular morphology under the microscope; thick fibromuscular bands; branching glandular vesicular; tubular structures	The papillary type or solid, tubulocystic, sieve-like type; abundant eosinophilic granulocytes, perinuclear halo	Cuboidal tumor cells, nested or tubular growth pattern. Characteristic morphology: the presence of vesicles or flocculent inclusions in the cytoplasm	*ALK*-rearranged RCC with *VCL* as a fusion gene: sickle-cell trait; eosinophilic granulocytic stroma; cytoplasmic lumen. Other *ALK*-rearranged RCC: similar to PRCC; consist of abundant intracellular and extracellular mucins; eosinophilic granuloplasm	At a high grade at the time of detection; infiltrative growth; sieve or reticular appearance
Ancillary test (IHC, FISH)	Positive: *PAX8* (100%); *TFE3* (95%); *CD10* (89%); achromatase (82%). Negative: *cytokeratin 7 (CK7); carbonic anhydrase 9 (CA9); GATA3*	Positive: *histone K; Melan-A* *TFEB*-amplified RCC: diffusely or patchily positive when tested for *TFEB* levels	Positive: *CK7; ELOC; CA9; CD10; ELOC* in the nucleus.	Positive: *PAX8*; succinate dehydrogenase B abnormal succinate semicarbonate (2SC)S-(2-succino)-cysteine. Negative: *FH; CK7; TFE3*	Positive: *PAX8*; epithelial membrane antigen (*EMA*). Negative: *SDHB; CK7; CD117; histone K; TFE3;* HMB45. *SDHA*-deficient RCC showed negativity for *SDHA*	Positive: *PAX7; CK10; AMACR; CD3; cytokeratin; ALK.* Negative: *carbonic anhydrase IX; TFE45; histone enzyme K; Melan A;* HMB45	Negative: *SMARCB1*
Oncological behavior and prognosis	May develop metastases within 20–30 years after diagnosis	*TFEB*-amplified RCC had higher tumor aggressiveness than *TFEB*-rearranged tumors. The 5-year survival rate for *TFEB*-amplified RCC was 48%	Has an aggressive oncological behavior	Have highly staged or distant metastases when diagnosed	Most cases are low grade and have a good prognosis with a low probability of metastasis	*ALK*-rearranged RCC with *VCL* as a fusion gene: no recurrence or distant metastasis. Other *ALK*-rearranged RCC: more aggressive clinical course	Often found at an advanced stage or with distant metastases; highly aggressive nature of the tumor. Average overall survival: 6–8 months

2. TFE3-Rearranged Renal Cell Carcinomas

Transcription factor binding to *IGHM* enhancer 3 (*TFE3*) is an important regulator of the immune system and has now been shown to cooperate with transcription factor EB (*TFEB*) to control and regulate carbohydrate and lipid metabolism and mitochondrial homeostasis [13]. The *TFE3/TFEB* rearrangement renal cell carcinoma is characterized by translocations involving the *TFE3* and *TFEB* genes. They are both derived from the microphthalmia transcription (*MiT*) family of heterotopic RCC according to the 2016 version of the WHO classification. The *MiT* subfamily of transcription factors includes *TFE3*, *TFEB*, *TFEC*, and *MITF* [14]. *TFE3*- and *TFEB*-rearranged RCC accounts for 1–4% of the newly diagnosed adult patients [15]. Recent studies have shown that *TFE3/TFEB*-rearranged RCC can be frequently detected in children [16]. In adults RCC patients, *TFE3* ectopic fusions with chaperone genes are more commonly seen [17], and there are no significant prognostic gender differences [15] (Figure 1). This ectopic fusion with a chaperone gene and the decreased immunity in adults *TFE3*-rearranged RCC patients cause them to have a potentially more aggressive course compared to the pediatric patients [16]. Current studies suggest that previous exposure to cytotoxic chemotherapy might be a predisposing factor [18].

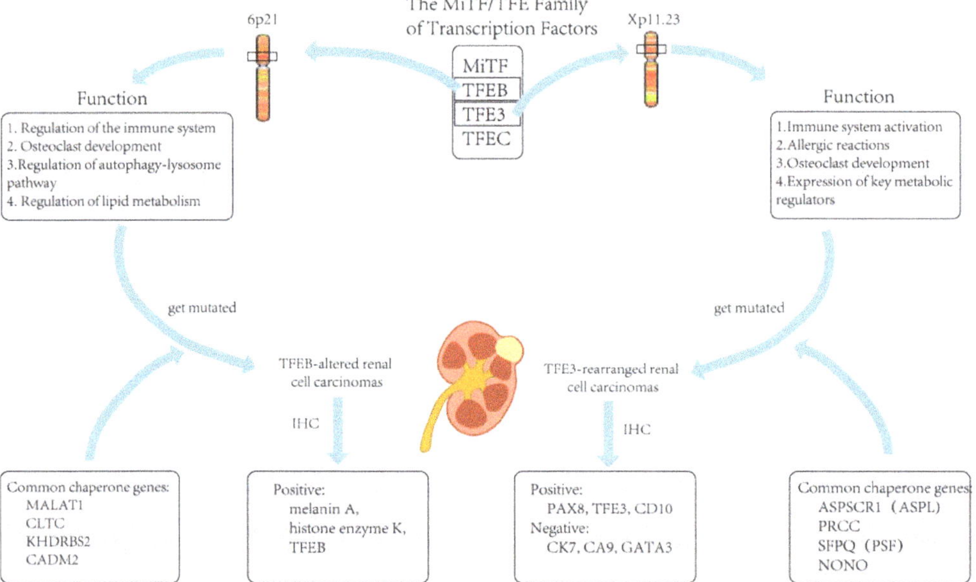

Figure 1. The role of *TFE3* in the organism and tumors caused by its mutation.

The list of chaperone genes has been growing and evolving, with more than a dozen having been reported [18]. The three most common translocations currently include a fusion of the *PRCC* and *TFE3* genes, a fusion of the *ASPL* (*ASPSCR1*) and *TFE3* genes, and a fusion of the *SFPQ* and *TFE3* genes [14]. In addition to this, there are also genes such as *CLTC*, *PARP14*, *RBM10*, *NONO*, and *MED15* that can be fused with ectopic *TFE3* [19]. However, current studies suggest that different chaperone genes may exhibit different oncological behaviors and tumor morphologies, and these features vary depending on the type of the involved chaperone genes [18]. For example, *TFE3* is more likely to exhibit lymph node metastasis when fused with *PRCC* than when fused with *ASPSCR3* [20].

In terms of histopathological morphology, the characteristics of *TFE3* fusion usually presents with transparent eosinophils, a papillary architecture, and psammoma bodies under the microscope [17,21]. However, due to chaperone genes, RCC with *TFE3* rearrangement may also resemble other types of RCC, including ccRCC, PRCC, and epithelioid vascular smooth muscle lipoma [14]. Therefore, attention should be paid and the impact of the genes that are fused with should be determined as much as possible both in the diagnosis and in the treatment of *TFE3*-rearranged renal cell carcinoma.

When facing *TFE3*-rearranged renal cell carcinoma, immunohistochemistry (IHC) is the most commonly used examination for diagnosis [13]. If IHC is not used at the time of diagnosis, a large proportion of TFE3-rearranged renal cell carcinomas are likely to be misdiagnosed as ccRCC [19]. For most other types of RCC, the positive IHC markers are *cytokeratin 7 (CK7)*, *carbonic anhydrase 9 (CA9)*, and *GATA3*. However, these are not expressed in *TFE3*-rearranged RCC and are usually positive for *histone K* [19]. In a recent review article, IHC data from nearly 400 cases of *TFE3*-rearranged RCC patients were analyzed, and the biomarkers with the highest probability of positivity were found to be *PAX8* (100%), *TFE3* (95%), *CD10* (89%), and *achromatase* (82%) [22]. However, *TFE3*-rearranged RCC did not always exhibit *TFE3* overexpression, and lower *TFE3* expression at the time of detection often resulted in false-positive or false-negative results; thus, this could limit the sensitivity and specificity of IHC for detecting *TFE3*-rearranged RCC [23]. In addition to this, the accuracy of IHC might be affected by the technique and be influenced by the formalin fixation time [18]. On the other hand, there has been no consensus or standardized guidelines regarding the judgment of *TFE3* staining results [18], and different pathologists might give completely opposite judgments if specimens show heterogeneous or focal staining. Lee HJ et al. used tissue specimens from 303 RCC patients for IHC testing and found that 23.2% of IHC-negative *TFE3* tumors were eventually diagnosed as *TFE3*-rearranged RCC [24]. Therefore, in clinical practice, a negative *TFE3* IHC result alone did not exclude the possibility of a *TFE3*-rearranged RCC case. Thus, in some cases, a combination of clinical presentation and other examination results might be needed.

The current literature suggests that the detection of *TFE3* gene rearrangements by fluorescence in situ hybridization (FISH) is more sensitive and advantageous in experimental manipulation than traditional IHC [23], and its results are more stable in formalin-fixed tissues [14]. Therefore, FISH is currently considered as the gold standard for the diagnosis of *TFE3*-rearranged RCC [23]. However, some chaperone genes, such as *NONO*, *RBM10*, and *GRIPAP1*, after fusing with *TFE3*, may not be detected by traditional FISH assays for significant *TFE3*-positive results. [18]. In addition, similar to IHC testing, the current standard definition of a positive FISH result varies widely among laboratories, from as low as 10% up to 30% [17]. These results suggest that, although the FISH test is currently the gold standard for the diagnosis of *TFE3*-rearranged RCC, in clinical practice, it should be carefully used together with other test results. For example, the previously mentioned IHC and FISH tests should be considered along with the option of gene probes or alternative molecular techniques [18]. In fact, FISH cannot provide information about fused genes, so in order to further confirm the diagnosis in clinical practice, RNA sequencing is often used to identify the gene involved in the translocation [17]. Recently, *TRIM63* determination by RNA in situ hybridization (RNA-ISH) was proposed as an alternative diagnostic tool for *TFE3*- and *TFEB*-rearranged RCC [25], but no strong evidence is available from in vitro studies.

Due to the rarity of *TFE3*-rearranged RCC and the fact that it has not been previously considered as a specific tumor subtype, there are no treatment recommendations for it to date [18]. Most previous treatment regimens are consistent with those for patients with ccRCC; however, due to recent developments in detection technology, its diagnosis has become more accurate, similarly to the detection of ccRCC. More importantly, drugs that normally treat ccRCC may not be effective against *TFE3*-rearranged RCC [16]. Additionally, Aldera AP et al. found that patients with *TFE3*-rearranged RCC may develop metastases within 20–30 years after diagnosis, so such patients may also need long-term clinical follow-up [26].

3. *TFEB*-Altered Renal Cell Carcinomas

As previously stated, *TFEB*-altered RCC has been included in the *MiTF*-translocated carcinoma family and *TFEB*-overexpressing renal tumors were initially identified in pediatric patients. Nowadays, with the availability of accurate examinations, more and more adult RCC patients are diagnosed with *TFEB*-altered RCC [27]. Nevertheless, the number of *TFEB*-altered RCC cases is still much lower than for *TFE3*-rearranged RCC [5]. There are two types of *TFEB*-altered RCC, including *TFEB*-rearranged RCC and *TFEB*-amplified RCC. The *TFEB* gene in *TFEB*-rearranged RCC is located on chromosome 6 and is most often translocated into chromosome 11, fusing with the *MALAT1* gene. Therefore, it was previously called t(6;11) RCC [19]. In the last few years, researchers have identified cases of RCC related to *TFEB* amplification, and after further testing and analysis, it was found that both genetic alteration patterns could co-exist in one case [5]. Due to the rarity of the disease, there are few studies on the distinction between different subtypes of *TFEB*-altered RCC, and current case studies show that the mean age of diagnosis for *TFEB*-amplified RCC is 62.5–64 years, while the mean age of diagnosis for *TFEB* translocated RCC is 32.8–34 years [28,29].

Similar to *TFE3*-rearranged RCC, *TFEB* can also be ectopically fused to chaperone genes [28] (Figure 1). Furthermore, for *TFEB*-amplified RCC, in addition to the possible elevated expression of *TFEB*, they are often accompanied by the amplification of other oncogenes, such as vascular endothelial growth factor A (*VEGFA*) and G1 S specific cyclin D3 (*CCND3*) [30]. It has been shown that these two genes are associated with aggressive oncological behavior [27], which would precisely explain the severe clinical symptoms and poor prognosis of patients with *TFEB*-amplified RCC.

Although *TFEB* genes are altered in *TFEB*-altered RCC, the characteristics of tumor growth vary considerably between different patterns of alteration. It has been widely reported that in *TFEB*-translocated RCC, the most commonly found morphology is a biphasic growth pattern consisting of large and small tumor cells [27], with smaller cells around the basement membrane-like structures. In addition to this, extensive hyalinization, a papillary architecture, and a clear cell morphology can be seen [31]. However, in *TFEB*-amplified RCC, this pattern is less common. Gupta S et al. investigated 37 patients with *TFEB*-altered RCC and found that nearly half of the patients had renal tubular structures and prominent cytoplasmic eosinophilia of tumor cells in their tumor specimens [27].

IHC and FISH are commonly used tests to detect *TFEB*-altered RCC; however, when assessing whether the *TFEB* gene is amplified or translocated, the markers used in the detection are quite similar. For *TFEB*-altered RCC, it has been shown that the staining results for both histone K and Melan-A are positive [31]. Similarly, Gupta S et al. and Wyvekens N et al. studied *TFEB*-amplified RCC and *TFEB*-translocated RCC, respectively, and they found that both types of tumors typically express melanin A and histone enzyme K. The difference was that tumor cells in *TFEB*-amplified RCC were usually diffusely or patchily positive when tested for *TFEB* levels [27]. However, there was also a subset of *TFEB*-amplified RCC that had lower *TFEB* expression levels than *TFEB*-translocated RCC [29]. Therefore, the type of *TFEB* gene alteration cannot be distinguished by a *TFEB*-specific assay alone. If a type of *TFEB* gene alteration is suspected, it should also be demonstrated using a FISH breakdown test or identified by RNA sequencing with a gene fusion examination [31]. In clinical practice, such detailed testing and diagnosis is not always necessary for all patients because of the very low incidence of the disease, the high cost of FISH, and the use of sequencing tests.

In addition to this, it has also been found that *TFEB*-amplified RCC exhibits a higher tumor aggressiveness than *TFEB*-rearranged tumors, and the 5-year survival rate for *TFEB*-amplified RCC is only 48% [32], while *TFEB*-translocated RCC progresses more slowly than *TFE3*-rearranged RCC. Therefore, in clinical practice, physicians should distinguish *TFEB*-amplified RCC from *TFEB*-translocated RCC. Since *TFEB*-altered RCC has often been previously diagnosed as ccRCC, its current treatment modality still differs little from the standard treatment for patients with ccRCC, which may also contribute to the poor prognostic outcome for patients with *TFEB*-altered RCC. We hope that more appropriate targeted drugs and treatment strategies for *TFEB*-altered RCC will become available in the future.

4. *Elongin C* (*ELOC*, Formerly *TCEB1*)-Mutated Renal Cell Carcinoma

Elongin C (*ELOC*) is a transcription factor in the human body and the product of this gene expression is *ELOC*, which is part of the *von Hippel–Lindau* (*VHL*) protein complex and is responsible for the ubiquitination of hypoxia inducible factor 1 alpha subunit (*HIF1α*) and its subsequent degradation [33]. Previous studies have shown that *HIF* can activate the transcription of a large number of oncogenes, leading to tumorigenesis [34]. *ELOC*-mutated RCC was classified as ccRCC in previous WHO classifications [5], accounting for 0.5% to 5% of ccRCC [35]. However, in recent years, *ELOC*-mutated RCC has been found to present as wild-type *VHL*, exhibiting somatic mutations in the *ELOC* gene and deletion of the alternative allele (8q21) [36]. In addition to this, the microscopic morphology of *ELOC*-mutated RCC also differs in many ways from ccRCC [34]. Therefore, in the latest WHO classification for RCC, it was assigned to the molecularly defined tumors as a separate pathological type. *ELOC*-mutated RCC is a rare form of RCC [33] that usually develops in middle-aged and elderly male patients, most of whom are around 50 years of age [34].

Unlike the previous tumor types, *ELOC*-mutated RCC can be seen under the microscope with a clear cellular morphology [34]. It is usually similar to ccRCC [37], which has a transparent cellular appearance [34]. This is one of the reasons why it was assigned to the ccRCC category in the previous WHO classification. However, recent studies have shown that *ELOC*-mutated RCC also have thick fibromuscular bands and branching glandular vesicular or tubular structures similar to the morphology of ccRCC [37,38], and these manifestations can be distinguished from ccRCC. When tested using IHC, *ELOC*-mutated RCC can show the same aspects as ccRCC in that it is positive for both *CA9* and *CD10* [39]. However, in the study by Wang Y et al., IHC testing was performed in four patients with *ELOC*-mutated RCC and it was found that they all showed strong positive expression for *CA9* and three patients showed positive results for *CK7* and *CD10*. In addition, the authors observed *ELOC* positivity localized only in the nucleus of all four patients [34]. Despite the small number of cases selected, this result might also indicate that *ELOC* positivity in the nucleus was a characteristic manifestation of *ELOC*-mutated RCC. Similarly, Shah RB et al. conducted a study including 21 RCC patients with *ELOC* mutations and found that 16 of them had IHC staining results expressing diffuse positivity for *CK7* [39]. In summary, in addition to observing the characteristic structure of *ELOC*-mutated RCC under a microscope, the use of IHC to detect *CK7*, *ELOC*, *CA9*, and *CD10* could further help to confirm the diagnosis.

Previous studies have shown that *ELOC*-mutated RCC tends to be inert compared to ccRCC [36], but recently there have been some case studies demonstrating that certain cases could exhibit an aggressive oncological behavior. For example, DiNatale RG et al. investigated clinical data from five patients with *ELOC*-mutated RCC and found that four of them had advanced tumors (stage III-IV) and four had developed distant metastases [33]. This aggressiveness might be related to oncogene activation due to mutations in *ELOC*. Since *ELOC*-mutated RCC was previously widely considered to be one type of ccRCC, the current treatment is largely consistent with the treatment guidelines for ccRCC.

5. Fumarate Hydratase-Deficient Renal Cell Carcinoma

Fumarate hydratase (*FH*) is an indispensable enzyme in the tricarboxylic acid cycle that produces cellular energy in the form of ATP through oxidative phosphorylation (*OXPHOS*) in mitochondria [40]. Mutations in the gene where *FH* is located can lead to fumarate accumulation, which not only causes an imbalance in the energy supply but also impairs the function of histones and DNA demethylases, thus causing abnormal gene expression [41]. Singh NP et al. analyzed the TCGA database and found that alterations in the *FH* gene were associated with the immune function of *PRCC* [42], in addition to the accumulation of metabolites, such as fumarate, which promote the expression of inflammatory factors and suppress the body's tumor immunity [43,44]. Fumarate hydratase (*FH*)-deficient RCC is a rare subtype of renal cancer that was considered a subtype of PRCC in the previous classification of RCC [40]. In the fifth edition of the WHO cancer classification, *FH*-deficient RCC has replaced hereditary leiomyomatosis and renal cell carcinoma (HLRCC) as a separate molecular subtype. It is characterized by germline mutations or somatic mutations in the *FH* gene, resulting in decreased expression of *FH* [45]. In addition to this, several studies have shown that methylation of genes, such as cyclin-dependent kinase inhibitor 2A (*CDKN2A*), O-6-methylguanine DNA methyltransferase (*MGMT*), adenomatous polyposis coli (*APC*), and tumor protein P53 (*TP53*), all of which are associated with tumorigenesis and progression, have been observed in *FH*-deficient RCC [45], which may explain the aggressiveness and poor prognostic outcome for patients observed in the clinic. HLRCC is an inherited syndrome caused by congenital mutations in *FH* gene and it is inherited in an autosomal dominant fashion. In clinical practice, HLRCC often presents as uterine tumors and smooth muscle tumors of the skin [5,40]. Indeed, it has long been shown that HLRCC increases the susceptibility to aggressive RCC [46]. However, there is no very precise treatment modality for patients with *FH*-deficient RCC.

Due to the rarity of *FH*-deficient RCC, the current knowledge of its disease characteristics and course is not very accurate. Yu YF et al. found that the mean age of onset was 36.7 years through a survey of 11 patients with *FH*-deficient RCC, which is lower than that of RCC patients without *FH* defects [47]. *FH*-deficient RCC could also exhibit many pathological structures, thus increasing its probability of being misdiagnosed [48]. Often, patients are much younger compared to other types of renal tumors when firstly diagnosed.

FH-deficient RCC can exhibit a variety of growth patterns and is, therefore, difficult to differentiate histologically [49]. The papillary type is the most common structure, and other common types include solid, tubulocystic, and sieve-like [47]. Microscopically, *FH*-deficient RCC also has characteristic histological manifestations, such as a papillary architecture with tubule cystic growth patterns, abundant eosinophilic granulocytes, and a perinuclear halo [40]. However, microscopic observation alone is not enough; more tests, such as IHC and imaging, are required to confirm the diagnosis [46]. In the clinical setting, genetic detection of mutations in *FH* is the gold standard for the diagnosis of *FH*-deficient RCC [50]. The imaging manifestations of *FH*-deficient RCC are very diverse, and it can present as a solid enhancing mass or as a mildly enhancing cystic mass, etc. These presentations cannot be distinguished from other types of RCC; therefore, diagnosis by imaging alone is incomplete [51]. Magnetic resonancespectroscopy (MRS) has also recently been proposed to be helpful in confirming the diagnosis of *FH*-deficient RCC. Wu G et al. used MRS in six patients with *FH*-deficient RCC and showed that the sensitivity, specificity, and accuracy were 69%, 100%, and 91%, respectively [52]. In IHC testing, the characteristic presentation of *FH*-deficient RCC is the lack of *FH* staining [45]; however, a recent study reported that there were isolated cases of *FH*-deficient RCC in which positive *FH* could still be detected [48]. Therefore, a positive result for *FH* does not completely exclude the possibility of *FH*-deficient RCC. In addition to detecting *FH*, studies in recent years suggested that some other biomarkers might play a key role in the detection of this disease. For example, *CK7* and *TFE3* usually show negative results, while *PAX8* and succinate dehydrogenase B abnormal succinate semicarbonate (2SC) S-(2-succino)-cysteine usually show positive results in the detection of patients with *FH*-deficient RCC [48,49,53].

Clinically, most *FH*-deficient RCC exhibit highly aggressive tumors, and patients are often found to have highly staged or distant metastases when they are diagnosed [54], with the most common sites of metastasis being the lymph nodes in the chest and abdomen, bone, and liver [55]. In addition, there is no clear standard treatment strategy for patients with *FH*-deficient RCC [45], and its highly aggressive course often makes treatment more difficult [46]. Most treatment stratigies for patients with *FH*-deficient RCC are quite similar to the treatment guidelines for patients withccRCC; however, due to the different pathogenesis and oncologic behavior, treatments that mimic ccRCC often result in an increased chance of distant metastasis and death for patients with *FH*-deficient RCC [53]. In the past years, several new drugs have been explored for the treatment of this disease, such as sunitinib, pazopanib and immune checkpoint inhibitors (ICIs), including ipilimumab and nivolumab [46]. However, the efficacy of these drugs is not yet supported by clear positive evidence. In a recent study comparing treatment outcomes in 55 patients with *FH*-deficient RCC, the analysis found that the treatment with ICIs in combination with tyrosine kinase inhibitor (TKI) may have a better clinical outcome compared to monotherapy [56]. In addition to this, Gleeson JP et al. analyzed 26 patients with *FH*-deficient RCC to assess the efficacy of combined treatment with vascular endothelial growth factor (*VEGF*) and mammalian target of rapamycin (*mTOR*), and the study demonstrated that the objective response rate of this combination therapy was 44% [46]. In addition to this, recent reports demonstrated that bevacizumab in combination with erlotinib had entered phase II clinical trials and was currently showing positive results [47]. In the future, more targeted agents and more standard treatments will be available to help patients with *FH*-deficient RCC.

6. Succinate Dehydrogenase-Deficient Renal Cell Carcinoma

Succinate dehydrogenase (*SDH*), a complex that functions in mitochondria, is composed of several subunits (*SDHA*, *SDHB*, *SDHC*, and *SDHD*) [57]. It plays an important role in cellular respiration and energy metabolism, catalyzing the conversion of succinate to fumarate [58]. In tumorigenesis, *SDH* is considered as a class of cancer suppressor gene [57], and current studies demonstrate that when the *SDH* gene germline is altered, it often results in the development of paragangliomas, gastrointestinal mesenchymal tumors, and pituitary adenomas [59,60]. In addition, *SDH*-deficient RCC has also been shown to be associated with *SDH* germline mutations, and by far the most commonly found are mutations in *SDHB*, while *SDHC*, *SDHA*, and *SDHD* mutations are rare. *SDH*-deficient RCC is rare, accounting for an estimated 0.05% to 0.2% of all RCC cases [61].

SDH-deficient RCC can be seen in a wide variety of age groups and, in a survey by Gill AJ et al., they found that the age of diagnosis of *SDH*-deficient RCC can range from 14 to 76 years and is predominate in male patients [61]. Unlike the previously described RCC, most *SDH*-deficient RCC cases are low grade and have a good prognosis with a low probability of metastasis [58]. However, some *SDH*-deficient RCC cases with high-grade nuclei, sarcomatoid changes, or coagulative necrosis can have an aggressive oncological behavior with a poor prognosis [61]. Therefore, in facing RCC patients with the above pathological features, an aggressive molecular diagnosis should be clarified and early therapeutic measures should be taken to improve the quality of life and life expectancy of these patients.

For *SDH*-deficient RCC, its tumor cells are usually cuboidal, with nested or tubular growth pattern. However, its most characteristic morphology compared to other RCCs is the presence of vesicles or flocculent inclusions in the cytoplasm [58], which is often due to the enlargement of mitochondria as a result of an altered respiratory chain [59]. In terms of IHC, the negative result of *SDHB* staining is currently considered important for the definitive confirmation of the diagnosis [61]. However, recent studies have shown that decreased *SDH* expression is also observed in some non-*SDH* germline-deficient tumors [62], a condition that may be somewhat misleading in IHC, and, therefore, it may be inaccurate to solely rely on the decreased *SDH* expression to make the diagnosis. *SDH*-deficient RCC usually shows negativity for *CK7*, *CD117*, histone K, *TFE3*, and *HMB45*, but positivity for biomarkers such as *PAX8* and epithelial membrane antigen (*EMA*) [58,61,63]. Another recent study indicated that tumor cells in *SDHA*-deficient RCC showed negativity for both *SDHA* and *SDHB*, while RCC caused by defects in the *SDHB*, *SDHC*, or *SDHD* genes only showed negativity for *SDHB* [64]. This is also a possible way to diagnose *SDH*-deficient RCC accurately.

Clinically, most *SDH*-deficient RCC patients present as low-grade tumors; however, in some rare cases, distant metastases may be present [61]. In this regard, most *SDH*-deficient RCC can usually be easily cured by surgical resection [59], and for early-stage tumors, even partial nephrectomy can be performed to preserve the kidney [58]. For patients with advanced-grade or with distant metastases, some studies have shown that targeted therapy with tyrosine kinase inhibitors, *VEGF*-targeted drugs, or *mTOR*-targeted drugs has shown positive therapeutic effects in patients with *SDH*-deficient RCC [65,66].

7. *ALK*-Rearranged Renal Cell Carcinomas

Anaplastic lymphoma kinase (*ALK*) is a membrane-associated tyrosine kinase that belongs to the insulin receptor family [67,68]. *ALK* functions to regulate cell proliferation and promote cell motility [69]. When *ALK* gene rearrangement occurs, it may lead to tumorigenesis. In 2011, two cases of *ALK*-rearranged RCC were first identified and diagnosed [70], and until now, it is still a very rare tumor [68], which accounts for 0.12–0.56% of all RCC cases [69]. Generally, a high expression of *ALK* can be observed in patients with *ALK*-rearranged RCC [71]. Similar to *TFE3*-rearraged RCC, it has many accompanying fusion genes. Various fusion genes have been identified in recent years, such as *VCL*, *TPM3*, *EML4*, *STRN*, and *HOOK1* [70], with renal tumors of *VCL* and *HOOK1* rearranged with *ALK* only described in pediatric patients [69].

Due to its rarity, there is no standard characteristic description of the clinical presentation of patients with *ALK*-rearranged RCC, which remains similar to PRCC and ccRCC in this sense [67]. *ALK*-rearranged RCC has many pathological manifestations, most of which display a shaped structure, in addition to solid and tubular patterns [70]. Among them, they can be roughly divided into two categories according to the morphology: One is *ALK*-rearranged RCC with *VCL* as a fusion gene, which occurs mostly in childhood and has a sickle-cell trait, eosinophilic granulocytic stroma, and cytoplasmic lumen [69,71]; the other category comprises other *ALK*-rearranged RCCs, most of which have a morphology similar to PRCC and also consist of abundant intracellular and extracellular mucins with eosinophilic granuloplasm [67,69]. In terms of IHC, the detection of *ALK* expressed in abundance in *ALK*-rearranged RCC using IHC has proven to be a valuable tool for the diagnosis of *ALK* [69]. In addition to this, several recent studies have found that the majority of *ALK*-rearranged RCC cases showed positive results for biomarkers such as *PAX7*, *CK10*, *AMACR*, *CD3*, and *cytokeratin*; negative results for biomarkers such as carbonic anhydrase IX, *TFE45*, histone enzyme K, Melan A, and *HMB45* [70,71]. These results can further help physicians to differentiate *ALK*-rearranged RCC from other types of RCC.

There is no standard treatment for patients with *ALK*-rearranged RCC; however, a recent study found that *ALK*-rearranged RCC with *VCL* as a fusion gene did not generally exhibit recurrence or distant metastasis [72], while *ALK*-rearranged RCC accompanied by other fusion genes showed a more aggressive clinical course [73]. As targeted agents continue to be developed, there is evidence that inhibitors of *ALK*, such as crizotinib and alectinib, can demonstrate efficacy in the treatment of nonsmall cell lung cancer and myofibroblastic tumors due to *ALK* rearrangements [74–76]. Although evidence for the treatment of *ALK*-rearranged RCC is still lacking, it is hoped that more clinical trials will be conducted in the future to demonstrate the efficacy of targeted agents for the treatment of patients with *ALK*-rearranged RCC.

8. *SMARCB1*-Deficient Renal Medullary Carcinoma

SWI/SNF-related, matrix-associated, actin-dependent regulator of chromatin subfamily B member 1 (*SMARCB1*) is a *SWI/SNF* protein complex that was considered to be a tumor suppressor in past studies and plays an important regulatory role in the organism [77]. In recent years, researchers have discovered that the *SMARCB1* gene is located on chromosome 22 and, when it is altered, *SMARCB1* expression is decreased or even absent [78], and a series of tumors are rapidly developed, such as malignant rhabdoid tumors of the central nervous system, renal medullary RCC, and epithelioid sarcoma [79]. In the 2022 edition of the WHO classification for RCC, this class of renal medullary carcinoma (RMC) with mutations in the *SMARCB1* gene is classified as a new molecular category called *SMARCB1*-

deficient RMC [5]. *SMARCB1*-deficient RMC is a rare cancer [80], which usually develops in patients with the sickle-cell trait (SCT) or sickle-cell disease (SCD).

SMARCB1-deficient RMC is an aggressive tumor that commonly affects males and is predominantly right sided [81]. It is often found at an advanced stage or with distant metastases, and recent studies have shown that *SMARCB1*-deficient RMC is also associated with the sickle-cell trait [82]. Specific symptoms are usually abdominal pain, hematuria, and weight loss [80], while distant metastases can be found in the renal lymph nodes, adrenal glands, lungs, and liver [83]. Due to the prevalence of *SMARCB1*-deficient RMC in children and adolescents and the aggressive nature of the tumor, early recognition and diagnosis are a priority for physicians.

In previous clinical practice, patients with *SMARCB1*-deficient RMC were often misdiagnosed as ccRCC [84]. With the advancement of detection technology in recent years, some characteristic manifestations of *SMARCB1*-deficient RMC have been gradually proposed. First, in addition to the previously mentioned clinical symptoms and prodromal nature during adolescence, *SMARCB1*-deficient RMC usually develops with SCT and SCD [83]. Secondly, the tumor is often already at a high grade at the time of detection, showing infiltrative growth and exhibiting a sieve or reticular appearance [85,86]. In addition to this, and most importantly, all *SMARCB1*-deficient RMC showed negative staining for *SMARCB1* when IHC for the detection of the *SMARCB1* protein was performed [83]. Therefore, when adolescent RCC patients with hematologic disorders such as SCT are identified in the clinic, physicians should perform IHC testing as early as possible to determine whether *SMARCB1*-deficient RMC is present.

Due to the rarity of the disease and the highly aggressive nature of the tumor, the current treatment options for *SMARCB1*-deficient RMC are not effective, with one study published in 2015 showing that the average overall survival of patients with *SMARCB1*-deficient RMC was only 6–8 months, with only one patient reaching 1 year [87]. Moreover, there is no standard treatment strategy for the disease. Due to the rapid progression of the disease, the predominant recommended treatment modality in the clinic is platinum-based chemotherapy [88]. In recent years, in addition to conventional treatments for kidney cancer, investigators have tried to explore the efficacy of various targeted agents for *SMARCB1*-deficient RMC. Examples include *VEGF* inhibitors, *mTOR* inhibitors (e.g., everolimus), etc. [83]; however, none of the patient outcomes have been very satisfactory. Immunosuppressive agents have been popular for oncology treatment, and Forrest SJ et al. tested 30 patients with *SMARCB1*-deficient RMC and found that 47% of them were positive for PD-L1 expression [89]. Furthermore, it has also been shown that, for *SMARCB1*-deficient RMC, differences in the tumor cell origin make it difficult for physicians to grasp the immune profile of the tumor [90]. Therefore, immunotherapy for *SMARCB1*-deficient RMC requires more in-depth studies in the future.

9. Conclusions

RCC is a common tumor that occurs mostly in men and most of them are low-grade tumors. However, in recent years, it has been discovered that RCC also has many specific molecular types, and the different molecular types may determine different clinical features and treatment outcomes. However, for many years, due to limited testing technology, many RCC patients were not diagnosed with a clear molecular type and most were managed according to the standard treatment protocol of ccRCC, resulting in poor outcomes and prognosis for many patients. This article presents the molecular types in the 2022 WHO classification of renal cancers, including the genetic alterations and clinical manifestations of each tumor type, followed by a summary of the current molecular testing results and current treatment status for each tumor type. Here, we suggest that urological clinicians should individualize the genetic level of testing when presented with RCC patients based on clinical manifestations and laboratory tests and should give targeted treatment after diagnosis. For certain congenital genetic defective RCCs, attention should also be paid to the effect of the genetic defect at other sites. However, because physicians did not previously pay much attention to the molecular types of kidney cancer, and because of the rarity of the onset of certain RCC types, the existing clinical studies are inevitably limited in terms of sample size, observation angle, and treatment bias, and there are still many inconsistent conclusions on the characteristic manifestations of molecular detection and clinical treatment criteria. In recent years, research on molecular detection technologies and targeted drugs or immune checkpoint inhibitors has progressed very rapidly, physicians' knowledge of the disease has become more and more mature, and significant progress has been made in the diagnosis and treatment of RCC. In the future, we hope that there will be more tests and detection standards for RCC in molecular science and effective drugs to help RCC patients have a better prognosis and higher quality of life.

Author Contributions: Conceptualization, G.Z.; original draft writing, X.H.; figures preparation and table making, C.T.; manuscript review and editing, G.Z.; funding acquisition, G.Z. All authors have read and agreed to the published version of the manuscript.

Funding: The Fundamental Research Funds for the Central Universities of China (No. xjj2018zyts34) and the Research Funds on Social Development from the Department of Science and Technology of Shaanxi Province of China (No. 2020SF-119) to Guodong Zhu are acknowledged.

Institutional Review Board Statement: Not applicable.

Informed Consent Statement: Not applicable.

Data Availability Statement: Not applicable.

Conflicts of Interest: The authors declare no conflict of interest.

References

1. Capitanio, U.; Montorsi, F. Renal cancer. *Lancet* **2016**, *10021*, 894–906. [CrossRef]
2. Dibajnia, P.; Cardenas, L.M.; Lalani, A.A. The emerging landscape of neo/adjuvant immunotherapy in renal cell carcinoma. *Hum. Vaccin. Immunother.* **2023**, *1*, 2178217. [CrossRef] [PubMed]
3. Bray, F.; Ferlay, J.; Soerjomataram, I.; Siegel, R.L.; Torre, L.A.; Jemal, A. Global cancer statistics 2018: GLOBOCAN estimates of incidence and mortality worldwide for 36 cancers in 185 countries. *CA Cancer J. Clin.* **2018**, *6*, 394–424. [CrossRef]
4. Motzer, R.J.; Jonasch, E.; Agarwal, N.; Alva, A.; Baine, M.; Beckermann, K.; Carlo, M.I.; Choueiri, T.K.; Costello, B.A.; Derweesh, I.H.; et al. Kidney Cancer, Version 3.2022, NCCN Clinical Practice Guidelines in Oncology. *J. Natl. Compr. Cancer Netw.* **2022**, *1*, 71–90. [CrossRef] [PubMed]
5. Moch, H.; Amin, M.B.; Berney, D.M.; Compérat, E.M.; Gill, A.J.; Hartmann, A.; Menon, S.; Raspollini, M.R.; Rubin, M.A.; Srigley, J.R.; et al. The 2022 World Health Organization Classification of Tumours of the Urinary System and Male Genital Organs-Part A: Renal, Penile, and Testicular Tumours. *Eur. Urol.* **2022**, *5*, 458–468. [CrossRef] [PubMed]
6. Gray, R.E.; Harris, G.T. Renal Cell Carcinoma: Diagnosis and Management. *Am. Fam. Physician* **2019**, *3*, 179–184.
7. Padala, S.A.; Barsouk, A.; Thandra, K.C.; Saginala, K.; Mohammed, A.; Vakiti, A.; Rawla, P.; Barsouk, A. Epidemiology of Renal Cell Carcinoma. *World J. Oncol.* **2020**, *3*, 79–87. [CrossRef] [PubMed]
8. Yong, C.; Stewart, G.D.; Frezza, C. Oncometabolites in renal cancer. *Nat. Rev. Nephrol.* **2020**, *3*, 156–172. [CrossRef]
9. Navani, V.; Heng., D.Y.C. Treatment Selection in First-line Metastatic Renal Cell Carcinoma-The Contemporary Treatment Paradigm in the Age of Combination Therapy: A Review. *JAMA Oncol.* **2022**, *8*, 292–299. [CrossRef] [PubMed]
10. Sandoval, J.; Esteller, M. Cancer epigenomics: Beyond genomics. *Curr. Opin. Genet. Dev.* **2012**, *1*, 50–55. [CrossRef] [PubMed]
11. Mitchell, T.J.; Turajlic, S.; Rowan, A.; Nicol, D.; Farmery, J.H.R.; O'Brien, T.; Martincorena, I.; Tarpey, P.; Angelopoulos, N.; Yates, L.R.; et al. TRACERx Renal Consortium. Timing the Landmark Events in the Evolution of Clear Cell Renal Cell Cancer: TRACERx Renal. *Cell* **2018**, *3*, 611–623.e17. [CrossRef]
12. Xing, T.; He, H. Epigenomics of clear cell renal cell carcinoma: Mechanisms and potential use in molecular pathology. *Chin. J. Cancer Res.* **2016**, *1*, 80–91.
13. Pastore, N.; Vainshtein, A.; Klisch, T.J.; Armani, A.; Huynh, T.; Herz, N.J.; Polishchuk, E.V.; Sandri, M.; Ballabio, A. TFE3 Regulates whole-body energy metabolism in cooperation with TFEB. *EMBO Mol. Med.* **2017**, *5*, 605–621. [CrossRef]
14. Argani, P. Translocation carcinomas of the kidney. *Genes. Chromosomes Cancer* **2022**, *5*, 219–227. [CrossRef]
15. Caliò, A.; Segala, D.; Munari, E.; Brunelli, M.; Martignoni, G. MiT Family Translocation Renal Cell Carcinoma: From the Early Descriptions to the Current Knowledge. *Cancers* **2019**, *8*, 1110. [CrossRef]
16. Sukov, W.R.; Hodge, J.C.; Lohse, C.M.; Leibovich, B.C.; Thompson, R.H.; Pearce, K.E.; Wiktor, A.E.; Cheville, J.C. TFE3 rearrangements in adult renal cell carcinoma: Clinical and pathologic features with outcome in a large series of consecutively treated patients. *Am. J. Surg. Pathol.* **2012**, *5*, 663–670. [CrossRef] [PubMed]

17. Akgul, M.; Williamson, S.R.; Ertoy, D.; Argani, P.; Gupta, S.; Caliò, A.; Reuter, V.; Tickoo, S.; Al-Ahmadie, H.A.; Netto, G.J.; et al. Diagnostic approach in TFE3-rearranged renal cell carcinoma: A multi-institutional international survey. *J. Clin. Pathol.* **2021**, *5*, 291–299. [CrossRef]
18. Kmeid, M.; Akgul, M. TFE3 Rearrangement and Expression in Renal Cell Carcinoma. *Int. J. Surg. Pathol.* **2022**, *1*, 10668969221108517. [CrossRef]
19. Caliò, A.; Marletta, S.; Brunelli, M.; Pedron, S.; Portillo, S.C.; Segala, D.; Bariani, E.; Gobbo, S.; Netto, G.; Martignoni, G. TFE3 and TFEB-rearranged renal cell carcinomas: An immunohistochemical panel to differentiate from common renal cell neoplasms. *Virchows Arch.* **2022**, *6*, 877–891. [CrossRef]
20. Argani, P.; Zhong, M.; Reuter, V.E.; Fallon, J.T.; Epstein, J.I.; Netto, G.J.; Antonescu, C.R. TFE3-Fusion Variant Analysis Defines Specific Clinicopathologic Associations Among Xp11 Translocation Cancers. *Am. J. Surg. Pathol.* **2016**, *6*, 723–737. [CrossRef] [PubMed]
21. Argani, P. MiT family translocation renal cell carcinoma. *Semin. Diagn. Pathol.* **2015**, *2*, 103–113. [CrossRef]
22. Tretiakova, M.S. Chameleon TFE3-translocation RCC and How Gene Partners Can Change Morphology: Accurate Diagnosis Using Contemporary Modalities. *Adv. Anat. Pathol.* **2022**, *3*, 131–140. [CrossRef]
23. Green, W.M.; Yonescu, R.; Morsberger, L.; Morris, K.; Netto, G.J.; Epstein, J.I.; Illei, P.B.; Allaf, M.; Ladanyi, M.; Griffin, C.A.; et al. Utilization of a TFE3 break-apart FISH assay in a renal tumor consultation service. *Am. J. Surg. Pathol.* **2013**, *8*, 1150–1163. [CrossRef]
24. Lee, H.J.; Shin, D.H.; Kim, S.Y.; Hwang, C.S.; Lee, J.H.; Park, W.Y.; Choi, K.U.; Kim, J.Y.; Lee, C.H.; Sol, M.Y.; et al. TFE3 translocation and protein expression in renal cell carcinoma are correlated with poor prognosis. *Histopathology* **2018**, *5*, 758–766. [CrossRef] [PubMed]
25. Wang, X.M.; Zhang, Y.; Mannan, R.; Skala, S.L.; Rangaswamy, R.; Chinnaiyan, A.; Su, F.; Cao, X.; Zelenka-Wang, S.; McMurry, L.; et al. TRIM63 is a sensitive and specific biomarker for MiT family aberration-associated renal cell carcinoma. *Mod. Pathol.* **2021**, *8*, 1596–1607. [CrossRef]
26. Aldera, A.P.; Ramburan, A.; John, J. TFE3-rearranged renal cell carcinoma with osseous metaplasia and indolent behaviour. *Urol. Case Rep.* **2022**, *42*, 102041. [CrossRef] [PubMed]
27. Gupta, S.; Argani, P.; Jungbluth, A.A.; Chen, Y.B.; Tickoo, S.K.; Fine, S.W.; Gopalan, A.; Al-Ahmadie, H.A.; Sirintrapun, S.J.; Sanchez, A.; et al. TFEB Expression Profiling in Renal Cell Carcinomas: Clinicopathologic Correlations. *Am. J. Surg. Pathol.* **2019**, *11*, 1445–1461. [CrossRef] [PubMed]
28. Argani, P.; Reuter, V.E.; Zhang, L.; Sung, Y.S.; Ning, Y.; Epstein, J.I.; Netto, G.J.; Antonescu, C.R. TFEB-amplified Renal Cell Carcinomas: An Aggressive Molecular Subset Demonstrating Variable Melanocytic Marker Expression and Morphologic Heterogeneity. *Am. J. Surg. Pathol.* **2016**, *11*, 1484–1495. [CrossRef]
29. Wyvekens, N.; Rechsteiner, M.; Fritz, C.; Wagner, U.; Tchinda, J.; Wenzel, C.; Kuithan, F.; Horn, L.C.; Moch, H. Histological and molecular characterization of TFEB-rearranged renal cell carcinomas. *Virchows Arch.* **2019**, *5*, 625–631. [CrossRef]
30. Williamson, S.R.; Grignon, D.J.; Cheng, L.; Favazza, L.; Gondim, D.D.; Carskadon, S.; Gupta, N.S.; Chitale, D.A.; Kalyana-Sundaram, S.; Palanisamy, N. Renal Cell Carcinoma with Chromosome 6p Amplification Including the TFEB Gene: A Novel Mechanism of Tumor Pathogenesis? *Am. J. Surg. Pathol.* **2017**, *3*, 287–298. [CrossRef] [PubMed]
31. Williamson, S.R.; Eble, J.N.; Palanisamy, N. Sclerosing TFEB-rearrangement renal cell carcinoma: A recurring histologic pattern. *Hum. Pathol.* **2017**, *62*, 175–179. [CrossRef] [PubMed]
32. Gupta, S.; Johnson, S.H.; Vasmatzis, G.; Porath, B.; Rustin, J.G.; Rao, P.; Costello, B.A.; Leibovich, B.C.; Thompson, R.H.; Cheville, J.C.; et al. TFEB-VEGFA (6p21.1) co-amplified renal cell carcinoma: A distinct entity with potential implications for clinical management. *Mod. Pathol.* **2017**, *7*, 998–1012. [CrossRef] [PubMed]
33. DiNatale, R.G.; Gorelick, A.N.; Makarov, V.; Blum, K.A.; Silagy, A.W.; Freeman, B.; Chowell, D.; Marcon, J.; Mano, R.; Sanchez, A.; et al. Putative Drivers of Aggressiveness in TCEB1-mutant Renal Cell Carcinoma: An Emerging Entity with Variable Clinical Course. *Eur. Urol. Focus.* **2021**, *2*, 381–389. [CrossRef] [PubMed]
34. Wang, Y.; Zhao, P.; Wang, L.; Wang, J.; Ji, X.; Li, Y.; Shi, H.; Li, Y.; Zhang, W.; Jiang, Y. Analysis of clinicopathological and molecular features of ELOC(TCEB1)-mutant renal cell carcinoma. *Pathol. Res. Pract.* **2022**, *235*, 153960. [CrossRef] [PubMed]
35. Brugarolas, J. Molecular genetics of clear-cell renal cell carcinoma. *J. Clin. Oncol.* **2014**, *18*, 1968–1976. [CrossRef]
36. Sato, Y.; Yoshizato, T.; Shiraishi, Y.; Maekawa, S.; Okuno, Y.; Kamura, T.; Shimamura, T.; Sato-Otsubo, A.; Nagae, G.; Suzuki, H.; et al. Integrated molecular analysis of clear-cell renal cell carcinoma. *Nat. Genet.* **2013**, *8*, 860–867. [CrossRef]
37. Shah, R.B. Renal Cell Carcinoma with Fibromyomatous Stroma-The Whole Story. *Adv. Anat. Pathol.* **2022**, *3*, 168–177. [CrossRef]
38. Hakimi, A.A.; Tickoo, S.K.; Jacobsen, A.; Sarungbam, J.; Sfakianos, J.P.; Sato, Y.; Morikawa, T.; Kume, H.; Fukayama, M.; Homma, Y.; et al. TCEB1-mutated renal cell carcinoma: A distinct genomic and morphological subtype. *Mod. Pathol.* **2015**, *6*, 845–853. [CrossRef]

39. Shah, R.B.; Stohr, B.A.; Tu, Z.J.; Gao, Y.; Przybycin, C.G.; Nguyen, J.; Cox, R.M.; Rashid-Kolvear, F.; Weindel, M.D.; Farkas, D.H.; et al. "Renal Cell Carcinoma with Leiomyomatous Stroma" Harbor Somatic Mutations of TSC1, TSC2, MTOR, and/or ELOC (TCEB1): Clinicopathologic and Molecular Characterization of 18 Sporadic Tumors Supports a Distinct Entity. *Am. J. Surg. Pathol.* **2020**, *5*, 571–581. [CrossRef]
40. Lindner, A.K.; Tulchiner, G.; Seeber, A.; Siska, P.J.; Thurnher, M.; Pichler, R. Targeting strategies in the treatment of fumarate hydratase deficient renal cell carcinoma. *Front. Oncol.* **2022**, *12*, 906014. [CrossRef]
41. Xiao, M.; Yang, H.; Xu, W.; Ma, S.; Lin, H.; Zhu, H.; Liu, L.; Liu, Y.; Yang, C.; Xu, Y.; et al. Inhibition of α-KG-dependent histone and DNA demethylases by fumarate and succinate that are accumulated in mutations of FH and SDH tumor suppressors. *Genes Dev.* **2012**, *12*, 1326–1338. [CrossRef]
42. Singh, N.P.; Vinod, P.K. Integrative analysis of DNA methylation and gene expression in papillary renal cell carcinoma. *Mol. Genet. Genom.* **2020**, *3*, 807–824. [CrossRef]
43. Arts, R.J.; Novakovic, B.; Ter Horst, R.; Carvalho, A.; Bekkering, S.; Lachmandas, E.; Rodrigues, F.; Silvestre, R.; Cheng, S.C.; Wang, S.Y.; et al. Glutaminolysis and Fumarate Accumulation Integrate Immunometabolic and Epigenetic Programs in Trained Immunity. *Cell Metab.* **2016**, *6*, 807–819. [CrossRef]
44. Ge, X.; Li, M.; Yin, J.; Shi, Z.; Fu, Y.; Zhao, N.; Chen, H.; Meng, L.; Li, X.; Hu, Z.; et al. Fumarate inhibits PTEN to promote tumorigenesis and therapeutic resistance of type2 papillary renal cell carcinoma. *Mol. Cell* **2022**, *7*, 1249–1260.e7. [CrossRef]
45. Sun, G.; Zhang, X.; Liang, J.; Pan, X.; Zhu, S.; Liu, Z.; Armstrong, C.M.; Chen, J.; Lin, W.; Liao, B.; et al. Integrated Molecular Characterization of Fumarate Hydratase-deficient Renal Cell Carcinoma. *Clin. Cancer Res.* **2021**, *6*, 1734–1743. [CrossRef]
46. Gleeson, J.P.; Nikolovski, I.; Dinatale, R.; Zucker, M.; Knezevic, A.; Patil, S.; Ged, Y.; Kotecha, R.R.; Shapnik, N.; Murray, S.; et al. Comprehensive Molecular Characterization and Response to Therapy in Fumarate Hydratase-Deficient Renal Cell Carcinoma. *Clin. Cancer Res.* **2021**, *10*, 2910–2919. [CrossRef]
47. Yu, Y.F.; He, S.M.; Wu, Y.C.; Xiong, S.W.; Shen, Q.; Li, Y.Y.; Yang, F.; He, Q.; Li, X.S. Clinicopathological features and prognosis of fumarate hydratase deficient renal cell carcinoma. *Beijing Da Xue Xue Bao Yi Xue Ban* **2021**, *4*, 640–646.
48. Lau, H.D.; Chan, E.; Fan, A.C.; Kunder, C.A.; Williamson, S.R.; Zhou, M.; Idrees, M.T.; Maclean, F.M.; Gill, A.J.; Kao, C.S. A Clinicopathologic and Molecular Analysis of Fumarate Hydratase-deficient Renal Cell Carcinoma in 32 Patients. *Am. J. Surg. Pathol.* **2020**, *1*, 98–110. [CrossRef]
49. Muller, M.; Guillaud-Bataille, M.; Salleron, J.; Genestie, C.; Deveaux, S.; Slama, A.; de Paillerets, B.B.; Richard, S.; Benusiglio, P.R.; Ferlicot, S. Pattern multiplicity and fumarate hydratase (FH)/S-(2-succino)-cysteine (2SC) staining but not eosinophilic nucleoli with perinucleolar halos differentiate hereditary leiomyomatosis and renal cell carcinoma-associated renal cell carcinomas from kidney tumors without FH gene alteration. *Mod. Pathol.* **2018**, *6*, 974–983.
50. Chen, Y.B.; Brannon, A.R.; Toubaji, A.; Dudas, M.E.; Won, H.H.; Al-Ahmadie, H.A.; Fine, S.W.; Gopalan, A.; Frizzell, N.; Voss, M.H.; et al. Hereditary leiomyomatosis and renal cell carcinoma syndrome-associated renal cancer: Recognition of the syndrome by pathologic features and the utility of detecting aberrant succination by immunohistochemistry. *Am. J. Surg. Pathol.* **2014**, *5*, 627–637. [CrossRef]
51. Nikolovski, I.; Carlo, M.I.; Chen, Y.B.; Vargas, H.A. Imaging features of fumarate hydratase-deficient renal cell carcinomas: A retrospective study. *Cancer Imaging* **2021**, *1*, 24. [CrossRef]
52. Wu, G.; Liu, G.; Wang, J.; Pan, S.; Luo, Y.; Xu, Y.; Kong, W.; Sun, P.; Xu, J.; Xue, W.; et al. MR Spectroscopy for Detecting Fumarate Hydratase Deficiency in Hereditary Leiomyomatosis and Renal Cell Carcinoma Syndrome. *Radiology* **2022**, *3*, 631–639. [CrossRef]
53. Liu, Y.; Dong, Y.; Gu, Y.; Xu, H.; Fan, Y.; Li, X.; Dong, L.; Zhou, L.; Yang, X.; Wang, C. GATA3 aids in distinguishing fumarate hydratase-deficient renal cell carcinoma from papillary renal cell carcinoma. *Ann. Diagn. Pathol.* **2022**, *60*, 152007. [CrossRef]
54. Grubb, R.L.; Franks, M.E.; Toro, J.; Middelton, L.; Choyke, L.; Fowler, S.; Torres-Cabala, C.; Glenn, G.M.; Choyke, P.; Merino, M.J.; et al. Hereditary leiomyomatosis and renal cell cancer: A syndrome associated with an aggressive form of inherited renal cancer. *J. Urol.* **2007**, *6*, 2074–2079. [CrossRef]
55. Choi, Y.; Keam, B.; Kim, M.; Yoon, S.; Kim, D.; Choi, J.G.; Seo, J.Y.; Park, I.; Lee, J.L. Bevacizumab Plus Erlotinib Combination Therapy for Advanced Hereditary Leiomyomatosis and Renal Cell Carcinoma-Associated Renal Cell Carcinoma: A Multicenter Retrospective Analysis in Korean Patients. *Cancer Res. Treat.* **2019**, *4*, 1549–1556. [CrossRef]
56. Xu, Y.; Kong, W.; Cao, M.; Wang, J.; Wang, Z.; Zheng, L.; Wu, X.; Cheng, R.; He, W.; Yang, B.; et al. Genomic Profiling and Response to Immune Checkpoint Inhibition plus Tyrosine Kinase Inhibition in FH-Deficient Renal Cell Carcinoma. *Eur. Urol.* **2023**, *2*, 163–172. [CrossRef]
57. Williamson, S.R.; Eble, J.N.; Amin, M.B.; Gupta, N.S.; Smith, S.C.; Sholl, L.M.; Montironi, R.; Hirsch, M.S.; Hornick, J.L. Succinate dehydrogenase-deficient renal cell carcinoma: Detailed characterization of 11 tumors defining a unique subtype of renal cell carcinoma. *Mod. Pathol.* **2015**, *1*, 80–94. [CrossRef]
58. Wang, G.; Rao, P. Succinate Dehydrogenase-Deficient Renal Cell Carcinoma: A Short Review. *Arch. Pathol. Lab. Med.* **2018**, *10*, 1284–1288. [CrossRef]
59. Tsai, T.H.; Lee, W.Y. Succinate Dehydrogenase-Deficient Renal Cell Carcinoma. *Arch. Pathol. Lab. Med.* **2019**, *5*, 643–647. [CrossRef]
60. Gill, A.J. Succinate dehydrogenase (SDH)-deficient neoplasia. *Histopathology* **2018**, *1*, 106–116. [CrossRef]
61. Gill, A.J.; Hes, O.; Papathomas, T.; Šedivcová, M.; Tan, P.H.; Agaimy, A.; Andresen, P.A.; Kedziora, A.; Clarkson, A.; Toon, C.W.; et al. Succinate dehydrogenase (SDH)-deficient renal carcinoma: A morphologically distinct entity: A clinicopathologic series of 36 tumors from 27 patients. *Am. J. Surg. Pathol.* **2014**, *12*, 1588–1602. [CrossRef]

62. Aggarwal, R.K.; Luchtel, R.A.; Machha, V.; Tischer, A.; Zou, Y.; Pradhan, K.; Ashai, N.; Ramachandra, N.; Albanese, J.M.; Yang, J.I.; et al. Functional succinate dehydrogenase deficiency is a common adverse feature of clear cell renal cancer. *Proc. Natl. Acad. Sci. USA* **2021**, *39*, e2106947118. [CrossRef]
63. Sun, A.; Liu, Z.; Wang, T.; Xing, J. Succinate dehydrogenase-deficient renal cell carcinoma: A case report and review of the literature. *Asian J. Surg.* **2021**, *4*, 692–693. [CrossRef]
64. Gill, A.J.; Pachter, N.S.; Chou, A.; Young, B.; Clarkson, A.; Tucker, K.M.; Winship, I.M.; Earls, P.; Benn, D.E.; Robinson, B.G.; et al. Renal tumors associated with germline SDHB mutation show distinctive morphology. *Am. J. Surg. Pathol.* **2011**, *10*, 1578–1585. [CrossRef]
65. Paik, J.Y.; Toon, C.W.; Benn, D.E.; High, H.; Hasovitz, C.; Pavlakis, N.; Clifton-Bligh, R.J.; Gill, A.J. Renal carcinoma associated with succinate dehydrogenase B mutation: A new and unique subtype of renal carcinoma. *J. Clin. Oncol.* **2014**, *6*, e10–e13. [CrossRef]
66. Linehan, W.M.; Ricketts, C.J. The metabolic basis of kidney cancer. *Semin. Cancer Biol.* **2013**, *1*, 46–55. [CrossRef]
67. Yu, W.; Wang, Y.; Jiang, Y.; Zhang, W.; Li, Y. Genetic analysis and clinicopathological features of ALK-rearranged renal cell carcinoma in a large series of resected Chinese renal cell carcinoma patients and literature review. *Histopathology* **2017**, *1*, 53–62. [CrossRef]
68. Jeanneau, M.; Gregoire, V.; Desplechain, C.; Escande, F.; Tica, D.P.; Aubert, S.; Leroy, X. ALK rearrangements-associated renal cell carcinoma (RCC) with unique pathological features in an adult. *Pathol. Res. Pract.* **2016**, *11*, 1064–1066. [CrossRef]
69. Gorczynski, A.; Czapiewski, P.; Korwat, A.; Budynko, L.; Prelowska, M.; Okon, K.; Biernat, W. ALK-rearranged renal cell carcinomas in Polish population. *Pathol. Res. Pract.* **2019**, *12*, 152669. [CrossRef]
70. Hang, J.F.; Chung, H.J.; Pan, C.C. ALK-rearranged renal cell carcinoma with a novel PLEKHA7-ALK translocation and metanephric adenoma-like morphology. *Virchows Arch.* **2020**, *6*, 921–929. [CrossRef]
71. Kuroda, N.; Sugawara, E.; Kusano, H.; Yuba, Y.; Yorita, K.; Takeuchi, K. A review of ALK-rearranged renal cell carcinomas with a focus on clinical and pathobiological aspects. *Pol. J. Pathol.* **2018**, *2*, 109–113. [CrossRef]
72. Debelenko, L.V.; Daw, N.; Shivakumar, B.R.; Huang, D.; Nelson, M.; Bridge, J.A. Renal cell carcinoma with novel VCL-ALK fusion: New representative of ALK-associated tumor spectrum. *Mod. Pathol.* **2011**, *3*, 430–442. [CrossRef] [PubMed]
73. Sukov, W.R.; Hodge, J.C.; Lohse, C.M.; Akre, M.K.; Leibovich, B.C.; Thompson, R.H.; Cheville, J.C. ALK alterations in adult renal cell carcinoma: Frequency, clinicopathologic features and outcome in a large series of consecutively treated patients. *Mod. Pathol.* **2012**, *11*, 1516–1525. [CrossRef] [PubMed]
74. Ou, S.H.; Bazhenova, L.; Camidge, D.R.; Solomon, B.J.; Herman, J.; Kain, T.; Bang, Y.J.; Kwak, E.L.; Shaw, A.T.; Salgia, R.; et al. Rapid and dramatic radiographic and clinical response to an ALK inhibitor (crizotinib, PF02341066) in an ALK translocation-positive patient with non-small cell lung cancer. *J. Thorac. Oncol.* **2010**, *12*, 2044–2046. [CrossRef] [PubMed]
75. Kimura, H.; Nakajima, T.; Takeuchi, K.; Soda, M.; Mano, H.; Iizasa, T.; Matsui, Y.; Yoshino, M.; Shingyoji, M.; Itakura, M.; et al. ALK fusion gene positive lung cancer and 3 cases treated with an inhibitor for ALK kinase activity. *Lung Cancer* **2012**, *1*, 66–72. [CrossRef]
76. Pal, S.K.; Bergerot, P.; Dizman, N.; Bergerot, C.; Adashek, J.; Madison, R.; Chung, J.H.; Ali, S.M.; Jones, J.O.; Salgia, R. Responses to Alectinib in ALK-rearranged Papillary Renal Cell Carcinoma. *Eur. Urol.* **2018**, *1*, 124–128. [CrossRef]
77. Wang, X.; Haswell, J.R.; Roberts, C.W. Molecular pathways: SWI/SNF (BAF) complexes are frequently mutated in cancer—mechanisms and potential therapeutic insights. *Clin. Cancer Res.* **2014**, *1*, 21–27. [CrossRef]
78. Msaouel, P.; Malouf, G.G.; Su, X.; Yao, H.; Tripathi, D.N.; Soeung, M.; Gao, J.; Rao, P.; Coarfa, C.; Creighton, C.J.; et al. Comprehensive Molecular Characterization Identifies Distinct Genomic and Immune Hallmarks of Renal Medullary Carcinoma. *Cancer Cell* **2020**, *5*, 720–734.e13. [CrossRef]
79. Agaimy, A. The expanding family of SMARCB1(INI1)-deficient neoplasia: Implications of phenotypic, biological, and molecular heterogeneity. *Adv. Anat. Pathol.* **2014**, *6*, 394–410. [CrossRef]
80. Hong, A.L.; Tseng, Y.Y.; Wala, J.A.; Kim, W.J.; Kynnap, B.D.; Doshi, M.B.; Kugener, G.; Sandoval, G.J.; Howard, T.P.; Li, J.; et al. Renal medullary carcinomas depend upon SMARCB1 loss and are sensitive to proteasome inhibition. *Elife* **2019**, *8*, e44161. [CrossRef]
81. Al-Daghmin, A.; Gaashan, M.; Haddad, H. Atypical presentation of renal medullary carcinoma: A case report and review of the literature. *Urol. Case Rep.* **2018**, *22*, 8–10. [CrossRef] [PubMed]
82. Holland, P.; Merrimen, J.; Pringle, C.; Wood, L.A. Renal medullary carcinoma and its association with sickle cell trait: A case report and literature review. *Curr. Oncol.* **2020**, *1*, e53–e56. [CrossRef] [PubMed]
83. Su, Y.; Hong, A.L. Recent Advances in Renal Medullary Carcinoma. *Int. J. Mol. Sci.* **2022**, *13*, 7097. [CrossRef] [PubMed]
84. Beckermann, K.E.; Sharma, D.; Chaturvedi, S.; Msaouel, P.; Abboud, M.R.; Allory, Y.; Bourdeaut, F.; Calderaro, J.; de Cubas, A.A.; Derebail, V.K.; et al. Renal Medullary Carcinoma: Establishing Standards in Practice. *J. Oncol. Pract.* **2017**, *7*, 414–421. [CrossRef]
85. Lopez-Beltran, A.; Cheng, L.; Raspollini, M.R.; Montironi, R. SMARCB1/INI1 Genetic Alterations in Renal Medullary Carcinomas. *Eur. Urol.* **2016**, *6*, 1062–1064. [CrossRef]
86. Scarpelli, M.; Mazzucchelli, R.; Lopez-Beltran, A.; Cheng, L.; De Nictolis, M.; Santoni, M.; Montironi, R. Renal cell carcinoma with rhabdoid features and loss of INI1 expression in an individual without sickle cell trait. *Pathology* **2014**, *7*, 653–655. [CrossRef]
87. Iacovelli, R.; Modica, D.; Palazzo, A.; Trenta, P.; Piesco, G.; Cortesi, E. Clinical outcome and prognostic factors in renal medullary carcinoma: A pooled analysis from 18 years of medical literature. *Can. Urol. Assoc. J.* **2015**, *3–4*, E172–E177. [CrossRef]

88. Wiele, A.J.; Surasi, D.S.; Rao, P.; Sircar, K.; Su, X.; Bathala, T.K.; Shah, A.Y.; Jonasch, E.; Cataldo, V.D.; Genovese, G.; et al. Efficacy and Safety of Bevacizumab Plus Erlotinib in Patients with Renal Medullary Carcinoma. *Cancers* **2021**, *9*, 2170. [CrossRef]
89. Forrest, S.J.; Al-Ibraheemi, A.; Doan, D.; Ward, A.; Clinton, C.M.; Putra, J.; Pinches, R.S.; Kadoch, C.; Chi, S.N.; DuBois, S.G.; et al. Genomic and Immunologic Characterization of INI1-Deficient Pediatric Cancers. *Clin. Cancer Res.* **2020**, *12*, 2882–2890. [CrossRef]
90. Ngo, C.; Postel-Vinay, S. Immunotherapy for SMARCB1-Deficient Sarcomas: Current Evidence and Future Developments. *Biomedicines* **2022**, *3*, 650. [CrossRef]

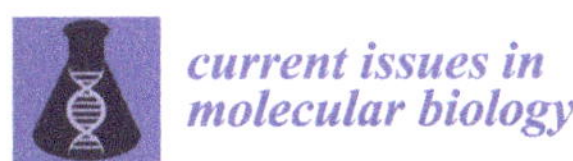

current issues in molecular biology

Article

Genomic Fabrics of the Excretory System's Functional Pathways Remodeled in Clear Cell Renal Cell Carcinoma

Dumitru Andrei Iacobas [1,*], Ehiguese Alade Obiomon [1] and Sanda Iacobas [2]

1 Personalized Genomics Laboratory, Undergraduate Medical Academy, Prairie View A&M University, Prairie View, TX 77446, USA; eobiomon3@pvamu.edu
2 Department of Pathology, New York Medical College, Valhalla, NY 10595, USA; sandaiacobas@gmail.com
* Correspondence: daiacobas@pvamu.edu

Abstract: Clear cell renal cell carcinoma (ccRCC) is the most frequent form of kidney cancer. Metastatic stages of ccRCC reduce the five-year survival rate to 15%. In this report, we analyze the ccRCC-induced remodeling of the five KEGG-constructed excretory functional pathways in a surgically removed right kidney and its metastasis in the chest wall from the perspective of the Genomic Fabric Paradigm (GFP). The GFP characterizes every single gene in each region by these independent variables: the average expression level (AVE), relative expression variability (REV), and expression correlation (COR) with each other gene. While the traditional approach is limited to only AVE analysis, the novel REV analysis identifies the genes whose correct expression level is critical for cell survival and proliferation. The COR analysis determines the real gene networks responsible for functional pathways. The analyses covered the pathways for aldosterone-regulated sodium reabsorption, collecting duct acid secretion, endocrine and other factor-regulated sodium reabsorption, proximal tubule bicarbonate reclamation, and vasopressin-regulated water reabsorption. The present study confirms the conclusion of our previously published articles on prostate and kidney cancers that even equally graded cancer nodules from the same tumor have different transcriptomic topologies. Therefore, the personalization of anti-cancer therapy should go beyond the individual, to his/her major cancer nodules.

Keywords: *ADCY6*; aldosterone-regulated sodium reabsorption; *AP2A1*; *AVP*; collecting duct acid secretion; *CREB3L4*; endocrine and other factor-regulated sodium reabsorption; *ESR1*; proximal tubule bicarbonate reclamation; vasopressin-regulated water reabsorption

check for **updates**

Citation: Iacobas, D.A.; Obiomon, E.A.; Iacobas, S. Genomic Fabrics of the Excretory System's Functional Pathways Remodeled in Clear Cell Renal Cell Carcinoma. *Curr. Issues Mol. Biol.* **2023**, *45*, 9471–9499. https://doi.org/10.3390/cimb45120594

Academic Editor: Stergios Boussios

Received: 19 October 2023
Revised: 18 November 2023
Accepted: 21 November 2023
Published: 24 November 2023

1. Introduction

Limits of the Gene Biomarker Paradigm in Cancer Diagnostics and Therapy

Cancer is a major cause of death worldwide and is likely the most funded and researched group of lethal diseases. Depending on the tumor localization, size, and metastatic stage, treatment options in specialized clinics may include surgery, chemotherapy, radiation therapy, hormone therapy, bone marrow transplantation, targeted therapy, and immunotherapy [1]. For smaller tumors, at early stages, thermal ablation offers a low-risk and minimally invasive solution [2,3]. Nevertheless, despite all the academic and industry efforts, we still do not have an efficient answer to cancer, suggesting the need for a novel approach.

According to the American Cancer Society, 52,360 men and 29,440 women are expected to be diagnosed with kidney and pelvis cancer in 2023, out of whom 9920 men and 4970 women may die from this disease [4]. The prevalence of kidney cancer is strongly dependent on age (most diagnosed people are over 65 years old), sex (twice more frequent in men than in women), and race (African Americans, American Indians, and Alaska Natives are affected in higher percentages than other races). When the cancer is localized only in the kidney, the 5-year survival rate is good (93%), however, it declines rapidly (15%)

when the cancer spreads to the lungs, brain, or bones [4]. The vast majority of kidney cancers are clear cell subtypes of Renal Cell Carcinoma (ccRCC), characterized by high inter- and intra-tumor heterogeneity and strong crosstalk with the cellular microenvironment [5].

A very dynamic and promising avenue is provided by gene therapy as an alternative to kidney transplantation [6]. As of 5 October 2023, PubMed lists 143,107 articles for "cancer gene therapy" published from 1966 onward, of which 12,818 were published in 2021 alone. The majority of these articles looked for gene biomarkers whose altered sequence and expression level were supposedly responsible for triggering cancerization and whose restoration allegedly provides the cure. In most publications, the biomarkers were identified by comparing sequencing (e.g., [7–9] and/or transcription (e.g., [10–12]) data in tissues collected from cancer-stricken and healthy people.

A potentially effective yet insufficiently exploited tool for both diagnostics and therapy is quantifying and managing the amount of cancer cell-secreted microRNAs in blood and urine. Certain miRNAs have been shown to alter the expression of oncogenic or tumor-suppressive genes, thus regulating the proliferation of cancer cells. Owing to accessibility, the dosing and manipulation of the amounts of selected types of urine miRNAs was proposed as an excellent non-invasive instrument for cancer detection and management [13].

The potency of gene therapy was also tested on standard human cancer cell cultures (e.g., [14–17]), but the relevance of the experimental results from the cell culture to the cancer reality is disputable. Nonetheless, as recently summarized [18], the non-malignant cells and molecular factors from the tumor microenvironment play "crucial roles" in the development of ccRCC. Thus, when taken from their natural environment and plated in a homo-cellular culture, the cancer cells will adapt their gene expression profiles to the new conditions. Therefore, interpreting the results from homo-cellular culture as valid for the hetero-cellular tissue is disputable. We have proven that the transcriptome of one cell type changes significantly in the proximity of another cell type by profiling mouse cortical astrocytes and immortalized precursor oligodendrocytes when plated separately or co-cultured in insert systems [19].

However, what are the real predictive values of the gene biomarkers for cancer diagnosis and therapy? The 38.0 release (31 August 2023) of the NIH-National Cancer Institute GDC Data Portal [20] containing genomic data collected from 88,991 cancer cases in 68 primary sites, reported a total of 2,903,037 mutations located in 22,588 genes. Importantly, the Portal reported mutations in almost all genes affecting each of the 68 primary sites. Table 1 summarizes the GDC data for 14 primary sites by presenting the number of mutated genes found in the investigated cases, how many of the mutated genes are protein coding, and the total number of mutations detected thus far for each site.

Table 1. Numbers of mutated genes in 14 primary sites (data from [20]). Note that the number of mutated genes in the listed individual sites represents from 86.94% (prostate cancer) to 95.07% (bone marrow cancer) of the 22,588 mutated genes reported in all 88,991 cancer cases located in all 68 primary sites.

Primary Site	# of Cases	# of Genes	Protein Coding	# of Mutations	Primary Site	# of Cases	# of Genes	Protein Coding	# of Mutations
Bladder	1725	20,183	19,692	114,662	Lung	12,262	21,318	19,790	443,974
Bone marrow	11,027	21,474	19,705	163,756	Ovary	3381	20,266	19,673	64,142
Brain	1452	20,343	19,729	93,128	Pancreas	2776	19,874	19,502	36,676
Breast	9121	20,454	19,727	113,777	Prostate	2387	19,638	19,402	27,468
Colorectal	8140	21,060	19,794	337,634	Skin	2893	20,739	19,770	353,213
Head & neck	2792	20,535	19,712	116,274	Stomach	1631	20,336	19,739	182,493
Kidney	3501	20,129	19,631	65,471	Uterus	2803	21,471	19,781	769,622

Moreover, almost every single gene was found to be mutated in at least one case from each of the 68 primary sites. For instance, with respect to the reported cancer cases from

Table 1, the titin (*TTN*) appeared to be mutated in the following percentages of reported cases: 12.41 of bladder cases, 2.75 of bone marrow, 14.47 of brain, 2.99 of breast, 5.05 of colorectal, 10.28 of head and neck, 5.71 of kidneys, 6.88 of lungs, 2.17 of pancreas, 2.60 of prostate, 13.46 of skin, 15.51 of stomach, and 12.21 of uterus cases. The percentages include all 13,073 distinct somatic mutations observed for this gene in 4512 out of the 88,991 cases included in the portal database. Thus, not only do none of the distinct mutations, but all kinds of altered sequences of *TTN* as a whole do not exhibit statistically significant sensitivity and/or specificity for a particular form of cancer. The same lack of significance is carried by all other "regular suspects", like tumor protein p53 (*TP53*), with 1341 mutations identified in 4934 cases across 47 out of 82 projects, or *KRAS* (1500 cases, 125 distinct mutations identified in 43 projects), or *PTEN* (1228 cases, 846 mutations across 38 projects). The most frequently mutated gene in kidney cancer is the von Hippel-Lindau tumor suppressor (*VHL*) which was detected in 342 (9.77%) out of 3501 cases. However, the *VHL* was also found to be mutated in 50 cases of cancer in other organs.

An excellent recent review indicated that one possible explanation for the unsatisfactory conventional anti-cancer therapy is its targeting of the somatic tumor cells instead of the cancer stem cells (CSC), "assumed to be responsible for tumor recurrence and metastasis" [21]. Therefore, targeting the CSC-signaling pathways might offer a much better alternative than attacking the cancer-specific surface proteins.

Mimicking a human cancer phenotype in genetically engineered animals (e.g., [22–25]) provides disputable etiologies showing that together with the manipulated gene, hundreds of other genes are regulated, as reported in many studies, ours included (e.g., [26]). The set of significantly regulated genes in the tissues of genetically engineered animals, with respect to their wild-type counterparts, depends on the profiled tissue (e.g., [27]), silencing method used (e.g., [28]), and the genetic background (e.g., [29]).

Owing to the unrepeatable combination of favoring factors (some of them changing in time) of race, sex, age, medical history, diet, climate, exposure to toxins, stress, and other external stimuli, each human is a DYNAMIC UNIQUE. This dynamic unicity requires a time-sensitive personalized therapeutic approach.

Some very important factors, as of yet still neglected in many published papers and public repositories, include the tumor's genomic, transcriptomic, and proteomic heterogeneity [30–33]. Thus, histopathologically distinct cancer nodules from the same tumor most frequently have different characteristics. Therefore, the best REFERENCE for cancer-related genomic alterations of an individual is not the tissue of the average healthy person of the same race, sex, and age group, but rather the quasi-normal tissue surrounding his/her cancer nodules [34]. With this reference in mind, the true goal of anti-cancer therapy is to restore what is considered normal for that person, hence the need for a personalized approach.

While the diagnostic value of the gene biomarkers is disputable, let us see whether the restoration of the correct sequence and/or expression level of the biomarkers can provide the therapeutic answer for cancer. Since the biomarkers are selected from the most frequently altered genes in cancer patients, it means that their sequences and/or expression levels are poorly protected by cellular homeostatic mechanisms like the minor players in cell life. Therefore, their restoration might be of little consequence.

It is very surprising (and disappointing) that almost all gene expression studies neglect about 99.99% of the information provided by the high throughput transcriptomic platforms (RNA-sequencing, Agilent microarray, Affimetrix, Illumina BeadChip arrays, etc.), which will be presented in the Results section below. The traditional analysis considers ONLY the expression levels of the quantified genes whose comparison between conditions tells what gene was significantly up-/down-regulated (according to the arbitrarily introduced cut-off for the absolute fold change) or turned on/off. The genes are eventually clustered according to their similar behaviors across conditions (e.g., [35,36]), but similar regulation does not necessarily mean that the clustered genes are interacting with each other (they may have an upstream common regulator or transcription factor).

Using publicly available software (based on text mining the peer-reviewed literature) such as Ingenuity [37], DAVID [38], and KEGG [39], the regulated genes might be organized into functional pathways. However, the topology of the pathways constructed by such software has three major flaws: universality, rigidity, and unicity. They are universal in that they do not discriminate with respect to the strain/race, sex, age, hormonal activity, etc., and even with respect to the tissue, such as those of the Ca^{2+}- and other signaling pathways. They are considered to be rigid for not changing in response to aging, medical treatment, external stimuli, and the progressions of a disease or other dynamic influencing factors. Finally, each constructed pathway has unique wiring for the genes and not a spectrum of several possible gene circuits. If two simple elements like hydrogen and carbon can combine in so many ways to form an unlimited variety of hydrocarbons, how could one assume that tens of much more complicated units (the genes) network in only a single way to accomplish a particular task? Therefore, we have used KEGG-constructed pathways only for illustrative purposes and the coordination analysis to determine the real gene networking.

Because of the above-discussed deficiencies of the biomarker approach, we switched our research from the biomarker to the Genomic Fabric Paradigm (GFP, [40]) approach. The GFP incorporates the traditional analysis of gene expression regulation while considering two additional classes of independent descriptors, and so offers the most theoretically possible comprehensive characterization of the transcriptome and personalized solutions for cancer gene therapy. The two additional transcriptomic descriptors of individual genes, the Relative Expression Variation and the Expression Correlation with each other gene, can be determined using the gene expression profiles of biological replicas without supplementary experimental costs. Through the use of the two additional groups of descriptors, the GFP approach increases by four orders of magnitude the amount of transcriptomic information extracted from a high throughput (ng RNA-sequencing or microarray) gene expression platform.

The present study complements a previously published article [40], with GFP analyses of the remodeling of the five KEGG-constructed excretion system's functional pathways in the kidney and chest wall regions of a 74-year-old man affected by metastatic clear cell renal cell carcinoma (ccRCC).

2. Materials and Methods

2.1. The Best Choice of Tissue Samples

Nevertheless, for statistical significance, a transcriptomic study should profile several biological replicas of the compared conditions. Most authors use three biological replicas, but four is (in our view) the best compromise between getting enough statistical relevance and the errors resulting from the inherent technical noise of the profiling method. In the case of solid tumors, the most reasonable choice is to take a point biopsy from the center of a cancer nodule (or each cancer nodule, if there are more) and another one from the surrounding (almost normal) tissue, split each biopsy into four parts, and profile separately the resulted quarters. Thus, the reference for the patient's cancer and the aim of the therapy is no longer the abstract, racially blind, ageless, and sexless model of the human body but rather his/her own normal tissue for his/her race, age, and sex. This procedure is standard in our lab and was used in investigations of surgically removed tumors from kidney [40], thyroid [41], and prostate [42] cancer patients.

In this study, we re-analyzed transcriptomic data from the surgically removed right kidney affected by ccRCC Fuhrman grade 3 (two primary cancer nodules, denoted as PTA and PTB in the renal medula) and its metastasis in the chest wall (CWM). The gene expression profiles of the three cancer nodules were compared to those of the quasi-normal surrounding kidney tissue (NOR). Data were obtained using Agilent-026652 Whole Human Genome Microarrays 4 × 44K v2 and are publicly accessible [43].

2.2. Data Filtering and Normalization

The hybridized microarray spots with a foreground fluorescence less than twice the background in one biological replica profiled with microarrays are eliminated from the analysis of all samples to be compared owing to the non-negligible technical noise. With this filtering, every profiled sample from all conditions to be compared was reduced to the same number of distinct transcripts, here "N = 13,314". The background-subtracted forward fluorescence of the microarray hybridized spot(s) with transcript "i" from the biological replica "k" (k = 1, 2, 3, 4) of condition "c", "$a_i^{(c;k)}$" were normalized to the expression of the median gene for that profiled sample. This normalization strategy makes comparable the expression profiles of all samples, with $a_i^{(c;k)} > 1$ indicating genes with a higher than median expression level and $a_i^{(c;k)} < 1$ genes with a lower than median expression level. For the analyzed microarray experiment, $a_i^{(c;k)}$ was the sum of the net fluorescence of all spots probing redundantly transcript "i" in the biological replica "k" of condition "c" (see Equation (A1) in Appendix A).

2.3. Independent Characteristics of Gene Expression
2.3.1. Normalized Average Expression Level

The filtered and normalized expression values of each transcript "i" were averaged over the biological replicas of each condition "c" resulting in $AVE_i^{(c)}$. $AVE_i^{(c)}$ is the genomic measure that everybody in the field uses to determine whether that transcript abundance was up-/down-regulated or turned on/off when comparing cancer with healthy samples. Thus, the GFP includes but is not reduced to the traditional gene expression analysis.

2.3.2. Relative Expression Variability

When properly selected (as quarters of point biopsies), the biological replicas may be considered as different instances of the same system subjected to distinct local (not significantly regulating) conditions. This applies to ccRCC samples owing to the strong crosstalk of cancer cells within the non-uniform microenvironment [5,18]. Thus, we can add as an independent feature of the transcript "i" in condition "c" the Relative Expression Variability, "$REV_i^{(c)}$", computed as the mid-chi-square (χ^2) interval estimate of the coefficient of variation for "n = 4" biological replicas and "v_i" spots probing redundantly the transcript "i" (Equation (A2) in Appendix A). The REV provides an indirect estimate of the strength of the cellular homeostatic mechanisms to control the transcript abundances, with the smallest REV indicating the most stably expressed (i.e., the most controlled) gene and the largest REV pointing to the most variably expressed (i.e., the least controlled) gene. Since more control means more energy spent by the cell, it is natural to assume that the right expressions of the most controlled genes were more important for the cell's survival and/or proliferation in the multicellular tissue. As such, the REV analysis tells the investigator firsthand about the cell priorities.

2.3.3. Expression Coordination

Genes are not single but team players in cell life. Considering the high efficiency of the cellular phenomena, we have introduced the Postulate of the Transcriptomic Stoichiometry (PTS) [44] as an extension to gene networking in functional pathways of Proust's Law of Definite Proportions from chemistry [45]. The PTS states that: *in any steady-state condition, expressions of genes whose encoded products are part of a functional pathway are coordinated to ensure the maximum efficiency of that functional pathway.* This means that the involved genes are set to produce the transcripts at the right abundance proportions. The PTS and the coordination analysis can be used to determine the real gene network responsible for a particular pathway in a given condition.

The most difficult question is how the genes are networked: in pairs (e.g., agonist-antagonist), in "ménage à trois" (e.g., agonist, antagonist, and a modulator of both), or in more complex gene inter-coordination clusters? To answer this question, we adopted a formalism of correlation functions similar to that used to describe the structure of simple liquids (like liquid argon) [46]. Thus, the configuration function "F" of "N" distinct genes is considered as a superposition of virtual configurations in which the genes are: independently expressed ("f_1"), coordinately expressed in pairs ("f_2"), coordinately expressed in triplets ("f_3"), and so on until all "N" genes are coordinately expressed in a cluster of all genes ("f_N"). As shown in Appendix A, the contributions of the distribution functions of higher than pair-correlated genes can be neglected, so that the configuration function can be approximated with a distribution of independently expressed genes and a distribution of coordinately expressed gene pairs (Equation (A3) in Appendix A).

The problem is now to select the suitable algorithm for gene pairing. There are several weighted and unweighted types of correlation algorithms (e.g., [47–50]) aiming to identify the interconnected genes based on their co-regulation determined from the meta-analysis of genomic data on healthy and cancer-affected populations. By contrast, our aim is to determine the gene network in a particular condition (normal or cancerous) of only one individual, so none of these cluster analyses are suitable for our endeavor.

The simplest gene pairing is completed using the Pearson pair-wise correlation coefficient (hereafter denoted as "COR" instead of the traditional "r") between the $\log_2$ of the normalized expressions of two genes ("i" and "j") in the four replicas of the same condition "c" (shown in Equation (A4) in Appendix A). Although Marbach et al. [51] have shown that Pearson's correlation coefficient is not the strongest way to determine gene networks, it is accurate enough when taking into account the technical noise of the gene expression platform. With four biological replicas, two genes are ($p < 0.05$) significantly synergistically expressed (i.e., a positive correlation) if $COR \geq 0.951$, antagonistically expressed (negative correlation) if $COR \leq -0.95$, and independently expressed (null correlation) if $|COR| \leq 0.05$. For microarrays probing the same transcript with two spots (i.e., 8 paired values), the $p < 0.05$ significance cut-off for synergism/antagonism is $|COR| \geq 0.71$, for three spots (12 paired values) it is $|COR| \geq 0.58$, and so on, with the Pearson cut-off decreasing when the number of probing spots increases [52].

In [53], we defined the coordination score "$COORD$" with p-value "p" of a pathway "Γ" in condition "$c = NOR, PTA, PTB, CWM$" as:

$$COORD_\Gamma^{(c)}(p) \equiv SYN_\Gamma^{(c)} + ANT_\Gamma^{(c)} - IND_\Gamma^{(c)} \tag{1}$$

where "$SYN/ANT/IND$" are the percentages of all gene pairs from the pathway "Γ" that are synergistically/antagonistically/independently expressed with a statistical significance (p-value) "p" in condition "c".

The $COORD$ score indicates the ($p > 0.05$) statistically significant influence of that gene on all other genes.

2.3.4. Topology of the Transcriptome and the Gene Master Regulator

The "REV" and "COR" can be used to determine the Gene Commanding Height (GCH) that establishes the importance hierarchy of the genes in each region. The top gene (highest GCH) is termed the Gene Master Regulator (GMR) of that region. The GMR is the highly protected gene (i.e., low REV, meaning it is critical for cell survival) that also has the strongest influence on the expression of other genes through expression coordination [54,55] (Equation (A5) in Appendix A). In all our cancer genomics studies to date ([34,40–42,54,55]), we found that cancer and normal cells from the same tumor are controlled by different GMRs. Moreover, the GMR gene of the cancer cells has low GCH in the normal cells. Therefore, silencing the GMR of the cancer cells is expected to selectively kill the cancer cells from the tissue with very little influence on the normal cells.

<u>Note:</u> For the GFP users, the transcriptome is no longer a chaotic collection of transcripts, but a multi-dimensional hierarchized mathematical entity subjected to dynamic sets of expression variability and expression correlations of its components.

2.4. Transcriptome Alteration in Cancer

2.4.1. Measures of Expression Regulation

We use the GFP to compare the transcriptome of a cancer nodule with that of the surrounding normal tissue in a tumor. Thus, in addition to identifying what genes were significantly up-/down regulated in cancer, we determine also how much of the homeostatic control of the transcript abundance was altered for each gene or group of genes, and how the gene networks were remodeled.

The traditional analysis considers a gene as significantly regulated in cancerous tissue compared to normal tissue if the expression ratio "x" (negative for down-regulation) has an absolute fold-change $|x|$ larger than an arbitrarily introduced cut-off (most frequently 1.5). In most cases, it also adds the condition that the p-value of the heteroscedastic t-test of mean expressions is less than 0.05. However, the cut-off fold-change could be too stringent for very stably expressed genes (leading to false negatives) or too lax for very unstably expressed ones (introducing false positives). Therefore, we use for each gene the cut-off of the absolute fold-change, "CUT", considering the combined contributions of its biological variability and the technical noises of the platform in the two profiled conditions. Thus, the gene "i" is considered significantly regulated in "cancer" with respect to the normal tissue (NOR), if the absolute fold-change exceeds the respective "CUT" and the p-value is less than 0.05 (Equation (A6) in Appendix A).

In many studies, transcriptome alteration is presented as percentages of the significantly up- and down-regulated genes. Nonetheless, the percentage of presentation implicitly assumes that all regulated genes are uniform (+1 or −1) contributors to the overall transcriptome alteration while neglecting the contributions of the not significantly regulated genes. A more accurate measure of the expression regulation is the Weighted Individual (gene) Regulation "WIR", whose absolute values can be averaged for the genes included in a functional pathway "T" as the Weighted Pathway Regulation "WPR". In addition to being applied to all genes (including those not significantly regulated), WIR takes also into account the absolute expression change and the statistical significance of the regulation (as shown in Equations (A7) and (A8) in Appendix A).

2.4.2. Regulation of the Control of Transcript Abundance

In a previous paper [56], we defined the Relative Expression Control, "REC," of gene "i" in condition (here region) "c" so that positive RECs indicate more than the median-controlled genes and negative values indicate less than the median-controlled genes in that condition (shown as Equation (A9) in Appendix A). As shown below (Figure 1), the ccRCC altered the REVs of individual genes and as a consequence, their hierarchy (illustrated in Figure 2 below). The difference between the REV inverses in a cancer nodule and those in the surrounding normal tissue (Equation (A10) in Appendix A) indicates how much the ccRCC altered the transcript abundance control.

2.4.3. Regulation of Expression Coordination

The regulation of expression coordination can be computed with regard to a single gene, a group of genes (like those involved in a particular pathway), or all quantified genes. In this report, the regulation of the expression coordination is limited to the five KEGG-constructed excretory system pathways (shown in Equation (A11) in Appendix A). Positive regulation values indicate an overall increase in the expression coordination while negative values indicate an overall decrease in the expression coordination of gene "i" with expressions of all genes from the reference pathway "T".

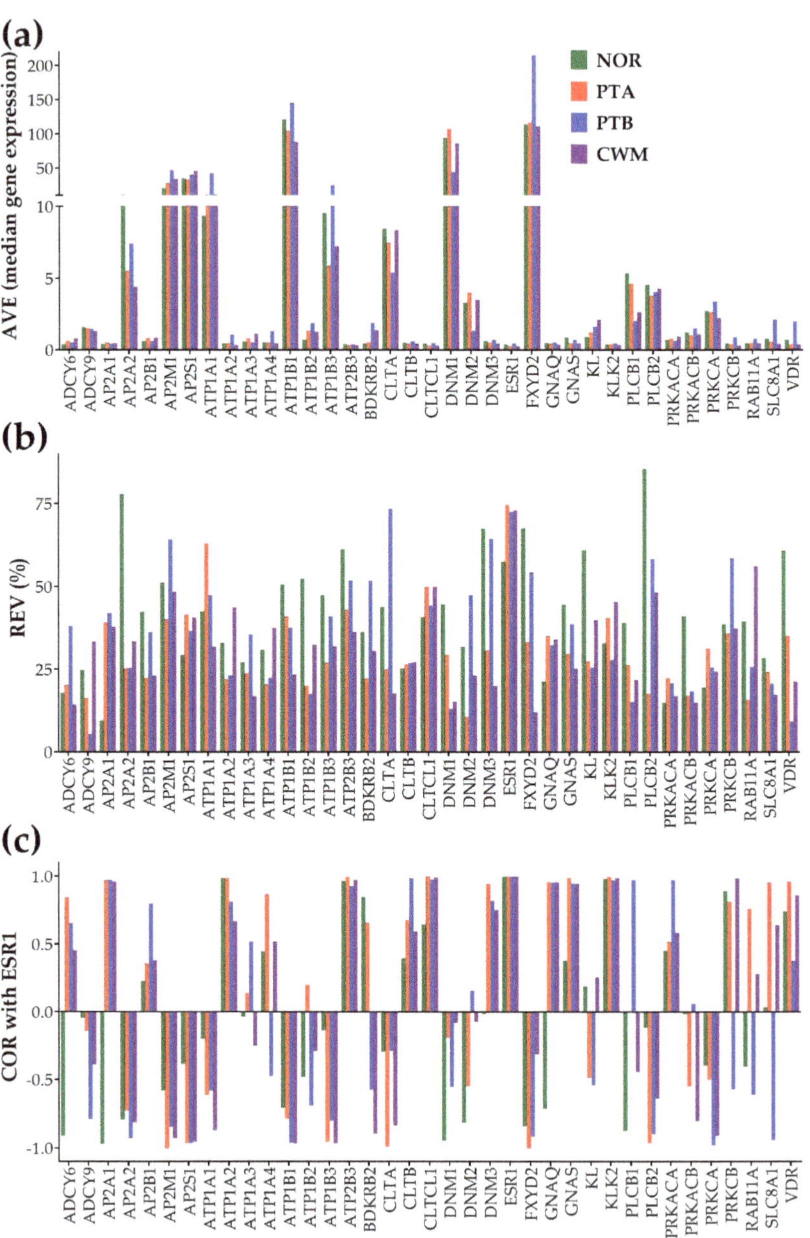

Figure 1. Independence of: (**a**) AVE, (**b**) REV, and (**c**) COR with *ESR1* characteristics for 37 genes involved in the KEGG-constructed pathway of the "Endocrine and other factor-regulated calcium absorption". The $COR_{ESR1,ESR1} = 1$ values in all conditions validate the coordination analysis. Observe the differences in all three gene characteristics between the two equally graded and located close to each other nodules of PTA and PTB.

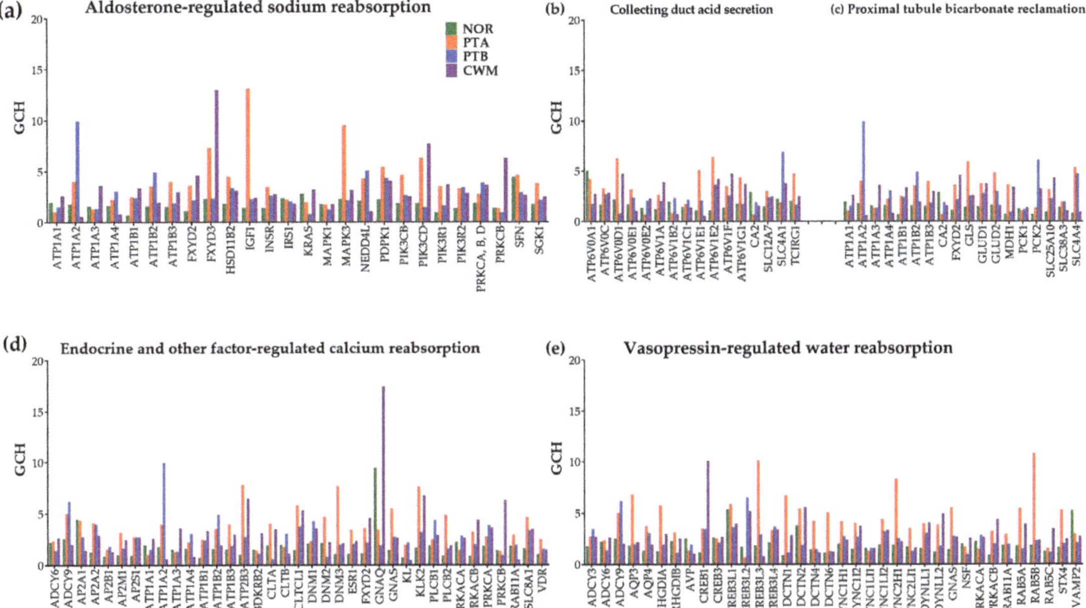

Figure 2. Gene Commanding Height (GCH) scores of the genes involved in the five KEGG-constructed excretory pathways: (**a**) Aldosterone-regulated sodium reabsorption, (**b**) Collecting duct acid secretion, (**c**) Proximal tubule bicarbonate reclamation, (**d**) Endocrine and other factor-regulated calcium reabsorption, and (**e**) Vasopressin-regulated water reabsorption.

2.4.4. The Transcriptomic Distance

Nonetheless, the most comprehensive measure of the transcriptome alteration should include all changes: in expression level, expression control, and expression coordination with all other genes. The Transcriptomic Distance of an Individual gene, "*TDI*", is the Euclidian distance from the origin of the 3D space whose orthogonal axes are: WIR, the regulation of the expression control, and the regulation of expression coordination (shown in Equation (A12) in Appendix A).

2.5. Functional Pathways

In this report, we used our experimental data from the profiled samples of a man with metastatic ccRCC [43] to determine the topology and the ccRCC-induced remodeling of the transcriptomes associated with the five KEGG-constructed functional pathways of the excretory system. The analyzed pathways were: (ALDO) hsa04960 Aldosterone-regulated sodium reabsorption [57], (COLL) hsa04966 Collecting duct acid secretion [58], (ENDO) hsa04961 Endocrine and other factor-regulated calcium reabsorption [59], (PROX) hsa04964 Proximal tubule bicarbonate reclamation [60], and (VASO) hsa04962 Vasopressin-regulated water reabsorption [61].

3. Results

3.1. The Global Picture

Expressions of 13,314 unigenes were adequately quantified, normalized, and organized in functional pathways in all 16 samples. However, not all excretory genes identified by the KEGG were analyzed. The missing genes were the ones that: (i) were not expressed in one of the four profiled regions, (ii) had no microarray spot to be probed by, or (iii) were probed by spots excluded from the analysis because of corrupted pixels.

Nonetheless, the functional pathways of the excretory system were not mutually exclusive, some genes being counted in two (e.g., *ADCY9* in ENDO and VASO) or even three excretory pathways (e.g., *ATP1A1* in ALDO, ENDO, and PROX). Therefore, the total number of the quantified distinct excretory genes was 108.

3.2. Independent Characteristics of Gene Expression

Figure 1 illustrates the independence of the three types of expression characteristics for 37 genes involved in the KEGG-constructed pathway of the "Endocrine and other factor—regulated calcium reabsorption" [59] using our microarray data on Fuhrman grade 3 metastatic ccRCC samples [38]. For the COR analysis, we chose to represent the expression correlation with the estrogen receptor 1 (*ESR1*), owing to the kidney being one of "the most estrogen-responsive, not reproductive organs in the body" [62].

Of note are the differences between the normal tissue and the cancer nodules not only in the average expression levels of certain genes (AVE, as expected and reported by the traditional analysis), but also in the relative expression variability (REV) and correlation (COR) with other genes (here with *ESR1*). For instance, the significant expression antagonism ($COR^{(NOR)} = -0.96624$, $p = 0.0338$) of *AP2A1* (adaptor-related protein complex 2, alpha 1 subunit) with *ESR1* in the normal tissue is switched into expression synergism in each of the three cancer nodules. We obtained the following correlation values (and statistical significances) between *ESR1* and *AP2A1* in the cancer samples: $COR^{(PTA)} = 0.9607$ (p-val = 0.03093), $COR^{(PTB)} = 0.970449$ (p-val = 0.02955), and $COR^{(CWM)} = 0.958562$ (p-val = 0.04144). This result means that in the normal tissue, when the expression of *ESR1* goes up, that of *AP2A1* goes down, and when *ESR1* goes down, *AP2A1* goes up. By contrast, in cancer, the expression of *ESR1* oscillates in-phase with the expression of *AP2A1*, so that expressions of both genes go up or down simultaneously. It is interesting to also note the differences in all three characteristics of the individual genes between the nodules of PTA and PTB.

The independence of the three characteristics of the individual genes collected from a regular gene expression experiment with four biological replicas of each histopathologically distinct region is visually evident. Therefore, each of the three types of analyses brings non-redundant information and is worth taking into account.

3.3. ccRCC Changed the Gene Hierarchy

Figure 2 presents the GCH scores for all quantified genes selected by the KEGG as involved in the five excretory functional pathways. Of note are again the substantial inter-regional differences in the genes' GCH scores, indicating distinct gene hierarchies within the corresponding pathways. For instance, the GCH of the *IGF1* (insulin-like growth factor 1 (somatomedin C)) from the "Aldosterone-regulated sodium reabsorption" pathway increased from 1.41 in NOR to 13.11 ($9.28\times$) in PTA, although it remains practically unchanged in PTB (2.25) and in CWM (2.38). The *CREB3L2* (cAMP responsive element binding protein 3-like 2) from the "Vasopressin-regulated water reabsorption" pathway exhibited a GCH increase of $9.57\times$ in PTB compared to PTA. The GCH of the *ATP1A2* (ATPase, Na$^+$/K$^+$ transporting, alpha 2 polypeptide) decreased by $16.85\times$ in CWM with respect to PTB.

3.4. Measures of Individual Gene Regulation

Figure 3 illustrates the six types of quantifying measures of the ccRCC-induced alterations of the genes from the KEGG-constructed pathway hsa04961 of "Endocrine and other factor-regulated calcium reabsorption". The six measure types are: the uniform $+1/-1$ for significant regulation, the expression ratio (negative for the down-regulation), the WIR (Weighted Individual (gene) Regulation, negative for down-regulation), the regulation of the transcription control, the regulation of the expression correlation with all other genes from the pathway, and the TDI (Transcriptomic Distance of Individual gene).

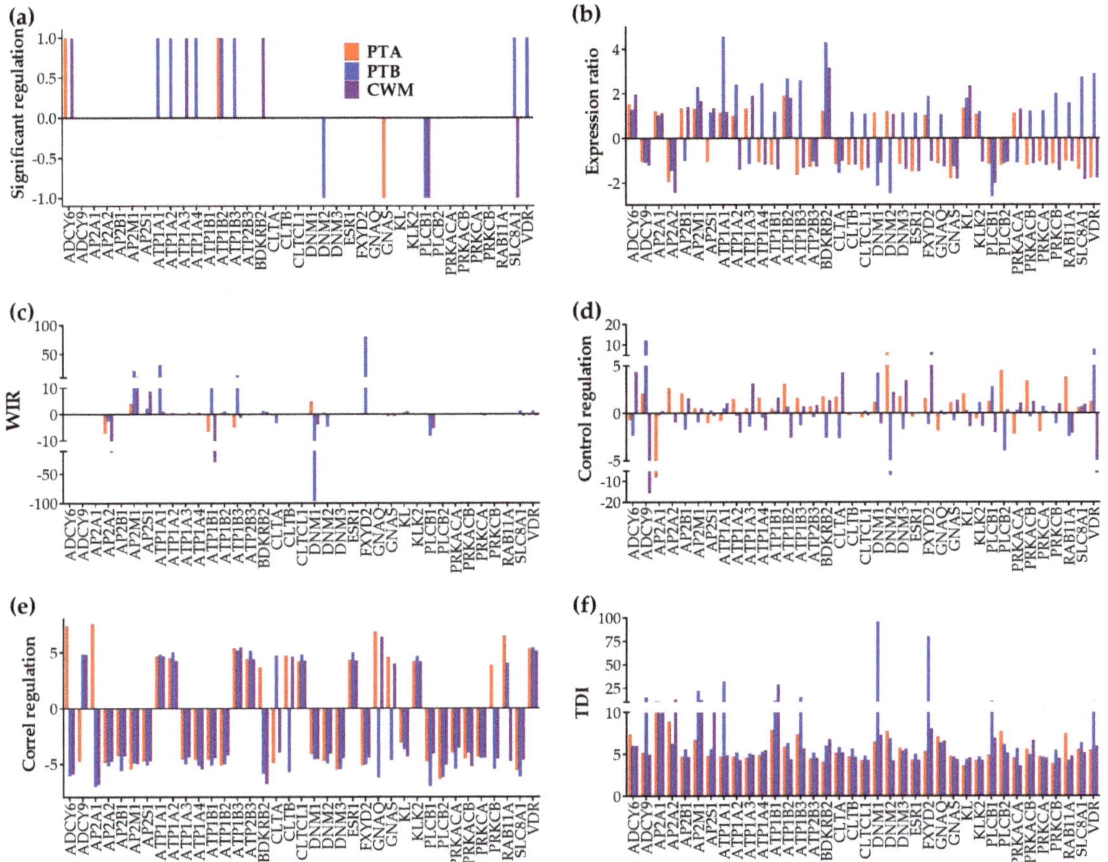

Figure 3. Six measures of individual gene regulation in the KEGG-constructed pathway of ENDO (Endocrine and other factor-regulated calcium reabsorption). (**a**) Uniform (+1/−1 for significant up-/down regulation). (**b**) Expression ratio × (negative for down-regulation). (**c**) Weighted individual (gene) regulation (WIR, negative for down-regulation). (**d**) Regulation of transcript abundance control mechanisms (negative for decreased control). (**e**) Regulation of expression coordination (with respect to every other gene of the pathway, negative for reduced correlation); (**f**) Transcriptomic distance of individual (gene) (here with respect to all its partners within the pathway). Observe that all but the uniform measure takes into account the contributions of every single gene.

There are notable differences among the three cancer nodules in all six measures. For instance, *SLC8A1* (solute carrier family 8 (sodium/calcium exchanger), member 1), which was not regulated in PTA, was significantly up-regulated in PTB (x = 2.75) and significantly down-regulated in CWM (x = −1.84). Only one gene, *ATP1B2* (ATPase, Na⁺/K⁺ transporting, beta 2 polypeptide), was significantly upregulated in both PTA (x = 1.91) and PTB (x = 2.67), but not in CWM. *PLCB1* (phospholipase C, beta 1 (phosphoinositide-specific) was the single gene significantly down-regulated in PTB (x = −2.60) and CWM (x = −2.01), but not in PTA.

As defined, the WIR takes larger absolute values for highly expressed genes in the reference (here NOR) region, sometimes even larger for not significantly regulated genes than for significantly regulated ones. For instance, *DNM2* (dynamin2), which is a significantly down-regulated gene in PTB (x$^{(PTB→NOR)}$ = −2.47, CUT$^{(PTB→NOR)}$ = 1.81,

$p^{(\text{PTB}\to\text{NOR})} = 0.0242$), had a $\text{WIR}^{(\text{PTB}\to\text{NOR})} = 4.73$. The *DNM2* contribution to the transcriptomic alteration in PTB was substantially overpassed by that of the not statistically significantly regulated *DNM1* (dynamin1: $\text{WIR}^{(\text{PTB}\to\text{NOR})} = 95.74$, $x^{(\text{PTB}\to\text{NOR})} = -2.12$, $\text{CUT}^{(\text{PTB}\to\text{NOR})} = 1.66$, $p^{(\text{PTB}\to\text{NOR})} = 0.0969 > 0.05$). The reason for this is that in NOR, $AVE_{DNN1} = 94.30 >> 3.30 = AVE_{DNN2}$. Nonetheless, by considering the whole change in the expression level, the WIR tells the investigator much more about the contribution of a gene to the transcriptome expression alteration than the uniform $+1/-1$ up-/down regulation.

Overall, the median expression control 100/<REV> increased from 2.58 in NOR to 4.27 in PTA, 2.92 in PTB, and 3.96 in CWM. In the illustrated pathway, we found genes with a substantial increase in expression control and genes with a substantial decrease in control in cancer. The most substantial increase of the expression control was for *ADCY9* (adenylate cyclase 9) in PTB ($\Delta REC^{(PTB\to NOR)} = 12.38$). It is remarkable that the control of *ADCY9* had a modest change in PTA ($\Delta REC^{(PTA\to NOR)} = 2.08$) but the largest decrease of all in CWM ($\Delta REC^{(CWM\to NOR)} = -15.50$). The largest decrease in PTA was exhibited by *AP2A1* (adaptor-related protein complex 2, alpha 1 subunit) with $\Delta REC^{(PTA\to NOR)} = -7.97$, but with insignificant changes in the other nodules with $\Delta REC^{(PTB\to NOR)} = -0.17$, and $\Delta REC^{(CWM\to NOR)} = 0.26$. Interestingly, there are genes (e.g., *DNM2*) whose control increased with respect to NOR in one cancer nodule (PTA, $\Delta REC^{(PTA\to NOR)} = 6.2$) but decreased in the equally ranked other nodule from the same tumor ($\Delta REC^{(PTB\to NOR)} = -7.2$), indicating a shift in the cell's priorities. A substantial shift in cell priorities occurred also for *VDR* (vitamin D (1,25-dihydroxyvitamin D3) receptor) between the nodules of PTB ($\Delta REC^{(PTB\to NOR)} = 7.92$) and CWM ($\Delta REC^{(CWM\to NOR)} = -6.07$).

We found genes, such as *ADCY6*, with an increased overall correlation with the other pathways' genes with respect to NOR in one nodule ($\Delta COR^{(PTA\to NOR)} = 7.39$) but decreased in the other nodules ($\Delta COR^{(PTB\to NOR)} = -6.04$, $\Delta COR^{(PTB\to NOR)} = -5.92$), indicating a profound remodeling of the gene networking. Interestingly, *ADCY6* was significantly up-regulated in PTA ($x = 1.54$) and CWM ($x = 1.96$), but not in PTB ($x = 1.29$). A very similar behavior, except that it was not significantly regulated in any of the three cancer nodules, was exhibited by *AP2A1* ($\Delta COR^{(PTA\to NOR)} = 7.60$, $\Delta COR^{(PTB\to NOR)} = -7.02$, $\Delta COR^{(PTB\to NOR)} = -6.84$).

Nevertheless, the most comprehensive measure that incorporates the changes in all three types of characteristics is the transcriptomic distance of an individual gene (TDI) from its $AVE^{(NOR)}$, $REV^{(NOR)}$, and $COR^{(NOR)}$ (with all other genes within the pathway) in the normal tissue. From the TDI perspective, *DNM1* ($\text{TDI}^{(PTB\to NOR)} = 95.95$ in PTB) followed by *FXYD2* ($\text{TDI}^{(PTB\to NOR)} = 80.21$ in PTB) were the most altered genes within this set. Interestingly, with all differences at the individual gene level, the median TDIs of the three nodules were close to each other (5.72 for PTA, 5.73 for PTB, and 5.37 for CWM).

3.5. Overall Regulation of the Excretory Pathways

Table 2 presents the overall gene expression alterations of the five KEGG-constructed excretory pathways as percentages of up- and down-regulation out of the quantified genes in each of the cancer nodules. Table 2 presents also the WPRs of the analyzed pathways.

With 23.08% in PTA and CWM, and 30.77% in PTB total percentage of up- and down-regulated genes, the ALDO appears as the most altered pathway. However, from the more comprehensive WPR perspective, the COLL is the most altered of the five pathways in all three cancer nodules.

Interestingly, all significantly regulated genes are up for PROX in all three cancer nodules, indicating a major activation of this pathway in ccRCC. The results on the ALDO are intriguing: while there were equal numbers of up- and down-regulated genes (11.54% of 26) in both PTA and CWM, in PTB all 30.77% significantly regulated genes were over-expressed. This means that the ALDO was balanced in PTA and CWM, but strongly activated in PTB. The ALDO regulomes (sets of significantly regulated genes in this pathway) of the three cancer nodules are different, with only one gene, *PIK3R2* (phosphoinositide-3-kinase, regulatory subunit 2 (beta)), being significantly up-regulated in all three nodules.

Table 2. Number of quantified out of number of included genes in the five KEGG-constructed functional pathways of the excretory system, their percentages, and overall weighted regulations in each cancer nodule. ALDO = Aldosterone-regulated sodium reabsorption, COLL = Collecting duct acid secretion, ENDO = Endocrine and other factor-regulated calcium reabsorption, PROX = Proximal tubule bicarbonate reclamation, and VASO = Vasopressin-regulated water reabsorption.

		PTA			PTB			CWM		
Path	Genes	%Up	%Down	WPR	%Up	%Down	WPR	%Up	%Down	WPR
ALDO	26/37	11.54	11.54	1.12	30.77	0.00	8.19	11.54	11.54	2.32
COLL	16/27	6.25	12.50	5.20	12.50	0.00	16.36	12.50	0.00	9.69
ENDO	37/53	5.41	2.70	0.88	18.92	5.41	7.68	8.11	5.41	2.15
PROX	18/23	16.67	0.00	0.96	38.89	0.00	9.04	11.11	0.00	2.12
VASO	36/44	8.33	11.11	0.69	11.11	5.56	0.93	2.78	8.33	0.77

Results from Table 2 show that even when closely located and with equal pathology grades, cancer nodules from the same tumor (PTA and PTB) may exhibit different gene alterations, questioning the validity of meta-analyses comparing ccRCC patients with healthy counterparts.

3.6. False Hits

Figure 4 presents the excretory genes that would have been considered as significantly regulated in the traditional analysis ($|x^{(\text{cancer}\to\text{NOR})}| > 1.5$) but were identified by our cut-off criterion as false positive hits ($|x^{(\text{cancer}\to\text{NOR})}| \leq \text{CUT}^{(\text{cancer}\to\text{NOR})}$). In contrast, the *CREB3L4* (cAMP responsive element binding protein 3-like 4) was identified as a false negative hit ($|x^{(\text{PTA}\to\text{NOR})}| < 1.50$) in PTA because $|x^{(\text{PTA}\to\text{NOR})}| = 1.40 > \text{CUT}^{(\text{PTA}\to\text{NOR})} = 1.38$). In consequence, the false positive hits were eliminated, and the false negative hits were included in the "excretory regulomes" of the three cancer nodules.

GeneName	PTA	CUT	PTB	CUT	CWM	CUT
AP2A2	-1.94	2.16				
AP2M1					1.67	2.00
AQP4					-1.79	2.26
ATP1B2					1.80	1.87
ATP1B3	-1.63	1.77				
ATP6V1E1	-1.58	1.92				
CLTA			-1.56	2.21		
CREB3L4	-1.40	1.38				
DYNC2LI1					1.60	2.69
GLS	-1.54	2.07				
GLUD2	-1.62	1.66	-1.57	1.65	-1.59	1.71
IGF1	-1.59	1.72				
INSR					-1.86	2.03
KL			1.80	1.93		
MDH1			1.51	2.05		
NEDD4L	-1.51	1.60				
PCK2	-1.69	2.19			-1.76	2.27
PIK3CB			1.55	1.99		
RAB11A			1.59	1.67		
SGK1			1.61	1.93		
SLC38A3			1.53	1.75		
SLC4A4					1.73	1.89
STX4					-1.55	1.80
TCIRG1					1.64	2.00
VDR	-1.77	1.99			-1.78	1.91

Figure 4. Excretory genes identified by our absolute fold-change criterion ($|x^{(\text{cancer}\to\text{NOR})}| > \text{CUT}^{(\text{cancer}\to\text{NOR})}$) as false positive hits (red accent 2 lighter 60% background) and false negative hits (light blue background) in the traditional analysis of gene expression. PTA/PTB/CWM = expression ratio (negative for down-regulation) in the indicated cancer nodule with respect to the normal tissue (NOR). CUT = absolute fold-change cut-off to consider a gene as significantly regulated.

3.7. Location of the Regulated Genes in the Excretory System's Functional Pathways

See Figure 5 for the hsa04962 (VASO) "Vasopressin-regulated water reabsorption" and Figures S1–S4 from Supplementary Material for the other four KEGG-constructed excretory system pathways present for every profiled cancer nodule and the localizations of the regulated genes. Of note are the inter-nodule differences in the subsets of the regulated genes.

We found that although the VASO gene of *AVP* (arginine vasopressin) was not significantly regulated in any of the three cancer nodules, expressions of several other genes were significantly altered (though not in the same way) in all profiled regions. For instance, the *AQP3* (aquaporin 3 (Gill blood group)) was found as down-regulated in PTA ($x^{(\mathrm{PTA}\to\mathrm{NOR})} = -2.808$) and CWM ($x^{(\mathrm{CWM}\to\mathrm{NOR})} = -5.846$) but up-regulated in PTB ($x^{(\mathrm{PTB}\to\mathrm{NOR})} = 2.034$).

The opposite regulations of the *AQP3* in the two closely located and equally pathologically graded nodules of PTA and PTB is another argument to consider the "transcriptomic signature" unreliable for ccRCC [63]. The GDC data Portal of the National Cancer Institute reports four cases of kidney cancer (three females and one male) where *AQP3* was found to be mutated.

Unfortunately, the important *AQP2* (aquaporin 2) and *AVPR2* (arginine vasopressin receptor 2) were not quantified in this experiment.

Only two excretory genes were similarly regulated in all three cancer nodules. The VASO gene *CREB3L4* (cAMP responsive element binding protein 3-like 4) was down-regulated: $x^{(\mathrm{PTA}\to\mathrm{NOR})} = -1.40$ (CUT $^{(\mathrm{PTA}\to\mathrm{NOR})} = 1.38$, $p^{(\mathrm{PTA}\to\mathrm{NOR})} = 0.040$); $x^{(\mathrm{PTB}\to\mathrm{NOR})} = -1.74$ (CUT$^{(\mathrm{PTB}\to\mathrm{NOR})} = 1.63$, $p^{(\mathrm{PTB}\to\mathrm{NOR})} = 0.031$), $x^{(\mathrm{CWM}\to\mathrm{NOR})} = -1.95$ (CUT$^{(\mathrm{CWM}\to\mathrm{NOR})} = 1.48$, $p^{(\mathrm{CWM}\to\mathrm{NOR})} = 0.003$). The down-regulation of the *CREB3L4* gene in PTA would have been considered not significant in the traditional analysis requiring $|x^{(\mathrm{PTA}\to\mathrm{NOR})}| > 1.50$, but was identified as significant by our algorithm that requires the absolute fold-change to exceed the cut-off value computed for that gene in the compared samples.

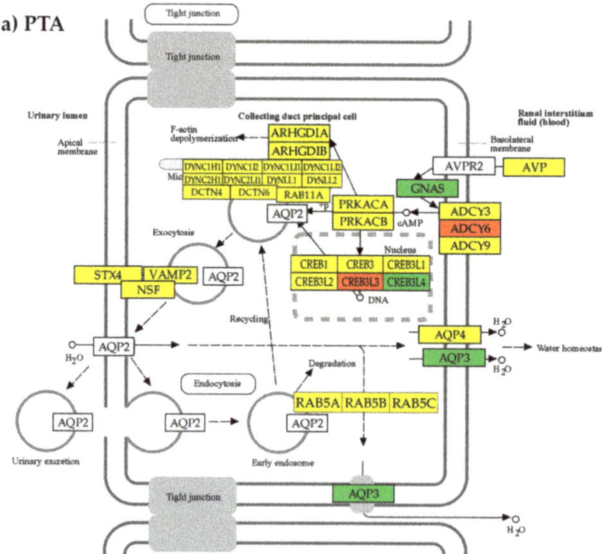

Figure 5. *Cont.*

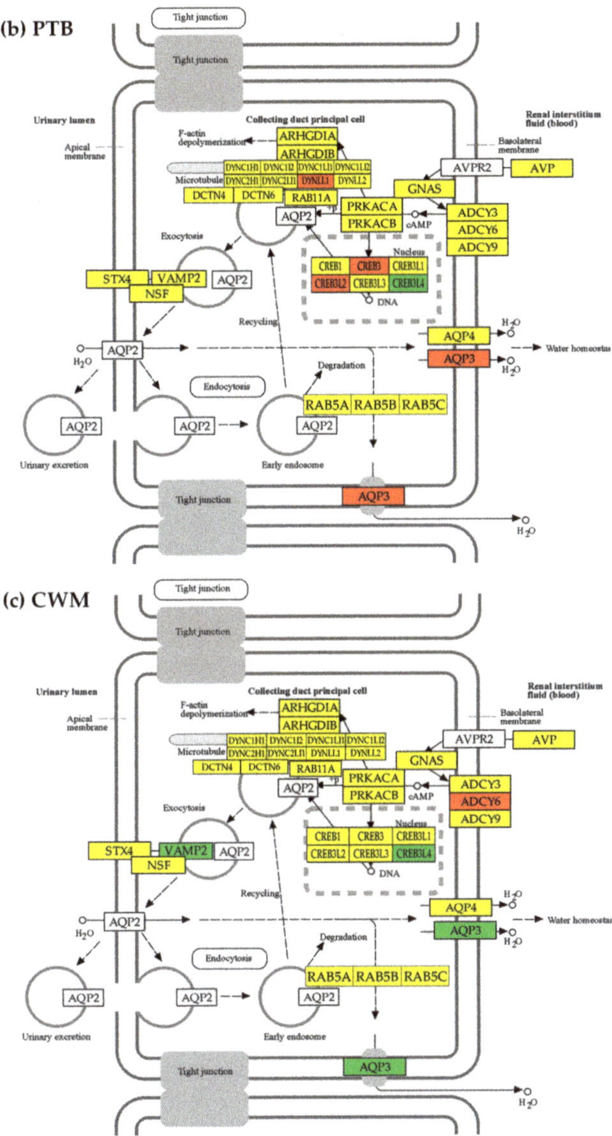

Figure 5. The regulated genes from the KEGG-constructed pathway of hsa04962 "Vasopressin-regulated water reabsorption" in the three cancer nodules with respect to the surrounding normal (NOR) tissue in the right kidney: (**a**) PTA, (**b**) PTB, and (**c**) CWM. Red/green background of the gene symbol indicates significant up-/down-regulation, yellow background indicates not statistically significant regulation, while blank background indicates that that gene was not quantified. Significantly regulated genes: *ADCY6* (adenylate cyclase 6), *AQP3*, *CREB3* (cAMP responsive element binding protein 3), *CREB3L2/3/4* (cAMP responsive element binding protein 3-like 2/3/4), *DCTN1/2* (dynactin 1/2), *DYNC2LI1* (dynein, cytoplasmic 2, light intermediate chain 1), *GNAS* (GNAS complex locus), and *VAMP2* (vesicle-associated membrane protein 2 (synaptobrevin 2)). Note the differences among the three nodules including that *AQP3* is down-regulated in PTA and CWM, but up-regulated in PTB.

In contrast, the ALDO gene *PIK3R2* (phosphoinositide-3-kinase, regulatory subunit 2 (beta)) was upregulated in all three cancer nodules: $x^{(\text{PTA}\rightarrow\text{NOR})} = 3.31$ ($\text{CUT}^{(\text{PTA}\rightarrow\text{NOR})} = 1.89$, $p^{(\text{PTA}\rightarrow\text{NOR})} = 0.013$), $x^{(\text{PTB}\rightarrow\text{NOR})} = 1.85$ ($\text{CUT}^{(\text{PTB}\rightarrow\text{NOR})} = 1.82$, $p^{(\text{PTB}\rightarrow\text{NOR})} = 0.046$), $x^{(\text{CWM}\rightarrow\text{NOR})} = 3.30$ ($\text{CUT}^{(\text{CWM}\rightarrow\text{NOR})} = 1.90$, $p^{(\text{CWM}\rightarrow\text{NOR})} = 0.032$).

Two genes were oppositely regulated in the nodules of PTB and CWM: the ALDO gene *SFN* (stratifin; $x^{(\text{PTB}\rightarrow\text{NOR})} = 2.03$, $x^{(\text{CWM}\rightarrow\text{NOR})} = -1.95$) and the ENDO gene *SLC8A1* (solute carrier family 8 (sodium/calcium exchanger), member 1; $x^{(\text{PTB}\rightarrow\text{NOR})} = 2.75$, $x^{(\text{CWM}\rightarrow\text{NOR})} = -1.84$).

Three genes were similarly regulated in PTA and PTB: the *ATP1B2* (ATPase, Na$^+$/K$^+$ transporting, beta 2 polypeptide; $x^{(\text{PTA}\rightarrow\text{NOR})} = 1.91$, $x^{(\text{PTB}\rightarrow\text{NOR})} = 2.67$), the *DCTN2* (dynactin 2 (p50); $x^{(\text{PTA}\rightarrow\text{NOR})} = -1.51$, $x^{(\text{PTB}\rightarrow\text{NOR})} = -2.06$), and the *SLC4A4* (solute carrier family 4 (sodium bicarbonate cotransporter), member 4; $x^{(\text{PTA}\rightarrow\text{NOR})} = 2.19$, $x^{(\text{PTB}\rightarrow\text{NOR})} = 2.57$).

Three regulated genes in PTA were similarly regulated in CWM: the *ADCY6* (adenylate cyclase 6; $x^{(\text{PTA}\rightarrow\text{NOR})} = 1.52$, $x^{(\text{CWM}\rightarrow\text{NOR})} = 1.96$), the *KRAS* (Kirsten rat sarcoma viral oncogene homolog; $x^{(\text{PTA}\rightarrow\text{NOR})} = 2.21$, $x^{(\text{CWM}\rightarrow\text{NOR})} = 2.77$), and the *PIK3CD* (phosphatidylinositol-4,5-bisphosphate 3-kinase, catalytic subunit delta; $x^{(\text{PTA}\rightarrow\text{NOR})} = -1.65$, $x^{(\text{CWM}\rightarrow\text{NOR})} = -1.97$).

Two genes were similarly regulated in PTB and CWM: the *CA2* (carbonic anhydrase II; $x^{(\text{PTB}\rightarrow\text{NOR})} = 9.94$, $x^{(\text{CWM}\rightarrow\text{NOR})} = 3.44$) and the *PLCB1* (phospholipase C, beta 1 (phosphoinositide-specific), $x^{(\text{PTB})} = -2.60$, $x^{(\text{CWM})} = -2.01$).

3.8. Tumor Heterogeneity of the Transcriptomic Networks

Figure 6 presents the ($p < 0.05$) significant synergism, antagonism, and independence among the genes from the hsa04961 pathway of ENDO (Endocrine and other factor-regulated calcium reabsorption) [59]. It is interesting to observe that the percentage of the synergistic pairs increased from 12.28% in NOR to 26.90% in PTA, 20.76% in PTB, and 16.96% in CWM. The percentage of the antagonistic pairs increased from 9.65% in NOR to 21.92% in PTA, 20.76% in PTB, and 16.08% in CWM, while that of the independently expressed pairs decreased from 12.28% in NOR to 4.09% in PTA, 4.68% in PTB, and 6.43% in CWM. Altogether, the coordination score COORD = %synergistic + %antagonistic − %independent increased from 9.65% in NOR to 44.74% in PTA, 36.84% in PTB, and 26.61% in CWM. These results indicate a substantial ccRCC-triggered increase in the inter-coordination of the genes involved in this pathway.

Of note are again the substantial differences between the PTA and PTB regions, indicating distinct wiring of the genes in the functional network. For instance, there are 7 genes (*AP2S1*, *ATP1A2*, *ATP2B3*, *CLTCL1*, *DNM3*, *ESR1*, and *PLCB2*) whose significant correlations with the sodium/calcium exchanger *SLC8A1* are opposite in the two kidney nodules, indicating major differences in the gene networking. Thus, the *ATP1A2*, *ATP2B3*, *CLTCL1*, *DNM3*, and *ESR1* are synergistically expressed with the *SLC8A1* in PTA but antagonistically expressed with the *SLC8A1* in PTB, while the *AP2S1* and *PLCB2* are antagonistically expressed with the *SLC8A1* in PTA but synergistically expressed with the *SLC8A1* in PTB.

Figure 7 presents the ($p < 0.05$) statistically significant synergism, antagonism, and independence of the quantified excretory genes with *AVP* in all four regions profiled. In the KEGG-constructed "Vasopressin-regulated water reabsorption" (VASO) pathway, *AVP* is directly connected to *AVPR2* (arginine vasopressin receptor 2; not quantified in the experiment) and indirectly connected to *GNAS*. Using the COR analysis, we found that in the normal kidney *AVP* is significantly connected to the *CREB3L1*, *CREB3L4*, *DYNC1H1*, and *RAB5A*.

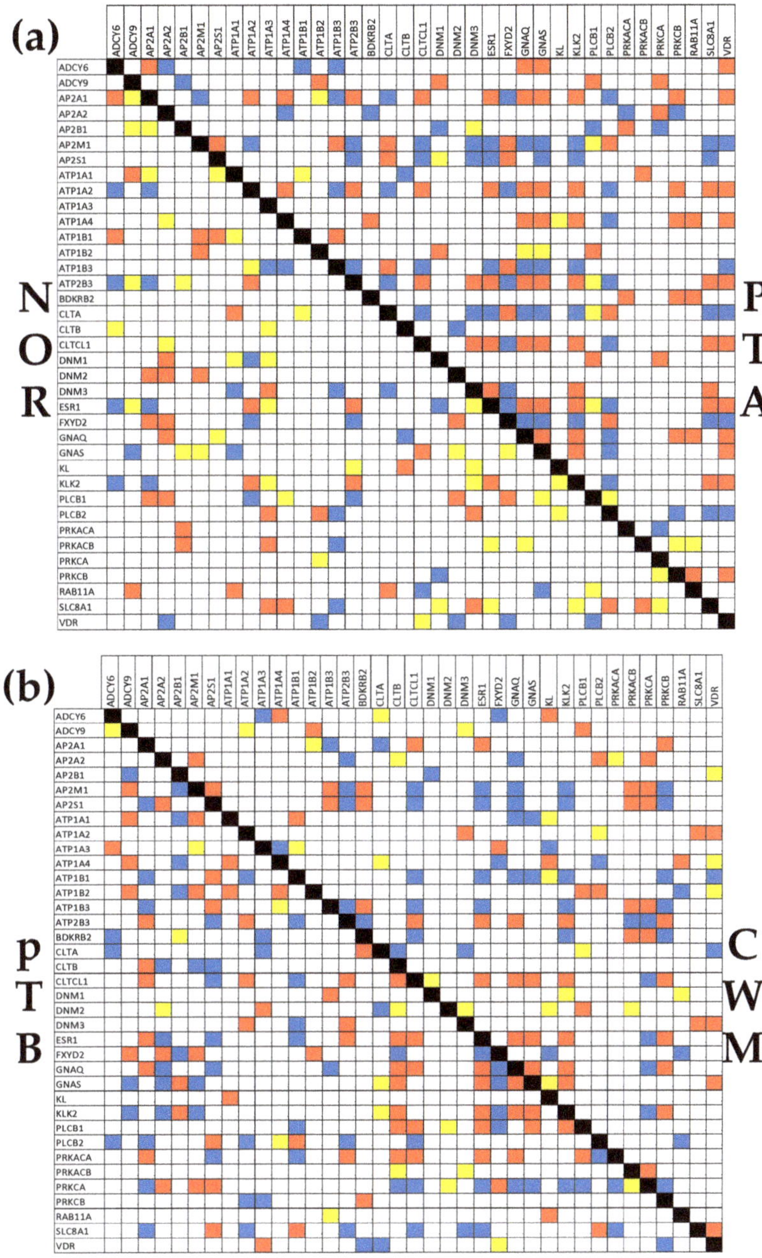

Figure 6. ($p < 0.05$) significant synergism, antagonism, and independence among the genes responsible for the Endocrine and other factor-regulated calcium reabsorption. (**a**) Significant gene expression correlations in NOR and PTA. (**b**) Significant gene expression correlations in PTB and CWM. A red/blue/yellow square indicates significant synergism/antagonism/independence of the genes labeling the intersecting row and column, while a blank square means a lack of statistical significance of the expression correlation.

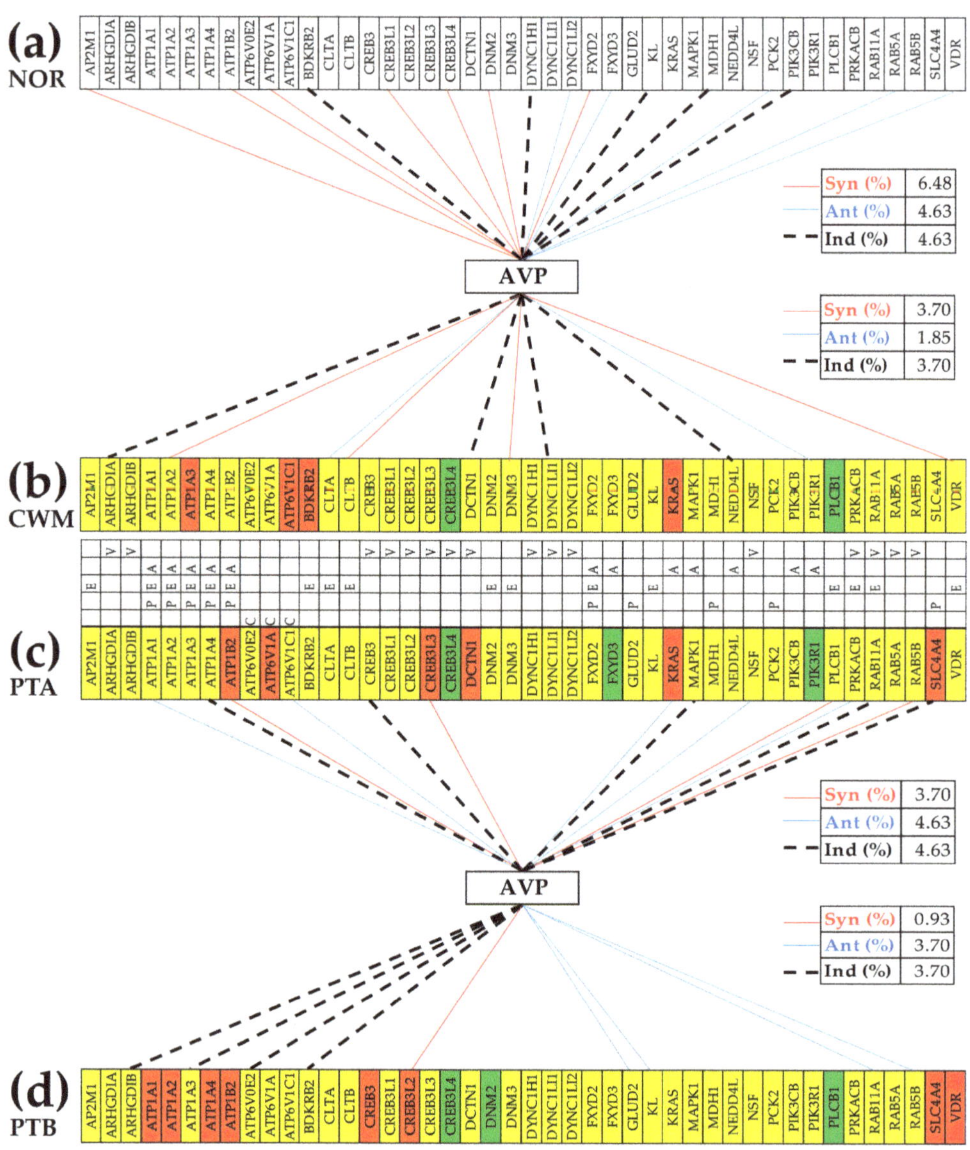

Figure 7. Statistically significant synergism, antagonism, and independence of excretory genes with arginine vasopressin (*AVP*) in all four profiled regions. (**a**) Significant expression correlation partners of *AVP* in NOR. (**b**) Significant expression correlation partners of *AVP* in CWM. (**c**) Significant expression correlation partners of *AVP* in PTA. (**d**) Significant expression correlation partners of *AVP* in PTB. A continuous red/blue line indicates synergism/antagonism, while a dashed black line indicates significant independence. Letters A, C, E, P, and V indicate the pathway affiliations of the genes: A = Aldosterone-regulated sodium reabsorption, C = Collecting duct acid secretion, E = Endocrine and other factor-regulated calcium reabsorption, P = Proximal tubule bicarbonate reclamation, and V = Vasopressin-regulated water reabsorption. Note: only the genes with significant correlations in at least one region were included in the figure.

The cancer reconfigures the VASO networks, so that in PTA, the *AVP* is significantly connected to the *CREB3L3, NSF, PRKACB,* and *RAB5B*. In PTB, the *AVP* is connected to the *CREB3L2, RAB11A,* and *RAB5B*, while in CWM it is connected to no hsa04962 gene. Remarkably, the hsa04962 genes that also have significant independence with respect to *AVP* are: *DYNC1H1* in NOR, *CREB3* and *RAB1A* in PTA, *ARHGDIB* in PTB and *ARHGDIA, DCTN1,* and *DYNC1H1* in CWM.

4. Discussion

This study continues the analyses from a previous article [40] on four samples collected from the chest wall metastasis (CWM), two primary tumor regions (PTA and PTB), and the surrounding normal tissue (NOR) in the right kidney of a man with metastatic ccRCC. In [40], we analyzed the ccRCC impact on cyclins, cyclin kinases, and the functional pathways of apoptosis, chemokine and VEGF signaling, oxidative phosphorylation, basal transcription factors, and RNA polymerase, while here we focused on the five KEGG-constructed functional pathways of the excretory system.

An important finding of the previous paper [40], reconfirmed in the present study, was that even equally Fuhrman-graded (3) and closely located cancer nodules from the same kidney exhibit substantial transcriptomic differences and are under the command of distinct gene master regulators. Considerably transcriptomic differences between equally graded cancer nodules were also reported by us in two prostate cancer studies [34,42]. Importantly, tumor heterogeneity, present even in patient-derived RCC organoids [64], is not limited to the genes' mutations [65] or expression levels [66], but encompasses also the control of the transcripts' abundances and the way the genes are interconnected in functional pathways. Tumor heterogeneity and the unrepeatable set of cancer-favoring factors characterizing each human make the "transcriptomic signature" unreliable (e.g., [67,68]), and even of the integrated proteogenomic characterization [69] derived from comparing genomic data from large populations of cancer and healthy people is made unreliable. Instead, it imposes the normal tissue surrounding the cancer nodule(s) in the tumor as the most trustable reference.

The novel findings on the alterations of the mechanisms controlling the transcript abundances and the remodeling of the functional pathways were possible by adopting the Genomic Fabric Paradigm (GFP) and the use of its mathematically advanced algorithms and software (detailed in [54]). The GFP improves the traditional gene expression analysis by replacing the arbitrarily introduced (1.5×) cut-off for the absolute fold-change of a gene to be considered as significantly regulated with a cut-off computed for each gene in the compared conditions that considers both biological variability and technical noise. While incorporating an improved version of the traditional analysis, the GFP goes further by also analyzing the gene expression control and inter-coordination. These additional characteristics provide as much supplementary information as going from knowing the height of a mountain to featuring it on a 3D scale model. The independence of the three types of expression characteristics was illustrated here for the genes involved in the KEGG-constructed pathway of the "Endocrine and other factor-regulated calcium absorption". The independence of the three types of expression characteristics can be proven for any other gene subset (principle discussed in [53]). For instance, in previously published cancer genomics papers, we proved the independence of genes involved in: chemokine signaling [40], apoptosis [55], evading apoptosis [42], and mTOR signaling [34].

The gene networks constructed with the COR analysis do not have the same faults of universality, unicity, and rigidity as those built with the KEGG and other software. The COR-determined networks are not universal but instead depend on the race, age, and other personal characteristics of the patient that modulate his/her gene expression, as we proved in [34] for two prostate cancer patients and two standard human prostate cancer cell lines. They are not unique, with cancer nodule(s) and normal tissue exhibiting different gene wiring, and certainly not rigid, but remodeling during aging, the progression of the disease, and in response to treatment and other external stimuli. Thus, the Postulate of

Transcriptomic Stoichiometry (PTS, [44]) also extends Dalton's Law of Multiple Proportions from chemistry [45] to gene networks.

This analysis of the expression correlation among the genes of the "Endocrine and other factor-regulated calcium reabsorption" revealed an interesting partnership with the estrogen receptor 1 (*ESR1*), which is a tumor driver and drug-targeting factor in cancer [70], and regulator of age-related mitochondrial dysfunction and inflammation [71]. Thus, while its expression synergisms with the *ATP2B3* (ATPase, Ca^{++} transporting, plasma membrane 3) and *KLK2* (kallikrein-related peptidase 2) in NOR are not modified by ccRCC, the antagonistic expression with the *AP2A1* is switched to a synergistic one in all three cancer nodules. Because the *AP2A1* is important in vesicle formation and intracellular membrane trafficking [72], our result indicates that cancer switched the type of estrogen influence on the intracellular transport phenomena. Interestingly, the altered expression of *ESR1* was associated with anti-cancer drug resistance, and non-coding events were identified in the regulatory hotspot of *AP2A1* in various metastatic cancers [73].

We found that the excretory genes rank much lower than the GMRs identified for each region in [40]. Thus, the *GNAQ* (guanine nucleotide-binding protein (G protein), q polypeptide), the top gene involved in the "Endocrine and other factor-regulated calcium reabsorption" in both NOR ($GCH^{(NOR)} = 9.46$) and CWM ($GCH^{(CWM)} = 17.43$), is far below the GMRs of these regions: the *DAPK3* (death-associated protein kinase 3, $GCH^{(NOR)} = 30.31$) and the *ALG13* (UDP-N-acetylglucosaminyltransferase subunit, $GCH^{(CWM)} = 82.95$). The *IGF1* (insulin-like growth factor 1 (somatomedin C), $GCH^{(PTA)} = 13.11$) from the pathway of the "Aldosterone-regulated sodium reabsorption" is far below the *TASOR* (transcription activation suppressor, $GCH^{(PTA)} = 63.97$) in PTA. The *ATP1A2* (ATPase, Na^+/K^+ transporting, alpha 2 polypeptide, $GCH^{(PTB)} = 9.95$) from the pathway of the "Proximal tubule bicarbonate reclamation" is below the *FAM27C* (family with sequence similarity 27, member C, $GCH^{(PTB)} = 57.19$) in PTB. The much lower GCH scores mean that although the excretory pathways were strongly regulated, no excretory gene is an efficient target for gene therapy against any of the three cancer nodules.

The substantial decrease in the homeostatic control of *AP2A1* expression in PTA but not in PTB and CWM indicates that, although not significantly regulated ($x^{(PTA \to NOR)} = 1.22 < 1.57 = CUT^{(PTA \to NOR)}$, $x^{(PTB \to NOR)} = 1.04 < 1.61 = CUT^{(PTB \to NOR)}$, $x^{(CWM \to NOR)} = 1.13 < 1.55 = CUT^{(CWM \to NOR)}$), the *AP2A1* lost its importance for the cell homeostasis in PTA while keeping it at NOR level in the other nodules. The most interesting case for the expression control analysis was that of *ADCY9*, a biomarker for glioma [74], lung [75], and hepatocellular carcinoma [76], as well as colorectal [77], bladder [78], and pancreatic cancers [79]. The *ADCY9* exhibited the largest increase of the homeostatic control in PTB and the largest decrease in CWM among all profiled excretory genes, while its control in PTA remained the same as in the normal tissue. These results indicate that while cancer had no effect on the homeostatic control of the *ADCY9* transcript abundance in one region (PTA) of the tumor, it made the right *ADCY9* expression critical in another region (PTB), and totally irrelevant in the metastasis nodule (CWM). Since the expression of *ADCY9* is normally four times more controlled than the median kidney gene, we predict that the expression manipulation of *ADCY9* will wholly jeopardize the PTB cells, and have a similar (high) negative effect on both normal and PTA cells, while stimulating the proliferation of CWM cells. Thus, at least for this patient, targeting *ADCY9* could be both beneficial and harmful.

The increased overall correlation of the *ADCY6* and *AP2A1* with all other genes from the "Endocrine and other factor-regulated calcium reabsorption" pathway (Figure 3e) in PTA, but the decrease in the other two cancer nodules, indicates substantially different gene networks. Thus, according to Figure 6, in NOR, the *AP2A1* has four statistically significant synergistically (*ADCY6*, *DNM2*, **FXYD2**, *PLCB1*) and four antagonistically (**ATP1A2**, **ATP2B3**, **ESR1**, **KLK2**) expressed partners, while in PTA it has eleven synergistic (*ADCY6*, *ATP1A2*, *ATP1A4*, **ATP2B3**, <u>*CLTCL1*</u>, <u>*ESR1*</u>, *GNAQ*, *GNAS*, **KLK2**, *PRKCB*, *VDR*) and five antagonistic (*ATP2M1*, <u>*ATP1B3*</u>, *CLTA*, **FXYD2**) partners (bolded symbols indicate the genes that switched the expression correlation type in PTA with respect to NOR). The partnerships of *AP2A1* were partially similar in PTB: it had synergism with <u>*ATP2B3*</u>, *CLTB*, <u>*CLTCL1*</u>, <u>*ESR1*</u>, and <u>*GNAQ*</u>, and antagonism with *AP2S1*, *ATP1B1*, <u>*ATP1B3*</u>, *PLCB2*, *PRKCA*, and *SLC8A1* (underlined symbols indicate common partners in PTA and PTB).

Because *ADCY6* is a promising target in cancer therapy [80], it is important to know how it relates to other excretory genes. Thus, in normal tissue (NOR) it is synergistically expressed with *AP2A1* and *ATP1B1*, and antagonistically expressed with *ATP1A2*, *ATP2B3*, *ESR1*, and *KLK2*, as well as independently expressed with *CLTB*. In PTA, the synergism with *AP2A1* is preserved but that with *ATP1B1* is switched to antagonism. PTA adds synergism with *GNAQ*, *GNAS*, and *VDR* and antagonism with *AP2A2* and *ATP1B3*. The overall coordination of *ADCY6* with other genes from the pathway is diminished in PTB (synergism with *ATP1A3* and antagonism with *BDKRB2*, *CLTA*, and *PLCB2*) and CWM (synergism with *ATP1A4* and *KL*, and antagonism with *ATP1A3* and *FXYD2*). Therefore, for this patient, manipulating the expression of *ADCY6* would have different effects on his cancer nodules.

The down-regulation of *AQP3* (x = −3.889) was reported recently [81] by performing a meta-analysis of public datasets from the publicly accessible databases of ONCOMINE [82] and UALCAN [83]. In the cited study, the authors compared the expressions of all aquaporin encoding genes (*AQP1/2/3/4/5/6/7/8/9/10/11*) in 533 tumor cases with 72 normal cases, using the uniform fold-change threshold of 1.50× and *p*-value < 0.01, to decide whether a gene was significantly regulated. Since no web resource specifies the exact locations in the kidney from which the samples have been collected, nor whether the tumors contained more cancer nodules, the authors implicitly assumed homogeneous gene expressions all over the tumor. In contrast, our results (*AQP3* down-regulated in PTA and CWM but up-regulated in PTB) indicate that ccRCC tumors do not exhibit uniform but rather heterogeneous gene expressions, and therefore the meta-analyses comparing cancer cases with healthy cases have little biological relevance beyond the statistics exercise. Instead, the best reference for cancer nodules is the normal tissue still present in the tumor.

Figure 7 shows distinct statistically significant coordination partners of *AVP*, the neuropeptide hormone arginine vasopressin, which is a very important regulator of kidney salt and water homeostasis [84]. Our analysis detailed also how *AVP*-dependent water reabsorption regulates the cAMP signaling pathway [85,86] through expression correlation with genes shared by the cAMP signaling pathway with the excretory pathways. Thus, in NOR, AVP is synergistically expressed with *ATP1B2*, *CREB3L1*, *CREB3L4*, and *FXYD2*, has no antagonistic partners, and is independently expressed with *PIK3CB*. In PTA, *AVP* is synergistically expressed with *ATP1B2* and *CREB3L3*, antagonistically expressed with *ATP1A1* and *PRKACB*, and independently expressed with *ATP1A4*, *CREB3*, and *MAPK1*. In PTB, *AVP* is synergistically expressed with *CREB3L2* and independently expressed with *ATP1A3*, while in CWM it is synergistically expressed with *ATP1A2* and antagonistically expressed with *PIK3R1*. *AVP* is also involved in the KEGG-constructed pathways of hsa04270 "Vascular smooth muscle contraction", and the very important hsa04020 "Calcium signaling pathway" and hsa04072 "Phospholipase signaling pathway".

Interestingly, *CREB3L4*, known for its role in proliferating prostate cancer cells [87], was significantly down-regulated in all kidney cancer nodules (x$^{(PTA)}$ = −1.40, x$^{(PTB)}$ = −1.74, x$^{(CWM)}$ = −1.95). Out of 18 regulated excretory genes in PTA and twenty-one in PTB, only five were similarly regulated (*ATP1B2*, *CREB3L4*, *DCTN2*, *PIK3R2*, *SLC4A4*) and one (*AQ3*) was oppositely regulated.

Altogether, the differences in gene expression regulation and remodeling of the excretory pathways among the three cancer nodules indicate that the bio-assays used to identify the ccRCC presence by the regulation of certain gene biomarkers might not be always valid. For instance, the *CA9* (carbonic anhydrase IX) that should be expressed only in ccRCC [88,89] was found by us to be expressed not only in the cancer nodules (*AVE*$^{(PTA)}$ = 2.11, *AVE*$^{(PTB)}$ = 0.27, *AVE*$^{(CWM)}$ = 2.65) but also in the NOR (*AVE*$^{(NOR)}$ = 2.76). Moreover, the *CA9* exhibited a significant down-regulation in PTB (x$^{(PTB \to NOR)}$ = −10.37).

Nonetheless, adoption of the GFP increases by 3–4 orders of magnitude the computational effort, thus opening the field of genomics to artificial intelligence applications [90] that have the power to combine practically unlimited amounts of histo-pathological, imaging and "omic" information [91].

5. Conclusions

Although based on a single metastatic ccRCC case, our study shows that adding the analyses of transcriptomic control and expression inter-coordination of the genes provides a significantly deeper understanding of cancer-related transcriptomic alterations. Going from traditional unidimensional gene expression analysis to the GFP is like going from determining only the numbers of electronic items of each type needed to fix a complex robot, to knowing also how to wire these items and what voltages to apply to each of them. Thus, the GFP not only incorporates the traditional gene expression analysis but complements it with a detailed examination of the expression control and gene inter-coordination, altogether providing a complete characterization of the transcriptome.

The unrepeatable combination of cancer-favoring factors among humans questions the validity of the cancer biomarkers derived from meta-analyses of large genomic datasets. The intra-tumor transcriptomic heterogeneity makes the transcriptomic signatures even less reliable. Therefore, we believe that the personalization of cancer gene therapy should go beyond the patient to the primary cancer clones present in his/her tumor. The progress of gene-editing technology will soon allow the industry to produce silencing constructs for all genes at reasonable prices so that personalized cancer gene therapy will become affordable.

Supplementary Materials: The following supporting information can be downloaded at: https://www.mdpi.com/article/10.3390/cimb45120594/s1, Figure S1: Regulated genes in the KEGG-constructed functional pathway hsa04960 Aldosterone-regulated sodium reabsorption, Figure S2: Regulated genes in the KEGG-constructed functional pathway hsa04966 Collecting duct acid secretion, Figure S3: Regulated genes in the KEGG-constructed functional pathway hsa04961 Endocrine and other factor-regulated calcium reabsorption, Figure S4: Regulated genes in the KEGG-constructed functional pathway hsa04964 Proximal tubule bicarbonate reclamation.

Author Contributions: Conceptualization, D.A.I.; methodology, D.A.I. and S.I.; software, D.A.I.; validation, D.A.I. and S.I.; formal analysis, D.A.I.; investigation, S.I.; resources, D.A.I.; data curation, D.A.I.; writing—original draft preparation, D.A.I.; writing—review and editing, E.A.O.; visualization, D.A.I. and E.A.O.; supervision, D.A.I.; project administration, D.A.I.; funding acquisition, D.A.I. All authors have read and agreed to the published version of the manuscript.

Funding: This research received no external funding.

Institutional Review Board Statement: The experimental dataset was retrieved from an investigation conducted according to the guidelines of the Declaration of Helsinki. At the time of the experiment, the study was part of D.A. Iacobas' project approved by the Institutional Review Boards (IRB) of the New York Medical College's (NYMC) and Westchester Medical Center's (WMC) Committees for Protection of Human Subjects. The approved IRB (L11,376 from 2 October 2015) granted access to frozen cancer specimens from the WMC Pathology Archives and depersonalized pathology reports, waiving the patient's informed consent.

Informed Consent Statement: The approved IRB granted access to frozen cancer specimens from the WMC Pathology Archives and depersonalized pathology reports, thus waiving the patient's informed consent.

Data Availability Statement: Transcriptomic data are publicly available at https://www.ncbi.nlm.nih.gov/geo/query/acc.cgi?acc=GSE72304 (accessed on 26 October 2023).

Conflicts of Interest: The authors declare no conflict of interest.

Appendix A

1. Normalized gene expression levels in the biological replica "k" in condition "c":

$$\forall c = PTA, PTB, CWM, NOR \; \& \; k = 1 \div 4 \;\; \alpha_i^{(c;k)} \equiv \frac{a_i^{(c;k)}}{\left\langle a_j^{(c;k)} \right\rangle_{j=1 \div N}} \Rightarrow \left\langle \alpha_j^{(c;k)} \right\rangle_{j=1 \div N} = 1 \tag{A1}$$

$a_i^{(c;k)}$ is the sum of the net fluorescences of all microarray spots probing gene "i" in the biological replica "k" of condition "c".

2. Relative expression variation:

$$REV_i^{(c)}(\alpha) \equiv \frac{\sigma_i^{(c)}}{2AVE_i^{(c)}} \left(\sqrt{\frac{r_i}{\chi^2(\beta;r)}} + \sqrt{\frac{r_i}{\chi^2(1-\beta;r)}} \right) \times 100\% \,, \; where:$$
$$\sigma_i^{(c)} = sdev\left(\alpha_i^{(c;k)}\right)_{k=1 \div 4}, \; \beta \; (usually \; \beta = 0.05) \; is \; the \; probability, \tag{A2}$$
$$r_i \; is \; the \; number \; of \; degrees \; of \; freedom, \; r_i = nv_i - 1$$

and χ^2 is the chi-square score with β probability for r degrees of freedom.

3. Transcriptome configuration function:

$$F(1,2,\ldots,N) = A_1 \underbrace{\prod_{i=1}^{N} f_1(i)}_{independent} + A_2 \underbrace{\prod_{i>j=1}^{N} f_2(i,j)}_{pair-wise} + A_3 \underbrace{\prod_{i>j>k=1}^{N} f_3(i,j,k)}_{3-genes \; clusters} + \ldots + A_N \underbrace{f_N(1,2,\ldots,N)}_{all \; genes \; cluster}$$

where $A_1, \ldots, A_N$ are the probabilities of each configuration of gene clustering, and $f_2, f_3, \ldots, f_N$ are distribution functions symmetrical to the permutation(s) of the genes. The above expansion satisfies the following conditions:

probability condition : $\sum_{p=1}^{N} A_p = 1$

norm conditions :

$$\int F(1,2,\ldots,N)dV = 1 \,, \; \forall p = 1 \div N \; \rightarrow \; \int_V \left(\prod_{i_1>i_2>\ldots>i_p=1}^{N} f_p(i_1, i_2, \ldots, i_p) \right) dV = 1$$

A 3-gene cluster can be approximated with the superposition of three paired genes:

$$\prod_{i>j>k}^{N} f_3(i,j,k) \simeq \prod_{i>j>k}^{N} f_2(i,j) f_2(j,k) f_2(k,i)$$

A 4-gene cluster can be approximated with four 3-gene clusters:

$$f_4(i,j,k,l) = f_3(i,j,k) f_3(j,k,l) f_3(k,l,i) f_3(l,i,j)$$

that can be further approximated with six squared distributions of paired genes:

$$f_4(i,j,k,l) = \underbrace{(f_2(i,j) f_2(j,k) f_2(k,i))}_{f_3(i,j,k)} \times \underbrace{(f_2(j,k) f_2(k,l) f_2(l,j))}_{f_3(j,k,l)} \times \underbrace{(f_2(k,l) f_2(l,i) f_2(i,k))}_{f_3(k,l,i)}$$
$$\times \underbrace{(f_2(i,j) f_2(j,l) f_2(l,i))}_{f_3(l,i,j)} = f_2^2(i,j) f_2^2(j,k) f_2^2(k,i) f_2^2(k,l) f_2^2(l,j) f_2^2(l,i)$$

and so on, giving the recurrence relation:

$$\forall p \geq 3 \ \& \ I_p = \text{a set of } p \text{ genes}$$

$$f_p(I_p) \simeq \prod_{\substack{\text{all combinations of p genes} \\ \text{in sets of } p-1 \text{ genes}}} f_{p-1}(I_{p-1}) \simeq \prod_{i>j\in I_p} f_2^{p-2}(i,j)$$

From the normalization condition, it results that:

$$\forall p \geq 3 \ \wedge \ \forall q > r = 1 \div N , \ \prod_{i>j\in I_p} f_2^{p-2}(i,j) << \prod_{i>j\in I_p} f_2(i,j)$$

Therefore, one can neglect the contributions of clusters with more than 2 genes:

$$F(1,2,\ldots,N) \approx A_1 \underbrace{\prod_{i=1}^{N} f_1(i)}_{\text{independent}} + A_2 \underbrace{\prod_{i>j=1}^{N} f_2(i,j)}_{\text{pair−wise}} \tag{A3}$$

4. **Pair-wise correlation of gene expression levels:**

$$COR_{i,j}^{(c)} \equiv \frac{\sum\limits_{k=1}^{4} \left(\log_2 \alpha_i^{(c;k)} - \log_2 AVE_i^{(c)}\right)\left(\log_2 \alpha_j^{(c;k)} - \log_2 AVE_j^{(c)}\right)}{\sqrt{\sum\limits_{k=1}^{4}\left(\log_2 \alpha_i^{(c;k)} - \log_2 AVE_i^{(c)}\right)^2}\sqrt{\sum\limits_{k=1}^{4}\left(\log_2 \alpha_j^{(c;k)} - \log_2 AVE_j^{(c)}\right)^2}} \tag{A4}$$

5. **Gene Commanding Height:**

$$GCH_i^{(c)} \equiv \frac{\langle REV \rangle^{(c)}}{REV_i^{(c)}} \times exp\left(\overline{4COR_{i,j}^{(c)^2}}\right) \tag{A5}$$

6. **Statistically significant regulation of the expression level:**

$$Abs\left(x_i^{(cancer \rightarrow NOR)}\right) \geq CUT_i^{(cancer \rightarrow NOR)} = 1 + \sqrt{2\left(\left(\frac{REV_i^{(cancer)}}{100}\right)^2 + \left(\frac{REV_i^{(NOR)}}{100}\right)^2\right)} \ \& \ p_i^{(cancer \rightarrow NOR)} < 0.05$$

$$\text{where}: x_i^{(cancer \rightarrow NOR)} = \begin{cases} \dfrac{AVE_i^{(cancer)}}{AVE_i^{(NOR)}} & if \quad AVE_i^{(cancer)} \geq AVE_i^{(NOR)} \\[2ex] -\dfrac{AVE_i^{(NOR)}}{AVE_i^{(cancer)}} & if \quad AVE_i^{(cancer)} < AVE_i^{(NOR)} \end{cases}, \ cancer = PTA, PTB, CWM \tag{A6}$$

7. **Weighted Individual (Gene) Regulation:**

$$WIR_i^{(cancer \rightarrow NOR)} \equiv AVE_i^{(NOR)} \frac{x_i^{(cancer \rightarrow NOR)}}{Abs\left(x_i^{(cancer \rightarrow NOR)}\right)} \underbrace{\left(Abs\left(x_i^{(cancer \rightarrow NOR)}\right) - 1\right)}_{\text{absolute fold−change}} \underbrace{\left(1 - p_i^{(cancer \rightarrow NOR)}\right)}_{\text{confidence}} \tag{A7}$$

8. **Weighted Pathway Regulation:**

$$WPR_\Gamma^{(cancer \rightarrow NOR)} = \overline{Abs\left(WIR_i^{(cancer \rightarrow NOR)}\right)}\Big|_{i \in \Gamma} \tag{A8}$$

9. **Relative Expression Control:**

$$REC_i^{(condition)} \equiv \left(\frac{\left\langle REV^{(condition)} \right\rangle}{REV_i^{(condition)}} - 1 \right) \times 100\%, \text{ where } \left\langle REV^{(condition)} \right\rangle \text{ is the median } REV \tag{A9}$$

10. **Regulation of the Expression Control:**

$$\Delta REC_i^{(cancer \to NOR)} \equiv \frac{const}{REV_i^{(cancer)}} - \frac{const}{REV_i^{(NOR)}}, \; const = \text{calibration constant} \tag{A10}$$

In this study *const* = 100.

11. **Regulation of the Expression Coordination:**

$$\Delta COR_{i,\Gamma}^{(cancer \to NOR)} \equiv \frac{\sum\limits_{j \in \Gamma} \left(COR_{i,j}^{(cancer)} - COR_{i,j}^{(NOR)} \right)}{Abs \left(\sum\limits_{j \in \Gamma} \left(COR_{i,j}^{(cancer)} - COR_{i,j}^{(NOR)} \right) \right)} \sum\limits_{j \in \Gamma} \left(COR_{i,j}^{(cancer)} - COR_{i,j}^{(NOR)} \right)^2 \tag{A11}$$

12. **Transcriptomic Distance:**

$$TDI_{i,\Gamma}^{(cancer \to NOR)} \equiv \sqrt{ \underbrace{\left(WIR_i^{(cancer \to NOR)} \right)^2}_{\substack{\text{regulation of} \\ \text{expression level}}} + \underbrace{\left(\Delta REC_i^{(cancer \to NOR)} \right)^2}_{\substack{\text{regulation of} \\ \text{transription control}}} + \underbrace{\sum\limits_{j \in \Gamma} \left(COR_{i,j}^{(cancer)} - COR_{i,j}^{(NOR)} \right)^2}_{\substack{\text{regulation of} \\ \text{expression coordination}}} } \tag{A12}$$

References

1. Cancer Treatment Options at Houston Methodist Organization. Available online: https://www.houstonmethodist.org/cancer/treatment-options/ (accessed on 3 May 2023).
2. Pecoraro, A.; Campi, R.; Marchioni, M. European Association of Urology Young Academic Urologists Renal Cancer Working Group. Techniques and outcomes of percutaneous tumour ablation for small renal masses. *Curr. Opin. Urol.* **2023**, *33*, 360–366. [CrossRef] [PubMed]
3. Lanza, C.; Carriero, S.; Ascenti, V.; Tintori, J.; Ricapito, F.; Lavorato, R.; Biondetti, P.; Angileri, S.A.; Piacentino, F.; Fontana, F.; et al. Percutaneous Application of High Power Microwave Ablation with 150 W for the Treatment of Tumors in Lung, Liver, and Kidney: A Preliminary Experience. *Technol. Cancer Res. Treat.* **2023**, *22*, 15330338231185277. [CrossRef] [PubMed]
4. Key Statistics about Kidney Cancer. Available online: https://www.cancer.org/cancer/kidney-cancer.html (accessed on 6 September 2023).
5. Mieville, V.; Griffioen, A.W.; Benamran, D.; Nowak-Sliwinska, P. Advanced in vitro models for renal cell carcinoma therapy design. *Biochim. Biophys. Acta (BBA)-Rev. Cancer* **2023**, *1878*, 188942. [CrossRef] [PubMed]
6. Dahle, D.O.; Skauby, M.; Langberg, C.W.; Brabrand, K.; Wessel, N.; Midtvedt, K. Renal Cell Carcinoma and Kidney Transplantation: A Narrative Review. *Transplantation* **2022**, *106*, e52–e63. [CrossRef]
7. Gong, Y.; Pang, H.; Yu, Z.; Wang, X.; Li, P.; Zhang, Q. Construction of inflammatory associated risk gene prognostic model of NSCLC and its correlation with chemotherapy sensitivity. *Ann. Med.* **2023**, *55*, 2200034. [CrossRef]
8. Hara, Y.; Shiba, N.; Yoshida, K.; Yamato, G.; Kaburagi, T.; Shiraishi, Y.; Ohki, K.; Shiozawa, Y.; Kawamura, M.; Kawasaki, H.; et al. *TP53* and *RB1* alterations characterize poor prognostic subgroups in pediatric acute myeloid leukemia. *Genes Chromosomes Cancer* **2023**, *62*, 412–422. [CrossRef]
9. Cancel-Tassin, G.; Koutros, S. Use of genomic markers to improve epidemiologic and clinical research in urology. *Curr. Opin. Urol.* **2023**, *33*, 414–420. [CrossRef]
10. Wu, F.; Ning, H.; Sun, Y.; Wu, H.; Lyu, J. Integrative exploration of the mutual gene signatures and immune microenvironment between benign prostate hyperplasia and castration-resistant prostate cancer. *Aging Male* **2023**, *26*, 2183947. [CrossRef]

11. Yerukala Sathipati, S.; Tsai, M.J.; Shukla, S.K.; Ho, S.Y. Artificial intelligence-driven pan-cancer analysis reveals miRNA signatures for cancer stage prediction. *HGG Adv.* **2023**, *4*, 100190. [CrossRef]

12. Yang, L.; Yang, M.; Cui, C.; Long, X.; Li, Y.; Dai, W.; Lang, T.; Zhou, Q. The myo-inositol biosynthesis rate-limiting enzyme ISYNA1 suppresses the stemness of ovarian cancer via Notch1 pathway. *Cell. Signal.* **2023**, *107*, 110688. [CrossRef]

13. Aveta, A.; Cilio, S.; Contieri, R.; Spena, G.; Napolitano, L.; Manfredi, C.; Franco, A.; Crocerossa, F.; Cerrato, C.; Ferro, M.; et al. Urinary MicroRNAs as Biomarkers of Urological Cancers: A Systematic Review. *Int. J. Mol. Sci.* **2023**, *24*, 10846. [CrossRef] [PubMed]

14. Huang, L.; Shao, J.; Xu, X.; Hong, W.; Yu, W.; Zheng, S.; Ge, X. WTAP regulates autophagy in colon cancer cells by inhibiting FLNA through N6-methyladenosine. *Cell Adhes. Migr.* **2023**, *17*, 1–13. [CrossRef] [PubMed]

15. Yang, R.Y.; Tan, J.Y.; Liu, Z.; Shen, X.L.; Hu, Y.J. Lappaol F regulates the cell cycle by activating CDKN1C/p57 in human colorectal cancer cells. *Pharm. Biol.* **2023**, *61*, 337–344. [CrossRef]

16. Yavuz, M.; Takanlou, L.S.; Avcı, Ç.B.; Demircan, T. A selective androgen receptor modulator, S4, displays robust anti-cancer activity on hepatocellular cancer cells by negatively regulating PI3K/AKT/mTOR signalling pathway. *Gene* **2023**, *869*, 147390. [CrossRef] [PubMed]

17. Ishiguro, M.; Fukushige, T.; Iwasaki, H. Establishment and Characterization of a TFE3-rearranged Renal Cell Carcinoma Cell Line (FU-UR-2) with the PRCC-TFE3 Fusion Transcript. *Anticancer Res.* **2023**, *43*, 3463–3470. [CrossRef] [PubMed]

18. Lasorsa, F.; Rutigliano, M.; Milella, M.; Ferro, M.; Pandolfo, S.D.; Crocetto, F.; Tataru, O.S.; Autorino, R.; Battaglia, M.; Ditonno, P.; et al. Cellular and Molecular Players in the Tumor Microenvironment of Renal Cell Carcinoma. *J. Clin. Med.* **2023**, *12*, 3888. [CrossRef]

19. Iacobas, D.A.; Iacobas, S.; Stout, R.F.; Spray, D.C. Cellular Environment Remodels the Genomic Fabrics of Functional Pathways in Astrocytes. *Genes* **2020**, *11*, 520. [CrossRef]

20. NIH-National Cancer Institute Genomic Data Commons Data Portal. Available online: https://portal.gdc.cancer.gov/ (accessed on 20 June 2023).

21. Lasorsa, F.; Rutigliano, M.; Milella, M.; Ferro, M.; Pandolfo, S.D.; Crocetto, F.; Autorino, R.; Battaglia, M.; Ditonno, P.; Lucarelli, G. Cancer Stem Cells in Renal Cell Carcinoma: Origins and Biomarkers. *Int. J. Mol. Sci.* **2023**, *24*, 13179. [CrossRef]

22. Sarkar, O.S.; Donninger, H.; Al Rayyan, N.; Chew, L.C.; Stamp, B.; Zhang, X.; Whitt, A.; Li, C.; Hall, M.; Mitchell, R.A.; et al. Monocytic MDSCs exhibit superior immune suppression via adenosine and depletion of adenosine improves efficacy of immunotherapy. *Sci. Adv.* **2023**, *9*, eadg3736. [CrossRef]

23. Liu, C.L.; Huang, W.C.; Cheng, S.P.; Chen, M.J.; Lin, C.H.; Chang, S.C.; Chang, Y.C. Characterization of Mammary Tumors Arising from MMTV-PyVT Transgenic Mice. *Curr. Issues Mol. Biol.* **2023**, *45*, 4518–4528. [CrossRef]

24. Aggen, D.H.; Ager, C.R.; Obradovic, A.Z.; Chowdhury, N.; Ghasemzadeh, A.; Mao, W.; Chaimowitz, M.G.; Lopez-Bujanda, Z.A.; Spina, C.S.; Hawley, J.E.; et al. Blocking IL1 Beta Promotes Tumor Regression and Remodeling of the Myeloid Compartment in a Renal Cell Carcinoma Model: Multidimensional Analyses. *Clin. Cancer Res.* **2021**, *27*, 608–621. [CrossRef] [PubMed]

25. Ding, W.Y.; Kuzmuk, V.; Hunter, S.; Lay, A.; Hayes, B.; Beesley, M.; Rollason, R.; Hurcombe, J.A.; Barrington, F.; Masson, C.; et al. Adeno-associated virus gene therapy prevents progression of kidney disease in genetic models of nephrotic syndrome. *Sci. Transl. Med.* **2023**, *15*, eabc8226. [CrossRef] [PubMed]

26. Iacobas, D.A.; Iacobas, S.; Urban-Maldonado, M.; Spray, D.C. Sensitivity of the brain transcriptome to connexin ablation. *Biochim. Biophys. Acta (BBA)-Biomembr.* **2005**, *1711*, 183–196. [CrossRef] [PubMed]

27. Iacobas, D.A.; Iacobas, S.; Spray, D.C. Connexin-dependent transcellular transcriptomic networks in mouse brain. *Prog. Biophys. Mol. Biol.* **2007**, *94*, 168–184. [CrossRef] [PubMed]

28. Iacobas, D.A.; Iacobas, S.; Urban-Maldonado, M.; Scemes, E.; Spray, D.C. Similar transcriptomic alterations in Cx43 knock-down and knock-out astrocytes. *Cell Commun. Adhes.* **2008**, *15*, 195–206. [CrossRef] [PubMed]

29. Iacobas, S.; Iacobas, D.A.; Spray, D.C.; Scemes, E. The connexin43 transcriptome during brain development: Importance of genetic background. *Brain Res.* **2012**, *1487*, 131–139. [CrossRef]

30. Li, F.; Aljahdali, I.A.M.; Zhang, R.; Nastiuk, K.L.; Krolewski, J.J.; Ling, X. Kidney cancer biomarkers and targets for therapeutics: Survivin (BIRC5), XIAP, MCL-1, HIF1α, HIF2α, NRF2, MDM2, MDM4, p53, KRAS and AKT in renal cell carcinoma. *J. Exp. Clin. Cancer Res.* **2021**, *40*, 254. [CrossRef]

31. Tu, S.-M.; Zhang, M.; Wood, C.G.; Pisters, L.L. Stem Cell Theory of Cancer: Origin of Tumor Heterogeneity and Plasticity. *Cancers* **2021**, *13*, 4006. [CrossRef]

32. Li, H.; Zhao, S.; Fan, H.Y.; Li, Y.; Wu, X.P.; Miao, Y.P. The Effect of Histogram Analysis of DCE-MRI Parameters on Differentiating Renal Tumors. *Clin. Lab.* **2023**, *69*, 2201–2207. [CrossRef]

33. Brady, L.; Kriner, M.; Coleman, I.; Morrissey, C.; Roudier, M.; True, L.D.; Gulati, R.; Plymate, S.R.; Zhou, Z.; Birditt, B.; et al. Inter and intra-tumor heterogeneity of metastatic prostate cancer determined by digital spatial gene expression profiling. *Nat. Commun.* **2021**, *12*, 1426. [CrossRef]

34. Iacobas, S.; Iacobas, D.A. Personalized 3-Gene Panel for Prostate Cancer Target Therapy. *Curr. Issues Mol. Biol.* **2022**, *44*, 360–382. [CrossRef] [PubMed]

35. de Vries-Brilland, M.; Rioux-Leclercq, N.; Meylan, M.; Dauvé, J.; Passot, C.; Spirina-Menand, E.; Flippot, R.; Fromont, G.; Gravis, G.; Geoffrois, L.; et al. Comprehensive analyses of immune tumor microenvironment in papillary renal cell carcinoma. *J. Immunother. Cancer* **2023**, *11*, e006885. [CrossRef] [PubMed]

36. Gui, Z.; Du, J.; Wu, N.; Shen, N.; Yang, Z.; Yang, H.; Wang, X.; Zhao, N.; Zeng, Z.; Wei, R.; et al. Immune regulation and prognosis indicating ability of a newly constructed multi-genes containing signature in clear cell renal cell carcinoma. *BMC Cancer* **2023**, *23*, 649. [CrossRef] [PubMed]

37. Qiagen Ingenuity Pathway Analysis. Available online: https://digitalinsights.qiagen.com/products-overview/discovery-insights-portfolio/analysis-and-visualization/qiagen-ipa/ (accessed on 12 July 2023).

38. DAVID Functional Annotation Bioinformatics Microarray Analysis. Available online: https://david.ncifcrf.gov (accessed on 12 July 2023).

39. Kyoto Encyclopedia of Genes and Genomes. Available online: https://www.kegg.jp/kegg/pathway.html (accessed on 12 July 2023).

40. Iacobas, D.A.; Mgbemena, V.; Iacobas, S.; Menezes, K.M.; Wang, H.; Saganti, P.B. Genomic fabric remodeling in metastatic clear cell renal cell carcinoma (ccRCC): A new paradigm and proposal for a personalized gene therapy approach. *Cancers* **2020**, *12*, 3678. [CrossRef] [PubMed]

41. Iacobas, D.A.; Tuli, N.; Iacobas, S.; Rasamny, J.K.; Moscatello, A.; Geliebter, J.; Tiwari, R.M. Gene master regulators of papillary and anaplastic thyroid cancer phenotypes. *Oncotarget* **2018**, *9*, 2410–2424. [CrossRef] [PubMed]

42. Iacobas, S.; Iacobas, D.A. A Personalized Genomics Approach of the Prostate Cancer. *Cells* **2021**, *10*, 1644. [CrossRef] [PubMed]

43. Remodeling of Major Genomic Fabrics and Their Interplay in Metastatic Clear Cell Renal Carcinoma. Available online: https://www.ncbi.nlm.nih.gov/geo/query/acc.cgi?acc=GSE72304 (accessed on 6 October 2023).

44. Iacobas, D.A.; Iacobas, S.; Lee, P.R.; Cohen, J.E.; Fields, R.D. Coordinated Activity of Transcriptional Networks Responding to the Pattern of Action Potential Firing in Neurons. *Genes* **2019**, *10*, 754. [CrossRef] [PubMed]

45. Ebbing, D.; Gammon, S.D. *General Chemistry—Standalone Book*; Student Ed.; Cengage Learning: Boston, MA, USA, 2015; pp. 88–100.

46. Hansen, J.-P.; McDonald, I.R. Chapter 7—Time-dependent Correlation and Response Functions. In *Theory of Simple Liquids*, 4th ed.; Academic Press: London, UK, 2013; pp. 265–310.

47. Eisen, M.; Spellman, P.; Brown, P.; Botstein, D. Cluster analysis and display of genome-wide expression patterns. *Proc. Natl. Acad. Sci. USA* **1998**, *95*, 14863–14868. [CrossRef]

48. Butte, A.J.; Tamayo, P.; Slonim, D.; Golub, T.R.; Kohane, I.S. Discovering functional relationships between RNA expression and chemotherapeutic susceptibility using relevance networks. *Proc. Natl. Acad. Sci. USA* **2000**, *97*, 12182–12186. [CrossRef]

49. Horvath, S.; Dong, J. Geometric Interpretation of Gene Coexpression Network Analysis. *PLoS Comput. Biol.* **2008**, *4*, e1000117. [CrossRef]

50. Oldham, M.C.; Langfelder, P.; Horvath, S. Network methods for describing sample relationships in genomic datasets: Application to Huntington's disease. *BMC Syst. Biol.* **2012**, *6*, 63. [CrossRef] [PubMed]

51. Marbach, D.; Costello, J.C.; Küner, R.; Vega, N.M.; Prill, R.J.; Camacho, D.M.; Allison, K.R.; Aderhold, A.; Bonneau, R.; Chen, Y.; et al. Wisdom of the crowds for robust gene network inference. *Nat. Methods* **2012**, *9*, 796–804. [CrossRef] [PubMed]

52. P Value from Pearson (R) Calculator. Available online: https://www.socscistatistics.com/pvalues/pearsondistribution.aspx (accessed on 1 September 2023).

53. Mathew, R.; Iacobas, S.; Huang, J.; Iacobas, D.A. Metabolic Deregulation in Pulmonary Hypertension. *Curr. Issues Mol. Biol.* **2023**, *45*, 4850–4874. [CrossRef]

54. Iacobas, S.; Ede, N.; Iacobas, D.A. The Gene Master Regulators (GMR) Approach Provides Legitimate Targets for Personalized, Time-Sensitive Cancer Gene Therapy. *Genes* **2019**, *10*, 560. [CrossRef] [PubMed]

55. Iacobas, D.A. Biomarkers, Master Regulators and Genomic Fabric Remodeling in a Case of Papillary Thyroid Carcinoma. *Genes* **2020**, *11*, 1030. [CrossRef]

56. Iacobas, D.A.; Xi, L. Theory and Applications of the (Cardio) Genomic Fabric Approach to Post-Ischemic and Hypoxia-Induced Heart Failure. *J. Pers. Med.* **2022**, *12*, 1246. [CrossRef]

57. Aldosterone-Regulated Sodium Reabsorption. Available online: https://www.genome.jp/pathway/hsa04960 (accessed on 1 September 2023).

58. Collecting Duct Acid Secretion. Available online: https://www.genome.jp/pathway/hsa04966 (accessed on 1 September 2023).

59. Endocrine and Other Factor-Regulated Calcium Reabsorption. Available online: https://www.genome.jp/pathway/hsa04961 (accessed on 1 September 2023).

60. Proximal Tubule Bicarbonate Reclamation. Available online: https://www.genome.jp/kegg-bin/show_pathway?hsa04964 (accessed on 9 January 2023).

61. Vasopressin-Regulated Water Reabsorption. Available online: https://www.genome.jp/kegg-bin/show_pathway?hsa04962 (accessed on 1 September 2023).

62. Thomas, W.; Harvey, B.J. Estrogen-induced signalling and the renal contribution to salt and water homeostasis. *Steroids* **2023**, *199*, 109299. [CrossRef]

63. Xu, W.H.; Xu, Y.; Tian, X.; Anwaier, A.; Liu, W.R.; Wang, J.; Zhu, W.K.; Cao, D.L.; Wang, H.K.; Shi, G.H.; et al. Large-scale transcriptome profiles reveal robust 20-signatures metabolic prediction models and novel role of G6PC in clear cell renal cell carcinoma. *J. Cell. Mol. Med.* **2020**, *24*, 9012–9027. [CrossRef]

64. Li, Z.; Xu, H.; Yu, L.; Wang, J.; Meng, Q.; Mei, H.; Cai, Z.; Chen, W.; Huang, W. Patient-derived renal cell carcinoma organoids for personalized cancer therapy. *Clin. Transl. Med.* **2022**, *12*, e970. [CrossRef]

65. Gerlinger, M.; Rowan, A.J.; Horswell, S.; Math, M.; Larkin, J.; Endesfelder, D.; Gronroos, E.; Martinez, P.; Matthews, N.; Stewart, A.; et al. Intratumor heterogeneity and branched evolution revealed by multiregion sequencing. *N. Engl. J. Med.* **2012**, *366*, 883–892, Erratum in: *N. Engl. J. Med.* **2012**, *367*, 976. [CrossRef]

66. Muto, Y.; Dixon, E.E.; Yoshimura, Y.; Wu, H.; Omachi, K.; Ledru, N.; Wilson, P.C.; King, A.J.; Eric Olson, N.; Gunawan, M.G.; et al. Defining cellular complexity in human autosomal dominant polycystic kidney disease by multimodal single cell analysis. *Nat. Commun.* **2022**, *13*, 6497. [CrossRef]

67. Jiang, Y.; Wang, Y.; Wang, Z.; Zhang, Y.; Hou, Y.; Wang, X. Anoikis-related genes signature development for clear cell renal cell carcinoma prognosis and tumor microenvironment. *Sci. Rep.* **2023**, *13*, 18909. [CrossRef]

68. Chen, W.; Lin, W.; Wu, L.; Xu, A.; Liu, C.; Huang, P. A Novel Prognostic Predictor of Immune Microenvironment and Therapeutic Response in Kidney Renal Clear Cell Carcinoma based on Necroptosis-related Gene Signature. *Int. J. Med. Sci.* **2022**, *19*, 377–392. [CrossRef]

69. Clark, D.J.; Dhanasekaran, S.M.; Petralia, F.; Pan, J.; Song, X.; Hu, Y.; da Veiga Leprevost, F.; Reva, B.; Lih, T.M.; Clinical Proteomic Tumor Analysis Consortium; et al. Integrated Proteogenomic Characterization of Clear Cell Renal Cell Carcinoma. *Cell* **2019**, *179*, 964–983.e31, Erratum in *Cell* **2020**, *180*, 207. [CrossRef]

70. Li, L.; Tan, H.; Zhou, J.; Hu, F. Predicting response of immunotherapy and targeted therapy and prognosis characteristics for renal clear cell carcinoma based on m1A methylation regulators. *Sci. Rep.* **2023**, *13*, 12645. [CrossRef]

71. Wang, X.X.; Myakala, K.; Libby, A.E.; Krawczyk, E.; Panov, J.; Jones, B.A.; Bhasin, K.; Shults, N.; Qi, Y.; Krausz, K.W.; et al. Estrogen-Related Receptor Agonism Reverses Mitochondrial Dysfunction and Inflammation in the Aging Kidney. *Am. J. Pathol.* **2023**, *193*, 1969–1987. [CrossRef]

72. Adamopoulos, P.G.; Kontos, C.K.; Diamantopoulos, M.A.; Scorilas, A. Molecular cloning of novel transcripts of the adaptor-related protein complex 2 alpha 1 subunit (AP2A1) gene, using Next-Generation Sequencing. *Gene* **2018**, *678*, 55–64. [CrossRef]

73. Pleasance, E.; Titmuss, E.; Williamson, L.; Kwan, H.; Culibrk, L.; Zhao, E.Y.; Dixon, K.; Fan, K.; Bowlby, R.; Jones, M.R.; et al. Pan-cancer analysis of advanced patient tumors reveals interactions between therapy and genomic landscapes. *Nat. Cancer* **2020**, *1*, 452–468. [CrossRef]

74. Zhang, G.; Xi, M.; Li, Y.; Wang, L.; Gao, L.; Zhang, L.; Yang, Z.; Shi, H. The ADCY9 genetic variants are associated with glioma susceptibility and patient prognosis. *Genomics* **2021**, *113*, 706–716. [CrossRef] [PubMed]

75. Tang, Y.; Wang, T.; Zhang, A.; Zhu, J.; Zhou, T.; Zhou, Y.L.; Shi, J. ADCY9 functions as a novel cancer suppressor gene in lung adenocarcinoma. *J. Thorac. Dis.* **2023**, *15*, 1018–1035. [CrossRef] [PubMed]

76. Chao, X.; Jia, Y.; Feng, X.; Wang, G.; Wang, X.; Shi, H.; Zhao, F.; Jiang, C. A Case-Control Study of ADCY9 Gene Polymorphisms and the Risk of Hepatocellular Carcinoma in the Chinese Han Population. *Front. Oncol.* **2020**, *10*, 1450. [CrossRef] [PubMed]

77. Li, H.; Liu, Y.; Liu, J.; Sun, Y.; Wu, J.; Xiong, Z.; Zhang, Y.; Li, B.; Jin, T. Assessment of ADCY9 polymorphisms and colorectal cancer risk in the Chinese Han population. *J. Gene Med.* **2021**, *23*, e3298. [CrossRef] [PubMed]

78. Chen, Q.; Cai, L.; Liang, J. Construction of prognosis model of bladder cancer based on transcriptome. *J. Zhejiang Univ. (Med. Sci.)* **2022**, *51*, 79–86. (In English) [CrossRef]

79. Lee, Y.H.; Gyu Song, G. Genome-wide pathway analysis in pancreatic cancer. *J. Buon* **2015**, *20*, 1565–1575. [PubMed]

80. Guo, R.; Liu, T.; Shasaltaneh, M.D.; Wang, X.; Imani, S.; Wen, Q. Targeting Adenylate Cyclase Family: New Concept of Targeted Cancer Therapy. *Front. Oncol.* **2022**, *12*, 829212. [CrossRef] [PubMed]

81. Wang, H.; Zhang, W.; Ding, Z.; Xu, T.; Zhang, X.; Xu, K. Comprehensive exploration of the expression and prognostic value of AQPs in clear cell renal cell carcinoma. *Medicine* **2022**, *101*, e29344. [CrossRef] [PubMed]

82. Oncomine Solutions For Next-Generation Sequencing. Available online: http://www.oncomine.org (accessed on 1 October 2023).

83. The University of ALabama at Birmingham CANcer Data Analysis Portal. Available online: http://ualcan.path.uab.edu (accessed on 1 October 2023).

84. Sinha, S.; Dwivedi, N.; Tao, S.; Jamadar, A.; Kakade, V.R.; Neil, M.O.; Weiss, R.H.; Enders, J.; Calvet, J.P.; Thomas, S.M.; et al. Targeting the vasopressin type-2 receptor for renal cell carcinoma therapy. *Oncogene* **2020**, *39*, 1231–1245. [CrossRef]

85. Baltzer, S.; Bulatov, T.; Schmied, C.; Krämer, A.; Berger, B.T.; Oder, A.; Walker-Gray, R.; Kuschke, C.; Zühlke, K.; Eichhorst, J.; et al. Aurora Kinase A Is Involved in Controlling the Localization of Aquaporin-2 in Renal Principal Cells. *Int. J. Mol. Sci.* **2022**, *23*, 763. [CrossRef]

86. KEGG-Constructed cAMP Signaling Pathway. Available online: https://www.genome.jp/pathway/hsa04024+109 (accessed on 1 October 2023).

87. Kim, T.H.; Park, J.M.; Kim, M.Y.; Ahn, Y.H. The role of CREB3L4 in the proliferation of prostate cancer cells. *Sci. Rep.* **2017**, *7*, 45300. [CrossRef]

88. Giménez-Bachs, J.M.; Salinas-Sánchez, A.S.; Serrano-Oviedo, L.; Nam-Cha, S.H.; Rubio-Del Campo, A.; Sánchez-Prieto, R. Carbonic anhydrase IX as a specific biomarker for clear cell renal cell carcinoma: Comparative study of Western blot and immunohistochemistry and implications for diagnosis. *Scand. J. Urol. Nephrol.* **2012**, *46*, 358–364. [CrossRef]

89. Tostain, J.; Li, G.; Gentil-Perret, A.; Gigante, M. Carbonic anhydrase 9 in clear cell renal cell carcinoma: A marker for diagnosis, prognosis and treatment. *Eur. J. Cancer* **2010**, *46*, 3141–3148. [CrossRef]

90. Ferro, M.; Falagario, U.G.; Barone, B.; Maggi, M.; Crocetto, F.; Busetto, G.M.; Giudice, F.D.; Terracciano, D.; Lucarelli, G.; Lasorsa, F.; et al. Artificial Intelligence in the Advanced Diagnosis of Bladder Cancer-Comprehensive Literature Review and Future Advancement. *Diagnostics* **2023**, *13*, 2308. [CrossRef]

91. Ferro, M.; Crocetto, F.; Barone, B.; Del Giudice, F.; Maggi, M.; Lucarelli, G.; Busetto, G.M.; Autorino, R.; Marchioni, M.; Cantiello, F.; et al. Artificial intelligence and radiomics in evaluation of kidney lesions: A comprehensive literature review. *Ther. Adv. Urol.* **2023**, *15*, 17562872231164803. [CrossRef]

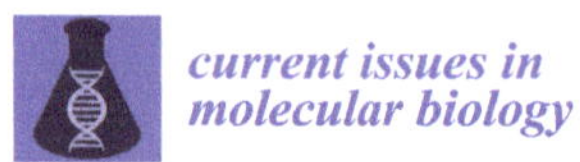

current issues in molecular biology

MDPI

Article

Potential Diagnostic Biomarker Detection for Prostate Cancer Using Untargeted and Targeted Metabolomic Profiling

Diana Nitusca [1,2], Carmen Socaciu [3], Andreea Iulia Socaciu [4], Ioan Ovidiu Sirbu [1,2], Razvan Bardan [5,6], Alin Adrian Cumpanas [5,6], Edward Seclaman [1,2] and Catalin Marian [1,2,*]

1   Department of Biochemistry and Pharmacology, Victor Babes University of Medicine and Pharmacy, Pta Eftimie Murgu Nr. 2, 300041 Timisoara, Romania; nitusca.diana@umft.ro (D.N.); ovidiu.sirbu@umft.ro (I.O.S.); eseclaman@umft.ro (E.S.)
2   Center for Complex Networks Science, Victor Babes University of Medicine and Pharmacy, Pta Eftimie Murgu Nr. 2, 300041 Timisoara, Romania
3   BIODIATECH, Research Center for Applied Biotechnology in Diagnosis and Molecular Therapy, 400478 Cluj-Napoca, Romania; csocaciudac@gmail.com
4   Department of Occupational Health, Iuliu Hateganu University of Medicine and Pharmacy, Str. Victor Babes Nr. 8, 400347 Cluj-Napoca, Romania; andreeaiso@gmail.com
5   Department of Urology, Victor Babes University of Medicine and Pharmacy, Pta Eftimie Murgu Nr. 2, 300041 Timisoara, Romania; razvan.bardan@umft.ro (R.B.); cumpanas.alin@umft.ro (A.A.C.)
6   Urology Clinic, Timisoara Emergency County Hospital, 300723 Timisoara, Romania
*   Correspondence: cmarian@umft.ro

Abstract: Prostate cancer (PCa) remains one of the leading causes of cancer mortality in men worldwide, currently lacking specific, early detection and staging biomarkers. In this regard, modern research focuses efforts on the discovery of novel molecules that could represent potential future non-invasive biomarkers for the diagnosis of PCa, as well as therapeutic targets. Mounting evidence shows that cancer cells express an altered metabolism in their early stages, making metabolomics a promising tool for the discovery of altered pathways and potential biomarker molecules. In this study, we first performed untargeted metabolomic profiling on 48 PCa plasma samples and 23 healthy controls using ultra-high-performance liquid chromatography coupled with electrospray ionization quadrupole time-of-flight mass spectrometry (UHPLC-QTOF-[ESI+]-MS) for the discovery of metabolites with altered profiles. Secondly, we selected five molecules (L-proline, L-tryptophan, acetylcarnitine, lysophosphatidylcholine C18:2 and spermine) for the downstream targeted metabolomics and found out that all the molecules, regardless of the PCa stage, were decreased in the PCa plasma samples when compared to the controls, making them potential biomarkers for PCa detection. Moreover, spermine, acetylcarnitine and L-tryptophan had very high diagnostic accuracy, with AUC values of 0.992, 0.923 and 0.981, respectively. Consistent with other literature findings, these altered metabolites could represent future specific and non-invasive candidate biomarkers for PCa detection, which opens novel horizons in the field of metabolomics.

Keywords: metabolomics; prostate cancer; biomarkers; diagnosis

Citation: Nitusca, D.; Socaciu, C.; Socaciu, A.I.; Sirbu, I.O.; Bardan, R.; Cumpanas, A.A.; Seclaman, E.; Marian, C. Potential Diagnostic Biomarker Detection for Prostate Cancer Using Untargeted and Targeted Metabolomic Profiling. *Curr. Issues Mol. Biol.* **2023**, *45*, 5036–5051. https://doi.org/10.3390/cimb45060320

Academic Editor: Dumitru A. Iacobas

Received: 26 April 2023
Revised: 27 May 2023
Accepted: 3 June 2023
Published: 8 June 2023

1. Introduction

One of the leading malignancies affecting men worldwide (of all races and ethnic groups) is prostate cancer (PCa). Globally, although this disease has a rapidly growing incidence (268,490 estimated new cases in 2022 alone in the US, according to the National Institutes of Health (NIH), it has a relatively high survival rate and favorable progression [1,2]. It typically affects middle-aged men (between 45 and 60 years), with risk factors including race, family history, genetic and environmental factors such as diet, smoking, obesity, chemical exposure, sexually-transmitted diseases and vasectomy [3].

The specific reasons behind the rise in PCa incidence over the years remain to be fully elucidated; modern research indicates that current diagnostic strategies are either highly invasive or lack proper specificity. This is the case for plasma prostate-specific antigen (PSA, ng/mL) screening, which, on the one hand, has helped reduce PCa-specific mortality over the last 30 years, but, on the other hand, has led to major overdiagnosis and over-treatment when used indiscriminately, due to its lack of specificity (being increased in other benign PCa-related diseases, such as prostatitis, benign prostatic hyperplasia, etc.) [4]. Other diagnostic methods include digital rectal examination (DRE), which is usually at risk for clinician subjectivity, magnetic resonance imaging (MRI) and the gold standard, transperineal prostate biopsy, which is highly invasive and leads to potential hospitalization due to infection, antibiotic use and the overall patient burden [5,6].

Therefore, in recent years, modern research has focused on the discovery of novel putative biomarkers that could improve the diagnosis and treatment of PCa, given the numerous publications dealing with this issue over the last decade. Such biomarkers belong to various classes of biological compounds, including proteins and other metabolites [7–12]. These molecules can now be easily identified and characterized (qualitatively and quantitatively) using a plethora of mass spectrometry-based techniques, generally designed as metabolomics approaches (referring to the scientific study of cellular processes involving small molecule metabolite profiles and their chemical fingerprints) [13]. In this regard, metabolomics is considered to represent the apogee of "omic" (genomic, transcriptomic and proteomic) technologies and helps in finding a detailed understanding of various biochemical events inside cells and their relationship with each other. Different metabolomic studies have been reported in biomarker discovery on various diseases in recent years (including PCa and other cancers) as a new powerful technology and a dynamic field for global comprehension of biological systems. Commonly applied techniques in metabolomic analysis are mass spectrometry (MS)-based techniques, including gas chromatography–mass spectrometry (GC–MS), liquid chromatography–mass spectrometry (LC–MS) or magnetic resonance spectroscopy (MRS) [14–16].

Since cancer cells are characterized by altered metabolic pathways, the determination of low-molecular-weight metabolites in biological fluids (via liquid biopsy) could represent a novel, minimally invasive diagnostic method associated with high diagnostic potential and specificity, reduced invasiveness, increased compliance and a low overall burden for patients suffering from this disease [17].

Recently, it was found that free AA and lipid profiles vary depending on the type of cancer and its stage [18]. Examples of AA that were previously found to be dysregulated in PCa samples when compared to controls include alanine, arginine, uracil, glutamate (from tissue specimens) and valine, taurine and leucine (from both tissue and urine samples) [19]. In addition, one systematic review conducted by Kdadra et al. (2019) revealed that, in general, the vast majority of metabolites with altered profiles between PCa patients and controls belong to either the lipids class (lysophosphatidylcholine 18:2, decanoilcarnitine, stearate, docosadienoate), AAs (methionine, glutamine, isoleucine, arginine, leucine), amines (ethanolamine, sarcosine) and/or their derivatives [20]. Moreover, a laboratory-developed test called Prostarix (performed by the CLIA lab at Metabolon, Inc. and offered through Boswick Laboratories) is commercially available for use in the decision to perform initial or repeat biopsies for DRE-negative patients with moderately high PSA values. This test includes metabolomic profiling for sarcosine, glycine, alanine and glutamate, and it is performed from urine samples. However, Prostarix has not been approved by the FDA yet, and the urine samples need to be collected immediately following a vigorous DRE [21].

Therefore, in this study, we performed an untargeted metabolomics analysis on 48 plasma samples from diagnosed PCa patients and on 23 plasma samples from healthy controls using ultra-high-performance liquid chromatography coupled with electrospray ionization quadrupole time-of-flight-mass spectrometry (UHPLC-QTOF-[ESI+]-MS) to discover potential biomarkers for PCa detection. Next, based on this study and other literature findings, we selected five molecules for our downstream targeted metabolomics study: two free AAs (L-proline, L-tryptophan) and other endogenous molecules, such as acetylcarnitine, lysophosphatidylcholine C18:2 and spermine. Our aim was to identify potential biomarkers for a more minimally invasive diagnosis and staging of PCa.

2. Materials and Methods

2.1. Study Subjects and Clinical Data

This study comprised a total number of 71 subjects (48 PCa patients and 23 healthy controls). Blood was collected from each individual at the Urology Clinic of the Clinical Emergency County Hospital in Timisoara, Romania, following their agreement to participate in the study (all the subjects filled in a written IRB informed consent for the use of their biological specimens). The study was approved by the Ethics Committee of the participating institutions, in accordance with the 1964 Declaration of Helsinki and its later amendments, venous blood was collected in specific EDTA collection tubes; then, the plasma was separated and kept at $-80\ ^\circ$C until further use. The patients were diagnosed with prostate adenocarcinoma following transrectal biopsies for histopathological diagnosis of PCa. The control volunteers had no prostate disease and normal PSA values (<4 ng/mL), verified by chemiluminescent microparticle immunoassay (Abbott Diagnostics, Chicago, IL, USA), and their blood samples were drawn at the same clinic.

2.2. Sample Preparation

A volume of 0.8 mL mix of methanol and acetonitrile (2:1 v/v) was added to 0.2 mL of blood plasma. The mixture was vortexed to precipitate the proteins, ultrasonicated for 5 min and kept for 24 h at $-20\ ^\circ$C. The supernatant was collected after centrifugation at 12,500 rpm for 10 min (4 $^\circ$C) and filtered through nylon filters (0.2 μm). Finally, the supernatant was placed in glass micro vials and transferred to the autosampler of the ultra-high-performance liquid chromatograph (UHPLC) before injection.

Quality control (QC) samples were also obtained by mixing the plasma samples to provide a representative "mean" sample and to validate the method. The QC samples were injected at the beginning and end of every 10th injection while analyzing the study samples in the UHPLC.

2.3. UHPLC-QTOF-ESI+-MS Analysis

The metabolic profiling was performed by ultra-high-performance liquid chromatography coupled with electrospray ionization quadrupole time-of-flight-mass spectrometry (UHPLC-QTOF-ESI+-MS) using a ThermoFisher Scientific UHPLC Ultimate 3000 instrument equipped with a quaternary pump, Dionex delivery system and MS detection equipment with MaXis Impact (Bruker Daltonics, Billerica, MA, USA). The metabolites were separated on an Acclaim C18 column (5 μm, 2.1 $\times$ 100 mm with a pore size of 30 nm (Thermo Scientific, Waltham, MA, USA) at 28 $^\circ$C. The mobile phase consisted of 0.1% formic acid in water (A) and 0.1% formic acid in acetonitrile (B). The elution time was set at 20 min. The flow rate was set at 0.3 mL/min. The gradient for the plasma samples was: 90 to 85% A (0–3 min), 85–50% A (3–6 min), 50–30% (6–8 min), 30–5% (8–12 min) and, afterward, increased to 90% at 20 min. The volume of injected extract was 5 μL, and the column temperature was 25 $^\circ$C. Several QC samples obtained from each group were used in parallel to calibrate the separations. Different volumes from a solution of 2 mg/mL Doxorubicin hydrochloride (MW = 580) were added in parallel to the QC samples as an internal standard (IS).

The ESI+-MS calibration was done with sodium formiate and the applied parameters were: capillary voltage of 3500 V, nebulizing gas pressure of 2.8 bar, drying gas flow of 12 L/min and drying temperature of 300 $^\circ$C. The m/z values for the molecules to be separated were set between 50 and 600 Da. The control of the instrument and the data processing were done using the specific software TofControl 3.2, HyStar 3.2, Data Analysis 4.2 (Bruker, Daltonics, MA, USA) and Chromeleon, respectively.

2.4. Data Processing and Statistical Analysis

The Bruker software Data Analysis 4.2, attached to the instrument, was used to process the acquired data. By using the peak dissect algorithm, details of the separated molecules were obtained. By using the algorithm Find Molecular Features (FMF), we generated the first advanced bucket matrix, which included, for each m/z value, the retention time, the peak area, the peak intensity and the signal/noise (S/N) ratio. From the total ion chromatogram, using specific algorithms, total ion chromatograms (TICs) and base peak chromatograms (BPCs) were obtained. The number of separated molecules ranged between 1800 and 3000 in all the plasma samples.

The matrices representing the peak intensity = f(m/z value) for each sample were stored in an Excel file.

In the first step, we eliminated the molecules having retention times below 0.8 min (dead volume of the HPLC column), the molecules with S/N values < 5 (noise elimination), molecules with m/z values over 550 Da and minor molecules and residues with peak intensities under 1000 units. The number of molecules selected for the statistics decreased to 540 and the range of m/z values was between 100 and 550 nm.

In the second step, the alignment of common molecules (with the same m/z value) in all the samples, using the online software from www.bioinformatica.isa.cnr.it/NEAPOLIS, was performed, keeping for the final matrix the molecules common in more than 80% of the samples. The final matrices contained the m/z values versus the peak MS peak intensity for each molecule and each sample.

In some cases, no peak intensities were found, meaning that the molecule was not present, or it was present at intensities below 1000 (traces).

The Excel matrix (.xlsx) was converted to a .csv file, which was introduced in the Metaboanalyst 5.0 platform for multivariate and univariate analysis (https://www.metabo analyst.ca/MetaboAnalyst/ModuleView.xhtml (accessed on 1 November 2022)).

The untargeted metabolomic analysis was performed by:

- Multivariate analysis comparing the control group with the whole patients' group, based on the final matrix.csv for each type of sample. Discriminations between these two groups were represented by the fold change, volcano test, PatternHunter analysis, partial least squares discriminant analysis (PLS-DA), sparse PLSDA (sPLS-DA) and variable importance in the projection (VIP) values including cross-validation parameters. Then, the random forest-based prediction test was applied, and the calculations of p-values were performed by t-tests. The heat maps of correlations were also built. Finally, using the biomarker analysis, the receiver operating curves (ROCs) and the values of the areas under the ROC curves (AUCs) were obtained, and the m/z values were ranked according to their sensitivity/specificity.
- Univariate analysis, which allowed us to compare the subgroups P1-P4 with the control group. The statistical analysis was conducted for each type of sample using one-way ANOVA, PLS-DA and sPLS-DA score analysis, PatternHunter, random forest and heat maps.

The results were presented graphically, and the putative biomarkers of differentiation were identified. The identification of molecules, based on their m/z value and the retention time, was conducted in agreement with our own database and other international databases for metabolomics: the Human Metabolome Database (http://www.hmdb.ca, accessed on 1 November 2022), Lipid Maps (http://www.lipidmaps.org, accessed on 1 November 2022), PubChem (https://pubchem.ncbi.nlm.nih.gov, accessed on 1 November 2022) and the Heidelberg Database (https://www.msomics.com, accessed on 1 November 2022).

For the statistical analysis of targeted metabolomics, we selected the matrices which included the above-mentioned five metabolites (m/z values vs. MS peak intensity, as .csv file) and applied the Metaboanalyst 5.0 platform for multivariate and univariate analysis (https://www.metaboanalyst.ca, accessed on 1 November 2022).

Differences between the two groups (the control versus the PCa group) were first analyzed using the multivariate analysis by fold change, PCA and PLS-DA score plots including VIP values. Volcano plots were generated with the log2 fold change values and Bonferroni-adjusted *p* values. The value of $p < 0.05$ was defined as statistically significant.

To test the discriminatory capacity of each metabolite, we performed a receiver operating characteristic (ROC) analysis. For the AUC values higher than 0.8, the metabolite was considered to have a very high prediction of disease and can be used as a candidate biomarker for further study.

In the second step, the one-way ANOVA univariate analysis tested the discrimination between the controls and the subgroups of PCa patients (PI, PII, PIII and PIV). The PCA and PLS-DA score plots including VIP values, cross-validation parameters, as well the mean decrease accuracy scores by random forest analysis, were performed.

According to the data collected from the untargeted metabolomics, we targeted five specific molecules in the plasma, and their MS peak intensities were compared in the different groups of patients, e.g., between groups C and P. The mean values of peak intensity (PI) and their standard deviations SD were calculated for these specific biomarkers, selected by the untargeted metabolomics and in agreement with the recent literature data. For quantitative analysis, the calibration curves were built with pure standards.

2.5. Quantitative Evaluation

2.5.1. Preparation of Calibration Solution and Quality Control (QC) Samples

The stock solutions of the five potential biomarkers, L-proline 10 mM, spermine 1 mM, acetylcarnitine 1 mM, L-Tryptophan 5 mM, lysophosphatidylcholine (LPC, 18:2) 2 mM, were dissolved in ultra-pure water. The stock solutions were successively diluted in the mix of methanol:acetonitrile 2:1 to obtain the series of working solutions at different concentration levels for external calibration. In parallel, volumes of 0.3 mL of QC deproteinated samples were spiked with different volumes of standard solutions.

2.5.2. Method Validation

According to the "Guidance for Industry: Bioanalytical Method Validation" recommended by the US Food and Drug Administration (FDA), the UHPLC-QTOF-ESI+-MS method was validated to evaluate the linearity, specificity, precision, accuracy, the limit of detection (LOD) and the limit of quantification (LOQ). Two calibration curves were generated: an external standard calibration curve (1), made by diluting standard solutions in the mobile phase, and an internal standard curve (2), whose linearity was determined for the QC samples spiked with different volumes of standard solutions. The mean peak area of three replicate measurements at each concentration was calculated.

2.5.3. Limit of Detection (LOD) and Limit of Quantification (LOQ)

The LOD was the lowest concentration of analyte in the test sample that could be reliably distinguished from zero to signal/noise ratio ≥ 10. The LOQ was the lowest concentration of analyte that could be determined with acceptable repeatability and trueness (signal/noise ratio ≥ 10 and SD values $\leq 40\%$).

2.6. Chemical Reagents

Pure standards of five targeted metabolites, namely L-proline, spermine, acetylcarnitine, L-tryptophan and lysophosphatidylcholine (LPC, 18:2) were purchased, as follows:

- L-proline (from Amino Acid Standard H product #20088, Thermo Scientific, Waltham, MA, USA). MW = 115.
- spermine (>97% product S3256, Sigma-Aldrich Chemie GmbH, St. Louis, MO, USA). MW = 202.
- o-acetyl-L-carnitine hydrochloride (J6153606; Alfa Aesar by Thermo Scientific, Waltham, MA, USA). MW = 203.
- L-tryptophan (>98.5% (T) (HPLC) (T0541;TCI Chemicals, Portland, OR, USA). MW = 204.

- lyso L-α-Lysophosphatidylcholine-LPC(18:2) from bovine brain (CAS nr. 9008-30-4) Sigma-Aldrich Chemie GmbH, St. Louis, MO, USA. MW = 519.

Other reagents such as HPLC-grade methanol, formic acid and acetonitrile were purchased from Sigma-Aldrich Chemie GmbH (St. Louis, MO, USA) and Thermo Fisher Scientific (Waltham, MA, USA). Ultrahigh-purity water was prepared by a Millipore-Q Water Purification System (Millipore, Darmstadt, Germany).

The instruments used in this study included a vortex mixer, Minicentrifuge Eppendorf (Thermo Fisher Scientific, Waltham, MA, USA) and UHPLC-QTOF-MS (Bruker GmbH, Bremen, Germany).

3. Results

3.1. Subjects' Clinicopathological Data

A total number of 71 male subjects were included in this study, of whom 48 had histopathologically confirmed PCa (different stages, different PSA levels), and 23 were age-matched healthy controls. Their data are presented in Table 1.

Table 1. Characteristics of the subjects included in the present study.

Characteristics	N = 48 (Patients)	%	N = 23 (Controls)	%
Age (mean ± SD)	63 (±5.49)		54 (±1.29)	
Prostate-specific antigen—PSA (ng/mL)				
<4	0	0.00	23	100.00
4–10	28	58.33	0	0.00
>10	20	41.66	0	0.00
Prostatic volume (mL) mean ± SD	44 (±19.83)			
PSA density (ng/mL^2) mean ± SD	0.28 (±0.15)			
Gleason score				
6	14	29.17		
7–8	32	66.67		
9	2	4.16		
AJCC stage				
I	5	4.16		
II-III	39	87.5		
IV	4	8.33		

Regarding demographics, the mean age of the patients was 63 years, while the mean age of the healthy controls was 54 years. More than half of the patients had PSA levels of 4–10 ng/mL (58.33%), while all the controls had PSA levels under 4 ng/mL (100%). The majority of tumors had Gleason scores of 7 or 8 (66.67%). We included in our study patients with different AJCC stages (I to IV), with the vast majority (81.25%) having either stage 2 or 3.

3.2. Untargeted Metabolomic Profiling

3.2.1. Metabolite Identification

Following the discovery phase of our study, via the untargeted plasma profiling, we retrieved a total number of 296 molecules with molecular weights below 550 Da. These molecules were subsequently identified with putative names using different international databases, such as the Human Metabolome Database (http://www.hmdb.ca, accessed on 1 November 2022), Lipid Maps (http://www.lipidmaps.org, accessed on 1 November 2022), PubChem (https://pubchem.ncbi.nlm.nih.gov, accessed on 1 November 2022) and Heidelberg Database (https://www.msomics.com, accessed on 1 November 2022) and are shown in Table S1 (Supplementary Material). The vast majority of the metabolites identified were either AAs (such as L-proline, L-tryptophan, L-valine, L-threonine, L-aspartic acid, ornithine, L-lysine, L-glutamic acid, etc.) or products derived from AA metabolism. Other metabolites were amines (spermine, glutamine, serotonin) or, in general, nitrogen-containing compounds, while a substantial part was made of metabolites derived from lipid metabolism (phospholipids, lysophospholipids and fatty acids). A small part was metabolites of nucleotide structure, such as uridine, guanosine and inosine. This preliminary data is consistent with other literature findings regarding free AAs and lipid profiles.

3.2.2. Multivariate Untargeted Metabolomics Statistical Analysis of the 71 Plasma Samples

Following identification, we performed the statistical analysis. Volcano plots were generated to show the significant molecules which are increased or decreased between the groups: controls, n = 23, and patients, n = 48. The volcano plot (Figure 1) shows the molecules (*m/z* values) that have significantly lower or higher intensities in the patients' group when compared to the control group.

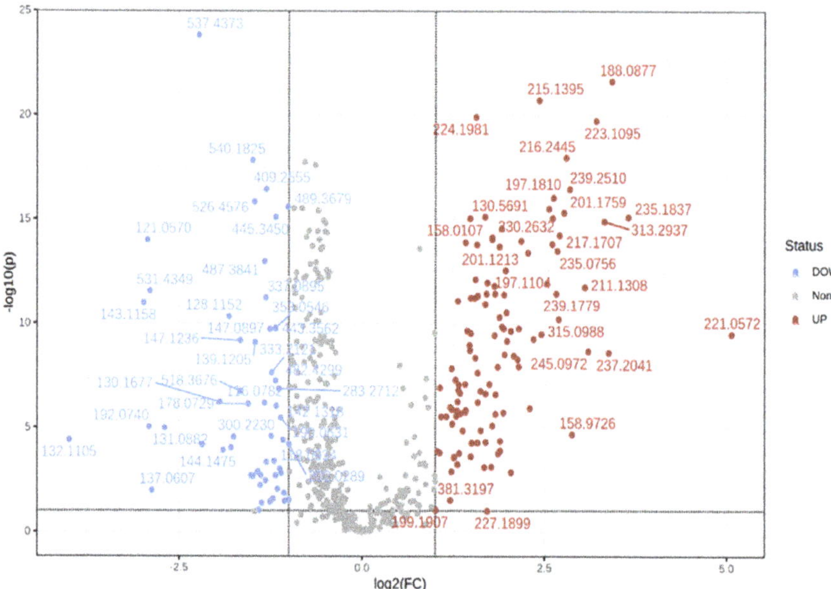

Figure 1. Volcano plot: red-marked *m/z* values are decreased in the patients' group compared to the control group; blue-marked *m/z* values are increased in the patients' group compared to the control group.

Next, a discrimination analysis was performed, where the partial least squares-discriminant analysis (PLS-DA) and sparse PLS-DA plots were generated. In general, they showed a good differentiation of metabolic patterns between the two groups (patients and controls). The sparse PLS-DA (sPLS-DA) algorithm reduces the number of feature variables (metabolites) and shows a more robust and easy-to-interpret model. One can control the "sparseness" of the model by adjusting the number of components in the model and the number of variables within each component (Figure 2a,b, respectively).

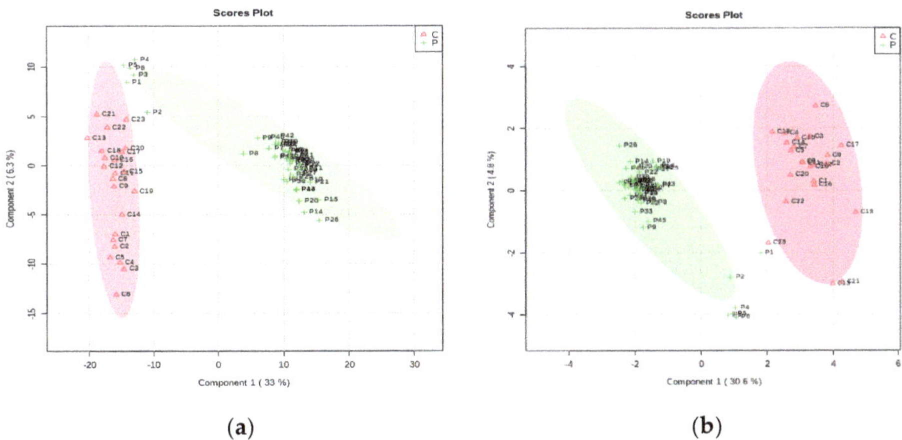

Figure 2. (**a**, **left**) PLS-DA plot with a covariance of 39.3% and (**b**, **right**) sPLS-DA plot with a covariance of 35.2% showing the discrimination between patient and control groups.

Then, we performed a biomarker analysis which included the analysis of the receiver operating characteristic (ROC) curve with the corresponding area under the curve values (AUC), and the metabolites were ranked based on the AUC values. The most representative molecules to discriminate between the control group and the patients'group, and to be considered putative biomarkers (with AUC values higher than 0.7), are shown in Table 2.

Table 2. The ranking of the first molecules based on AUC values higher than 0.7.

m/z	AUC	*p*-Value	Identification
188.0877	0.9973	2.43×10^{-22}	N1-Acetylspermidine
537.4373	0.9955	1.52×10^{-24}	Stearyl Stearate & Isomers
335.298	0.9873	3.07×10^{-16}	Docosatrienoic Acid C22:3
326.301	0.9846	1.86×10^{-18}	N-Myristoyl Proline
215.1395	0.9828	1.973×10^{-21}	Methyl lauric acid
540.1825	0.9828	1.525×10^{-18}	Hexacosanoyl Carnitine
223.1095	0.9792	1.92×10^{-20}	L-Cystathionine
489.3679	0.9792	2.59×10^{-16}	Cytidine 5'-Diphosphocholine
331.0208	0.9774	2.56×10^{-18}	Deoxycorticosterone
300.223	0.9765	2.66×10^{-5}	D-Sphingosine
526.4576	0.9728	1.45×10^{-16}	LPS(18:0)
239.251	0.9701	3.66×10^{-17}	5-Oxo-Tetradecadienoic Acid (C14:2)
409.2555	0.9701	3.64×10^{-17}	Ursocholic/Muricholic Acid Acid
205.1072	0.9692	3.16×10^{-16}	L-Tryptophan
227.1899	0.9692	0.099	Carnosine
419.2139	0.9683	7.21×10^{-15}	Palmitoyl Glucuronide
181.9605	0.9674	3.11×10^{-16}	Tyrosine
224.1981	0.9674	1.26×10^{-20}	N-Acetyl-Tyrosine

In addition, a discrimination analysis was also performed for the patients' group, according to their AJCC stage (I-IV), and for the controls, and the PLS-DA plot which resulted from the ANOVA analysis can be seen in Figure 3.

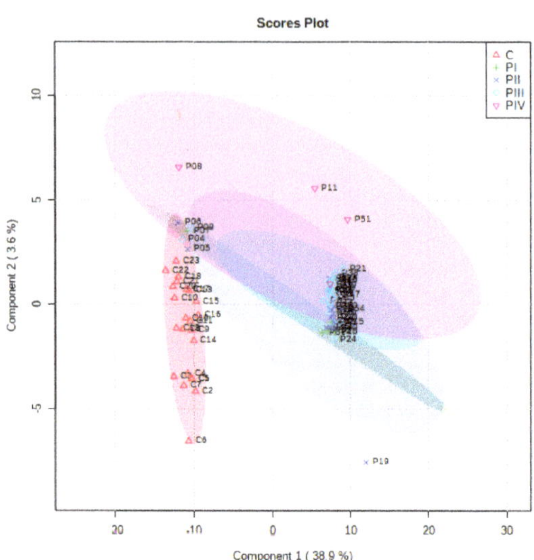

Figure 3. PLS-DA score plots showing the differences between controls and patients (subgroups PI-PIV).

From the PLS-DA analysis, the VIP scores were calculated and the ranking of the first 15 molecules to be considered responsible for the discrimination is presented in Figure 4.

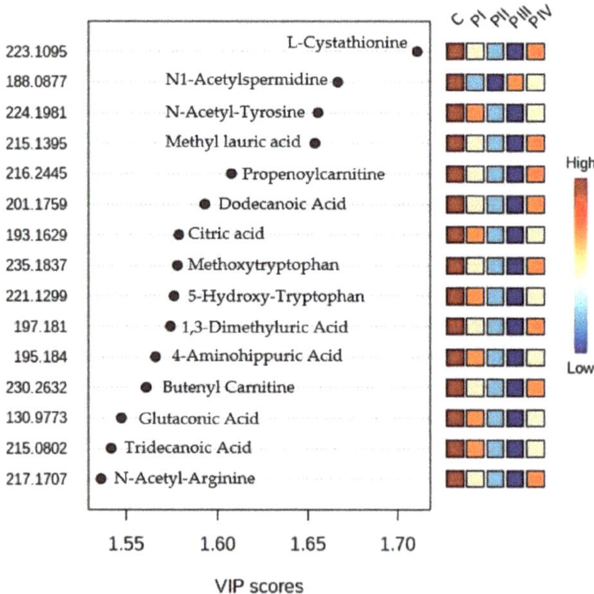

Figure 4. Ranking of top 15 molecules with higher VIP scores, according to PLS-DA analysis (their identification can be made by using Table S1).

3.3. Targeted Metabolomic Profiling

According to our previous untargeted analysis and consistent with other literature findings, we selected five different molecules (L-proline, spermine, acetylcarnitine,

L-tryptophan and lysophosphatidylcholine LPC 18:2) for our downstream targeted study. Even though these molecules were not generally among the top-ranked metabolites found in our untargeted study, we chose them based on the available literature findings regarding the three main metabolisms affected in PCa: AAs, phosphatidylcholines and carnitines.

3.3.1. Calibrations and Validation Parameters

The determination of linear ranges (calibration curves and equations including R^2 values), the limit of detection (LOD) and the limit of quantification (LOQ) of each standard are given in Table 3. The correlation coefficients (R^2) were higher than 0.898 for all the standards in their linear range, showing good linear relationships within linear ranges. All the LOD values were in the range of 0.3–4 µM, and the LOQ values were in the range of 0.9–5.5 µM.

Table 3. Validation parameters (linear range, curve equation, the correlation coefficients (R^2), limit of detection (LOD) and limit of quantification (LOQ) for each of the five molecules selected as potential biomarkers.

Molecule (*m/z*)	Linear Range µM	Curve Equation	R^2	LOD µM	LOQ µM
L-proline (116.1142)	10–100	y = 11290x − 119623	0.898	4.0	5.5
Spermine (203.0655)	0.5–5	y = 498406x + 51799	0.997	0.1	0.3
Acetylcarnitine (204.1369)	1–5	y = 36813x − 3879.2	0.994	0.2	0.8
L-tryptophan (205.1072)	4–40	y = 19944x − 7756	0.998	0.8	1.0
LPC 18:2 (520.3663)	1–10	y = 172161x + 91874	0.965	0.3	0.9

The validation of the LC–MS method for the quantitative evaluation of the metabolites was conducted using controlled additions of the internal standard (doxorubicin—DOXO) and each of the five pure standards to the quality control (QC) extracts.

To the same volume of QC extracts (0.3 mL), we added 0.2 mL of the five standard solutions (50 µM L-proline, 5 µM spermine, acetylcarnitine or LPC and 20 µM L-tryptophan) and of the internal standard (DOXO 2 mg/mL = 3.44 µM) with known concentrations of each metabolite. Table 4 shows the initial concentrations of metabolites after mixing with the QC extract and the measured concentrations after the LC–MS analysis. The recovery percentage was calculated as a measure of the method's reproducibility.

Table 4. The recovery percentage (%) was calculated from the measured concentrations of internal standard (IS) and each metabolite (pure standard) compared to their initial concentrations, after addition to QC extract.

Metabolite	Initial Concentration (mM)	Measured Concentration (mM)	Recovery (%)
L-proline	20	17.5	87.5
Spermine	2	1.82	91.0
Acetylcarnitine	2	1.88	94.0
L-tryptophan	8	7.55	94.3
LPC(18:2)	2	1.85	92.5
IS (DOXO)	1.4	1.25	89.3

3.3.2. Quantitative Determination

Unpaired *t*-tests showed that there were statistically significant differences between the patients (all stages) and controls for all five molecules analyzed. Spermine, acetylcarnitine and L-tryptophan showed *p* values below 0.0001, while L-proline and LPC 18:2 had *p* values of 0.02 and 0.01, respectively. All five molecules had significantly lower concentrations in the patients' group (all stages) when compared to the control group, as seen in Figure 5.

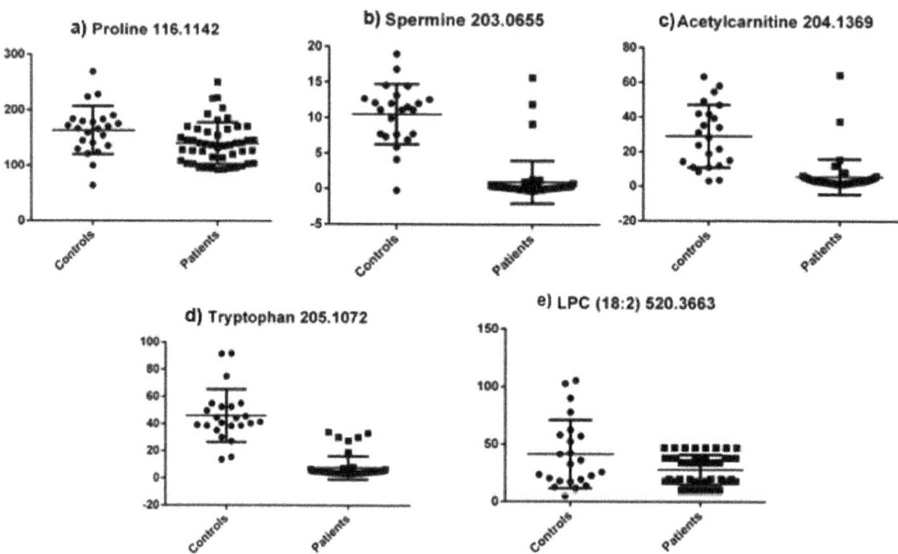

Figure 5. Metabolite concentrations (μM) between all patients and controls for the five selected molecules in the targeted analysis.

Next, we performed ROC curve analyses for the aforementioned molecules, and the obtained data can be seen in Table 5.

Table 5. ROC curve analysis for the molecules selected in the targeted metabolomic profiling.

Metabolite	*m*/*z*	AUC Value	95% CI	*p* Value
L-proline	116.1142	0.682	0.548–0.816	0.01
Spermine	203.0655	0.922	0.832–1.01	<0.0001
Acetylcarnitine	204.1369	0.923	0.855–0.991	<0.0001
L-tryptophan	205.1072	0.981	0.957–1.005	<0.0001
LPC (18:2)	520.3663	0.608	0.453–0.764	0.140

According to this data, the molecule with the highest diagnostic potential is represented by L-tryptophan (AUC value of 0.981), suggesting very good diagnostic potential for PCa detection. The same is true for acetylcarnitine and spermine (AUC values of 0.923 and 0.922, respectively), while L-proline and LPC 18:2 showed only a moderate diagnostic value, both having AUC values below 0.7. The ROC curves for these five molecules can be seen in Figure 6.

The quantitative evaluation, based on the curve equations for each of the five biomarkers, is presented in Table 6.

All the metabolites had decreased levels in the patients' group, regardless of their stage, when compared to the control group. For L-proline, we observed a gradual decrease in concentration from stage I to stage III, which slightly increased in stage IV. In addition, spermine levels decreased more than 10 times in the patients' group (all stages) compared to the control group, having the most significant change of all metabolites. For acetylcarnitine, there was a gradual increase in concentration, which was directly proportional to stages II-IV, while the reverse occurred for L-tryptophan (stages I-III), which had a similar profile to L-proline.

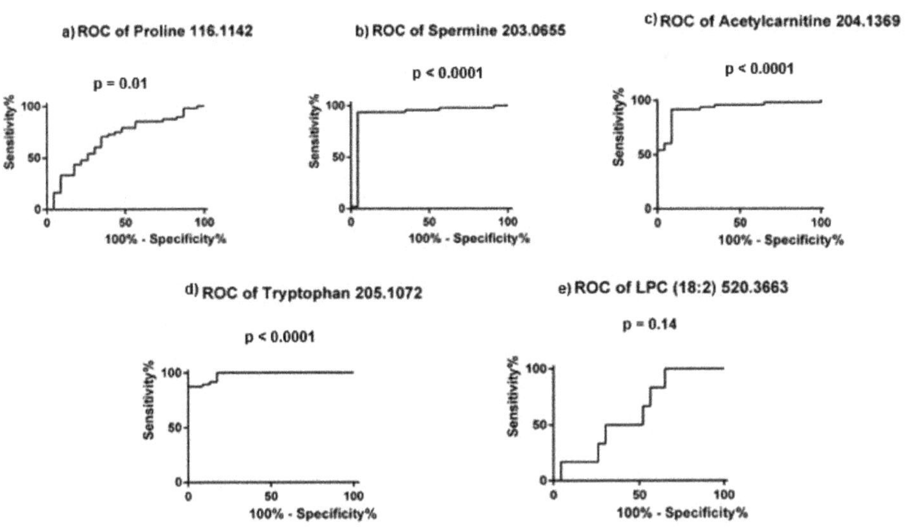

Figure 6. ROC curves for the five molecules selected in the targeted analysis.

Table 6. The mean values of concentrations (μM) of the five potential biomarkers targeted in this study for the control group, patients' group (all) and subgroups, PI, PII, PIII and PIV (according to their AJCC stage). The percentage of standard deviations (SD, %) for each group is also presented SD (%) = 100 × SD/mean value).

Molecule	Controls		Patients (All Stages)		PI		PII		PIII		PIV	
	Mean (μM)	SD (%)	Mean (μM)	SD (%)	Mean (μM)	SD (%)	Mean (μM)	SD (%)	Mean (μM)	SD (%)	Mean (μM)	SD (%)
L-proline	163.580	43.45	140.36	37.57	152.08	24.57	134.93	30.98	140.36	45.68	147.46	27.59
Spermine	10.484	4.22	0.98	3.04	3.05	7.05	0.06	0.37	1.19	2.99	0.87	0.10
Acetylcarnitine	29.020	18.03	5.65	10.27	4.10	4.56	3.49	1.69	6.29	12.98	12.53	16.70
L-tryptophan	45.908	19.39	7.63	8.45	9.67	11.58	7.60	9.45	6.16	5.67	13.68	13.65
LPC (18:2)	41.274	29.77	27.93	12.93	28.18	12.34	28.34	13.95	27.09	12.90	30.77	14.24

In addition, to evaluate the staging potential of these five molecules, ordinary one-way ANOVA tests were performed on the values of concentration (μM) for the controls and for the patients grouped according to their AJCC stage (I, II, III and IV). Statistically significant differences were observed for spermine, acetylcarnitine and L-tryptophan ($p < 0.0001$), but not for L-proline ($p = 0.209$) or LPC 18:2 ($p = 0.159$).

4. Discussion

In the present study, we performed an untargeted and targeted metabolomic analysis of human plasma samples from PCa-diagnosed patients compared to healthy controls using UHPLC–MS. The aim was to discover and validate potential biomarkers for the early detection and staging of this urologic malignancy in a minimally invasive fashion (from plasma samples). In the targeted analysis, we performed calibrations and validation analyses to evaluate the concentrations and the changes in the levels of the five potential biomarkers targeted: L-proline, L-tryptophan, spermine, acetylcarnitine and lysophosphatidylcholine (18:2). We found out that, regardless of the PCa stage, all the selected metabolites were significantly decreased in the PCa plasma samples when compared to the healthy controls, making these molecules potential future biomarkers for PCa diagnosis.

In this regard, spermine is a polyamine involved in cellular metabolism, having ornithine AA as a precursor. Our findings regarding decreased levels of L-proline in PCa patients' plasma samples vs. healthy controls are consistent with other literature findings, as Bentrad et al. (2019) described that the concentration of spermine was 7–34 times lower in PCa samples when compared to healthy individuals and even to patients with benign prostate hyperplasia [22]. In contrast, one study proposed spermine (among other metabolites) as a potential biomarker in distinguishing aggressive from indolent PCa, as spermine levels were decreased in the aggressive phenotype [23].

Acetylcarnitine, another potential biomarker that we found to be decreased in PCa plasma in our study, is an endogenous acetylated form of L-carnitine, a source that releases (by esterase hydrolysis) L-carnitine, which is known to be involved in the transport of fatty acids into the mitochondria for β-oxidation and ATP generation. In PCa, in particular, acetylcarnitine was shown to downregulate different pathways involved in angiogenesis (such as VEGF and CXCL8) and invasion (via the downregulation of CXCR4/CXCL12 and MMP-9), playing, therefore, a protective role against PCa development. The same study conducted by Baci et al. (2019) found that acetylcarnitine induces apoptosis of PCa cells, reduced cancer cell proliferation, and halted proinflammatory cytokines and chemokines production, such as TNF-α, IFN-γ and CCL2, CXCL12 and receptor CXCR4, respectively, therefore altering migration, invasion and adhesion processes. The authors, therefore, propose acetylcarnitine as a novel repurposed dietary supplement for chemoprevention in PCa [24].

L-tryptophan is an essential α-AA with an indole side chain, used in the biosynthesis of proteins. In PCa, in our study, we found out that L-tryptophan was significantly decreased in all the patients' samples compared to the controls, resulting in the highest diagnostic accuracy (AUC value of 0.981). One study revealed that aromatic AAs (including L-tryptophan) have altered metabolisms and play harmonizing roles in PCa and that their modifications could be used as potential biomarkers in the early detection of this disease [25].

Another AA that we chose to investigate in our targeted metabolomic profiling is L-proline, a proteinogenic, secondary, non-essential AA, synthesized endogenously from L-glutamate. It is generally used in the biosynthesis of proteins, although it does not contain the free amino group, NH2. However, the exact role of L-proline metabolism in PCa is yet to be fully elucidated. Nonetheless, a very recent study (2022) proposes prostate-specific membrane antigen (PSMA) as a promising future therapeutic target against PCa, as it demonstrates that this protein modulates PCa progression by regulating arginine and L-proline biosynthesis (its depletion promoted the synthesis of these two AA and inhibited androgen receptor expression) [26].

In addition, lysophosphatidylcholines are minor phospholipids from cell membranes and are found in blood plasma (8–12%), including different acyl groups from saturated or unsaturated fatty acids. In this regard, Li et al. (2021) demonstrated that high LPC levels in urine are associated with PCa development, therefore proposing LPCs as novel biomarkers for the detection of this malignancy [27]. Moreover, Zhou et al. (2012) revealed that PCa progression is positively correlated with the expression level of the enzyme that catalyzes the remodeling of phosphatidylcholine (PC) reaction, namely lysophosphatidylcholine acyltransferase 1 (LPCAT1), and this correlation was independent of age, race and PSA levels of PCa patients, proposing LPCAT1 and LPC as novel biomarkers for PCa detection [28].

Moreover, significant associations between the lipid profile and malignancy were validated, and the phospholipid composition was characteristically altered in the tissues of patients who responded to androgen receptor inhibition. In this regard, a study of 40 prostate tissues and 40 healthy controls using two imaging methods showed, following the analysis of phosphatidylcholine, lysophosphatidylcholine, sphingomyelin and phosphatidylethanolamine classes, that prostate tumors are correlated with increased fatty acid synthesis and lipid oxidation. Phosphatidylcholine variants PC 16:0/16:1, PC 16:0/18:2, PC 18:0/22:5, PC 18:1/18:2, PC 18:1/20:0 and PC 18:1/20:4 showed the highest discriminatory power between the two tested groups, suggesting that lipidomics may represent an alternative future diagnostic strategy for PCa [29].

Nonetheless, it is clear that no single analytical method can accommodate the chemical diversity of the entire metabolome; thus, a multi-platform approach can provide a more comprehensive understanding of metabolic changes. Therefore, the metabolomic profile obtained by combining the MS approach with UHPLC techniques could obtain, with greater accuracy, more precise information regarding the mechanisms of PCa and could help in the much earlier detection of cancer, given the fact that the alteration of the cellular metabolism occurs in the early stages of the tumorigenesis process.

Taken together, our findings suggest that molecules belonging to the AA, lipid and polyamine classes might represent novel candidate biomarkers for PCa early detection and staging, using minimally invasive, *bona fide* tools such as metabolomics (and/or lipidomics).

Our study has some limitations, however, that mainly arise from the small population size used in this research (71 subjects). Another important characteristic that could not be adjusted for with the available blood samples from our participants was the 9-year age mismatch between the patients and the controls, which is known to possibly affect metabolite levels. In addition, our analyses do not adjust for other covariates, due to the limited information available regarding the participants (no information on smoking, history of diabetes, and BMI). Moreover, the ROC analyses from the targeted profiling were performed on the same population, as additional blood samples from a separate cohort were not available for this study. Another limitation of our study arises from the fact that, currently, there is no available scientific background to support our hypotheses. At the moment, there are no metabolomic case-control studies performed on acetylcarnitine, L-tryptophan, L-proline or LPC (18:2) in relation to PCa diagnosis and staging alone, in spite of the fact that their metabolism has been studied in tumorigenesis, in general, and in other human diseases [30–34]. Therefore, assessing their true diagnostic value as biomarkers for PCa detection and staging remains an issue in dire need of further extensive research in order to avoid providing a false diagnosis. Hence, our results are relatively preliminary and should be viewed in the larger context of biomarker discovery for PCa.

Supplementary Materials: The following supporting information can be downloaded at: https://www.mdpi.com/article/10.3390/cimb45060320/s1, Table S1. Metabolite identification with putative names via HMDB, PubChem, Lipid Maps and Heidelberg databases.

Author Contributions: Conceptualization, C.M. and D.N.; methodology, C.S. and A.I.S.; software, C.S. and A.I.S.; validation, C.S. and C.M.; formal analysis, C.S. and A.I.S.; investigation, C.S., R.B. and A.A.C.; resources, D.N.; data curation, C.S., R.B. and A.A.C.; writing—original draft preparation, D.N. and C.S.; writing—review and editing, C.M., C.S., R.B., E.S. and I.O.S.; visualization, C.M., R.B., I.O.S., E.S. and A.A.C.; supervision, C.M., I.O.S. and E.S.; project administration, D.N.; funding acquisition, D.N. All authors have read and agreed to the published version of the manuscript.

Funding: This research was funded by an internal doctoral research grant (No. 25332/22.12.2021) from the University of Medicine and Pharmacy "Victor Babes", Timisoara.

Institutional Review Board Statement: The study was conducted in accordance with the Declaration of Helsinki and approved by the Ethics Committee of the University of Medicine and Pharmacy "Victor Babes", Timisoara (proto-col code no. 9/13.05.2014 extended by code no. 33_2017 and protocol code No. 71/2021 (2023)). The study was also approved by the Ethics Committee of the Clinical Emergency County Hospital in Timisoara, code no. 71/05.08.2014.

Informed Consent Statement: Informed consent was obtained from all the subjects involved in the study. Written informed consent was obtained from the patients to publish this paper.

Data Availability Statement: All data is available in the manuscript.

Conflicts of Interest: The authors declare no conflict of interest.

References

1. Sekhoacha, M.; Riet, K.; Motloung, P.; Gumenku, L.; Adegoke, A.; Mashele, S. Prostate Cancer Review: Genetics, Diagnosis, Treatment Options, and Alternative Approaches. *Molecules* **2022**, *27*, 5730. [CrossRef] [PubMed]
2. Cancer Stat Facts: Prostate Cancer. Available online: https://seer.cancer.gov/statfacts/html/prost.html (accessed on 5 January 2023).
3. American Cancer Sociecty (ACS). Prostate Cancer Risk Factors. Available online: https://www.cancer.org/cancer/prostate-cancer/causes-risks-prevention/risk-factors.html (accessed on 5 January 2023).
4. Rodriguez, J.F.; Eggener, S.E. Prostate Cancer and the Evolving Role of Biomarkers in Screening and Diagnosis. *Radiol. Clin. N. Am.* **2018**, *56*, 187–196. [CrossRef] [PubMed]
5. Ortner, G.; Tzanaki, E.; Rai, B.P.; Nagele, U.; Tokas, T. Transperineal prostate biopsy: The modern gold standard to prostate cancer diagnosis. *Turk. J. Urol.* **2021**, *47* (Suppl. S1), S19–S26. [CrossRef] [PubMed]
6. Bach, C.; Pisipati, S.; Daneshwar, D.; Wright, M.; Rowe, E.; Gillatt, D.; Persad, R.; Koupparis, A. The status of surgery in the management of high-risk prostate cancer. *Nat. Rev. Urol.* **2014**, *11*, 342–351. [CrossRef]
7. Velonas, V.M.; Woo, H.H.; dos Remedios, C.G.; Assinder, S.J. Current status of biomarkers for prostate cancer. *Int. J. Mol. Sci.* **2013**, *14*, 11034–11060. [CrossRef]
8. Spratlin, J.L.; Serkova, N.J.; Eckhardt, S.G. Clinical applications of metabolomics in oncology: A review. *Clin. Cancer Res.* **2009**, *15*, 431–440. [CrossRef] [PubMed]
9. Sreekumar, A.; Poisson, L.M.; Rajendiran, T.M.; Khan, A.P.; Cao, Q.; Yu, J.; Laxman, B.; Mehra, R.; Lonigro, R.J.; Li, Y.; et al. Metabolomic profiles delineate potential role for sarcosine in prostate cancer progression. *Nature* **2009**, *457*, 910–914. [CrossRef]
10. Trock, B.J. Application of metabolomics to prostate cancer. *Urol Oncol.* **2011**, *29*, 572–581. [CrossRef]
11. Kim, Y.; Kislinger, T. Novel approaches for the identification of biomarkers of aggressive prostate cancer. *Genome Med.* **2013**, *5*, 56. [CrossRef]
12. Rigau, M.; Olivan, M.; Garcia, M.; Sequeiros, T.; Montes, M.; Colás, E.; Llauradó, M.; Planas, J.; de Torres, I.; Morote, J.; et al. The Present and Future of Prostate Cancer Urine Biomarkers. *Int. J. Mol. Sci.* **2013**, *14*, 12620–12649. [CrossRef]
13. Idle, J.R.; Gonzalez, F.J. Metabolomics. *Cell Metab.* **2007**, *6*, 348–351. [CrossRef] [PubMed]
14. Zhang, A.; Sun, H.; Yan, G.; Wang, P.; Wang, X. Mass spectrometry-based metabolomics: Applications to biomarker and metabolic pathway research. *Biomed. Chromatogr.* **2016**, *30*, 7–12. [CrossRef] [PubMed]
15. Zhang, T.; Watson, D.G.; Wang, L.; Abbas, M.; Murdoch, L.; Bashford, L.; Ahmad, I.; Lam, N.-Y.; Ng, N.C.F.; Leung, H.Y. Application of Holistic Liquid Chromatography-High Resolution Mass Spectrometry Based Urinary Metabolomics for Prostate Cancer Detection and Biomarker Discovery. *PLoS ONE* **2013**, *8*, e65880. [CrossRef] [PubMed]
16. Alonso, A.; Marsal, S.; Julià, A. Analytical methods in untargeted metabolomics: State of the art in 2015. *Front. Bioeng. Biotechnol.* **2015**, *3*, 23. [CrossRef] [PubMed]
17. Ahmad, F.; Cherukuri, M.K.; Choyke, P.L. Metabolic reprogramming in prostate cancer. *Br. J. Cancer* **2021**, *125*, 1185–1196. [CrossRef]
18. Thapar, R.; Titus, M.A. Recent Advances in Metabolic Profiling And Imaging of Prostate Cancer. *Curr. Metab.* **2014**, *2*, 53–69. [CrossRef]
19. Lima, A.R.; Pinto, J.; Amaro, F.; Bastos, M.L.; Carvalho, M.; Guedes de Pinho, P. Advances and Perspectives in Prostate Cancer Biomarker Discovery in the Last 5 Years through Tissue and Urine Metabolomics. *Metabolites* **2021**, *11*, 181. [CrossRef]
20. Kdadra, M.; Höckner, S.; Leung, H.; Kremer, W.; Schiffer, E. Metabolomics Biomarkers of Prostate Cancer: A Systematic Review. *Diagnostics* **2019**, *9*, 21. [CrossRef]
21. Sartori, D.A.; Chan, D.W. Biomarkers in prostate cancer: What's new? *Curr. Opin. Oncol.* **2014**, *26*, 259–264. [CrossRef]
22. Bentrad, V.V.; Gogol, S.V.; Zaletok, S.P.; Vitruk, Y.V.; Stakhovskyi, E.O.; Grechko, B.O. Urinary spermine level as novel additional diagnostic marker of prostate cancer. *Ann. Oncol.* **2019**, *30*, 123. [CrossRef]
23. Giskeødegård, G.F.; Bertilsson, H.; Selnæs, K.M.; Wright, A.J.; Bathen, T.F.; Viset, T.; Halgunset, J.; Angelsen, A.; Gribbestad, I.S.; Tessem, M.-B. Spermine and Citrate as Metabolic Biomarkers for Assessing Prostate Cancer Aggressiveness. *PLoS ONE* **2013**, *8*, e62375. [CrossRef] [PubMed]
24. Baci, D.; Bruno, A.; Cascini, C.; Gallazzi, M.; Mortara, L.; Sessa, F.; Pelosi, G.; Albini, A.; Noonan, D.M. Acetyl-L-Carnitine downregulates invasion (CXCR4/CXCL12, MMP-9) and angiogenesis (VEGF, CXCL8) pathways in prostate cancer cells: Rationale for prevention and interception strategies. *J. Exp. Clin. Cancer Res.* **2019**, *38*, 464. [CrossRef] [PubMed]
25. Akbari, Z.; Dijojin, R.T.; Zamani, Z.; Hosseini, R.H.; Arjmand, M. Aromatic amino acids play a harmonizing role in prostate cancer: A metabolomics-based cross-sectional study. *Int. J. Reprod. Biomed.* **2021**, *19*, 741–750. [CrossRef] [PubMed]
26. Hong, X.; Mao, L.; Xu, L.; Hu, Q.; Jia, R. Prostate-specific membrane antigen modulates the progression of prostate cancer by regulating the synthesis of arginine and proline and the expression of androgen receptors and Fos proto-oncogenes. *Bioengineered* **2022**, *13*, 995–1012. [CrossRef]
27. Li, X.; Nakayama, K.; Goto, T.; Kimura, H.; Akamatsu, S.; Hayashi, Y.; Fujita, K.; Kobayashi, T.; Shimizu, K.; Nonomura, N.; et al. High level of phosphatidylcholines/lysophosphatidylcholine ratio in urine is associated with prostate cancer. *Cancer Sci.* **2021**, *112*, 4292–4302. [CrossRef]
28. Zhou, X.; Lawrence, T.J.; He, Z.; Pound, C.R.; Mao, J.; Bigler, S.A. The expression level of lysophosphatidylcholine acyltransferase 1 (LPCAT1) correlates to the progression of prostate cancer. *Exp. Mol. Pathol.* **2012**, *92*, 105–110. [CrossRef]

29. Buszewska-Forajta, M.; Pomastowski, P.; Monedeiro, F.; Walczak-Skierska, J.; Markuszewski, M.; Matuszewski, M.; Markuszewski, M.J.; Buszewski, B. Lipidomics as a Diagnostic Tool for Prostate Cancer. *Cancers* **2021**, *13*, 2000. [CrossRef]
30. Geng, P.; Qin, W.; Xu, G. Proline metabolism in cancer. *Amino Acids* **2021**, *53*, 1769–1777. [CrossRef]
31. Phang, J.M.; Liu, W.; Hancock, C.N.; Fischer, J.W. Proline metabolism and cancer: Emerging links to glutamine and collagen. *Curr. Opin. Clin. Nutr. Metab. Care* **2015**, *18*, 71–77. [CrossRef]
32. Platten, M.; Nollen, E.A.A.; Röhrig, U.F.; Fallarino, F.; Opitz, C.A. Tryptophan metabolism as a common therapeutic target in cancer, neurodegeneration and beyond. *Nat. Rev. Drug. Discov.* **2019**, *18*, 379–401. [CrossRef]
33. Dambrova, M.; Makrecka-Kuka, M.; Kuka, J.; Vilskersts, R.; Nordberg, D.; Attwood, M.M.; Smesny, S.; Sen, Z.D.; Guo, A.C.; Oler, E.; et al. Acylcarnitines: Nomenclature, Biomarkers, Therapeutic Potential, Drug Targets, and Clinical Trials. *Pharmacol. Rev.* **2022**, *74*, 506–551. [CrossRef] [PubMed]
34. Law, S.H.; Chan, M.L.; Marathe, G.K.; Parveen, F.; Chen, C.H.; Ke, L.Y. An Updated Review of Lysophosphatidylcholine Metabolism in Human Diseases. *Int. J. Mol. Sci.* **2019**, *6*, 1149. [CrossRef] [PubMed]

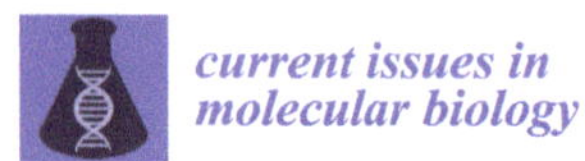

Review

Interaction and Collaboration of SP1, HIF-1, and MYC in Regulating the Expression of Cancer-Related Genes to Further Enhance Anticancer Drug Development

Kotohiko Kimura, Tiffany L. B. Jackson and Ru Chih C. Huang *

Department of Biology, Johns Hopkins University, 3400 N. Charles Street, Baltimore, MD 21218-2685, USA
* Correspondence: rhuang@jhu.edu; Tel.: +1-410-516-5181; Fax: +1-410-516-5213

Abstract: Specificity protein 1 (SP1), hypoxia-inducible factor 1 (HIF-1), and MYC are important transcription factors (TFs). SP1, a constitutively expressed housekeeping gene, regulates diverse yet distinct biological activities; MYC is a master regulator of all key cellular activities including cell metabolism and proliferation; and HIF-1, whose protein level is rapidly increased when the local tissue oxygen concentration decreases, functions as a mediator of hypoxic signals. Systems analyses of the regulatory networks in cancer have shown that SP1, HIF-1, and MYC belong to a group of TFs that function as master regulators of cancer. Therefore, the contributions of these TFs are crucial to the development of cancer. SP1, HIF-1, and MYC are often overexpressed in tumors, which indicates the importance of their roles in the development of cancer. Thus, proper manipulation of SP1, HIF-1, and MYC by appropriate agents could have a strong negative impact on cancer development. Under these circumstances, these TFs have naturally become major targets for anticancer drug development. Accordingly, there are currently many SP1 or HIF-1 inhibitors available; however, designing efficient MYC inhibitors has been extremely difficult. Studies have shown that SP1, HIF-1, and MYC modulate the expression of each other and collaborate to regulate the expression of numerous genes. In this review, we provide an overview of the interactions and collaborations of SP1, HIF1A, and MYC in the regulation of various cancer-related genes, and their potential implications in the development of anticancer therapy.

Keywords: hypoxia-inducible factor 1; SP1; MYC; cancer

Citation: Kimura, K.; Jackson, T.L.B.; Huang, R.C.C. Interaction and Collaboration of SP1, HIF-1, and MYC in Regulating the Expression of Cancer-Related Genes to Further Enhance Anticancer Drug Development. *Curr. Issues Mol. Biol.* **2023**, *45*, 9262–9283. https://doi.org/10.3390/cimb45110580

Academic Editor: Dumitru A. Iacobas

Received: 28 September 2023
Revised: 9 November 2023
Accepted: 11 November 2023
Published: 17 November 2023

1. Introduction: Specificity Protein 1, Hypoxia-Inducible Factor-1, and MYC as Master Regulators of Cancer

Recent progress in systems biology has shown that several specific factors are participants of a network that function as master regulators of cancer [1–3]. Wilson and Volker Filipp investigated complementary omics in human cancer, and discovered a close teamwork of transcriptional and epigenomic machinery, which is tightly connected and comprises histone lysine demethylase 3A, basic helix-loop-helix factors, MYC, hypoxia-inducible factor 1 alpha (HIF1A), and sterol regulatory element-binding transcription factor 1, as well as differentiation factors such as activator protein 1, myogenic differentiation 1, specificity protein 1 (SP1), Meis homeobox 1, zinc finger E-box-binding homeobox 1, and ETS like-1 protein (ETS1) [1]. Cao et al. [2] showed that 10 long non-coding RNA (lncRNA)-transcription factor (TF) pairs including four glycolysis-related lncRNAs (FTX, long intergenic non-protein coding RNA 472, proteasome 20S subunit alpha 3 antisense RNA 1, and small nucleolar RNA host gene 14) and six TFs (forkhead box protein P1, SP1, MYC, FOX-M1, hypoxia-inducible factor 1 alpha [HIF1A], and FOS) are involved in the progression of human lung adenocarcinoma. Malik et al. [3] discovered, using a statistical method called CoMEx (Combined score of DNA Methylation and Expression) to assess differentially expressed and methylated genes/microRNAs (miRNAs) between human seminoma and normal tissues, two hub miRNAs (miR-182-5p and miR-338-3p), five hub

TFs (ETS1, HIF1A, hepatocyte nuclear factor-1 alpha, MYC, and SP1), and three hub genes (*cadherin 1, C-X-C chemokine receptor type 4,* and *Snail family transcriptional repressor 1*) in the seminoma-specific regulatory network. Interestingly, in all of these studies, three TFs, namely SP1, HIF1A, and MYC, were among the factors that participated in the cancer regulatory network. In addition, many studies have shown that SP1, HIF1A, and MYC are often upregulated in cancer [4–9]. Together, these data suggest that SP1, HIF1A, and MYC have crucial roles in cancer development, and that interfering with their activity could negatively impact cancer development and progressions. For this reason, enormous efforts have been undertaken to develop inhibitors for SP1, HIF1A, and MYC. Accordingly, numerous inhibitors of SP1 or HIF1A have been developed [10–14]; however, designing MYC inhibitors has been extremely difficult [15]. Nevertheless, all of the inhibitors against SP1, HIF1A, or MYC can be considered potential anticancer drugs due to the nature of these TFs as master regulators of cancer.

2. What Are SP1, HIF-1, and MYC, and How Do These TFs Benefit Cancer

SP1, HIF-1, and MYC are three major TFs that play important roles as master regulators of cancer, so the next question is—what are these TFs and how do they benefit cancer as regulators of gene expression?

2.1. SP1: Housekeeping Gene That Regulates Biological Activities

SP1 is a ubiquitous TF from the Sp/Krüppel-like family (KLF) of TFs, which are the major forms of zinc finger DNA-binding proteins [16]. The defining feature of SP1-like/KLF proteins is a highly conserved DNA-binding domain (>65% sequence identity among family members) at the C-terminus that has three tandem Cys2His2 (C2H2) zinc finger motifs [17]. Likewise, SP1 contains three highly homologous C2H2 regions [18,19], which exhibit direct binding to DNA at the C-terminal regions of the protein, thus enhancing gene transcription [20]. By contrast, the N-terminal regions of the proteins are more divergent [21]. SP1 has four unstructured domains A, B, C, and D, starting from the C-terminal region of the protein. The two main transactivating domains of SP1 are A and B, which are capable of direct interaction with the components of transcription machinery such as TATA-binding protein (TBP) and TBP-associated factor 4 [22]. The C domain is not indispensable but is highly charged and supports DNA binding and transactivation. The D domain, also known as the C-terminal region of SP1, has multimeric domains and is responsible for the binding of consensus sequences such as 5′-(G/T) GGGCGG(G/A)(G/A)(G/T)-3′ (the sequences are referred to as the GC box) [23]. The N-terminal region of SP1 is a small inhibitory domain, which mainly regulates the functions of domains A and B and is linked to the A domain with a serine/threonine-rich region [22]. The transcriptional activity and stability of SP1 are influenced by its post-translational modifications. SP1 undergoes acetylation, SUMOylation, ubiquitination, and glycosylation after its translation [24,25]. Acetylation of SP1 takes place in the DNA-binding domain [26]. Glycosylation occurs at the O-GlcNAc linkages at the serine and threonine residues in SP1, which can either induce or suppress DNA binding and transcription [27]. SUMOylation, occurring in the Lys16 region, controls the transcription of SP1 by instigating alterations in the chromatin structure, making the DNA inaccessible for transcription [28]. The proteasomal degradation of SP1 is carried out by ubiquitination, where the β-transducin repeat-containing protein (TCRP) ubiquitin ligase complex interacts with SP1 through the DSG (Asp-Ser-Gly) destruction box (β-TCRP binding motif) within the C-terminus of SP1 [29]. SP1 is critical for early embryonic development [30,31], but its expression decreases with age and there is evidence that the transformation of normal cells to cancer cells is associated with the upregulation of SP1, SP3, and SP4 [10,32]. Functional studies have demonstrated that the SP-like family of TFs regulates various genes responsible for cancer-related cellular mechanisms; SP1, SP3, and SP4 are also non-oncogene addiction (NOA) genes and thus are important drug targets [33]. NOA genes are essential for supporting the stress-burdened phenotype of tumors and thus are vital for their survival. The most important functional role of SP1 in normal cells is the regulation of cell cycle and cellular reprogramming [5]. Since cell proliferation and differentiation are the most active during the developmental stage of organisms, SP1 plays critical roles during early developmental stages perhaps for this reason [30,31]. This also indicates that SP1 is still an essential component of cellular mechanisms during adulthood although less so compared with during developmental stages.

2.2. HIF-1: Functions as a Mediator of Hypoxic Signals

HIF-1 is the most important factor involved in the cellular response to hypoxia [34,35]. The broad impact of HIF-1 on cell biology is reflected in the total number of hypoxic target genes, which is estimated to be approximately 1–2% of all human genes [36]. HIF-1 plays important roles in energy metabolism and angiogenesis, especially in cancer progression [34]. It is composed of two subunits, HIF1A and HIF1B (aryl hydrocarbon receptor nuclear translocator). Among these two subunits, only HIF1A is activated under hypoxia and HIF1B is not regulated by oxygen [35]. The dual functional protein apurinic/apyrimidinic endonuclease 1 is an enzyme in DNA base excision repair but also works as a redox factor to maintain HIF1A in the reduced state that is necessary for its transcriptional function [35]. In the presence of oxygen, prolyl hydroxylase hydroxylates HIF1A and hydroxylated HIF1A binds to the tumor suppressor von Hippel–Lindau protein (pVHL), a component of the E3 ubiquitin ligase complex. This interaction causes HIF1A to become ubiquitinated and targeted to the proteasome, where it is degraded. However, under hypoxia, HIF1A is not degraded by the proteasome since prolyl hydroxylase is not functional, so HIF1A dimerizes with HIF1B and binds to the hypoxia response element (HRE) in the promoters of target genes, initiating the expression of genes that promote adaptation to hypoxia [35]. HIF1A as well as the more cell-specific HIF2A are important regulators of the hypoxic response. Although both HIF1A and HIF2A are highly conserved at the protein level, share a similar domain structure, heterodimerize with HIF1B (HIF-2 is formed by the assembly of HIF2A and HIF1B), and bind to the same DNA sequence (the HRE), their effects on the expression of various genes differ [37].

2.3. SP1 and HIF-1

The importance of HIF-1 and SP1 in cancer development is beyond dispute. In fact, it has been shown that both HIF-1 and SP1 are involved in every aspect of cancer-related cellular mechanisms. For instance, both SP1 and HIF-1 play important roles in the regulation of cancer metabolism in carbohydrates [34,38–41] and lipids [42–44]. Both are involved in anticancer immunity via regulation of immune-related cells [45–51]; the tumor microenvironment (TME)/oncometabolites [52–58]; and transforming growth factor beta, which regulates the immune system [59–64]. SP1 promotes tumor angiogenesis via activation of vascular endothelial growth factor (VEGF), epidermal growth factor receptor (EGFR), and VEGF receptor 3 (VEGFR3) [65–67], whereas HIF-1 is a master regulator of angiogenesis, participating in vasculature formation by synergistic correlations with other proangiogenic factors such as VEGF, placental growth factor, and angiopoietins [68]. In addition, SP1 plays an important role in each of the crucial events of metastasis, namely, adhesion, invasion, migration, and angiogenesis [65–67,69–71]. Both SP1 and HIF-1 are also involved in the regulation of cellular stress mechanisms as mediators of the protection of cancer cells against various stresses [72–74].

2.4. MYC: A Master Regulator of Cellular Activity

MYC is a transcription factor that belongs to the basic helix-loop-helix-leucine zipper (bHLHZip) family and regulates cell growth, differentiation, metabolism, and cell death. Thus, MYC functions as a master regulator of major cellular functions [75–78]. Studies using knockout mice have shown that MYC is particularly important for cell growth (accumulation of the body mass) and is indispensable during the period of both embryogenesis and adulthood [79]. c-MYC is the prototype member of the MYC family, which also includes N-MYC and L-MYC proteins in mammalian cells. All three members of the

MYC family are highly homologous but distributed differently. c-MYC is ubiquitous and highly abundant in proliferating cells, whereas N-MYC and L-MYC display more restricted expression at distinct stages of cell and tissue development. MYC proteins exist within the MYC/MAX/MXD network. To fold and become transcriptionally active, c-MYC must first heterodimerize with MAX, a process governed by the coiling of their bHLHZip domains. Once dimerized, the c-MYC/MAX complex acts as a master transcriptional regulator by binding via its basic region to the specific DNA consensus sequence 5′-CANNTG-3′. Due to the multifunctional activities of MYC in cellular functions, cancers with MYC activation elicit many of the important hallmarks essential for autonomous neoplastic growth. In fact, MYC aberrations or upregulation of MYC-related pathways occur in many cancers. In preclinical animal models, MYC inactivation can result in sustained tumor regression, a phenomenon that has been attributed to oncogene addiction [80].

Recently, it was shown that MYC overexpression leads to increased chromatin interactions at super-enhancers and MYC-binding sites [81]. This shows the importance of MYC overexpression in the regulation of cancer development and suggests that super-enhancers might be a potential target for anticancer therapy. Recent studies have also demonstrated that MYC signaling can enable tumor cells to dysregulate the TME and evade the host immune response [82]. Due to the importance of MYC as a regulator of both cancer and the TME, MYC inhibitors may be a holy grail of anticancer drugs. For this reason, many therapeutic agents that directly target MYC are under development; however, to date, their clinical efficacy remains to be demonstrated partially due to the extreme difficulty of developing efficient MYC inhibitors specifically targeted for cancer therapy [15,83]. In this regard, Omomyc, a newly developed MYC inhibitor is more specific in targeting MYC-related genes responsible for cancer development than other MYC inhibitors, might provide insights into how to target MYC for cancer therapy [84].

3. Interactions among SP1, HIF-1, and MYC with One Another and Other TFs

3.1. Modulation of SP1, HIF-1, and MYC Activities

SP1, HIF-1, and MYC modulate the expression of numerous genes as major TFs. However, these TFs do not work independently and are in fact under the regulation of many other cellular components. For example, SP1, HIF-1, and MYC can interact and modulate the activities of each other. Figure 1 shows the promoters of human *SP1, HIF1A,* and *MYC* genes [85–92]. 'SP1' and 'HRE' in the figure indicate the locations of SP1 and the HRE consensus sequences, respectively. The SP1 consensus sequences are usually the GC boxes, whereas the HIF-1 consensus sequences (of the HRE) usually contain the nucleotide residues '5′-RCGTG-3′. The *SP1* promoter contains numerous SP1 consensus sequences as well as NF-Y and E2F consensus sequences. SP1 binds to NF-Y and E2F consensus sequences as well as SP1 consensus sequences in the *SP1* promoter [86,87]. These data suggest that SP1 can autoregulate its transcriptional activity. In addition to these consensus sequences, there is an HRE in the *SP1* promoter (Figure 1) [85] to which HIF-1 binds and stimulates the transcriptional activity of the *SP1* promoter [85]. It has been shown that the mRNA and protein levels of SP1 are decreased by silencing HIF1A in human cultured esophageal squamous cell carcinoma cells, whereas overexpression of HIF1A significantly increases these levels [85]. These data indicate that HIF-1 upregulates SP1 through its binding to the HRE.

There are numerous SP1 consensus sequences in the *HIF1A* gene promoter, which suggests that SP1 can induce *HIF1A* gene expression (Figure 1) [88,89]; however, no definitely active HRE has been found in the *HIF1A* promoter to date (Figure 1) [88,89]. These data confirm that the HRE in the *SP1* gene promoter contributes to the induction of *SP1* gene transcription by HIF1A, and SP1 consensus sequences in the *HIF1A* gene promoter contribute to the induction of *HIF1A* gene transcription by SP1 [85–89]. Thus, there may be a positive activation feedback loop of HIF-1 and SP1, and hypoxia-mediated induction of HIF-1 may trigger the activation of both SP1 and HIF-1 until normoxia deactivates HIF-1. On the other hand, in the promoter of the *MYC* gene, there is one SP1 consensus sequence located upstream of the transcription start site (TSS) and two located downstream of the TSS [90–92]. Meanwhile, there are two E2F consensus sequences in the *SP1* gene promoter, to which MYC and SP1 can bind (Figure 1). Although, as discussed later, many studies have shown that SP1 and MYC collaborate in the transcriptional regulation of various genes, to date, there has been no definitive study showing that SP1 directly regulates transcription of the *MYC* gene [90–92] or that MYC directly regulates transcription of the *SP1* gene.

SP1 promoter

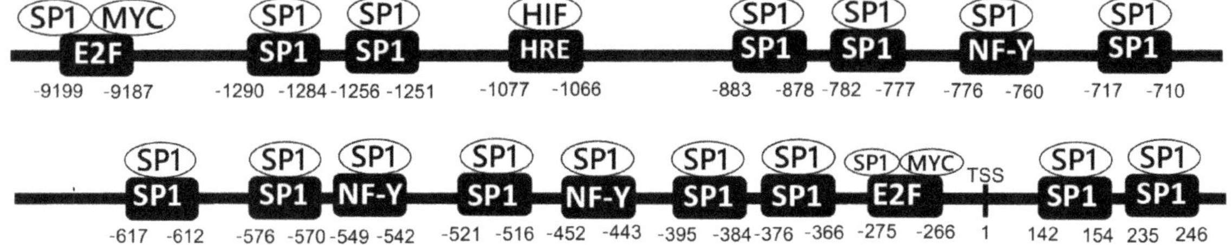

HIF1A promoter

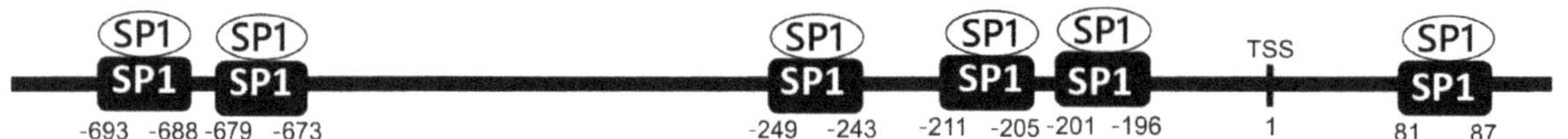

MYC promoter

Figure 1. The promoter structures of human *SP1*, *HIF1A*, and *MYC* genes. The consensus sequences and their potential binding proteins are shown in each promoter. SP1 binds to SP1 consensus sequences (GC box) as well as NF-Y and E2F consensus sequences. HIF-1 binds to the HRE. MYC, similarly to SP1, binds to E2F. Myc-associated zinc finger protein (MAZ) is an important regulatory protein associated with *MYC* gene expression and binds to MAZ consensus sequences [90,92]. The nucleotide numbers are numbered from the transcription start site (TSS). The TSS for *MYC* gene promoter is for the P1 promoter [90,92].

3.2. Effect of HIF-1 on SP1 Gene Expression and Vice Versa

Expression of the *SP1* gene can be upregulated by HIF-1 transcriptionally by the binding of HIF-1 to its consensus sequences in the *SP1* gene promoter, as described in Section 3.1 [85]. This is shown schematically in Figure 2A. Meanwhile, Figure 2B–D schematically shows how *HIF1A* expression is regulated by SP1, using several examples. Insulin increases *HIF1A* promoter activity by reactive oxygen species (ROS) via SP1 in murine 3T3-L1 preadipocytes [93]. *HIF1A* transcription is downregulated by protein arginine methyltransferase 1 (PRMT1), a protein whose transcription is regulated by SP1 in human Hela cervical carcinoma and human HEK293T embryonic kidney cells [94]. In the former example, SP1 is activated by phosphoinositide 3-kinase/protein kinase C via ROS, and then induces *HIF1A* transcription (Figure 2B). In the latter example, the suppression of PRMT1, which prevents the recruitment of SP1/SP3 to the *HIF1A* gene promoter, allows SP1/SP3 to activate the transcription of HIF1A (Figure 2C). In both cases, SP1 directly induces transcriptional activity of the *HIF1A* gene via its binding to SP1 consensus sequences in the *HIF1A* gene promoter (Figure 1) and upregulates expression of the *HIF1A* gene [77–89]. Meanwhile, SP1 can indirectly regulate HIF1A expression by modulating the gene expression of histone deacetylase 4 (*HDAC4*) in rat cardiomyocytes (Figure 2D) [95]. SP1 upregulates the activity of the *HDAC4* gene promoter, thereby promoting deacetylation and impairing the secretion of high mobility group box 1 in mouse intestinal epithelial cells [96]. Likewise, HDAC4 can prevent the acetylation of HIF1A, thereby stabilizing the protein in human pVHL-null kidney cancer cell lines [97]. In this way, SP1 upregulates HIF1A expression either directly by activating *HIF1A* gene expression via binding to the *HIF1A* gene promoter (Figure 2B,C) or indirectly by stabilizing HIF1A protein via modulation of *HDAC4* gene expression (Figure 2D). Either way, SP1 increases the activity of HIF1A. Unlike the *HIF1A* gene, the *HIF1B* gene is constitutively expressed [98]. Therefore, the activity of HIF-1, which is composed of HIF1A and HIF1B, is regulated by adjusting the mRNA and protein levels of HIF1A in cells as well as by modulating the levels of co-activators for HIF-1 [37].

The mechanism of SP1 activation by HIF1A

The mechanism of HIF1A activation by SP1

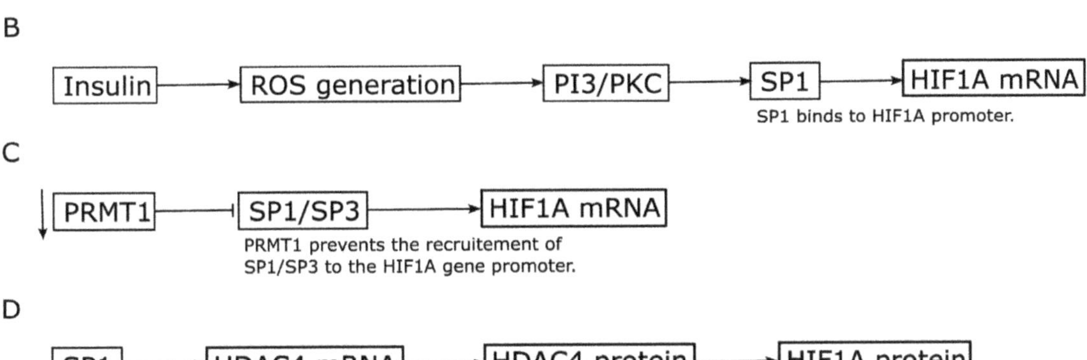

Figure 2. Schematic of the mechanism of activation of SP1 by HIF1A and that of HIF1A by SP1. (**A**) The mechanism of SP1 activation by HIF1A. (**B–D**) The mechanism of HIF1A activation by SP1. While HIF1A regulates SP1 expression via binding to the promoter of a *SP1* gene, SP1 can regulate HIF1A expression at both the mRNA level (**B,C**) and protein level (**D**). PKC: protein kinase C.

3.3. Effects of HIF-1 Compared to the Effects of SP1 on MYC Gene Activities

While HIF-1 induces *SP1* gene expression, it inhibits the activity of MYC (without affecting *MYC* gene expression) [85,99–101]. Since activation of MYC is usually associated with cell growth, MYC activities must be suppressed under hypoxia, which is a condition unsuitable for rapid cell growth due to a lack of oxygen, which is required for efficient biological energy production. Thus, under hypoxia, MYC activity is inhibited by HIF1A as an adaptive response that promotes cell survival under low oxygen conditions. Since there is no HRE in the *MYC* gene promoter (Figure 1), HIF1A is unlikely to inhibit transcription of the *MYC* gene by directly binding to the *MYC* promoter. However, there are several mechanisms by which HIF can inhibit MYC activity. First, HIF1A can antagonize MYC transcriptional activity at MYC target genes by interfering with MYC binding to protein partners. For instance, HIF1A binds to MAX and disrupts MYC/MAX complexes, leading to reduced cyclin D2 expression, induction of p21 (CDKN1A), and G1 phase cell cycle arrest in human pVHL-null kidney cancer cell lines [102]. Meanwhile, under hypoxia, HIF-1 can induce MAX interactor 1, dimerization protein, which inhibits the transcriptional activity of MYC by competing for MAX and represses MYC target genes [103] such as peroxisome proliferator-activated receptor gamma coactivator 1-beta in human pVHL-null kidney cancer cell lines [104] or ornithine decarboxylase in multiple human cancer cell lines [105]. Second, HIF1A directly inhibits MYC transcriptional activity by DNA-binding site competition. For instance, HIF1A displaces MYC binding from the promoter of cyclin-dependent kinase inhibitor 1A (*CDKN1A, p21cip1*) and upregulates the expression of p21 (CDKN1A) in human HCT116 colorectal carcinoma cell line [106]. HIF1A also competes against MYC for binding to SP1, a coactivator of MYC, at the promoters of MYC target genes such as MutS homolog 2 (*MSH2*), *MSH6*, and nibirin, which encode DNA repair proteins, in human HCT116 colorectal carcinoma cell line [107,108] and the E-type prostanoid receptor in human HCA-7 colon cancer cell line [109]. Third, several studies have shown that HIF-1A promotes proteasomal degradation of MYC under chronic hypoxia conditions [104,110–112].

In contrast to HIF1A, HIF2A promotes MYC activity [37]. Overexpression of HIF2A enhances SP1 activity and promotes MYC-driven interleukin 8 expression in human microvascular endothelial cells [113]. HIF2A also enhances MYC activity in human pVHL-null kidney cancer cell lines and primary mouse embryo fibroblasts [102,114]. Consistently, HIF2A deletion has been shown to reduce MYC transcription in human pVHL-null kidney cancer cells implanted in mice [115]. HIF2A promotes MYC activity by stabilizing the MYC/MAX complex [104,116]. Importantly, HIF2A-induced stabilization of MYC/MAX heterodimer is much stronger than HIF1A-mediated degradation of MYC in human cancer cells [104,110–112], leading to MYC activation under hypoxia [116].

MYC upregulates HIF1A proteins although there are no MYC consensus sequences in the promoter of the *HIF1A* gene (Figure 1) [117–130]. The post-transcriptional regulation of HIF1A is responsible for the induction of HIF1A via MYC. For example, transient knockdown of MYC downregulates HIF1A protein levels in multiple human myeloma cells [117]. Overexpression of MYC in human colon cancer and esophageal cancer cells promotes the expression of HIF1A at the post-transcriptional level [118,119]. Overexpression of MYC significantly stabilizes HIF1A and enhances HIF1A accumulation under both normoxic and hypoxic conditions in human normal immortalized mammary epithelial cells and breast cancer cells [120]. Accumulation of HIF1A by MYC leads to the induction of HIF1A targets and is required for MYC-induced anchorage-independent cell growth and proliferation [120]. Mechanistically, MYC prevents HIF1A degradation by reducing HIF1A binding to the pVHL complex, although it increases the level of pVHL complex components [120]. Further, MYC promotes pVHL SUMOylation while repressing its ubiquitination, thereby inhibiting HIF1A ubiquitination and proteasomal degradation [121]. Besides hypoxia, HIF1A expression can be increased via oxygen-independent mechanisms under certain normoxic conditions such as ROS and nitrogen species. MYC increases mitochondrial oxidative phosphorylation and generation of ROS [122]. An increased level of ROS in the mitochondria leads to the stabilization and accumulation of HIF1A by inhibiting prolyl hydroxylase under normoxic conditions [122,123].

Recently, it was shown that MYC induces HIF2A expression as well. MYC has been shown to preferentially bind to the *HIF2A* gene promoter in mouse Sca1C+ cancer stem cells (CSCs) in a MYC-driven mouse T-cell leukemia model and the equivalent ATP-binding cassette superfamily G member 2+ CSC population in human acute lymphoblastic lymphoma, and activate HIF2A expression [124]. HIF2A regulates stem cell function by inducing the expression of octamer-binding transcription factor 4 [125] and AlkB homolog 5, an m6A demethylase that demethylates *Nanog* mRNA and increases Nanog expression [126]. In fact, the stem cell factors Nanog and SRY-box 2 facilitate MYC-mediated induction of HIF2A, playing a critical role in stem cell renewal and tumor stemness [127].

To date, there is limited literature on the effect of SP1 on *MYC* gene expression or the effect of MYC on *SP1* gene expression. However, Parisi et al. [128] identified a functionally distinct signature for strong dual MYC/SP1 sites in various gene promoters. This finding indicates that although SP1 and MYC do not greatly influence each other's expression transcriptionally or post-transcriptionally, there is a distinct mechanism by which they collaborate to regulate the transcription of specifically selected sets of target genes regulated by both SP1 and MYC.

Overall, these data suggest that there is a positive activation loop of HIF-1 (HIF1A) and SP1, which mostly occurs through induction of the transcriptional activity of the *HIF1A* gene via SP1 and that of the *SP1* gene via HIF-1 (Figure 2). HIF-1 negatively regulates MYC through post-transcriptional mechanisms, and MYC activates HIF-1 through post-transcriptional mechanism. Interestingly, unlike HIF-1 and MYC, there is a positive activation loop of HIF-2 and MYC, which occurs via the combination of both transcriptional and post-transcriptional mechanisms. By contrast, there does not seem to be a direct effect of SP1 on *MYC* transcription or of MYC on *SP1* transcription, although SP1 and MYC collaborate to transcriptionally regulate their target genes.

4. Collaboration of SP1, HIF-1, and MYC in Transcriptional Regulation of Their Target Genes

SP1, HIF-1, and MYC interact with each other either transcriptionally or post-transcript ionally and modulate the activity of each other, which demonstrates that there is some collaboration of these TFs in the execution of their activities. However, since SP1, HIF-1, and MYC are first and foremost TFs, their more important collaborations take place when these TFs modulate transcription of their target genes.

Many studies have investigated the mechanisms underlying how SP1 and HIF-1 collaborate in transcriptional regulation of their target genes. One example is the detailed study of the effect of SP1 and HIF-1 on the promoter activity of the human erythropoietin receptor gene [129]. That study showed that the binding of SP1 and HIF-1 to their binding sites in the promoter additively increases the transcriptional activity of the promoter. Another example is the detailed study on regulation of the human retinoic acid receptor-related orphan receptor alpha 4 (*RORalpha*) gene by the interaction between HIF-1 and SP1 [130]. In that case, it was shown that the binding sites for HIF-1 and SP1 in the promoter of this gene are situated closely to each other, and that HIF-1 functionally interacts with SP1 [130]. It was also shown that the HIF2A/SP1/HDAC4 network is involved in transcriptional activation of the human coagulation factor VII gene promoter [131]. Although HIF2A instead of HIF1A is involved in this case, these data suggest that the complex network of HIF1A/HIF2A/SP1/HDAC4 exists, as there is a link between SP1 and HIF1A via HDAC4 (Figure 2D) [95].

The collaboration of HIF-1 and MYC in transcriptional regulation of their target genes has already been described in the previous section. As aforementioned, since HIF-1 and MYC do not modify the expression of each other transcriptionally, the interaction between HIF-1 and MYC occurs either post-transcriptionally (HIF-1 usually suppresses MYC while MYC usually activates HIF-1) or through their collaboration to regulate the expression of their target genes. As an example of HIF-1 modulating the MYC-regulated transcription of genes, for instance, HIF-1 inhibits MYC-dependent induction of the transcriptional activity of the human *CDKN1A* gene promoter via a HIF1A–MYC mechanism [106]. This involves functional antagonism of the transcription repressor MYC via protein–protein interactions. This mechanism is independent of HIF1A DNA binding and transcriptional activity; instead, HIF1A displaces MYC from binding to the *CDKN1A* promoter. A similar mechanism also works for regulation of the human *MSH2* gene promoter [107]. In this case, neither HIF1A nor MYC binds directly to the *MSH2* promoter. Rather, both HIF1A and MYC discretely interact with the constitutively bound TF SP1 on the *MSH2* promoter, whereas HIF1A dominates SP1 binding in hypoxia by competing with MYC. As a result, SP1 acts as a molecular switch by recruiting HIF1A for the hypoxic repression of *MSH2*. This mechanism is a good example of how HIF-1 can suppress rather than induce gene expression under hypoxia. In addition, this mechanism also shows the diversity of how HIF-1, SP1, and MYC collaborate to control the transcriptional activity.

There is no evidence to suggest that SP1 and MYC directly affect the transcription of each other. However, the collaboration of SP1 and MYC in the regulation of their target genes has been well described in the literature [132–137]. Among the genes whose transcription is regulated by the collaboration of MYC and SP1, there are various genes involved in the regulation of CSCs such as telomerase reverse transcriptase (*TERT*), *BMI1*, cluster of differentiation 133 (*CD133*), and *CD147* [134–137]. These genes are often upregulated in cancer. In fact, most of the genes involved in the regulation of CSCs are regulated by HIF-1 as well [138–141]. Hence, these data indicate that the genes involved in the regulation of CSCs are in most cases regulated by SP1, HIF-1, and MYC. Since CSCs possess 'stemness' properties, which are reflected in their capacity to self-renew and generate differentiated cells that contribute to tumor heterogeneity [142,143], the contribution of CSCs has fundamental importance in the development of cancer; therefore, the eradication of CSCs is crucial for the success of anticancer therapy. As aforementioned, SP1, HIF-1, and MYC are all participants of cancer regulatory networks. The fact that the genes involved in the regulation of CSCs are all controlled by SP1, HIF-1, and MYC indicates that the very reason why these TFs are important participants of cancer regulatory networks might be because they regulate CSCs.

Figure 3 shows the promoters of the human *TERT*, *BMI1*, *CD133*, and *CD147* genes, which contain consensus sequences for SP1, HIF-1, and MYC [136,138–141,144–146]. Either SP1, HIF-1, or MYC bind to their respective consensus sequences in the promoters of these genes and induce transcriptional activity [136,138–141,144–146]. The promoter of the *CD133* gene is very complex, and HIF1A binds to the ETS-binding sites rather than the HRE [136,141]. Interestingly, there are only a small number of HREs, whereas there are often clusters of SP1-binding sites. The HRE and SP1-binding sites are often situated closely together, which suggests that SP1 and HIF-1 collaborate to regulate the transcription of these promoters in a similar manner to that of the *RAR Related Orphan Receptor A* gene promoter [130]. The consensus sequences of the HRE are similar to those of the E-box to which MYC binds [147–149]. Hence, the HRE (or E-box) can provide the point of interaction between HIF-1 and MYC. Regarding the *TERT*, *BMI1*, and *CD147* genes, MYC binds to the HRE (or E-box) in their promoters and controls expression of these genes [137,150,151], whereas in the case of the *CD133* gene, MYC binds somewhere in the vicinity of the CpG islands (which have SP1 consensus sequences) in its promoter and controls expression of the gene [136]. As seen by the great structural differences between the *CD133* gene promoter and those of others, there is some diversity in how SP1, HIF-1, or MYC controls the genes involved in the regulation of stem cells.

Based on the current knowledge about the transcription factors SP1, HIF-1, and MYC, the following conclusions can be drawn. First, as described in Section 2, all HIF-1, SP1, and MYC are deeply involved in cancer-related cellular mechanisms including metabolism, angiogenesis, anticancer immunity, and regulation of TME. Importantly, as described in Section 3, HIF-1 and SP1 usually induce the expression of each other while HIF-1 suppresses the expression of MYC and MYC induces that of HIF-1. This indicates that HIF-1 and SP1 can cooperatively activate cancer-related cellular mechanisms while the relationship between HIF-1 and MYC regarding the regulation of cancer-related cellular mechanisms can be variable depending on the context. Second, the CSC-related genes, which have fundamental importance in oncogenesis, are all positively regulated by HIF-1, SP1, and MYC at the transcriptional level (Figure 3). Overall, these results suggest that inhibitors for

HIF-1 and SP1 likely induce anticancer effects in cooperation by suppressing the activity of cancer-related cellular mechanisms (including the mechanisms underlying CSC regulation) while using MYC inhibitors as anticancer drugs requires some cautions.

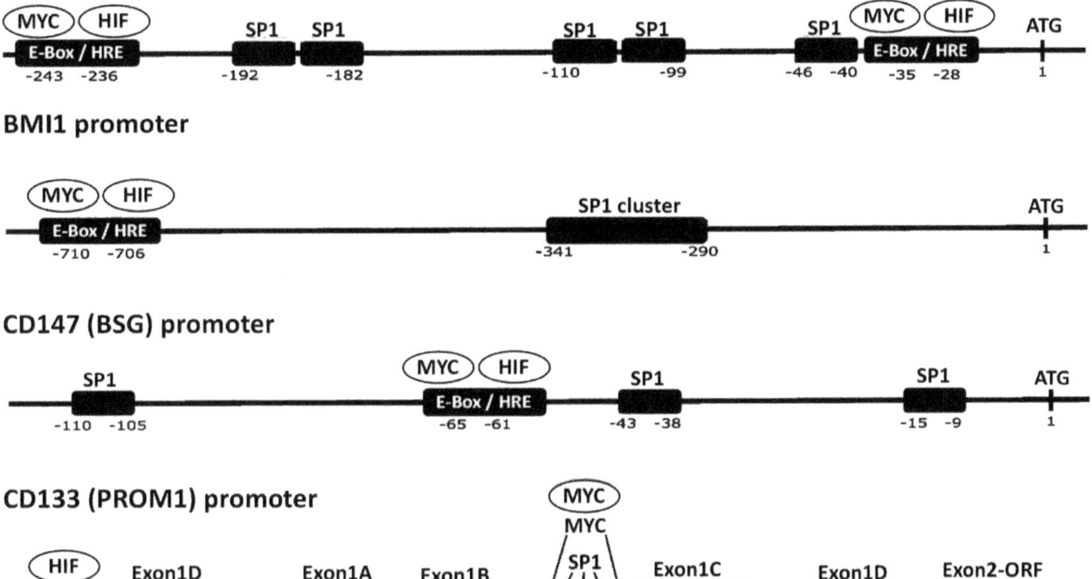

Figure 3. Locations of SP1 and HIF-1 consensus sequences in the promoters of genes which are involved in the regulation of CSCs. Schematic of the locations of SP1 and HIF-1 consensus sequences in the promoters of human *TERT*, *BMI1*, *CD147*, and *CD133* genes, which are all involved in the regulation of CSCs. The nucleotide numbers are numbered from the ATG (translation initiation) sites. HIF-1 binds to the ETS consensus sequence in the *CD133* promoter, whereas HIF-1 binds to the HRE in other promoters. The nucleotide sequences of the HRE are similar to those of the E-Box, which is a consensus sequence for MYC binding. The objects with descriptions of 'MYC' or 'HIF' in the circles indicate the TFs. The MYC TF, designated by the 'MYC' in the circle, binds to the E-Box or the MYC consensus sequences in the vicinity of CpG islands (SP1 consensus sequences), whereas the HIF-1 TF, designated by the 'HIF' in the circle, binds to the HRE or the ETS.

5. Implications of the Interactions and Collaborations of SP1, HIF-1, and MYC in the Development of Anticancer Drugs, and Their Future Perspectives

5.1. SP1, HIF-1, and MYC as Targets for Potential Anticancer Therapies

There are many lines of evidence indicating that SP1, HIF-1, and MYC play important roles in the development of cancer [1–9]. Hence, these TFs must be among the major targets of potential anticancer therapies. Many inhibitors of SP1 and HIF-1 have been developed [11,12]; however, currently an adequately efficient inhibitor of MYC is not available due to the extreme difficulty of designing MYC inhibitors specifically targeted against cancer [15]. Thus, it is unknown how well putative MYC inhibitors work as anticancer agents. In the end, it might turn out that a putative MYC inhibitor is key for efficiently eliminating cancer. On the other hand, a putative MYC inhibitor might have strong side effects due to its broad influence over the normal essential activities of its target genes. In fact, to some extent, this has been suggested by the results of clinical trials of mithramycin, which is considered to be an SP1 inhibitor but also inhibits transcription of the *MYC* gene [152]. Mithramycin is rarely used as an anticancer drug due to its strong side effects [153]. Although it is not known whether these side effects are derived from the activity of mithramycin in inhibiting *MYC* transcription, this result suggests that MYC inhibition might cause too many side effects.

The design of Omomyc may be used as a guide for the development of effective MYC inhibitors [84]. The action of Omomyc is different from that of many other MYC inhibitors, which have the ability to reduce *MYC* expression by gene knockout or RNA interference. Omomyc selectively targets MYC protein interactions: it binds C-MYC and N-MYC, MAX and MIZ-1, but does not bind MAD or select HLH proteins. Specifically, it prevents MYC binding to promoter E-boxes and the transactivation of target genes while retaining MIZ-1-dependent binding to promoters and transrepression. Clinical trials to date have indicated that the side effects of Omomyc are mild and well tolerated unlike many other MYC inhibitors [154]. In addition, recent findings about the important roles of MYC overexpression in activating super-enhancers, which are often deregulated in cancer, suggest that the points of interactions between super-enhancers and MYC could be more specific targets for designing anticancer drugs than MYC proteins [81]. Recent progress in MYC-related research indicates that MYC inhibitors that are more specifically targeted against cancer-related genes than conventional MYC inhibitors, might be developed in the future.

Currently, there are no anticancer drugs with strong enough activity to eradicate cancer in many patients. One way to compensate for this limitation of anticancer drugs is to combine several drugs to strengthen the anticancer activity of each single drug. Since inhibitors of SP1 and HIF-1 are already available, it is feasible to combine them. As described in Section 2.3, there are many genes that play crucial roles in cancer development and whose expression is regulated by both SP1 and HIF-1. In addition, it was also shown that the genes involved in the regulation of CSCs, which have fundamental importance in cancer development, are regulated by HIF-1, SP1, and MYC (Figure 3). The degree of dependency of the transcription of these genes on either SP1 or HIF-1 can differ from one gene to the other. However, it can be expected that inhibiting the binding of both HIF-1 and SP1 to the promoters of genes whose expression is regulated by both HIF-1 and SP1, should lead to stronger reduction of the expression of these genes than inhibiting binding of either HIF-1 or SP1 alone. Furthermore, since SP1 and HIF1A induce the expression of each other as mentioned above (in other words, positive feedback regulation works for the expression of SP1 and HIF1A; Figure 2), inhibiting the activity of both SP1 and HIF-1 can be expected to downregulate the expression of both TFs far more efficiently than inhibiting the activity of either TF alone. Therefore, it makes sense to combine SP1 inhibitors with HIF-1 inhibitors as potential drugs to be used for combinational anticancer therapy. This concept is shown as a scheme to illustrate what could potentially happen when cancers are treated with a combination of both HIF-1 and SP1 inhibitors (Figure 4). Recently, it was shown that the interaction of SP1 and HIF1A modulated the behavior of cancer cells in a hypoxic microenvironment and promoted cancer development [155]. These data suggest that combination treatment of SP1 and HIF1A inhibitors could effectively disrupt the interaction of SP1 and HIF1A and inhibit cancer development in a hypoxic microenvironment, and support the benefit of using combination treatment of SP1 and HIF1A inhibitors.

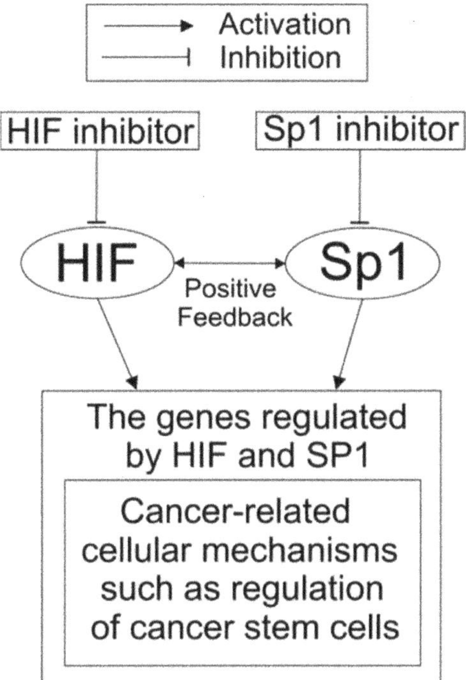

Figure 4. Scheme to illustrate what could happen when cancers are treated with a combination of HIF-1 and SP1 inhibitors. There are intricate mechanisms underlying the collaboration of HIF-1 and SP1 to regulate cancer-related cellular mechanisms. HIF-1 and SP1 usually induce the expression of each other (as described in Section 3) and activate their target genes in corporation (as described in Section 4). Therefore, it is expected that combination treatment of HIF and SP1 inhibitors can specifically suppress the activity of cancer-related cellular mechanisms.

5.2. Tetramethyl-O-Nordihydroguaiaretic Acid as an Anticancer Drug

Tetramethyl-O-nordihydroguaiaretic acid (M_4N) is an anticancer drug candidate that has been studied for many years [156–172]. This compound reversibly inhibits the binding of SP1 to its DNA consensus sequence thereby functioning as an SP1 inhibitor unlike mithramycin which is an irreversible inhibitor of SP1 [156]. M_4N also decreases HIF1A content in cancer cells under hypoxia [159], indicating that it functions as a dual inhibitor of both SP1 and HIF1A. Importantly, M_4N essentially has no strong side effects [171], suggesting that the reversible inhibition of SP1 and HIF1A bindings to their target genes might not cause strong toxicity. M_4N also downregulates the expression of proteins, such as B-cell lymphoma 2/adenovirus E1B 19kDa protein-interacting protein 3, X-linked inhibitor of apoptosis protein, survivin, VEGF, and lactate dehydrogenase A [161,163,164,169,170], and suppresses energy metabolism [162,168]. The function of these proteins as well as energy metabolism are associated with the activities of SP1 and HIF-1 [34,38–44]. M_4N can markedly induce cellular stress in cancer cells [163,169]. The activity of SP1 and HIF-1 in protecting cells from cellular stress has been previously described [72–74], and thus M_4N as an SP1 and HIF-1 dual inhibitor likely suppresses this activity. However, the stress induced by M_4N does not seem to be strong enough to induce significant tumoricidal activity, as M_4N treatment alone has not been shown to induce anticancer activity strong enough to eliminate cancer (although it slows cancer growth) in various xenograft mice and many patients [163,165,166,169,171]. The most compelling data showing the efficacy of M_4N as an anticancer drug is a clinical trial that was conducted in India in the 1990s treating patients with oral squamous cell carcinoma [172].

However, it might be sufficient to make cancer cells vulnerable against a second anticancer drug that is able to induce strong cytocidal effects, as shown by the synergistic induction of anticancer activity with the combination treatment of M_4N with a second anticancer drug in various xenograft mouse experiments [163,165,166,169]. It was also shown that combination treatment with M_4N suppresses energy metabolism and oncometabolites in addition to inducing strong cellular stress, making it a potentially impressive anticancer treatment [163,169]. In addition, M_4N induces interleukin 21 and enhances B cell-mediated humoral immunity in tumor-bearing mice [173]. It has also been shown that M_4N can downregulate oncometabolites such as lactate and 2-hydroxyglutarate, which suppress activity of the immune system in cancer cells [169], and thus can activate immune reactions via the downregulation of oncometabolites. Overall, these data showed that M_4N could make the TME unfavorable for cancer cells to thrive due to its effect to activate the immune system. The roles of SP1 and HIF-1 in regulating the immune system has already been described in the literature [45–64], which suggests that M_4N induces its effects on the immune system via its activity as a dual inhibitor of SP1 and HIF-1.

5.3. A Possible Usage of the Combination Treatment of HIF-1 and SP1 Inhibitors as an Anticancer Therapy

Since M_4N, which inhibits both SP1 and HIF-1, does not have strong enough anticancer activity to eliminate cancer in most patients [171], combination treatment of SP1 and HIF-1 inhibitors is also likely not able to induce strong enough anticancer activity to eliminate cancer. However, the results showing that M_4N can induce anticancer activity synergistically in combination with a second anticancer drug [163,165,166,169] suggest that the combination treatment of SP1 and HIF-1 inhibitors could significantly strengthen the anticancer activity of a third anticancer drug. Other than TFs such as SP1, HIF-1, and MYC, there are various additional factors, particularly lncRNAs and miRNAs [174], which play important roles in the development of cancer, as predicted by systems analyses of the regulatory networks in cancer [1–3]. All these factors can naturally be targets of anticancer drugs. To increase the anticancer efficacy of HIF-1 or SP1 inhibitors, combination treatments of HIF-1 and/or SP1 inhibitors with inhibitors of these other factors (e.g., certain lncRNAs and miRNAs as well as MYC), which function as parts of the cancer regulatory networks, may need to be considered as well. In this regard, Omomyc with its selective targeting against MYC-related genes is a particularly attractive candidate to combine with HIF and SP1 dual inhibitors such as M_4N [84].

5.4. Roles of HIF-1, SP1, and MYC in Normal Cells and the Potential Toxicity of Anticancer Drugs Targeted on These Transcription Factors

One of many reasons why anticancer drugs do not work well in the clinics is that most anticancer drugs have strong toxicity to normal cells as well as having an effect on cancer cells. As a result, cancer patients treated with these drugs often experience strong side effects and cannot tolerate the drugs until cancers are totally eliminated. Therefore, good anticancer drugs need to specifically target cancer cells without causing strong toxicity to normal cells. Keeping that in mind, the question becomes: how potentially toxic are anticancer drugs that target HIF-1, SP1, or MYC for normal cells?

First, HIF-1 is a master regulator of hypoxic signals so that it is required mainly under hypoxic conditions only [34,35]. Thus, although activation of HIF-mediated signaling mechanisms results in various benefits to cancer cells and the upregulation of HIF-1 is often found in various tumors, only a moderate amount of the expression of HIF-1 is presumably needed to maintain the health of normal cells. For this reason, it would not cause too many difficulties to develop safe and effective inhibitors for HIF-1 as anticancer drugs. Second, SP1 is involved in important normal cellular functions such as cell proliferation and cellular differentiation [5]. However, importantly, although SP1 is essential for normal tissues during embryogenesis and adulthood, it is less important during adulthood than embryogenesis [5,30,31]. Meanwhile, SP1 is often upregulated in cancers [10,32,33]. This suggests that SP1 inhibitors with only moderate inhibitory activities could possibly incur negative impact on cancer cells without inducing strong adverse effects on normal cells. As mentioned above, mithramycin, a well-known SP1 inhibitor with some activity to inhibit MYC, has strong side effects [152,153]. However, a newly developed mithramycin analogue EC-8042 showed fewer side effects than mithramycin [175], which suggested that there were ways to improve safety of SP1 inhibitors. Moreover, M_4N which only reversibly inhibits SP1 binding to SP1-target genes showed few side effects if any [171]. Overall, this suggests that safe and effective SP1 inhibitors without strong side effects are feasible. Finally, MYC is involved in some of the most important functions in normal cells [75–78]. Therefore, it would be very difficult to eliminate all side effects from any potential MYC inhibitors. In this sense, Omomyc with its selective targeting against MYC-related genes can potentially be a good start to develop safe and effective MYC inhibitors as anticancer drugs [84]. Largely, this analysis suggests that there are potential methods to develop safe and effective combination treatments with HIF inhibitors and SP1 inhibitors.

6. Conclusions and Future Directions

HIF-1, SP1, and MYC, which function as master regulators of cancer, interact with each other and modulate the expression of many genes whose functions are associated with the development and maintenance of cancer, such as the genes involved in the regulation of proliferation, CSCs, metabolism, angiogenesis, stress response, and metastasis (as described in Sections 1 and 2). In addition, SP1, HIF-1, or MYC modulates the TME, such as the immune system and the production of oncometabolites, in such a way to facilitate tumor development (as discussed in Section 2). For this reason, inhibitors of HIF-1, SP1, and MYC should be able to work as anticancer agents. Currently, many inhibitors of HIF-1 and SP1 are available, although there are no efficient inhibitors of MYC available due to the extreme difficulty in developing them [15]. In this sense, Omomyc might become a breakthrough for the future development of MYC inhibitors [84]. Although many inhibitors of either HIF-1 or SP1 have some anticancer activity, none of them have activity strong enough to eradicate cancer in the majority of patients. This has led to the idea to combining HIF-1 inhibitors with SP1 inhibitors to improve the anticancer activity of each drug.

It has been shown that M_4N [156–172], a newly developed anticancer drug candidate that inhibits both HIF-1 and SP1, modulates the expression of various genes whose promoters are controlled by SP1 and HIF-1. M_4N also suppresses energy metabolism in cancer and induces humoral immunity in the TME. Clinical trials of M_4N have shown some anticancer activity, but this activity is not strong enough to eradicate cancer in most patients. However, combination treatments of M_4N with various second drugs have been shown to synergistically improve anticancer activity in xenograft mouse studies. The conclusions obtained from the data on M_4N suggest that, while a combination treatment of an HIF-1 inhibitor with an SP1 inhibitor might improve the anticancer efficacy of the drugs to some extent without curing most cancer patients, a combination treatment of HIF-1 and SP1 inhibitors with a third appropriately selected anticancer drug might significantly improve the anticancer activity of the third drug.

In addition, as discussed above, it is realistic to think developing safe and effective inhibitors for HIF-1 and SP1 is achievable. In fact, M_4N with activities as a dual inhibitor for HIF-1 and SP1 does not cause strong side effects [171]. Since we have not tested the safety and efficacy of the combination treatment of M_4N with the second anticancer drug for human patients yet, we do not know how safe and effective this combination treatment can be. Therefore, clinical trials of M_4N (or the combined treatment of HIF-1 and SP1 inhibitors) with various second anticancer drugs are urgently needed. Although it has been shown by numerous mouse xenograft studies that M_4N can synergistically induce anticancer activity with many different second anticancer drugs, the optimal selection of these second drugs has not been established yet. As described, Omomyc is potentially an interesting choice as a second drug [84]. This line of studies also needs to be carried out together with the clinical trials of the combination treatments suggested above.

Author Contributions: Conceptualization, K.K. and R.C.C.H.; methodology, K.K. and R.C.C.H.; validation, K.K. and R.C.C.H.; formal analysis, K.K. and R.C.C.H.; resources, K.K. and T.L.B.J.; data curation, K.K. and R.C.C.H.; writing—original draft preparation K.K. and R.C.C.H.; writing—review and editing, K.K., T.L.B.J. and R.C.C.H.; supervision, R.C.C.H.; project administration, K.K., T.L.B.J. and R.C.C.H.; funding acquisition, R.C.C.H. All authors have read and agreed to the published version of the manuscript.

Funding: This study was supported by grants from the National Institutes of Health (R01DE12165), 806 Biocuremedical, LLC, and the Dorothy Yen Trust (P 690-C25-2407) to RCH. The funders had no role in conceptualization, design of the study, interpretation of the data, decision to publish, or preparation of the manuscript.

Institutional Review Board Statement: Not applicable.

Informed Consent Statement: Not applicable.

Data Availability Statement: Not applicable.

Conflicts of Interest: The authors have declared that no conflict of interest exists. However, for full disclosure, we acknowledge that Ru Chih C. Huang is a principal inventor of several Johns Hopkins University patents on M_4N.

References

1. Wilson, S.; Filipp, F.V. A network of epigenomic and transcriptional cooperation encompassing an epigenomic master regulator in cancer. *NPJ Syst. Biol. Appl.* **2018**, *1*, 24. [CrossRef] [PubMed]
2. Cao, P.; Zhao, B.; Xiao, Y.; Hu, S.; Kong, K.; Han, P.; Yue, J.; Deng, Y.; Zhao, Z.; Wu, D.; et al. Understanding the Critical Role of Glycolysis-Related lncRNAs in Lung Adenocarcinoma Based on Three Molecular Subtypes. *Biomed Res. Int.* **2022**, *2022*, 7587398. [CrossRef] [PubMed]
3. Mallik, S.; Qin, G.; Jia, P.; Zhao, Z. Molecular signatures identified by integrating gene expression and methylation in non-seminoma and seminoma of testicular germ cell tumours. *Epigenetics* **2021**, *16*, 162–176. [CrossRef]
4. Sankpal, U.T.; Maliakal, P.; Bose, D.; Kayaleh, O.; Buchholz, D.; Basha, R. Expression of specificity protein transcription factors in pancreatic cancer and their association in prognosis and therapy. *Curr. Med. Chem.* **2012**, *19*, 3779–3786. [CrossRef] [PubMed]
5. Vellingiri, B.; Iyer, M.; Devi Subramaniam, M.; Jayaramayya, K.; Siama, Z.; Giridharan, B.; Narayanasamy, A.; Abdal Dayem, A.; Cho, S.G. Understanding the Role of the Transcription Factor Sp1 in Ovarian Cancer: From Theory to Practice. *Int. J. Mol. Sci.* **2020**, *21*, 1153. [CrossRef] [PubMed]
6. Gao, Y.; Gan, K.; Liu, K.; Xu, B.; Chen, M. SP1 Expression and the Clinicopathological Features of Tumors: A Meta-Analysis and Bioinformatics Analysis. *Pathol. Oncol. Res.* **2021**, *27*, 581998. [CrossRef]
7. Abu, E.L.; Maaty, M.A.; Terzic, J.; Keime, C.; Rovito, D.; Lutzing, R.; Yanushko, D.; Parisotto, M.; Grelet, E.; Namer, I.J.; et al. Hypoxia-mediated stabilization of HIF1A in prostatic intraepithelial neoplasia promotes cell plasticity and malignant progression. *Sci. Adv.* **2022**, *8*, eabo2295. [CrossRef]
8. Wu, Q.; You, L.; Nepovimova, E.; Heger, Z.; Wu, W.; Kuca, K.; Adam, V. Hypoxia-inducible factors: Master regulators of hypoxic tumor immune escape. *J. Hematol. Oncol.* **2022**, *15*, 77. [CrossRef]
9. Mahauad-Fernandez, W.D.; Felsher, D.W. The Myc and Ras Partnership in Cancer: Indistinguishable Alliance or Contextual Relationship? *Cancer Res.* **2020**, *80*, 3799–3802. [CrossRef]
10. Vizcaíno, C.; Mansilla, S.; Portugal, J. Sp1 transcription factor: A long-standing target in cancer chemotherapy. *Pharmacol. Ther.* **2015**, *152*, 111–124. [CrossRef]
11. Safe, S.; Abbruzzese, J.; Abdelrahim, M.; Hedrick, E. Specificity Protein Transcription Factors and Cancer: Opportunities for Drug Development. *Cancer Prev. Res.* **2018**, *11*, 371–382. [CrossRef] [PubMed]
12. Semenza, G.L. Evaluation of HIF-1 inhibitors as anticancer agents. *Drug Discov. Today* **2007**, *12*, 853–859. [CrossRef] [PubMed]
13. Joshi, E.; Pandya, M.; Desai, U. Current Drugs and Their Therapeutic Targets for Hypoxia-inducible Factors in Cancer. *Curr. Protein Pept. Sci.* **2023**, *24*, 447–464. [CrossRef]
14. Shirai, Y.; Chow, C.C.T.; Kambe, G.; Suwa, T.; Kobayashi, M.; Takahashi, I.; Harada, H.; Nam, J.M. An Overview of the Recent Development of Anticancer Agents Targeting the HIF-1 Transcription Factor. *Cancers* **2021**, *13*, 2813. [CrossRef]
15. Llombart, V.; Mansour, M.R. Therapeutic targeting of "undruggable: MYC. *EBioMedicine* **2022**, *75*, 103756. [CrossRef] [PubMed]
16. Szpirer, J.; Szpirer, C.; Riviere, M.; Levan, G.; Marynen, P.; Cassiman, J.J.; Wiese, R.; DeLuca, H.F. The Sp1 transcription factor gene (SP1) and the 1,25-dihydroxyvitamin D3 receptor gene (VDR) are colocalized on human chromosome arm 12q and rat chromosome 7. *Genomics* **1991**, *11*, 168–173. [CrossRef]
17. Kaczynski, J.; Cook, T.; Urrutia, R. Sp1- and Krüppel-like transcription factors. *Genome Biol.* **2003**, *4*, 206. [CrossRef]

18. Narayan, V.A.; Kriwacki, R.W.; Caradonna, J.P. Structures of zinc finger domains from transcription factor Sp1. Insights into sequence–specific protein–DNA recognition. *J. Biol. Chem.* **1997**, *272*, 7801–7809. [CrossRef]
19. Oka, S.; Shiraishi, Y.; Yoshida, T.; Ohkubo, T.; Sugiura, Y.; Kobayashi, Y. NMR structure of transcription factor Sp1 DNA binding domain. *Biochemistry* **2004**, *43*, 16027–16035. [CrossRef]
20. Nagaoka, M.; Shiraishi, Y.; Sugiura, Y. Selected base sequence outside the target binding site of zinc finger protein Sp1. *Nucleic Acids Res.* **2001**, *29*, 4920–4929. [CrossRef]
21. Cook, T.; Gebelein, B.; Belal, M.; Mesa, K.; Urrutia, R. Three conserved transcriptional repressor domains are a defining feature of the TIEG subfamily of Sp1-like zinc finger proteins. *J. Biol. Chem.* **1999**, *274*, 29500–29504. [CrossRef] [PubMed]
22. Kadonag, J.T.; Courey, A.J.; Ladika, J.; Tjian, R. Distinct regions of Sp1 modulate DNA binding and transcriptional activation. *Science* **1988**, *242*, 1566–1570. [CrossRef] [PubMed]
23. Kadonaga, J.T.; Jones, K.A.; Tjian, R. Promoter specific activation of RNA polymerase II transcription by Sp1. *Trends Biochem. Sci.* **1986**, *11*, 20–23. [CrossRef]
24. Suske, G. The Sp-family of transcription factors. *Gene* **1999**, *238*, 291–300. [CrossRef] [PubMed]
25. Tan, N.; Khachigian, L. Sp1 phosphorylation and its regulation of gene transcription. *Mol. Cell. Biol.* **2009**, *29*, 2483–2488. [CrossRef]
26. Suzuki, T.; Kimura, A.; Nagai, R.; Horikoshi, M. Regulation of interaction of the acetyltransferase region of p300 and the DNA-binding domain of Sp1 on and through DNA binding. *Genes Cells* **2000**, *5*, 29–41. [CrossRef]
27. Jackson, S.P.; Tjian, R. O-glycosylation of eukaryotic transcription factors: Implications for mechanisms of transcriptional regulation. *Cell* **1988**, *55*, 125–133. [CrossRef]
28. Stielow, B.; Sapetschnig, A.; Wink, C.; Kruger, I.; Suske, G. SUMO-modified Sp3 represses transcription by provoking local heterochromatic gene silencing. *EMBO Rep.* **2008**, *9*, 899–906. [CrossRef]
29. Wei, S.; Chuang, H.C.; Tsai, W.C.; Yang, H.C.; Ho, S.R.; Paterson, A.J.; Kulp, S.K.; Chen, C.S. Thiazolidinediones mimic glucose starvation in facilitating Sp1 degradation through the up-regulation of beta-transducin repeat-containing protein. *Mol. Pharmacol.* **2009**, *76*, 47–57. [CrossRef]
30. Marin, M.; Karis, A.; Visser, P.; Grosveld, F.; Philipsen, S. Transcription factor Sp1 is essential for early embryonic development but dispensable for cell growth and differentiation. *Cell* **1997**, *89*, 619–628. [CrossRef]
31. Gilmour, J.; Assi, S.A.; Jaegle, U.; Kulu, D.; van de Werken, H.; Clarke, D.; Westhead, D.R.; Philipsen, S.; Bonifer, C. A crucial role for the ubiquitously expressed transcription factor Sp1 at early stages of hematopoietic specification. *Development* **2014**, *141*, 2391–2401. [CrossRef] [PubMed]
32. Li, L.; Davie, J.R. The role of Sp1 and Sp3 in normal and cancer cell biology. *Ann. Anat.-Anat. Anz.* **2010**, *192*, 275–283. [CrossRef]
33. Hjaltelin, J.X.; Izarzugaza, J.M.G.; Jensen, L.J.; Russo, F.; Westergaard, D.; Brunak, S. Identification of hyper-rewired genomic stress non-oncogene addiction genes across 15 cancer types. *NPJ Syst. Biol. Appl.* **2019**, *5*, 27. [CrossRef] [PubMed]
34. Wicks, E.E.; Semenza, G.L. Hypoxia-inducible factors: Cancer progression and clinical translation. *J. Clin. Investig.* **2022**, *132*, e159839. [CrossRef]
35. Harris, A.L. Hypoxia—A key regulatory factor in tumour growth. *Nat. Rev. Cancer* **2002**, *2*, 38–47. [CrossRef] [PubMed]
36. Mazure, N.M.; Brahimi-Horn, M.C.; Berta, M.A.; Benizri, E.; Bilton, R.L.; Dayan, F.; Ginouvès, A.; Berra, E.; Pouysségur, J. HIF-1: Master and commander of the hypoxic world. A pharmacological approach to its regulation by siRNAs. *Biochem. Pharmacol.* **2004**, *68*, 971–980. [CrossRef] [PubMed]
37. Loboda, A.; Jozkowicz, A.; Dulak, J. HIF-1 and HIF-2 transcription factors--similar but not identical. *Mol. Cells* **2010**, *29*, 435–442. [CrossRef]
38. Archer, M.C. Role of sp transcription factors in the regulation of cancer cell metabolism. *Genes Cancer* **2011**, *2*, 712–719. [CrossRef]
39. Beishline, K.; Azizkhan-Clifford, J. Sp1 and the 'hallmarks of cancer'. *FEBS J.* **2015**, *282*, 224–258. [CrossRef]
40. Semenza, G.L. HIF-1 mediates metabolic responses to intratumoral hypoxia and oncogenic mutations. *J. Clin. Investig.* **2013**, *123*, 3664–3671. [CrossRef]
41. Paredes, F.; Williams, H.C.; San Martin, A. Metabolic adaptation in hypoxia and cancer. *Cancer Lett.* **2021**, *502*, 133–142. [CrossRef] [PubMed]
42. Zhu, J.; Sun, Y.; Luo, J.; Wu, M.; Li, J.; Cao, Y. Specificity protein 1 regulates gene expression related to fatty acid metabolism in goat mammary epithelial cells. *Int. J. Mol. Sci.* **2015**, *16*, 1806–1820. [CrossRef] [PubMed]
43. Shimano, H.; Sato, R. SREBP-regulated lipid metabolism: Convergent physiology—Divergent pathophysiology. *Nat. Rev. Endocrinol.* **2017**, *13*, 710–730. [CrossRef] [PubMed]
44. Mylonis, I.; Simos, G.; Paraskeva, E. Hypoxia-Inducible Factors and the Regulation of Lipid Metabolism. *Cells* **2019**, *8*, 214. [CrossRef]
45. Leone, R.D.; Powell, J.D. Metabolism of immune cells in cancer. *Nat. Rev. Cancer* **2020**, *20*, 516–531. [CrossRef]
46. Ho, P.-C.; Bihuniak, J.D.; Macintyre, A.N.; Staron, M.; Liu, X.; Amezquita, R.; Tsui, Y.-C.; Cui, G.; Micevic, G.; Perales, J.C.; et al. Phosphoenolpyruvate is a metabolic checkpoint of anti- tumor T cell responses. *Cell* **2015**, *162*, 1217–1228. [CrossRef]
47. Chang, C.-H.; Qiu, J.; O'Sullivan, D.; Buck, M.D.; Noguchi, T.; Curtis, J.D.; Chen, Q.; Gindin, M.; Gubin, M.M.; van der Windt, G.J.W.; et al. Metabolic competition in the tumor microenvironment is a driver of cancer progression. *Cell* **2015**, *162*, 1229–1241. [CrossRef]

48. Doedens, A.L.; Stockmann, C.; Rubinstein, M.P.; Liao, D.; Zhang, N.; DeNardo, D.G.; Coussens, L.M.; Karin, M.; Goldrath, A.W.; Johnson, R.S. Macrophage expression of hypoxia-inducible factor-1 alpha suppresses T-cell function and promotes tumor progression. *Cancer Res.* **2010**, *70*, 7465–7475. [CrossRef]

49. Takeda, N.; O'Dea, E.L.; Doedens, A.; Kim, J.W.; Weidemann, A.; Stockmann, C.; Asagiri, M.; Simon, M.C.; Hoffmann, A.; Johnson, R.S. Differential activation and antagonistic function of HIF-{alpha} isoforms in macrophages are essential for NO homeostasis. *Genes Dev.* **2010**, *24*, 491–501. [CrossRef]

50. Noman, M.Z.; Desantis, G.; Janji, B.; Hasmim, M.; Karray, S.; Dessen, P.; Bronte, V.; Chouaib, S. PD-L1 is a novel direct target of HIF-1α, and its blockade under hypoxia enhanced MDSC-mediated T cell activation. *J. Exp. Med.* **2014**, *211*, 781–790. [CrossRef]

51. Lee, J.H.; Elly, C.; Park, Y.; Liu, Y.-C. E3 ubiquitin ligase VHL regulates hypoxia-inducible factor-1α to maintain regulatory T cell stability and suppressive capacity. *Immunity* **2015**, *42*, 1062–1074. [CrossRef]

52. Yang, L.; Yang, C. Oncometabolites in Cancer: Current Understanding and Challenges. *Cancer Res.* **2021**, *81*, 2820–2823. [CrossRef]

53. Böttcher, M.; Renner, K.; Berger, R.; Mentz, K.; Thomas, S.; Cardenas-Conejo, Z.E.; Dettmer, K.; Oefner, P.J.; Mackensen, A.; Kreutz, M.; et al. D-2-hydroxyglutarate interferes with HIF-1α stability skewing T-cell metabolism towards oxidative phosphorylation and impairing Th17 polarization. *Oncoimmunology* **2018**, *7*, e1445454. [CrossRef] [PubMed]

54. Fischer, K.; Hoffmann, P.; Voelkl, S.; Meidenbauer, N.; Ammer, J.; Edinger, M.; Gottfried, E.; Schwarz, S.; Rothe, G.; Hoves, S.; et al. Inhibitory effect of tumor cell–derived lactic acid on human T cells. *Blood* **2007**, *109*, 3812–3819. [CrossRef] [PubMed]

55. Brand, A.; Singer, K.; Koehl, G.E.; Kolitzus, M.; Schoenhammer, G.; Thiel, A.; Matos, C.; Bruss, C.; Klobuch, S.; Peter, K.; et al. LDHA- associated lactic acid production blunts tumor immunosurveillance by T and NK cells. *Cell Metab.* **2016**, *24*, 657–671. [CrossRef] [PubMed]

56. Labadie, B.W.; Bao, R.; Luke, J.J. Reimagining IDO Pathway inhibition in cancer immunotherapy via downstream focus on the tryptophan–kynurenine–aryl hydrocarbon axis. *Clin. Cancer Res.* **2019**, *25*, 1462–1471. [CrossRef] [PubMed]

57. Liu, Y.; Liang, X.; Dong, W.; Fang, Y.; Lv, J.; Zhang, T.; Fiskesund, R.; Xie, J.; Liu, J.; Yin, X.; et al. Tumor- repopulating cells induce PD-1 expression in CD8⁺ T cells by transferring kynurenine and AhR activation. *Cancer Cell* **2018**, *33*, 480–494.e7. [CrossRef]

58. Schuettengruber, B.; Doetzlhofer, A.; Kroboth, K.; Wintersberger, E.; Seiser, C. Alternate activation of two divergently transcribed mouse genes from a bidirectional promoter is linked to changes in histone modification. *J. Biol. Chem.* **2003**, *278*, 1784–1793. [CrossRef]

59. Sanjabi, S.; Oh, S.A.; Li, M.O. Regulation of the Immune Response by TGF-β: From Conception to Autoimmunity and Infection. *Cold Spring Harb. Perspect. Biol.* **2017**, *9*, a022236. [CrossRef]

60. Kim, S.J.; Glick, A.; Sporn, M.B.; Roberts, A.B. Characterization of the promoter region of the human transforming growth factor-beta 1 gene. *J. Biol. Chem.* **1989**, *264*, 402–408. [CrossRef]

61. Qureshi, H.Y.; Sylvester, J.; El Mabrouk, M.; Zafarullah, M. TGF-beta-induced expression of tissue inhibitor of metalloproteinases-3 gene in chondrocytes is mediated by extracellular signal-regulated kinase pathway and Sp1 transcription factor. *J. Cell. Physiol.* **2005**, *203*, 345–352. [CrossRef]

62. Wang, J.P.; Wen, M.H.; Chen, Y.T.; Lee, H.H.; Chiang, E.R.; Lee, Y.T.; Liu, C.L.; Chen, T.H.; Hung, S.C. Trichostatin A inhibits TGF-β1 induced in vitro chondrogenesis of hMSCs through Sp1 suppression. *Differentiation* **2011**, *81*, 119–126. [CrossRef]

63. Datta, P.K.; Blake, M.C.; Moses, H.L. Regulation of plasminogen activator inhibitor-1 expression by transforming growth factor-beta-induced physical and functional interactions between smads and Sp1. *J. Biol. Chem.* **2000**, *275*, 40014–40019. [CrossRef]

64. Fajardo, O.A.; Thompson, K.; Parapuram, S.K.; Liu, S.; Leask, A. Mithramycin reduces expression of fibro-proliferative mRNAs in human gingival fibroblasts. *Cell Prolif.* **2011**, *44*, 166–173. [CrossRef]

65. Su, F.; Geng, J.; Li, X.; Qiao, C.; Luo, L.; Feng, J.; Dong, X.; Lv, M. SP1 promotes tumor angiogenesis and invasion by activating VEGF expression in an acquired trastuzumab-resistant ovarian cancer model. *Oncol. Rep.* **2017**, *38*, 2677–2684. [PubMed]

66. Xu, K.; Shu, H.K. EGFR activation results in enhanced cyclooxygenase-2 expression through p38 mitogen-activated protein kinase-dependent activation of the Sp1/Sp3 transcription factors in human gliomas. *Cancer Res.* **2007**, *67*, 6121–6129. [CrossRef] [PubMed]

67. Hertel, J.; Hirche, C.; Wissmann, C.; Ebert, M.P.; Höcker, M. Transcription of the vascular endothelial growth factor receptor-3 (VEGFR3) gene is regulated by the zinc finger proteins Sp1 and Sp3 and is under epigenetic control: Transcription of vascular endothelial growth factor receptor 3. *Cell Oncol.* **2014**, *37*, 131–145. [CrossRef] [PubMed]

68. Zimna, A.; Kurpisz, M. Hypoxia-Inducible Factor-1 in Physiological and Pathophysiological Angiogenesis: Applications and Therapies. *Biomed Res. Int.* **2015**, *2015*, 549412. [CrossRef]

69. Nam, E.H.; Lee, Y.; Zhao, X.F.; Park, Y.K.; Lee, J.W.; Kim, S. ZEB2-Sp1 cooperation induces invasion by upregulating cadherin-11 and integrin a5 expression. *Carcinogenesis* **2014**, *35*, 302–314. [CrossRef]

70. Kolesnikoff, N.; Attema, J.L.; Roslan, S.; Bert, A.G.; Schwarz, Q.P.; Gregory, P.A.; Goodall, G.J. Specificity protein 1 (Sp1) maintains basal epithelial expression of the miR-200 family: Implications for epithelial-mesenchymal transition. *J. Biol. Chem.* **2014**, *289*, 11194–11205. [CrossRef]

71. Qian, Y.; Yao, W.; Yang, T.; Yang, Y.; Liu, Y.; Shen, Q.; Zhang, J.; Qi, W.; Wang, J. aPKC-i/P-Sp1/Snail signaling induces epithelial-mesenchymal transition and immunosuppression in cholangiocarcinoma. *Hepatology* **2017**, *66*, 1165–1182. [CrossRef] [PubMed]

72. Chhunchha, B.; Fatma, N.; Bhargavan, B.; Kubo, E.; Kumar, A.; Singh, D.P. Specificity protein, Sp1-mediated increased expression of Prdx6 as a curcumin-induced antioxidant defense in lens epithelial cells against oxidative stress. *Cell Death Dis.* **2011**, *2*, e234. [CrossRef] [PubMed]
73. Dauer, P.; Gupta, V.K.; McGinn, O.; Nomura, A.; Sharma, N.S.; Arora, N.; Giri, B.; Dudeja, V.; Saluja, A.K.; Banerjee, S. Inhibition of Sp1 prevents ER homeostasis and causes cell death by lysosomal membrane permeabilization in pancreatic cancer. *Sci. Rep.* **2017**, *7*, 1564. [CrossRef] [PubMed]
74. Akman, M.; Belisario, D.C.; Salaroglio, I.C.; Kopecka, J.; Donadelli, M.; De Smaele, E.; Riganti, C. Hypoxia, endoplasmic reticulum stress and chemoresistance: Dangerous liaisons. *J. Exp. Clin. Cancer Res.* **2021**, *40*, 28. [CrossRef] [PubMed]
75. Duffy, M.J.; O'Grady, S.; Tang, M.; Crown, J. MYC as a target for cancer treatment. *Cancer Treat. Rev.* **2021**, *94*, 102154. [CrossRef]
76. Das, S.K.; Lewis, B.A.; Levens, D. MYC: A complex problem. *Trends Cell Biol.* **2023**, *33*, 235–246. [CrossRef]
77. Madden, S.K.; de Araujo, A.D.; Gerhardt, M.; Fairlie, D.P.; Mason, J.M. Taking the Myc out of cancer: Toward therapeutic strategies to directly inhibit c-Myc. *Mol. Cancer* **2021**, *20*, 3. [CrossRef]
78. Cencioni, C.; Scagnoli, F.; Spallotta, F.; Nasi, S.; Illi, B. The "Superoncogene" Myc at the Crossroad between Metabolism and Gene Expression in Glioblastoma Multiforme. *Int. J. Mol. Sci.* **2023**, *24*, 4217. [CrossRef]
79. Pirity, M.; Blanck, J.K.; Schreiber-Agus, N. Lessons learned from Myc/Max/Mad knockout mice. *Curr. Top. Microbiol. Immunol.* **2006**, *302*, 205–234. [CrossRef]
80. Felsher, D.W. MYC Inactivation Elicits Oncogene Addiction through Both Tumor Cell-Intrinsic and Host-Dependent Mechanisms. *Genes Cancer* **2010**, *1*, 597–604. [CrossRef]
81. See, Y.X.; Chen, K.; Fullwood, M.J. MYC overexpression leads to increased chromatin interactions at super-enhancers and MYC binding sites. *Genome Res.* **2022**, *32*, 629–642. [CrossRef]
82. Dhanasekaran, R.; Deutzmann, A.; Mahauad-Fernandez, W.D.; Hansen, A.S.; Gouw, A.M.; Felsher, D.W. The MYC oncogene—The grand orchestrator of cancer growth and immune evasion. *Nat. Rev. Clin. Oncol.* **2022**, *19*, 23–36. [CrossRef]
83. Karadkhelkar, N.M.; Lin, M.; Eubanks, L.M.; Janda, K.D. Demystifying the Druggability of the MYC Family of Oncogenes. *J. Am. Chem. Soc.* **2023**, *145*, 3259–3269. [CrossRef]
84. Savino, M.; Annibali, D.; Carucci, N.; Favuzzi, E.; Cole, M.D.; Evan, G.I.; Soucek, L.; Nasi, S. The action mechanism of the Myc inhibitor termed Omomyc may give clues on how to target Myc for cancer therapy. *PLoS ONE* **2011**, *6*, e22284. [CrossRef] [PubMed]
85. Hu, X.; Lin, J.; Jiang, M.; He, X.; Wang, K.; Wang, W.; Hu, C.; Shen, Z.; He, Z.; Lin, H.; et al. HIF-1α Promotes the Metastasis of Esophageal Squamous Cell Carcinoma by Targeting SP1. *J. Cancer* **2020**, *11*, 229–240. [CrossRef] [PubMed]
86. Nicolás, M.; Noé, V.; Ciudad, C.J. Transcriptional regulation of the human Sp1 gene promoter by the specificity protein (Sp) family members nuclear factor Y (NF-Y) and E2F. *Biochem. J.* **2003**, *371 Pt 2*, 265–275. [CrossRef]
87. Nicolás, M.; Noé, V.; Jensen, K.B.; Ciudad, C.J. Cloning and characterization of the 5'-flanking region of the human transcription factor Sp1 gene. *J. Biol. Chem.* **2001**, *276*, 22126–22132. [CrossRef] [PubMed]
88. Iyer, N.V.; Leung, S.W.; Semenza, G.L. The human hypoxia-inducible factor 1alpha gene: HIF1A structure and evolutionary conservation. *Genomics* **1998**, *52*, 159–165. [CrossRef]
89. Kato, A.; Ng, S.; Thangasamy, A.; Han, H.; Zhou, W.; Raeppel, S.; Fallon, M.; Guha, S.; Ammanamanchi, S. A potential signaling axis between RON kinase receptor and hypoxia-inducible factor-1 alpha in pancreatic cancer. *Mol. Carcinog.* **2021**, *60*, 734–745. [CrossRef]
90. Liu, J.; Levens, D. Making myc. *Curr. Top. Microbiol. Immunol.* **2006**, *302*, 1–32. [CrossRef]
91. Geltinger, C.; Hörtnagel, K.; Polack, A. TATA box and Sp1 sites mediate the activation of c-myc promoter P1 by immunoglobulin kappa enhancers. *Gene Expr.* **1996**, *6*, 113–127.
92. Wierstra, I.; Alves, J. The c-myc promoter: Still MysterY and challenge. *Adv. Cancer Res.* **2008**, *99*, 113–333. [CrossRef]
93. Biswas, S.; Mukherjee, R.; Tapryal, N.; Singh, A.K.; Mukhopadhyay, C.K. Insulin regulates hypoxia-inducible factor-1α transcription by reactive oxygen species sensitive activation of Sp1 in 3T3-L1 preadipocyte. *PLoS ONE* **2013**, *8*, e62128. [CrossRef] [PubMed]
94. Lafleur, V.N.; Richard, S.; Richard, D.E. Transcriptional repression of hypoxia-inducible factor-1 (HIF-1) by the protein arginine methyltransferase PRMT1. *Mol. Biol. Cell* **2014**, *25*, 925–935. [CrossRef] [PubMed]
95. Wu, M.; Huang, Z.; Huang, W.; Lin, M.; Liu, W.; Liu, K.; Li, C. microRNA-124-3p attenuates myocardial injury in sepsis via modulating SP1/HDAC4/HIF-1α axis. *Cell Death Discov.* **2022**, *8*, 40. [CrossRef] [PubMed]
96. Liu, Z.M.; Wang, X.; Li, C.X.; Liu, X.Y.; Guo, X.J.; Li, Y.; Chen, Y.L.; Ye, H.X.; Chen, H.S. SP1 Promotes HDAC4 Expression and Inhibits HMGB1 Expression to Reduce Intestinal Barrier Dysfunction, Oxidative Stress, and Inflammatory Response after Sepsis. *J. Innate Immun.* **2022**, *14*, 366–379. [CrossRef]
97. Geng, H.; Harvey, C.T.; Pittsenbarger, J.; Liu, Q.; Beer, T.M.; Xue, C.; Qian, D.Z. HDAC4 protein regulates HIF1α protein lysine acetylation and cancer cell response to hypoxia. *J. Biol. Chem.* **2011**, *286*, 38095–38102. [CrossRef]
98. Baddela, V.S.; Sharma, A.; Michaelis, M.; Vanselow, J. HIF1 driven transcriptional activity regulates steroidogenesis and proliferation of bovine granulosa cells. *Sci. Rep.* **2020**, *10*, 3906. [CrossRef]
99. Jeong, J.K.; Park, S.Y. Transcriptional regulation of specific protein 1 (SP1) by hypoxia-inducible factor 1 alpha (HIF-1α) leads to PRNP expression and neuroprotection from toxic prion peptide. *Biochem. Biophys. Res. Commun.* **2012**, *429*, 93–98. [CrossRef]

100. Culver, C.; Melvin, A.; Mudie, S.; Rocha, S. HIF-1α depletion results in SP1-mediated cell cycle disruption and alters the cellular response to chemotherapeutic drugs. *Cell Cycle* **2011**, *10*, 1249–1260. [CrossRef]

101. Li, Y.; Sun, X.X.; Qian, D.Z.; Dai, M.S. Molecular Crosstalk Between MYC and HIF in Cancer. *Front. Cell Dev. Biol.* **2020**, *8*, 590576. [CrossRef]

102. Gordan, J.D.; Bertout, J.A.; Hu, C.J.; Diehl, J.A.; and Simon, M.C. HIF-2alpha promotes hypoxic cell proliferation by enhancing c-myc transcriptional activity. *Cancer Cell* **2007**, *11*, 335–347. [CrossRef] [PubMed]

103. Conacci-Sorrell, M.; McFerrin, L.; Eisenman, R.N. An overview of MYC and its interactome. *Cold Spring Harb. Perspect. Med.* **2014**, *4*, a014357. [CrossRef] [PubMed]

104. Zhang, H.; Gao, P.; Fukuda, R.; Kumar, G.; Krishnamachary, B.; Zeller, K.I.; Dang, C.V.; Semenza, G.L. HIF-1 inhibits mitochondrial biogenesis and cellular respiration in VHL-deficient renal cell carcinoma by repression of C-MYC activity. *Cancer Cell* **2007**, *11*, 407–420. [CrossRef] [PubMed]

105. Corn, P.G.; Ricci, M.S.; Scata, K.A.; Arsham, A.M.; Simon, M.C.; Dicker, D.T.; El-Deiry, W.S. Mxi1 is induced by hypoxia in a HIF-1-dependent manner and protects cells from c-Myc-induced apoptosis. *Cancer Biol. Ther.* **2005**, *4*, 1285–1294. [CrossRef]

106. Koshiji, M.; Kageyama, Y.; Pete, E.A.; Horikawa, I.; Barrett, J.C.; Huang, L.E. HIF-1alpha induces cell cycle arrest by functionally counteracting Myc. *EMBO J.* **2004**, *23*, 1949–1956. [CrossRef]

107. Koshiji, M.; To, K.K.; Hammer, S.; Kumamoto, K.; Harris, A.L.; Modrich, P.; Huang, L.E. HIF-1alpha induces genetic instability by transcriptionally downregulating MutSalpha expression. *Mol. Cell.* **2005**, *17*, 793–803. [CrossRef]

108. To, K.K.; Sedelnikova, O.A.; Samons, M.; Bonner, W.M.; Huang, L.E. The phosphorylation status of PAS-B distinguishes HIF-1alpha from HIF-2alpha in NBS1 repression. *EMBO J.* **2006**, *25*, 4784–4794. [CrossRef]

109. Seira, N.; Yamagata, K.; Fukushima, K.; Araki, Y.; Kurata, N.; Yanagisawa, N.; Mashimo, M.; Nakamura, H.; Regan, J.W.; Murayama, T.; et al. Cellular density-dependent increases in HIF-1alpha compete with c-Myc to down-regulate human EP4 receptor promoter activity through Sp-1-binding region. *Pharmacol. Res. Perspect.* **2018**, *6*, e00441. [CrossRef]

110. Wong, W.J.; Qiu, B.; Nakazawa, M.S.; Qing, G.; Simon, M.C. MYC degradation under low O$_2$ tension promotes survival by evading hypoxia-induced cell death. *Mol. Cell. Biol.* **2013**, *33*, 3494–3504. [CrossRef]

111. Li, Q.; Kluz, T.; Sun, H.; Costa, M. Mechanisms of c-myc degradation by nickel compounds and hypoxia. *PLoS ONE* **2009**, *4*, e8531. [CrossRef] [PubMed]

112. Zarrabi, A.J.; Kao, D.; Nguyen, D.T.; Loscalzo, J.; Handy, D.E. Hypoxia-induced suppression of c-Myc by HIF-2alpha in human pulmonary endothelial cells attenuates TFAM expression. *Cell. Signal.* **2017**, *38*, 230–237. [CrossRef] [PubMed]

113. Florczyk, U.; Czauderna, S.; Stachurska, A.; Tertil, M.; Nowak, W.; Kozakowska, M.; Poellinger, L.; Jozkowicz, A.; Loboda, A.; Dulak, J. Opposite effects of HIF-1alpha and HIF-2alpha on the regulation of IL-8 expression in endothelial cells. *Free Radic. Biol. Med.* **2011**, *51*, 1882–1892. [CrossRef]

114. Thaysen, J.; Boisen, A.; Hansen, O.; Bouwstra, S. HIF-alpha effects on c-Myc distinguish two subtypes of sporadic VHL-deficient clear cell renal carcinoma. *Cancer Cell* **2008**, *14*, 435–446. [CrossRef]

115. Hoefflin, R.; Harlander, S.; Schäfer, S.; Metzger, P.; Kuo, F.; Schönenberger, D.; Adlesic, M.; Peighambari, A.; Seidel, P.; Chen, C.-Y.; et al. HIF-1alpha and HIF-2alpha differently regulate tumour development and inflammation of clear cell renal cell carcinoma in mice. *Nat. Commun.* **2020**, *11*, 4111. [CrossRef] [PubMed]

116. Xue, G.; Yan, H.-L.; Zhang, Y.; Hao, L.-Q.; Zhu, X.-T.; Mei, Q.; Sun, S.-H. c-Myc-mediated repression of miR-15-16 in hypoxia is induced by increased HIF-2alpha and promotes tumor angiogenesis and metastasis by upregulating FGF2. *Oncogene* **2015**, *34*, 1393–1406. [CrossRef]

117. Zhang, J.; Sattler, M.; Tonon, G.; Grabher, C.; Lababidi, S.; Zimmerhackl, A.; Raab, M.S.; Vallet, S.; Zhou, Y.; Cartron, M.-A.; et al. Targeting angiogenesis via a c-Myc/hypoxia-inducible factor-1alpha dependent pathway in multiple myeloma. *Cancer Res.* **2009**, *69*, 5082–5090. [CrossRef]

118. Chen, C.; Cai, S.; Wang, G.; Cao, X.; Yang, X.; Luo, X.; Feng, Y.; Hu, J. c-Myc enhances colon cancer cell-mediated angiogenesis through the regulation of HIF-1alpha. *Biochem. Biophys. Res. Commun.* **2013**, *430*, 505–511. [CrossRef]

119. Weili, Z.; Zhikun, L.; Jianmin, W.; Qingbao, T. Knockdown of USP28 enhances the radiosensitivity of esophageal cancer cells via the c-Myc/hypoxia inducible factor-1 alpha pathway. *J. Cell. Biochem.* **2019**, *120*, 201–212. [CrossRef]

120. Doe, M.R.; Ascano, J.M.; Kaur, M.; Cole, M.D. Myc posttranscriptionally induces HIF1 protein and target gene expression in normal and cancer cells. *Cancer Res.* **2012**, *72*, 949–957. [CrossRef]

121. Fu, R.; Chen, Y.; Wang, X.-P.; An, T.; Tao, L.; Zhou, Y.-X.; Huang, Y.-J.; Chen, B.-A.; Li, Z.-Y.; You, Q.-D.; et al. Wogonin inhibits multiple myeloma-stimulated angiogenesis via c-Myc/VHL/HIF-1alpha signaling axis. *Oncotarget* **2016**, *7*, 5715–5727. [CrossRef]

122. Lee, K.-M.; Giltnane, J.M.; Balko, J.M.; Schwarz, L.J.; Guerrero-Zotano, A.L.; Hutchinson, K.E.; Nixon, M.J.; Estrada, M.V.; Sánchez, V.; Sanders, M.E.; et al. MYC and MCL1 Cooperatively Promote Chemotherapy Resistant Breast Cancer Stem Cells via Regulation of Mitochondrial Oxidative Phosphorylation. *Cell Metab.* **2017**, *26*, 633–647 e637. [CrossRef] [PubMed]

123. Patten, D.A.; Lafleur, V.N.; Robitaille, G.A.; Chan, D.A.; Giaccia, A.J.; Richard, D.E. Hypoxia-inducible factor-1 activation in nonhypoxic conditions: The essential role of mitochondrial-derived reactive oxygen species. *Mol. Biol. Cell.* **2010**, *21*, 3247–3257. [CrossRef] [PubMed]

124. Das, B.; Pal, B.; Bhuyan, R.; Li, H.; Sarma, A.; Gayan, S.; Talukdar, J.; Sandhya, S.; Bhuyan, S.; Gogoi, G.; et al. MYC Regulates the HIF2α Stemness Pathway via *Nanog* and *Sox2* to Maintain Self-Renewal in Cancer Stem Cells versus Non-Stem Cancer Cells. *Cancer Res.* **2019**, *79*, 4015–4025. [CrossRef]

125. Covello, K.L.; Kehler, J.; Yu, H.; Gordan, J.D.; Arsham, A.M.; Hu, C.-J.; Labosky, P.A.; Simon, M.C.; Keith, B. HIF-2alpha regulates Oct-4: Effects of hypoxia on stem cell function, embryonic development, and tumor growth. *Genes Dev.* **2006**, *20*, 557–570. [CrossRef]
126. Zhang, C.; Samanta, D.; Lu, H.; Bullen, J.W.; Zhang, H.; Chen, I.; He, X.; Semenza, G.L. Hypoxia induces the breast cancer stem cell phenotype by HIF-dependent and ALKBH5-mediated m^{6}A-demethylation of NANOG mRNA. *Proc. Natl. Acad. Sci. USA* **2016**, *113*, E2047–E2056. [CrossRef]
127. Soleymani Abyaneh, H.; Gupta, N.; Alshareef, A.; Gopal, K.; Lavasanifar, A.; Lai, R. Hypoxia Induces the Acquisition of Cancer Stem-like Phenotype Via Upregulation and Activation of Signal Transducer and Activator of Transcription-3 (STAT3) in MDA-MB-231, a Triple Negative Breast Cancer Cell Line. *Cancer Microenviron.* **2018**, *11*, 141–152. [CrossRef] [PubMed]
128. Parisi, F.; Wirapati, P.; Naef, F. Identifying synergistic regulation involving c-Myc and sp1 in human tissues. *Nucleic Acids Res.* **2007**, *35*, 1098–1107. [CrossRef]
129. Su, T.; Liu, P.; Ti, X.; Wu, S.; Xue, X.; Wang, Z.; Dioum, E.; Zhang, Q. HIF1α, EGR1 and SP1 co-regulate the erythropoietin receptor expression under hypoxia: An essential role in the growth of non-small cell lung cancer cells. *Cell Commun. Signal.* **2019**, *17*, 152. [CrossRef] [PubMed]
130. Miki, N.; Ikuta, M.; Matsui, T. Hypoxia-induced activation of the retinoic acid receptor-related orphan receptor alpha4 gene by an interaction between hypoxia-inducible factor-1 and Sp1. *J. Biol. Chem.* **2004**, *279*, 15025–15031. [CrossRef]
131. Koizume, S.; Ito, S.; Miyagi, E.; Hirahara, F.; Nakamura, Y.; Sakuma, Y.; Osaka, H.; Takano, Y.; Ruf, W.; Miyagi, Y. HIF2α-Sp1 interaction mediates a deacetylation-dependent FVII-gene activation under hypoxic conditions in ovarian cancer cells. *Nucleic Acids Res.* **2012**, *40*, 5389–5401. [CrossRef] [PubMed]
132. Zhang, Y.; Chen, H.-X.; Zhou, S.-Y.; Wang, S.-X.; Zheng, K.; Xu, D.-D.; Liu, Y.-T.; Wang, X.-Y.; Yan, H.-Z.; Zhang, L.; et al. Sp1 and c-Myc modulate drug resistance of leukemia stem cells by regulating survivin expression through the ERK-MSK MAPK signaling pathway. *Mol. Cancer* **2015**, *14*, 56. [CrossRef]
133. Liu, H.; Zhou, M.; Luo, X.; Zhang, L.; Niu, Z.; Peng, C.; Ma, J.; Peng, S.; Zhou, H.; Xiang, B.; et al. Transcriptional regulation of BRD7 expression by Sp1 and c-Myc. *BMC Mol. Biol.* **2008**, *9*, 111. [CrossRef] [PubMed]
134. Kyo, S.; Takakura, M.; Taira, T.; Kanaya, T.; Itoh, H.; Yutsudo, M.; Ariga, H.; Inoue, M. Sp1 cooperates with c-Myc to activate transcription of the human telomerase reverse transcriptase gene (hTERT). *Nucleic Acids Res.* **2000**, *28*, 669–677. [CrossRef] [PubMed]
135. Wang, H.B.; Liu, G.H.; Zhang, H.; Xing, S.; Hu, L.J.; Zhao, W.F.; Xie, B.; Li, M.Z.; Zeng, B.H.; Li, Y.; et al. Sp1 and c-Myc regulate transcription of BMI1 in nasopharyngeal carcinoma. *FEBS J.* **2013**, *280*, 2929–2944. [CrossRef]
136. Gopisetty, G.; Xu, J.; Sampath, D.; Colman, H.; Puduvalli, V.K. Epigenetic regulation of CD133/PROM1 expression in glioma stem cells by Sp1/myc and promoter methylation. *Oncogene* **2013**, *32*, 3119–3129. [CrossRef]
137. Kong, L.M.; Liao, C.G.; Zhang, Y.; Xu, J.; Li, Y.; Huang, W.; Zhang, Y.; Bian, H.; Chen, Z.N. A regulatory loop involving miR-22, Sp1, and c-Myc modulates CD147 expression in breast cancer invasion and metastasis. *Cancer Res.* **2014**, *74*, 3764–3778. [CrossRef]
138. Nishi, H.; Nakada, T.; Kyo, S.; Inoue, M.; Shay, J.W.; Isaka, K. Hypoxia-inducible factor 1 mediates upregulation of telomerase (hTERT). *Mol. Cell. Biol.* **2004**, *24*, 6076–6083. [CrossRef] [PubMed]
139. Du, R.; Xia, L.; Ning, X.; Liu, L.; Sun, W.; Huang, C.; Wang, H.; Sun, S. Hypoxia-induced Bmi1 promotes renal tubular epithelial cell-mesenchymal transition and renal fibrosis via PI3K/Akt signal. *Mol. Biol. Cell* **2014**, *25*, 2650–2659. [CrossRef]
140. Ke, X.; Fei, F.; Chen, Y.; Xu, L.; Zhang, Z.; Huang, Q.; Zhang, H.; Yang, H.; Chen, Z.; Xing, J. Hypoxia upregulates CD147 through a combined effect of HIF-1α and Sp1 to promote glycolysis and tumor progression in epithelial solid tumors. *Carcinogenesis* **2012**, *33*, 1598–1607. [CrossRef]
141. Ohnishi, S.; Maehara, O.; Nakagawa, K.; Kameya, A.; Otaki, K.; Fujita, H.; Higashi, R.; Takagi, K.; Asaka, M.; Sakamoto, N.; et al. Hypoxia-inducible factors activate CD133 promoter through ETS family transcription factors. *PLoS ONE* **2013**, *8*, e66255. [CrossRef]
142. Barbato, L.; Bocchetti, M.; Di Biase, A.; Regad, T. Cancer Stem Cells and Targeting Strategies. *Cells* **2019**, *8*, 926. [CrossRef] [PubMed]
143. Batlle, E.; Clevers, H. Cancer stem cells revisited. *Nat. Med.* **2017**, *23*, 1124–1134. [CrossRef]
144. Kyo, S.; Takakura, M.; Fujiwara, T.; Inoue, M. Understanding and exploiting hTERT promoter regulation for diagnosis and treatment of human cancers. *Cancer Sci.* **2008**, *99*, 1528–1538. [CrossRef] [PubMed]
145. Cheng, D.; Zhao, Y.; Wang, S.; Jia, W.; Kang, J.; Zhu, J. Human Telomerase Reverse Transcriptase (hTERT) Transcription Requires Sp1/Sp3 Binding to the Promoter and a Permissive Chromatin Environment. *J. Biol. Chem.* **2015**, *290*, 30193–30203. [CrossRef]
146. Kong, L.M.; Liao, C.G.; Fei, F.; Guo, X.; Xing, J.L.; Chen, Z.N. Transcription factor Sp1 regulates expression of cancer-associated molecule CD147 in human lung cancer. *Cancer Sci.* **2010**, *101*, 1463–1470. [CrossRef] [PubMed]
147. Tang, J.; Luo, Z.; Zhou, G.; Song, C.; Yu, F.; Xiang, J.; Li, G. Cis-regulatory functions of overlapping HIF-1alpha/E-box/AP-1-like sequences of CD164. *BMC Mol. Biol.* **2011**, *12*, 44. [CrossRef] [PubMed]
148. Kokkonen, N.; Ulibarri, I.F.; Kauppila, A.; Luosujärvi, H.; Rivinoja, A.; Pospiech, H.; Kellokumpu, I.; Kellokumpu, S. Hypoxia upregulates carcinoembryonic antigen expression in cancer cells. *Int. J. Cancer* **2007**, *121*, 2443–2450. [CrossRef]
149. Xu, D.; Popov, N.; Hou, M.; Wang, Q.; Björkholm, M.; Gruber, A.; Menkel, A.R.; Henriksson, M. Switch from Myc/Max to Mad1/Max binding and decrease in histone acetylation at the telomerase reverse transcriptase promoter during differentiation of HL60 cells. *Proc. Natl. Acad. Sci. USA* **2001**, *98*, 3826–3831. [CrossRef]

150. Khattar, E.; Tergaonkar, V. Transcriptional Regulation of Telomerase Reverse Transcriptase (TERT) by MYC. *Front. Cell Dev. Biol.* **2017**, *5*, 1. [CrossRef]
151. Ramaiah, M.J.; Vaishnave, S. BMI1 and PTEN are key determinants of breast cancer therapy: A plausible therapeutic target in breast cancer. *Gene* **2018**, *678*, 302–311. [CrossRef]
152. Hardenbol, P.; Van Dyke, M.W. In vitro inhibition of c-myc transcription by mithramycin. *Biochem. Biophys. Res. Commun.* **1992**, *185*, 553–558. [CrossRef] [PubMed]
153. Holstege, C.P. Mithramycin. In *Encyclopedia of Toxicology*, 3rd ed.; Wexler, P., Ed.; Academic Press: Cambridge, MA, USA, 2014; pp. 347–348, ISBN 9780123864550. [CrossRef]
154. Beaulieu, M.E.; Jauset, T.; Massó-Vallés, D.; Martínez-Martín, S.; Rahl, P.; Maltais, L.; Zacarias-Fluck, M.F.; Casacuberta-Serra, S.; Serrano Del Pozo, E.; Fiore, C.; et al. Intrinsic cell-penetrating activity propels Omomyc from proof of concept to viable anti-MYC therapy. *Sci. Transl. Med.* **2019**, *11*, eaar5012. [CrossRef] [PubMed]
155. Wu, D.; Chen, T.; Zhao, X.; Huang, D.; Huang, J.; Huang, Y.; Huang, Q.; Liang, Z.; Chen, C.; Chen, M.; et al. HIF1α-SP1 interaction disrupts the circ-0001875/miR-31-5p/SP1 regulatory loop under a hypoxic microenvironment and promotes non-small cell lung cancer progression. *J. Exp. Clin. Cancer Res.* **2022**, *41*, 156. [CrossRef] [PubMed]
156. Gnabre, J.N.; Brady, J.N.; Clanton, D.J.; Ito, Y.; Dittmer, J.; Bates, R.B.; Huang, R.C. Inhibition of human immunodeficiency virus type 1 transcription and replication by DNA sequence-selective plant lignans. *Proc. Natl. Acad. Sci. USA* **1995**, *92*, 11239–11243. [CrossRef]
157. Gnabre, J.N.; Ito, Y.; Ma, Y.; Huang, R.C. Isolation and anti-HIV-1 lignans from Larreatridentata by counter-current chromatography. *J. Chromatogr. A* **1996**, *719*, 353–364. [CrossRef]
158. Gnabre, J.N.; Huang, R.C.; Bates, R.B.; Burns, J.J.; Caldera, S.; Malcomson, M.E.; McClure, K.J. Characterization of Anti-HIV lignans from Larreatridentata. *Tetrahedron* **1995**, *51*, 12203–12210. [CrossRef]
159. Hwu, J.R.; Tseng, W.N.; Gnabre, J.; Giza, P.; Huang, R.C.C. Antiviral activities of methylated nordihydroguaiaretic acids. 1. Synthesis, structure identification, and inhibition of Tat regulated HIV transactivation. *J. Med. Chem.* **1998**, *41*, 2994–3000. [CrossRef]
160. Chen, H.; Teng, L.; Li, J.N.; Park, R.; Mold, D.E.; Gnabre, J.; Hwu, J.R.; Tseng, W.N.; Huang, R.C. Antiviral activities of methylated nordihydroguaiaretic acids. 2. Targeting herpes simplex virus replication by the mutation insensitive transcription inhibitor tetra-O-methyl-NDGA. *J. Med. Chem.* **1998**, *41*, 3001–3007. [CrossRef]
161. Heller, J.D.; Kuo, J.; Wu, T.C.; Kast, W.M.; Huang, R.C. Tetra-O-methyl nordihydroguaiaretic acid induces G2 arrest in mammalian cells and exhibits tumoricidal activity in vivo. *Cancer Res.* **2001**, *61*, 5499–5504.
162. Park, R.; Chang, C.C.; Liang, Y.C.; Chung, Y.; Henry, R.A.; Lin, E.; Mold, D.E.; Huang, R.C. Systemic treatment with tetra-O-methyl nordihydroguaiaretic acid suppresses the growth of human xenograft tumors. *Clin. Cancer Res.* **2005**, *11*, 4601–4609. [CrossRef] [PubMed]
163. Kimura, K.; Huang, R.C. Tetra-O-Methyl Nordihydroguaiaretic Acid Broadly Suppresses Cancer Metabolism and Synergistically Induces Strong Anticancer Activity in Combination with Etoposide, Rapamycin and UCN-01. *PLoS ONE* **2016**, *11*, e0148685. [CrossRef] [PubMed]
164. Huang, R.C.; Kimura, K. Methods for Inhibition of BNIP3 and Prevention and Treatment of Ischemia Reperfusion Injury by Tetra-O-Methyl Nordihydroguaiaretic Acid. U.S. 9456995B2, 18 July 2013.
165. Huang, R.C.; Kimura, K. Suppression of Cancer Growth and Metastasis Using Nordihydroguaiaretic Acid Derivatives with Metabolic Modulators. U.S. 2011/0014192 A1, 20 January 2011.
166. Huang, R.C.; Kimura, K. Suppression of Cancer Growth and Metastasis Using Nordihydroguaiaretic Acid Derivatives with 7-hydroxystaurosporine. U.S. 2013/0053335 A1, 28 February 2013.
167. Huang, R.C.; Mold, D.; Ruland, C.; Liang, Y.C.; Chun, J.H. Compositions Comprising NDGA Derivatives and Sorafenib and Their Use in Treatment of Cancer. PCT/US13/24595, 4 June 2013; U.S. 2015/0018302 A1, 15 January 2015.
168. Chang, C.C.; Liang, Y.C.; Klutz, A.; Hsu, C.I.; Lin, C.F.; Mold, D.E.; Chou, T.C.; Lee, Y.C.; Huang, R.C.C. Reversal of multidrug resistance by two nordihydroguaiaretic acid derivatives, M₄N and maltose-M₃N, and their use in combination with doxorubicin or paclitaxel. *Cancer Chemother. Pharmacol.* **2006**, *58*, 640–653. [CrossRef]
169. Kimura, K.; Chun, J.H.; Lin, Y.; Liang, Y.; Jackson, T.L.B.; Huang, R.C.C. Tetra-*O*-methyl-nordihydroguaiaretic acid inhibits energy metabolism and synergistically induces anticancer effects with temozolomide on LN229 glioblastoma tumors implanted in mice while preventing obesity in normal mice that consume high-fat diets. *PLoS ONE* **2023**, *18*, e0285536. [CrossRef] [PubMed]
170. Chang, C.C.; Heller, J.D.; Kuo, J.; Huang, R.C. Tetra-O-methyl nordihydroguaiaretic acid induces growth arrest and cellular apoptosis by inhibiting Cdc2 and survivin expression. *Proc. Natl. Acad. Sci. USA* **2004**, *101*, 13239–13244. [CrossRef]
171. Grossman, S.A.; Ye, X.; Peereboom, D.; Rosenfeld, M.R.; Mikkelsen, T.; Supko, J.G.; Desideri, S.; Adult Brain Tumor Consortium. Phase I study of terameprocol in patients with recurrent high-grade glioma. *Neuro-Oncology* **2012**, *14*, 511–517. [CrossRef]
172. Nair, M.K.; Pandey, M.; Huang, R.C.C. A Single Group Study into the Effect of Intralesional Tetra-O-Methyl Nordihydroguaiaretic Acid (M₄N) in Oral Squamous Cell Carcinoma. *World J. Med. Res.* **2019**, *8*, 1–10.
173. Huang, R.C.; Chun, J.; Lin, Y.; Liang, Y.; Liao, K.W.; Jackson, T.; Lai, C.; Mold, D. Formulations of Terameprocol and Temozolomide and Their Use in Stimulation of Humoral Immunity in Tumors. PCT/US2020/062288, 25 November 2020; WO/2021/108601, 3 June 2021.

174. Ratti, M.; Lampis, A.; Ghidini, M.; Salati, M.; Mirchev, M.B.; Valeri, N.; Hahne, J.C. MicroRNAs (miRNAs) and Long Non-Coding RNAs (lncRNAs) as New Tools for Cancer Therapy: First Steps from Bench to Bedside. *Target Oncol.* **2020**, *15*, 261–278. [CrossRef]
175. Tornin, J.; Martinez-Cruzado, L.; Santos, L.; Rodriguez, A.; Núñez, L.E.; Oro, P.; Hermosilla, M.A.; Allonca, E.; Fernández-García, M.T.; Astudillo, A.; et al. Inhibition of SP1 by the mithramycin analog EC-8042 efficiently targets tumor initiating cells in sarcoma. *Oncotarget* **2016**, *7*, 30935–30950. [CrossRef]

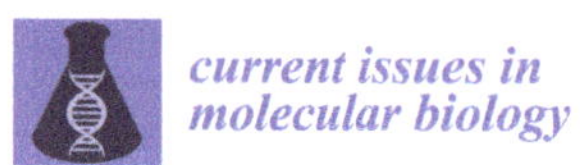

current issues in
molecular biology

Review

The Potential of Dendritic Cell Subsets in the Development of Personalized Immunotherapy for Cancer Treatment

Anna Valerevna Gorodilova [1], Kristina Viktorovna Kitaeva [1], Ivan Yurevich Filin [1], Yuri Pavlovich Mayasin [1], Chulpan Bulatovna Kharisova [1], Shaza S. Issa [2], Valeriya Vladimirovna Solovyeva [1] and Albert Anatolyevich Rizvanov [1,*]

[1] Institute of Fundamental Medicine and Biology, Kazan Federal University, 420008 Kazan, Russia; anagorodilova@yandex.ru (A.V.G.); krvkitaeva@kpfu.ru (K.V.K.); ivyfilin@kpfu.ru (I.Y.F.); mayasin_yuriy@mail.ru (Y.P.M.); harisovachulpan@gmail.com (C.B.K.); vavsoloveva@kpfu.ru (V.V.S.)

[2] Department of Genetics and Biotechnology, St. Petersburg State University, 199034 St. Petersburg, Russia; st103070@student.spbu.ru

* Correspondence: albert.rizvanov@kpfu.ru; Tel.: +7-905-316-7599

Abstract: Since the discovery of dendritic cells (DCs) in 1973 by Ralph Steinman, a tremendous amount of knowledge regarding these innate immunity cells has been accumulating. Their role in regulating both innate and adaptive immune processes is gradually being uncovered. DCs are proficient antigen-presenting cells capable of activating naive T-lymphocytes to initiate and generate effective anti-tumor responses. Although DC-based immunotherapy has not yielded significant results, the substantial number of ongoing clinical trials underscores the relevance of DC vaccines, particularly as adjunctive therapy or in combination with other treatment options. This review presents an overview of current knowledge regarding human DCs, their classification, and the functions of distinct DC populations. The stepwise process of developing therapeutic DC vaccines to treat oncological diseases is discussed, along with speculation on the potential of combined therapy approaches and the role of DC vaccines in modern immunotherapy.

Keywords: dendritic cells; immunotherapy; clinical trials; tumor-associated antigen; cancer

check for updates

Citation: Gorodilova, A.V.; Kitaeva, K.V.; Filin, I.Y.; Mayasin, Y.P.; Kharisova, C.B.; Issa, S.S.; Solovyeva, V.V.; Rizvanov, A.A. The Potential of Dendritic Cell Subsets in the Development of Personalized Immunotherapy for Cancer Treatment. *Curr. Issues Mol. Biol.* **2023**, *45*, 8053–8070. https://doi.org/10.3390/cimb45100509

Academic Editor: Dumitru A. Iacobas

Received: 29 August 2023
Revised: 27 September 2023
Accepted: 30 September 2023
Published: 1 October 2023

1. Introduction

Dendritic cells (DCs) are proficient antigen-presenting cells (APCs) that play a pivotal role in initiating an adaptive immune response [1]. Among the diverse range of APCs in the human body, such as macrophages and B-lymphocytes, DCs are considered the most efficient in capturing antigens at the site of infection and delivering them to secondary lymphoid organs, where T-cell clustering takes place for subsequent antigen presentation and the activation of effector cells [2].

DCs are usually referred to as a link between innate and adaptive immunity. As part of the innate immune system, DCs contribute to the initiation of inflammatory processes while also playing a crucial role in activating the acquired immune response by presenting antigens on major histocompatibility complex (MHC) molecules [3]. Modern sequencing technologies have enabled the gradual characterization of the diversity of human DC subsets, with the determination of their exceptional role in shaping the immune response.

The main concept behind creating a vaccine based on DCs is to utilize their ability to activate and enhance the immune response against specific antigens [4]. Mature antigen-loaded DCs are capable of activating the immune system and directing it towards fighting tumors. Therefore, a DC-based vaccine allows us to increase the immune response against a specific antigen and improves the treatment effectiveness. One of the main advantages of this type of therapy is its low toxicity compared to other methods and its use as a safe adjuvant treatment method [5].

2. Biology of Dendritic Cells

Among other populations of immune cells, DCs are identified by their high expression of MHC II molecules and CD11c, which are considered necessary for their primary functions of antigen capture and subsequent processing in complex with MHC molecules [6]. To carry out the processes of the modulation of the immune response, DCs must migrate to the site of inflammation along the gradient of chemokine concentration. DCs express the chemokine receptors C-C type 5 and C-C type 6. The signaling axes CCR5-CCL5 and CCR6-CCL20, which include CCR5 and CCR6 and their ligands expressed by the tumor microenvironment (TME), are important for the successful recruitment of DCs to the TME [7]. DCs, which penetrate into the TME under cytokine gradient conditions, are capable of producing cytokines, which induce the migration and modulate the action of lymphocytes [8–10]. Additionally, DCs are able to effectively recruit NK cells into the TME [11] and activate them, particularly through the production of cytokines CXCL9 and CXCL10 [12].

DCs exist in two distinct physiological states in the human body. In tissues, DCs in a steady state or immature condition display low levels of costimulatory molecules and are incapable of activating naive T-lymphocytes [13]. Immature DCs also exhibit high endocytic capacity, high levels of adhesion molecules for tissue localization, and low levels of immune-stimulatory cytokines. Antigen capture processes performed by DCs are diverse and involve mechanisms such as phagocytosis [14], receptor-mediated endocytosis (lectin-dependent endocytosis, Toll-like receptor-mediated endocytosis) [15,16], and macropinocytosis [17].

The recognition of pathogen-associated molecular patterns (PAMPs) or damage-associated molecular patterns (DAMPs) serves as a stimulus for DCs to transition into a mature state. This transition is accompanied by changes in the expression of costimulatory molecules, integrin, and chemokine receptors, as well as the suppression of adhesion molecule expression [18]. All these processes contribute to the migration of DCs from the initial tissue site to the secondary lymphoid organs for the presentation of endogenous peptides via MHC I to CD8$^+$ T-cells, and exogenous peptides in complex with MHC II to CD4$^+$ T-lymphocytes [19]. In addition to antigen presentation, DCs can interact with T-cells through protein factors and costimulatory molecules such as CD80, CD86, OX40 ligand (OX40L), and CD70 [13,20].

3. Dendritic Cell Populations in the Human Body

DC populations in the human body demonstrate complex phenotypic and functional heterogeneity, which accounts for their broad functionality and substantial role in modulating immune responses. The following major subsets of DCs are observed in the human body: plasmacytoid DCs (pDCs), monocyte-derived DCs (moDCs), and conventional dendritic cells (cDCs) (Figure 1). These populations share a common myeloid precursor and are distinguished from other immune cells by their high expression of MHC II and CD11c molecules. However, during ontogenesis, they exhibit distinct repertoires of surface markers [6], which will be further discussed below. Interestingly, DC development is influenced by the microenvironment, especially in non-lymphohematopoietic tissue (lungs, skin, etc.), which additionally emphasizes their functional plasticity [21].

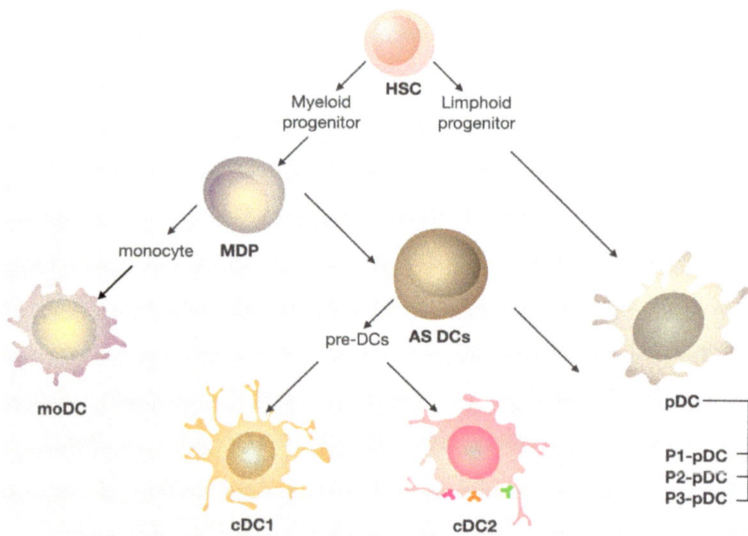

Figure 1. The scheme of the ontogenesis of DC populations. Myeloid and lymphoid precursors develop from hematopoietic stem cells. From the myeloid precursor, the macrophage DC progenitor (MDP) develops. The MDP further differentiates into monocytes and AXL[+] SIGLEC6[+] cells (AS DCs). AS DCs are capable of giving rise to both pre-DCs and pDC lineages. Pre-DCs are the precursors of cDC1 and cDC2. The pDC population, which originates from the lymphoid precursor, is also considered heterogeneous, with three distinct groups identified: P1-pDC, P2-pDC, P3-pDC.

3.1. Conventional Dendritic Cells (cDCs)

cDCs are considered to play a crucial role in the activation of naive T-cells. Their main function is to capture and degrade protein antigens and present them as peptides in complex with MHC class I or II molecules [22]. cDCs are predominantly located in non-lymphoid tissues, especially in barrier tissues, which are the main sites of pathogen entry, where they perform antigen capture functions [23].

There are two main subsets of cDCs, cDC1 and cDC2, with each performing specific functions. The cDC1 subset is characterized by chemokine receptor XCR1 expression, with its ligand XCL1 secreted on the surfaces of CD8[+] T-cells [24]. Among the distinctive set of specific markers, CLEC9A, involved in the uptake of necrotic cells; CD141, a cell adhesion molecule; and CADM1, CD103, CD8α, and BDCA-3, whose role is not fully understood, should be noted [25]. This subset is characterized by the expression of transcription factors BATF3 and IRF8 [26]. cDC1s are also major producers of IL-12, which is necessary for the differentiation of various T-cell populations [27]. cDC1s are also specialized in activating CD8[+] naive T-cells through cross-presentation via MHC I molecules, playing a crucial role in anti-tumor and antiviral immune responses [24,28]. The presence of Toll-like receptor 3 (TLR3), an endosomal protein, is characteristic of cDC1s, and it is necessary for the recognition of double-stranded RNA, an intermediate in the replication of many viruses [29].

Subpopulation cDC2 is predominantly localized in secondary lymphoid tissue and is characterized by the expression of CD11b, CD1c, BDCA-1, and Dectin-1 surface markers [30,31]. Unlike cDC1, this subset neither expresses TLR3 nor is a major producer of IL-12. Physiologically, cDC2s induce the activation of naive CD4[+] cells [30,32]. For example, in the induction of Th17 cells, cDC2s express IL-23 and IL-6 molecules, although the exact activation mechanisms are still being elucidated [18].

Using RNA sequencing technologies, it is possible to identify new subsets of DCs from those described above. The diversity of the precursors of the aforementioned populations at different stages of differentiation is increasingly taken into account. A. Villani's group identified a population of DC precursors—AS DCs—characterized by the expression of AXL, SIGLEC1, and SIGLEC6 antigens, capable of giving rise to both cDC and pDC lineages [33]. P. See and others have identified a population of pre-DCs with an immunophenotypic profile of $CD123^+$ $CD33^+$ $CD45RA^+$. Pre-DCs appear to be a later stage of differentiation from AS DCs and are precursors of cDC1 and cDC2, but not pDCs. Due to the presence of common markers, pre-DCs may be mistakenly classified as pDCs. Therefore, the secretion of IL-12 and activation of T-cells in pDCsare likely due to the presence of a small population of classical DC precursors. Pre-DCs differentiate towards $CADM1^+$ $CD1c^-$ pre-cDC1 and CADM1-$CD1c^+$ pre-cDC2 [34].

3.2. Plasmacytoid Dendritic Cells (pDCs)

PDCs are characterized by the production of type I interferons (IFN-α), which is attributed to the endoplasmic reticulum and Golgi apparatus in their cellular structure [35]. Among the expressed receptors, CD11c, CD33, CD11b, and CD13 are absent, while GMDP, CD123 (IL-3R), and CD45RA are observed. Receptors involved in IFN-α production include CD303 (CLEC4C; BDCA-2), CD304 (neuropilin; BDCA-4), CD85k (ILT3), CD85g (ILT7), Fc ε R1, BTLA, CD358, and CD300A [36–38]. pDCs are primarily involved in the detection of viral infections and do not play a significant role in stimulating naive T-cells. pDCs express TLR7 and TLR9 in order to recognize nucleic acid molecules, which are key receptors in recognizing endosomal patterns [39–41].

The production of IFN-α is not the only function of this subset. Alculumbre et al. were able to characterize three subsets of pDCs from the general population based on the expression of the costimulatory and inhibitory molecules PD-L1 and CD80. After removing AS DCs that could have influenced the experimental results, the multifunctional property of pDCs was determined. P1-pDCs (PD-L1$^+$CD80$^-$) displayed a pronounced plasmacytoid morphology, and they are the main producers of IFN-α. P3-pDCs (PD-L1$^-$CD80$^+$) acquire a dendritic morphology and adaptive immune functions. P2-pDCs (PD-L1$^+$CD80$^+$) promote T-cell activation and differentiation towards Th2 cells.

3.3. Monocyte-Derived Dendritic Cells (moDCs)

MoDCs are derived from monocytes, and they undergo functional changes in inflammatory foci. Similar to other DC populations, moDCs can also transport antigens to lymphoid tissue [42]. However, the process of differentiation can be significantly influenced by the TME conditions, such as increased lactate levels, a byproduct of tumor cell metabolism, which has a negative effect on the increase in moDCs in the TME [43]. In addition, high levels of lactate also negatively affect the production of anti-inflammatory cytokines and the cytotoxic properties of T-cells and NK cells [44,45]. MoDCs can be easily obtained from human peripheral blood monocytes by laboratory generation [46]. Experiments show that moDCs are weaker stimulators of T-cell activation; nevertheless, upon migration to the lymph nodes, moDCs can transfer captured antigens from the periphery and present them to resident DCs in the lymphoid organs. This cell population exhibits heterogeneous expression of markers including CD13, CD33, CD11b, CD11c, CD172a, S100A8/9, CCR2, CD1c, CD1a, Fc εR1, IRF4, and ZBTB46. Similar to classical DCs, they express CD11c and MHC II [47]. moDCs are effective and perform their functions efficiently in vivo, by interacting with cDCs and responding to the microenvironment's cytokine profile. However, in vitro studies of moDC-based vaccines have often yielded disappointing results, which can vary depending on the utilized activation and differentiation methods [48].

4. In Vitro DC Vaccine Design

4.1. DC Differentiation

DC vaccine development is a multi-step process, with its efficacy influenced by factors such as the culture conditions, antigen selection, and additional parameters. The first step in DC vaccine development is obtaining a sufficient number of cells for further manipulations.

Current approaches to generating DCs mainly focus on in vitro differentiation from CD14$^+$ or CD34$^+$ monocyte precursors [49–52] (Figure 2). This direction is driven by the fact that DCs, circulating in the body, are a small population of immune cells, and isolating them in the required quantity to achieve therapeutic effects is rather challenging.

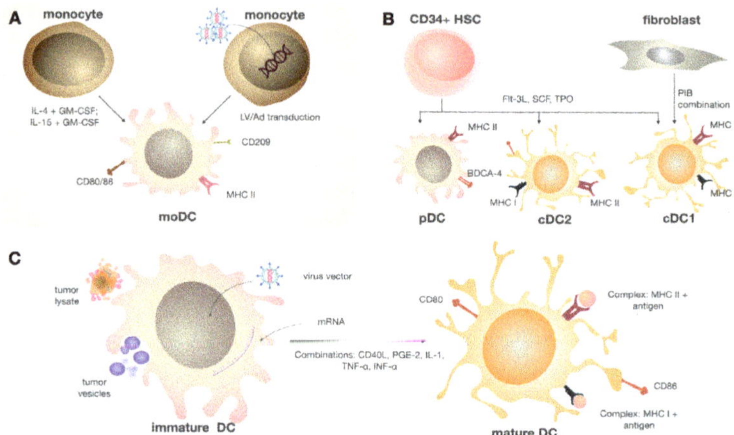

Figure 2. DCs can be created by various methods in vitro. (**A**) MoDCs are obtained through directed differentiation from CD14$^+$ monocytes using various cytokine combinations, such as IL-4, GM-CSF, IFN-α, IL-15, etc. Resulting cells express several surface markers that are characteristic of moDCs and necessary for antigen presentation [53]. (**B**) A combination of three main subsets of DCs—cDC1, cDC2, and pDC—can be obtained from HSC CD34$^+$ using FMS-like tyrosine kinase 3 ligand (FLT3L), thrombopoietin (TPO), and stem cell factor (SCF). A population similar in marker composition to cDC1 can be obtained through the direct reprogramming of fibroblasts transduced with the transcription factor set PU.1 + IRF8 + BATF3 (PIB) [54]. (**C**) The process of antigen internalization and DC activation occurs, where the antigens can be tumor vesicles, inactivated tumor cells, or the lysates of tumor cells. Antigen capture occurs, mainly through receptor-mediated phagocytosis mechanisms (lectin-dependent endocytosis, Toll-like receptor endocytosis, and macropinocytosis). One method involves transduction, where the DNA sequence encodes for antigens. An electroporation procedure is used to internalize mRNA molecules. In addition to antigen processing in complex with MHC molecules, a combination of activating molecules serves as a stimulus for maturation under in vitro conditions. The activated (mature) state of DCs is characterized by changes in the expression of costimulatory molecules (CD80, CD86) and integrin and chemokine receptors (CCR7), as well as the suppression of adhesion molecule expression.

It is worth noting that ex-vivo-generated cells exhibit transcriptional profile differences compared to their in vivo counterparts [55,56]. This could potentially be the cause of the limited therapeutic efficacy observed in DC vaccines utilizing these strategies. Nevertheless, many preclinical and clinical trials have demonstrated the ability of generated DCs to activate T-cells and secrete anti-inflammatory cytokines, such as IL-12 [57].

4.1.1. DCs Derived from Monocytes (moDCs)

MoDCs are derived through the directed differentiation of monocytes isolated from the peripheral blood mononuclear cell fraction (PBMC), obtained from whole blood or leukapheresis [58,59] (Figure 3). Isolation is commonly performed using plastic adherence, positive selection using antibody-coated magnetic beads, or flow cytometry. This method represents the most commonly used approach among published articles [60–64]. CD14$^+$ monocytes are differentiated into immature DCs over several days, alongside various factors, with the combination of IL-4 and granulocyte–macrophage colony-stimulating factor (GM-CSF) being the "gold standard" [65,66]. However, other combinations of different cytokines have also been tested. DCs differentiated in the presence of GM-CSF and IL-15 have shown to be the most effective inducers of Th17 responses [67]. The combination of GM-CSF and IFN-α also contributes to the activation of effector CD8$^+$ lymphocytes and Th1 cells [68]. Subsequently, immature moDCs are loaded with antigen and matured using a set of factors. After 1–2 days, mature moDCs present as cells loaded with tumor-associated antigen (TAA), which are then cryopreserved and thawed as needed [59]. Although this method is considered to be time-consuming, it is, however, being currently used in several medical institutions.

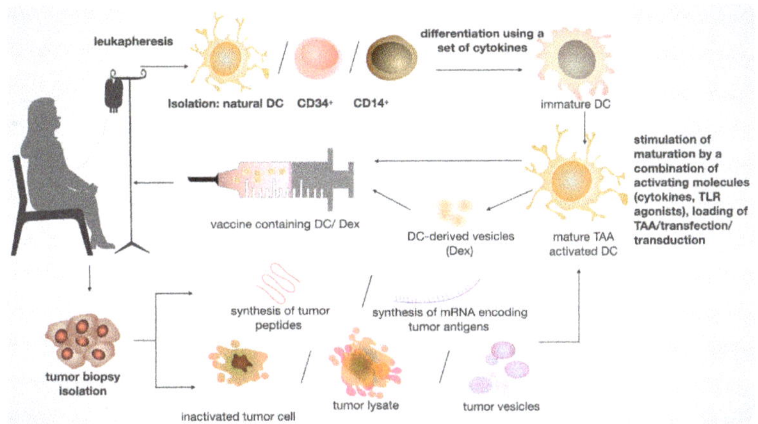

Figure 3. The main stages of creating a personalized DC-based vaccine for cancer treatment. Natural DCs, CD14$^+$ monocytes, or CD34$^+$ are isolated from leukapheresis material. CD14$^+$ and CD34$^+$ cells are differentiated into immature DCs. For TAA, tumor material is isolated, which can be used to construct the necessary antigen. Mature DCs are obtained using a combination of activating molecules and loading TAA. The ready-made injection consists of mature TAA-activated DCs or vesicles obtained from activated DCs (Dex).

The use of autologous DCs is considered a priority approach, as it avoids the immune rejection of the cells. Despite the explicit advantages of autologous therapy, allogeneic DC therapy also has been tested in several clinical trials [69]. Allogeneic DCs represent an attractive material, as the donor's immune system is not compromised due to oncological conditions. However, this approach can be challenging given that it requires careful donor selection.

4.1.2. DCs Derived from CD34$^+$ Progenitors

Another approach to generating DCs involves differentiating CD34$^+$ hematopoietic stem progenitor cells (HSPCs). By combining specific factors, certain populations of DCs can be obtained. For example, in several studies, the generation of cDC1s, the most efficient in antigen cross-presentation, has been demonstrated using a combination of recombinant FLT3L, SCF, GM-CSF, and IL-4 [70,71]. Generated cells exhibited the phenotype of true cDC1s: CD141$^+$ CLEC9A$^+$ XCR1$^+$. However, obtaining an adequate number of cells to achieve therapeutic effects and multiple infusions remains a challenge [72]. Some studies have reported an up to 20-fold increase in the yield of generated cDC1s when co-cultured with HSPCs and the OP9 cell line, compared to classical methods [73,74].

4.1.3. Genetic Reprogramming in DCs

Another approach to obtaining DCs involves genetic reprogramming, which, in theory, can allow the generation of the desired DC population, depending on the designed genetic cassette. Using a lentiviral vector encoding GM-CSF, IL-4, and melanoma-associated antigen (TRP2), it has been possible to differentiate CD14$^+$ monocytes into moDCs loaded with the tumor antigen TRP2 from melanoma [75]. A similar approach has been used to generate induced cDC1s from fibroblasts by transducing a lentiviral vector that induces the expression of key transcription factors for cDC1s: PU.1, IRF8, and BATF3 [76,77].

4.2. DC Maturation

The next rational step in creating a DC vaccine is the process of DC maturation. This process involves antigen loading and the activation of DCs using factors that influence their physiological processes. There is currently no consensus on the optimal composition of activating molecules. These can include cytokines such as TNF-α, IFN-γ, Toll-like receptor (TLR) agonists (e.g., LPS), and agonistic recombinant proteins (e.g., CD40L). It should be noted that IFN-α is included in the cytokine combination in order to mimic viral infection [78]. Several approaches have been developed for the ex vivo generation of therapeutic DC vaccines, involving the acquisition of TAA. Such approaches include synthesized tumor peptides [79], full-length proteins, tumor heat shock proteins [80], autologous tumor cells (lysates and inactivated cells) [81,82], the introduction of mRNA encoding tumor antigens [83], and tumor vesicles [84]. All antigen-loading methods have shown efficacy under in vitro conditions, considering the combination with maturation factors. The use of tumor lysates is considered the most common approach to loading tumor cells [85,86]. A significant advantage of this method is enabling the loading of DCs with a polyantigen complex, including neo-antigens, specific to the patient's tumor, due to the cellular heterogeneity of the malignancy.

An alternative approach to loading DCs would be the in vivo delivery of antigens to DCs, using liposomes [87], genetic vectors [88], or the fusion of antigens with monoclonal antibodies [89]. In this case, liposomes serve as a delivery method, for example, of RNA molecules encoding tumor antigens. This method leads to the activation of DCs in situ through endocytosis mechanisms [87]. Adeno-associated viruses carrying an antigen sequence can also act as viral delivery agents [88]. Liposomes conjugated with antibodies on their surfaces can also serve as an alternative antigen delivery method, which can enhance specific binding to target cells [89]. A proper combination of antigens and activating molecules, which do not induce immunosuppression or immune tolerance, would ensure optimal DC maturation and the subsequent priming of T-cells.

4.3. Vesicles Derived from Dendritic Cells

It is known that extracellular vesicles (EVs) released by immune cells modulate cell interactions in the TME [90]. For example, EVs can inhibit tumor growth, stimulate an immune response against malignant cells, and improve the infiltration of other immune cells into the TME [91]. Additionally, EVs can transmit information between immune cells, allowing them to coordinate their actions and enhance the immune response against the tumor [92]. Therefore, one of the new approaches to oncotherapy is the use of vesicles obtained from dendritic cells (Dex), which possess the properties of stem cells and have several advantages. The utilization of Dex as an alternative antigen delivery method, and their involvement in the in vivo activation of effector cells, is currently considered a promising direction in immunotherapy [93].

The process of Dex production involves all previous stages of DC creation, with the final step being the isolation of these structures from the supernatant through sequential centrifugation. Dex is naturally secreted by DCs and, similar to the parent cell, possesses a bilayer lipid membrane with a characteristic repertoire of protein molecules, including MHC I and MHC II, as well as costimulatory molecules such as CD80 and CD86, necessary for the interaction and activation of CD8$^+$ and CD4$^+$ T-cells [94]. Additionally, the Dex membrane expresses intercellular adhesion molecule 1 (ICAM-1) [45]. Research also indicates that Dex contains various cytoplasmic proteins and microRNAs [95]. As extracellular structures, Dex are less susceptible to tumor immunosuppressive mechanisms, suggesting a potentially more effective T-cell response [96]. Moreover, due to their stable configuration resembling exosomes and other extracellular vesicles, Dex can be stored frozen for at least six months [97].

4.4. Adaptive Transfer of DC Vaccines

In the context of in vivo conditions, DCs, following successful activation, migrate to the lymph nodes, via chemokine gradient, in order to interact with T-cells. Therefore, choosing a delivery method for autologous or allogeneic DC injections is crucial in achieving the required therapeutic effects. Intradermal injections of labeled DCs have shown that only approximately 2–4% of DCs migrate to the draining lymph nodes. Remaining cells perish at the injection site and are subsequently eliminated by immune cells (macrophages). However, W. Joost Lesterhuis et al. conducted a study comparing different methods of DC administration. In the study, it was demonstrated that the subcutaneous injection of DCs leads to the higher induction of anti-tumor properties in T-cells [98]. These findings can be explained by the fact that only the most mature and differentiated cells reach the lymph nodes. In intranodal injection, DCs are delivered directly to the site of interaction with lymphocytes. Nonetheless, this method does not achieve an optimal response, compared to subcutaneous injection, which may be attributed to the administration of activated and non-activated DCs, as well as non-viable vaccine cells [99]. Direct injection into the lymph nodes eliminates the loss of non-migratory cells; however, this approach requires precise manipulation control.

5. Application of DC Vaccines in Cancer Therapy

Currently, a wealth of preclinical research results has accumulated and been published, demonstrating the anti-tumor potential of DC vaccines. Despite progression in DC development from different precursors, and the utilization of TAA and activation factor combinations, as mentioned earlier, DC vaccines have shown discouraging results in clinical practice. Undoubtedly, one of DC vaccines' advantages is the rare occurrence of third–fourth-grade adverse effects, as demonstrated in numerous clinical trials. Most side effects are minimal and characterized by first–second-grade symptoms, such as weakness, irritation at the injection site, and flu-like symptoms [100]. Toxic effects of the third–fourth grade have been reported in some published clinical trials and are likely to be associated with the therapeutic combination used [64].

5.1. Preclinical Studies

The diversity of preclinical studies aimed at assessing the effectiveness of DC-based vaccines demonstrates the current and future directions of development in this field. We consider several types of studies that reflect general trends in DC vaccine development.

Accumulating data suggest that DC vaccines, based on cDC1s, are more effective in priming T-cell responses, compared to similar moDC therapies. In their study, Stephen Ferris et al. evaluated the induction of T-cell responses by moDCs of bone marrow origin and generated cDC1s using mouse models. The study compared the ability of moDCs and cDC1s to directly prime T-cells in lymphatic vessels, without natural DCs' involvement. *Irf8*$^{+}$*32*$^{-/-}$ mice, which lack endogenous cDC1s, were used for this purpose. Although the cross-presentation of antigens was demonstrated in in vitro models, the authors concluded that moDC injection in *Irf8*$^{+}$*32*$^{-/-}$ mouse xenotransplant models did not lead to tumor-specific responses without the involvement of cDCs [101]. The results of preclinical studies, involving various combinations of DC-based therapies, are presented in Table 1.

Shin-Wha Lee et al. found that a therapy approach using CD8α^{+} DCs, induced from HSCs, similar to the population of human CD141^{+} DCs, not only promotes tumor regression but also contributes to a higher level of immune-stimulating cells such as CD4^{+}, CD8^{+}, and CD11c^{+}, as well as a lower level of the immunosuppressive cytokine IL-10 at lower therapeutic doses, compared to moDC therapy in a mouse model [102]. This, therefore, emphasizes the relevance of developing therapeutic vaccines based on cDCs.

The current preclinical development of cDC-based vaccines aims not only to demonstrate differences, but also to optimize the protocol of obtaining cDCs from HSCs with maximal yields and a complete immunophenotypic marker set. Yuanzhi Bian et al. established the synergistic role of IFN-γ and TLR agonists in activating an immortalized mouse DC cell line (JAWSII (ATCC® CRL-11904™, Manassas, VA, USA). Their research established a significant difference in the percentage of activated DCs treated with IFN-γ+poly I:C (polyinosinic-polycytidylic acid, a TLR agonist), compared to DCs treated with IFN-γ alone, indicating IFN-γ involvement in TLR signaling pathways upon their co-administration [103]. Mariana Oliveira et al. found that inhibiting signal transmission through WASp and Arp2/3, using a small molecule called CK666, promotes cross-presentation by reducing phagosomal acidification, resulting in antigen release into the cytoplasm. As a result, TAA presentation is mediated by MHC I molecules instead of MHC II molecules, leading to a higher level of proliferation of specific CD8$^+$ T-cells in vitro and in vivo and the prolonged survival of mice receiving CK666-treated DCs [104].

Table 1. Results of some preclinical studies based on DC therapy.

Therapy	Model for Research	Results	References
Comparison of moDC-based therapy and cDC1-based therapy	*Irf8$^+$32$^{-/-}$*, *Batf3$^{-/-}$* mice C57BL/6, CD45.2$^+$ *Irf8$^+$32$^{-/-}$*, mice with subcutaneous methylcholanthrene (MCA)-induced fibrosarcoma injections	Lack of tumor-specific response in the therapy of moDCs in *Irf8$^+$32$^{-/-}$* mice	[101]
Comparison of CD8α$^+$DC-based therapy and moDC-based therapy	C57BL/6 mice with orthotopic model of ID8 cancer	Reduced volume of ascites in both groups Decreased level of regulatory T-cells (Treg), IL-10, increased expression of CD3, CD4, CD8, and CD11c markers in the CD8α$^+$ DC group	[102]
DC-based therapy + inhibitor Arp2/3 CK666	CD45.2 WT, OT-I and CD45.1 (Ly5.1) mice	The combination of DCs and CK666 inhibitor led to a reduction in phagosomal acidification and an increase in CD8$^+$ T-cell proliferation, compared to the control group	[104]
bmDC therapy + DNA vaccine	Human MUC1 transgenic mice	Tumor regression was observed only in mice receiving therapy with bmDCs + DNA vaccine	[105]
DC-based vaccine + αPD-1	C3H/HeJ mice by transplanting murine MBT-2 bladder cancer cells	In the group "DC + αPD-1", there was higher survival, IFN-γ production, and frequency of CD8$^+$ and CD4$^+$ T-cells in the spleen	[106]

Retno Murwanti et al. presented data on combined therapy using a DNA vaccine along with DCs derived from bone marrow (bmDCs). The DNA vaccine targeting MUC1, a tumor-associated antigen, and autologous bmDCs was tested as a monotherapy in human MUC1 transgenic mice with colorectal tumors. However, results showed that tumor regression was only achieved through the combination of the DNA vaccine and bmDCs. Accordingly, the authors highlighted bmDCs' crucial contribution in enhancing the anti-tumor immune response in combination with a DNA vaccine designed for a specific target molecule [105].

Soyeon Lim et al. found that the combination of a DC vaccine loaded with lysate and an anti-PD-1 antibody (αPD-1) had a more pronounced therapeutic effect, compared to mice receiving either DC therapy alone or αPD-1 in a mouse model of bladder cancer. In the study, increased secretion of IFN-γ and splenocyte cytotoxicity were also found in mice [106]. Felipe Cezar de Mato tested the influence of peptides, isolated from spider venom, on the modulation of mouse DCs in vitro, based on previous data on the cytotoxic properties of these peptides on glioblastoma cells. The study results showed statistically significant differences in the expression of the costimulatory molecule CD86, when using the peptide and tumor cell lysate, compared to a DC + lysate, as well as the increased secretion of some proinflammatory cytokines [107].

5.2. Clinical Studies

To date, a multitude of clinical trials have used different DC vaccines, with the majority of them utilizing monocyte-derived DCs for loading with antigens [60,62,108,109], while only a few highlight the use of neoantigens [61]. Some studies employ DCs of natural myeloid origin [110]; however, the method of generating DCs from monocytes is considered the most commonly encountered. Selective clinical trials will be discussed further, as they reflect general trends in vaccine development and demonstrate patient outcomes.

5.2.1. DC Progenitor-Based Therapy

At the end of 2022, data from phase 3 clinical trials of therapy, under the registered trademark DCVax-L—autologous mature DCs loaded with tumor lysate for the treatment of glioma (NCT00045968)—were published. PBMCs were obtained by leukapheresis and then cultured in the presence of GM-CSF and IL-4 cytokines. Antigen loading was performed using a tumor lysate obtained from tumor resection [111]. The phase 3 trials involved 331 patients, of whom 232 received DCVax-L injections, along with standard temozolomide treatment, and 99 patients were in the placebo group. The study aimed to analyze the vaccine effectiveness and its impact on the survival of patients with newly diagnosed glioblastoma (nGBM) or its recurrent form (rGBM). The study, which started in 2007, showed a statistically significant increase in patient survival for both NDGB and RGB patients. The median overall survival (mOS) for 232 NDGB patients receiving DCVax-L was 19.3 (95% confidence interval (CI) 17.5–21.3) months, compared to 16.5 (95% CI 16.0–17.5) months in the control group (98% CI 0.00–0.94). Among 64 RGB patients receiving DCVax-L, the mOS was 13.2 (95% CI 9.7–16.8) months after recurrence, compared to 7.8 (95% CI 7.2–8.2) months in the control group (98% CI) [60]. Clinical trial results based on personalized DC vaccines are presented in Table 2.

Nicholas J Vogelzang and colleagues demonstrated the results of phase 3 of a completed VIABLE study. The study tested the efficacy of autologous DCs, combined with docetaxel and prednisone, compared to a placebo group receiving only standard chemotherapy (NCT02111577), for 1182 men with metastatic castration-resistant prostate cancer. The vaccine contained moDCs loaded with inactivated human prostate adenocarcinoma cell line cells (LNCaP). Despite promising results from phase 1/2 clinical trials, prior to VIABLE, the mOS was 23.9% in the treatment group and 24.3% in the placebo group, indicating no difference in overall survival between the two groups [112]. The authors also highlighted a lack of adverse events (AE) in patients. Such results were attributed to the terminal stage of cancer in many patients, and the fact that the mOS rate in patients who had been receiving abiraterone or enzalutamide prior to the trial was significantly lower, indicating possible drug resistance [63].

Trials involving DC transduction for antigen processing are of particular interest. A phase 1 clinical trial (NCT01730118) tested the therapeutic efficacy of autologous DCs, transduced with an adenovirus expressing human epidermal growth factor receptor 2 (HEP2), for patients with metastatic solid tumors overexpressing HEP2. These patients had either progressive disease after standard treatment or no evidence of disease after tumor resection. Thirty-three patients were included in the study and divided into several groups, with each receiving different amounts of DCs. One patient achieved a complete response (CR), one achieved a partial response (PR), and five patients achieved disease stabilization (DS). Significantly, the study showed no AE with transduced DC therapy [109]. In the study NCT01826877, DCs delivering antigens via an adenoviral vector were also tested. No significant results were obtained in the study, with only one patient completing the treatment and achieving DS after 27 months [108]. Lisa H. Butterfield and colleagues investigated the effect of transduced DNA vaccines on patients with melanoma, in combination with intravenously administered IFN-α. The vaccine was constructed against three commonly known melanoma antigens (tyrosinase, MART-1, and MAGE-A6), in order to stimulate polyclonal CD8$^+$ and CD4$^+$ responses. Two CR, eight cases of stable disease (SD), and 14 cases of disease progression (PD) were recorded. In 51% of patients (18 out of 35), first–fourth-grade AE were observed; however, second–fourth-grade AE were likely associated with the use of IFN-α. Most participants showed an enhancement in tumor-specific CD8$^+$ and CD4$^+$ T-cell responses, which did not correlate with increased levels of IL-12. Moreover, it is worth noting that adding IFN-α did not improve the immune or clinical response in this trial [113].

As an example of a clinical trial based on a neoantigen DNA vaccine, the study by Zhenyu Ding et al. (NCT02956551) can be considered. For 12 patients with non-small cell lung cancer, DCs loaded with neoantigen peptides were used as immunotherapy in combination with cyclophosphamide treatment. In order to obtain neoantigens, the study authors performed RNA sequencing, along with the whole exome sequencing of DNA from patient biopsy material. The method of obtaining monocyte-derived DCs was chosen quite classically (leukapheresis + IL-4 and GM-CSF). Along with DC vaccination, patients received maintenance therapy, such as ipilimumab, radiation therapy, or chemotherapy, as the disease progressed. In the latest published trial report, AE 1–2 were recorded, an objective response (OR) was achieved in 25% of cases, and it was also mentioned that personalized neoantigen vaccines have the ability to induce a T-cell response [61].

Koichi Mitsuya and colleagues used polarized α-type 1 DCs in a clinical trial as a vaccine against multifocal glioblastoma. DCs were activated with a cytokine cocktail and a complex of synthetic peptides. Combining the vaccine with standard therapy showed an anti-tumor effect, with an mOS of 19.0 months. The study also found significant IL-12 secretion by α-type DCs, which correlated with increased levels of IFN-γ, potentially inducing the IFN-γ-mediated stimulation of effector T-cells. The long-term follow-up of patients (up to 6 years) revealed better outcomes for those receiving DC therapy, compared to the control group [114].

5.2.2. Therapy Based on Natural DCs

Clinical trial NCT02574377 evaluated immunological effects in patients with stage 3 resected melanoma, who had undergone autologous myeloid-derived cDC2 or pDC therapy. Cells were isolated by apheresis and loaded with melanoma-associated peptides. No third-grade or higher AE were observed, and antigen-specific CD8$^+$ T-cells were present in 80% of patients, and CD4$^+$ T-cells in 64%, following the first vaccine injection. Median disease-free survival (mDFS) was 19.4 months, and 6 out of 15 patients showed no recurrence at the end of the study [110]. In 2023, a phase 1/2 clinical trial of a neoadjuvant DC vaccine for ovarian cancer treatment was launched, with the selection of natural cDC1 as the most effective DC population for antigen cross-presentation as a therapy basis (NCT05773859)

5.2.3. Therapy Based on DC-Derived Vesicles (Dex)

Dex vaccines have been tested in a series of clinical trials. In phase 1 of the Dex clinical trial, those obtained from autologous DCs were administered to patients with metastatic melanoma. Despite the promising method, only 2 out of 15 participants developed a PR, while SD was observed in two patients. However, no AE were identified, and 8 out of 13 individuals showed increased effector function of NK cells [115]. In another phase 1 clinical trial, autologous Dex therapy loaded with MAGE tumor antigens demonstrated MAGE-specific T-cell responses in 33% of patients with non-small cell lung cancer, as well as an increase in NK lytic activity in 16% of patients [116]. After the phase 2 study using Dex-stimulated IFN-γ, Dex was shown to enhance anti-tumor immunity in the NK cells of patients with advanced non-small cell lung cancer (NCT01159288). The results indicate that the Dex vaccine promotes the increased effector function of NKp30-NK cells, which is closely related to high levels of MHC II expression and IFN-γ content. No T-cell response was detected in patients. One patient developed third-grade hepatotoxicity, while, in other cases, AE did not exceed the second grade [117].

Table 2. Results from clinical trials using dendritic-cell-based therapeutic vaccines.

Therapy	Type of Cancer	Participants	Efficiency, %	References
Neoantigen-primed DC vaccine + cyclophosphamide	Non-small cell lung cancer	12	25 OR 75 DCR 100 AE (1–2)	NCT02956551
Adenoviral transduced autologous human epidermal growth factor receptor (AdHER)/neu DC vaccine	Metastatic solid tumors characterized by HER2/Neu expression	33	3 CR 3 PR 15 SD 100 AE	[116]
Autologous DCs transduced with AdGMCA9 (DC-AdGMCAIX)	Metastatic renal cell carcinoma	11	45 AE (1–2) 9 SD	NCT01826877
Autologous DCs pulsed with tumor lysate antigen (DCVax®-L)	Newly diagnosed glioblastoma (NDG); recurrent glioblastoma (RG)	232—NDG 64—RG	1,5 AE 19.3 mOS (months), NDG 13.2 mOS (months), RG	NCT00045968
IKKβ-matured RNA-transfected DC vaccine + immune checkpoint blockade (ICB)	Metastatic uveal melanoma	12	No published results	NCT04335890
Autologous DCs + docetaxel + prednisone	Metastatic castration-resistant prostate cancer	1182	23.9 mOS (months) 24.3 mOS (months, placebo group)	NCT02111577
Alpha-type-1 polarized DC-based vaccination	Newly diagnosed high-grade glioma	16	19 mOS (months)	[114]
DCs transduced with MART-1, tyrosinase, and MAGE-A6+ IFN-α	Melanoma	35	5.7 PR 23 SD 40 PD 51.4 AE 36 mOS (months)	NCT01622933
Therapy based on natural autologous cDC2 and pDCs	Melanoma	15	19.4 mDFS (months) 100 AE (1–2)	NCT02574377
Exosomes derived from autologous moDCs	Metastatic melanoma	15	13 PR 13 SD	[58]
IFN-γ-exosomes derived from DC (IFN-γ- Dex)	Non-small cell lung cancer	22	15 mOS (months) 19 AE (1–3) 32 SD (for 4 months)	NCT01159288

OR—objective response, DCR—disease control rate, AE—adverse events, CR—complete response, PR—partial response, SD—stable disease, mOS—median overall survival, PD—progressive disease, mDFS—median disease-free survival.

6. Conclusions

Despite the discouraging results of DC-based immunotherapy approaches, addressing the challenge of enhancing the clinical response remains relevant. Vaccines may be a proper treatment choice for individuals with tumors non-responsive to CAR-T therapy, immune checkpoint inhibitors, or monoclonal antibodies and may also be suitable for maintaining remission. The analysis of different populations of DCs in the human body provides a clear understanding of their contribution to the mechanisms of immune function—in particular, anti-tumor processes. Among preclinical studies, there is a trend towards using generated cDCs and pDCs as immunotherapy, demonstrating promising results. However, obtaining sufficient quantities of DCs from precursors in vitro to achieve a clinical response in patients remains an open question.

In conclusion, it should be noted that numerous clinical trials aiming to test combined treatment options are currently either in the patient recruitment stage or have not been completed yet. Although DC-based vaccines face a number of difficulties, recent advances in the development of new DC vaccination schemes as adjuvant therapy will make a significant contribution to therapy, especially for solid tumors. It is likely that the new generation of DC vaccines based on mRNA loading and combined with various immune adjuvants will lead to a significant T-cell response and a reduction in the influence of the TME. An alternative and promising therapy option is the direct generation of cDC1 from the patient's blood as the most effective APC for an anti-tumor response. The most important steps in improving DC vaccines are optimizing the methods of differentiating DCs from precursor cells, selecting the most effective method of loading, and combining DC vaccines with other forms of therapy, which may contribute to increasing their effectiveness and achieving the best clinical outcome for cancer patients.

Author Contributions: Conceptualization, A.A.R. and V.V.S.; writing—original draft preparation, A.V.G., Y.P.M. and C.B.K.; writing—review and editing, K.V.K., V.V.S. and S.S.I.; visualization, A.V.G. and I.Y.F.; supervision, A.A.R. All authors have read and agreed to the published version of the manuscript.

Funding: This research was funded by the Russian Science Foundation grant, grant number 22-24-20018.

Institutional Review Board Statement: Not applicable.

Informed Consent Statement: Not applicable.

Data Availability Statement: Authors can confirm that all relevant data used to support the findings of this study are including within the article.

Acknowledgments: This work is part of the Kazan Federal University Strategic Academic Leadership Program (PRIORITY-2030).

Conflicts of Interest: The authors declare no conflict of interest.

References

1. Steinman, R.M.; Witmer, M.D. Lymphoid Dendritic Cells Are Potent Stimulators of the Primary Mixed Leukocyte Reaction in Mice. *Proc. Natl. Acad. Sci. USA* **1978**, *75*, 5132–5136. [CrossRef] [PubMed]
2. Stumbles, P.A.; Himbeck, R.; Frelinger, J.A.; Collins, E.J.; Lake, R.A.; Robinson, B.W.S. Cutting Edge: Tumor-Specific CTL Are Constitutively Cross-Armed in Draining Lymph Nodes and Transiently Disseminate to Mediate Tumor Regression Following Systemic CD40 Activation. *J. Immunol.* **2004**, *173*, 5923–5928. [CrossRef] [PubMed]
3. Filin, I.Y.; Kitaeva, K.V.; Rutland, C.S.; Rizvanov, A.A.; Solovyeva, V.V. Recent Advances in Experimental Dendritic Cell Vaccines for Cancer. *Front. Oncol.* **2021**, *11*, 730824. [CrossRef] [PubMed]
4. Okada, H.; Kalinski, P.; Ueda, R.; Hoji, A.; Kohanbash, G.; Donegan, T.E.; Mintz, A.H.; Engh, J.A.; Bartlett, D.L.; Brown, C.K.; et al. Induction of CD8$^+$ T-Cell Responses against Novel Glioma-Associated Antigen Peptides and Clinical Activity by Vaccinations with α-Type 1 Polarized Dendritic Cells and Polyinosinic-Polycytidylic Acid Stabilized by Lysine and Carboxymethylcellulose in Patients with Recurrent Malignant Glioma. *J. Clin. Oncol.* **2011**, *29*, 330–336. [CrossRef]
5. Zhu, P.; Li, S.-Y.; Ding, J.; Fei, Z.; Sun, S.-N.; Zheng, Z.-H.; Wei, D.; Jiang, J.; Miao, J.-L.; Li, S.-Z.; et al. Combination Immunotherapy of Glioblastoma with Dendritic Cell Cancer Vaccines, Anti-PD-1 and Poly I:C. *J. Pharm. Anal.* **2023**, *13*, 616–624. [CrossRef]
6. Vermaelen, K.; Pauwels, R. Accurate and Simple Discrimination of Mouse Pulmonary Dendritic Cell and Macrophage Populations by Flow Cytometry: Methodology and New Insights. *Cytom. A* **2004**, *61*, 170–177. [CrossRef]
7. Bell, D.; Chomarat, P.; Broyles, D.; Netto, G.; Harb, G.M.; Lebecque, S.; Valladeau, J.; Davoust, J.; Palucka, K.A.; Banchereau, J. In Breast Carcinoma Tissue, Immature Dendritic Cells Reside within the Tumor, Whereas Mature Dendritic Cells Are Located in Peritumoral Areas. *J. Exp. Med.* **1999**, *190*, 1417–1426. [CrossRef]
8. Tuna, H.; Avdiushko, R.G.; Sindhava, V.J.; Wedlund, L.; Kaetzel, C.S.; Kaplan, A.M.; Bondada, S.; Cohen, D.A. Regulation of the Mucosal Phenotype in Dendritic Cells by PPARγ: Role of Tissue Microenvironment. *J. Leukoc. Biol.* **2014**, *95*, 471–485. [CrossRef]
9. Collin, M.; Bigley, V. Human Dendritic Cell Subsets: An Update. *Immunology* **2018**, *154*, 3–20. [CrossRef]
10. Zhang, Y.; Guan, X.; Jiang, P. Cytokine and Chemokine Signals of T-Cell Exclusion in Tumors. *Front. Immunol.* **2020**, *11*, 594609. [CrossRef]
11. Persson, C.M.; Chambers, B.J. Plasmacytoid Dendritic Cell-Induced Migration and Activation of NK Cells in Vivo. *Eur. J. Immunol.* **2010**, *40*, 2155–2164. [CrossRef]

12. Jacobs, B.; Gebel, V.; Heger, L.; Grèze, V.; Schild, H.; Dudziak, D.; Ullrich, E. Characterization and Manipulation of the Crosstalk Between Dendritic and Natural Killer Cells Within the Tumor Microenvironment. *Front. Immunol.* **2021**, *12*, 670540. [CrossRef] [PubMed]
13. Trombetta, E.S.; Mellman, I. Cell Biology of Antigen Processing in Vitro and in Vivo. *Annu. Rev. Immunol.* **2005**, *23*, 975–1028. [CrossRef] [PubMed]
14. Hoffmann, E.; Kotsias, F.; Visentin, G.; Bruhns, P.; Savina, A.; Amigorena, S. Autonomous Phagosomal Degradation and Antigen Presentation in Dendritic Cells. *Proc. Natl. Acad. Sci. USA* **2012**, *109*, 14556–14561. [CrossRef] [PubMed]
15. Platt, C.D.; Ma, J.K.; Chalouni, C.; Ebersold, M.; Bou-Reslan, H.; Carano, R.A.D.; Mellman, I.; Delamarre, L. Mature Dendritic Cells Use Endocytic Receptors to Capture and Present Antigens. *Proc. Natl. Acad. Sci. USA* **2010**, *107*, 4287–4292. [CrossRef]
16. Bonifaz, L.C.; Bonnyay, D.P.; Charalambous, A.; Darguste, D.I.; Fujii, S.-I.; Soares, H.; Brimnes, M.K.; Moltedo, B.; Moran, T.M.; Steinman, R.M. In Vivo Targeting of Antigens to Maturing Dendritic Cells via the DEC-205 Receptor Improves T Cell Vaccination. *J. Exp. Med.* **2004**, *199*, 815–824. [CrossRef]
17. Norbury, C.C.; Chambers, B.J.; Prescott, A.R.; Ljunggren, H.G.; Watts, C. Constitutive Macropinocytosis Allows TAP-Dependent Major Histocompatibility Complex Class I Presentation of Exogenous Soluble Antigen by Bone Marrow-Derived Dendritic Cells. *Eur. J. Immunol.* **1997**, *27*, 280–288. [CrossRef]
18. Hilligan, K.L.; Ronchese, F. Antigen Presentation by Dendritic Cells and Their Instruction of CD4[+] T Helper Cell Responses. *Cell Mol. Immunol.* **2020**, *17*, 587–599. [CrossRef]
19. MacNabb, B.W.; Tumuluru, S.; Chen, X.; Godfrey, J.; Kasal, D.N.; Yu, J.; Jongsma, M.L.M.; Spaapen, R.M.; Kline, D.E.; Kline, J. Dendritic Cells Can Prime Anti-Tumor CD8[+] T Cell Responses through Major Histocompatibility Complex Cross-Dressing. *Immunity* **2022**, *55*, 982–997.e8. [CrossRef]
20. Ukyo, N.; Hori, T.; Yanagita, S.; Ishikawa, T.; Uchiyama, T. Costimulation through OX40 Is Crucial for Induction of an Alloreactive Human T-Cell Response. *Immunology* **2003**, *109*, 226–231. [CrossRef]
21. Heidkamp, G.F.; Sander, J.; Lehmann, C.H.K.; Heger, L.; Eissing, N.; Baranska, A.; Lühr, J.J.; Hoffmann, A.; Reimer, K.C.; Lux, A.; et al. Human Lymphoid Organ Dendritic Cell Identity Is Predominantly Dictated by Ontogeny, Not Tissue Microenvironment. *Sci. Immunol.* **2016**, *1*, eaai7677. [CrossRef]
22. Bowman-Kirigin, J.A.; Desai, R.; Saunders, B.T.; Wang, A.Z.; Schaettler, M.O.; Liu, C.J.; Livingstone, A.J.; Kobayashi, D.K.; Durai, V.; Kretzer, N.M.; et al. The Conventional Dendritic Cell 1 Subset Primes CD8[+] T Cells and Traffics Tumor Antigen to Drive Antitumor Immunity in the Brain. *Cancer Immunol. Res.* **2023**, *11*, 20–37. [CrossRef]
23. Villar, J.; Segura, E. Decoding the Heterogeneity of Human Dendritic Cell Subsets. *Trends Immunol.* **2020**, *41*, 1062–1071. [CrossRef] [PubMed]
24. Granot, T.; Senda, T.; Carpenter, D.J.; Matsuoka, N.; Weiner, J.; Gordon, C.L.; Miron, M.; Kumar, B.V.; Griesemer, A.; Ho, S.-H.; et al. Dendritic Cells Display Subset and Tissue-Specific Maturation Dynamics over Human Life. *Immunity* **2017**, *46*, 504–515. [CrossRef] [PubMed]
25. Schreibelt, G.; Tel, J.; Sliepen, K.H.E.W.J.; Benitez-Ribas, D.; Figdor, C.G.; Adema, G.J.; de Vries, I.J.M. Toll-like Receptor Expression and Function in Human Dendritic Cell Subsets: Implications for Dendritic Cell-Based Anti-Cancer Immunotherapy. *Cancer Immunol. Immunother.* **2010**, *59*, 1573–1582. [CrossRef] [PubMed]
26. Edelson, B.T.; Kc, W.; Juang, R.; Kohyama, M.; Benoit, L.A.; Klekotka, P.A.; Moon, C.; Albring, J.C.; Ise, W.; Michael, D.G.; et al. Peripheral CD103+ Dendritic Cells Form a Unified Subset Developmentally Related to CD8α[+] Conventional Dendritic Cells. *J. Exp. Med.* **2010**, *207*, 823–836. [CrossRef] [PubMed]
27. Garris, C.S.; Arlauckas, S.P.; Kohler, R.H.; Trefny, M.P.; Garren, S.; Piot, C.; Engblom, C.; Pfirschke, C.; Siwicki, M.; Gungabeesoon, J.; et al. Successful Anti-PD-1 Cancer Immunotherapy Requires T Cell-Dendritic Cell Crosstalk Involving the Cytokines IFN-γ and IL-12. *Immunity* **2018**, *49*, 1148–1161.e7. [CrossRef]
28. Mair, F.; Liechti, T. Comprehensive Phenotyping of Human Dendritic Cells and Monocytes. *Cytom. A* **2021**, *99*, 231–242. [CrossRef]
29. Mishra, G.P.; Jha, A.; Ahad, A.; Sen, K.; Sen, A.; Podder, S.; Prusty, S.; Biswas, V.K.; Gupta, B.; Raghav, S.K. Epigenomics of Conventional Type-I Dendritic Cells Depicted Preferential Control of TLR9 versus TLR3 Response by NCoR1 through Differential IRF3 Activation. *Cell Mol. Life Sci.* **2022**, *79*, 429. [CrossRef]
30. Binnewies, M.; Mujal, A.M.; Pollack, J.L.; Combes, A.J.; Hardison, E.A.; Barry, K.C.; Tsui, J.; Ruhland, M.K.; Kersten, K.; Abushawish, M.A.; et al. Unleashing Type-2 Dendritic Cells to Drive Protective Antitumor CD4[+] T Cell Immunity. *Cell* **2019**, *177*, 556–571.e16. [CrossRef]
31. Guilliams, M.; Ginhoux, F.; Jakubzick, C.; Naik, S.H.; Onai, N.; Schraml, B.U.; Segura, E.; Tussiwand, R.; Yona, S. Dendritic Cells, Monocytes and Macrophages: A Unified Nomenclature Based on Ontogeny. *Nat. Rev. Immunol.* **2014**, *14*, 571–578. [CrossRef] [PubMed]
32. Patente, T.A.; Pinho, M.P.; Oliveira, A.A.; Evangelista, G.C.M.; Bergami-Santos, P.C.; Barbuto, J.A.M. Human Dendritic Cells: Their Heterogeneity and Clinical Application Potential in Cancer Immunotherapy. *Front. Immunol.* **2018**, *9*, 3176. [CrossRef] [PubMed]
33. Villani, A.-C.; Satija, R.; Reynolds, G.; Sarkizova, S.; Shekhar, K.; Fletcher, J.; Griesbeck, M.; Butler, A.; Zheng, S.; Lazo, S.; et al. Single-Cell RNA-Seq Reveals New Types of Human Blood Dendritic Cells, Monocytes, and Progenitors. *Science* **2017**, *356*, eaah4573. [CrossRef]

34. See, P.; Dutertre, C.-A.; Chen, J.; Günther, P.; McGovern, N.; Irac, S.E.; Gunawan, M.; Beyer, M.; Händler, K.; Duan, K.; et al. Mapping the Human DC Lineage through the Integration of High-Dimensional Techniques. *Science* **2017**, *356*, eaag3009. [CrossRef] [PubMed]

35. Hernández, S.S.; Jakobsen, M.R.; Bak, R.O. Plasmacytoid Dendritic Cells as a Novel Cell-Based Cancer Immunotherapy. *Int. J. Mol. Sci.* **2022**, *23*, 11397. [CrossRef]

36. Dzionek, A.; Fuchs, A.; Schmidt, P.; Cremer, S.; Zysk, M.; Miltenyi, S.; Buck, D.W.; Schmitz, J. BDCA-2, BDCA-3, and BDCA-4: Three Markers for Distinct Subsets of Dendritic Cells in Human Peripheral Blood. *J. Immunol.* **2000**, *165*, 6037–6046. [CrossRef]

37. Ju, X.; Zenke, M.; Hart, D.N.J.; Clark, G.J. CD300a/c Regulate Type I Interferon and TNF-α Secretion by Human Plasmacytoid Dendritic Cells Stimulated with TLR7 and TLR9 Ligands. *Blood* **2008**, *112*, 1184–1194. [CrossRef]

38. Bao, M.; Liu, Y.-J. Regulation of TLR7/9 Signaling in Plasmacytoid Dendritic Cells. *Protein Cell* **2013**, *4*, 40–52. [CrossRef]

39. Yun, T.J.; Igarashi, S.; Zhao, H.; Perez, O.A.; Pereira, M.R.; Zorn, E.; Shen, Y.; Goodrum, F.; Rahman, A.; Sims, P.A.; et al. Human Plasmacytoid Dendritic Cells Mount a Distinct Antiviral Response to Virus-Infected Cells. *Sci. Immunol.* **2021**, *6*, eabc7302. [CrossRef]

40. Das, A.; Chauhan, K.S.; Kumar, H.; Tailor, P. Mutation in Irf8 Gene (Irf8R294C) Impairs Type I IFN-Mediated Antiviral Immune Response by Murine pDCs. *Front. Immunol.* **2021**, *12*, 758190. [CrossRef]

41. Deb, P.; Dai, J.; Singh, S.; Kalyoussef, E.; Fitzgerald-Bocarsly, P. Triggering of the cGAS-STING Pathway in Human Plasmacytoid Dendritic Cells Inhibits TLR9-Mediated IFN Production. *J. Immunol.* **2020**, *205*, 223–236. [CrossRef] [PubMed]

42. Sprangers, S.; de Vries, T.J.; Everts, V. Monocyte Heterogeneity: Consequences for Monocyte-Derived Immune Cells. *J. Immunol. Res.* **2016**, *2016*, 1475435. [CrossRef] [PubMed]

43. Goenka, A.; Khan, F.; Verma, B.; Sinha, P.; Dmello, C.C.; Jogalekar, M.P.; Gangadaran, P.; Ahn, B. Tumor Microenvironment Signaling and Therapeutics in Cancer Progression. *Cancer Commun.* **2023**, *43*, 525–561. [CrossRef] [PubMed]

44. Husain, Z.; Seth, P.; Sukhatme, V.P. Tumor-Derived Lactate and Myeloid-Derived Suppressor Cells. *Oncoimmunology* **2013**, *2*, e26383. [CrossRef]

45. Xia, J.; Miao, Y.; Wang, X.; Huang, X.; Dai, J. Recent Progress of Dendritic Cell-Derived Exosomes (Dex) as an Anti-Cancer Nanovaccine. *Biomed. Pharmacother.* **2022**, *152*, 113250. [CrossRef]

46. Chometon, T.Q.; da Silva Siqueira, M.; Anna, J.C.S.; Almeida, M.R.; Gandini, M.; de Almeida Nogueira, A.C.M.; Antas, P.R.Z. A Protocol for Rapid Monocyte Isolation and Generation of Singular Human Monocyte-Derived Dendritic Cells. *PLoS ONE* **2020**, *15*, e0231132. [CrossRef]

47. Eguíluz-Gracia, I.; Bosco, A.; Dollner, R.; Melum, G.R.; Lexberg, M.H.; Jones, A.C.; Dheyauldeen, S.A.; Holt, P.G.; Bækkevold, E.S.; Jahnsen, F.L. Rapid Recruitment of CD14(+) Monocytes in Experimentally Induced Allergic Rhinitis in Human Subjects. *J. Allergy Clin. Immunol.* **2016**, *137*, 1872–1881.e12. [CrossRef] [PubMed]

48. Perez, C.R.; De Palma, M. Engineering Dendritic Cell Vaccines to Improve Cancer Immunotherapy. *Nat. Commun.* **2019**, *10*, 5408. [CrossRef]

49. Kalantari, T.; Kamali-Sarvestani, E.; Zhang, G.-X.; Safavi, F.; Lauretti, E.; Khedmati, M.-E.; Rostami, A. Generation of Large Numbers of Highly Purified Dendritic Cells from Bone Marrow Progenitor Cells after Co-Culture with Syngeneic Murine Splenocytes. *Exp. Mol. Pathol.* **2013**, *94*, 336–342. [CrossRef]

50. Luo, X.; Balan, S.; Arnold-Schrauf, C.; Dalod, M. In Vitro Generation of Human Cross-Presenting Type 1 Conventional Dendritic Cells (cDC1s) and Plasmacytoid Dendritic Cells (pDCs). *Methods Mol. Biol.* **2023**, *2618*, 133–145. [CrossRef] [PubMed]

51. Jin, D.; Sprent, J. GM-CSF Culture Revisited: Preparation of Bulk Populations of Highly Pure Dendritic Cells from Mouse Bone Marrow. *J. Immunol.* **2018**, *201*, 3129–3139. [CrossRef] [PubMed]

52. van Eck van der Sluijs, J.; van Ens, D.; Thordardottir, S.; Vodegel, D.; Hermens, I.; van der Waart, A.B.; Falkenburg, J.H.F.; Kester, M.G.D.; de Rink, I.; Heemskerk, M.H.M.; et al. Clinically Applicable CD34+-Derived Blood Dendritic Cell Subsets Exhibit Key Subset-Specific Features and Potently Boost Anti-Tumor T and NK Cell Responses. *Cancer Immunol. Immunother.* **2021**, *70*, 3167–3181. [CrossRef]

53. Nielsen, M.C.; Andersen, M.N.; Møller, H.J. Monocyte Isolation Techniques Significantly Impact the Phenotype of Both Isolated Monocytes and Derived Macrophages in Vitro. *Immunology* **2020**, *159*, 63–74. [CrossRef]

54. Pires, C.F.; Rosa, F.F.; Kurochkin, I.; Pereira, C.-F. Understanding and Modulating Immunity With Cell Reprogramming. *Front. Immunol.* **2019**, *10*, 2809. [CrossRef]

55. Lundberg, K.; Albrekt, A.-S.; Nelissen, I.; Santegoets, S.; de Gruijl, T.D.; Gibbs, S.; Lindstedt, M. Transcriptional Profiling of Human Dendritic Cell Populations and Models—Unique Profiles of in Vitro Dendritic Cells and Implications on Functionality and Applicability. *PLoS ONE* **2013**, *8*, e52875. [CrossRef] [PubMed]

56. Carpentier, S.; Vu Manh, T.-P.; Chelbi, R.; Henri, S.; Malissen, B.; Haniffa, M.; Ginhoux, F.; Dalod, M. Comparative Genomics Analysis of Mononuclear Phagocyte Subsets Confirms Homology between Lymphoid Tissue-Resident and Dermal XCR1(+) DCs in Mouse and Human and Distinguishes Them from Langerhans Cells. *J. Immunol. Methods* **2016**, *432*, 35–49. [CrossRef]

57. Carreno, B.M.; Becker-Hapak, M.; Huang, A.; Chan, M.; Alyasiry, A.; Lie, W.-R.; Aft, R.L.; Cornelius, L.A.; Trinkaus, K.M.; Linette, G.P. IL-12p70-Producing Patient DC Vaccine Elicits Tc1-Polarized Immunity. *J. Clin. Investig.* **2013**, *123*, 3383–3394. [CrossRef]

58. Lopes, A.M.M.; Michelin, M.A.; Murta, E.F.C. Monocyte-Derived Dendritic Cells from Patients with Cervical Intraepithelial Lesions. *Oncol. Lett.* **2017**, *13*, 1456–1462. [CrossRef] [PubMed]

59. Vopenkova, K.; Mollova, K.; Buresova, I.; Michalek, J. Complex Evaluation of Human Monocyte-Derived Dendritic Cells for Cancer Immunotherapy. *J. Cell Mol. Med.* **2012**, *16*, 2827–2837. [CrossRef] [PubMed]
60. Liau, L.M.; Ashkan, K.; Brem, S.; Campian, J.L.; Trusheim, J.E.; Iwamoto, F.M.; Tran, D.D.; Ansstas, G.; Cobbs, C.S.; Heth, J.A.; et al. Association of Autologous Tumor Lysate-Loaded Dendritic Cell Vaccination With Extension of Survival Among Patients With Newly Diagnosed and Recurrent Glioblastoma: A Phase 3 Prospective Externally Controlled Cohort Trial. *JAMA Oncol.* **2023**, *9*, 112–121. [CrossRef]
61. Ding, Z.; Li, Q.; Zhang, R.; Xie, L.; Shu, Y.; Gao, S.; Wang, P.; Su, X.; Qin, Y.; Wang, Y.; et al. Personalized Neoantigen Pulsed Dendritic Cell Vaccine for Advanced Lung Cancer. *Signal Transduct. Target. Ther.* **2021**, *6*, 26. [CrossRef]
62. Koch, E.A.T.; Schaft, N.; Kummer, M.; Berking, C.; Schuler, G.; Hasumi, K.; Dörrie, J.; Schuler-Thurner, B. A One-Armed Phase I Dose Escalation Trial Design: Personalized Vaccination with IKKβ-Matured, RNA-Loaded Dendritic Cells for Metastatic Uveal Melanoma. *Front. Immunol.* **2022**, *13*, 785231. [CrossRef] [PubMed]
63. Vogelzang, N.J.; Beer, T.M.; Gerritsen, W.; Oudard, S.; Wiechno, P.; Kukielka-Budny, B.; Samal, V.; Hajek, J.; Feyerabend, S.; Khoo, V.; et al. Efficacy and Safety of Autologous Dendritic Cell-Based Immunotherapy, Docetaxel, and Prednisone vs Placebo in Patients With Metastatic Castration-Resistant Prostate Cancer: The VIABLE Phase 3 Randomized Clinical Trial. *JAMA Oncol.* **2022**, *8*, 546–552. [CrossRef]
64. Vreeland, T.J.; Clifton, G.T.; Hale, D.F.; Chick, R.C.; Hickerson, A.T.; Cindass, J.L.; Adams, A.M.; Bohan, P.M.K.; Andtbacka, R.H.I.; Berger, A.C.; et al. A Phase IIb Randomized Controlled Trial of the TLPLDC Vaccine as Adjuvant Therapy After Surgical Resection of Stage III/IV Melanoma: A Primary Analysis. *Ann. Surg. Oncol.* **2021**, *28*, 6126–6137. [CrossRef] [PubMed]
65. Sallusto, F.; Lanzavecchia, A. Efficient Presentation of Soluble Antigen by Cultured Human Dendritic Cells Is Maintained by Granulocyte/Macrophage Colony-Stimulating Factor plus Interleukin 4 and Downregulated by Tumor Necrosis Factor Alpha. *J. Exp. Med.* **1994**, *179*, 1109–1118. [CrossRef]
66. Syme, R.; Bajwa, R.; Robertson, L.; Stewart, D.; Glück, S. Comparison of CD34 and Monocyte-Derived Dendritic Cells from Mobilized Peripheral Blood from Cancer Patients. *Stem Cells* **2005**, *23*, 74–81. [CrossRef] [PubMed]
67. Harris, K.M. Monocytes Differentiated with GM-CSF and IL-15 Initiate Th17 and Th1 Responses That Are Contact-Dependent and Mediated by IL-15. *J. Leukoc. Biol.* **2011**, *90*, 727–734. [CrossRef]
68. Mohty, M.; Vialle-Castellano, A.; Nunes, J.A.; Isnardon, D.; Olive, D.; Gaugler, B. IFN-α Skews Monocyte Differentiation into Toll-like Receptor 7-Expressing Dendritic Cells with Potent Functional Activities. *J. Immunol.* **2003**, *171*, 3385–3393. [CrossRef]
69. Flörcken, A.; Kopp, J.; van Lessen, A.; Movassaghi, K.; Takvorian, A.; Jöhrens, K.; Möbs, M.; Schönemann, C.; Sawitzki, B.; Egerer, K.; et al. Allogeneic Partially HLA-Matched Dendritic Cells Pulsed with Autologous Tumor Cell Lysate as a Vaccine in Metastatic Renal Cell Cancer: A Clinical Phase I/II Study. *Hum. Vaccines Immunother.* **2013**, *9*, 1217–1227. [CrossRef]
70. Balan, S.; Ollion, V.; Colletti, N.; Chelbi, R.; Montanana-Sanchis, F.; Liu, H.; Vu Manh, T.-P.; Sanchez, C.; Savoret, J.; Perrot, I.; et al. Human XCR1+ Dendritic Cells Derived in Vitro from CD34+ Progenitors Closely Resemble Blood Dendritic Cells, Including Their Adjuvant Responsiveness, Contrary to Monocyte-Derived Dendritic Cells. *J. Immunol.* **2014**, *193*, 1622–1635. [CrossRef]
71. Proietto, A.I.; Mittag, D.; Roberts, A.W.; Sprigg, N.; Wu, L. The Equivalents of Human Blood and Spleen Dendritic Cell Subtypes Can Be Generated in Vitro from Human CD34(+) Stem Cells in the Presence of Fms-like Tyrosine Kinase 3 Ligand and Thrombopoietin. *Cell Mol. Immunol.* **2012**, *9*, 446–454. [CrossRef] [PubMed]
72. Bedke, N.; Swindle, E.J.; Molnar, C.; Holt, P.G.; Strickland, D.H.; Roberts, G.C.; Morris, R.; Holgate, S.T.; Davies, D.E.; Blume, C. A Method for the Generation of Large Numbers of Dendritic Cells from CD34+ Hematopoietic Stem Cells from Cord Blood. *J. Immunol. Methods* **2020**, *477*, 112703. [CrossRef]
73. Kirkling, M.E.; Cytlak, U.; Lau, C.M.; Lewis, K.L.; Resteu, A.; Khodadadi-Jamayran, A.; Siebel, C.W.; Salmon, H.; Merad, M.; Tsirigos, A.; et al. Notch Signaling Facilitates In Vitro Generation of Cross-Presenting Classical Dendritic Cells. *Cell Rep.* **2018**, *23*, 3658–3672.e6. [CrossRef] [PubMed]
74. Balan, S.; Arnold-Schrauf, C.; Abbas, A.; Couespel, N.; Savoret, J.; Imperatore, F.; Villani, A.-C.; Vu Manh, T.-P.; Bhardwaj, N.; Dalod, M. Large-Scale Human Dendritic Cell Differentiation Revealing Notch-Dependent Lineage Bifurcation and Heterogeneity. *Cell Rep.* **2018**, *24*, 1902–1915.e6. [CrossRef] [PubMed]
75. Sundarasetty, B.S.; Chan, L.; Darling, D.; Giunti, G.; Farzaneh, F.; Schenck, F.; Naundorf, S.; Kuehlcke, K.; Ruggiero, E.; Schmidt, M.; et al. Lentivirus-Induced "Smart" Dendritic Cells: Pharmacodynamics and GMP-Compliant Production for Immunotherapy against TRP2-Positive Melanoma. *Gene Ther.* **2015**, *22*, 707–720. [CrossRef]
76. Rosa, F.F.; Pires, C.F.; Kurochkin, I.; Ferreira, A.G.; Gomes, A.M.; Palma, L.G.; Shaiv, K.; Solanas, L.; Azenha, C.; Papatsenko, D.; et al. Direct Reprogramming of Fibroblasts into Antigen-Presenting Dendritic Cells. *Sci. Immunol.* **2018**, *3*, eaau4292. [CrossRef]
77. Rosa, F.F.; Pires, C.F.; Kurochkin, I.; Halitzki, E.; Zahan, T.; Arh, N.; Zimmermannová, O.; Ferreira, A.G.; Li, H.; Karlsson, S.; et al. Single-Cell Transcriptional Profiling Informs Efficient Reprogramming of Human Somatic Cells to Cross-Presenting Dendritic Cells. *Sci. Immunol.* **2022**, *7*, eabg5539. [CrossRef]
78. Hervas-Stubbs, S.; Mancheño, U.; Riezu-Boj, J.-I.; Larraga, A.; Ochoa, M.C.; Alignani, D.; Alfaro, C.; Morales-Kastresana, A.; Gonzalez, I.; Larrea, E.; et al. CD8 T Cell Priming in the Presence of IFN-α Renders CTLs with Improved Responsiveness to Homeostatic Cytokines and Recall Antigens: Important Traits for Adoptive T Cell Therapy. *J. Immunol.* **2012**, *189*, 3299–3310. [CrossRef]

79. Rosalia, R.A.; Quakkelaar, E.D.; Redeker, A.; Khan, S.; Camps, M.; Drijfhout, J.W.; Silva, A.L.; Jiskoot, W.; van Hall, T.; van Veelen, P.A.; et al. Dendritic Cells Process Synthetic Long Peptides Better than Whole Protein, Improving Antigen Presentation and T-Cell Activation. *Eur. J. Immunol.* **2013**, *43*, 2554–2565. [CrossRef]
80. Binder, R.J.; Anderson, K.M.; Basu, S.; Srivastava, P.K. Cutting Edge: Heat Shock Protein Gp96 Induces Maturation and Migration of CD11c+ Cells in Vivo. *J. Immunol.* **2000**, *165*, 6029–6035. [CrossRef]
81. Palucka, A.K.; Ueno, H.; Connolly, J.; Kerneis-Norvell, F.; Blanck, J.-P.; Johnston, D.A.; Fay, J.; Banchereau, J. Dendritic Cells Loaded with Killed Allogeneic Melanoma Cells Can Induce Objective Clinical Responses and MART-1 Specific CD8+ T-Cell Immunity. *J. Immunother.* **2006**, *29*, 545–557. [CrossRef]
82. Geskin, L.J.; Damiano, J.J.; Patrone, C.C.; Butterfield, L.H.; Kirkwood, J.M.; Falo, L.D. Three Antigen-Loading Methods in Dendritic Cell Vaccines for Metastatic Melanoma. *Melanoma Res.* **2018**, *28*, 211–221. [CrossRef] [PubMed]
83. De Keersmaecker, B.; Claerhout, S.; Carrasco, J.; Bar, I.; Corthals, J.; Wilgenhof, S.; Neyns, B.; Thielemans, K. TriMix and Tumor Antigen mRNA Electroporated Dendritic Cell Vaccination plus Ipilimumab: Link between T-Cell Activation and Clinical Responses in Advanced Melanoma. *J. Immunother. Cancer* **2020**, *8*, e000329. [CrossRef] [PubMed]
84. Chulpanova, D.S.; Kitaeva, K.V.; James, V.; Rizvanov, A.A.; Solovyeva, V.V. Therapeutic Prospects of Extracellular Vesicles in Cancer Treatment. *Front. Immunol.* **2018**, *9*, 1534. [CrossRef] [PubMed]
85. Rojas-Sepúlveda, D.; Tittarelli, A.; Gleisner, M.A.; Ávalos, I.; Pereda, C.; Gallegos, I.; González, F.E.; López, M.N.; Butte, J.M.; Roa, J.C.; et al. Tumor Lysate-Based Vaccines: On the Road to Immunotherapy for Gallbladder Cancer. *Cancer Immunol. Immunother.* **2018**, *67*, 1897–1910. [CrossRef] [PubMed]
86. Aerts, J.G.J.V.; de Goeje, P.L.; Cornelissen, R.; Kaijen-Lambers, M.E.H.; Bezemer, K.; van der Leest, C.H.; Mahaweni, N.M.; Kunert, A.; Eskens, F.A.L.M.; Waasdorp, C.; et al. Autologous Dendritic Cells Pulsed with Allogeneic Tumor Cell Lysate in Mesothelioma: From Mouse to Human. *Clin. Cancer Res.* **2018**, *24*, 766–776. [CrossRef]
87. Kranz, L.M.; Diken, M.; Haas, H.; Kreiter, S.; Loquai, C.; Reuter, K.C.; Meng, M.; Fritz, D.; Vascotto, F.; Hefesha, H.; et al. Systemic RNA Delivery to Dendritic Cells Exploits Antiviral Defence for Cancer Immunotherapy. *Nature* **2016**, *534*, 396–401. [CrossRef]
88. Vaccination with an Adenoviral Vector Encoding the Tumor Antigen Directly Linked to Invariant Chain Induces Potent CD4+ T-cell-independent CD8+ T-cell-mediated Tumor Control—Sorensen—2009—European Journal of Immunology—Wiley Online Library. Available online: https://onlinelibrary.wiley.com/doi/full/10.1002/eji.200939543 (accessed on 24 September 2023).
89. Tacken, P.J.; Figdor, C.G. Targeted Antigen Delivery and Activation of Dendritic Cells in Vivo: Steps towards Cost Effective Vaccines. *Semin. Immunol.* **2011**, *23*, 12–20. [CrossRef]
90. Xie, F.; Zhou, X.; Fang, M.; Li, H.; Su, P.; Tu, Y.; Zhang, L.; Zhou, F. Extracellular Vesicles in Cancer Immune Microenvironment and Cancer Immunotherapy. *Adv. Sci.* **2019**, *6*, 1901779. [CrossRef]
91. Zippoli, M.; Ruocco, A.; Novelli, R.; Rocchio, F.; Miscione, M.S.; Allegretti, M.; Cesta, M.C.; Amendola, P.G. The Role of Extracellular Vesicles and Interleukin-8 in Regulating and Mediating Neutrophil-Dependent Cancer Drug Resistance. *Front. Oncol.* **2022**, *12*, 947183. [CrossRef]
92. Ding, Y.-N.; Ding, H.-Y.; Li, H.; Yang, R.; Huang, J.-Y.; Chen, H.; Wang, L.-H.; Wang, Y.-J.; Hu, C.-M.; An, Y.-L.; et al. Photosensitive Small Extracellular Vesicles Regulate the Immune Microenvironment of Triple Negative Breast Cancer. *Acta Biomater.* **2023**, *167*, 534–550. [CrossRef]
93. Hodge, A.L.; Baxter, A.A.; Poon, I.K.H. Gift Bags from the Sentinel Cells of the Immune System: The Diverse Role of Dendritic Cell-Derived Extracellular Vesicles. *J. Leukoc. Biol.* **2022**, *111*, 903–920. [CrossRef]
94. Pitt, J.M.; André, F.; Amigorena, S.; Soria, J.-C.; Eggermont, A.; Kroemer, G.; Zitvogel, L. Dendritic Cell–Derived Exosomes for Cancer Therapy. *J. Clin. Investig.* **2016**, *126*, 1224–1232. [CrossRef] [PubMed]
95. Montecalvo, A.; Larregina, A.T.; Shufesky, W.J.; Stolz, D.B.; Sullivan, M.L.G.; Karlsson, J.M.; Baty, C.J.; Gibson, G.A.; Erdos, G.; Wang, Z.; et al. Mechanism of Transfer of Functional microRNAs between Mouse Dendritic Cells via Exosomes. *Blood* **2012**, *119*, 756–766. [CrossRef] [PubMed]
96. Zuo, B.; Zhang, Y.; Zhao, K.; Wu, L.; Qi, H.; Yang, R.; Gao, X.; Geng, M.; Wu, Y.; Jing, R.; et al. Universal Immunotherapeutic Strategy for Hepatocellular Carcinoma with Exosome Vaccines That Engage Adaptive and Innate Immune Responses. *J. Hematol. Oncol.* **2022**, *15*, 46. [CrossRef] [PubMed]
97. Andre, F.; Escudier, B.; Angevin, E.; Tursz, T.; Zitvogel, L. Exosomes for Cancer Immunotherapy. *Ann. Oncol.* **2004**, *15* (Suppl. S4), 141–144. [CrossRef]
98. Lesterhuis, W.J.; de Vries, I.J.M.; Schreibelt, G.; Lambeck, A.J.A.; Aarntzen, E.H.J.G.; Jacobs, J.F.M.; Scharenborg, N.M.; van de Rakt, M.W.M.M.; de Boer, A.J.; Croockewit, S.; et al. Route of Administration Modulates the Induction of Dendritic Cell Vaccine-Induced Antigen-Specific T Cells in Advanced Melanoma Patients. *Clin. Cancer Res.* **2011**, *17*, 5725–5735. [CrossRef]
99. Edele, F.; Dudda, J.C.; Bachtanian, E.; Jakob, T.; Pircher, H.; Martin, S.F. Efficiency of Dendritic Cell Vaccination against B16 Melanoma Depends on the Immunization Route. *PLoS ONE* **2014**, *9*, e105266. [CrossRef]
100. Filin, I.Y.; Solovyeva, V.V.; Kitaeva, K.V.; Rutland, C.S.; Rizvanov, A.A. Current Trends in Cancer Immunotherapy. *Biomedicines* **2020**, *8*, 621. [CrossRef]
101. Ferris, S.T.; Ohara, R.A.; Ou, F.; Wu, R.; Huang, X.; Kim, S.; Chen, J.; Liu, T.-T.; Schreiber, R.D.; Murphy, T.L.; et al. cDC1 Vaccines Drive Tumor Rejection by Direct Presentation Independently of Host cDC1. *Cancer Immunol. Res.* **2022**, *10*, 920–931. [CrossRef]
102. Lee, S.-W.; Lee, H.; Lee, K.-W.; Kim, M.-J.; Kang, S.W.; Lee, Y.-J.; Kim, H.; Kim, Y.-M. CD8α+ Dendritic Cells Potentiate Antitumor and Immune Activities against Murine Ovarian Cancers. *Sci. Rep.* **2023**, *13*, 98. [CrossRef] [PubMed]

103. Bian, Y.; Walter, D.L.; Zhang, C. Efficiency of Interferon-γ in Activating Dendritic Cells and Its Potential Synergy with Toll-like Receptor Agonists. *Viruses* **2023**, *15*, 1198. [CrossRef] [PubMed]
104. Oliveira, M.M.S.; D'Aulerio, R.; Yong, T.; He, M.; Baptista, M.A.P.; Nylén, S.; Westerberg, L.S. Increased Cross-Presentation by Dendritic Cells and Enhanced Anti-Tumour Therapy Using the Arp2/3 Inhibitor CK666. *Br. J. Cancer* **2023**, *128*, 982–991. [CrossRef] [PubMed]
105. Murwanti, R.; Denda-Nagai, K.; Sugiura, D.; Mogushi, K.; Gendler, S.J.; Irimura, T. Prevention of Inflammation-Driven Colon Carcinogenesis in Human MUC1 Transgenic Mice by Vaccination with MUC1 DNA and Dendritic Cells. *Cancers* **2023**, *15*, 1920. [CrossRef] [PubMed]
106. Lim, S.; Park, J.-H.; Chang, H. Enhanced Anti-Tumor Immunity of Vaccine Combined with Anti-PD-1 Antibody in a Murine Bladder Cancer Model. *Investig. Clin. Urol.* **2023**, *64*, 74–81. [CrossRef] [PubMed]
107. de Mato, F.C.; Barreto, N.; Cordeiro, G.; Munhoz, J.; Bonfanti, A.P.; da Rocha-E-Silva, T.A.A.; Sutti, R.; Cruz, P.B.M.; Sanches, L.R.; Bombeiro, A.L.; et al. Isolated Peptide from Spider Venom Modulates Dendritic Cells In Vitro: A Possible Application in Oncoimmunotherapy for Glioblastoma. *Cells* **2023**, *12*, 1023. [CrossRef]
108. Faiena, I.; Comin-Anduix, B.; Berent-Maoz, B.; Bot, A.; Zomorodian, N.; Sachdeva, A.; Said, J.; Cheung-Lau, G.; Pang, J.; Macabali, M.; et al. A Phase I, Open-Label, Dose-Escalation, and Cohort Expansion Study to Evaluate the Safety and Immune Response to Autologous Dendritic Cells Transduced With AdGMCA9 (DC-AdGMCAIX) in Patients With Metastatic Renal Cell Carcinoma. *J. Immunother.* **2020**, *43*, 273–282. [CrossRef]
109. Maeng, H.M.; Moore, B.N.; Bagheri, H.; Steinberg, S.M.; Inglefield, J.; Dunham, K.; Wei, W.-Z.; Morris, J.C.; Terabe, M.; England, L.C.; et al. Phase I Clinical Trial of an Autologous Dendritic Cell Vaccine Against HER2 Shows Safety and Preliminary Clinical Efficacy. *Front. Oncol.* **2021**, *11*, 789078. [CrossRef]
110. Bloemendal, M.; Bol, K.F.; Boudewijns, S.; Gorris, M.A.J.; de Wilt, J.H.W.; Croockewit, S.A.J.; van Rossum, M.M.; de Goede, A.L.; Petry, K.; Koornstra, R.H.T.; et al. Immunological Responses to Adjuvant Vaccination with Combined CD1c+ Myeloid and Plasmacytoid Dendritic Cells in Stage III Melanoma Patients. *Oncoimmunology* **2022**, *11*, 2015113. [CrossRef]
111. Polyzoidis, S.; Ashkan, K. DCVax®-L—Developed by Northwest Biotherapeutics. *Hum. Vaccines Immunother.* **2014**, *10*, 3139–3145. [CrossRef]
112. Podrazil, M.; Horvath, R.; Becht, E.; Rozkova, D.; Bilkova, P.; Sochorova, K.; Hromadkova, H.; Kayserova, J.; Vavrova, K.; Lastovicka, J.; et al. Phase I/II Clinical Trial of Dendritic-Cell Based Immunotherapy (DCVAC/PCa) Combined with Chemotherapy in Patients with Metastatic, Castration-Resistant Prostate Cancer. *Oncotarget* **2015**, *6*, 18192–18205. [CrossRef] [PubMed]
113. Butterfield, L.H.; Vujanovic, L.; Santos, P.M.; Maurer, D.M.; Gambotto, A.; Lohr, J.; Li, C.; Waldman, J.; Chandran, U.; Lin, Y.; et al. Multiple Antigen-Engineered DC Vaccines with or without IFNα to Promote Antitumor Immunity in Melanoma. *J. Immunother. Cancer* **2019**, *7*, 113. [CrossRef]
114. Mitsuya, K.; Akiyama, Y.; Iizuka, A.; Miyata, H.; Deguchi, S.; Hayashi, N.; Maeda, C.; Kondou, R.; Kanematsu, A.; Watanabe, K.; et al. Alpha-Type-1 Polarized Dendritic Cell-Based Vaccination in Newly Diagnosed High-Grade Glioma: A Phase II Clinical Trial. *Anticancer Res.* **2020**, *40*, 6473–6484. [CrossRef] [PubMed]
115. Escudier, B.; Dorval, T.; Chaput, N.; André, F.; Caby, M.-P.; Novault, S.; Flament, C.; Leboulaire, C.; Borg, C.; Amigorena, S.; et al. Vaccination of Metastatic Melanoma Patients with Autologous Dendritic Cell (DC) Derived-Exosomes: Results of Thefirst Phase I Clinical Trial. *J. Transl. Med.* **2005**, *3*, 10. [CrossRef] [PubMed]
116. Morse, M.A.; Garst, J.; Osada, T.; Khan, S.; Hobeika, A.; Clay, T.M.; Valente, N.; Shreeniwas, R.; Sutton, M.A.; Delcayre, A.; et al. A Phase I Study of Dexosome Immunotherapy in Patients with Advanced Non-Small Cell Lung Cancer. *J. Transl. Med.* **2005**, *3*, 9. [CrossRef] [PubMed]
117. Besse, B.; Charrier, M.; Lapierre, V.; Dansin, E.; Lantz, O.; Planchard, D.; Le Chevalier, T.; Livartoski, A.; Barlesi, F.; Laplanche, A.; et al. Dendritic Cell-Derived Exosomes as Maintenance Immunotherapy after First Line Chemotherapy in NSCLC. *Oncoimmunology* **2016**, *5*, e1071008. [CrossRef] [PubMed]

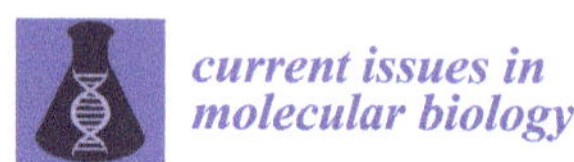
current issues in molecular biology

Article

Targeted Delivery of Chimeric Antigen Receptor into T Cells via CRISPR-Mediated Homology-Directed Repair with a Dual-AAV6 Transduction System

Pablo D. Moço [iD], Omar Farnós [iD], David Sharon and Amine A. Kamen *[iD]

Department of Bioengineering, McGill University, Montreal, QC H3A 0E9, Canada;
pablo.moco@mail.mcgill.ca (P.D.M.)
* Correspondence: amine.kamen@mcgill.ca; Tel.: +1-(514)-398-5775

Abstract: CAR-T cell therapy involves genetically engineering T cells to recognize and attack tumour cells by adding a chimeric antigen receptor (CAR) to their surface. In this study, we have used dual transduction with AAV serotype 6 (AAV6) to integrate an anti-CD19 CAR into human T cells at a known genomic location. The first viral vector expresses the Cas9 endonuclease and a guide RNA (gRNA) targeting the T cell receptor alpha constant locus, while the second vector carries the DNA template for homology-mediated CAR insertion. We evaluated three gRNA candidates and determined their efficiency in generating indels. The AAV6 successfully delivered the CRISPR/Cas9 machinery in vitro, and molecular analysis of the dual transduction showed the integration of the CAR transgene into the desired location. In contrast to the random integration methods typically used to generate CAR-T cells, targeted integration into a known genomic locus can potentially lower the risk of insertional mutagenesis and provide more stable levels of CAR expression. Critically, this method also results in the knockout of the endogenous T cell receptor, allowing target cells to be derived from allogeneic donors. This raises the exciting possibility of "off-the-shelf" universal immunotherapies that would greatly simplify the production and administration of CAR-T cells.

Keywords: adeno-associated viral vectors; T cells; genetic engineering; CRISPR/Cas9; homology-directed repair; chimeric antigen receptor

Citation: Moço, P.D.; Farnós, O.; Sharon, D.; Kamen, A.A. Targeted Delivery of Chimeric Antigen Receptor into T Cells via CRISPR-Mediated Homology-Directed Repair with a Dual-AAV6 Transduction System. *Curr. Issues Mol. Biol.* **2023**, *45*, 7705–7720. https://doi.org/10.3390/cimb45100486

Academic Editor: Dumitru A. Iacobas

Received: 26 August 2023
Revised: 18 September 2023
Accepted: 20 September 2023
Published: 22 September 2023

1. Introduction

CAR-T cell therapy is a type of immunotherapy that involves the genetic engineering of a patient's T cells to express a chimeric antigen receptor (CAR) on their surface. This synthetic receptor artificially redirects the specificity of the T cells to tumour cells [1]. Such receptors have an extracellular recognition domain derived from an antibody and can recognize any antigen to which it has an affinity. Their intracellular domain can combine signals from the receptor complex, such as CD3ζ and T cell co-stimulatory molecules [2]. Once the target antigen is recognized, the CAR-T cell is activated, releasing cytokines and enzymes that kill the target tumour cell. CAR-T cells recognize molecules on the surfaces of tumour cells independently of the major histocompatibility complex (MHC), making the antitumour response more effective [3]. This therapy has shown great promise in early clinical trials, leading to its approval by the US Food and Drug Administration for treating certain types of leukemia and lymphoma [1]. The production process, however, can take several weeks and requires specialized equipment and trained personnel [4], leading to expensive therapy, with overall treatment costs surpassing 1 million US dollars [5].

Recombinant adeno-associated viruses (AAVs) are small, non-pathogenic viruses packaging a 4.7-kb-long ssDNA genome that have been extensively studied for their potential use in gene therapy [6]. For the generation of recombinant AAVs, the *Rep* and *Cap* genes, which encode for proteins related to viral genome replication and capsid formation and the assembly of the wildtype AAV, are replaced with the transgene of interest flanked by inverted terminal repeats (ITRs). One of the characteristics of recombinant AAV vectors is their ability to deliver and express transgenes stably without integration into the host genome because these vectors do not retain the *Rep* gene [7]. In most cases, the AAV genome remains in an episomal state, which confers low toxicity and allows for the expression of transgenes for long periods of time in non-dividing cells [8]. However, the lack of integration can also limit the durability and effectiveness of AAV gene therapies. On the other hand, the AAV-delivered ssDNA genome can serve as a template for gene targeting via homologous recombination [9–11]. While still not completely understood, this method is considered more efficient than other conventional approaches, such as plasmid transfection. Possible explanations include more efficient nuclear delivery of the transgene, the fact that the AAV genome is single-stranded, and the effect of the ITRs on recruiting the cellular repair machinery [9,12–14]. Additionally, the introduction of a double-strand break (DSB) into the host genome by endonucleases improves the frequency of this AAV-mediated recombination [14–17].

Five types of nuclease proteins can be used to make specific changes to the genome by creating a DSB in the DNA: Zinc-finger nucleases (ZFNs), transcription activator-like effector nucleases (TALENs), homing endonucleases, meganucleases, and CRISPR RNA-guided nucleases such as Cas9. These proteins have been employed to integrate the CAR transgene into T cells at specific sites, as reviewed by Dabiri et al. [18]. The CRISPR/Cas9 system has made it possible to introduce DSBs at specific genomic loci with high precision. This allows for targeting the AAV genome to a specific genomic locus, where it can integrate into the host genome via homology-directed repair. This approach can increase the stability and expression of transgenes delivered by AAVs, such as the chimeric antigen receptor [19]. In the case of CAR-T cell therapies, the use of AAVs and CRISPR/Cas9 can enable the site-specific integration of CAR transgenes into the host genome, leading to more durable and effective CAR-T cells [20]. Moreover, the CRISPR/Cas9 system can knock out endogenous genes in T cells, such as the T cell receptor, to generate "universal" CAR-T cells that are potentially more affordable and accessible [4,21].

Current CAR-T cell products are generated via the transduction of lentivirus (Kymriah, Breyanzi, Abecma, and Carvykti) or γ-retrovirus (Yescarta and Tecartus) due to the achievement of high rates of transduction and long-term expression of the CAR transgene. The use of these vectors is deemed safe [22], but their variable transgene integration [23,24] poses the risk of insertional oncogenesis [25–28] and of clonal expansion [29]. Additionally, the transgene integration can occur in sites with distinct transcriptional activity, resulting in variable CAR expression and inadequate therapeutic outcomes [30]. Targeting the CAR transgene to the T-cell receptor alpha constant (*TRAC*) locus alleviates these risks and leads to CAR-T cells with increased antitumour efficacy while exhibiting a reduced exhaustion profile, as previously reported [20]. Generating a potential universal CAR-T cell product is another benefit of targeting the *TRAC* [31]. By knocking out the endogenous TCR, the risk of graft versus host disease (GvHD) is eliminated as the recipient's immune system does not recognize the alloantigen [32,33]. The endogenous MHC can also be knocked out to generate "off-the-shelf" allogeneic CAR-T cells by targeting the beta-2 microglobulin (*B2M*) gene, reducing the risk of graft rejection [32]. In contrast to the current labour-intensive and time-consuming manufacturing process of CAR-T cell products [4], which increases its price and limits its application and availability to patients [5], the manufacturing of universal CAR-T cells has the potential to lower the costs and increase the accessibility of the product to patients whose T cells have been depleted due to cancer or previous treatments [4,34]. Additional targets for knockout can include *PD1*, for the generation of exhaustion-resistant CAR-T cells [35,36], and some target antigens that are also expressed on the surfaces of T cells, such as CD7 and CD33, for the reduction of on-target/off-tumour toxicity [37,38]. Here, we describe an alternative method for the site-specific integration of CAR into T cells via dual transduction with AAV6, a serotype extensively used for homology-mediated transgene integration in different human cells, including induced pluripotent stem cells [39,40], hematopoietic stem cells [17,41–43], and, most importantly, the immune cells NK and T cells [20,36,44–48]. The first viral vector encodes for the CRISPR/Cas9 machinery, while the second vector carries the DNA template for transgene integration via homology-directed repair. Simultaneous endogenous TCR and CAR integration disruption could be produced from allogeneic donors as an "off-the-shelf" therapy. This study shows a proof of principle for dual-AAV6-mediated site-specific CAR-T cell generation.

2. Materials and Methods

2.1. Cell Culture

HEK293SF-3F6 (HEK293SF) cells were kindly provided by the National Research Council Canada. Jurkat cells were kindly provided by Prof. Mohammad-Ali Jenabian (Immuno-Virology Lab, Department of Biological Sciences and CERMO-FC Research Centre, Université du Québec à Montréal (UQAM), Montréal, Québec, Canada). HEK293SF cells were maintained in serum-free suspension cultures at 37 °C, 5% CO_2, and 75% relative humidity in a shaker incubator (Infors HT, Bottmingen-Basel, Switzerland) at 135 rpm speed of agitation. The medium for both cell maintenance and viral vector production was HyCell TransFx-H (Cytiva Life Sciences, Marlborough, MA, USA) supplemented with 0.1% *w/v* of Kolliphor P188 (Sigma-Aldrich, Burlington, MA, USA) and 4 mM GlutaMAX (Gibco, Billings, MT, USA). Jurkat cells were maintained in RPMI-1640 (Cytiva Life Sciences, Marlborough, MA, USA) supplemented with 10% *v/v* fetal bovine serum (FBS, Cytiva Life Sciences, Marlborough, MA, USA) and 1% *v/v* penicillin–streptomycin (pen-strep, Gibco, Billings, MT, USA) at 37 °C, 5% CO_2, and 75% relative humidity in a static incubator (Panasonic, Tokyo, Japan).

2.2. Guide RNAs and Plasmids

The following guide RNAs targeting the *TRAC* locus were evaluated: (1) ataggcagaca-gacttgtca; (2) gtctctcagctggtacacgg; and (3) tacacggcagggtcagggtt. Each guide RNA (gRNA) was synthesized as 2 complementary DNA strands with BsaI overhangs at the 5′ end. The oligo DNAs were annealed and inserted into the plasmid pX601-AAV-CMV::NLS-SaCas9-NLS-3xHA-bGHpA;U6::BsaI-sgRNA (a gift from Feng Zhang, Addgene plasmid #61591; http://n2t.net/addgene:61591, accessed on 7 April 2021; RRID:Addgene_61591), which encodes for the SaCas9, via restriction cloning at the BsaI site [49]. The three plasmids containing each of the three gRNAs were named pX601-AAV-SaCas9-TRAC#1, pX601-AAV-SaCas9-TRAC#2, and pX601-AAV-SaCas9-TRAC#3, respectively. The plasmid containing the DNA template for homology-directed repair was designed to encode for the gene of the anti-CD19 chimeric antigen receptor (FMC63-28Z receptor protein) and an EGFP marker based on the plasmid pSLCAR-CD19-BBz [50], flanked by a 150-bp homology arm upstream of the template (left homology arm, LHA) and a 500-bp homology arm downstream of the template (right homology arm, LHA). The ITRs flank the entire construction for packaging into an AAV vector. The construct was synthesized by BioBasic (Markham, ON, Canada) and cloned into a pUC57 vector. The final plasmid was named pUC57_AAV_anti-CD19_CAR_HDR.

2.3. Analysis of Double-Strand Break Efficiency

HEK293SF cells were transfected with each of the three plasmids (1 µg/mL) to evaluate double-strand break (DSB) efficiency. Non-transfected cells were used as the negative control. Ninety-six hours post-transfection, the genomic DNA of the cells was extracted using a PureLink™ Genomic DNA Mini Kit (Invitrogen, Waltham, MA, USA), as per the manufacturer's instruction. The CRISPR/Cas9 target site was amplified via PCR using primers targeting the *TRAC* locus (forward: 5′-GCCAACATACCATAAACCTCCC-3′; reverse: 5′-GGACTGCCAGAACAAGGCTC-3′). The amplicons were sequenced by Sanger sequencing at Genome Québec (Montréal, QC, Canada), and the DSB efficiency was evaluated by Tracking of Indels by Decomposition (TIDE) [51]. The TIDE software (http://shinyapps.datacurators.nl/tide/, accessed on 15 September 2022) utilizes quantitative sequence trace data from two standard capillary sequencing reactions. These sequence traces are subjected to analysis through a custom decomposition algorithm, identifying the major induced mutations (insertions and deletions (indels)) at the intended editing site and quantifying their occurrence within the cell population to determine the overall efficiency.

2.4. Viral Vector Production

Adeno-associated viral vectors serotype 6 packaging the SaCas9 and gRNA #1 (AAV6-SaCas9) or the CAR (AAV6-CAR) genomes were produced by triple transient transfection as previously described [52], where the transgene plasmids used were pX601-AAV-SaCas9-TRAC#1 and pUC57_AAV_anti-CD19_CAR_HDR, respectively. After viral harvest, the cell lysate was clarified using a 1.2/0.5-μm Optiscale capsule (Millipore Sigma, Burlington, MA, USA). The clarified lysate was then subjected to single-step affinity capture chromatography using a commercially available 5-mL prepacked immunoaffinity resin column, POROS™ CaptureSelect™ AAVX (ThermoFisher Scientific, Waltham, MA, USA). The affinity purification process was conducted on the ÄKTA Avant25 FPLC system (Cytiva Life Sciences, Marlborough, MA, USA). The affinity-resin-bound AAVs were eluted in 0.1 M glycine (pH 2.5) and immediately neutralized by adding 10% v/v of neutralization buffer containing 1 M Tris, 20 mM $MgCl_2$, and 20% sucrose (pH 8.8). The purified AAV6 vectors were further concentrated using 100-KDa Amicon® Ultra-5 centrifugal filter units (Millipore Sigma, Burlington, MA, USA) and stored at −80 °C. The AAV6 vector titers were determined by ddPCR analysis as previously described [52], using the following pair of primers: 5′-GGCCAGATTCAGGATGTGCT-3′ (forward) and 5′-CATCATCCCCAGAAGCGTGT-3′ (reverse), for AAV6-SaCas9; 5′-GAGGAAACGGGGCAGAAAGA-3′ (forward) and 5′-GGCCTTCCTGAGGGTTCTTC-3′ (reverse), for AAV6-CAR.

2.5. Analysis of AAV6-Delivered CRISPR/Cas9

The efficiency of the AAV6-delivered SaCas9 and gRNA #1 to be expressed and cleave the target DNA was evaluated by transducing HEK293SF cells at a multiplicity of infection (MOI) of 10^3 and co-infection with E1/E3-deleted adenovirus type 5. RNA was extracted from the cells 24 and 48 h post-transduction. The extraction was done using the Aurum™ Total RNA Mini Kit (Bio-Rad, Hercules, CA, USA). cDNA was synthesized using the iScript™ cDNA Synthesis Kit (Bio-Rad, Hercules, CA, USA), per protocol. Oligo(dT) primers were used to prepare the sample for the detection of SaCas9 expression, and gene-specific primers (5′-CGCCAACAAGTTGACGAGAT-3′ and 5′-GGCAGACAGACTTGTCAGTTTTA-3′) were used to prepare the sample for the detection of gRNA expression. The cDNAs were amplified via PCR using primers for SaCas9 (5′-GGCCAGATTCAGGATGTGCT-3′ and 5′-CATCATCCCCAGAAGCGTGT-3′) and gRNA #1 (same pair of primers used for cDNA synthesis). The size of the amplicons was analyzed via agarose gel electrophoresis. At 24 and 48 h post-transduction, samples were collected to detect the expressed SaCas9 protein. Following cell lysis, the samples were run on polyacrylamide gel electrophoresis. After blotting the samples to a nitrocellulose membrane, the membrane was blocked and incubated with 0.5 μg/mL of anti-SaCas9 monoclonal antibody (11C12, GenScript, Piscataway, NJ, USA) overnight. Peroxidase goat anti-mouse antibody (IgG (H + L) Jackson ImmunoResearch, West Grove, PA, USA, RRID:AB_2307346) was used as the secondary antibody. Genomic DNA was also extracted from the cells 48 h post-transfection. As previously described, the *TRAC* locus was amplified by PCR, and the material was sequenced and analyzed via agarose gel electrophoresis and TIDE.

2.6. Dual-AAV Transduction of Jurkat Cells

Jurkat cells were resuspended at a density of 0.5×10^6 cells/mL in RPMI supplemented with 10% FBS and 1% penicillin–streptomycin and seeded into wells of a 48-well plate. The cells were transduced with AAV6-SaCas9 and AAV6-CAR at medium and high MOI. For medium MOI, 1×10^5 genome-containing particles (viral genomes, VG) of each vector were added per cell. For high MOI, 2×10^6 VG of AAV6-SaCas9 and 7×10^6 VG of AAV6-CAR were added per cell. The transduced cells were incubated at 37 °C, 5% CO_2, and 75%

relative humidity in a static incubator, as previously described. At 48 h post-transduction, the cells for the high MOI group were resuspended in a fresh medium.

2.7. Genomic Analysis of Modified Jurkat Cells

Extraction of genomic DNA from up to 5×10^5 Jurkat cells was done at 48 and 96 h post-transduction for the medium and high MOI groups, respectively. Non-transduced cells were used as controls. Genomic DNA was extracted as described above and used as a template for PCR using primers for the *TRAC* locus. The size of the PCR products was analyzed by agarose gel electrophoresis, and the amplicons were sequenced for TIDE analysis. Quantitation of gel bands was done using Image Lab 6.1 (Bio-Rad, Hercules, CA, USA) using a 1-kb Plus DNA Ladder (New England BioLabs, Ipswich, MA, USA) as a standard curve. Purified PCR products were used for the evaluation of the site-specific insertion of the transgene. The PCR reaction contained one primer positioned upstream of the 3′ junction (TRAC-Forward: 5′-GCCAACATACCATAAACCTCCC-3′) and one downstream, inside the transgene (CAR-Reverse: 5′-GGCCTTCCTGAGGGTTCTTC-3′). The PCR product was run on an agarose gel, and the 2.2-kb band was extracted using the QIAquick Gel Extraction Kit (QIAGEN, Germantown, MD, USA). The extracted DNA was analyzed by Sanger sequencing and enzymatic digestion with KpnI-HF (New England Biolabs, Ipswich, MA, USA).

2.8. Analysis of CRISPR/Cas9 Off-Targets

The bioinformatics-based tool COSMID (CRISPR Off-target Sites with Mismatches, Insertions, and Deletions, https://crispr.bme.gatech.edu/, accessed on 14 September 2022) [53] was used to generate a list of predicted off-target sites. Genomic DNA from modified cells was extracted and used as the template for PCR with primers flanking the predicted off-target sites. The amplicons were sequenced, and TIDE was used to analyze the double-strand break.

3. Results

3.1. Guide RNA Design, Cloning, and Selection

Three gRNAs were designed to target the first exon of the *TRAC* locus (Figure 1A). Single-strand DNA oligos with complementing sequences corresponding to these gRNAs and a BsaI overhang were synthesized. The complementing strands were annealed and used as templates for cloning into the BsaI site of a plasmid encoding the *Staphylococcus aureus* Cas9 (SaCas9) and the gRNA scaffold flanked by AAV inverted terminal repeats (ITRs) for packaging into AAV vectors (Figure 1B).

The first step was identifying the most efficient gRNA for the targeting and performance of the double-strand break of the *TRAC* locus. For this, the highly permissible cell line HEK293SF was transfected with the plasmids encoding the SaCas9 and one of the gRNAs. Genomic DNA was extracted from the cells 96 h post-transfection, and the CRISPR/Cas9 target site was amplified via PCR. The material was sequenced, and the efficiency of the double-strand break was analyzed by TIDE (Table 1 and Figure 1C). The plasmid containing gRNA #1 (pX601-AAV-SaCas9-TRAC#1) resulted in the highest efficiency (21.7%) and was selected to be used in the following experiments.

Table 1. Double-strand break efficiency of tested guide RNAs.

gRNA Sequence	Efficiency (%)
#1 ataggcagacagacttgtca	21.7
#2 gtctctcagctggtacacgg	12.1
#3 tacacggcagggtcagggtt	2

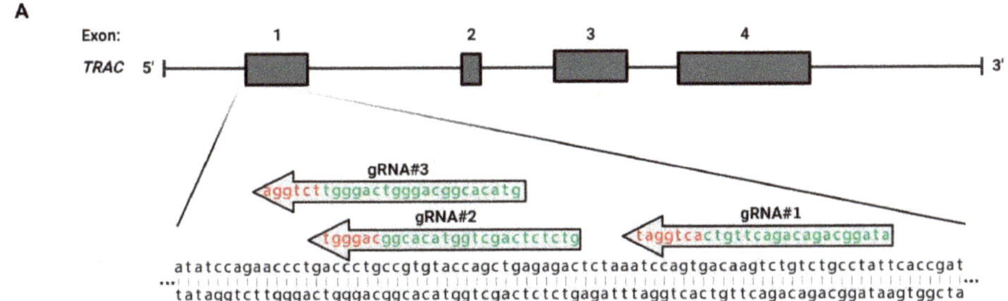

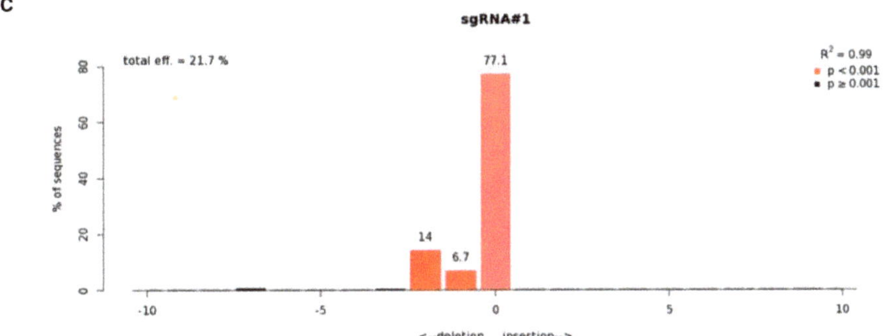

Figure 1. Targeting of *TRAC* locus via AAV6-delivered SaCas9. (**A**) Location of the three designed guide RNAs (gRNAs, in green) and their respective PAM sequences (in red) for targeting of the first exon of the T cell receptor alpha constant (*TRAC*) locus. The four exons of the *TRAC* locus are identified by the numbers 1 to 4. (**B**) AAV-packaged genome encoding for the SaCas9 and one gRNA against the *TRAC* locus. CMV, cytomegalovirus promoter and enhancer; NLS, nuclear localization signal (5′: from SV40 (simian virus 40) large T antigen; 3′: from nucleoplasmin); SaCas9, Cas9 endonuclease from the *Staphylococcus aureus* type II CRISPR/Cas system; 3xHA, three tandem HA epitope tags; pA, bovine growth hormone polyadenylation signal; U6, U6 promoter; gRNA, guide RNA targeting the *TRAC* locus; Scaff, guide RNA scaffold for the *S. aureus* CRISPR/Cas9 system; ITR, inverted terminal repeat. (**C**) Assessment of genome editing by sequence trace decomposition (TIDE) of gRNA #1. Different bar colours indicate different *p*-values: bright red, $p < 0.001$; black, $p \geq 0.001$.

3.2. Evaluation of Viral Delivery in HEK293SF Cells

The adeno-associated viral vector packaging the SaCas9 and gRNA #1 (subsequently referred to as AAV6-SaCas9) was produced via the triple transient transfection of HEK293SF cells, purified via affinity chromatography, and concentrated using 100-kDa centrifugal filter units. We first verified the ability of the AAV6-SaCas9 to deliver the CRISPR/Cas9 system and successfully cut the genome at the target site. HEK293SF cells were transduced with AAV6-SaCas9 (MOI ~ 10^3) in the presence of a human adenovirus serotype 5 (Ad5) at MOI 5. The Ad5 was added to the transduction assay to accelerate and enhance transgene expression [54–56]. At 24 and 48 h post-transduction, it was possible to detect the expression of the SaCas9 mRNA and gRNA via agarose gel electrophoresis, with band sizes corresponding to the expected PCR amplicon, respectively 166 and 90 bp (Figure 2A). The Western blot from cell lysate samples showed that the SaCas9 protein was successfully expressed in the cells transduced with AAV6-SaCas9 (Figure 2B). The next step evaluated the DSB efficiency in genomic DNA extracted from the cells 48 h post-transduction. The target site at the *TRAC* locus was amplified via PCR, and sequencing data were analyzed via TIDE (Figure 2C). Differently from the transfection of HEK293SF cells with pX601-AAV-SaCas9-TRAC#1, the transduction with AAV6-SaCas9 resulted in 2.4% indel efficiency, which could be explained by the MOI of AAV6 used.

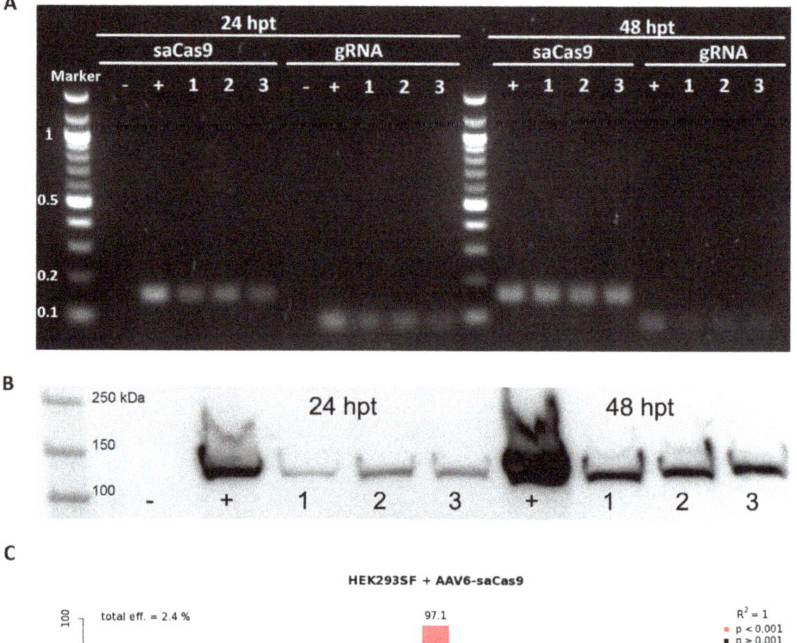

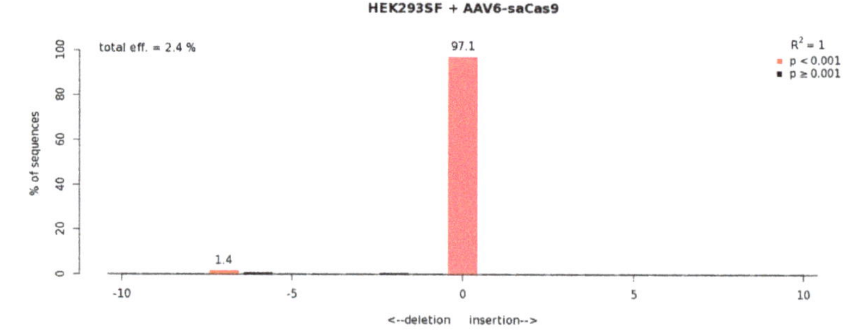

Figure 2. Evaluation of AAV6 delivery of CRISPR/Cas9 system in HEK293SF cells. Samples were collected 24 or 48 h post-transduction (hpt). Negative (−) control is a sample from non-transduced cells; positive (+) control is a sample from cells transfected with a plasmid encoding for SaCas9 and gRNA (pX601-AAV-SaCas9-TRAC#1); 1–3 indicate the replicates. (**A**) Expression of SaCas9 messenger RNA and gRNA detected via agarose gel electrophoresis after RT-PCR. Marker in kb. (**B**) Detection of SaCas9 protein by Western blot. (**C**) Assessment of genome editing by sequence trace decomposition. Different bar colours indicate different *p*-values: bright red, $p < 0.001$; black, $p \geq 0.001$.

3.3. Design of the DNA Template for Homology-Directed Repair

The third-generation anti-CD19 chimeric antigen was chosen to be incorporated into the DNA template for homology-directed repair. This CAR contains the variable region sequences of the FMC63 monoclonal antibody, which recognizes human CD19 and contains both the CD28 and 4-1BB co-stimulatory signalling domains (Figure 3A). The DNA template also incorporates an EGFP marker and homology arms to the *TRAC* locus (Figure 3B). The total length of the template DNA was 4.0 kb, which allowed sufficient space for the addition of the AAV ITRs. The whole construct was synthesized and cloned into the pUC57 plasmid (pUC57_AAV_anti-CD19_CAR_HDR). This plasmid was used for packaging the DNA template into the AAV6 vectors (subsequently referred to as AAV6-CAR).

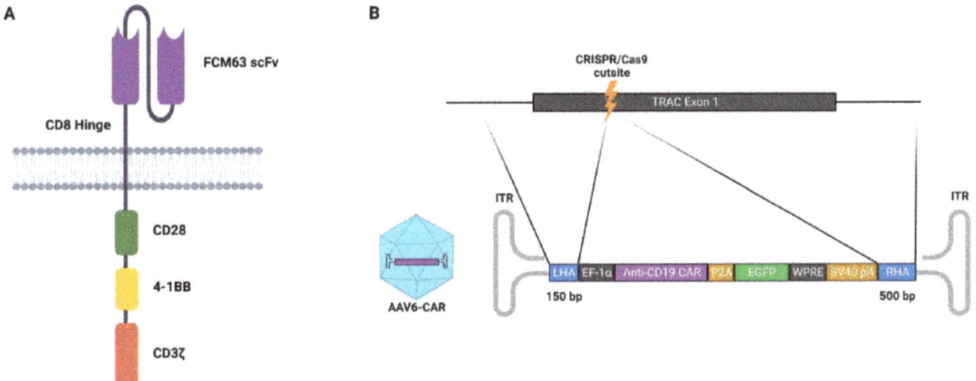

Figure 3. Anti-CD19 chimeric antigen receptor. (**A**) Third-generation chimeric antigen receptor comprising FCM63 single-chain variable fragment (scFv) specific for CD19, CD8 hinge, co-stimulatory domains CD28 and 4-1BB, and signalling domain CD3ζ. (**B**) Top, *TRAC* locus exon 1; bottom, AAV6 containing the CAR cassette flanked by homology arms. LHA, 150-bp left homology arm; EF-1α, core promoter for human elongation factor EF-1α; anti-CD19 CAR, third-generation anti-CD19 chimeric antigen receptor; P2A, 2A self-cleaving peptide; EGFP, enhanced green fluorescent protein; WPRE, woodchuck hepatitis virus posttranscriptional regulatory element; SV40_PA, SV40 (simian virus 40) polyadenylation signal; RHA, 500-bp right homology arm; ITR, inverted terminal repeat.

3.4. Production of Adeno-Associated Viral Vectors and Site-Specific Modification of Jurkat Cells via Dual-AAV Transduction

In order to obtain a sufficient amount of viral vectors for the dual-AAV transduction of the Jurkat cells, a 2-L production was carried out for each vector (AAV6-SaCas9 and AAV6-CAR). The cells, which contained the viral vectors, were pelleted and concentrated by a factor of 4 before viral harvest. The viral titers obtained, expressed as genome-containing viral particles, were 7.7×10^9 and 1.5×10^{10} VG/mL for AAV6-SaCas9 and AAV6-CAR, respectively. These viral titers were incompatible with the high-MOI transduction of the Jurkat cells; thus, the viral vector preparations underwent purification via affinity capture chromatography and concentration through a 100-KDa cut-off centrifugal filter unit. As shown in Table 2, the overall vector recovery was around 40–50%. Post-concentration, the viral titer was 1.7×10^{12} and 3.9×10^{12} VG/mL for AAV6-SaCas9 and AAV6-CAR, respectively. These post-concentration titers were compatible with high-MOI transduction, and the viral vectors were used in the subsequent experiments.

Table 2. Production of AAV6 vectors.

Vector	Step	Total VG	% Step Recovery	% Overall Recovery
AAV6-SaCas9	Harvest	4.34×10^{12}	100	100
	Purification	1.91×10^{12}	44.08	-
	Concentration	1.76×10^{12}	91.75	40.47
AAV6-CAR	Harvest	8.65×10^{12}	100	100
	Purification	8.17×10^{12}	94.40	-
	Concentration	3.99×10^{12}	48.89	46.15

The dual-transduction of Jurkat cells (Figure 4) was first conducted at a medium MOI, where the cells were transduced with both AAV6 vectors at an MOI of 10^5 VG/cells. The transgene insertion was evaluated via analysis of the gDNA 48 h post-transduction; however, it was not possible to detect the integration in the target locus (Figure 4A), and a TIDE analysis of the sequencing traces resulted in a total DSB efficiency of 1% (Figure 4C). Molecular analysis of the target locus did not result in the detection of transgene integration either. The low efficiency in the DSB and failure to detect transgene integration could have resulted from inefficient transduction due to a still not appropriate MOI or a premature analysis of the genomic modification. The incubation time of only 48 h may have been insufficient for the formation of the complementary second strand of the single-stranded AAV genome, which is needed for the expression of the transgene. It is plausible that this incubation was too brief to allow for the expression of the nuclease, which is essential for genome cleavage. Therefore, a longer incubation period may be necessary.

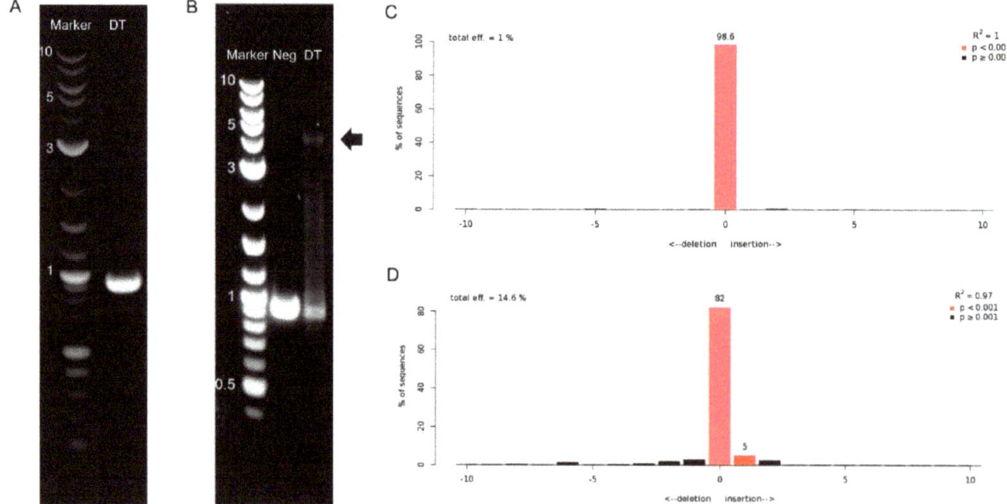

Figure 4. Insertion of anti-CD19 CAR into the *TRAC* locus via dual AAV6 transduction. (**A**) Detection of transgene insertion is not detected in medium-MOI transduction 48 h post-transduction. (**B**) Detection of transgene insertion via PCR amplification of the *TRAC* locus 96 h post-high-MOI transduction. DNA marker in kilobases (kb); Neg, non-transduced Jurkat cells; DT, dual transduced Jurkat cells. Black arrow indicates amplicon with the expected transgene size (~4.3 kb). (**C**) Assessment of genome editing by sequence trace decomposition of medium-MOI transduction. (**D**) Assessment of genome editing by sequence trace decomposition of high-MOI transduction. Different bar colours indicate different *p*-values: bright red, $p < 0.001$; black, $p \geq 0.001$.

A second transduction was conducted at a high MOI to account for the previously observed problems. AAV6-SaCas9 and AAV6-CAR were added to the cells at MOI 2×10^6 and 7×10^6 VG/cells, respectively. Furthermore, the genomic analysis was done 96 h post-transduction, allowing more time for the expression and function of the nuclease, increasing the possibility of detecting both genomic cleavage and homology-directed repair. Following a PCR using primers flanking the target region, a band corresponding to the expected size of the inserted transgene (4.3 kb) was detected via agarose gel electrophoresis (Figure 4B). This band was calculated to have a relative abundance of 10.9% via densitometric analysis. Moreover, the absolute quantitation of the bands resulted in an abundance of 9.1% for the 4.3-kb band. The TIDE analysis of this high-MOI dual transduction resulted in 14.6% of DSB efficiency (Figure 4D). It is important to note that TIDE can only detect small indels of up to 50 base pairs. Thus, this result does not consider transgene insertion via homology-directed repair. These results suggest that approximately 10% of the genomic material analyzed contained the insertion.

To further confirm the insertion of the transgene into the specific CRISPR/Cas9 cut site, the amplified product from the TRAC-flanking PCR was purified and used as the template for a second PCR ("nested"). A nested PCR was done using the same forward primer flanking the *TRAC* locus and a reverse primer that annealed inside the transgene, more specifically in the CAR region, resulting in the product shown in Figure 5A. As expected, a band of 2.2 kb was obtained (Figure 5B). This band was then extracted and digested with KpnI, generating approximately 550- and 1650-bp (Figure 5C) fragments corresponding to the expected digested DNA. The 2.2-kb band from the undigested fragment was also visible, as well as a 0.9-kb band, the same size as the unmodified *TRAC* locus also visible in Figure 4B. The material from the 2.2-kb band was also sequenced, confirming the insertion of the transgene template at the *TRAC* locus (Figure 5D).

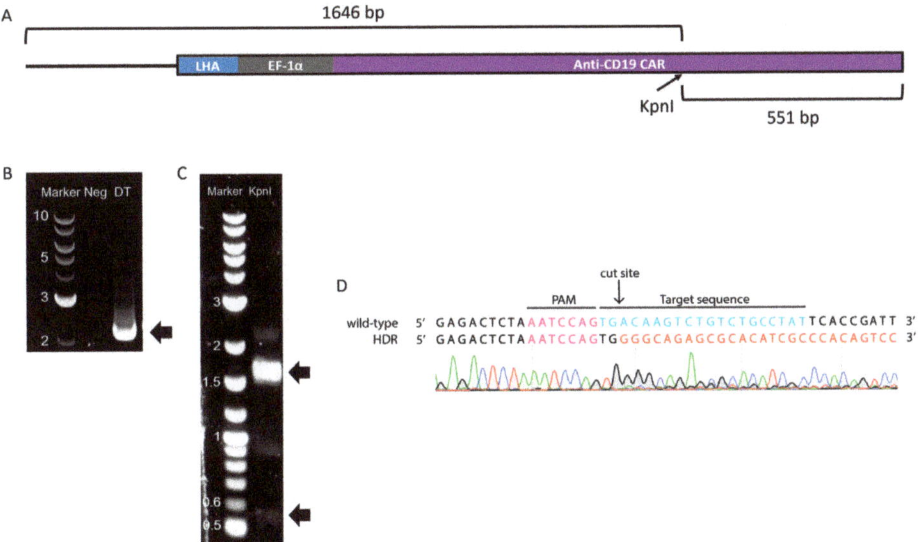

Figure 5. Molecular analysis of site-specific homology-mediated knock-in of anti-CD19 CAR gene. (**A**) Schematic representation of the PCR product containing the inserted DNA template with a KpnI site located within the CAR gene. (**B**) Agarose gel electrophoresis of the amplicons from non-transduced cells (Neg) and dual-transduced (DT) cells, resulting in a DNA band of 2.2 kb (black arrow). (**C**) Agarose gel electrophoresis of the 2.2-kb fragment digested with KpnI. Black arrows indicate the expected fragments with sizes 1.6 and 0.5 kb. The 2.2-kb band from the undigested material is also visible. DNA marker in kilobases (kb). (**D**) Sequence alignment of the 5′-end of genome-edited cells. Magenta, PAM site; blue, gRNA binding site; and red, integrated donor fragment.

3.5. Off-Target Evaluation

Finally, to evaluate the specificity of the chosen gRNA, COSMID detected 22 sites in the genome that diverged from the expected target site by less than four base pairs. The predicted sites were ranked based on similarity to the target sequence. Table 3 shows six of the top predicted sites evaluated for the occurrence of double-strand breaks in transfected HEK293SF cells and the dual-transduced Jurkat cells. Indel frequencies were low for all the potential off-target sites. In the HEK293SF cells, the off-target sites where activity was the highest (OT1 and OT6) were mutated in 1.5% of the cells; however, they were not located in known gene coding regions. The second-highest indel frequency (OT4, 1.3%) was detected in the gene *CACBN2*. The third-highest indel frequency was detected in a pseudogene (*LPAL2*). Three off-target sites failed to be sequenced for the indel analysis in Jurkat cells (OT1, OT4, and OT10), due to inadequate template quality/concentration or due to the presence of double sequences. No indel was identified for the remaining analyzed sites, corresponding to OT3, OT6, and OT8. These results show that the SaCas9 coupled with the chosen gRNA induces DNA breaks at the TRAC locus with a high specificity profile and without significant off-target effects.

Table 3. Analyzed off-target sites.

Name	Target Sequence	Gene	Product	% Indel HEK293SF	Jurkat
gRNA #1	*ATAGGCAGACAGACTTGTCACTGGAT*	*TRAC*	*T-cell receptor*	*21.7*	*14.6*
OT1	AAGTAAGACAGACTTGTCAGTGGGT	LOC112267962	Long non-coding RNA	1.5	-
OT3	ATAGTCACACAGACTGTCACTGAAT	LINC01035	Long non-coding RNA	0.1	0
OT4	ATAGGAGACAGGATTGTCAAGGGAT	CACNB2	Voltage-gated calcium channel	1.3	-
OT6	ATAGAGCAGACAGACAGGTCAAAGAAT	-	-	1.5	0
OT8	ATGGCAGAGAGACTTGCCATTGGAT	LPAL2	Lipoprotein(a) 2 pseudogene	1.1	0
OT10	GTAGTCAGACAGACTTGTATTGGAT	GTDC1	Glycosyltransferase-like	0.2	-

Analysis of desired target site is italicized.

4. Discussion

To avoid the use of the Cas9 from *Streptococcus pyogenes* (SpCas9) and, consequently, having to split the CRISPR/Cas9 machinery into two AAV vectors [57,58], we decided to utilize the smaller Cas9 from *Staphylococcus aureus* (SaCas9) in order to package both nuclease and gRNA into a single AAV vector [49]. The SaCas9 efficiency is comparable to that of SpCas9, and it is able to mediate in vivo genome editing [59–61]. Moreover, depending on the PAM sequence, SaCas9 presents higher cleavage activity than SpCas9 and FnCpf1 [62]. From the three gRNAs targeting the first exon of the *TRAC* locus tested in this study, we achieved gene editing efficiency similar to results previously reported for other targets and cell types [61].

The AAV serotype 6 was chosen due to its high efficiency in transducing T cells compared to other serotypes [44]. However, the MOI required to effectively transduce cells is considered high, reaching values around 10^6 VG/cell [42,44]. This could explain the low indel efficiencies obtained when the CRISPR/Cas9 machinery was delivered via AAV6 at a low MOI to HEK293SF and medium MOI to Jurkat cells. The genome modification efficiency increased significantly when Jurkat cells were transduced at a high MOI, as reported by other researchers [63,64].

The DNA template design had to consider the ~4.7-kb packaging limit of the AAV vector. The transgene containing the anti-CD19 CAR and EGFP marker spanned over 3.4 kb, with the ITRs accounting for an additional 400 bp. This implies around 900 bp in designing the homology arms. We then decided to evaluate the shortest homology arms necessary for a successful HDR, as determined by Hirata and Russell [65]. Thus, the homology arm upstream of the transgene (left homology arm, LHA) contained a

150-bp homologous to nucleotide sequence upstream of the cut site in the *TRAC* locus. Downstream of the transgene, a 500-bp homology arm was included. Both the MOI and the length of the homology arms are essential factors for the effectiveness of gene targeting by AAV vectors [16,65]. Homology arms ranging from 300 to 985 bp have been described for the AAV6-mediated targeting of CAR into the *TRAC* locus [46,47] and around 600 bp for other genomic locations [45,48]. The HDR efficiency obtained in our study (~10%) is similar to the one reported by Sather et al. [45] while using megaTAL, but slightly inferior (2–5 times lower) to those achieved with homing nucleases or CRISPR [20,36,46]. Future studies should optimize the MOI and the length and position of the homology arms in order to improve the HDR efficiency obtained in this study and to allow the future use of these vectors in a therapeutic setting [44]. Increasing the length of the homology arms would be possible while still within the AAV packaging limit by removing the fluorescent marker, EGFP, or the exogenous promoter, EF1-α, from the transgene cassette. Fluorescent markers are helpful during pre-clinical research but are not desired for clinical applications. By removing the exogenous promoter and controlling the expression of the CAR transgene by the endogenous *TRAC* promoter, stable and robust expression can be achieved, similar to the physiological TCR [34].

In this work, we report the effective combination of CRISPR and AAV6 delivery of a DNA template for the targeted insertion of CAR into the *TRAC* locus. Similar methods have been published previously for the site-specific integration of CAR [20,36,48]. In these reports, however, the CRISPR machinery is delivered as ribonucleoproteins (RNPs) via electroporation of the cells, which limits their approaches to in vitro or ex vivo genome modification. Because our method exploits an AAV6 vector to deliver the CRISPR/Cas9 machinery, it opens up the possibility for the in vivo generation of CAR-T cells. In vivo delivery of RNPs can be achieved through the use of lipid-based nanocarriers [66], polymer-based nanoparticles [67,68], peptide nanoparticles [69,70], or inorganic nanoparticles [71]. These methods, however, have limited efficiency compared to AAV vectors [72]. In vivo delivery of CRISPR-based therapies via AAV vectors has been reported in several research types with wildtype and disease-model animals [73]. Nawaz et al. [74] recently reported the AAV-mediated in vivo generation of CAR-T cells. In addition, in mice, dual-AAV delivery has been successfully reported for the in vivo HDR-mediated correction of metabolic liver disease and modification of mitotic and postmitotic neurons [64,75].

Low cell recovery was observed after 48 h from the transduction, which prevented us from confirming the expression of the CAR transgene inserted into the TRAC locus. This could have resulted from genotoxicity due to the high MOI needed for efficient transduction [76]. Further investigation is necessary to increase the recovery of the modified cells. Other studies have employed a similarly high MOI without reporting toxicities; however, different transduction regimens were used [20,44,45]. For instance, Wang et al. [44] replaced the culture medium 16 h after adding the viral vectors to halt the transduction. On the other hand, we decided to maintain the viral vector in the culture medium in an attempt to enhance transduction and further improve the chances for genome modification. Other approaches to improve cell recovery include the optimization of the seeding density, medium supplementation with higher amounts of fetal bovine serum, supplementation with IL-2, and/or cell activation via CD3 and CD28 receptors [77]. Another possible cause of the observed toxicity is off-target genome modification by the CRISPR/Cas9 system. However, only low percentages of off-target effects were identified. Most occurred in low-risk locations, such as non-coding sequences or pseudogenes. The only off-target identified in a known gene occurred in *CACBN2*, which is dominantly expressed in the retina (https: //www.proteinatlas.org/ENSG00000165995-CACNB2, accorded on 17 February 2023) and should have a limited impact in T cells. The SaCas9 has been characterized as inducing very low percentages of off-target indels [49]. However, it is not possible to disregard that a non-identified off-target resulted in decreased cell viability over time. Hence, a deeper screening of off-target activity is essential in improving the safety and efficacy of the method here presented [73].

To ensure a thorough understanding of the effectiveness of the dual-AAV6 transduction method, it is necessary to conduct further analyses to evaluate and characterize the modified cells. This includes examining the expression of the CAR protein and the endogenous TCR on the surfaces of the transduced cells through flow cytometric analysis. Successful CAR integration into the *TRAC* locus should replace the expression of the endogenous TCR with the anti-CD19 CAR. These two proteins can also be used for cell sorting and enrichment of the culture for TCR$^-$CAR$^+$ cells. Evaluating the functionality of anti-CD19 CAR-T cells can be achieved by co-cultivating them with CD19-expressing target cells. Although Jurkat cells cannot induce the death of target cells, they are activated by the target antigen, which can be detected by the expression of CD69 on the Jurkat cell surface [50]. Furthermore, conducting a complete immunophenotypic analysis of the CAR-T cells can provide a better understanding of the cells' phenotype, activation, and functional profile [78].

5. Conclusions

These results serve as a proof of concept that dual-AAV6 transduction can be employed for the site-specific integration of anti-CD19 CAR into T cells, with no significant off-target effects. Potential applications of the method reported here include the generation of ex vivo "off-the-shelf" universal immunotherapy. Further research is needed to improve the HDR efficiency, assess CAR protein expression and TCR knockout, and evaluate CAR functionality against target cells. This dual-AAV transduction process broadens the possibility for the in vivo generation of CAR-T cells but is not limited to this transgene and could be exploited for other gene therapy applications.

Author Contributions: Conceptualization: P.D.M., O.F., A.A.K.; Investigation: P.D.M.; Formal analysis: P.D.M., O.F., D.S.; Writing—original draft: P.D.M.; Writing—review and editing: P.D.M., O.F., D.S., A.A.K.; Supervision: A.A.K., Funding acquisition: A.A.K. All authors have read and agreed to the published version of the manuscript.

Funding: P.D.M. was funded by a doctoral scholarship from the Fonds de Recherche du Québec—Santé (FRQS), #284444. A.A.K. is partially funded through Canada Research Chair CRC-240394.

Institutional Review Board Statement: Not applicable.

Informed Consent Statement: Not applicable.

Data Availability Statement: The original contributions presented in the study are included in the article; further inquiries can be directed to the corresponding author.

Conflicts of Interest: The authors declare no conflict of interest.

References

1. Lu, J.; Jiang, G. The journey of CAR-T therapy in hematological malignancies. *Mol. Cancer* **2022**, *21*, 194. [CrossRef] [PubMed]
2. Finney, H.M.; Lawson, A.D.G.; Bebbington, C.R.; Weir, A.N.C.; Christopher, R.; Weir, A.N.C.; Bebbington, C.R.; Weir, A.N.C. Chimeric receptors providing both primary and costimulatory signaling in T cells from a single gene product. *J. Immunol.* **1998**, *161*, 2791–2797. [CrossRef] [PubMed]
3. Brentjens, R.J.; Santos, E.; Nikhamin, Y.; Yeh, R.; Matsushita, M.; La Perle, K.; Quintás-Cardama, A.; Larson, S.M.; Sadelain, M. Genetically targeted T cells eradicate systemic acute lymphoblastic leukemia xenografts. *Clin. Cancer Res.* **2007**, *13*, 5426–5435. [CrossRef] [PubMed]
4. Lin, H.; Cheng, J.; Mu, W.; Zhou, J.; Zhu, L. Advances in Universal CAR-T Cell Therapy. *Front. Immunol.* **2021**, *12*, 744823. [CrossRef] [PubMed]
5. Lin, J.K.; Lerman, B.J.; Barnes, J.I.; Boursiquot, B.C.; Tan, Y.J.; Robinson, A.Q.L.; Davis, K.L.; Owens, D.K.; Goldhaber-Fiebert, J.D. Cost Effectiveness of Chimeric Antigen Receptor T-Cell Therapy in Relapsed or Refractory Pediatric B-Cell Acute Lymphoblastic Leukemia. *J. Clin. Oncol.* **2018**, *36*, 3192–3202. [CrossRef] [PubMed]
6. Kuzmin, D.A.; Shutova, M.V.; Johnston, N.R.; Smith, O.P.; Fedorin, V.V.; Kukushkin, Y.S.; van der Loo, J.C.M.; Johnstone, E.C. The clinical landscape for AAV gene therapies. *Nat. Rev. Drug Discov.* **2021**, *20*, 173–174. [CrossRef]
7. McCarty, D.M.; Young, S.M., Jr.; Samulski, R.J. Integration of adeno-associated virus (AAV) and recombinant AAV vectors. *Annu. Rev. Genet.* **2004**, *38*, 819–845. [CrossRef]

8. Dhungel, B.P.; Bailey, C.G.; Rasko, J.E.J. Journey to the Center of the Cell: Tracing the Path of AAV Transduction. *Trends Mol. Med.* **2021**, *27*, 172–184. [CrossRef]
9. Russell, D.W.; Hirata, R.K. Human gene targeting by viral vectors. *Nat. Genet.* **1998**, *18*, 325–330. [CrossRef]
10. Vasileva, A.; Jessberger, R. Precise hit: Adeno-associated virus in gene targeting. *Nat. Rev. Microbiol.* **2005**, *3*, 837–847. [CrossRef]
11. Khan, I.F.; Hirata, R.K.; Russell, D.W. AAV-mediated gene targeting methods for human cells. *Nat. Protoc.* **2011**, *6*, 482. [CrossRef]
12. Hirsch, M.L. Adeno-associated virus inverted terminal repeats stimulate gene editing. *Gene Ther.* **2015**, *22*, 190–195. [CrossRef]
13. Barzel, A.; Paulk, N.K.; Shi, Y.; Huang, Y.; Chu, K.; Zhang, F.; Valdmanis, P.N.; Spector, L.P.; Porteus, M.H.; Gaensler, K.M.; et al. Promoterless gene targeting without nucleases ameliorates haemophilia B in mice. *Nature* **2015**, *517*, 360–364. [CrossRef] [PubMed]
14. Gaj, T.; Epstein, B.E.; Schaffer, D.V. Genome Engineering Using Adeno-associated Virus: Basic and Clinical Research Applications. *Mol. Ther.* **2016**, *24*, 458–464. [CrossRef] [PubMed]
15. Porteus, M.H.; Cathomen, T.; Weitzman, M.D.; Baltimore, D. Efficient gene targeting mediated by adeno-associated virus and DNA double-strand breaks. *Mol. Cell. Biol.* **2003**, *23*, 3558–3565. [CrossRef]
16. Miller, D.G.; Petek, L.M.; Russell, D.W. Adeno-associated virus vectors integrate at chromosome breakage sites. *Nat. Genet.* **2004**, *36*, 767–773. [CrossRef] [PubMed]
17. Dever, D.P.; Bak, R.O.; Reinisch, A.; Camarena, J.; Washington, G.; Nicolas, C.E.; Pavel-Dinu, M.; Saxena, N.; Wilkens, A.B.; Mantri, S.; et al. CRISPR/Cas9 β-globin gene targeting in human haematopoietic stem cells. *Nature* **2016**, *539*, 384–389. [CrossRef]
18. Dabiri, H.; Safarzadeh Kozani, P.; Habibi Anbouhi, M.; Mirzaee Godarzee, M.; Haddadi, M.H.; Basiri, M.; Ziaei, V.; Sadeghizadeh, M.; Hajizadeh Saffar, E. Site-specific transgene integration in chimeric antigen receptor (CAR) T cell therapies. *Biomark. Res.* **2023**, *11*, 67. [CrossRef]
19. Moço, P.D.; Aharony, N.; Kamen, A. Adeno-Associated Viral Vectors for Homology-Directed Generation of CAR-T Cells. *Biotechnol. J.* **2020**, *15*, 1900286. [CrossRef]
20. Eyquem, J.; Mansilla-Soto, J.; Giavridis, T.; van der Stegen, S.J.; Hamieh, M.; Cunanan, K.M.; Odak, A.; Gonen, M.; Sadelain, M. Targeting a CAR to the TRAC locus with CRISPR/Cas9 enhances tumour rejection. *Nature* **2017**, *543*, 113–117. [CrossRef]
21. Guo, Y.; Tong, C.; Su, L.; Zhang, W.; Jia, H.; Liu, Y.; Yang, Q.; Wu, Z.; Wang, Y.; Han, W. CRISPR/Cas9 genome-edited universal CAR T cells in patients with relapsed/refractory lymphoma. *Blood Adv.* **2022**, *6*, 2695–2699. [CrossRef] [PubMed]
22. McGarrity, G.J.; Hoyah, G.; Winemiller, A.; Andre, K.; Stein, D.; Blick, G.; Greenberg, R.N.; Kinder, C.; Zolopa, A.; Binder-Scholl, G.; et al. Patient monitoring and follow-up in lentiviral clinical trials. *J. Gene. Med.* **2013**, *15*, 78–82. [CrossRef]
23. Schroder, A.R.; Shinn, P.; Chen, H.; Berry, C.; Ecker, J.R.; Bushman, F. HIV-1 integration in the human genome favors active genes and local hotspots. *Cell* **2002**, *110*, 521–529. [CrossRef]
24. Mitchell, R.S.; Beitzel, B.F.; Schroder, A.R.; Shinn, P.; Chen, H.; Berry, C.C.; Ecker, J.R.; Bushman, F.D. Retroviral DNA integration: ASLV, HIV, and MLV show distinct target site preferences. *PLoS Biol.* **2004**, *2*, E234. [CrossRef] [PubMed]
25. Themis, M.; Waddington, S.N.; Schmidt, M.; von Kalle, C.; Wang, Y.; Al-Allaf, F.; Gregory, L.G.; Nivsarkar, M.; Themis, M.; Holder, M.V.; et al. Oncogenesis following delivery of a nonprimate lentiviral gene therapy vector to fetal and neonatal mice. *Mol. Ther.* **2005**, *12*, 763–771. [CrossRef] [PubMed]
26. Howe, S.J.; Mansour, M.R.; Schwarzwaelder, K.; Bartholomae, C.; Hubank, M.; Kempski, H.; Brugman, M.H.; Pike-Overzet, K.; Chatters, S.J.; de Ridder, D.; et al. Insertional mutagenesis combined with acquired somatic mutations causes leukemogenesis following gene therapy of SCID-X1 patients. *J. Clin. Investig.* **2008**, *118*, 3143–3150. [CrossRef]
27. Hacein-Bey-Abina, S.; Garrigue, A.; Wang, G.P.; Soulier, J.; Lim, A.; Morillon, E.; Clappier, E.; Caccavelli, L.; Delabesse, E.; Beldjord, K.; et al. Insertional oncogenesis in 4 patients after retrovirus-mediated gene therapy of SCID-X1. *J. Clin. Investig.* **2008**, *118*, 3132–3142. [CrossRef]
28. Schlimgen, R.; Howard, J.; Wooley, D.; Thompson, M.; Baden, L.R.; Yang, O.O.; Christiani, D.C.; Mostoslavsky, G.; Diamond, D.V.; Duane, E.G.; et al. Risks Associated With Lentiviral Vector Exposures and Prevention Strategies. *J. Occup. Environ. Med.* **2016**, *58*, 1159–1166. [CrossRef]
29. Aiuti, A.; Biasco, L.; Scaramuzza, S.; Ferrua, F.; Cicalese, M.P.; Baricordi, C.; Dionisio, F.; Calabria, A.; Giannelli, S.; Castiello, M.C.; et al. Lentiviral hematopoietic stem cell gene therapy in patients with Wiskott-Aldrich syndrome. *Science* **2013**, *341*, 1233151. [CrossRef]
30. Ellis, J. Silencing and variegation of gammaretrovirus and lentivirus vectors. *Hum. Gene Ther.* **2005**, *16*, 1241–1246. [CrossRef]
31. Torikai, H.; Reik, A.; Liu, P.Q.; Zhou, Y.; Zhang, L.; Maiti, S.; Huls, H.; Miller, J.C.; Kebriaei, P.; Rabinovich, B.; et al. A foundation for universal T-cell based immunotherapy: T cells engineered to express a CD19-specific chimeric-antigen-receptor and eliminate expression of endogenous TCR. *Blood* **2012**, *119*, 5697–5705. [CrossRef]
32. Liu, X.; Zhang, Y.; Cheng, C.; Cheng, A.W.; Zhang, X.; Li, N.; Xia, C.; Wei, X.; Liu, X.; Wang, H. CRISPR-Cas9-mediated multiplex gene editing in CAR-T cells. *Cell Res.* **2017**, *27*, 154–157. [CrossRef]
33. Anwer, F.; Shaukat, A.A.; Zahid, U.; Husnain, M.; McBride, A.; Persky, D.; Lim, M.; Hasan, N.; Riaz, I.B. Donor origin CAR T cells: Graft versus malignancy effect without GVHD, a systematic review. *Immunotherapy* **2017**, *9*, 123–130. [CrossRef]
34. Dimitri, A.; Herbst, F.; Fraietta, J.A. Engineering the next-generation of CAR T-cells with CRISPR-Cas9 gene editing. *Mol. Cancer* **2022**, *21*, 78. [CrossRef] [PubMed]
35. Ren, J.; Zhang, X.; Liu, X.; Fang, C.; Jiang, S.; June, C.H.; Zhao, Y. A versatile system for rapid multiplex genome-edited CAR T cell generation. *Oncotarget* **2017**, *8*, 17002–17011. [CrossRef]

36. Dai, X.; Park, J.J.; Du, Y.; Kim, H.R.; Wang, G.; Errami, Y.; Chen, S. One-step generation of modular CAR-T cells with AAV-Cpf1. *Nat. Methods* **2019**, *16*, 247–254. [CrossRef]

37. Gomes-Silva, D.; Srinivasan, M.; Sharma, S.; Lee, C.M.; Wagner, D.L.; Davis, T.H.; Rouce, R.H.; Bao, G.; Brenner, M.K.; Mamonkin, M. CD7-edited T cells expressing a CD7-specific CAR for the therapy of T-cell malignancies. *Blood* **2017**, *130*, 285–296. [CrossRef] [PubMed]

38. Borot, F.; Wang, H.; Ma, Y.; Jafarov, T.; Raza, A.; Ali, A.M.; Mukherjee, S. Gene-edited stem cells enable CD33-directed immune therapy for myeloid malignancies. *Proc. Natl. Acad. Sci. USA* **2019**, *116*, 11978–11987. [CrossRef] [PubMed]

39. Martin, R.M.; Ikeda, K.; Cromer, M.K.; Uchida, N.; Nishimura, T.; Romano, R.; Tong, A.J.; Lemgart, V.T.; Camarena, J.; Pavel-Dinu, M.; et al. Highly Efficient and Marker-free Genome Editing of Human Pluripotent Stem Cells by CRISPR-Cas9 RNP and AAV6 Donor-Mediated Homologous Recombination. *Cell Stem Cell* **2019**, *24*, 821–828. [CrossRef]

40. Fu, Y.-W.; Dai, X.-Y.; Wang, W.-T.; Yang, Z.-X.; Zhao, J.-J.; Zhang, J.-P.; Wen, W.; Zhang, F.; Oberg, K.C.; Zhang, L.; et al. Dynamics and competition of CRISPR–Cas9 ribonucleoproteins and AAV donor-mediated NHEJ, MMEJ and HDR editing. *Nucleic Acids Res.* **2021**, *49*, 969–985. [CrossRef]

41. Ellis, B.L.; Hirsch, M.L.; Barker, J.C.; Connelly, J.P.; Steininger, R.J., 3rd; Porteus, M.H. A survey of ex vivo/in vitro transduction efficiency of mammalian primary cells and cell lines with Nine natural adeno-associated virus (AAV1-9) and one engineered adeno-associated virus serotype. *Virol. J.* **2013**, *10*, 74. [CrossRef] [PubMed]

42. Wang, J.; Exline, C.M.; DeClercq, J.J.; Llewellyn, G.N.; Hayward, S.B.; Li, P.W.-L.; Shivak, D.A.; Surosky, R.T.; Gregory, P.D.; Holmes, M.C.; et al. Homology-driven genome editing in hematopoietic stem and progenitor cells using ZFN mRNA and AAV6 donors. *Nat. Biotechnol.* **2015**, *33*, 1256. [CrossRef] [PubMed]

43. De Ravin, S.S.; Reik, A.; Liu, P. Q.; Li, L.; Wu, X.; Su, L.; Raley, C.; Theobald, N.; Choi, U.; Song, A.H.; et al. Targeted gene addition in human CD34+ hematopoietic cells for correction of X-linked chronic granulomatous disease. *Nat. Biotechnol.* **2016**, *34*, 424–429. [CrossRef] [PubMed]

44. Wang, J.; DeClercq, J.J.; Hayward, S.B.; Li, P.W.; Shivak, D.A.; Gregory, P.D.; Lee, G.; Holmes, M.C. Highly efficient homology-driven genome editing in human T cells by combining zinc-finger nuclease mRNA and AAV6 donor delivery. *Nucleic Acids Res.* **2015**, *44*, e30. [CrossRef] [PubMed]

45. Sather, B.D.; Romano Ibarra, G.S.; Sommer, K.; Curinga, G.; Hale, M.; Khan, I.F.; Singh, S.; Song, Y.; Gwiazda, K.; Sahni, J.; et al. Efficient modification of CCR5 in primary human hematopoietic cells using a megaTAL nuclease and AAV donor template. *Sci. Transl. Med.* **2015**, *7*, 307ra156. [CrossRef] [PubMed]

46. MacLeod, D.T.; Antony, J.; Martin, A.J.; Moser, R.J.; Hekele, A.; Wetzel, K.J.; Brown, A.E.; Triggiano, M.A.; Hux, J.A.; Pham, C.D.; et al. Integration of a CD19 CAR into the TCR Alpha Chain Locus Streamlines Production of Allogeneic Gene-Edited CAR T Cells. *Mol. Ther.* **2017**, *25*, 949–961. [CrossRef]

47. Hale, M.; Lee, B.; Honaker, Y.; Leung, W.H.; Grier, A.E.; Jacobs, H.M.; Sommer, K.; Sahni, J.; Jackson, S.W.; Scharenberg, A.M.; et al. Homology-Directed Recombination for Enhanced Engineering of Chimeric Antigen Receptor T Cells. *Mol. Ther. Methods Clin. Dev.* **2017**, *4*, 192–203. [CrossRef]

48. Naeimi Kararoudi, M.; Likhite, S.; Elmas, E.; Yamamoto, K.; Schwartz, M.; Sorathia, K.; de Souza Fernandes Pereira, M.; Sezgin, Y.; Devine, R.D.; Lyberger, J.M.; et al. Optimization and validation of CAR transduction into human primary NK cells using CRISPR and AAV. *Cell Rep. Methods* **2022**, *2*, 100236. [CrossRef]

49. Ran, F.A.; Cong, L.; Yan, W.X.; Scott, D.A.; Gootenberg, J.S.; Kriz, A.J.; Zetsche, B.; Shalem, O.; Wu, X.; Makarova, K.S.; et al. In vivo genome editing using Staphylococcus aureus Cas9. *Nature* **2015**, *520*, 186–191. [CrossRef]

50. Bloemberg, D.; Nguyen, T.; Maclean, S.; Zafer, A.; Gadoury, C.; Gurnani, K.; Chattopadhyay, A.; Ash, J.; Lippens, J.; Harcus, D.; et al. A High-Throughput Method for Characterizing Novel Chimeric Antigen Receptors in Jurkat Cells. *Mol. Ther. Methods Clin. Dev.* **2020**, *16*, 238–254. [CrossRef]

51. Brinkman, E.K.; Chen, T.; Amendola, M.; van Steensel, B. Easy quantitative assessment of genome editing by sequence trace decomposition. *Nucleic Acids Res.* **2014**, *42*, e168. [CrossRef] [PubMed]

52. Joshi, P.R.H.; Bernier, A.; Moço, P.D.; Schrag, J.; Chahal, P.S.; Kamen, A. Development of a Scalable and Robust Anion-Exchange Chromatographic Method for Enriched Recombinant Adeno-Associated Virus Preparations in Genome Containing Vector Capsids of Serotypes- 5, 6, 8, and 9. *Mol. Ther. Methods Clin. Dev.* **2021**, *21*, 341–356. [CrossRef]

53. Cradick, T.J.; Qiu, P.; Lee, C.M.; Fine, E.J.; Bao, G. COSMID: A Web-based Tool for Identifying and Validating CRISPR/Cas Off-target Sites. *Mol. Ther.-Nucleic Acids* **2014**, *3*, e214. [CrossRef] [PubMed]

54. Song, L.; Samulski, R.J.; Hirsch, M.L. Adeno-Associated Virus Vector Mobilization, Risk Versus Reality. *Hum. Gene Ther.* **2020**, *31*, 1054–1067. [CrossRef] [PubMed]

55. Xiao, W.; Warrington, K.H.; Hearing, P.; Hughes, J.; Muzyczka, N. Adenovirus-Facilitated Nuclear Translocation of Adeno-Associated Virus Type 2. *J. Virol.* **2002**, *76*, 11505–11517. [CrossRef] [PubMed]

56. Fisher, K.J.; Gao, G.P.; Weitzman, M.D.; DeMatteo, R.; Burda, J.F.; Wilson, J.M. Transduction with recombinant adeno-associated virus for gene therapy is limited by leading-strand synthesis. *J. Virol.* **1996**, *70*, 520. [CrossRef]

57. Swiech, L.; Heidenreich, M.; Banerjee, A.; Habib, N.; Li, Y.; Trombetta, J.; Sur, M.; Zhang, F. In vivo interrogation of gene function in the mammalian brain using CRISPR-Cas9. *Nat. Biotechnol.* **2015**, *33*, 102–106. [CrossRef]

58. Chew, W.L.; Tabebordbar, M.; Cheng, J.K.W.; Mali, P.; Wu, E.Y.; Ng, A.H.M.; Zhu, K.; Wagers, A.J.; Church, G.M. A multifunctional AAV–CRISPR–Cas9 and its host response. *Nat. Methods* **2016**, *13*, 868. [CrossRef]

59. Pan, X.; Philippen, L.; Lahiri, S.K.; Lee, C.; Park, S.H.; Word, T.A.; Li, N.; Jarrett, K.E.; Gupta, R.; Reynolds, J.O.; et al. In Vivo Ryr2 Editing Corrects Catecholaminergic Polymorphic Ventricular Tachycardia. *Circ. Res.* **2018**, *123*, 953–963. [CrossRef]
60. Wang, Y.; Wang, B.; Xie, H.; Ren, Q.; Liu, X.; Li, F.; Lv, X.; He, X.; Cheng, C.; Deng, R.; et al. Efficient Human Genome Editing Using SaCas9 Ribonucleoprotein Complexes. *Biotechnol. J.* **2019**, *14*, 1800689. [CrossRef]
61. Weng, S.; Gao, F.; Wang, J.; Li, X.; Chu, B.; Wang, J.; Yang, G. Improvement of muscular atrophy by AAV–SaCas9-mediated myostatin gene editing in aged mice. *Cancer Gene Ther.* **2020**, *27*, 960–975. [CrossRef] [PubMed]
62. Xie, H.; Tang, L.; He, X.; Liu, X.; Zhou, C.; Liu, J.; Ge, X.; Li, J.; Liu, C.; Zhao, J.; et al. SaCas9 Requires 5′-NNGRRT-3′ PAM for Sufficient Cleavage and Possesses Higher Cleavage Activity than SpCas9 or FnCpf1 in Human Cells. *Biotechnol. J.* **2018**, *13*, e1700561. [CrossRef] [PubMed]
63. Kumar, N.; Stanford, W.; de Solis, C.; Aradhana; Abraham, N.D.; Dao, T.J.; Thaseen, S.; Sairavi, A.; Gonzalez, C.U.; Ploski, J.E. The Development of an AAV-Based CRISPR SaCas9 Genome Editing System That Can Be Delivered to Neurons in vivo and Regulated via Doxycycline and Cre-Recombinase. *Front. Mol. Neurosci.* **2018**, *11*, 413. [CrossRef] [PubMed]
64. Yang, Y.; Wang, L.; Bell, P.; McMenamin, D.; He, Z.; White, J.; Yu, H.; Xu, C.; Morizono, H.; Musunuru, K.; et al. A dual AAV system enables the Cas9-mediated correction of a metabolic liver disease in newborn mice. *Nat. Biotechnol.* **2016**, *34*, 334–338. [CrossRef]
65. Hirata, R.K.; Russell, D.W. Design and packaging of adeno-associated virus gene targeting vectors. *J. Virol.* **2000**, *74*, 4612–4620. [CrossRef]
66. Cheng, Q.; Wei, T.; Farbiak, L.; Johnson, L.T.; Dilliard, S.A.; Siegwart, D.J. Selective organ targeting (SORT) nanoparticles for tissue-specific mRNA delivery and CRISPR-Cas gene editing. *Nat. Nanotechnol.* **2020**, *15*, 313–320. [CrossRef]
67. Chen, G.; Abdeen, A.A.; Wang, Y.; Shahi, P.K.; Robertson, S.; Xie, R.; Suzuki, M.; Pattnaik, B.R.; Saha, K.; Gong, S. A biodegradable nanocapsule delivers a Cas9 ribonucleoprotein complex for in vivo genome editing. *Nat. Nanotechnol.* **2019**, *14*, 974–980. [CrossRef]
68. Wan, T.; Chen, Y.; Pan, Q.; Xu, X.; Kang, Y.; Gao, X.; Huang, F.; Wu, C.; Ping, Y. Genome editing of mutant KRAS through supramolecular polymer-mediated delivery of Cas9 ribonucleoprotein for colorectal cancer therapy. *J. Control. Release* **2020**, *322*, 236–247. [CrossRef]
69. Lostale-Seijo, I.; Louzao, I.; Juanes, M.; Montenegro, J. Peptide/Cas9 nanostructures for ribonucleoprotein cell membrane transport and gene edition. *Chem. Sci.* **2017**, *8*, 7923–7931. [CrossRef]
70. Krishnamurthy, S.; Wohlford-Lenane, C.; Kandimalla, S.; Sartre, G.; Meyerholz, D.K.; Theberge, V.; Hallee, S.; Duperre, A.M.; Del'Guidice, T.; Lepetit-Stoffaes, J.P.; et al. Engineered amphiphilic peptides enable delivery of proteins and CRISPR-associated nucleases to airway epithelia. *Nat. Commun.* **2019**, *10*, 4906. [CrossRef]
71. Lee, K.; Conboy, M.; Park, H.M.; Jiang, F.; Kim, H.J.; Dewitt, M.A.; Mackley, V.A.; Chang, K.; Rao, A.; Skinner, C.; et al. Nanoparticle delivery of Cas9 ribonucleoprotein and donor DNA in vivo induces homology-directed DNA repair. *Nat. Biomed. Eng.* **2017**, *1*, 889–901. [CrossRef] [PubMed]
72. Behr, M.; Zhou, J.; Xu, B.; Zhang, H. In vivo delivery of CRISPR-Cas9 therapeutics: Progress and challenges. *Acta Pharm. Sin. B* **2021**, *11*, 2150–2171. [CrossRef] [PubMed]
73. Lau, C.H.; Suh, Y. In vivo genome editing in animals using AAV-CRISPR system: Applications to translational research of human disease. *F1000Res* **2017**, *6*, 2153. [CrossRef] [PubMed]
74. Nawaz, W.; Huang, B.; Xu, S.; Li, Y.; Zhu, L.; Yiqiao, H.; Wu, Z.; Wu, X. AAV-mediated in vivo CAR gene therapy for targeting human T-cell leukemia. *Blood Cancer J.* **2021**, *11*, 119. [CrossRef] [PubMed]
75. Nishiyama, J.; Mikuni, T.; Yasuda, R. Virus-Mediated Genome Editing via Homology-Directed Repair in Mitotic and Postmitotic Cells in Mammalian Brain. *Neuron* **2017**, *96*, 755–768. [CrossRef] [PubMed]
76. Dudek, A.M.; Porteus, M.H. Answered and Unanswered Questions in Early-Stage Viral Vector Transduction Biology and Innate Primary Cell Toxicity for Ex-Vivo Gene Editing. *Front. Immunol.* **2021**, *12*, 660302. [CrossRef]
77. Ghaffari, S.; Torabi-Rahvar, M.; Omidkhoda, A.; Ahmadbeigi, N. Impact of various culture conditions on ex vivo expansion of polyclonal T cells for adoptive immunotherapy. *APMIS* **2019**, *127*, 737–745. [CrossRef]
78. de Azevedo, J.T.C.; Mizukami, A.; Moco, P.D.; Malmegrim, K.C.R. Immunophenotypic Analysis of CAR-T Cells. In *Chimeric Antigen Receptor T Cells*; Swiech, K., Malmegrim, K.C.R., Picanco-Castro, V., Eds.; Methods in Molecular Biology; Humana: New York, NY, USA, 2020; pp. 195–201.

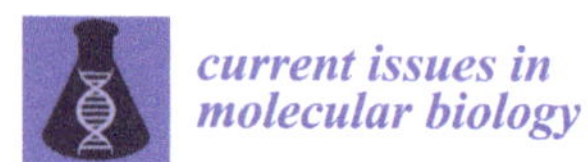

current issues in molecular biology

Article

Identification of a Novel ERK5 (MAPK7) Inhibitor, MHJ-627, and Verification of Its Potent Anticancer Efficacy in Cervical Cancer HeLa Cells

Jeonghye Hwang [1], Hyejin Moon [2], Hakwon Kim [2] and Ki-Young Kim [1,3,*]

1. Department of Genetics and Biotechnology, Kyung Hee University, Yongin 17104, Republic of Korea; hjh7956@khu.ac.kr
2. Department of Applied Chemistry, Global Center for Pharmaceutical Ingredient Materials, Kyung Hee University, Yongin 17104, Republic of Korea; hjm6719@naver.com (H.M.); hwkim@khu.ac.kr (H.K.)
3. Graduate School of Biotechnology, Kyung Hee University, Yongin 17104, Republic of Korea
* Correspondence: kiyoung@khu.ac.kr; Tel.: +82-312012633

Abstract: Extracellular signal-regulated kinase 5 (ERK5), a member of the mitogen-activated protein kinase (MAPK) family, is involved in key cellular processes. However, overexpression and upregulation of ERK5 have been reported in various cancers, and ERK5 is associated with almost every biological characteristic of cancer cells. Accordingly, ERK5 has become a novel target for the development of anticancer drugs as inhibition of ERK5 shows suppressive effects of the deleterious properties of cancer cells. Herein, we report the synthesis and identification of a novel ERK5 inhibitor, MHJ-627, and verify its potent anticancer efficacy in a yeast model and the cervical cancer HeLa cell line. MHJ-627 successfully inhibited the kinase activity of ERK5 (IC$_{50}$: 0.91 μM) and promoted the mRNA expression of tumor suppressors and anti-metastatic genes. Moreover, we observed significant cancer cell death, accompanied by a reduction in mRNA levels of the cell proliferation marker, proliferating cell nuclear antigen (*PCNA*), following ERK5 inhibition due to MHJ-627 treatment. We expect this finding to serve as a lead compound for further identification of inhibitors for ERK5-directed novel approaches for oncotherapy with increased specificity.

Keywords: ERK5; MAPK7; BMK1; ERK5 inhibitor; anticancer; oncotherapy; imidazolium

check for updates

Citation: Hwang, J.; Moon, H.; Kim, H.; Kim, K.-Y. Identification of a Novel ERK5 (MAPK7) Inhibitor, MHJ-627, and Verification of Its Potent Anticancer Efficacy in Cervical Cancer HeLa Cells. *Curr. Issues Mol. Biol.* **2023**, *45*, 6154–6169. https://doi.org/10.3390/cimb45070388

Academic Editor: Dumitru A. Iacobas

Received: 29 June 2023
Revised: 19 July 2023
Accepted: 21 July 2023
Published: 24 July 2023

1. Introduction

Extracellular signal-regulated kinase 5 (ERK5), also termed big mitogen-activated protein kinase 1 (BMK1) and mitogen-activated protein kinase 7 (MAPK7), belongs to the mitogen-activated protein kinase (MAPK) family, which mainly consists of four subfamilies in mammalian cells: ERK1/2, c-Jun-N-terminal kinases (JNK)1/2/3, p38α/β/γ/δ, and ERK5 [1]. In MAPK signaling cascades, three kinds of kinase are consecutively activated: a MAPK kinase kinase (MAPKKK), a MAPK kinase (MAPKK) and a MAP kinase (MAPK). In the ERK5 signaling pathway, MEKK2/3 are activated by various extracellular stimuli such as mitogens, cytokines, and cellular stresses, and they subsequently phosphorylate and activate MEK5 [2,3]. Once activated, MEK5 activates the apical kinase ERK5 by phosphorylating the T-E-Y motif in the activation loop within the ERK5 kinase domain [4,5]. Activated ERK5 then modulates a wide array of key cellular processes such as cell survival, proliferation, differentiation, angiogenesis, and apoptosis [6]. Structurally, the unique behavior of ERK5 among MAPK members is due to its extended C-terminal non-catalytic domain, which contains a transcriptional activation domain [7,8]. While other conventional MAP kinases transmit signals to downstream molecules mainly by phosphorylation, ERK5 can regulate downstream targets in two ways: direct substrate phosphorylation through its N-terminal kinase domain and transcriptional activation through autophosphorylation on its C-terminal non-kinase domain [9]. Thus, ERK5 is able to translocate to the nucleus and directly control gene expression by activating transcription factors [10].

With lines of accumulating research, overexpression and upregulation of ERK5 have been reported in various cancers, and ERK5 is widely implicated in the biological characteristics of cancer cells [11,12]. Moreover, ERK5 inhibition has been shown to suppress cancer cell proliferation, especially HeLa cells, and to induce tumor cell death in various tumor types [13–16]. Accordingly, ERK5 has emerged as a potential novel therapeutic target for overcoming malignancies and suppressing the deleterious actions of cancer cells [17].

However, to date, there has been a lack of high throughput screening systems to detect changes in ERK5 activity in animal cells due to its complex networks and cross-talk of signaling pathways [18]. Thus, we established a simple and time-saving yeast model system which could be utilized as a primary ERK5 inhibitor screening procedure to select putative ERK5 inhibitors among various candidate compounds we had (other compounds not mentioned), based on the well-established homologous pathway in yeast. Mpk1 (Slt2) in the CWI (Cell Wall Integrity) pathway that is functionally homologous to the ERK5 in humans is found in the yeast *Saccharomyces cerevisiae* [19]. It was demonstrated by Truman et al. that the expression of human ERK5 in Mpk1-defective yeasts is capable of rescuing diverse phenotypes attributable to the loss of native Mpk1 and therefore yeast Mpk1 is a functional homologue of human ERK5 [19]. The C-terminal domain of Mpk1 also possesses a transcriptional activating potential like ERK5, not to mention the striking sequence similarity within the N-terminal domain (49.7%) [19,20]. Since it is known that Mpk1 activates Rlm1 transcription factor by directly phosphorylating it and Rlm1 subsequently activates the transcription of *MLP1*, the decrease in *MLP1* expression shown by the β-galactosidase reporter can be interpreted as inhibited catalytic activity of Mpk1 (ERK5 homologue). This model system enables us to easily observe the alteration in Mpk1 activity using the *MLP1-lacZ* reporter plasmid. Moreover, there are two types of transcriptional regulatory pathways of ERK5, one reliant on the kinase domain and the other on the transcriptional activation domain. Since it is demonstrated by Jung et al. that Mpk1-Rlm1-*MLP1* pathway is mediated by the catalytic action of Mpk1, this model system makes it easier to achieve our goal to develop a kinase inhibitor of ERK5 [21].

Therefore, we ultimately aimed to develop a potential anticancer drug candidate for ERK5 inhibition through a series of experiments in a yeast model and the cervical cancer HeLa cell line. We report the synthesis and identification of a novel ERK5 inhibitor, MHJ-627, and verify its potent anticancer efficacy.

2. Materials and Methods

2.1. Instruments and Chemicals

All chemical reagents were purchased from Acros Organics (Brookline, MA, USA), Alfa Aesar (Haverhill, MA, USA), Sigma-Aldrich (St. Louis, MO, USA) or Tokyo Chemical Industry (Tokyo, Japan) and were used as received. The progress of reactions was monitored through thin-layer chromatography (TLC, silica gel 60 F254; Merck, Darmstadt, Germany). Melting points (m.p.) were determined on a Barnstead Electrothermal 9100 instrument and were uncorrected. ^{1}H and ^{13}C NMR spectra were recorded on a JEOL JNM-ECZ400S (Tokyo, Japan). NMR solvent was purchased from Cambridge Isotope Laboratories, Inc. (Andover, MA, USA) and spectra are referenced relative to the chemical shift of tetramethylsilane (TMS) as an internal standard. Chemical shifts (δ) are reported in parts-per-million (ppm), and coupling constants (J) are reported in Hertz (Hz). High-resolution mass spectroscopy was performed with a JEOL JMS-700 mass spectrometer. 1-(1,4-Bis(isopentyloxy)naphthalen-2-yl)-2-bromoethanone (2) and 1-isopentyl-1*H*-benzo[*d*]imidazole (3) were prepared as previously described [22,23].

Synthesis of 3-(2-(1,4-bis(isopentyloxy)naphthalen-2-yl)-2-oxoethyl)-1-isopentyl-1*H*-benzo[*d*]imidazol-3-ium bromide, MHJ-627: a solution of 1-(1,4-bis(isopentyloxy)naphthalen-2-yl)-2-bromoethanone (2, 0.10 g, 0.24 mml, 1.0 eq) and 1-isopentyl-1*H*-benzo[*d*]imidazole (3, 0.045 g, 0.24 mmol, 1.0 eq) in acetonitrile (4.8 mL, 0.05 M) was stirred at reflux for 24 h. After the reaction was complete, the product was concentrated in vacuo and recrystallized in ether to yield MHJ-627 as an ivory solid. Yield: 80%. m.p.: 195.6–197.3 °C. ^{1}H NMR

(400 MHz, $(CD_3)_2SO$) δ (ppm): 0.97–1.00 (m, 18H), 1.61–1.71 (m, 1H), 1.75–1.80 (m, 2H), 1.84–1.97 (m, 6H), 4.21 (t, *J* = 6.4 Hz, 2H), 4.25 (t, *J* = 6.4 Hz, 2H), 4.65 (t, *J* = 7.6 Hz, 2H), 6.26 (s, 2H), 7.23 (s, 1H), 7.67–7.81 (m, 4H), 8.09 (dd, *J* = 1.2, 7.6 Hz, 1H), 8.17 (dd, *J* = 1.2, 7.6 Hz, 1 H), 8.19–8.28 (m, 2H), and 9.82 (s, 1H). ^{13}C NMR (100 MHz, $(CD_3)_2SO$) δ (ppm): 22.10, 22.40, 22.60, 24.59, 24.78, 24.98, 37.14, 37.25, 38.48, 45.23, 55.83, 66.55, 75.51, 102.06, 113.67, 114.41, 122.30, 123.43, 123.68, 126.52, 126.68, 127.83, 128.25, 129.00, 129.05, 130.65, 131.96, 143.39, 150.58, 151.41, and 190.94. HRMS (FAB+ mode) m/z Calcd. for $C_{34}H_{45}N_2O_3$ [M-Br]+ 529.3430, found 529.3433.

A commercialized ERK5 inhibitor, XMD8-92 (S7525), and a MEK1/2 inhibitor, U0126 (S1102), were purchased from Selleck Chemicals (Houston, TX, USA).

2.2. Yeast Strains, Plasmids, Growth Conditions, and Transformation

The *S. cerevisiae* strain BY4742 was grown in a YEPD medium containing 2% Bacto peptone, 1% Bacto yeast extract, and 2% glucose at 30 °C in a shaking incubator [24]. *Escherichia coli* DH5α was used to distribute the plasmids. For selective growth, they were grown in a Luria Bertani (LB) medium containing 1% Bacto-tryptone, 0.5% Bacto-yeast extract, 1% NaCl, and 100 μg/mL Ampicillin at 37 °C in a shaking incubator. The plasmid used to transform the yeast contains an *MLP1* promoter followed by *lacZ*, which makes it possible to detect the expression level of *MLP1* through β-galactosidase expression [24–26]. Yeasts were transformed with *MLP1-lacZ*-containing plasmids using the standard lithium acetate-PEG method. Yeast transformants were cultured in a synthetic defined (SD) medium without uracil (SD-Ura) at 23 °C in a shaking incubator for 18 h until the exponential growth phase and then moved to a YEPD medium and cultured for 18 h to produce enough cells for the experiment [27]. For the ONPG assay, yeast cells were adjusted to OD_{600} = 1.0 with YEPD medium.

2.3. Animal Cell Lines and Culture

Human cervical cancer cell line HeLa cells (Korean Cell Line Bank, Seoul, Republic of Korea) were selected since they are commonly used in the study of ERK5 due to their ability to provide a clear observation of ERK5 activity [28,29]. It is known that negative regulation of ERK5 induces apoptosis in HeLa cells since ERK5 activity is necessary for survival of HeLa cells [16]. Cells were routinely cultured in Dulbecco's Modified Eagle's Medium (DMEM) high glucose, supplemented with 10% (*vol/vol*) fetal bovine serum (FBS) and 1% (*vol/vol*) penicillin/streptomycin. HeLa cells were grown at 37 °C in a humidified incubator with 5% CO_2 [30].

2.4. β-Galactosidase Reporter Assay

Yeast cells bearing *MLP1-lacZ* reporter plasmids were prepared as described above. For this, 3 mL of cells (OD_{600} = 1.0) were treated with 15 μL of compounds. Yeast cells were prepped via centrifugation and resuspended in 250 μL of breaking buffer (100 mM Tris-HCl pH = 8, 1 mM dithiothreitol, and 20% glycerol), with 100 μL of glass beads of 0.4–0.6 mm in diameter [31]. Yeast cells were homogenized via a bead beater to extract proteins. After 6 cycles of bead beating, samples were clarified via centrifugation at 12,000 RPM for 15 min at 4 °C. The Bradford method was used to measure protein concentration. Here, 100 μL of protein extracts containing 15 ug of proteins were mixed with 900 μL of Z buffer (60 mM $Na_2HPO_4 \cdot 7H_2O$, 40 mM $NaH_2PO_4 \cdot H_2O$, 10 mM KCl, 1 mM $MgSO_4 \cdot 7H_2O$, 50 mM 2-Mercaptoethanol, and pH = 7.0). After 5 min at 28 °C, 200 μL of O-nitrophenyl-β-D-galactopyranoside (ONPG) solution (4 mg/mL in Z buffer) was added. The reaction was conducted at 28 °C in a water bath for 3 h until the mixture obtained a pale-yellow color. The reaction was terminated by adding 500 μL of Na_2CO_3 solution. To measure the degree of ONPG hydrolysis by β-galactosidase, optical density was measured at 420 nm using a spectrophotometer [32]. All the procedures are based on Rose and Botstein's method [26,33,34]. Miller unit was calculated as follows.

$$\frac{OD_{420} \times 1.7}{0.0045 \times \text{protein concentration (mg/mL)} \times \text{protein extract volume (mL)} \times \text{time (m)}}$$

2.5. In Vitro Kinase Assay

An in vitro kinase assay was conducted to determine the inhibition of ERK5 kinase activity caused by MHJ-627 at concentrations of 5 μM, 1 μM, 0.1 μM, and 0 μM. The kinase assay was performed using Z'-LYTE™ Kinase Assay Kit—Ser/Thr 4 Peptide (PV3177; Thermo Fisher Scientific, Waltham, MA, USA) following the manufacturer's instruction. For this, 9 ng of ERK5 (ab126913; Abcam, Eugene, OR, USA) was used per one kinase reaction, and 100 μM ATP was used to drive the kinase reaction [35]. Fluorescence intensity was detected with a Varioskan™ LUX multimode microplate reader (VL0000D0; Thermo Fisher Scientific, Waltham, MA, USA).

2.6. Transient Transfection and qRT-PCR-Based Luciferase Reporter Assay

To measure the activity of AP-1, which is activated by ERK5, HeLa cells were transfected with pGL4.44 plasmid [*luc2P*/AP1 RE/Hygro] containing six copies of an AP-1 response element (AP1 RE), which drives transcription of the luciferase reporter gene *luc2P* (Photinus pyralis), using a LipofectamineTM 3000 reagent (Invitrogen, Waltham, MA, USA) according to the manufacturer's protocol [36]. After 24 h of transfection, the cells were seeded at a density of 3×10^5 cells per well in a 6-well plate. After 24 h, the cells were treated with MHJ-627 at concentrations of 5 μM, 1 μM, 0.1 μM, and 0 μM, as well as with XMD8-92 (positive control) at a concentration of 5 μM. To measure the mRNA expression level of luciferase, quantitative real-time PCR was conducted.

2.7. Quantitative Real-Time PCR Analysis

HeLa cells were seeded at a density of 3.0×10^5 cells per well of 6-well plates in 2 mL of serum-containing DMEM and were further cultured for 24 h for attachment. Then, various concentration (5 μM, 1 μM, and 0.1 μM) of MHJ-627 dissolved in 2 mL of serum-free DMEM were added, and cells were further cultured for 24 h. After 24 h of treatment, total RNA was isolated using a Trizol reagent (Thermo Fisher Scientific, Waltham, MA, USA) according to the manufacturer's protocol and reverse-transcribed to cDNA [37]. qRT-PCR was carried out using 2X SybrGreen Real-Time PCR Master Mix (Biofact, Daejeon, Republic of Korea). A housekeeping gene, *GAPDH*, served as an endogenous control [38]. Sequences of the primers used are listed in Table 1. $2^{-\Delta\Delta Cq}$ was calculated in duplicate, and an average of the two values was used to analyze expression of the genes [39].

Table 1. List of primers used in quantitative real-time PCR analysis.

Gene	Primer Sequence (5′ to 3′)	References
GAPDH	F: GTGAAGGTCGGAGTCAACG R: TGAGGTCAATGAAGGGGTC	[37]
PCNA	F: AACCTCACCAGTATGTCCAA R: ACTTTCTCCTGGTTTGGTG	[40]
DDIT4	F: GTGGAGGTGGTTTGTGTATC R: CACCCCTTGCTACTCTTAC	This study
CXCL1	F: AAAGCTTGCCTCAATCCTGC R: CTTCAGGAACAGCCACCAGT	This study
KLF4	F: CCAATTACCCATCCTTCCTG R: CGATCGTCTTCCCCTCTTTG	This study
NR4A1	F: GCTTCATGCCAGCATTATGG R: GTTCGGACAACTTCCTTCAC	This study

Table 1. *Cont.*

Gene	Primer Sequence (5′ to 3′)	References
RORα	F: AGGCTCGCTAGAGGTGGTGTT R: TGAGAGTCAAAGGCACGGC	This study
PTPRC	F: CTTCAGTGGTCCCATTGTGGTG R: CCACTTTGTTCTCGGCTTCCAG	This study
CCL5	F: TCATTGCTACTGCCCTCTGC R: TACTCCTTGATGTGGGCACG	This study
ICAM1	F: AGCGGCTGACGTGTGCAGTAAT R: TCTGAGACCTCTGGCTTCGTCA	This study
SIGLEC1	F: ACCTGGAGGAAACTGACAGTGG R: CTCAGTGTCACTGCCTGTCCTT	This study
luc2P	F: CTTTTGCAGCCCTTTCTTGC R: CTTTTGCAGCCCTTTCTTGC	This study

2.8. Western Blot Analysis

HeLa cells were seeded at a density of 3.0×10^5 cells per well of 6-well plates in 2 mL of serum-containing DMEM and were further cultured for 24 h for attachment. Then, various concentrations (5 µM, 1 µM, 0.1 µM, and 0 µM) of MHJ-627 dissolved in 2 mL of serum-free DMEM were applied to cells and further cultured for 24 h. After 24 h treatment, cells were lysed in radio-immunoprecipitation assay (RIPA) buffer containing 150 mM sodium chloride, 1% Triton X-100, 0.5% sodium deoxycholate, 0.1% sodium dodecyl sulfate (SDS), 50 mM Tris (pH 8.0), and a complete protease inhibitor cocktail (BIOMAX, Seoul, Republic of Korea). Protein concentration was determined using the BCA protein assay kit (TaKaRa, San Jose, CA, USA) according to the manufacturer's protocol. An equal amount of protein (10 µg/lane) was separated using sodium dodecyl sulfate-polyacrylamide gel electrophoresis (SDS-PAGE), transferred to a polyvinylidene difluoride (PVDF) membrane, and blocked with 5% BSA and 5% skim milk in a TBST buffer (20 mM Tris-HCl, 150 mM NaCl, and 0.1% Tween 20, pH 7.6) [37]. The membranes were probed with primary antibodies against GAPDH (sc-25778), ERK5 (sc-398015), and phospho-ERK5 (sc-135760) (Santa Cruz Biotechnology, Inc., Dallas, TX, USA) at 4 °C overnight. Membranes were then incubated with secondary antibodies for one hour at room temperature. Protein bands were developed with an ECL reagent and detected using a UVITEC imaging system equipment (UVITEC, Cambridge, UK) [40,41]. Relative protein expression from the Western blot data was determined using ImageJ.

2.9. Cytotoxicity Assay

Cytotoxicity, the ability of compounds to kill cancer cells, was evaluated via methylthiazol tetrazolium (MTT) assay. HeLa cells were seeded at a density of 1.0×10^4 cells per well of 96-well plates in 100 µL serum-containing DMEM and were further cultured for 24 h for settlement as described during cell culture [42,43]. Subsequently, 100 µL of serum-free DMEM compounds with various concentrations was added to each well, and cells received 24 h compound exposures. The reason for serum starvation was to eliminate the possibility of serum affecting the results of the assay and to only observe the effects of the treated compounds [44–47]. MHJ-627, at concentrations of 100 µM, 50 µM, 10 µM, 5 µM, 1 µM, 0.1 µM, and 0 µM, was added to the cells. Compounds were dissolved in 100% dimethyl sulfoxide (DMSO) at the original concentration of 10 mM. In order to prevent the dilution of DMSO from interfering with the results, dilution proceeded while maintaining the same percentage of DMSO in the treatment. After 24 h of compound exposures, MTT solution (5 mg/mL) was diluted 10 times in serum-free DMEM and then added to the wells after suction. Then, the plate was further maintained at 37 °C in the incubator for 3 h. Briefly, 100 µL of DMSO was added to each well after suction in order to dissolve the formazan crystals, and the plate was wrapped in aluminum foil to avoid light and gently shaken on an orbital shaker for 30 more minutes [48]. Absorption values at 540 nm and 570 nm were measured via a microplate spectrophotometer (BioTek Instruments, Winooski, VT, USA). Survival of untreated cells was regarded as a negative control and set as 100%. Then, survival of treated cells was calculated as a percentage of negative control [49]. As MTT showed that most of the cells were dead at 10 μM of MHJ-627, 5 μM was set as the maximum concentration in other experiments.

2.10. Statistical Analysis

All of the experiments were performed in duplicate and independently repeated at least 3 times. All the data are presented as the mean $\pm$ standard deviation (SD). Statistically significant differences were analyzed using two-tailed t test when only two groups were compared and one-way ANOVA with Dunnett's post hoc test using 0 μM as a control when more than two groups were compared [50]. $p < 0.05$ was considered statistically significant. All statistical analyses were conducted using GraphPad Prism software program version 5.0 (Graphpad Software, La Jolla, CA, USA).

3. Results and Discussion

3.1. MHJ-627 Compound Synthesis

In our previous study (Figure 1), we synthesized various naphthalene-2-acyl thiazolium salts by combining the structures of 1,4-dialkoxynaphthalene and thiazole and evaluated their potential as AGE (advanced glycation end products) breakers [51]. The 1,4-dialkoxynaphthalene moiety played a significant role in their pharmacological activity. Subsequently, we replaced thiazole with an imidazole ring to produce 1,4-dialkoxynaphthalene-2-acyl imidazolium salt derivatives, which exhibited antifungal activity [22]. We further confirmed that combination with the 1,4-dialkoxynaphthalene moiety served as a good pharmacophore. After screening the activity of several 1,4-dialkoxynapthalene imidazolium salts, it was confirmed that MHJ-627 is a potent ERK5 inhibitor (Figure 1).

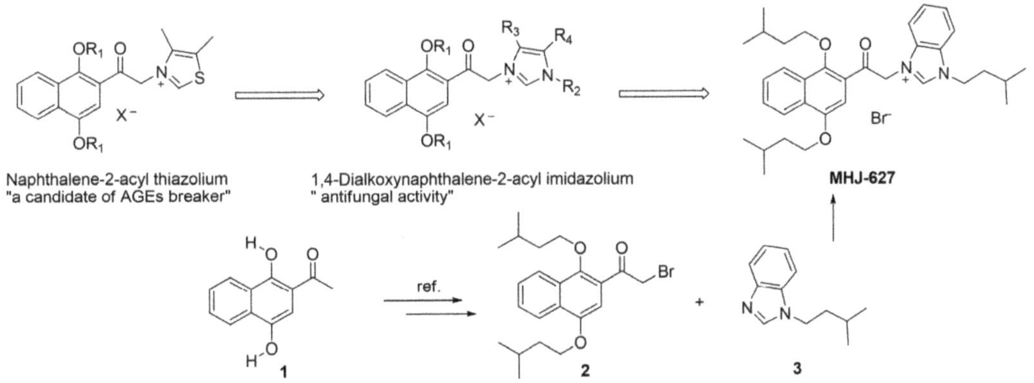

Figure 1. Design and synthesis of the new 1,4-dialkoxynaphthalen-2-acyl imidazolium salt, MHJ-627.

The synthesis of MHJ-627 was carried out as follows: a key intermediate acyl bromide 2 was synthesized from a commercially available starting compound 1 using a known process [22] and subsequently reacted with benzimidazole 3 to produce the desired compound, MHJ-627.

3.2. MHJ-627 Suppressed the Catalytic Activity of Mpk1 to Activate Rlm1 Transcription Factor and Attenuated the Expression of MLP1

As a primary putative ERK5 inhibitor screening procedure among various candidate compounds (other compounds not mentioned), we evaluated the ability of MHJ-627 to reduce the kinase activity of Mpk1, a functional homologue of human ERK5 [19], by examining the expression of *MLP1*, a target gene of Mpk1, through a transcriptional reporter assay using an *MLP1-lacZ* reporter plasmid in a yeast model system [24]. It is known that Rlm1 transcription factor activated by the kinase activity of Mpk1 promotes the transcription of *MLP1* (Figure 2a) [24]. MHJ-627 significantly suppressed *MLP1* expression by 66% compared to the control treated with DMSO only (Figure 2b). Since Rlm1 regulation is already demonstrated to be dependent on kinase activity of Mpk1 [21,24], this result implies that MHJ-627 impaired the kinase activity of Mpk1 to phosphorylate Rlm1 transcription factor and consequently inhibited *MLP1* expression, suggesting that MHJ-627 may also inhibit the kinase activity of human ERK5.

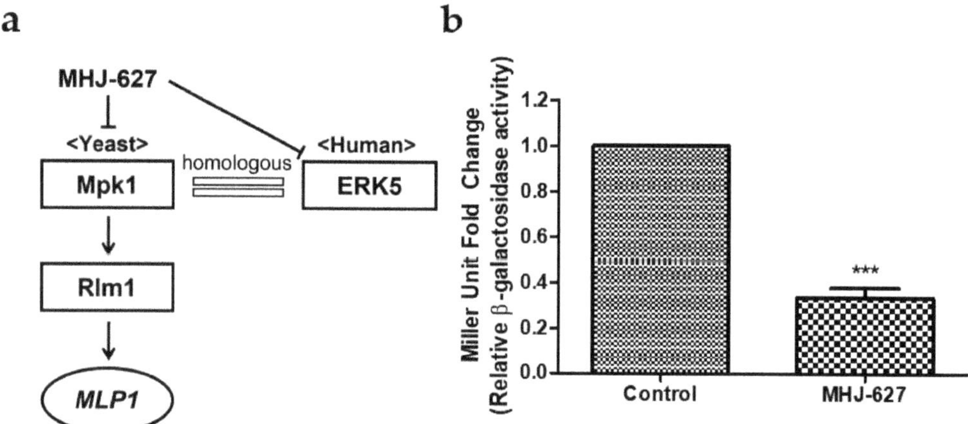

Figure 2. MHJ-627 suppressed the kinase activity of Mpk1 and attenuated *MLP1* expression in an *S. cerevisiae* model. (**a**) Schematic representation of Mpk1 regulation in the *S. cerevisiae* model system, which is functionally homologous to the human ERK5. Inactivation of Mpk1 activity results in downregulated transcriptional activity of Rlm1 transcription factor and subsequent decrease in *MLP1* expression; (**b**) effect of MHJ-627 on expression of *MLP1* measured by β-galactosidase activity. Yeasts were transformed with *MLP1-lacZ* reporter plasmid and treated with 15 μL of DMSO (control) and MHJ-627 in 3 mL of media. The data were calibrated to the control value (DMSO control = 1). Data are presented as mean ± SD. Each experiment was performed in duplicate and repeated at least three times. Two-tailed unpaired Student's *t* test (*** $p < 0.001$) was used for significance.

3.3. MHJ-627 Inhibited the Kinase Activity of Human ERK5 In Vitro

To further verify the capability of MHJ-627 to inhibit the kinase activity of human ERK5, a FRET-based in vitro kinase assay was carried out [52]. Relative kinase activity of ERK5 dropped to 0.58 at 0.1 μM, 0.49 at 1 μM, and 0.44 at 5 μM, which means MHJ-627 exhibited inhibitory activity by 42% at 0.1 μM, 51% at 1 μM, and 56% at 5 μM, respectively (Figure 3). This dose-dependent decrease in kinase activity according to the concentration of MHJ-627 shows that MHJ-627 also impairs the kinase activity of human ERK5 (IC$_{50}$: 0.91 μM), as we expected from the previous yeast screening. XMD8-92, a commercialized ERK5 inhibitor, was used as a positive control and showed an inhibition rate of 56% at 5 μM, which means that MHJ-627 and XMD8-92 exhibit similar inhibitory activity at 5 μM in vitro [33,52]. Since we confirmed that most of the cells were severely affected at 10 μM of MHJ-627 in the MTT assay, 5 μM was set as the maximum concentration in this assay.

3.4. MHJ-627 Suppressed the Activity of ERK5 and Impaired AP-1 Activity in HeLa Cells

To further examine the ability of MHJ-627 to inhibit the kinase activity of human ERK5 in cells, activation of activator protein-1 (AP-1), a downstream transcription factor of ERK5, was measured via a luciferase reporter [36]. HeLa cells were transfected with a plasmid bearing an AP-1 response element followed by a luciferase reporter gene and the mRNA level of luciferase was measured using quantitative real-time PCR (qRT-PCR). As the AP-1 transcription factor is a downstream target of ERK5, even though it is also a downstream of ERK1/2, it is often used to evaluate the alteration in ERK5 activity in cells [53]. Since ERK5 is often dysregulated in cancers, AP-1 is also found in a hyperactivated form in tumor cells [54,55]. The mRNA of luciferase transcribed by the AP-1 transcription factor decreased in a dose-dependent manner (Figure 4), signifying hindered AP-1 activation by ERK5 following MHJ-627 and XMD8-92 treatment [56]. This result suggests that MHJ-627 successfully inhibits ERK5 both in vitro and at the cell level.

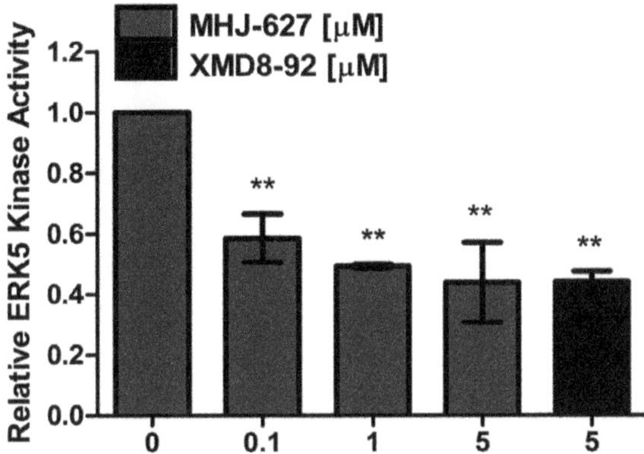

Figure 3. MHJ-627 reduced the kinase activity of human ERK5 in vitro. Relative ERK5 kinase activity following MHJ-627 treatment was measured via in vitro kinase assay. Kinase activity of ERK5 was reduced dose-dependently, supporting ERK5-inhibitory activity of MHJ-627 in vitro. Relative ERK5 kinase activity of the 0 μM control was set as 1. Data are presented as mean ± SD. Each experiment was performed in duplicate and repeated at least three times. One-way ANOVA (** $p < 0.01$) was used for significance. All values were compared to the 0 μM control value to determine the significance.

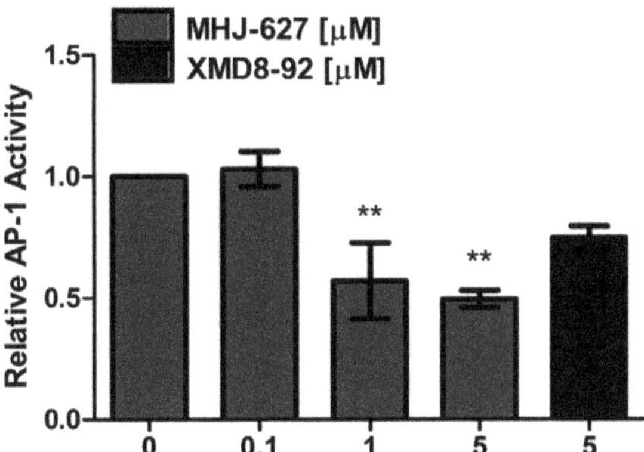

Figure 4. MHJ-627 suppressed ERK5 kinase activity to activate AP-1 transcription factor. To determine the ability of ERK5 to activate the transcription factor AP-1, luciferase reporter plasmid was transformed into HeLa cells and qRT-PCR was conducted to measure the mRNA level of luciferase after 24 h compound treatment. There was a decrease in luciferase mRNA levels, indicating reduced activity of AP-1 possibly caused by suppressed activity of ERK5 to activate AP-1. Relative AP-1 activity of the 0 μM control was set as 1. Data are presented as mean ± SD. Each experiment was performed in duplicate and repeated at least three times. One-way ANOVA (** $p < 0.01$) was used for significance. All values were compared to the 0 μM control value to determine the significance.

3.5. ERK5 Inhibition by MHJ-627 Modified the mRNA Expression of Genes Regulated by ERK5

qRT-PCR was performed to assess the expression of genes that are previously reported to be regulated or influenced by ERK5 in the gene expression analyses of ERK5 signaling though they are not the established direct targets of ERK5 [57]. As previously reported, downregulation of proliferating cell nuclear antigen (PCNA) expression is a known effect of ERK5 inhibition or ablation. Consistently, we observed a decrease in mRNA expression of *PCNA*, which is involved in DNA replication and repair machinery (Figure 5a) [58–60]. In contrast, as illustrated by previous literature, the mRNA level of DNA damage-inducible transcript 4 (*DDIT4*), which acts as a negative regulator of the mammalian target of rapamycin (mTOR) pathway, was elevated (Figure 5b) [57,61]. Furthermore, we observed increases in mRNA expression of the genes that can be categorized into two groups based on the function of the proteins they encode: transcription factors and immune-related proteins. The mRNA expression of KLF transcription factor 4 (*KLF4*), nuclear receptor subfamily 4 group A member 1 (*NR4A1*) and retinoic acid receptor-related orphan receptor-alpha (*RORα*), which act as transcription factors, was upregulated (Figure 5c). There was an increase in mRNA expression of protein tyrosine phosphatase receptor type C (*PTPRC*), C-C motif chemokine ligand 5 (*CCL5*), intercellular adhesion molecule 1 (*ICAM1*), sialic acid binding Ig like lectin 1 (*SIGLEC1*), and C-X-C motif chemokine ligand 1 (*CXCL1*), which all are related to immunity (Figure 5d). Since an activation of immune cells and immune responses following ERK5 inhibition has been reported, we speculate that this increase in expression is due to a feedback loop of the signal transduction pathway [11]. The decrease in *PCNA* mRNA and increase in *DDIT4* and *CXCL1* mRNA, which have been reported to occur when ERK5 is inhibited, are evidence that MHJ-627 effectively targets ERK5 [58].

Especially, PCNA, which is distinctly considered a cell proliferation marker due to its accumulation in late G1 and S phases, is strongly suggested to be involved in cell survival and tumorigenesis [62,63]. Considering previous knowledge that the degradation of PCNA inhibits cancer proliferation in vitro and in vivo, a dose-dependent decrease in *PCNA* mRNA levels may be indicative of the anticancer efficacy of MHJ-627 [64]. Therefore, in future study, we will conduct in-depth study on how ERK5 inhibition downregulates PCNA and verify if PCNA could be a direct target of ERK5. *KLF4* is suggested to act as a tumor suppressor, and its expression is often downregulated in some types of cancer, including cervical cancer, colorectal cancer, and lung cancer [65–67]. Particularly, in cervical cancer, previous study has shown the inactivation of *KLF4* as a tumor suppressor [68]. Similarly, *RORα* is a potential tumor suppressor, and its downregulation, which is related to tumor progression, is often observed in cancers [69,70]. Nuclear receptor 4A1 (*NR4A1*) is proposed to be downregulated in metastatic tumors and to play a protective role against metastasis [71,72]. Taken together, these results suggest that MHJ-627-induced ERK5 inhibition contributes to establishing a suitable environment to overcome malignancies by promoting the expression of some tumor suppressors and anti-metastatic genes which we assumed to be an outcome of targeting overexpressed ERK5 in cancers. However, in the cases of *KLF4*, *NR4A1*, and *ICAM1*, the trend of alteration in mRNA expression when treated with MHJ-627 was different from the positive control, increasing in the MHJ-627 treatment while decreasing in the positive control treatment [73]. Since the ERK5-inhibitory effect of MHJ-627 was already demonstrated via in vitro kinase assay in Figure 3, this result indicates that the mechanism governing ERK5 inhibition of these compounds may be somewhat different and needs to be further investigated in a follow-up study to identify the precise mechanism of MHJ-627's inhibition of ERK5 activity.

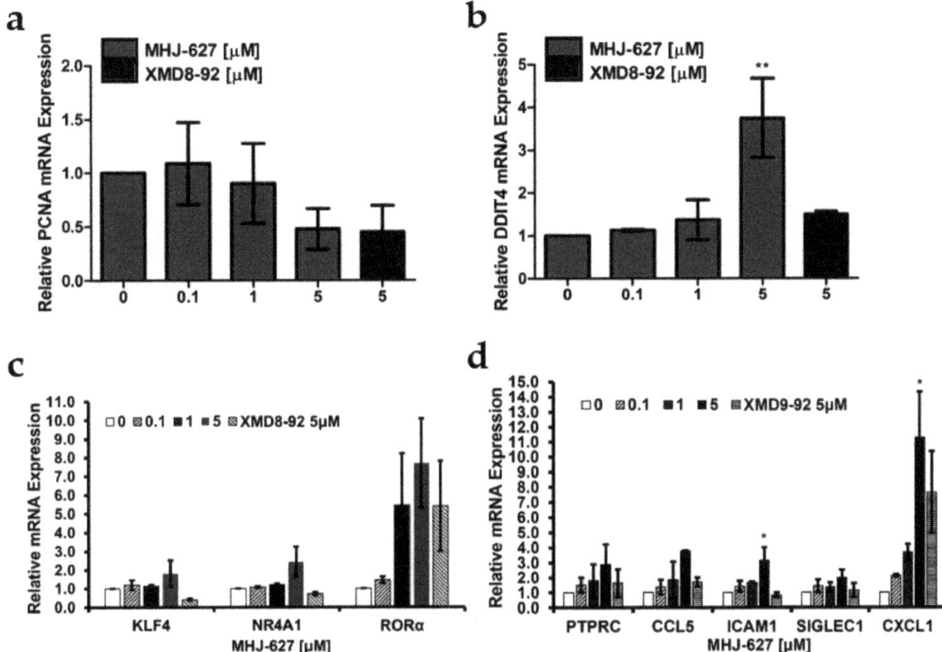

Figure 5. Alteration in mRNA expression pattern of the genes influenced by ERK5 after MHJ-627 treatment. (**a**) Decrease in mRNA expression of *PCNA*, which is a cell proliferation marker; (**b**) increase in mRNA expression of *DDIT4*, which is reported to increase when ERK5 is inhibited; (**c**) increase in mRNA expression of genes that encode transcription factors; (**d**) increase in mRNA expression of genes that encode immune-related proteins. Relative mRNA expression of genes influenced by ERK5 was measured via qRT-PCR analysis after 24 h compound treatment in HeLa cells. Relative mRNA expression of the 0 μM control was set as 1. Data are presented as mean ± SD. Each experiment was performed in duplicate and repeated at least three times. One-way ANOVA (* $p < 0.05$, ** $p < 0.01$) was used for significance. All values were compared to the 0 μM control value to determine the significance.

3.6. MHJ-627 Paradoxically Increased ERK5 Expression Possibly due to the Stimulatory Crosstalk of the ERK1/2 Pathway

To determine whether MHJ-627 affects the protein expression levels of ERK5 and pERK5, Western blot analysis was conducted. MHJ-627 paradoxically appeared to elevate ERK5 expression and phosphorylation, and so did the positive control, XMD8-92 (Figure 6a–c). However, even though the protein expression and phosphorylation of ERK5 increased, previous experimental results from Figures 3 and 4 have already shown that the actual activity of ERK5 was successfully inhibited as expected.

Since crosstalk and feedback loop mechanisms of other signaling pathways have been suggested as the most challenging problem in developing kinase inhibitors, we assumed that the elevations in ERK5 expression and phosphorylation may be attributed partly to the stimulatory crosstalk and compensatory action of the PI3K-AKT pathway or the ERK1/2 pathway [74,75]. Therefore, we examined the effect of the ERK1/2 pathway by treatment with 5 μM of an MEK1/2 inhibitor, U0126, which inhibits the activation of ERK1/2, together with various concentrations of MHJ-627 [76]. As expected, protein expression and phosphorylation of ERK5 among the lanes showed no difference (Figure 6d), suggesting that the previous increase in expression was due to the compensatory action of the ERK1/2 pathway, at least in part. Nevertheless, the precise mechanism is still unknown and is yet to be identified.

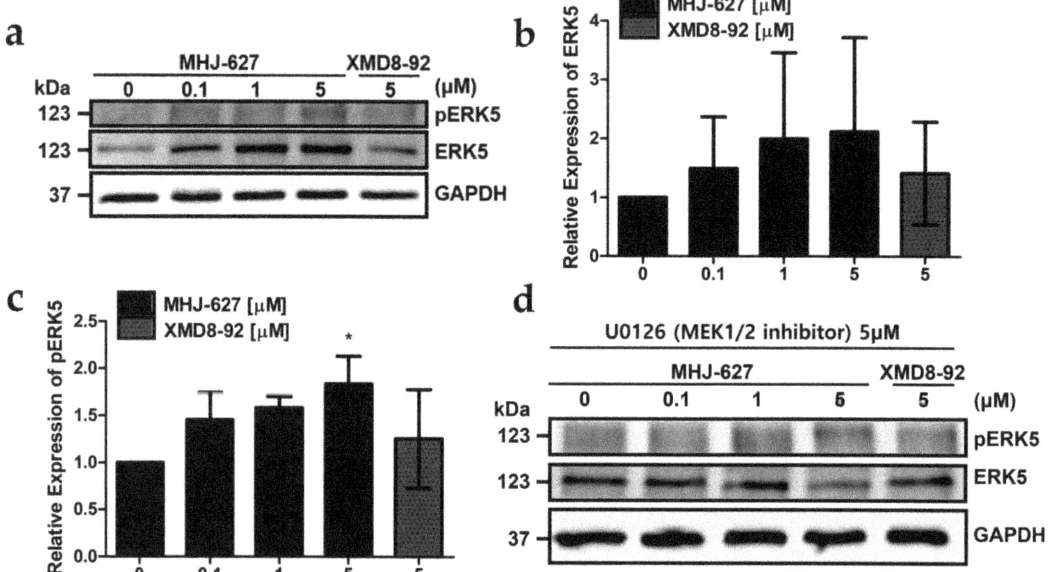

Figure 6. MHJ-627 paradoxically upregulated the expression and phosphorylation of ERK5, possibly due to the stimulatory crosstalk of the ERK1/2 pathway. (**a**) Western blot image depicting the elevations in ERK5 protein expression and phosphorylation. Effect of MHJ-627 on the protein expression and phosphorylation of ERK5 was measured via Western blot analysis after HeLa cells were treated with compounds for 24 h; (**b**) quantitation of Western blot showing a paradoxical increase in ERK5 expression; (**c**) quantitation of Western blot showing a trend of increase in ERK5 phosphorylation; (**d**) the increase in ERK5 expression and phosphorylation was due to the compensatory action of ERK1/2. GAPDH was used as a loading control. Relative protein expression of the 0 µM control was set as 1. Western blot data were quantified using ImageJ software. Data are presented as mean ± SD. Each experiment was performed in duplicate and repeated at least three times. One-way ANOVA (* $p < 0.05$) was used for significance. All values were compared to the 0 µM control value to determine the significance.

3.7. MHJ-627 Showed Anti-Proliferative Effect in the Human Cervical Cancer HeLa Cells

To measure the anticancer efficacy of MHJ-627, an MTT assay was conducted. HeLa cells were treated with the indicated concentration of MHJ-627 for 24 h and 48 h. XMD8-92 was used as a positive control. HeLa cells treated with XMD8-92 showed a significant decline in cell viability and showed an anti-proliferative effect of 16.9% after 24 h and of 22.7% after 48 h at 5 µM treatment, providing evidence for the possible anticancer efficacy of ERK5 inhibition (Figure 7a). The viability of HeLa cells significantly decreased in a dose-dependent manner after MHJ-627 treatment (Figure 7b). Especially, MHJ-627 exhibited anti-proliferative effect of 61% after 24 h (IC$_{50}$: 2.45 µM) and 94.2% after 48 h at 5 µM treatment. Almost every cancer was severely affected at a concentration of 10 µM in both the 24 h and 48 h treatments. The fact that MHJ-627 significantly exhibited higher cytotoxicity in HeLa cells confirms the higher anticancer efficacy of MHJ-627, which possibly resulted from the stronger ERK5-inhibitory activity, since negative regulation of ERK5 is known to induce apoptosis in HeLa cells [16]. This result casts a new light on the promise that MHJ-627 may serve as a more potent ERK5 inhibitor than the ones previously identified.

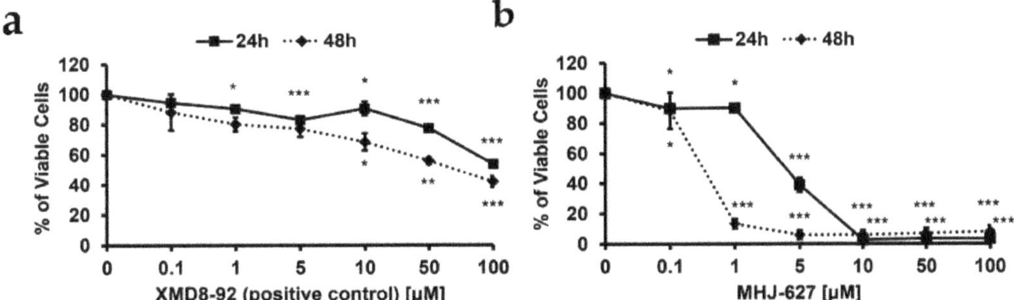

Figure 7. MHJ-627 had an anti-proliferative effect through inhibition of ERK5. (**a**) Effect of XMD8-92 on viability of HeLa cells showed that inhibition of ERK5 exhibited an anti-proliferative effect for HeLa cells; (**b**) effect of MHJ-627 on viability of HeLa cells showed its potent anti-proliferative efficacy. Cell viability was determined via MTT assay after 24 h and 48 h compound treatments with indicated concentration. XMD8-92 was used as a positive control. MHJ-627 showed higher cytotoxicity compared to the positive control, suggesting its potent anticancer efficacy and ERK5-inhibitory activity since ERK5 activity is necessary for the survival of HeLa cells. (0 µM control = 100%). Data are presented as mean ± SD. Each experiment was performed in duplicate and repeated at least three times. One-way ANOVA (* $p < 0.05$, ** $p < 0.01$, *** $p < 0.001$) was used for significance. All values were compared to the 0 µM control value to determine the significance.

4. Conclusions

ERK5 is a rising therapeutic target to combat cancer since its overexpression and dysregulation have been reported in various types of cancer [11,12]. However, despite its pivotal involvement in tumorigenesis, most previous works have focused on ERK1/2. In this study, we synthesized a novel ERK5 inhibitor, MHJ-627, and verified its potent anticancer property in the cervical cancer HeLa cells. MHJ-627 successfully impaired the kinase activity of ERK5 to produce significant anticancer efficacy accompanying upregulation of tumor suppressors and anti-metastatic genes, suggesting MHJ-627 as a promising ERK5 inhibitor.

There is no doubt that inhibition of ERK5 is a promising novel way to combat cancer [12,17]. Moreover, in the case of ERK1/2 inhibition, where extensive studies have been carried out, the compensatory elevation in the ERK5 pathway has conferred resistance to the ERK1/2 therapy [77,78]. Development of an ERK5 inhibitor for a combination therapy with ERK1/2 inhibitors may contribute to overcoming this resistance [75,79]. In this study, we focused on confirming ERK5-inhibitory activity of MHJ-627 which was also identified in our in silico simulation model that MHJ-627 actually binds to an ATP-binding pocket of ERK5. In future studies, our next goal is to utilize MHJ-627 as a lead compound and modify it to be a more potent ERK5 inhibitor with increased specificity to ERK5 that can exhibit a powerful anticancer efficacy. Further study with improved ERK5 inhibitor will also include ERK5 knock-out and knock-down models to clearly demonstrate its specific ERK5-inhibitory efficacy.

Author Contributions: Conceptualization, H.K. and K.-Y.K.; methodology, J.H. and K.-Y.K.; validation, J.H., H.K. and K.-Y.K.; formal analysis, J.H.; synthesis, H.M.; investigation, J.H.; resources, H.K.; data curation, J.H.; writing—original draft preparation, J.H.; writing—review and editing, H.K. and K.-Y.K.; visualization, J.H.; supervision, K.-Y.K.; project administration, K.-Y.K.; funding acquisition, H.K. All authors have read and agreed to the published version of the manuscript.

Funding: This research was funded by the GRRC Program of Gyeonggi province [GRRC-KyungHee-2023(B01)], Republic of Korea.

Institutional Review Board Statement: Not applicable.

Informed Consent Statement: Not applicable.

Data Availability Statement: All data generated or analyzed during this study are included in this published article.

Acknowledgments: This research was supported by the Signal Transduction and Transcription Regulation Laboratory (STTRL) of Kyung Hee university. I would like to express gratitude to all the members of STTRL for their continuous support and helpful advice.

Conflicts of Interest: The authors declare no conflict of interest.

References

1. Morrison, D.K. MAP kinase pathways. *Cold Spring Harb. Perspect. Biol.* **2012**, *4*, a011254. [CrossRef] [PubMed]
2. Nguyen, D.; Lemos, C.; Wortmann, L.; Eis, K.; Holton, S.J.; Boemer, U.; Moosmayer, D.; Eberspaecher, U.; Weiske, J.; Lechner, C.; et al. Discovery and characterization of the potent and highly selective (Piperidin-4-yl)pyrido[3,2-*d*]pyrimidine based in vitro probe BAY-885 for the kinase ERK5. *J. Med. Chem.* **2019**, *62*, 928–940. [CrossRef]
3. Hayashi, M.; Lee, J.D. Role of the BMK1/ERK5 signaling pathway: Lessons from knockout mice. *J. Mol. Med.* **2004**, *82*, 800–808. [CrossRef]
4. Hoang, V.T.; Yan, T.J.; Cavanaugh, J.E.; Flaherty, P.T.; Beckman, B.S.; Burow, M.E. Oncogenic signaling of MEK5-ERK5. *Cancer Lett.* **2017**, *392*, 51–59. [CrossRef] [PubMed]
5. Mody, N.; Campbell, D.G.; Morrice, N.; Peggie, M.; Cohen, P. An analysis of the phosphorylation and activation of extracellular-signal-regulated protein kinase 5 (ERK5) by mitogen-activated protein kinase kinase 5 (MKK5) in vitro. *Biochem. J.* **2003**, *372*, 567–575. [CrossRef]
6. Zhang, W.; Liu, H.T. MAPK signal pathways in the regulation of cell proliferation in mammalian cells. *Cell Res.* **2002**, *12*, 9–18. [CrossRef]
7. Nithianandarajah-Jones, G.N.; Wilm, B.; Goldring, C.E.; Müller, J.; Cross, M.J. ERK5: Structure, regulation and function. *Cell. Signal.* **2012**, *24*, 2187–2196. [CrossRef] [PubMed]
8. Buschbeck, M.; Ullrich, A. The unique C-terminal tail of the mitogen-activated protein kinase ERK5 regulates its activation and nuclear shuttling. *J. Biol. Chem.* **2005**, *280*, 2659–2667. [CrossRef]
9. Nishimoto, S.; Nishida, E. MAPK signalling: ERK5 versus ERK1/2. *EMBO Rep.* **2006**, *7*, 782–786. [CrossRef]
10. Morimoto, H.; Kondoh, K.; Nishimoto, S.; Terasawa, K.; Nishida, E. Activation of a C-terminal transcriptional activation domain of ERK5 by autophosphorylation. *J. Biol. Chem.* **2007**, *282*, 35449–35456. [CrossRef]
11. Stecca, B.; Rovida, E. Impact of ERK5 on the hallmarks of cancer. *Int. J. Mol. Sci.* **2019**, *20*, 1426. [CrossRef]
12. Monti, M.; Celli, J.; Missale, F.; Cersosimo, F.; Russo, M.; Belloni, E.; Di Matteo, A.; Lonardi, S.; Vermi, W.; Ghigna, C.; et al. Clinical significance and regulation of ERK5 expression and function in cancer. *Cancers* **2022**, *14*, 348. [CrossRef]
13. Gavine, P.R.; Wang, M.; Yu, D.; Hu, E.; Huang, C.; Xia, J.; Su, X.; Fan, J.; Zhang, T.; Ye, Q.; et al. Identification and validation of dysregulated MAPK7 (ERK5) as a novel oncogenic target in squamous cell lung and esophageal carcinoma. *BMC Cancer* **2015**, *15*, 454. [CrossRef]
14. Shukla, A.; Miller, J.M.; Cason, C.; Sayan, M.; MacPherson, M.B.; Beuschel, S.L.; Hillegass, J.; Vacek, P.M.; Pass, H.I.; Mossman, B.T. Extracellular signal-regulated kinase 5: A potential therapeutic target for malignant mesotheliomas. *Clin. Cancer Res.* **2013**, *19*, 2071–2083. [CrossRef] [PubMed]
15. Kato, Y.; Tapping, R.I.; Huang, S.; Watson, M.H.; Ulevitch, R.J.; Lee, J.D. Bmk1/Erk5 is required for cell proliferation induced by epidermal growth factor. *Nature* **1998**, *395*, 713–716. [CrossRef]
16. Zheng, F.; Zhang, J.; Luo, S.; Yi, J.; Wang, P.; Zheng, Q.; Wen, Y. miR-143 is associated with proliferation and apoptosis involving ERK5 in HeLa cells. *Oncol. Lett.* **2016**, *12*, 3021–3027. [CrossRef]
17. Simões, A.E.; Rodrigues, C.M.; Borralho, P.M. The MEK5/ERK5 signalling pathway in cancer: A promising novel therapeutic target. *Drug Discov. Today* **2016**, *21*, 1654–1663. [CrossRef]
18. Braicu, C.; Buse, M.; Busuioc, C.; Drula, R.; Gulei, D.; Raduly, L.; Rusu, A.; Irimie, A.; Atanasov, A.G.; Slaby, O.; et al. A Comprehensive review on MAPK: A promising therapeutic target in cancer. *Cancers* **2019**, *11*, 1618. [CrossRef]
19. Truman, A.W.; Millson, S.H.; Nuttall, J.M.; King, V.; Mollapour, M.; Prodromou, C.; Pearl, L.H.; Piper, P.W. Expressed in the yeast *Saccharomyces cerevisiae*, human ERK5 is a client of the Hsp90 chaperone that complements loss of the Slt2p (Mpk1p) cell integrity stress-activated protein kinase. *Eukaryot. Cell* **2006**, *5*, 1914–1924. [CrossRef] [PubMed]
20. Soler, M.; Plovins, A.; Martín, H.; Molina, M.; Nombela, C. Characterization of domains in the yeast MAP kinase Slt2 (Mpk1) required for functional activity and in vivo interaction with protein kinases Mkk1 and Mkk2. *Mol. Microbiol.* **1995**, *17*, 833–842. [CrossRef] [PubMed]
21. Jung, U.S.; Sobering, A.K.; Romeo, M.J.; Levin, D.E. Regulation of the yeast Rlm1 transcription factor by the Mpk1 cell wall integrity MAP kinase. *Mol. Microbiol.* **2002**, *46*, 781–789. [CrossRef] [PubMed]
22. Lee, J.; Kim, J.G.; Lee, H.; Lee, T.H.; Kim, K.Y.; Kim, H. Antifungal activity of 1,4-Dialkoxynaphthalen-2-Acyl imidazolium salts by inducing apoptosis of pathogenic *Candida* spp. *Pharmaceutics* **2021**, *13*, 312. [CrossRef]

23. Lee, H.; Jeon, Y.; Moon, H.; Lee, E.H.; Lee, T.H.; Kim, H. Synthesis of 1, 4-Dialkoxynaphthalene-Based Imidazolium salts and their cytotoxicity in cancer cell lines. *Int. J. Mol. Sci.* **2023**, *24*, 2713. [CrossRef]
24. Kim, K.Y.; Truman, A.W.; Levin, D.E. Yeast Mpk1 mitogen-activated protein kinase activates transcription through Swi4/Swi6 by a noncatalytic mechanism that requires upstream signal. *Mol. Cell. Biol.* **2008**, *28*, 2579–2589. [CrossRef] [PubMed]
25. Guarente, L. Yeast promoters and *lacZ* fusions designed to study expression of cloned genes in yeast. *Meth. Enzymol.* **1983**, *101*, 181–191. [CrossRef]
26. Rose, M.; Botstein, D. Construction and use of gene fusions to *lacZ* (beta-galactosidase) that are expressed in yeast. *Meth. Enzymol.* **1983**, *101*, 167–180. [CrossRef]
27. Stahl, G.; Salem, S.N.; Chen, L.; Zhao, B.; Farabaugh, P.J. Translational accuracy during exponential, postdiauxic, and stationary growth phases in *Saccharomyces cerevisiae*. *Eukaryot. Cell* **2004**, *3*, 331–338. [CrossRef]
28. Cude, K.; Wang, Y.; Choi, H.J.; Hsuan, S.L.; Zhang, H.; Wang, C.Y.; Xia, Z. Regulation of the G2-M cell cycle progression by the ERK5-NFkappaB signaling pathway. *J. Cell Biol.* **2007**, *177*, 253–264. [CrossRef]
29. Lin, E.C.; Amantea, C.M.; Nomanbhoy, T.K.; Weissig, H.; Ishiyama, J.; Hu, Y.; Sidique, S.; Li, B.; Kozarich, J.W.; Rosenblum, J.S. ERK5 kinase activity is dispensable for cellular immune response and proliferation. *Proc. Natl. Acad. Sci. USA.* **2016**, *113*, 11865–11870. [CrossRef]
30. Siano, G.; Caiazza, M.C.; Ollà, I.; Varisco, M.; Madaro, G.; Quercioli, V.; Calvello, M.; Cattaneo, A.; Di Primio, C. Identification of an ERK inhibitor as a therapeutic drug against Tau aggregation in a New Cell-Based Assay. *Front. Cell. Neurosci.* **2019**, *13*, 386. [CrossRef]
31. Arias, P.; Díez-Muñiz, S.; García, R.; Nombela, C.; Rodríguez-Peña, J.M.; Arroyo, J. Genome-wide survey of yeast mutations leading to activation of the yeast cell integrity MAPK pathway: Novel insights into diverse MAPK outcomes. *BMC Genom.* **2011**, *12*, 390. [CrossRef] [PubMed]
32. Becker, J.M. *Biotechnology: A Laboratory Course*, 1st ed.; Academic Press: Cambridge, MA, USA, 1990; pp. 129–132.
33. Kang, C.; Kim, J.S.; Kim, C.Y.; Kim, E.Y.; Chung, H.M. The pharmacological inhibition of ERK5 enhances apoptosis in acute myeloid leukemia cells. *Int. J. Stem Cell* **2018**, *11*, 227–234. [CrossRef] [PubMed]
34. Zhao, C.; Jung, U.S.; Garrett-Engele, P.; Roe, T.; Cyert, M.S.; Levin, D.E. Temperature-induced expression of yeast *FKS2* is under the dual control of protein kinase C and calcineurin. *Mol. Cell. Biol.* **1998**, *18*, 1013–1022. [CrossRef]
35. Myers, S.M.; Bawn, R.H.; Bisset, L.C.; Blackburn, T.J.; Cottyn, B.; Molyneux, L.; Wong, A.C.; Cano, C.; Clegg, W.; Harrington, R.W.; et al. High-throughput screening and hit validation of extracellular-related kinase 5 (ERK5) inhibitors. *ACS Comb. Sci.* **2016**, *18*, 444–455. [CrossRef] [PubMed]
36. Gomez, N.; Erazo, T.; Lizcano, J.M. ERK5 and cell proliferation: Nuclear localization is what matters. *Front. Cell Dev. Biol.* **2016**, *4*, 105. [CrossRef]
37. Nguyen, A.T.; Kim, K.Y. Inhibition of proinflammatory cytokines in *Cutibacterium acnes*-induced inflammation in HaCaT cells by using *Buddleja davidii* aqueous extract. *Int. J. Inflam.* **2020**, *2020*, 8063289. [CrossRef]
38. Yamada, Y.; Watanabe, Y.; Zhang, J.; Haraoka, J.; Ito, H. Changes in cortical and cerebellar bcl-2 mRNA levels in the developing hydrocephalic rat (LEW-HYR) as measured by a real time quantified RT-PCR. *Neuroscience* **2002**, *114*, 165–171. [CrossRef]
39. Livak, K.J.; Schmittgen, T.D. Analysis of relative gene expression data using real-time quantitative PCR and the 2(-Delta Delta C(T)) Method. *Methods* **2001**, *25*, 402–408. [CrossRef]
40. Kim, M.; Kim, J.G.; Kim, K.Y. *Trichosanthes kirilowii* Extract Promotes Wound Healing through the Phosphorylation of ERK1/2 in Keratinocytes. *Biomimetics* **2022**, *7*, 154. [CrossRef]
41. Kim, M.; Kim, J.; Shin, Y.K.; Kim, K.Y. Gentisic acid stimulates Keratinocyte proliferation through ERK1/2 phosphorylation. *Int. J. Med. Sci.* **2020**, *17*, 626–631. [CrossRef]
42. Wang, Y.; Yu, H.; Zhang, J.; Gao, J.; Ge, X.; Lou, G. Hesperidin inhibits HeLa cell proliferation through apoptosis mediated by endoplasmic reticulum stress pathways and cell cycle arrest. *BMC Cancer* **2015**, *15*, 682. [CrossRef]
43. Huang, L.; Huang, Q.Y.; Huang, H.Q. The evidence of HeLa cell apoptosis induced with tetraethylammonium using proteomics and various analytical methods. *J. Biol. Chem.* **2014**, *289*, 2217–2229. [CrossRef] [PubMed]
44. Pirkmajer, S.; Chibalin, A.V. Serum starvation: Caveat emptor. *Am. J. Physiol. Cell Physiol.* **2011**, *301*, C272–C279. [CrossRef] [PubMed]
45. Mbeunkui, F.; Fodstad, O.; Pannell, L.K. Secretory protein enrichment and analysis: An optimized approach applied on cancer cell lines using 2D LC-MS/MS. *J. Proteome Res.* **2006**, *5*, 899–906. [CrossRef] [PubMed]
46. Lambert, K.; Pirt, S.J. Growth of human diploid cells (strain MRC-5) in defined medium; replacement of serum by a fraction of serum ultrafiltrate. *J. Cell. Sci.* **1979**, *35*, 381–392. [CrossRef] [PubMed]
47. Colzani, M.; Waridel, P.; Laurent, J.; Faes, E.; Rüegg, C.; Quadroni, M. Metabolic labeling and protein linearization technology allow the study of proteins secreted by cultured cells in serum-containing media. *J. Proteome Res.* **2009**, *8*, 4779–4788. [CrossRef] [PubMed]
48. Park, S.C.; Kim, J.G.; Shin, Y.K.; Kim, K.Y. Antimicrobial activity of 4-hydroxyderricin, sophoraflavanone G, acetylshikonin, and kurarinone against the bee pathogenic bacteria *Paenibacillus larvae* and *Melissococcus plutonius*. *J. Apic. Res.* **2021**, *60*, 118–122. [CrossRef]
49. Gao, L.; Fei, J.; Zhao, J.; Li, H.; Cui, Y.; Li, J. Hypocrellin-loaded gold nanocages with high two-photon efficiency for photothermal/photodynamic cancer therapy in vitro. *ACS Nano* **2012**, *6*, 8030–8040. [CrossRef]

50. Kim, J.; Shin, Y.K.; Kim, K.Y. Promotion of Keratinocyte proliferation by Tracheloside through ERK1/2 stimulation. *Evid. Based Complement. Altern. Med.* **2018**, *2018*, 4580627. [CrossRef]

51. Samsuzzaman, M.; Lee, J.H.; Moon, H.; Lee, J.; Lee, H.; Lim, Y.; Park, M.G.; Kim, H.; Kim, S.Y. Identification of a potent NAFLD drug candidate for controlling T2DM-mediated inflammation and secondary damage in vitro and in vivo. *Front. Pharmacol.* **2022**, *13*, 943879. [CrossRef]

52. Yang, Q.; Deng, X.; Lu, B.; Cameron, M.; Fearns, C.; Patricelli, M.P.; Yates, J.R., 3rd; Gray, N.S.; Lee, J.D. Pharmacological inhibition of BMK1 suppresses tumor growth through promyelocytic leukemia protein. *Cancer Cell.* **2010**, *18*, 258–267. [CrossRef] [PubMed]

53. Cook, S.J.; Tucker, J.A.; Lochhead, P.A. Small molecule ERK5 kinase inhibitors paradoxically activate ERK5 signalling: Be careful what you wish for. *Biochem. Soc. Trans.* **2020**, *48*, 1859–1875. [CrossRef]

54. Eferl, R.; Wagner, E.F. AP-1: A double-edged sword in tumorigenesis. *Nat. Rev. Cancer* **2003**, *3*, 859–868. [CrossRef]

55. Ozanne, B.W.; Spence, H.J.; McGarry, L.C.; Hennigan, R.F. Invasion is a genetic program regulated by transcription factors. *Curr. Opin. Genet. Dev.* **2006**, *16*, 65–70. [CrossRef] [PubMed]

56. Elkins, J.M.; Wang, J.; Deng, X.; Pattison, M.J.; Arthur, J.S.; Erazo, T.; Gomez, N.; Lizcano, J.M.; Gray, N.S.; Knapp, S. X-ray crystal structure of ERK5 (MAPK7) in complex with a specific inhibitor. *J. Med. Chem.* **2013**, *56*, 4413–4421. [CrossRef]

57. Schweppe, R.E.; Cheung, T.H.; Ahn, N.G. Global gene expression analysis of ERK5 and ERK1/2 signaling reveals a role for HIF-1 in ERK5-mediated responses. *J. Biol. Chem.* **2006**, *281*, 20993–21003. [CrossRef]

58. Yang, X.; Zhong, D.; Gao, W.; Liao, Z.; Chen, Y.; Zhang, S.; Zhou, H.; Su, P.; Xu, C. Conditional ablation of MAPK7 expression in chondrocytes impairs endochondral bone formation in limbs and adaptation of chondrocytes to hypoxia. *Cell Biosci.* **2020**, *10*, 103. [CrossRef]

59. Tubita, A.; Lombardi, Z.; Tusa, I.; Lazzeretti, A.; Sgrignani, G.; Papini, D.; Menconi, A.; Gagliardi, S.; Lulli, M.; Dello Sbarba, P.; et al. Inhibition of ERK5 elicits cellular senescence in melanoma via the cyclin-dependent kinase inhibitor p21. *Cancer Res.* **2022**, *82*, 447–457. [CrossRef]

60. Kelman, Z. PCNA: Structure, functions and interactions. *Oncogene* **1997**, *14*, 629–640. [CrossRef] [PubMed]

61. Sofer, A.; Lei, K.; Johannessen, C.M.; Ellisen, L.W. Regulation of mTOR and cell growth in response to energy stress by REDD1. *Mol. Cell. Biol.* **2005**, *25*, 5834–5845. [CrossRef]

62. Wang, L.; Kong, W.; Liu, B.; Zhang, X. Proliferating cell nuclear antigen promotes cell proliferation and tumorigenesis by up-regulating STAT3 in non-small cell lung cancer. *Biomed. Pharmacother.* **2018**, *104*, 595–602. [CrossRef]

63. Lu, E.M.; Ratnayake, J.; Rich, A.M. Assessment of proliferating cell nuclear antigen (PCNA) expression at the invading front of oral squamous cell carcinoma. *BMC Oral. Health* **2019**, *19*, 233. [CrossRef] [PubMed]

64. Chang, S.C.; Gopal, P.; Lim, S.; Wei, X.; Chandramohan, A.; Mangadu, R.; Smith, J.; Ng, S.; Gindy, M.; Phan, U.; et al. Targeted degradation of PCNA outperforms stoichiometric inhibition to result in programed cell death. *Cell Chem. Biol.* **2022**, *29*, 1601–1615.e7. [CrossRef] [PubMed]

65. Zammarchi, F.; Morelli, M.; Menicagli, M.; Di Cristofano, C.; Zavaglia, K.; Paolucci, A.; Campani, D.; Aretini, P.; Boggi, U.; Mosca, F.; et al. KLF4 is a novel candidate tumor suppressor gene in pancreatic ductal carcinoma. *Am. J. Clin. Pathol.* **2011**, *178*, 361–372. [CrossRef] [PubMed]

66. Zhao, W.; Hisamuddin, I.M.; Nandan, M.O.; Babbin, B.A.; Lamb, N.E.; Yang, V.W. Identification of *Krüppel-like factor 4* as a potential tumor suppressor gene in colorectal cancer. *Oncogene* **2004**, *23*, 395–402. [CrossRef]

67. Hu, W.; Hofstetter, W.L.; Li, H.; Zhou, Y.; He, Y.; Pataer, A.; Wang, L.; Xie, K.; Swisher, S.G.; Fang, B. Putative tumor-suppressive function of *Krüppel-like factor 4* in primary lung carcinoma. *Clin. Cancer Res.* **2009**, *15*, 5688–5695. [CrossRef]

68. Yang, W.T.; Zheng, P.S. Promoter hypermethylation of KLF4 inactivates its tumor suppressor function in cervical carcinogenesis. *PLoS ONE* **2014**, *9*, e88827. [CrossRef]

69. Xiong, G.; Xu, R. Retinoid orphan nuclear receptor alpha (RORα) suppresses the epithelial-mesenchymal transition (EMT) by directly repressing Snail transcription. *J. Biol. Chem.* **2022**, *298*, 102059. [CrossRef]

70. Du, J.; Xu, R. RORα, a potential tumor suppressor and therapeutic target of breast cancer. *Int. J. Mol. Sci.* **2012**, *13*, 15755–15766. [CrossRef]

71. Ramaswamy, S.; Ross, K.N.; Lander, E.S.; Golub, T.R. A molecular signature of metastasis in primary solid tumors. *Nat. Genet.* **2003**, *33*, 49–54. [CrossRef]

72. Alexopoulou, A.N.; Leao, M.; Caballero, O.L.; Da Silva, L.; Reid, L.; Lakhani, S.R.; Simpson, A.J.; Marshall, J.F.; Neville, A.M.; Jat, P.S. Dissecting the transcriptional networks underlying breast cancer: NR4A1 reduces the migration of normal and breast cancer cell lines. *Breast Cancer Res.* **2010**, *12*, R51. [CrossRef] [PubMed]

73. Sureban, S.M.; May, R.; Weygant, N.; Qu, D.; Chandrakesan, P.; Bannerman-Menson, E.; Ali, N.; Pantazis, P.; Westphalen, C.B.; Wang, T.C.; et al. XMD8-92 inhibits pancreatic tumor xenograft growth via a DCLK1-dependent mechanism. *Cancer Lett.* **2014**, *351*, 151–161. [CrossRef] [PubMed]

74. Mendoza, M.C.; Er, E.E.; Blenis, J. The Ras-ERK and PI3K-mTOR pathways: Cross-talk and compensation. *Trends Biochem. Sci.* **2011**, *36*, 320–328. [CrossRef] [PubMed]

75. Wang, G.; Zhao, Y.; Liu, Y.; Sun, D.; Zhen, Y.; Liu, J.; Fu, L.; Zhang, L.; Ouyang, L. Discovery of a novel dual-target inhibitor of ERK1 and ERK5 that induces regulated cell death to overcome compensatory mechanism in specific tumor types. *J. Med. Chem.* **2020**, *63*, 3976–3995. [CrossRef]

76. Wang, Z.Q.; Chen, X.C.; Yang, G.Y.; Zhou, L.F. U0126 prevents ERK pathway phosphorylation and interleukin-1beta mRNA production after cerebral ischemia. *Chin. Med. Sci. J.* **2004**, *19*, 270–275.
77. Tubita, A.; Tusa, I.; Rovida, E. Playing the Whack-A-Mole game: ERK5 activation emerges among the resistance mechanisms to RAF-MEK1/2-ERK1/2- targeted therapy. *Front. Cell Dev. Biol.* **2021**, *9*, 647311. [CrossRef]
78. de Jong, P.R.; Taniguchi, K.; Harris, A.R.; Bertin, S.; Takahashi, N.; Duong, J.; Campos, A.D.; Powis, G.; Corr, M.; Karin, M.; et al. ERK5 signalling rescues intestinal epithelial turnover and tumour cell proliferation upon ERK1/2 abrogation. *Nat. Commun.* **2016**, *7*, 11551. [CrossRef]
79. Cook, S.J.; Lochhead, P.A. ERK5 signalling and resistance to ERK1/2 pathway therapeutics: The path less travelled? *Front. Cell Dev. Biol.* **2022**, *10*, 839997. [CrossRef]

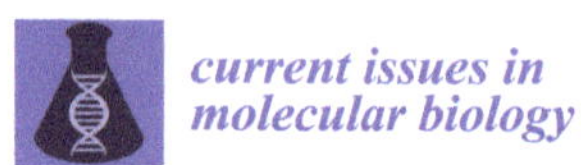
current issues in molecular biology

Article

T-Lymphocytes Activated by Dendritic Cells Loaded by Tumor-Derived Vesicles Decrease Viability of Melanoma Cells In Vitro

Ivan Yurevich Filin, Yuriy Pavlovich Mayasin, Chulpan Bulatovna Kharisova, Anna Valerevna Gorodilova, Daria Sergeevna Chulpanova, Kristina Viktorovna Kitaeva, Albert Anatolyevich Rizvanov * and Valeria Vladimirovna Solovyeva

Institute of Fundamental Medicine and Biology, Kazan Federal University, 420008 Kazan, Russia; ivyfilin@kpfu.ru (I.Y.F.); mayasin_yuriy@mail.ru (Y.P.M.); harisovachulpan@gmail.com (C.B.K.); anagorodilova@yandex.ru (A.V.G.); daschulpanova@kpfu.ru (D.S.C.); krvkitaeva@kpfu.ru (K.V.K.); vavsoloveva@kpfu.ru (V.V.S.)
* Correspondence: rizvanov@gmail.com; Tel.: +7-905-316-7599

Abstract: Immunotherapy represents an innovative approach to cancer treatment, based on activating the body's own immune system to combat tumor cells. Among various immunotherapy strategies, dendritic cell vaccines hold a special place due to their ability to activate T-lymphocytes, key players in cellular immunity, and direct them to tumor cells. In this study, the influence of dendritic cells processed with tumor-derived vesicles on the viability of melanoma cells in vitro was investigated. Dendritic cells were loaded with tumor-derived vesicles, after which they were used to activate T-cells. The study demonstrated that such modified T-cells exhibit high activity against melanoma cells, leading to a decrease in their viability. Our analysis highlights the potential efficacy of this approach in developing immunotherapy against melanoma. These results provide new prospects for further research and the development of antitumor strategies based on the mechanisms of T-lymphocyte activation using tumor-derived vesicles.

Keywords: dendritic cells; cytochalasin B; induced vesicles; antitumor vaccine; immunotherapy; GM-CSF; antigen presentation; cytotoxic effect

check for
updates

Citation: Filin, I.Y.; Mayasin, Y.P.; Kharisova, C.B.; Gorodilova, A.V.; Chulpanova, D.S.; Kitaeva, K.V.; Rizvanov, A.A.; Solovyeva, V.V. T-Lymphocytes Activated by Dendritic Cells Loaded by Tumor-Derived Vesicles Decrease Viability of Melanoma Cells In Vitro. *Curr. Issues Mol. Biol.* **2023**, *45*, 7827–7841. https://doi.org/10.3390/cimb45100493

Academic Editor: Dumitru A. Iacobas

Received: 18 August 2023
Revised: 15 September 2023
Accepted: 22 September 2023
Published: 26 September 2023

1. Introduction

Cancer cells within the body possess the ability to evade the antitumor response of the immune system due to angiogenesis, fibroblasts, and suppressor cytokines that attract corresponding immune cells to the tumor area, thereby collectively forming the tumor microenvironment (TME) [1]. TME plays a fundamental role in the tumor escaping from the immune system. The presence of cells in the TME, such as suppressor cells of myeloid origin, tumor-associated macrophages, and normal regulatory T-cells suppress the activity of infiltrating cytotoxic T-lymphocytes (CTLs), which are the main effector cells of the immune system. Immunosuppressive cytokines, such as interleukin (IL)-6, IL-10, transforming growth factor (TGF)-β, and indoleamine 2,3-dioxygenase (IDO), lead to immune tolerance, increasing the number of suppressor immune cells while reducing the number of CTLs [2]. Fibroblasts, in turn, are recruited and reprogrammed by tumor cells, after which they are able to produce a variety of growth signals thereby affecting tumor growth, invasion, and metastasis [3].

Combining traditional cancer treatment methods, including surgery, chemotherapy, and radiation therapy, with immunotherapy by specifically targeting the tumor and its TME can improve treatment response rates. Immunotherapy has several advantages due to its activation of the body's specific anti-tumor response. A therapeutic vaccine based on dendritic cells (DCs) may be a method that enhances the mechanism of a specific immune response. DCs have a crucial ability in that they cause the immune system to direct a specific immune response by internalizing foreign antigens and presenting them to CTLs. DCs are potent professional antigen-presenting cells due to the presence of the major histocompatibility complex (MHC)-II and costimulatory molecules (CD80/83/86) on their surface. Over the past two decades, DC vaccines have been shown in numerous clinical trials to be a safe therapy that can induce antitumor immunity. The most commonly observed adverse effects (AEs) include anemia, back pain, chills, fatigue, fever, headache, and nausea, generally not exceeding the 1st or 2nd grade and usually resolving within 1–2 days after vaccine administration [4]. There is currently an FDA-approved therapeutic vaccine, Sipuleucel-T, for the treatment of metastatic prostate cancer. This vaccine, based on DCs loaded with a recombinant hybrid protein (fusion of prostatic acid phosphatase (PAP) and granulocyte macrophage colony-stimulating factor (GM-CSF)), demonstrated a median improvement in overall survival of 4.1 months compared to the placebo group among men with metastatic castration-resistant prostate cancer [4]. The loading of DCs with tumor antigen is a critical component for an effective vaccine. There are several approaches to improve the efficiency of antigen presentation in order to induce a more potent immune response directed against tumor cells. The use of tumor-specific antigens (TSAs) is preferred for a more effective targeted antitumor response using unique tumor antigens. Autologous tumor mRNA, which can be delivered to DCs by electroporation [5] or viral vector [6], can be used as an antigen. However, isolating TSAs or utilizing their mRNA is energy and resource-intensive work. Moreover, TSAs for some tumors have not yet been identified. Tumor lysate is a reliable source of tumor antigens, especially for tumors whose TSAs have not been identified. In addition, the isolation of tumor lysates is a rapid and inexpensive method that can be personalized. Apart from tumor lysates, autologous tumor-derived vesicles can also serve as personalized sources of tumor antigens. Extracellular vesicles (EVs) are membrane-bound structures containing proteins and lipids as well as nuclear and mitochondrial components. EVs are able to fuse with recipient cells through endocytosis. EVs can be used to present antigens to immune system cells. However, natural EVs are not produced in sufficient amounts by tumor cells. There are alternative ways to produce EVs to overcome this problem; for example, cytochalasin B can be used. Cytochalasin B is a substance that causes disorganization of the cell cytoskeleton by preventing polymerization of its cellular structures, particularly actin filaments. Active shaking of cells treated with cytochalasin B leads to cell disintegration and formation of a large number of vesicles of different sizes. The use of cytochalasin B-induced membrane vesicles (CIMVs) may represent a novel approach of tumor antigen presentation to DCs. Furthermore, parental tumor cells from which CIMVs are derived can be modified with the cytokine gene *GM-CSF* for overexpression of the corresponding protein and subsequent immunomodulatory effects. To analyze the antigen-presenting and cytotoxic activity of immune cells, we used native CIMVs and CIMVs carrying GM-CSF derived from immortalized human melanoma M14 cells as tumor antigens to load monocyte-derived DCs (moDCs). We also loaded moDCs with CIMVs derived from triple-negative MDA-MB 231 breast cancer to compare the specific cytotoxic activity of T-cells after antigen presentation to moDCs. Also, we analyzed the cytotoxic effect of activated T-cells on human mesenchymal stem cells (MSCs).

2. Materials and Methods

2.1. Cultivation Conditions of Primary and Immortalized Human Cells

Mononuclear cells were isolated from human peripheral blood (PBMCs) of a healthy donor in a Ficoll density gradient (1.077 g/cm^3, Cat. #P052p, PanEco, Moscow, Russia) in accordance with approved ethical standards and current legislation (protocol approved by the Biomedical Ethics Committee of Kazan Federal University (#3 dated 23 March 2017)) Informed consent was obtained from a healthy donor. The obtained blood was diluted in Dulbecco's phosphate-buffered saline (DPBS) (Cat. #P060, PanEco, Moscow, Russia) (1:1) and then layered on Ficoll (1:1). Subsequently, centrifugation was performed at 1900 rpm for 20 min at room temperature. Mononuclear cells were collected, washed twice with

PBS, resuspended in complete RPMI-1640 medium (Catalog #C330p, PanEco, Moscow, Russia) that contained 10% fetal bovine serum (FBS) (SH30071.03, HyClone, Logan, UT, USA), 2 mM L-glutamine (Catalog #F032, PanEco, Moscow, Russia), and 5000 µg/mL penicillin-streptomycin mixture (Catalog #A063p, PanEco, Moscow, Russia). Samples were then cultured in a humid atmosphere at 37 °C and 5% CO_2. MSCs from human adipose tissue were isolated using a standard protocol [7].

The melanoma cell line M14 was obtained from the John Wayne Cancer Institute (JWCI) cell repository. Cells were cultured in a humid atmosphere at 37 °C and 5% CO_2 in complete RPMI 1640 medium (Catalog #C330p, PanEco, Moscow, Russia). Human embryonic kidney cells HEK293T (ATCC #CRL-3216) and human triple-negative breast cancer cells MDA-MB-231 (ATCC #HTB-26) were obtained from the American Type Culture Collection (ATCC, Manassas, VI, USA). Cells were cultured in a humid atmosphere at 37 °C and 5% CO_2 in complete DMEM with 4500 mg/l glucose (Catalog #C420p, PanEco, Russia). All cell lines were tested for the presence of mycoplasma.

2.2. Lentivirus Production

The donor plasmid encoding the human *GM-CSF* gene (CSF2/pENTR223) was obtained from the Harvard Plasmid Database (HsCD00365931). The *GM-CSF* gene was then recloned from the donor vector pENTR223 into the lentiviral expression vector pLX303 (AddGene #25897) using LR recombination (Catalog #11791020, Gateway™ LR Clonase™ II Enzyme mix, ThermoFisher Scientific, Waltham, MA, USA). Production of second-generation recombinant lentiviruses encoding *GM-CSF* was achieved by co-transfecting the packaging cell line HEK293T with three plasmids: the vector plasmid encoding *GM-CSF*, the packaging plasmid (psPAX2, AddGene #12260), and the envelope plasmid (pCMV-VSV-G, AddGene #8454) using calcium phosphate transfection. The culture medium was replaced with fresh medium 18 h later, followed by collection of the lentivirus-containing supernatant every 12 h for 48 h. The supernatant was stored at 4 °C during collection and then filtered through a 0.2 µm membrane (GVS filter technology, Morecambe, UK). Lentiviral particles were concentrated by ultracentrifugation for 2 h at 26,000 rpm at 4 °C.

2.3. Genetic Modification and Selection

The M14 cell line was transduced with the *GM-CSF* gene encoded by lentivirus in serum-free medium using 10 µg/mL protamine sulfate (#P4020, Sigma, St. Louis, MO, USA). Cells were seeded at 3×10^4 cells/well in a 6-well plate and incubated with lentivirus for 6 h. Transduced cells were selected using 7.5 µg/mL blasticidin S-HCl (#R21001, Gibco, Billings, MT, USA) for 7 days. Cell culture was tested for mycoplasma and showed negative results.

2.4. Quantitative Polymerase Chain Reaction (qPCR)

Native and modified cell samples were resuspended in 200 µL TRIzol™ reagent (Catalog #15596026, Invitrogen, Waltham, MA, USA), followed by total RNA extraction according to the manufacturer's recommendations. The cDNA synthesis was performed using GoScript Reverse Transcription System kit (Promega, Madison, WI, USA) according to the manufacturer's instructions. Nucleotide sequences of primers and fluorescent probes for 18S ribosomal RNA and *GM-CSF* genes were selected using GenScript Online Real-time PCR (TaqMan) Primer Design Tool (GenScript, Piscataway, NJ, USA). Primers and probes were synthesized by Litech (Moscow, Russia) (Table 1). Real-time polymerase chain reaction (qPCR) was performed according to TaqMan technology using MicroAmp 96-well plates (Bio-Rad, Hercules, CA, USA). For the amplification reaction, 1 µL of cDNA, 0.3 µL of primer and sample mix (at a final concentration of 300 nM), 4.7 µL of MilliQ H_2O, and 4 µL of 10x TaqMan buffer (Eurogen, Russia) were used. The final reaction volume was 10 µL. Amplification was performed on a CFX96 Touch™ Real-Time PCR Detection System (Bio-Rad, USA) using the following temperature profile: denaturation at 95 °C for 1 min;

44 cycles including denaturation at 95 °C for 30 s; primer annealing at 55 °C for 30 s; and elongation at 72 °C for 1 min.

Table 1. Nucleotide Sequences of Primers and Fluorescent Probes.

Gene	Forward Primer (5′–3′)	Reverse Primer (5′–3′)
18S	GCCGCTAGAGGTGAAATTCTTG	CATTCTTGGCAAATGCTTTCG
GM-CSF	GCGTCTCCTGAACCTGAGTA	CCCTGCTTGTACAGCTCCAG

2.5. Western Blot Analysis

Samples of native and modified cells (1×10^6 cells) were lysed using RIPA buffer (#89900, ThermoFisher Scientific, USA) containing a protease and phosphatase inhibitor mixture (#78444, ThermoFisher Scientific, USA). Protein concentration was determined using the Pierce™ BCA Protein Assay Kit (#23225, ThermoFisher Scientific, USA). Equal amounts of protein were denatured on a dry bath at 95 °C for 5 min, loaded and separated on 12% SDS-PAGE gels, and then transferred to PVDF membranes. The membranes were preblocked in 5% BCA solution (ThermoFisher Scientific, USA) and then incubated with primary antibodies at 4 °C overnight, followed by washing with Phosphate Buffered Saline with Tween™ 20 (#P1379, Sigma, USA) (PBS-T) and incubation with secondary antibodies at room temperature for 2 h. The membranes were then washed with PBS-T, and the target protein band was visualized using HRP (BioRad, Hercules, CA, USA) and analyzed using the ChemiDoc XRS+ system (BioRad Laboratories, Irvine, CA, USA). Primary rabbit polyclonal antibodies to GM-CSF (1:500) were purchased from Abcam, Cambridge, UK (ab300495). Secondary goat anti-rabbit IgG H&L (Alkaline Phosphatase) antibodies (1:2000) were purchased from Abcam, UK (ab97048).

2.6. Immunocytochemistry

Immunocytochemical analysis was performed to analyze GM-CSF protein synthesis in native M14 and M14-GM-CSF cells. Cells were seeded in 24-well plates (1×10^3 cells/well) on coverslips in complete RPMI-1640 medium. On the next day, the culture medium was aspirated, and cells were fixed with chilled methanol for 10 min at room temperature. After fixation, cells were washed with Tris-buffered saline (TBS) three times for 5 min each. The cells were then incubated with primary antibodies against GM-CSF (Cat. #ab54429, Abcam, Cambridge, UK, dilution 1:100 in TBS) for 1 h at room temperature. The stained cells were washed three times for 5 min each with TBS to remove unbound antibodies, and then incubated with secondary antibodies (Goat anti-Rabbit IgG (P&L) Fluorescein, Catalog #A102FN, American Qualex, San Clemente, CA, USA; dilution 1:1000 in TBS) for 1 h at room temperature, followed by washing. DAPI fluorescent dye (4′,6-diamidino-2-phenylindole; 1:50,000 dilution in TBS; Cat. #D1306, Invitrogen, USA) was used to stain the nucleus, then incubated for 5 min at room temperature, followed by washing. The coverslips were mounted on slides using mounting medium (ImmunoHistoMount, Catalog #sc-45086, Santa Cruz Biotechnology, Dallas, TX, USA). The analysis was conducted using a confocal microscope LSM 780 (Carl Zeiss, Jena, Germany) and ZEN (black edition) 2.3 software (Carl Zeiss, Germany).

2.7. Isolation of Cytochalasin-Induced Membrane Vesicles (CIMVs)

Cells were washed from FBS with PBS (Catalog #P060, PanEco, Russia). After counting, 2×10^6 cells were suspended in 10 mL serum-free medium containing cytochalasin B from Drechslera dematioidea (#C6762-5mg, Sigma-Aldrich, St. Louis, MO, USA) at a concentration of 10 µg/mL. Cells were incubated in a humid atmosphere for 30 min at 37 °C and 5% CO_2 with gentle shaking every 10 min. After incubation, cells were vortexed for 1 min, then centrifuged for 10 min at 700 rpm (Biosan LMC-3000 laboratory centrifuge). The supernatant was collected and centrifuged for 10 min at 1400 rpm; then, the supernatant was collected and filtered through a polyvinylidene difluoride (PVDF) membrane filter

(GVS filter technology, UK) with a pore size of 1 micron. After filtration, the supernatant was centrifuged for 30 min at 12,000 rpm. The resulting pellet containing CIMVs was collected. Protein concentration was determined using the Pierce BCA Protein Assay Kit (#23225, ThermoFisher Scientific, USA) according to the manufacturer's instructions.

2.8. Size Characterization of Cytochalasin-Induced Membrane Vesicles (CIMVs)

To determine the size of CIMVs, scanning electron microscopy was used. Isolated CIMVs, as described earlier, were deposited onto glass slides in the wells of a 96-well plate by centrifugation at 3000 rpm for 30 min at room temperature. The deposited CIMVs were fixed in 10% formalin for 15 min, dehydrated with ethanol increasing in concentration from 30% to absolute, and then air-dried for 24 h. Before visualization, the samples were coated with palladium using the Quorum T150ES sputter coater (Quorum Technologies Ltd., Lewes, UK), and analysis was performed using the SEM Merlin (Carl Zeiss, Germany). Additionally, flow cytometry (BD FACSAria III, BD Biosciences, San Jose, CA, USA) was used to measure the size of CIMVs using a mixture of calibration particles (0.22–0.45–0.88–1.34 µm) (Spherotech, Lake Forest, IL, USA). At least 50,000 events were acquired for each sample.

2.9. Assessment of Nuclear, Membrane, and Mitochondrial Components in CIMVs

To analyze the presence of nuclear, membrane, and mitochondrial components in CIMVs, native M14, M14-GM-CSF, and MDA-MB 231 cells were trypsinized and washed in PBS twice. Next, 5×10^5 cells were suspended in 1 mL of PBS and stained using 1 µL of DiO vital dye (Vybrant Multicolor Cell-Labeling Kit, Invitrogen, Waltham, MA, USA) for 10 min at room temperature and in the absence of light. Subsequently, the cells were washed twice and resuspended in 500 µL of PBS, followed by staining with 2 µL of 300 nM MitoTracker Red FM (Molecular probes, Invitrogen, Waltham, MA, USA) for 15 min at 37 °C. Afterward, the cells were again washed twice with PBS, and CIMVs were isolated as described earlier. These isolated CIMVs were fixed in cold methanol for 5 min at room temperature, then resuspended in 500 µL of PBS and stained using 0.5 µL of 7-aminoactinomycin D (7AAD) (Molecular probes, Invitrogen, Waltham, MA, USA) for 10 min at room temperature in the absence of light. The isolated CIMVs were analyzed using FACS Aria III (BD Biosciences, USA) and BD FACSDiva™ software version 7.0.

2.10. Assessment of Surface and Intracellular Markers in CIMVs

To analyze the expression of the cluster of differentiation (CD) markers typical for tumor cells on the surface of the CIMVs, the following antibodies were used: PE/Cyanine7 anti-human CD81 (#349512, Biolegend, San Diego, CA, USA), Alexa Fluor® 488 anti-human Hsp70 (#648004, Biolegend, USA), PerCP/Cyanine5.5 anti-human CD63 (#353020, Biolegend, USA), PE anti-human TSG101 (ab209927, abcam, USA), Alexa Fluor® 594 anti-human Calnexin (ab203439, abcam, USA). Briefly, CIMVs were isolated from 2×10^5 native M14, M14-GM-CSF, and MDA-MB 231 cells that were pre-fixed in cold methanol for 5 min at room temperature, washed once in PBS and stained with antibodies for 30 min at room temperature in the absence of light. Then, CIMVs were washed with PBS and analyzed using the FACSAria III flow cytometer (BD Biosciences, San Jose, CA, USA) and BD FACSDiva™ software version 7.0.

2.11. Generation of Dendritic Cells from Monocytes

Monocytes were isolated from healthy donor PBMCs by adhesion. They were seeded at a density of 10 million cells per well and incubated in a humid atmosphere at 37 °C and 5% CO_2 for 1.5 h. After this, non-adherent cells were removed along with the medium, and fresh complete RPMI-1640 medium (Catalog #C330p, PanEco, Russia) which contained 10% FBS, 2 mM L-glutamine and 1% autologous plasma was added. The generation of moDCs was carried out in 6-well plates. On day 0, isolated monocytes (5×10^6 cells/well) were cultured in 3 mL of complete RPMI-1640 medium supplemented with 250 IU/mL

IL-4 (Catalog #PSG040-10, SCI-Store, Moscow, Russia), and 800 IU/mL GM-CSF (Catalog #PSG030-10, SCI-Store, Russia) at 37 °C and 5% CO_2. On days 2 and 5, 1.5 mL of the medium was collected, centrifuged, and the cell pellet was resuspended in 1.5 mL of complete medium supplemented with double concentration of IL-4 and GM-CSF. This mixture was returned to the original culture. On day 8, 1.5 mL of the medium was collected, centrifuged, and the cell pellet was resuspended in 1.5 mL of medium supplemented with 2000 IU/mL IL-6 (Catalog #PSG180-10, SCI-Store, Russia), 400 IU/mL IL-1β (RPA563Hu01, Cloud-Clone Corp., Katy, Houston, TX, USA), 2000 IU/mL TNF-α (Catalog #PSG250-10, SCI-Store, Russia), and 2 μg/mL prostaglandin E2 (PGE2) (#900117P-5MG, Sigma-Aldrich, USA). Additionally, 15 μg/mL of CIMVs were added to the culture. The cells were then cultured for an additional 48 h. On day 9, the viability, yield, and absolute cell count of mature moDCs were determined by flow cytometry using specific surface markers (see Section 2.10). Subsequently, mature moDCs were re-seeded in 6-well plates for antigen presentation with fresh PBMCs.

2.12. Immunophenotyping of Monocyte-Derived Dendritic Cells after Differentiation from Monocytes

Sensitive conjugated antibodies were used to determine the target population of mature moDCs. Cells were removed from the culture plate and washed from the medium with PBS. The moDCs were then stained with conjugated antibodies: Pacific Blue™ anti-human CD3 (#300330, Biolegend, USA), Pacific Blue™ anti-human CD19 (#302232, Biolegend, USA), Pacific Blue™ anti-human CD56 (#362520, Biolegend, USA), Pacific Blue™ anti-human CD20 (#302328, Biolegend, USA), PerCP/Cyanine5.5 anti-human CD11c (#301624, Biolegend, USA), FITC antibody to human HLA-DR (#307604, Biolegend, USA), APC anti-human CD64 (#305014, Biolegend, USA), PE anti-human CD80 (#305208, Biolegend, USA), PE anti-human CD83 (#305308, Biolegend, USA) according to the manufacturer's instructions. Data were analyzed using FACS Aria III (BD Biosciences, USA) and BD FACSDiva™ software version 7.0.

2.13. Analysis of the Interaction between Cytochalasin B-Induced Membrane Vesicles and Monocyte-Derived Dendritic Cells

To avoid loss of CIMVs during staining, M14 cells were pre-stained with CellTracker™ Green CMFDA dye (#C7025, Invitrogen, USA) according to the manufacturer's instructions. After staining, CIMVs were isolated according to the previously described protocol. The moDCs were stained with the Vybrant™ Multicolor Cell-Labeling Kit (#V-22889, Thermo Fisher Scientific, USA) using DiD spectrum according to the manufacturer's instructions and seeded in a 12-well culture plate. The moDCs were incubated with CIMVs for 3 h in a humidified atmosphere at 37 °C and 5% CO_2. The moDCs were settled by centrifugation at 1400 rpm for 5 min. The cells were then washed from the growth culture medium with PBS and deposited at the bottom of the wells of the culture plate onto a coverslip. At this point, it is important to ensure that the coverslip remains at the bottom of the well. This procedure was followed by incubation for 30 min at room temperature for deposition and attachment of moDCs to the coverslips. After cell attachment, PBS was carefully removed, and cells were fixed with 500 μL of 10% formalin for 10 min at room temperature. The fixed cells were first washed once with 500 μL of PBS for 5 min and then permeabilized with 500 μL of 0.5% Triton X-100 for 10 min. The permeabilized cells were washed once more with 500 μL PBS for 5 min. The nuclei were stained with DAPI (dilution 1:10,000) for 10 min, then washed twice with PBS for 5 min to remove excess dye. For sample preparation, coverslips were carefully mounted on microscope slides using aqueous mounting medium (ab128982, Abcam, UK). Samples were analyzed by confocal microscopy using a LSM 780 confocal microscope and ZEN (black edition) 2.3 software (Carl Zeiss, Germany) at the KFU Interdisciplinary Center for Analytical Microscopy.

2.14. Analysis of Surface Markers of Peripheral Blood Mononuclear Cells after Culturing with Activated Mature Monocyte-Derived Dendritic Cells

The isolated PBMCs were seeded in a 6-well culture plate with activated mature moDCs at a ratio of 1:30. Native PBMCs were used as controls. Flow cytometry analysis was performed after 72 h of incubation. Also, some of these PBMCs were used for tumor cell apoptosis analysis after 72 h of co-culture, respectively. Conjugated antibody staining was performed to detect activated immune cell populations. Cells were divided into panels and stained with sensitive conjugated antibodies: FITC anti-human CD8a (#300906, Biolegend, USA), APC anti-human CD4 (#357408, Biolegend, USA), PE/Cy7 anti-human CD38 (#356608, Biolegend, USA), PE anti-human HLA-DR (#307605, Biolegend, USA), Brilliant Violet 421™ anti-human CD107a (LAMP-1) (#328626, Biolegend, USA), FITC anti-human CD3 (#300306, Biolegend, USA), Pacific Blue™ anti-human CD4 (#317429, Biolegend, USA), PE anti-human CD127 (IL-7Rα) (#351304, Biolegend, USA), PE/Cyanine7 anti-human CD25 (#356108, Biolegend, USA), PE anti-human CD196/CCR6 (#353410, Biolegend, USA), APC anti-human CD183/CXCR3 (#353708, Biolegend, USA), PerCP/Cyanine5.5 anti-human CD56/NCAM (#362506, Biolegend, USA) according to the manufacturer's instructions. Data were analyzed using FACS Aria III (BD Biosciences, USA) and BD FACSDiva™ software version 7.0.

2.15. Assessment of Tumor Cell Viability

Native M14 melanoma cells were seeded in a 12-well culture plate (5×10^4 cells per well) preliminarily 24 h before the apoptosis assay. MSCs cells were also seeded in the same amount to test for nonspecific cytotoxicity. Then, activated PBMCs (2.5×10^5 cells per well) were added to M14 cells and MSCs. M14 and MSC cells were harvested and washed with DPBS after 24 h. APC Annexin V Apoptosis Detection Kit with PI (#640932, Biolegend, USA) was used for apoptosis assay according to the manufacturer's instructions. Data were analyzed using FACS Aria III (BD Biosciences, USA) and BD FACSDiva™ software version 7.0.

2.16. Statistical Analysis

GraphPad Prism 8 software (GraphPad Software, San Diego, CA, USA) was used for statistical analysis. One way ANOVA analysis was used to compare independent groups by quantitative characteristics. Differences between groups were considered statistically significant at $p < 0.05$, $p < 0.0001$.

3. Results

3.1. Production of Genetically Modified Melanoma Cells Overexpressing GM-CSF

Genetically modified human melanoma M14 cells were obtained by lentiviral transduction. Confirmation of GM-CSF expression was demonstrated by qPCR and Western blot. Elevated levels of GM-CSF gene transcripts (78.3 gene copies per cell, standard deviation (SD) $\pm$ 10.12) were demonstrated in M14-GM-CSF compared to the absence of such in native M14 cells. Membrane vesicles were isolated from both native and genetically modified M14 cells using cytochalasin B treatment. The presence of GM-CSF protein in cells and CIMVs was confirmed by Western blot analysis (Figure 1B). The data obtained indicate that modified M14 tumor cells and CIMVs isolated from them contain GM-CSF protein with a molecular mass of 14 kDa. This protein was absent in native M14 tumor cells and isolated CIMVs. The presence of GM-CSF protein in M14 cells was also confirmed by immunofluorescence analysis (Figure 1A).

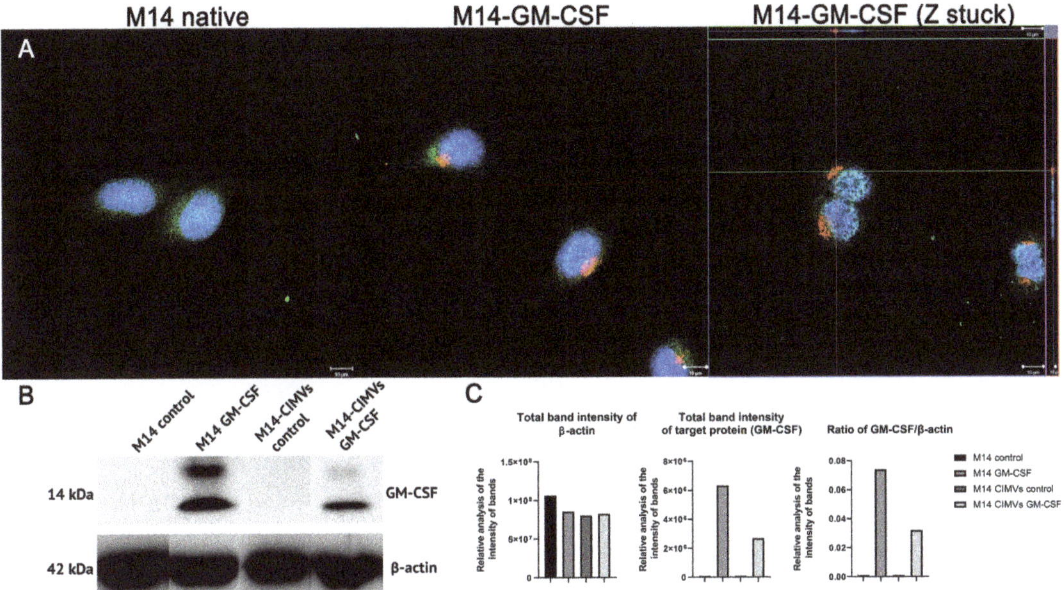

Figure 1. Characterization of native and genetically modified M14 melanoma cells. (**A**) Immunofluorescence analysis of GM-CSF synthesis in M14 (blue DAPI, green DIO, red monoclonal antibody (mAb) to GM-CSF) (Scale bar: 10 μm); (**B**) Western blot analysis of GM-CSF expression in M14 melanoma cells and CIMVs using β-actin as an internal control; (**C**) The relative analysis of the intensity of protein bands.

3.2. Preparation and Analysis of Induced Membrane Vesicles from Native and Genetically Modified Tumor Cells

Induced membrane vesicles were isolated from native and genetically modified M14 tumor cells as well as from MDA-MB 231 cells by treating the cells with cytochalasin B. It was shown that the size of isolated CIMVs from M14 averaged 240 nm (SD ± 70), from M14-GM-CSF averaged 250 nm (SD ± 50), and from MDA-MB 231 averaged 190 nm (SD ± 80) (Figure 2B).

Scanning electron microscopy (SEM) was used to examine the shape and size of CIMVs (Figure 2A). They had a characteristic spherical shape and varied in size. It is important to note that some deformation of vesicles was observed, which is a common phenomenon when preparing samples for SEM analysis and is not an artifact of the vesicle isolation process using cytochalasin B.

Furthermore, flow cytometry data indicated the presence of nuclear components in the following averages: M14 CIMVs—43.6% (SD ± 2.75), M14 CIMVs-GM-CSF—34.45% (SD ± 13.7), MDA-MB 231 CIMVs—39.4% (SD ± 9.5). Additionally, mitochondrial components were observed with the following averages: M14 CIMVs—66.8% (SD ± 5.15), M14 CIMVs-GM-CSF—51.5% (SD ± 17.1), MDA-MB 231 CIMVs—55.15% (SD ± 11.2). Cytoplasmic components were also present, with the following averages: M14 CIMVs—71.4% (SD ± 8.4), M14 CIMVs-GM-CSF—61% (SD ± 22.7), MDA-MB 231 CIMVs—59.5% (SD ± 15.22) (Figure 2C).

Upon staining vesicles for typical tumor biomarkers, the presence of surface receptors CD63, CD81, and TSG-101 was demonstrated. The averages for M14 CIMVs were 29% (SD ± 3), 40% (SD ± 3.95), and 9% (SD ± 0.46). The averages for M14 CIMVs-GM-CSF were 35.5% (SD ± 3), 52% (SD ± 2.76), and 8% (SD ± 0.78). And the averages for MDA-MB 231 CIMVs were 14.5% (SD ± 0.45), 17% (SD ± 0.7), and 3.5% (SD ± 0.6), respectively (Figure 2D). Additionally, the presence of intracellular receptors Calnexin, Hsp70, and TSG-101 was demonstrated. The averages for M14 CIMVs were 8.9% (SD ± 0.25),

0.77% (SD ± 0.06), and 7.8% (SD ± 0.3). The averages for M14 CIMVs-GM-CSF were 9.1% (SD ± 1.35), 0.73% (SD ± 0.67), and 2.77% (SD ± 1.7). And the averages for MDA-MB 231 CIMVs were 18.6% (SD ± 1.16), 3.47% (SD ± 0.75), and 23.2% (SD ± 0.92), respectively (Figure 2E).

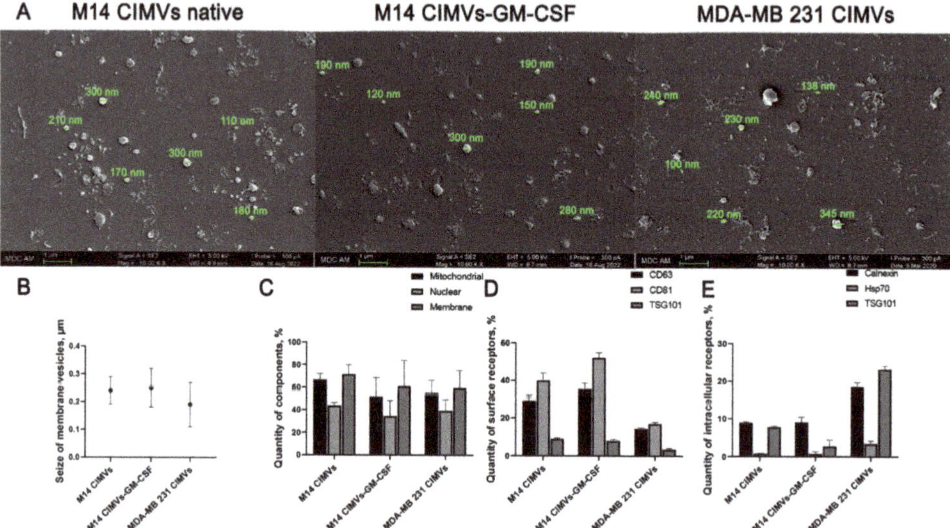

Figure 2. Analysis of the presence of intracellular components and the size of cytochalasin B-induced membrane vesicles isolated from native M14, M14-GM-CSF and MDA-MB 231. The size of induced membrane vesicles was determined by (**A**) scanning electron microscopy (scale bar: 1 μm) and (**B**) flow cytometry; (**C**) The presence of mitochondrial, nuclear and cytoplasmic components and the presence of (**D**) cell surface and (**E**) intracellular receptors in CIMVs were analyzed by flow cytometry.

3.3. Generation and Characterization of Dendritic Cells from the Population of Human Peripheral Blood Monocytes

As a result of targeted differentiation, rounded and dendritic morphology was observed in the moDCs generated (Figure 3).

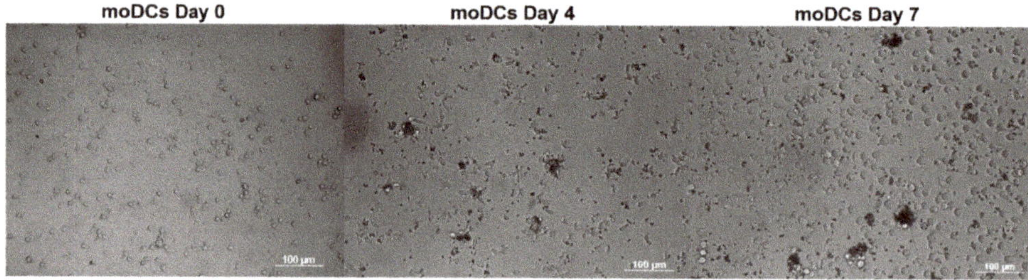

Figure 3. Mature monocyte-derived dendritic cells obtained after monocyte differentiation on the seventh day of cultivation (bright-field microscopy). Scale bar: 100 μm.

An immunophenotypic analysis was performed using flow cytometry with conjugated antibodies to confirm the phenotype of mature dendritic cells. As an example, the strategy for gating a population of moDCs from PBMCs used during flow cytometry is shown in Figure 4.

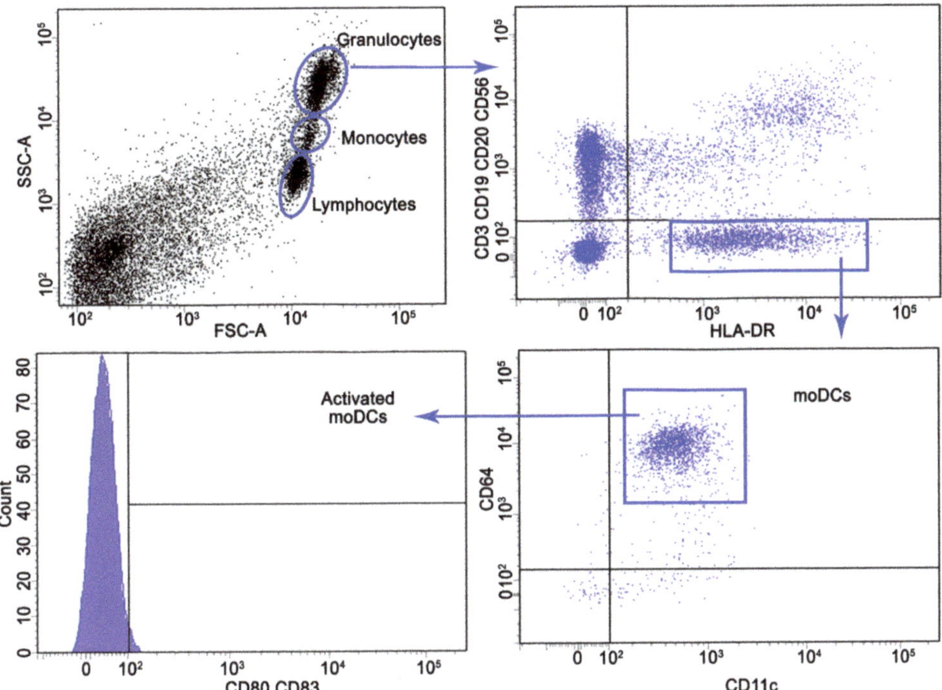

Figure 4. Gating strategy for the identification of CD3⁻CD19⁻CD20⁻CD56⁻HLA-DR⁺CD64⁺CD11c⁺ moDCs and activated moDCs (CD80⁺CD83⁺) population of PBMCs.

During the analysis of surface markers on mature moDCs, an increase of 65% in the population of moDCs differentiated from the monocyte population compared to control PBMCs was observed. Moreover, activated moDCs expressing CD80 and CD83 co-stimulatory domains loaded with M14 CIMVs, M14 CIMVs-GM-CSF and MDA-MB 231 CIMVs among differentiated moDCs were 56% (SD $\pm$ 7.24), 64% (SD $\pm$ 6.55) and 60% (SD $\pm$ 7.67), respectively (Figure 5A), while the number of activated moDCs among the population of DC in PBMCs culture was 38% (SD $\pm$ 3.59) (Figure 5A).

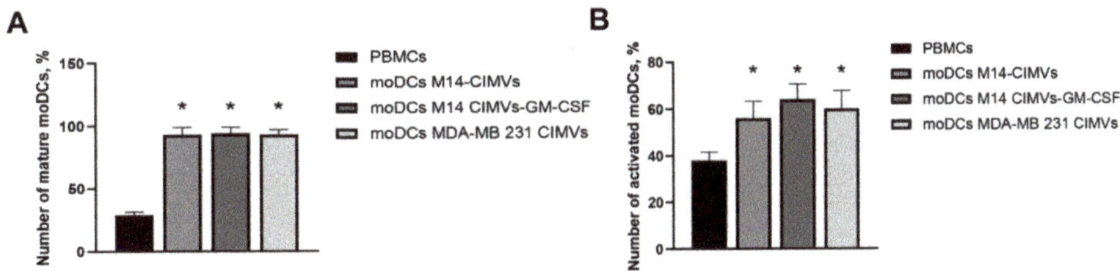

Figure 5. (**A**) Evaluation of differentiation efficiency of mature moDCs loaded with M14 CIMVs, M14 CIMVs-GM-CSF and MDA-MB 231 CIMVs; (**B**) Number of activated moDCs among the DC population. *—compared to the control group, * $p < 0.0001$.

3.4. Analysis of the Interaction between Induced Membrane Vesicles and Monocyte-Derived Dendritic Cells

To analyze the interaction of moDCs with tumor-derived CIMVs, moDCs prestained with vital dyes were added to the bottom of the wells of the culture plate for 3 h. The moDCs were then washed to remove any vesicles, seeded on coverslips, and prepared for confocal microscopy. As a result, a clear interaction between moDCs and CIMVs was demonstrated, characterized by the uptake of CIMVs by monocyte-derived dendritic cells through endocytosis or fusion processes (Figure 6).

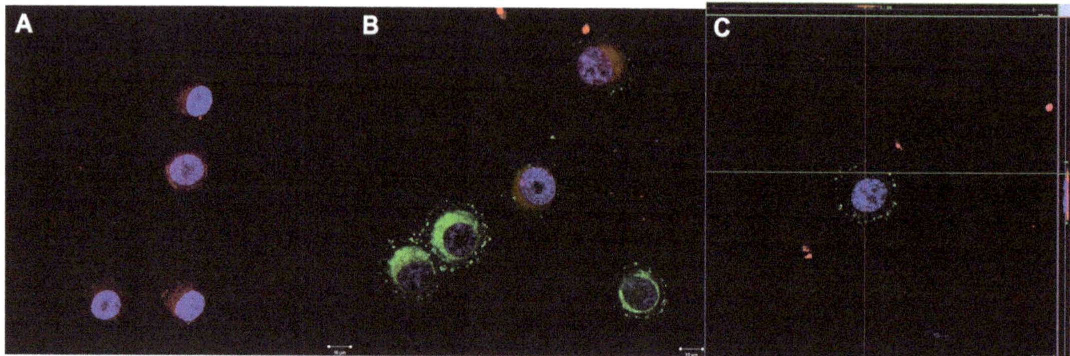

Figure 6. Confocal Microscopy (Scale 10 μm). (**A**) Control moDCs (Red DID, Blue DAPI); (**B**) moDCs (Red DID, Blue DAPI) after cultivation with CIMVs (Green DIO); (**C**) Z-stack moDCs (Red DID, Blue DAPI) after cultivation with CIMVs (Green DIO).

3.5. Analysis of Mononuclear Cell Activation and Cytotoxic Efficacy after Co-Culture with Activated Monocyte-Derived Dendritic Cells

An increase in the number of HLA-DR$^+$/CD38$^+$ cytotoxic T-lymphocytes (CTLs) was observed in all experimental samples (Figure 7A). However, co-culture of moDCs loaded with M14 CIMVs-GM-CSF with PBMCs led to a 29% increase (95% confidence interval (CI) of 25.03 to 32.57) in HLA-DR$^+$/CD38$^+$ CTLs compared to control PBMCs. Additionally, an increase in the number of T-helper 2 (Th2) cells was demonstrated in all experimental samples, showing a 10% increase (95% CI 6.87 to 11.39) compared to control cells (Figure 7B).

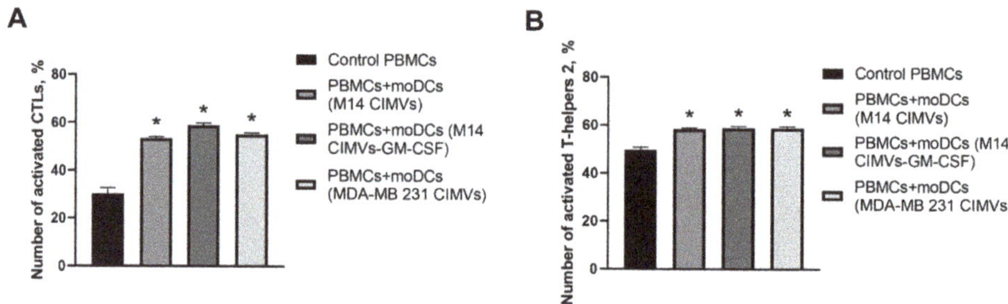

Figure 7. Comparative assessment of CTLs (**A**) and Th2 (**B**) activation by dendritic cells loaded with M14 CIMVs, M14 CIMVs-GM-CSF, and MDA-MB 231 CIMVs. *—compared to the control group, $p < 0.0001$.

To determine cytotoxic effectiveness towards M14 cells, unactivated PBMCs and PBMCs activated by DCs loaded with M14 CIMVs, M14 CIMVs-GM-CSF, and MDA-MB 231 CIMVs were added. Similarly, to assess nonspecific cytotoxicity, the same PBMCs were added to MSCs. PBMCs with tumor cells and MSCs were cultured for 24 h. Untreated M14 cells and MSCs were used as controls. It was observed that the presence of activated PBMCs (M14 CIMVs) and PBMCs (M14 CIMVs-GM-CSF) led to a reduction in the number of viable M14 cells by 10.5% (95% CI of 5.18 to 15.71) and 15.3% (95% CI of 10.07 to 20.60), respectively, compared to control cells. Furthermore, when PBMCs (MDA-MB 231 CIMVs) were added to M14 cells, a decrease in viable cell count was also observed by 14% (95%

CI of 7.52 to 20.65) compared to control M14 cells and by 11% (95% CI of 4.74 to 17.33) compared to coculture of M14 cells and unactivated PBMCs (Figure 8A). When unactivated and activated PBMCs were added to MSCs, no statistically significant differences were found compared to control MSCs (Figure 8B).

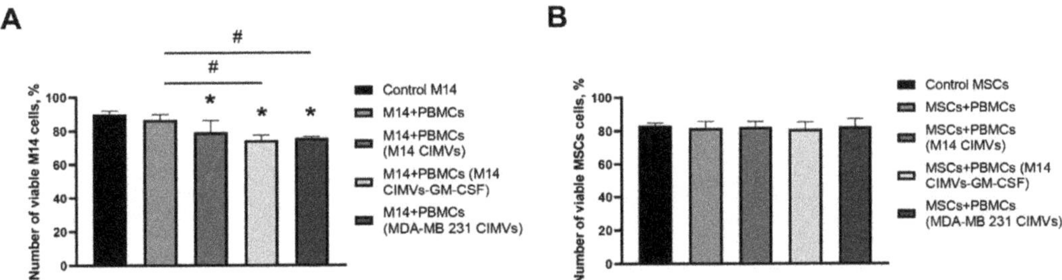

Figure 8. The number of viable M14 cells (**A**) and MSCs (**B**) after 24 h of cultivation with non-activated PBMCs or PBMCs activated by moDCs loaded with M14-CIMVs, M14-CIMVs-GM-CSF, and MDA-MB 231-CIMVs. *—compared to the control group, $p < 0.0001$; #—$p < 0.0001$.

4. Discussion

Under natural conditions, extracellular vesicles (EVs) serve as crucial transport mediators between cells, facilitating the transfer of various proteins, lipids, and nucleic acids. They can also regulate gene expression in recipient cells, as well as have internal effects on regulation of gene expression in parental cells [8]. Functioning as mediators of intercellular communication by carrying molecules from parent cells, EVs are considered promising tools for drug delivery and presentation of foreign antigens in cancer treatment [9]. However, obtaining a sufficient quantity of natural EVs for necessary research is a laborious task as these EVs are released by cells in limited amounts. In order to facilitate the isolation procedure and increase the yield of EVs, we treated tumor cells with cytochalasin B in our study.

Most commonly, the sizes of the membranous vesicles derived from tumor cells, including exosomes, range from 30 to 150 nm, while the size of the microvesicles ranges from 100 to 1000 nm [10–12]. In our study, the average size of isolated CIMVs from M14 cells was 240 nm, while CIMVs from M14-GM-CSF cells measured around 250 nm. The average size of the isolated CIMVs from MDA-MB 231 cells was approximately 190 nm. The CIMVs we investigated showed positive expression of typical EV biomarkers such as CD63, CD81, HSP70, TSG-101, and Calnexin. It is notable that MDA-MB 231 CIMVs had more intracellular markers than surface markers, while for M14 CIMVs, the pattern was reversed. The presence of these markers indicates that the studied CIMVs are tumorigenic [11,13]. Nuclear, mitochondrial, and lipid components were also analyzed. It's noted that tumor-derived exosomes contain a variety of genetic elements, such as DNA, mRNA, miRNA, proteins, and lipids [10]. The presence of mitochondrial components in CIMVs was, on average, from 51–66%, nuclear component was found in 34–44% of cases, and the membrane component of vesicles was from 59–71% of cases.

Dendritic cells were derived from human peripheral blood monocytes using growth factors such as GM-CSF and IL-4. A cytokine cocktail including PGE2, IL-1β, IL-6, and TNF-α was used to mature moDCs. This cytokine cocktail combining several protein factors leads to an improved immune response [2]. Immunophenotypic analysis showed high levels of moDC-specific receptor expression. These cells had a round dendritic morphology with reduced adhesion (Figure 3), which is consistent with data from the literature [14].

The protein GM-CSF was chosen as an immunomodulatory agent due to its ability to provide additional stimulation to moDCs. The moDCs activated in this way promote the differentiation of naive CD4$^+$ T-cells into T-helper subsets and also contribute to the development of CD8$^+$ T-cell responses [15–18]. Furthermore, this cytokine can serve as a potential immune adjuvant in inducing an antitumor response [18]. Thus, it was decided to use tumor cells overexpressing GM-CSF to isolate CIMVs. The presence of the protein in cells and vesicles was confirmed by immunofluorescence and Western blot analyses (Figure 1). In addition, CIMVs uptake by moDCs via endocytosis or fusion was shown by immunofluorescence analysis, which is correlated with literature data [19]. In our study, we found that a higher quantity of co-stimulatory molecules was present on the surface of moDCs loaded with CIMVs-GM-CSF, as well as an increase in the number of activated CTLs after antigen presentation by these moDCs. However, no statistically significant differences were observed among the experimental groups.

Cytotoxic CD8$^+$ T-cells play a critical role in anti-tumor immune responses [20]. Due to the ability of DCs to present tumor antigen to CD8$^+$ T-cells, tumor-derived CIMVs may be potential carriers of tumor-specific antigens. Tumor-derived exosomes from human melanoma cells promote the maturation of bone marrow derived DCs, resulting in the induction of proliferation of dendritic cell-processed T-cells [21]. By analyzing specific surface markers for CTLs and T-helpers, we observed that the culture of CD8$^+$ T-cells with moDCs loaded with M14 CIMVs, CIMVs-GM-CSF, or MDA-MB 231 CIMVs leads to a significant increase in activated CD38$^+$ HLA-DR$^+$ killer T-cells compared to the control group. Notably, a more significant increase in the number of CD38$^+$ HLA-DR$^+$ T-killers was in the group with moDCs loaded with M14 CIMVs-GM-CSF (2-fold). At the same time, an increase in Th2 number was observed in all experimental groups compared to the control (1.2 times). The transcriptional profile of DCs maturing under the influence of TNF-α/CD40L demonstrates that such DCs polarize T-cells towards a Th2 response [2]. It is known that a high expression of Th1 gene is associated with a favorable and non-recurrent outcome [22–25]. Meanwhile, a Th2 response has been linked to tumor immune evasion in mouse studies [26,27]. However, there are studies showing that the main role of Th2 cells in patient prognosis has not been confirmed [28]. It is known that the provision of an antitumor response by Th2 cells strongly depends on the type and stage of an individual tumor [29]. In particular, adoptive therapy using Th2 cells has already been shown to be effective in animal models of melanoma [30]. Moreover, there are studies showing that Th2 cells play an important role in inhibiting the progression of colon and pancreatic cancer in mice by recruiting eosinophils to tumors where they produce cytotoxic factors [31]. Thus, using the cytotoxic activity of eosinophils Th2 cells has great potential in the antitumor response [32].

In order to evaluate the efficiency of CD8$^+$ and CD4$^+$ T-cell activation, we performed cytotoxic activity assay. After 24 h of culturing activated PBMCs with M14 melanoma cells, we observed a decrease in the number of viable cells in all experimental samples compared to untreated M14 cells, except for the co-culture of M14 cells with unactivated PBMCs. Similar data indicate that human DCs loaded with apoptotic tumor cells and subsequently matured efficiently generated T cell-mediated antitumor responses in vitro [33]. Interestingly, PBMCs activated by moDCs-MDA-MB 231-CIMVs exhibited similar cytotoxic activity against M14 melanoma cells as PBMCs activated by moDCs-M14-CIMVs and moDCs-M14-CIMVs-GM-CSF. These findings suggest that CIMVs obtained from breast cancer and melanoma cells likely carry certain tumor-associated antigens (TAAs). It is also known that tumor-specific CTL clones can recognize breast cancer and melanoma cells [34]. Moreover, various members of the melanoma antigen (MAGE)-A and -C gene subfamilies are typically aberrantly activated in breast cancer [35]. However, co-culture of activated PBMCs with MSCs for 24 h showed no significant difference compared to untreated MSCs, indicating the absence of a nonspecific cytotoxic response. Thus, these results suggest the ability of moDCs loaded with tumor-derived CIMVs to activate CD8$^+$ T-killers, and the generated T-killers are able to kill human melanoma tumor cells in vitro.

5. Conclusions

In conclusion, this study demonstrates the potential of dendritic cell-based therapies using tumor-specific CIMVs to activate CD8$^+$ cytotoxic T-lymphocytes and initiate effective anti-tumor responses. These results reveal new evidence of complex interactions between immune cell populations and tumors, providing new potential opportunities to develop novel and targeted immunotherapeutic approaches to more effectively combat cancer. Furthermore, research is needed to fully explore the potential of these immune responses and develop personalized and efficient methods for cancer treatment.

Author Contributions: Conceptualization, I.Y.F., K.V.K. and V.V.S.; methodology, I.Y.F., K.V.K. and V.V.S.; formal analysis, Y.P.M., A.V.G. and C.B.K.; investigation, D.S.C. and I.Y.F. (flow cytometry analysis) and D.S.C. (confocal microscopy); writing—original draft preparation, I.Y.F.; writing—review and editing, V.V.S.; visualization, I.Y.F.; supervision, A.A.R. All authors have read and agreed to the published version of the manuscript.

Funding: This research was funded by the Russian Science Foundation grant No. 22-24-20018.

Institutional Review Board Statement: All manipulations were carried out in accordance with approved ethical standards and current legislation (the protocol was approved by the Committee on Biomedical Ethics of Kazan Federal University (No. 3, 23 March 2017).

Informed Consent Statement: An informed consent was obtained from all study participants.

Data Availability Statement: Authors can confirm that all relevant data used to support the findings of this study are including within the article.

Acknowledgments: This work is part of Kazan Federal University Strategic Academic Leadership Program (PRIORITY-2030).

Conflicts of Interest: The authors declare no conflict of interest.

References

1. Wang, Q.; Shao, X.; Zhang, Y.; Zhu, M.; Wang, F.X.; Mu, J.; Li, J.; Yao, H.; Chen, K. Role of Tumor Microenvironment in Cancer Progression and Therapeutic Strategy. *Cancer Med.* **2023**, *12*, 11149–11165. [CrossRef]
2. Sadeghzadeh, M.; Bornehdeli, S.; Mohahammadrezakhani, H.; Abolghasemi, M.; Poursaei, E.; Asadi, M.; Zafari, V.; Aghebati-Maleki, L.; Shanehbandi, D. Dendritic Cell Therapy in Cancer Treatment; the State-of-the-Art. *Life Sci.* **2020**, *254*, 117580. [CrossRef]
3. Boyle, S.T. RISING STARS: Hormonal Regulation of the Breast Cancer Microenvironment. *J. Mol. Endocrinol.* **2023**, *70*, e220174. [CrossRef]
4. Kantoff, P.W.; Higano, C.S.; Shore, N.D.; Berger, E.R.; Small, E.J.; Penson, D.F.; Redfern, C.H.; Ferrari, A.C.; Dreicer, R.; Sims, R.B. Sipuleucel-T Immunotherapy for Castration-Resistant Prostate Cancer. *N. Engl. J. Med.* **2010**, *363*, 411–422. [CrossRef]
5. Gilboa, E.; Vieweg, J. Cancer Immunotherapy with MRNA-transfected Dendritic Cells. *Immunol. Rev.* **2004**, *199*, 251–263. [CrossRef] [PubMed]
6. De Andrade Pereira, B.; Fraefel, C. Novel Immunotherapeutic Approaches in Targeting Dendritic Cells with Virus Vectors. *Discov. Med.* **2015**, *20*, 111–119. [PubMed]
7. Chulpanova, D.S.; Gilazieva, Z.E.; Kletukhina, S.K.; Aimaletdinov, A.M.; Garanina, E.E.; James, V.; Rizvanov, A.A.; Solovyeva, V.V. Cytochalasin B-Induced Membrane Vesicles from Human Mesenchymal Stem Cells Overexpressing IL2 Are Able to Stimulate CD8$^+$ T-Killers to Kill Human Triple Negative Breast Cancer Cells. *Biology* **2021**, *10*, 141. [CrossRef]
8. Lipps, G. Biological Properties of Extracellular Vesicles and Their Physiological Functions. *J. Extracell. Vesicles* **2015**, *4*, 27066.
9. Becker, A.; Thakur, B.K.; Weiss, J.M.; Kim, H.S.; Peinado, H.; Lyden, D. Extracellular Vesicles in Cancer: Cell-to-Cell Mediators of Metastasis. *Cancer Cell* **2016**, *30*, 836–848. [CrossRef]
10. Kok, V.C.; Yu, C.-C. Cancer-Derived Exosomes: Their Role in Cancer Biology and Biomarker Development. *Int. J. Nanomed.* **2020**, *15*, 8019–8036. [CrossRef]
11. Benecke, L.; Chiang, D.M.; Ebnoether, E.; Pfaffl, M.W.; Muller, L. Isolation and Analysis of Tumor-derived Extracellular Vesicles from Head and Neck Squamous Cell Carcinoma Plasma by Galectin-based Glycan Recognition Particles. *Int. J. Oncol.* **2022**, *61*, 133. [CrossRef] [PubMed]
12. Liu, J.; Ma, J.; Tang, K.; Huang, B. Therapeutic Use of Tumor Cell-Derived Extracellular Vesicles. *Extracell. Vesicles Methods Protoc.* **2017**, *1660*, 433–440.
13. Yoshioka, Y.; Konishi, Y.; Kosaka, N.; Katsuda, T.; Kato, T.; Ochiya, T. Comparative Marker Analysis of Extracellular Vesicles in Different Human Cancer Types. *J. Extracell. Vesicles* **2013**, *2*, 20424. [CrossRef] [PubMed]
14. Calmeiro, J.; Mendes, L.; Duarte, I.F.; Leitão, C.; Tavares, A.R.; Ferreira, D.A.; Gomes, C.; Serra, J.; Falcão, A.; Cruz, M.T. In-Depth Analysis of the Impact of Different Serum-Free Media on the Production of Clinical Grade Dendritic Cells for Cancer Immunotherapy. *Front. Immunol.* **2021**, *11*, 593363. [CrossRef]
15. Conti, L.; Gessani, S. GM-CSF in the Generation of Dendritic Cells from Human Blood Monocyte Precursors: Recent Advances. *Immunobiology* **2008**, *213*, 859–870. [CrossRef]

16. Ko, H.-J.; Brady, J.L.; Ryg-Cornejo, V.; Hansen, D.S.; Vremec, D.; Shortman, K.; Zhan, Y.; Lew, A.M. GM-CSF–Responsive Monocyte-Derived Dendritic Cells Are Pivotal in Th17 Pathogenesis. *J. Immunol.* **2014**, *192*, 2202–2209. [CrossRef]

17. Korniotis, S.; Saichi, M.; Trichot, C.; Hoffmann, C.; Amblard, E.; Viguier, A.; Grondin, S.; Noel, F.; Mattoo, H.; Soumelis, V. GM-CSF-Activated Human Dendritic Cells Promote Type 1 T Follicular Helper Cell Polarization in a CD40-Dependent Manner. *J. Cell Sci.* **2022**, *135*, jcs260298. [CrossRef]

18. Yan, W.-L.; Shen, K.-Y.; Tien, C.-Y.; Chen, Y.-A.; Liu, S.-J. Recent Progress in GM-CSF-Based Cancer Immunotherapy. *Immunotherapy* **2017**, *9*, 347–360. [CrossRef]

19. Pedrioli, G.; Paganetti, P. Hijacking Endocytosis and Autophagy in Extracellular Vesicle Communication: Where the inside Meets the Outside. *Front. Cell Dev. Biol.* **2021**, *8*, 595515. [CrossRef]

20. Raskov, H.; Orhan, A.; Christensen, J.P.; Gögenur, I. Cytotoxic CD8$^+$ T Cells in Cancer and Cancer Immunotherapy. *Br. J. Cancer* **2021**, *124*, 359–367. [CrossRef]

21. Marton, A.; Vizler, C.; Kusz, E.; Temesfoi, V.; Szathmary, Z.; Nagy, K.; Szegletes, Z.; Varo, G.; Siklos, L.; Katona, R.L. Melanoma Cell-Derived Exosomes Alter Macrophage and Dendritic Cell Functions in Vitro. *Immunol. Lett.* **2012**, *148*, 34–38. [CrossRef]

22. Knutson, K.L.; Disis, M. Tumor Antigen-Specific T Helper Cells in Cancer Immunity and Immunotherapy. *Cancer Immunol. Immunother.* **2005**, *54*, 721–728. [CrossRef] [PubMed]

23. Lu, X. Impact of IL-12 in Cancer. *Curr. Cancer Drug Targets* **2017**, *17*, 682–697. [CrossRef] [PubMed]

24. Haabeth, O.A.W.; Lorvik, K.B.; Hammarström, C.; Donaldson, I.M.; Haraldsen, G.; Bogen, B.; Corthay, A. Inflammation Driven by Tumour-Specific Th1 Cells Protects against B-Cell Cancer. *Nat. Commun.* **2011**, *2*, 240. [CrossRef] [PubMed]

25. Nonaka, K.; Saio, M.; Umemura, N.; Kikuchi, A.; Takahashi, T.; Osada, S.; Yoshida, K. Th1 Polarization in the Tumor Microenvironment Upregulates the Myeloid-Derived Suppressor-like Function of Macrophages. *Cell. Immunol.* **2021**, *369*, 104437. [CrossRef] [PubMed]

26. Osawa, E.; Nakajima, A.; Fujisawa, T.; Kawamura, Y.I.; Toyama-Sorimachi, N.; Nakagama, H.; Dohi, T. Predominant T Helper Type 2-inflammatory Responses Promote Murine Colon Cancers. *Int. J. Cancer* **2006**, *118*, 2232–2236. [CrossRef]

27. Ziegler, A.; Heidenreich, R.; Braumüller, H.; Wolburg, H.; Weidemann, S.; Mocikat, R.; Röcken, M. EpCAM, a Human Tumor-Associated Antigen Promotes Th2 Development and Tumor Immune Evasion. *Blood J. Am. Soc. Hematol.* **2009**, *113*, 3494–3502. [CrossRef]

28. Tosolini, M.; Kirilovsky, A.; Mlecnik, B.; Fredriksen, T.; Mauger, S.; Bindea, G.; Berger, A.; Bruneval, P.; Fridman, W.-H.; Pages, F. Clinical Impact of Different Classes of Infiltrating T Cytotoxic and Helper Cells (Th1, Th2, Treg, Th17) in Patients with Colorectal Cancer. *Cancer Res.* **2011**, *71*, 1263–1271. [CrossRef]

29. Schreiber, S.; Hammers, C.M.; Kaasch, A.J.; Schraven, B.; Dudeck, A.; Kahlfuss, S. Metabolic Interdependency of Th2 Cell-Mediated Type 2 Immunity and the Tumor Microenvironment. *Front. Immunol.* **2021**, *12*, 632581. [CrossRef]

30. Mattes, J.; Hulett, M.; Xie, W.; Hogan, S.; Rothenberg, M.E.; Foster, P.; Parish, C. Immunotherapy of Cytotoxic T Cell–Resistant Tumors by T Helper 2 Cells: An Eotaxin and STAT6-Dependent Process. *J. Exp. Med.* **2003**, *197*, 387–393. [CrossRef]

31. Jacenik, D.; Karagiannidis, I.; Beswick, E.J. Th2 Cells Inhibit Growth of Colon and Pancreas Cancers by Promoting Anti-Tumorigenic Responses from Macrophages and Eosinophils. *Br. J. Cancer* **2023**, *128*, 387–397. [CrossRef] [PubMed]

32. Ellyard, J.I.; Simson, L.; Parish, C.R. Th2-mediated Anti-tumour Immunity: Friend or Foe? *Tissue Antigens* **2007**, *70*, 1–11. [CrossRef] [PubMed]

33. Hoffmann, T.K.; Meidenbauer, N.; Dworacki, G.; Kanaya, H.; Whiteside, T.L. Generation of Tumor-Specific T Lymphocytes by Cross-Priming with Human Dendritic Cells Ingesting Apoptotic Tumor Cells. *Cancer Res.* **2000**, *60*, 3542–3549. [PubMed]

34. Wang, R.-F.; Johnston, S.L.; Zeng, G.; Topalian, S.L.; Schwartzentruber, D.J.; Rosenberg, S.A. A Breast and Melanoma-Shared Tumor Antigen: T Cell Responses to Antigenic Peptides Translated from Different Open Reading Frames. *J. Immunol.* **1998**, *161*, 3596–3606. [CrossRef]

35. Kunstič, T.T.; Debeljak, N.; Tacer, K.F. Heterogeneity in Hormone-Dependent Breast Cancer and Therapy: Steroid Hormones, HER2, Melanoma Antigens, and Cannabinoid Receptors. *Adv. Cancer Biol.-Metastasis* **2023**, *7*, 100086.

Article

A Comparative Analysis of NOX4 Protein Expression in Malignant and Non-Malignant Thyroid Tumors

Salma Fenniche [1,2,3,4], Mohamed Oukabli [5,6], Yassire Oubaddou [1], Hafsa Chahdi [5,6], Amal Damiri [5,6], Abir Alghuzlan [2,4], Abdelilah Laraqui [5], Nadia Dakka [1], Youssef Bakri [1], Corinne Dupuy [2,3,4] and Rabii Ameziane El Hassani [1,*]

1   Laboratory of Biology of Human Pathologies (BioPatH), Faculty of Sciences, Mohammed V University in Rabat, Rabat 1014, Morocco; salma.fenniche@um5r.ac.ma (S.F.); yassire.oubaddou@um5r.ac.ma (Y.O.); n.dakka@um5r.ac.ma (N.D.); y.bakri@um5r.ac.ma (Y.B.)
2   Gustave Roussy Cancer Campus, Pavillon de Recherche N°2, F-94805 Villejuif, France; abir.alghuzlan@gustaveroussy.fr (A.A.); corinne.dupuy@gustaveroussy.fr (C.D.)
3   Faculty of Medicine and Pharmacy, University Paris-Saclay, F-91400 Orsay, France
4   Unité Mixte de Recherche UMR9019 Centre National de la Recherche Scientifique, Pavillon de Recherche N°2, F-94805 Villejuif, France
5   Service of Anatomical Pathology, Military Hospital of Instruction Mohammed V (HMIMV-R), Rabat 1014, Morocco; oukablimohamed@yahoo.fr (M.O.); hchahdi168@gmail.com (H.C.); amaldamiripath@gmail.com (A.D.); loranjad@yahoo.fr (A.L.)
6   Faculty of Medicine and Pharmacy, Mohammed V University in Rabat, Rabat 10001, Morocco
*   Correspondence: r.ameziane@um5r.ac.ma or ame_rbi@yahoo.fr

Citation: Fenniche, S.; Oukabli, M.; Oubaddou, Y.; Chahdi, H.; Damiri, A.; Alghuzlan, A.; Laraqui, A.; Dakka, N.; Bakri, Y.; Dupuy, C.; et al. A Comparative Analysis of NOX4 Protein Expression in Malignant and Non-Malignant Thyroid Tumors. *Curr. Issues Mol. Biol.* **2023**, *45*, 5811–5823. https://doi.org/10.3390/cimb45070367

Academic Editor: Dumitru A. Iacobas

Received: 17 May 2023
Revised: 8 June 2023
Accepted: 14 June 2023
Published: 13 July 2023

Abstract: The comparative analysis of the expression of the reactive oxygen species-generating NADPH oxidase NOX4 from TCGA data shows that the NOX4 transcript is upregulated in papillary thyroid carcinomas (PTC)-BRAF^{V600E} tumors compared to PTC-BRAF^{wt} tumors. However, a comparative analysis of NOX4 at the protein level in malignant and non-malignant tumors is missing. We explored NOX4 protein expression by immunohistochemistry staining in malignant tumors (28 classical forms of PTC (C-PTC), 17 follicular variants of PTC (F-PTC), and three anaplastic thyroid carcinomas (ATCs)) and in non-malignant tumors (six lymphocytic thyroiditis, four Graves' disease, ten goiters, and 20 hyperplasias). We detected the BRAF^{V600E} mutation by Sanger sequencing and digital droplet PCR. The results show that NOX4 was found to be higher (score $\geq$ 2) in C-PTC (92.9%) compared to F-PTC (52.9%) and ATC (33.3%) concerning malignant tumors. Interestingly, all C-PTC-BRAF^{V600E} expressed a high score for NOX4 at the protein level, strengthening the positive correlation between the BRAF^{V600E} mutation and NOX4 expression. In addition, independent of the mutational status of BRAF, we observed that 90% of C-PTC infiltrating tumors showed high NOX4 expression, suggesting that NOX4 may be considered a complementary biomarker in PTC aggressiveness. Interestingly, NOX4 was highly expressed in non-malignant thyroid diseases with different subcellular localizations.

Keywords: papillary thyroid carcinomas; BRAF^{V600E} hot spot mutation; NADPH oxidase NOX4; biomarker

1. Introduction

Thyroid cancer is the most common endocrine disease, and papillary thyroid carcinomas (PTC) are the most frequent forms of thyroid cancer from the follicular origin (78–90%), followed by follicular thyroid carcinomas (FTC) and anaplastic thyroid carcinomas (ATC) [1–3]. An alarming worldwide increase in PTC incidence has been reported, and the molecular/histological heterogeneity of PTC complicates their management from both diagnostic and prognostic perspectives [2,4–9]. The BRAF^{V600E} mutation is detected in 28–90% of PTC tumors, indicating that this driver mutation is a common event in these tumors [1,3,10–16].

BRAF^{V600E} is a potent activator of the MAPK pathway and is often associated with aggressiveness, metastasis, radioiodine refractoriness, and mortality in thyroid cancer [13,15–18]. BRAF^{V600E} is exclusively detected in PTC (mainly in classical PTC 'C-PTC' and rarely in the follicular variant of PTC 'F-PTC') and ATC, but never in FTC and rarely in benign thyroid adenomas [2,19,20]. BRAF^{V600E} is useful as a prognostic and predictive biomarker in managing thyroid cancer. Regarding 10-year survival, PTC and FTC have survival rates of about 90–98%, whereas poorly differentiated thyroid carcinomas (PDTC) and ATC have survival rates of 50% and <10%, respectively [1]. Interestingly, high-grade tumors (PDTC and ATC) can arise from PTC and FTC through progressive genetic alterations and dedifferentiation [1,21], suggesting that thyroid tumor dedifferentiation can occur independently of the BRAF^{V600E} mutation because FTC is always BRAF^{wt}. In addition, the absence of a positive correlation between the BRAF^{V600E} mutation and thyroid tumor aggressiveness was reported [22–25]. An interesting classification of PTC has been proposed that groups PTC tumors independently of their molecular signature. PTC-BRAF^{V600E}-like tumors, including PTC-BRAF^{V600E} and PTC-BRAF^{wt}, share higher transcriptional MAPK pathway activity and higher dedifferentiation [7,26]. Consequently, exploring additional biomarkers other than BRAF^{V600E} could improve the management of PTC patients.

Reactive oxygen species (ROS) have been suspected of being involved in thyroid tumorigenesis for several years, and the role of ROS-generating NADPH oxidases has been analyzed in thyroid carcinomas [17,27,28]. Weyemi U et al. observed that NADPH oxidase 4 (NOX4) protein is overexpressed in thyroid cancer tissue (11 PTC) compared to normal adjacent tissue (eight NAT) [27]. Azouzi N et al. established a positive link between BRAF^{V600E} and NOX4 expression (at the mRNA level) by exploring about 500 PTC (BRAF^{V600E} versus BRAF^{wt}) from the TCGA database 'Genome Atlas' [17]. This study explored NOX4 expression at the protein level in 134 thyroid tissues (48 malignant tissues, 46 normal tissues surrounding the tumors, and 40 non-malignant tissues). Our results showed that the NOX4 protein score is higher in PTC, and we strengthened the link between NOX4 protein expression and BRAF^{V600E} mutation in PTC. Interestingly, and unlike the mutational status of BRAF, NOX4 expression is higher in C-PTC infiltrating tumors, highlighting a potential role of NOX4 as a marker of aggressiveness in PTC. Finally, we observed an overexpression of NOX4 in non-malignant diseases with different subcellular localization.

2. Materials and Methods

This study was approved by the Ethics Committee for Biomedical Research (CERB) of the Faculty of Medicine and Pharmacy in Rabat, with approval number 52/20.

2.1. Study Subjects

In this retrospective study, we collected 134 FFPE thyroid tissues from patients diagnosed at the Department of Anatomical Pathology in Military Hospital of Instruction Mohammed V in Rabat (HMIMV-R) between January 2015 and December 2021. Our cohort included 46 NAT, 28 classical forms of papillary thyroid carcinomas (C-PTC), 17 follicular variants of papillary thyroid carcinomas (F-PTC), three ATC, and 40 non-malignant thyroid tissues (including six lymphocytic thyroiditis, four Graves' disease, ten goiters, and 20 hyperplasias). Malignant and non-malignant thyroid tissues were histologically classified according to the World Health Organization (WHO) classification (Lloyd R.V. et al., 4th Edition, IARC: Lyon 2017) by an experienced anatomopathologist at the time of diagnosis. In addition, retrospective diagnosis confirmation was performed by experienced anatomopathologists from both HMIMV in Rabat (Morocco) and Gustave Roussy Cancer Institute (France). FFPE block selection was based on (1) cell sufficiency (more than 90% of tumor cells in thyroid tumors/absence of tumor cells in non-tumoral tissues) and (2) FFPE block availability.

The collected information from the medical files of the selected patients constituted a database for this study, providing all clinicopathological features of the patients.

2.2. NOX4 Immunohistochemistry Staining

Immunohistochemical staining of NOX4 was performed using rabbit polyclonal anti-NOX4 (ab154244, Abcam, Cambridge, United Kingdom), and Dako EnVision™ FLEX High pH 'LINK'/FLEX IHC Microscope Slides (Agilent, Santa Clara, California, United States) in an automaton (Autostainer Link 48; Agilent) according to the manufacturer's protocols. These antibodies (Abs), whose immunogen corresponds to human NOX4 (aa 252–564), are suitable for IHC staining of FFPE tissue (IHC-P). The validation of NOX4 staining was performed after testing several Ab dilutions (1:250, 1:500, 1:750 in EnVision™ FLEX Antibody Diluent) on normal human kidney tissue (FFPE Block) (Supplementary Figure S2). Counterstaining was performed by hematoxylin 'EnVision™ FLEX Hematoxylin'. Immunostaining of thyroid tissue without NOX4 Ab served as a negative control.

An experienced anatomopathologist from HMIMV in Rabat and Gustave Roussy Cancer Institute (France) assigned two scores to NOX4 protein expression. A score <2 corresponds to an absence or low expression of NOX4 protein, and a score $\geq$2 corresponds to a medium or high expression of NOX4 protein. As usually used by trained pathologist, the score of the staining intensity could be negative (0), weak (1), moderate (2), and strong (3) [29]. For our study, we classified scores <2 (including 0 and 1) and $\geq$2 (including 2 and 3).

In addition, the subcellular localization of NOX4 was noted. Slides were scanned with a NanoZoomer HT C9600 (Hamamatsu Photonics KK, Hamamatsu, Japan) digital scanner, using a ×40 objective. Scanned slides were uploaded on the CaloPix database (Tribvn Healthcare, Chatillon, France).

2.3. Mutational Analysis

$BRAF^{V600E}$ hotspot mutation was analyzed in 48 malignant thyroid tissues, including 28 C-PTCs, 17 F-PTCs, and three ATCs, using Sanger direct sequencing and digital droplet PCR after genomic DNA extraction.

The genomic DNA extraction was performed in 8 sections of 10 μm each, according to the manufacturer's 'QIAamp DNA FFPE Tissue Kit (Qiagen)' protocol, specially designed for purifying DNA from FFPE tissue sections. The quality and concentration of the extracted DNA were determined using the IMPLEN NanoPhotometer N60.

After genomic DNA extraction, PCRs were conducted using the HotStart Taq polymerase (Qiagen) with the following conditions: 95 °C for 15 min, followed by 42 cycles of 95 °C for 30 s, 53 °C for 30 s, and 72 °C for 45 s, and a final extension cycle of 72 °C for 10 min, all carried out in the Thermal Cycler (ProFlex PCR System; Thermo Fisher, Waltham, MA, USA). The forward and reverse primers for BRAF Exon 15 were (F) 5′-TCA TAA TGC TTG CTC TGA TAG GA-3′ and (R) 5′-GGC CAA AAA TTT AAT CAG TGG A-3′. The positive PCR products for $BRAF^{V600E}$, obtained after migration in 1.5% agarose gel electrophoresis, were purified using Exo-SAP before being bidirectionally sequenced with the Big Dye Terminator sequencing kit (Applied Biosystems, Foster City, CA, USA) at the UATRS Platform, CNRST, Rabat, Morocco. DNA sequences were analyzed with the SeqScape 2.7 software (Applied Biosystems, Waltham, MA, USA).

For digital droplet PCR, genomic DNA was extracted using the 'Maxwell® RSC DNA FFPE Kit' from Promega/'Maxwell® RSC Instruments (Cat.# AS4500, AS8500)' according to the manufacturer's protocols (ASB1450). The extracted genomic DNA was sequenced using Droplet Digital PCR 'ddPCR' at the platform of molecular medicine in Gustave Roussy institute by determining the copy number of $BRAF^{V600E}$ and $BRAF^{wt}$ in each sample.

2.4. Statistical Analysis

All statistical analyses were performed using GraphPad Prism 8. Fisher's exact and chi-square tests examined the associations between clinicopathological variables, $BRAF^{V600E}$, NOX4 protein expression, and NOX4 protein localization. A *p*-value of <0.05 was considered statistically significant for all analyses.

3. Results

3.1. Clinicopathological Characteristics of Patients

The analysis of the clinicopathologic data of our cohort revealed a female predominance (81%) (Table 1). The median age at diagnosis was 42.5 years, with a standard error of 2.4 years. Histologically, PTC comprise several histological subtypes (variants) that share specific nuclear features of PTC. We observed that C-PTC constituted the most frequent histological subtype (58.33%), followed by F-PTC (35.41%) and finally ATCs (6.26%) (Table 1). Based on aggressive features, we found that most thyroid cancers presented with a tumor size greater than 1 cm, while we noted the presence of vascular emboli in 15% of cases, capsular breach in 23% of cases, and lymph node metastases in 6.3% of cases (Table 1).

Table 1. Clinicopathological parameters of patients with thyroid tumors (n = 48).

Clinicopathological Parameters	n (%)
Gender	
Female	39 (81%)
Male	9 (19%)
Age	
Median ± standard error	42.5 ⊥ 2.4
Histological variant	
C-PTC	28/48 (58.33%)
F-PTC	17/48 (35.41%)
ATC	3/48 (6.26%)
Tumor size	
≤1 cm	7/48 (15%)
>1 cm	30/48 (63%)
unknown	11/48 (23%)
vascular emboli	
Presence	7/48 (15%)
Absence	41/48 (85%)
Capsular breach	
Presence	11/48 (23%)
Absence	37/48 (77%)
lymph node metastasis	
Presence	3/48 (6.3%)
Absence	34/48 (70.8%)
Unknown	11/48 (22.9%)

PTC: papillary thyroid carcinomas, C-PTC: classical form of papillary thyroid carcinomas, F-PTC: follicular variants of papillary thyroid carcinomas, ATC: anaplastic thyroid carcinomas. *Unknown*: absence of this information in both registers and medical files.

3.2. Clinicopathological Parameters of Each Histological Subtype of Thyroid Carcinomas

The comparative analysis between the histological type of 48 thyroid carcinomas (45 PTCs and three ATCs) and their clinicopathological characteristics revealed that 46.4% of C-PTC occurs in individuals aged ≥45 years old. However, 52.9% of F-PTCs occur in individuals aged <45 years old, while 66.7% of ATCs occur in individuals aged >45 years old (see Table 2). Regarding tumor size, we observed that most PTC tumors were >1 cm (64.3% for C-PTCs and 70.6% for F-PTC). As for vascular emboli and capsular breach, we found them, respectively, in 33.3% and 0% of ATCs, 17.9% and 32.1% of C-PTC, and 5.9% and 11.8% of F-PTC (see Table 2). However, we detected lymph node metastasis only in the classical form of PTC (10.7%) (Table 2).

Table 2. Clinicopathological parameters of each histological subtype of thyroid carcinomas ($n = 48$).

Clinicopathological Parameters	PTC ($n = 45$)	C-PTC ($n = 28$)	F-PTC ($n = 17$)	ATC ($n = 3$)	C-PTC-F-PTC Chi-Square/Fisher's Exact Test (p-Value < 0.05)
Age					
<45	19/45 (42.2%)	10/28 (35.7%)	9/17 (52.9%)	1/3 (33.3%)	0.1516
≥45	16/45 (35.6%)	13/28 (46.4%)	3/17 (17.7%)	2/3 (66.7%)	
unknown	10/45 (22.2%)	5/28 (17.9%)	5/17 (29.4%)	-	
Tumor size					
≤1 cm	7/45 (15.6%)	4/28 (14.3%)	3/17 (17.6%)	-	>0.9999
>1 cm	30/45 (66.7%)	18/28 (64.3%)	12/17 (70.6%)	-	
Unknown	8/45 (17.7%)	6/28 (21.4%)	2/17 (11.8%)	3 (100%)	
vascular emboli					
Presence	6/45 (13.3%)	5/28 (17.9%)	1/17 (5.9%)	1/3 (33.3%)	0.3846
Absence	39/45 (86.7%)	23/28 (82.1%)	16/17 (94.1%)	2/3 (66.7%)	
Capsular breach					
Presence	11/45 (24.4%)	9/28 (32.1%)	2/17 (11.8%)	0 (0%)	0.1647
Absence	34/45 (75.6%)	19/28 (67.9%)	15/17 (88.2%)	3/3 (100%)	
lymph node metastasis					
Presence	3/45 (6.6%)	3/28 (10.7%)	0 (0%)	0 (0%)	0.2432
Absence	34/45 (75.6%)	18/28 (64.2%)	16/17 (94.1%)	0 (0%)	
Unknown	8/45 (17.8%)	7/28 (25%)	1/17 (5.9%)	3/3 (100%)	

PTC: papillary thyroid carcinomas, *C-PTC*: classical form of papillary thyroid carcinomas, *F-PTC*: follicular variants of papillary thyroid carcinomas, *ATC*: anaplastic thyroid carcinomas. *Unknown*: absence of this information in both registers and medical files.

3.3. $BRAF^{V600E}$ Mutation in Thyroid Carcinomas

We analyzed the mutational profile of 48 thyroid carcinomas (45 PTC = 28 C-PTC + 17 F-PTC and three ATC) regarding the $BRAF^{V600E}$ hotspot mutation using Sanger direct sequencing and digital droplet PCR (ddPCR). The mutational profiles of 41 samples ($BRAF^{V600E}$ or $BRAF^{wt}$) were detected by Sanger direct sequencing, while ddPCR explored 7 samples. Genomic DNA extracted from FFPE block tissues is known to be extensively fragmented [30–32]. Therefore, seven samples whose results were not exploitable by Sanger direct sequencing were analyzed by ddPCR. (Figure S1 Supplementary Materials) shows an example of electropherogram sequences (Figure S1a Supplementary Materials) and an example of results from ddPCR (Figure S1b Supplementary Materials) for $BRAF^{V600E}$ mutation in C-PTC and $BRAF^{wt}$ in F-PTC. In this study, $BRAF^{V600E}$ mutation was found exclusively in the classical form of PTC (10 C-PTC-$BRAF^{V600E}$/28 C-PTC: 35.7%), while no mutation was detected in the follicular variant of PTC and ATC (see Figure 1). This percentage is about 20.83% in all thyroid carcinomas (10 C-PTC-$BRAF^{V600E}$/48; 48 = 28 C-PTC + 17 F-PTC + 3 ATC) and 22.22% in all papillary thyroid carcinomas (ten C-PTC-$BRAF^{V600E}$/45; 45 = 28 C-PTC + 17 F-PTC). Finally, we did not observe any statistically significant association between $BRAF^{V600E}$ mutation and the aggressiveness of thyroid carcinomas (see Table 3).

3.4. NOX4 Protein Expression in Human Malignant Thyroid Tissues

The antibody specificity and dilution choice (1:250) were validated in human kidney tissues (Figure S2) known to express NOX4 protein [33]. NOX4 protein expression was analyzed by immunohistochemistry staining in thyroid tumor tissues and their paired NAT (Figure 2a). Our results showed that 92.8% of C-PTC, 52.9% of F-PTC, and 33.3% of ATC express high levels of NOX4 protein (score ≥ 2) (see Figure 2b). No high NOX4 protein expression was observed in normal adjacent tissues (see Figure 2b). Notably, most FFPE thyroid tumor blocks contain both tumor areas and their normal adjacent tissues

NAT (42/46 NAT), enabling us to immunostain both tumor and normal adjacent tissues on the same slide (see Figure 2a). Importantly, all thyroid tumors (BRAF^{V600E}) overexpressed NOX4 protein (100%: 10 C-PTC/10-C-PTC), highlighting the positive correlation between BRAF^{V600E} mutation and NOX4 protein expression in PTC (p-value < 0.0001) (see Figure 2c). However, a non-negligible percentage of BRAF^{wt} thyroid tumors (68.4%: 26/38) also exhibited a high level of NOX4 expression (see Figure 2c).

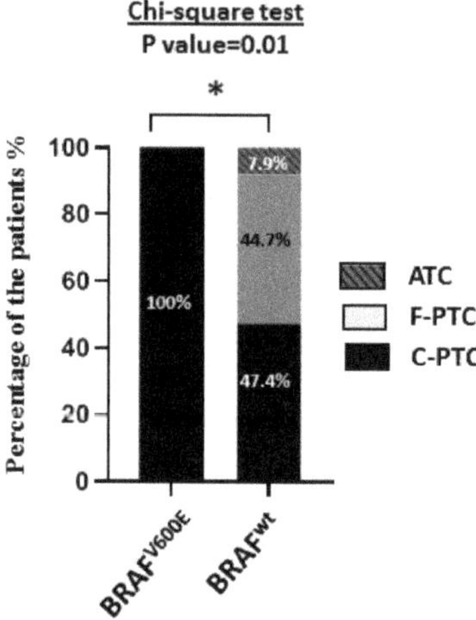

Figure 1. BRAF^{V600E} mutation in thyroid carcinomas. Association between BRAF^{V600E} mutation and histological type of thyroid carcinomas (n = 48). In total, 100% of thyroid carcinoma harboring BRAF^{V600E} are exclusively C-PTC. *ATC*: anaplastic thyroid carcinoma (n = 3). *C-PTC*: classical forms of papillary thyroid carcinoma (n = 28). *F-PTC*: follicular variants of papillary thyroid carcinoma (n = 17). BRAF^{V600E} (n = 10), BRAF^{wt} (n = 38). The statistical tests performed by GraphPad 8. * The statistical significance is affirmed by a p-value less than 0.05. *: p-value $\leq$ 0.05.

Table 3. BRAF^{V600E} mutation and C-PTC clinicopathological parameters (n = 28).

Clinicopathological Parameters	C-PTC BRAF^{V600E} (n = 10)	C-PTC BRAF^{wt} (n = 18)	Total	Chi-Square/Fisher's Exact Test (p-Value < 0.05)
Tumor size				
$\leq$1 cm	1/10 (10%)	2/18 (11.1%)	3	>0.9999
>1 cm	7/10 (70%)	12/18 (66.7%)	19	
unknown	2/10 (20%)	4/18 (22.2%)	6	
lymph node metastasis				
Presence	1/10 (10%)	2/18 (11.1%)	3	>0.9999
Absence	8/10 (80%)	10/18 (55.6%)	18	
unknown	1/10 (10%)	6/18 (33.3%)	7	
vascular emboli				
Presence	2/10 (20%)	3/18 (16.7%)	5	0.8253
Absence	8/10 (80%)	15/18 (83.3%)	23	
Capsular breach				
Presence	4/10 (40%)	5/18 (27.8%)	9	0.5070
Absence	6/10 (60%)	13/18 (72.2%)	19	

C-PTC: Classical form of papillary thyroid carcinomas. *Unknown*: absence of this information in both registers and medical files.

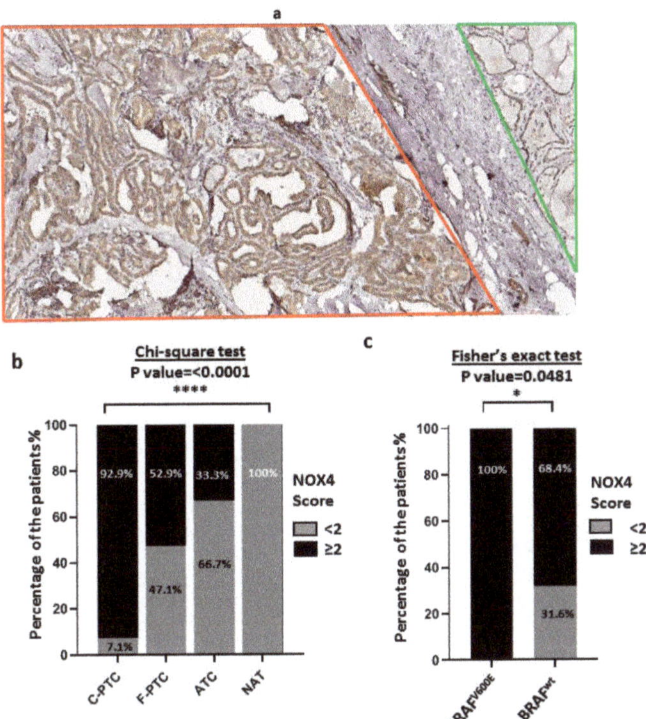

Figure 2. NOX4 protein expression in thyroid carcinomas. (**a**) Representative example of NOX4 protein expression in C-PTC. Red lines represent tumor tissue sections, and green lines represent normal adjacent tissue (NAT) sections on the same slide ($\times$10). There is a high expression level of NOX4 protein in C-PTC and a low expression level of NOX4 in its adjacent normal tissue (NAT). (**b**) Comparative analysis of NOX4 protein expression in human thyroid tumors ($n = 48$:28 C-PTC, 17 F-PTC, 3 ATC) compared to their normal adjacent tissues (46 NAT). Percentage calculated according to the number of each score <2 and ≥2. The score <2 represents an absence or very low and low expression of NOX4 protein ($n = 58$). The score ≥2 represents a middle or/and high expression level of NOX4 protein ($n = 36$). (**c**) Association between $BRAF^{V600E}$ mutation and NOX4 expression. In total, 100% of thyroid carcinoma harboring $BRAF^{V600E}$ mutation showed high expression level of NOX4 protein. $BRAF^{V600E}$ ($n = 10$), $BRAF^{wt}$ ($n = 38$). The statistical tests are performed by GraphPad 8. * The statistical significance is affirmed by a p-value under 0.05. *: p-value ≤ 0.05, ****: p-value ≤ 0.0001.

To investigate the involvement of NOX4 protein in thyroid tumor aggressiveness, we observed a statistically significant correlation between the overexpression of NOX4 protein and the presence of a capsular breach, indicating the association of NOX4 protein with the aggressiveness of thyroid carcinomas in our cohort (p-value = 0.0481) (see Table 4). Additionally, we evaluated NOX4 protein expression in infiltrating PTC tumors, characterized by the infiltration of tumor borders and consequently presenting a risk of metastasis initiation. We found that the majority of PTC infiltrating tumors exhibited a high level of NOX4 protein expression (83.3%: score ≥2 for all PTC and 90% for C-PTC: score ≥ 2) (see Figure 3a,b), highlighting a potential role of NOX4 protein in PTC aggressiveness. However, only 40% of infiltrating tumors (C-PTC) were $BRAF^{V600E}$ positive (see Figure 3c), while all these tumors overexpressed NOX4 (100% of C-PTC $BRAF^{V600E}$: score ≥ 2) (see Figure 3d). Additionally, 83.3% of infiltrating C-PTC ($BRAF^{wt}$) overexpressed NOX4 protein independently of the presence of the $BRAF^{V600E}$ mutation (Figure 3d) (Fisher's test: $p ≥ 0.999$ for $BRAF^{wt}$ infiltrating C-PTC versus $BRAF^{V600E}$ infiltrating C-PTC).

Table 4. NOX4 protein expression and clinicopathological parameters of thyroid tumors.

Clinicopathological Parameters	n (%)	NOX4 Overexpression (Score: ≥ 2)	NOX4 Low Expression (Score: <2)	Fisher's Exact Test (p-Value < 0.05)
Gender				
Female	39/48 (81%)	30/39 (76.9%)	9/39 (23.1%)	>0.9999
Male	9/48 (19%)	7/9 (77.8%)	2/9 (22.2%)	
Histological variant				
C-PTC	28/48 (58.33%)	26/28 (92.9%)	2/28 (7.1%)	
F-PTC	17/48 (35.41%)	9/17 (52.9%)	8/17 (47.1%)	<0.0001
ATC	3/48 (6.26%)	1/3 (33.3%)	2/3 (66.7%)	
Tumor size				
≤ 1 cm	7/48 (15%)	7/7 (100%)	0 (0%)	
>1 cm	30/48 (63%)	22/30 (73.3%)	8/30 (26.7%)	0.3079
unknown	11/48 (23%)	7/11 (63.4%)	4/11 (36.6%)	
vascular emboli				
Presence	7/48 (15%)	6/7 (85.7%)	1/7 (14.3%)	
Absence	41/48 (85%)	30/41 (73.2%)	11/41 (26.8%)	0.6621
Capsular breach				
Presence	11/48 (23%)	11/11 (100%)	0/11 (0%)	
Absence	37/48 (77%)	25/37 (67.6%)	12/37 (32.4%)	0.0442
lymph node metastasis				
Presence	3/48 (6.3%)	3/3 (100%)	0/3 (0%)	
Absence	34/48 (70.8%)	24/34 (70.6%)	10/34 (29.4%)	0.5483
unknown	11/48 (22.9%)	9/11 (81.8%)	2/11 (18.2%)	

PTC: papillary thyroid carcinomas, *C-PTC*: classical form of papillary thyroid carcinomas, *F-PTC*: follicular variant of papillary thyroid carcinomas, *ATC*: anaplastic thyroid carcinomas. *NOX4*: NADPH oxidase 4. *Unknown*: absence of this information in both registers and medical files.

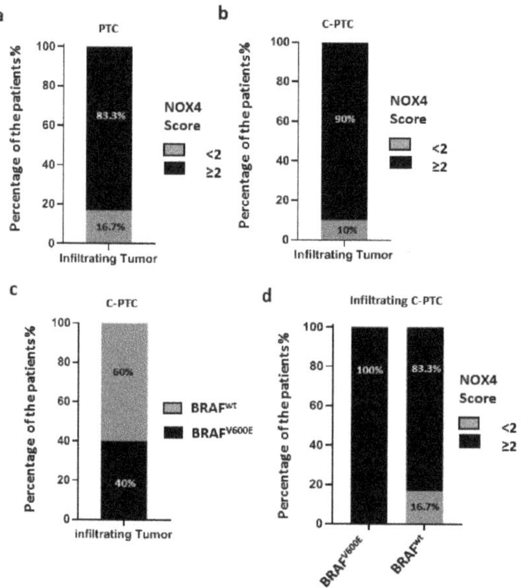

Figure 3. Correlation of infiltrating character followed by the limitation of the border of the tumor with NOX4 protein expression and BRAF^{V600E} mutation in PTC (n = 21: 17 c-PTC, 4 F-PTC). (**a**) In total, 83% (10/12) of papillary thyroid carcinoma with an infiltrating character show overexpression of NOX4 protein (n = 12). (**b**) C-PTC: high expression of NOX4 protein in infiltrating tumors (n = 10) (90% (9/10)) with a score ≥ 2. (**c**) C-PTC: 40% (4/10) of papillary thyroid carcinoma with an infiltrating character harbor BRAF^{V600E} mutation. (**d**) In total, 100% of infiltrating C-PTC harboring BRAF^{V600E} mutation overexpress NOX4 protein. Percentage calculated according to the total number of infiltrating tumors of each histological type.

3.5. NOX4 Protein Expression and Subcellular Localization in Non-Malignant Human Thyroid Tissues

We explored 40 non-malignant human thyroid disease tissues (six lymphocytic thyroiditis, four Graves' disease, ten goiters, and 20 hyperplasias) and observed that the protein level of NOX4 was higher in Graves' disease (100%: 4/4), goiters (80%: 8/10), and hyperplasias (70%: 14/20) (Figure 4a). Interestingly, we observed a different subcellular localization of NOX4 in non-malignant human thyroid tissues compared to malignant thyroid tissues (Figure 4c,d). NOX4 immunostaining revealed a perinuclear and cytoplasmic (intracellular) localization in both malignant thyroid tissues (100%: 48/48) and their normal adjacent tissues (100%: 46/46) (see Figure 4c). However, NOX4 localization was different in non-malignant thyroid tissues, being exclusively intracellular in both lymphocytic thyroiditis (100%: 6/6) and Graves' disease (100%: 4/4), nuclear and intracellular in goiters (100%: 10/10), and at different localizations in hyperplasias: intracellular (15%: 3/20), nuclear (35%: 7/20), and intracellular and nuclear (50%: 10/20) (Figure 4b).

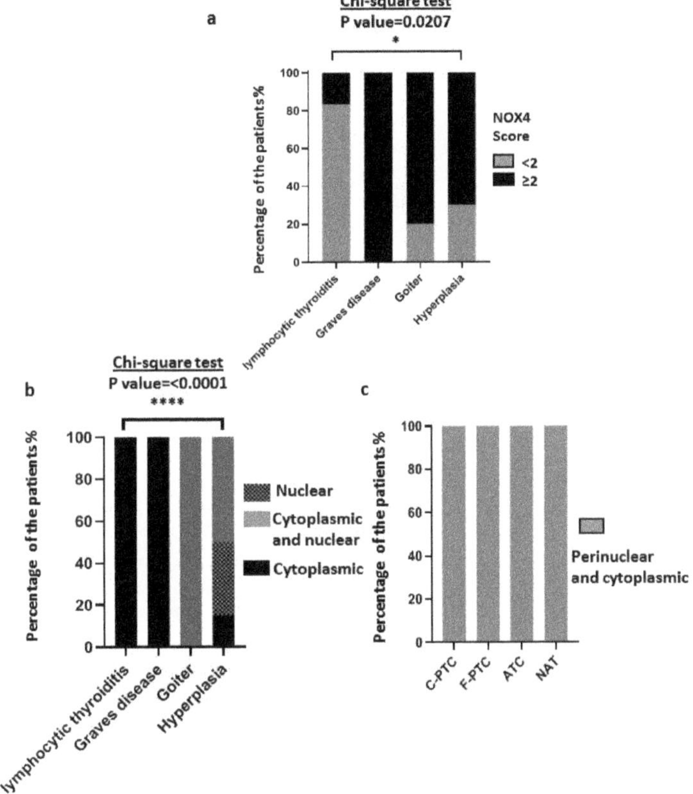

Figure 4. Immunohistochemical analysis of NOX4 protein expression and subcellular localization in thyroid carcinomas and non-malignant thyroid diseases. (**a**) NOX4 protein expression in 40 non-malignant thyroid diseases (six lymphocytic thyroiditis, four Graves' disease, ten goiters, and 20 hyperplasias. (**b**) Subcellular localization of NOX4 protein in 40 non-malignant thyroid diseases (six lymphocytic thyroiditis, four Graves' disease, ten goiters, and 20 hyperplasias). (**c**) Subcellular localization of NOX4 protein in 94 malignant thyroid tissues (28 c-PTC, 17 F-PTC, 3 ATC, and 46 NAT). The statistical tests performed by GraphPad 8. * The statistical significance is affirmed by a *p*-value less than 0.05. *: *p*-value $\leq$ 0.05, ****: *p*-value $\leq$ 0.0001.

4. Discussion

PTC is a heterogeneous group of thyroid cancer, and its molecular and histological diversity constitute a real challenge for managing these tumors. Molecular exploration of new and/or complementary biomarkers could improve the management of PTC patients. ROS are involved in both physiological and pathological processes in the thyroid, and human cells can produce ROS through various enzymes, including NADPH oxidases (NOX1, NOX2, NOX3, NOX4, NOX5, DUOX1, and DUOX2). Thyrocytes express three NADPH oxidases (DUOX1, DUOX2, and NOX4) with different subcellular localizations and functions. In the thyroid gland, DUOX2 contributes mainly to thyroid hormone synthesis by providing H_2O_2 to thyroid peroxidase (TPO), while ROS-generating NADPH oxidases DUOX1 and NOX4 are associated with oxidative DNA damage that could promote thyroid radio-carcinogenesis and oncogenes thyroid cancer dedifferentiation, respectively [17,27,33–35].

Importantly, unlike the other NADPH oxidases, the level of NOX4 protein is directly related to the level of ROS generation. Therefore, a high level of NOX4 may promote oxidative DNA damage and genomic instability in thyroid cells [33]. In this study, we analyzed NOX4 expression at the protein level by immunostaining in 134 thyroid tissues (48 thyroid carcinomas, 46 normal adjacent tissues, and 40 non-malignant thyroid tissues) to improve our understanding of the potential role of NOX4 as a predictive marker of aggressiveness in PTC. Overall, we observed a high level of NOX4 protein in thyroid cancer tissues compared to normal adjacent tissues (92.9% of C-PTC, 52.9% of F-PTC, and 33.3% of ATC and 0% of NAT), which is consistent with previous reports with fewer tissues [28,36] (Figure 2b).

We previously showed a positive correlation between NOX4 mRNA and BRAF[V600E] mutation in about 500 PTC from TCGA data [17], but this has not been established at the protein level in thyroid cancer tissues. In this study, all PTC-BRAF[V600E] overexpressed NOX4 protein (Figure 2c). However, 68.4% of PTC-BRAF[wt] also showed a high score of NOX4 protein (Figure 2c), emphasizing that NOX4 expression could be upregulated in thyroid cancer independently of BRAF mutational status. The TGF-β pathway plays a key role in thyroid tumorigenesis. The oncogenic effect of BRAF[V600E] in thyroid cells was shown to be mediated in part by TGF-β [37], and NOX4 is upregulated in thyroid cancer cells under the control of the BRAF[V600E]-TGF-β axis [17]. The tumor microenvironment, including cancer-associated fibroblasts (CAF) and myeloid cells such as macrophages, also produces TGF-β [38]. Therefore, the microenvironment may contribute to inducing NOX4 in tumoral cells.

To establish the link between NOX4 protein and the infiltrating ability of PTC, we explored the presence or absence of a frank capsular breach, as its presence indicates the ability of tumors to extend/invade and promote extrathyroidal extension. More than 90% of C-PTC infiltrating tumors overexpressed NOX4 protein (score $\geq$ 2), suggesting a role of NOX4 in thyroid tumor aggressiveness (Figure 3).

Certainly, the expression level of NOX4 protein in terms of ROS production can be related to the incidence of oxidative DNA damage in thyrocytes. Therefore, the cellular localization of NOX4 appears to be of great interest. Indeed, the nuclear and perinuclear localization of ROS-generating NOX4 can promote genomic instability associated with thyrocyte transformation. The perinuclear localization of NOX4 observed in malignant human thyroid tissues (Figure 4) is concordant with this, as genomic instability is a characteristic of malignant tumors. Interestingly, goiter and hyperplasia tissues with a high score of NOX4 protein also showed nuclear and perinuclear localization (Figure 4), questioning the potential role of NOX4 in the progression of these non-malignant diseases to thyroid cancer. Indeed, a high risk of cancer was reported in goiters (18%) [39], and thyroid hyperplasia is reported as the most frequent benign disease associated with PTC [40]. Graves' disease also shows a high level of NOX4 protein (Figure 4a), and this result can be explained by the constitutive activation of the TSH receptor (stimulating antibody, mutations) in Graves' disease tissue [41]. The upregulation of NOX4 by TSH has previously been reported [27].

5. Conclusions

Taken together, the main conclusion of our study is that NOX4 expression could be a potential co-marker of thyroid cancer aggressiveness. However, complementary studies are needed to explore this possibility further. Indeed, BRAF^{V600E} mutation is not often associated with thyroid aggressiveness [22–25]. Several isoforms of NOX4 exist (from GenCards and Aceview), and an antibody recognizing the conserved catalytic core of NOX4 cannot discriminate between the different isoforms in thyroid tissues. Therefore, identifying NOX4 isoforms associated with different diseases may improve diagnosis and/or prognosis.

Supplementary Materials: The following supporting information can be downloaded at: https://www.mdpi.com/article/10.3390/cimb45070367/s1, Figure S1: An example of electropherogram sequences (Figure S1a) and an example of results from ddPCR (Figure S1b) for BRAF^{V600E} mutation in C-PTC and BRAF^{wt} in F-PTC. Figure S2: NOX4 Immunoexpression in normal kidney cells. Validation of anti-NOX4 antibody (ab154244) using different dilutions test (1:250,1:500 and 1:750).

Author Contributions: Conceptualization, R.A.E.H.; methodology, S.F., Y.O., M.O. and A.A.; software, S.F. and Y.O.; validation, S.F., Y.O., H.C., A.D. and A.L.; formal analysis, Y.B., N.D., M.O., C.D. and R.A.E.H.; investigation, R.A.E.H., Y.B., M.O. and C.D.; resources, R.A.E.H. and M.O.; data curation, S.F. and Y.O.; writing—original draft preparation, R.A.E.H., S.F. and Y.O.; writing—review and editing, R.A.E.H., C.D., Y.B., M.O., S.F. and Y.O.; visualization, R.A.E.H.; supervision, R.A.E.H., C.D., Y.B., N.D. and M.O.; project administration, R.A.E.H.; funding acquisition, R.A.E.H., C.D. and Y.B. All authors have read and agreed to the published version of the manuscript.

Funding: Research funded by (1) Cancer Research Institute (IRC). www.irc.ma; Project: 767/AAmP2019. (2) PHC TOUBKAL N° TBK/20/113 N°CAMPUS: 43586YA (Maroc/France) (2020-2022); (3) Agence Nationale Des Plantes Médicinales Et Aromatiques (ANPMA'); and (4) doctoral scholarship to S. FENNICHE from CNRST, Maroc (Bourse d'excellence); (5): Publication fees supported by the "Cancer Research Institute IRC", Kingdom of Morocco. www.irc.ma.

Institutional Review Board Statement: This study was approved by the Ethics Committee for Biomedical Research (CERB) of the Faculty of Medicine and Pharmacy in Rabat, with approval number 52/20.

Informed Consent Statement: Not applicable.

Data Availability Statement: Archived datasets are not available due to the ethic restriction.

Acknowledgments: We thank Abderrahmane Al Bouzidi for his constructive contribution at the initiation of this research work.

Conflicts of Interest: The authors declare no conflict of interest.

References

1. Nikiforov, Y.E.; Nikiforova, M.N. Molecular Genetics and Diagnosis of Thyroid Cancer. *Nat. Rev. Endocrinol.* **2011**, *7*, 569–580. [CrossRef] [PubMed]
2. Hu, J.; Yuan, I.J.; Mirshahidi, S.; Simental, A.; Lee, S.C.; Yuan, X. Thyroid Carcinoma: Phenotypic Features, Underlying Biology and Potential Relevance for Targeting Therapy. *Int. J. Mol. Sci.* **2021**, *22*, 1950. [CrossRef] [PubMed]
3. Kaabouch, M.; Chahdi, H.; Azouzi, N.; Oukabli, M.; Rharrassi, I.; Boudhas, A.; Jaddi, H.; Ababou, M.; Dakka, N.; Boichard, A.; et al. BRAFV600E Hot Spot Mutation in Thyroid Carcinomas: First Moroccan Experience from a Single-Institution Retrospective Study. *Afr. Health Sci.* **2020**, *20*, 1849–1856. [CrossRef] [PubMed]
4. Agarwal, S.; Bychkov, A.; Jung, C.-K. Emerging Biomarkers in Thyroid Practice and Research. *Cancers* **2021**, *14*, 204. [CrossRef]
5. Aschebrook-Kilfoy, B.; Schechter, R.B.; Shih, Y.-C.T.; Kaplan, E.L.; Chiu, B.C.H.; Angelos, P.; Grogan, R.H. The Clinical and Economic Burden of a Sustained Increase in Thyroid Cancer Incidence. *Cancer Epidemiol. Biomark. Prev.* **2013**, *22*, 1252–1259. [CrossRef]
6. Kilfoy, B.A.; Zheng, T.; Holford, T.R.; Han, X.; Ward, M.H.; Sjodin, A.; Zhang, Y.; Bai, Y.; Zhu, C.; Guo, G.L.; et al. International Patterns and Trends in Thyroid Cancer Incidence, 1973–2002. *Cancer Causes Control* **2009**, *20*, 525–531. [CrossRef]
7. Agrawal, N.; Akbani, R.; Aksoy, B.A.; Ally, A.; Arachchi, H.; Asa, S.L.; Auman, J.T.; Balasundaram, M.; Balu, S.; Baylin, S.B.; et al. Integrated Genomic Characterization of Papillary Thyroid Carcinoma. *Cell* **2014**, *159*, 676–690. [CrossRef]
8. Giordano, T.J. Follicular Cell Thyroid Neoplasia: Insights from Genomics and The Cancer Genome Atlas Research Network. *Curr. Opin. Oncol.* **2016**, *28*, 1–4. [CrossRef]

9. Jung, C.K.; Little, M.P.; Lubin, J.H.; Brenner, A.V.; Wells, S.A.; Sigurdson, A.J.; Nikiforov, Y.E. The Increase in Thyroid Cancer Incidence during the Last Four Decades Is Accompanied by a High Frequency of BRAF Mutations and a Sharp Increase in RAS Mutations. *J. Clin. Endocrinol. Metab.* **2014**, *99*, E276–E285. [CrossRef]

10. Newfield, R.S.; Jiang, W.; Sugganth, D.X.; Hantash, F.M.; Lee, E.; Newbury, R.O. Mutational Analysis Using Next Generation Sequencing in Pediatric Thyroid Cancer Reveals BRAF and Fusion Oncogenes Are Common. *Int. J. Pediatr. Otorhinolaryngol.* **2022**, *157*, 111121. [CrossRef]

11. Nikita, M.E.; Jiang, W.; Cheng, S.-M.; Hantash, F.M.; McPhaul, M.J.; Newbury, R.O.; Phillips, S.A.; Reitz, R.E.; Waldman, F.M.; Newfield, R.S. Mutational Analysis in Pediatric Thyroid Cancer and Correlations with Age, Ethnicity, and Clinical Presentation. *Thyroid* **2016**, *26*, 227–234. [CrossRef]

12. Pérez-Fernández, L.; Sastre, J.; Zafón, C.; Oleaga, A.; Castelblanco, E.; Capel, I.; Galofré, J.C.; Guadalix-Iglesias, S.; De la Vieja, A.; Riesco-Eizaguirre, G. Validation of Dynamic Risk Stratification and Impact of BRAF in Risk Assessment of Thyroid Cancer, a Nation-Wide Multicenter Study. *Front. Endocrinol.* **2023**, *13*, 1071775. [CrossRef]

13. Kebebew, E.; Weng, J.; Bauer, J.; Ranvier, G.; Clark, O.H.; Duh, Q.-Y.; Shibru, D.; Bastian, B.; Griffin, A. The Prevalence and Prognostic Value of BRAF Mutation in Thyroid Cancer. *Ann. Surg.* **2007**, *246*, 466–471. [CrossRef]

14. Kimura, E.T.; Nikiforova, M.N.; Zhu, Z.; Knauf, J.A.; Nikiforov, Y.E.; Fagin, J.A. High Prevalence of BRAF Mutations in Thyroid Cancer: Genetic Evidence for Constitutive Activation of the RET/PTC-RAS-BRAF Signaling Pathway in Papillary Thyroid Carcinoma. *Cancer Res.* **2003**, *63*, 1454–1457.

15. Xing, M.; Alzahrani, A.S.; Carson, K.A.; Viola, D.; Elisei, R.; Bendlova, B.; Yip, L.; Mian, C.; Vianello, F.; Tuttle, R.M.; et al. Association Between BRAF V600E Mutation and Mortality in Patients with Papillary Thyroid Cancer. *JAMA* **2013**, *309*, 1493. [CrossRef]

16. Ulisse, S.; Baldini, E.; Lauro, A.; Pironi, D.; Tripodi, D.; Lori, E.; Ferent, I.C.; Amabile, M.I.; Catania, A.; Di Matteo, F.M.; et al. Papillary Thyroid Cancer Prognosis: An Evolving Field. *Cancers* **2021**, *13*, 5567. [CrossRef]

17. Azouzi, N.; Cailloux, J.; Cazarin, J.M.; Knauf, J.A.; Cracchiolo, J.; Al Ghuzlan, A.; Hartl, D.; Polak, M.; Carré, A.; El Mzibri, M.; et al. NADPH Oxidase NOX4 Is a Critical Mediator of BRAF V600E—Induced Downregulation of the Sodium/Iodide Symporter in Papillary Thyroid Carcinomas. *Antioxid. Redox Signal.* **2017**, *26*, 864–877. [CrossRef]

18. Ieni, A.; Vita, R.; Pizzimenti, C.; Benvenga, S.; Tuccari, G. Intratumoral Heterogeneity in Differentiated Thyroid Tumors: An Intriguing Reappraisal in the Era of Personalized Medicine. *J. Pers. Med.* **2021**, *11*, 333. [CrossRef]

19. Bangaraiahgari, R.; Panchangam, R.B.; Puthenveetil, P.; Mayilvaganan, S.; Bangaraiahgari, R.; Reddy Banala, R.; Karunakaran, P.; Md, R. Is There Adenoma-Carcinoma Sequence between Benign Adenoma and Papillary Cancer of Thyroid: A Genomic Linkage Study. *Ann. Med. Surg.* **2020**, *60*, 695–700. [CrossRef]

20. Park, J.Y.; Kim, W.Y.; Hwang, T.S.; Lee, S.S.; Kim, H.; Han, H.S.; Lim, S.D.; Kim, W.S.; Yoo, Y.B.; Park, K.S. BRAF and RAS Mutations in Follicular Variants of Papillary Thyroid Carcinoma. *Endocr. Pathol.* **2013**, *24*, 69–76. [CrossRef]

21. Baloch, Z.W.; Asa, S.L.; Barletta, J.A.; Ghossein, R.A.; Juhlin, C.C.; Jung, C.K.; LiVolsi, V.A.; Papotti, M.G.; Sobrinho-Simões, M.; Tallini, G.; et al. Overview of the 2022 WHO Classification of Thyroid Neoplasms. *Endocr. Pathol.* **2022**, *33*, 27–63. [CrossRef] [PubMed]

22. Ito, Y.; Yoshida, H.; Maruo, R.; Morita, S.; Takano, T.; Hirokawa, M.; Yabuta, T.; Fukushima, M.; Inoue, H.; Tomoda, C.; et al. BRAF Mutation in Papillary Thyroid Carcinoma in a Japanese Population: Its Lack of Correlation with High-Risk Clinicopathological Features and Disease-Free Survival of Patients. *Endocr. J.* **2009**, *56*, 89–97. [CrossRef] [PubMed]

23. Trovisco, V.; Soares, P.; Preto, A.; De Castro, I.V.; Lima, J.; Castro, P.; Máximo, V.; Botelho, T.; Moreira, S.; Meireles, A.M.; et al. Type and Prevalence of BRAF Mutations Are Closely Associated with Papillary Thyroid Carcinoma Histotype and Patients' Age but Not with Tumour Aggressiveness. *Virchows Arch.* **2005**, *446*, 589–595. [CrossRef] [PubMed]

24. Barbaro, D.; Incensati, R.M.; Materazzi, G.; Boni, G.; Grosso, M.; Panicucci, E.; Lapi, P.; Pasquini, C.; Miccoli, P. The BRAF V600E Mutation in Papillary Thyroid Cancer with Positive or Suspected Pre-Surgical Cytological Finding Is Not Associated with Advanced Stages or Worse Prognosis. *Endocrine* **2014**, *45*, 462–468. [CrossRef]

25. Gandolfi, G.; Sancisi, V.; Torricelli, F.; Ragazzi, M.; Frasoldati, A.; Piana, S.; Ciarrocchi, A. Allele Percentage of the BRAF V600E Mutation in Papillary Thyroid Carcinomas and Corresponding Lymph Node Metastases: No Evidence for a Role in Tumor Progression. *J. Clin. Endocrinol. Metab.* **2013**, *98*, E934–E942. [CrossRef]

26. Landa, I.; Ibrahimpasic, T.; Boucai, L.; Sinha, R.; Knauf, J.A.; Shah, R.H.; Dogan, S.; Ricarte-Filho, J.C.; Krishnamoorthy, G.P.; Xu, B.; et al. Genomic and Transcriptomic Hallmarks of Poorly Differentiated and Anaplastic Thyroid Cancers. *J. Clin. Investig.* **2016**, *126*, 1052–1066. [CrossRef]

27. Weyemi, U.; Caillou, B.; Talbot, M.; Ameziane-El-Hassani, R.; Lacroix, L.; Lagent-Chevallier, O.; Al Ghuzlan, A.; Roos, D.; Bidart, J.-M.; Virion, A.; et al. Intracellular Expression of Reactive Oxygen Species-Generating NADPH Oxidase NOX4 in Normal and Cancer Thyroid Tissues. *Endocr. Relat. Cancer* **2010**, *17*, 27–37. [CrossRef]

28. Muzza, M.; Pogliaghi, G.; Colombo, C.; Carbone, E.; Cirello, V.; Palazzo, S.; Frattini, F.; Gentilini, D.; Gazzano, G.; Persani, L.; et al. Oxidative Stress Correlates with More Aggressive Features in Thyroid Cancer. *Cancers* **2022**, *14*, 5857. [CrossRef]

29. Cizkova, K.; Foltynkova, T.; Gachechiladze, M.; Tauber, Z. Comparative Analysis of Immunohistochemical Staining Intensity Determined by Light Microscopy, ImageJ and QuPath in Placental Hofbauer Cells. *Acta Histochem. Cytochem.* **2021**, *54*, 21–29. [CrossRef]

30. Ofner, R.; Ritter, C.; Ugurel, S.; Cerroni, L.; Stiller, M.; Bogenrieder, T.; Solca, F.; Schrama, D.; Becker, J.C. Non-Reproducible Sequence Artifacts in FFPE Tissue: An Experience Report. *J. Cancer Res. Clin. Oncol.* **2017**, *143*, 1199–1207. [CrossRef]
31. Wong, S.Q.; Li, J.; Tan, A.Y.C.; Vedururu, R.; Pang, J.M.B.; Do, H.; Ellul, J.; Doig, K.; Bell, A.; Macarthur, G.A.; et al. Sequence Artefacts in a Prospective Series of Formalin-Fixed Tumours Tested for Mutations in Hotspot Regions by Massively Parallel Sequencing. *BMC Med. Genom.* **2014**, *7*, 23. [CrossRef]
32. Cazzato, G.; Caporusso, C.; Arezzo, F.; Cimmino, A.; Colagrande, A.; Loizzi, V.; Cormio, G.; Lettini, T.; Maiorano, E.; Scarcella, V.; et al. Formalin-Fixed and Paraffin-Embedded Samples for Next Generation Sequencing: Problems and Solutions. *Genes* **2021**, *12*, 1472. [CrossRef]
33. Ameziane-El-Hassani, R.; Schlumberger, M.; Dupuy, C. NADPH Oxidases: New Actors in Thyroid Cancer? *Nat. Rev. Endocrinol.* **2016**, *12*, 485–494. [CrossRef]
34. Ameziane El Hassani, R.; Buffet, C.; Leboulleux, S.; Dupuy, C. Oxidative Stress in Thyroid Carcinomas: Biological and Clinical Significance. *Endocr. Relat. Cancer* **2019**, *26*, R131–R143. [CrossRef]
35. Ameziane-El-Hassani, R.; Talbot, M.; de Souza Dos Santos, M.C.; Al Ghuzlan, A.; Hartl, D.; Bidart, J.-M.; De Deken, X.; Miot, F.; Diallo, I.; de Vathaire, F.; et al. NADPH Oxidase DUOX1 Promotes Long-Term Persistence of Oxidative Stress after an Exposure to Irradiation. *Proc. Natl. Acad. Sci. USA* **2015**, *112*, 5051–5056. [CrossRef]
36. Weyemi, U.; Lagente-Chevallier, O.; Boufraqech, M.; Prenois, F.; Courtin, F.; Caillou, B.; Talbot, M.; Dardalhon, M.; Al Ghuzlan, A.; Bidart, J.-M.; et al. ROS-Generating NADPH Oxidase NOX4 Is a Critical Mediator in Oncogenic H-Ras-Induced DNA Damage and Subsequent Senescence. *Oncogene* **2012**, *31*, 1117–1129. [CrossRef]
37. Riesco-Eizaguirre, G.; Rodríguez, I.; De la Vieja, A.; Costamagna, E.; Carrasco, N.; Nistal, M.; Santisteban, P. The BRAFV600E Oncogene Induces Transforming Growth Factor β Secretion Leading to Sodium Iodide Symporter Repression and Increased Malignancy in Thyroid Cancer. *Cancer Res.* **2009**, *69*, 8317–8325. [CrossRef]
38. Rabold, K.; Zoodsma, M.; Grondman, I.; Kuijpers, Y.; Bremmers, M.; Jaeger, M.; Zhang, B.; Hobo, W.; Bonenkamp, H.J.; de Wilt, J.H.W.; et al. Reprogramming of Myeloid Cells and Their Progenitors in Patients with Non-Medullary Thyroid Carcinoma. *Nat. Commun.* **2022**, *13*, 6149. [CrossRef]
39. Smith, J.J.; Chen, X.; Schneider, D.F.; Broome, J.T.; Sippel, R.S.; Chen, H.; Solórzano, C.C. Cancer after Thyroidectomy: A Multi-Institutional Experience with 1523 Patients. *J. Am. Coll. Surg.* **2013**, *216*, 571–577. [CrossRef]
40. de Carlos, J.; Ernaga, A.; Irigaray, A.; Pineda, J.J.; Echegoyen, A.; Salvador, P.; Anda, E. Incidentally Discovered Papillary Thyroid Microcarcinoma in Patients Undergoing Thyroid Surgery for Benign Disease. *Endocrine* **2022**, *77*, 325–332. [CrossRef]
41. Führer, D. Constitutive TSH Receptor Activation as a Hallmark of Thyroid Autonomy. *Endocrine* **2020**, *68*, 274–278. [CrossRef] [PubMed]

current issues in molecular biology

Article

Metabolic Silencing via Methionine-Based Amino Acid Restriction in Head and Neck Cancer

Anna Chiara Wünsch [1,†], Elena Ries [1,†], Sina Heinzelmann [1], Andrea Frabschka [1], Peter Christoph Wagner [1], Theresa Rauch [1], Corinna Koderer [1], Mohamed El-Mesery [2], Julian Manuel Volland [1], Alexander Christian Kübler [1], Stefan Hartmann [1] and Axel Seher [1,*]

[1] Department of Oral and Maxillofacial Plastic Surgery, University Hospital Wuerzburg, D-97070 Wuerzburg, Germany; wuenschanna@outlook.de (A.C.W.); elena.ries@icloud.com (E.R.); sina.heinzelmann@gmx.de (S.H.); andrea.frabschka@yahoo.de (A.F.); peterchristophwagner@t-online.de (P.C.W.); theresa.rauch98@web.de (T.R.); corinna.koderer@posteo.de (C.K.); julian_volland@gmx.de (J.M.V.); kuebler_a@ukw.de (A.C.K.); hartmann_s2@ukw.de (S.H.)

[2] Department of Biochemistry, Faculty of Pharmacy, Mansoura University, Mansoura 35516, Egypt; m_elmesery@mans.edu.eg

* Correspondence: seher_a@ukw.de; Tel.: +49-931-201-74841

† These authors contributed equally to this work.

Abstract: In recent years, various forms of caloric restriction (CR) and amino acid or protein restriction (AAR or PR) have shown not only success in preventing age-associated diseases, such as type II diabetes and cardiovascular diseases, but also potential for cancer therapy. These strategies not only reprogram metabolism to low-energy metabolism (LEM), which is disadvantageous for neoplastic cells, but also significantly inhibit proliferation. Head and neck squamous cell carcinoma (HNSCC) is one of the most common tumour types, with over 600,000 new cases diagnosed annually worldwide. With a 5-year survival rate of approximately 55%, the poor prognosis has not improved despite extensive research and new adjuvant therapies. Therefore, for the first time, we analysed the potential of methionine restriction (MetR) in selected HNSCC cell lines. We investigated the influence of MetR on cell proliferation and vitality, the compensation for MetR by homocysteine, the gene regulation of different amino acid transporters, and the influence of cisplatin on cell proliferation in different HNSCC cell lines.

Keywords: amino acid restriction; caloric restriction; methionine; HNSCC; SCCHN; cisplatin; amino acid transporter; SLC-family; cell vitality; low energy metabolism

Citation: Wünsch, A.C.; Ries, E.; Heinzelmann, S.; Frabschka, A.; Wagner, P.C.; Rauch, T.; Koderer, C.; El-Mesery, M.; Volland, J.M.; Kübler, A.C.; et al. Metabolic Silencing via Methionine-Based Amino Acid Restriction in Head and Neck Cancer. *Curr. Issues Mol. Biol.* **2023**, *45*, 4557–4573. https://doi.org/10.3390/cimb45060289

Academic Editor: Dumitru A. Iacobas

Received: 19 April 2023
Revised: 17 May 2023
Accepted: 22 May 2023
Published: 24 May 2023

1. Introduction

Head and neck squamous cell carcinoma (HNSCC) is one of the most common tumour types, with over 600,000 new cases diagnosed annually worldwide. With a 5-year survival rate of approximately 55%, the poor prognosis is constant and has not truly improved, despite extensive research and the introduction of new adjuvant therapies [1,2]. For this reason, the development of new strategies in cancer therapy is of enormous importance.

One of the essential characteristics of neoplastic cells is their unlimited proliferation, which results in a space-occupying lesion. For successful proliferation, two components must be available in sufficient-energy and mass. Energy is available as ATP/NADH, and mass is available as amino acids; the main components of cell mass [3] are enzymes, structural proteins, and precursors and intermediates of numerous metabolic molecules. In recent years, various forms of restriction have become increasingly important; among these strategies, the approach, which we refer to as "metabolic silencing", leads to the induction of so-called low energy metabolism (LEM). If one continuously limits energy in the form of caloric restriction (CR) [4] or mass by amino acid or protein restriction (AAR or PR) [5], the cell switches to an economical mode, which leads to the inhibition of proliferation and the induction of autophagy at the cellular level [6,7]. Such restrictions have extremely positive effects when applied permanently. The most important are the extension of lifespan and the reduction and prevention of age-associated diseases such as type II diabetes, cardiovascular diseases, and cancer. Surprisingly, this principle applies across species from yeast to Drosophila, rodents, and humans [4,5].

Moreover, the mechanisms can be explained very well. Complex networks such as hormones and growth factors, as well as the interactions between individual organs from the brain to the liver, are of great importance [8]. However, many of the decisive processes within the individual cell take place at the molecular level. A multitude of sensors continuously measure energy and mass. The ratio of AMP to ATP is measured by AMP kinase (AMPK) [9], the NAD+ level is measured via sirtuins [10], and the contents of selected amino acids (e.g., leucine, arginine, glutamine, serine, and methionine) are measured via various protein complexes. SAMTOR, for example, indirectly measures methionine content via the intermediate product S-adenosylmethionine (SAM) [11,12]. Vitamins and their intracellular availability can also have a significant influence on the proteome and its activity [13]. All these signals then transmit to the essential intracellular switching centre—the mechanistic target of rapamycin (mTOR). The sum of the signals mTOR receives determines whether mTOR actively promotes proliferation/growth or whether the cell switches to LEM by inhibiting proliferation and activating autophagy [11,14,15].

In principle, it is possible to exert a targeted influence on this network and induce LEM by means of metabolic silencing. Thus, a permanent reduction in calories by 10% to 50% or a reduction in the amount of protein is not always necessary; the restriction of a selected amino acid is sufficient to set in motion a process that is almost identical to the mechanisms at work in CR or PR. Due to the central position of the amino acid methionine, methionine restriction (MetR) is often used [16,17]. Although MetR is easy to implement in everyday laboratory work, e.g., by simply removing methionine from the cell culture medium, it is much more difficult to implement as a therapy since methionine must be either enzymatically degraded in vivo (by means of the bacterial enzyme methionase) or removed from ingested food [18,19]. Basically, MetR in vitro is only representative of energy or mass restriction, also known as metabolic silencing, which is a universal and methodical approach within tumour therapy. MetR shows the enormous potential of the AAR and allows molecular biological processes to be analysed in the laboratory, but it is in no way mandatory that this approach be implemented in vivo.

In principle, AAR and PR have great potential in cancer therapy. Levine et al. [20] impressively demonstrated the influence of PR on the progression of tumours in mice via in vivo experiments. Thus, in mice fed a lower amount of protein (4–7%) compared to the control group, the size of the developing tumours was drastically reduced (by approx. 80%), and in some of the mice (approx. 20%) the development of tumours was completely suppressed, although 10,000 aggressive neoplastic cells were inoculated [20].

In this work, for the first time, we analysed the potential of MetR to inhibit proliferation in selected HNSCC cell lines. In addition, the cell lines were tested for their methionine dependence. Although methionine itself is an essential amino acid, methionine can be regenerated at the cellular level from various intermediates, such as homocysteine (Hcy). Tumour cells very often lose this ability, which may enhance the effect of MetR as a tumour therapy [21]. We also tested the influence of MetR on cell vitality, the gene regulation of different amino acid transporters, and the influence of cisplatin on cell proliferation in different HNSCC cell lines.

2. Materials and Methods

2.1. Cell Culture

The murine fibroblast cell line L929 was purchased from the Leibniz Institute, DSMZ-German Collection of Microorganisms and Cell Cultures GmbH (Braunschweig, Germany). The human cell lines HeLa and HaCaT, and the HNSCC cell lines FaDu, Detroit562, SCC9, and SCC25 were purchased from the American Type Culture Collection (ATCC)(LGC Standards, Wesel, Germany). L929 and HeLa cells were cultured in RPMI 1640 medium (Gibco, Life Technologies; Darmstadt, Germany), HaCaT cells were cultured in DMEM (Gibco, Life Technologies), FaDu and Detroit562 cells were cultured in MEM alpha (Gibco, Life Technologies), and SCC9 and SCC25 cells were cultured in DMEM/F12 (Gibco, Life Technologies). All cell lines were cultured with 10% FCS (Sigma–Aldrich, Darmstadt, Germany) and 1% penicillin/streptomycin (P/S; 100 U/mL penicillin and 100 μg/mL streptomycin (Thermo Fisher Scientific, Darmstadt, Germany)). Media for SCC9 and SCC25 cell lines were additionally supplemented with 25 mM HEPES (Thermo Fisher) and 10 μL/L hydrocortisone (400 ng/mL) (Sigma–Aldrich). All cells were cultured at 37 °C in a humidified atmosphere containing 5% CO_2. For the AAR experiments, media lacking the amino acid methionine was used. For the controls (full medium), the amino acid methionine (Sigma–Aldrich, Darmstadt, Germany) was added at the concentrations indicated as follows: RPMI 1640 medium (Genaxxon, Ulm, Germany) 15 mg/L L-methionine; MEM alpha (Bio&Cell, Feucht, Germany) 15 mg/L L-methionine; DMEM (Gibco, Life Technologies) 30 mg/L L-methionine; and DMEM/F12 (Genaxxon) 17.25 mg/L L-methionine.

2.2. Crystal Violet Staining

Cells were seeded at 10,000 cells in 100 μL of culture medium per well of a 96-well plate and incubated overnight. The following day, the cells were incubated in a full or methionine-free (Met(-)) medium. For compensation experiments, DL-Hcy (Sigma–Aldrich) was used at 800 μM and 1.5 μM vitamin B12 (Sigma–Aldrich) in Met(-) medium. The number of measured values, the incubation period, and the number of experimental repetitions are mentioned in the corresponding figure legends. For staining, the supernatants were removed, and the cells in each well were incubated with 50 μL of crystal violet solution (1% crystal violet in 20% methanol; Carl Roth, Karlsruhe, Germany) for 10 min and subsequently washed five times with distilled water. The plates were dried for 2 h in the dark. For quantification, 100 μL of methanol was added to each well, and the plate was incubated for 10 min until the crystal violet was completely dissolved. The photometric absorbance was measured at 595 nm using a microplate reader (Tecan, Crailsheim, Germany). For data analysis, the experiments were repeated at the indicated times to calculate the mean values and standard deviations. The results were normalised to the untreated control (100%). The relative cell number values determined via the crystal violet assay with the stimulated probes (CV_S) were normalised to those of the untreated control (CV_C) ((CV_S/CV_C) = CV_R). To obtain the percentage values, the CV_R value was multiplied by 100 (RCN (%) = (CV_S/CV_C) × 100 = CV_R%).

2.3. ImageXpress Pico Automated Cell Imaging System—Digital Microscopy (Pico Assay)

Cells were seeded at 10,000 cells in 100 μL of culture medium per well of a 96-well plate and incubated overnight. The following day, the cells were incubated in a complete or methionine-free medium. The incubation time is stated in the corresponding figure legends. For staining, 10 μL of Hoechst staining solution (1:200 dilution of Hoechst 33342 (Thermo Fisher, Darmstadt, Germany) (10 mg/mL in H_2O) in medium) was added to each well. After a 20–30 min incubation period, wells were analysed with an ImageXpress Pico automated cell imaging system (Molecular Devices, San Jose, CA, USA) via automated digital microscopy. The cells were analysed with transmitted light and in the DAPI channel at 4× magnification. The complete area of every well was screened. The focus and exposure time were set via auto setup and controlled by analysing 3–4 test wells. Finally, every result was confirmed visually, and 95% of cells were counted and analysed.

2.4. Analysis of the Cell Progression Rate Using the Pico Assay

Cells were seeded at 10,000 cells in 100 μL of culture medium per well in a 96-well plate. After 24, 31, 48, 55, 72, 79, 96, and 103 h, cell numbers were measured with six values for every time point as described under the Pico Assay Section (2.3). The growth of a cell population can be described with the following formula:

$$N_t = N_0 \times 2^{(t \times f)}$$

(N_t = cell number at time t; N_0 = cell number at time 0; t = time in days (d); f = cell division frequency (1/d)).

To determine f, the formula is rearranged as follows:

$$f = \frac{\left(log\left(\frac{N_t}{N_0}\right)/\log(2) \right)}{t}$$

To obtain an overview, the measured values were first plotted as a simple diagram. From this, it was possible to see at what point the growth entered the plateau phase. From these values, the individual f values were calculated for the intermediate periods (e.g., $\Delta 24/32$, $\Delta 32/48$). The total value f was then calculated as the mean of the Δf values.

2.5. Live/Dead Assay

The cell lines L929, HeLa, HaCaT, Detroit562, FaDu, SCC9, and SCC25 were seeded at 10,000 cells/well on a 96-well plate. On the following day, the medium was removed, and the cells were stimulated with a full medium or Met(-) medium. Cells stimulated with 1 µM staurosporine (Seleckchem, Planegg, Germany) served as the death control in each case. After 6 h, 24 h, and 48 h, measurements were performed with the EarlyTox Live/Dead Assay Kit (Molecular Devices). For this, the medium was removed, and 100 µL/well of a staining solution (5 µg/mL Hoechst 33342 (company), 6 µM EthD-III, and 6 µM CAM in PBS (company)) was added per well. After 30 min of incubation, measurements were made using the "Cell Scoring: 3 Channels" program of the ImageXpress Pico Automated Cell Imaging System (Molecular Devices). To determine the absolute cell number, measurements were made in transmitted light and in the DAPI channel using Hoechst 33342, as well as via the FITC (CAM) and TexasRed (TRITC)/(EthD-III) channels over the entire area of the respective well at 4x magnification. Exposure time and focus were determined by auto settings. Subsequently, for better visualisation of the cells, the measurement was repeated with a selected area of the well (1.39 mm × 1.39 mm) at $10\times$ magnification, from which the images shown in this publication were selected. For each experimental condition, 2 wells were evaluated. The experiment was performed two times.

2.6. Semiquantitative RT–PCR

A total of 500,000 cells/well were seeded in a 6-well plate. The following day, cells were stimulated with or without methionine in the corresponding medium for 24 h and 72 h. RNA isolation was performed using an RNeasy Mini Kit (Qiagen, Hilden, Germany) according to the instructions provided by the manufacturer. One microgram of RNA was transcribed into cDNA using the QuantiTect Reverse Transcription Kit (Qiagen) according to the instructions provided by the manufacturer. Next, 20 ng cDNA was used in the PCR with 1.5 µL of the appropriate primer (QuantiTect Primer Assay, Qiagen) and 12.5 µL of a ready-to-use qPCR master mix (QuantiTect SYBR Green PCR Kit, Qiagen). The thermal cycling program was composed of initial denaturation at 95 °C for 15 min, 39 cycles at 95 °C for 15 s, 30 s at 54 °C, and 30 s at 72 °C. Triplicates for each data point were measured. The β-actin and GAPDH genes were used as internal controls (standard). The following primers from Qiagen were used: ACTB QT00095431, GAPDH QT00079247, SLC7A5 (LAT1) QT00089145, SLC7A11 QT00002674, SLC1A5 QT00083909, and SLC43a2 QT00027559. For analysis, the $2^{-\Delta\Delta Ct}$ method for relative quantification of gene expression was used, including error calculation (range) based on Livak and Schmittgen [22].

2.7. Statistical Analysis

Statistical analysis was performed using the GraphPad Prism program (version 6.04; GraphPad Software, San Diego, CA, USA). One-way ANOVA was used to compare and analyse data of different groups, followed by the Tukey–Kramer multiple comparison test (ns; nonsignificant; * $p < 0.05$, ** $p < 0.01$; *** $p < 0.001$).

3. Results

As already mentioned, the proliferation of neoplastic cells is an essential characteristic. For this reason, we analysed the influence of MetR on the proliferation of the established HNSCC cell lines FaDu, Detroit562, SCC9, and SCC25 in initial experiments. For comparison, the two human cell lines HeLa, generated from cervical carcinoma cells, and HaCaT, an immortalised keratinocyte line, were used. To assess the potential of AAR, the murine fibroblast cell line L929, which was established as a model system for MetR in earlier work by our research group, was also used [23].

3.1. Restriction of Methionine Inhibited the Proliferation of HNSCC Cell Lines

In each case, 10,000 cells/well were seeded on 96-well plates and stimulated the following day with a control medium or Met(-). At time points of 0 h, 24 h, 72 h, and 120 h, the cells were analysed by digital microscopy using the ImageXpress Pico automated cell imaging system (Figure 1a–g). Then, the DNA of the cell nuclei was stained with Hoechst 33342, allowing automated counting of the stained nuclei over the entire area of the well and, thus, determination of the absolute cell number. In HeLa cells (Figure 1a), MetR significantly inhibited proliferation at 24 h, while in HaCaT cells (Figure 1b), the effect was recognisable after 72 h. In essence, MetR had an antiproliferative effect in HNSCC cells that was observed at 24 h in FaDu cells (Figure 1c), and significant after 72 h in Detroit562 (Figure 1d), SCC9 (Figure 1e), and SCC25 cells (Figure 1f). Although MetR had a clear effect in all HNSCC cell lines analysed, a comparison of the results with those obtained in the murine cell line L929 (Figure 1g) clearly showed that the effect took more time and was less pronounced in these cell lines than that in the control. This became clear in the overview of the analysed cells (Figure 1h). Crystal violet staining, a simple and fast method, was used here for the analysis. In this case, data from the control group at the respective time were set as 100%, and data from the Met(-) group are shown relative to these baseline data. Although only relative cell numbers could be determined in this way, this experiment allowed a better representation of the dynamics of MetR. It became clear that the murine cell line (green line) reacted most effectively to MetR. The two human cell lines (black lines) showed almost equal efficiency, but the response in these cells was delayed compared to that of L929 cells. All HNSCC cells (red lines) responded to MetR, but after an initial decrease in proliferation, the cells entered a plateau phase again, indicating that they could at least partially or temporarily compensate for the effects of MetR.

3.2. Analysis of the Cell Progression Rate

The different speeds at which cells react to a MetR are influenced by many factors, from the possibility of storing methionine to regenerating the molecule via detours, e.g., autophagy. Another factor is the proliferation rate of the cell. As a simple example, if a cell line does not divide within the first 24 h after splitting in culture, no effect of MetR on the proliferation rate can be determined. In addition, the speed at which a cell divides plays a role; this speed is referred to here as the f value. The f value indicates the frequency with which a cell or a cell population divides/doubles with a simple numerical value. A value of 1 means that the population doubles once within one day. Thus, we determined the f value for the cell lines used, as previously published [24]. Three different f values were determined for further analysis. The value f_{total} indicates the division rate over the entire period investigated until cell division enters the saturation phase. In Figure 2, the measured values used for this analysis are marked in black, and the excluded measured values are marked in red. The value f_{start} indicates the division rate at the beginning of the experiment within the first 7 h, and f_{max} is the highest f value achieved within the measurement period.

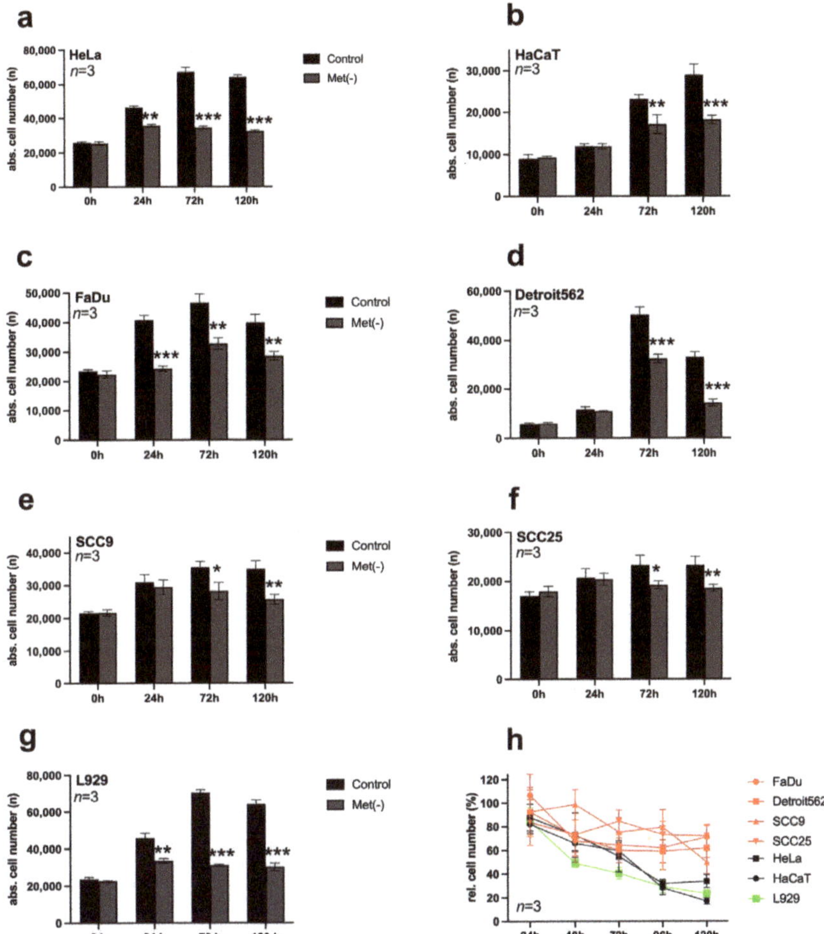

Figure 1. Analysis of proliferation in different cell lines subjected to restriction of the amino acid methionine. (**a–g**) In each case, 10,000 cells/well were seeded on 96-well plates and stimulated in triplicate on the following day with a control medium or Met(-) medium. Absolute cell counts were analysed at 0 h, 24 h, 72 h, and 120 h using digital microscopy with the ImageXpress Pico automated cell imaging system. A summary of the results from three experimental replicates ($n = 3$) is shown in each case. (**h**) Summary of the proliferation analysis based on relative cell number counted using crystal violet staining. In each case, 10,000 cells/well were seeded on 96-well plates and stimulated the following day with a control medium (set as 100%) or Met(-) medium. Cells were analysed at 24 h, 48 h, 72 h, 96 h, and 120 h. A summary of the results from three experimental replicates ($n = 3$) is shown in each case (* $p < 0.05$, ** $p < 0.01$; *** $p < 0.001$).

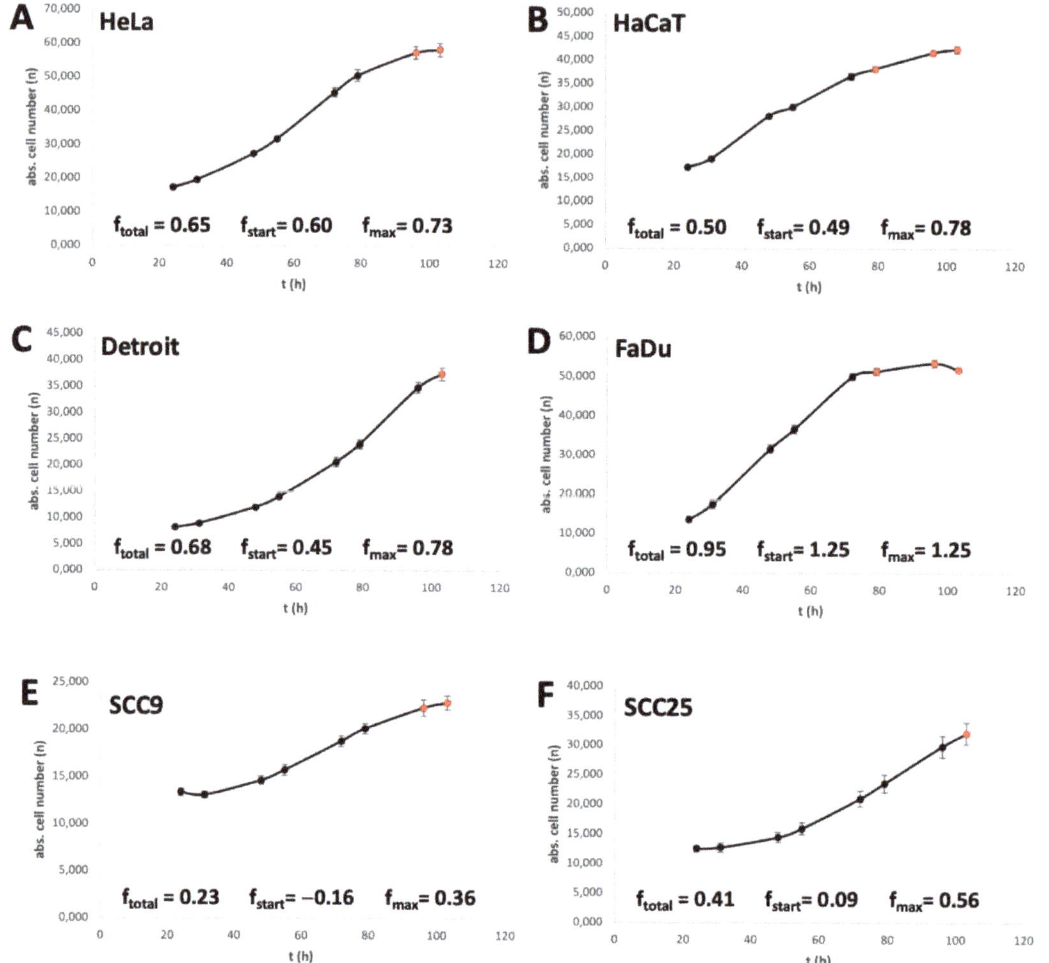

Figure 2. Analysis of the cell progression rate. (**A–F**) Cells were seeded at 10,000 cells/well in a 96-well plate. After 24, 31, 48, 55, 72, 79, 96, and 103 h, cell numbers were measured as described in Section 2 under Pico Assay (2.3). The summarised results (*n* = 3) are shown in Figure 2.

It is easy to see that the f values and the response rate to MetR correlate within a certain range. This is particularly clear for the cell lines SCC9 and SCC25, which have the lowest f values (Figure 2E,F) and the lowest response rate to MetR (Figure 1e,f). On the other hand, FaDu, with the highest f values, also has the corresponding best MetR response rate. In HaCaT and Detroit cells, the low MetR sensitivity seems to be due to the slow f_{start} values (0.49 and 0.45) at the beginning of proliferation. Here, the cells have to go until the amino acid restriction is reflected in the division behaviour. On the other hand, the HeLa cell line starts proliferating relatively quickly and responds to MetR at an early stage (Figures 1a and 2A).

3.3. HNSCC Cell Lines Are Mainly Methionine Dependent

Methionine is one of the most important amino acids, partly because of its roles as an amino acid required for the initiation of protein expression and a central molecule in metabolism. Thus, methionine can generally be regenerated by cells from various intermediates, such as SAM or Hcy, to a limited extent [25]. However, tumour cells often lose this ability [21]. For this reason, the methionine dependence of the selected cell lines was analysed in a compensation experiment. In each case, 10,000 cells/well were seeded on 96-well plates and stimulated on the following day with control medium, Met(-) medium, or Met(-) medium to which 800 µM Hcy had been added (Met(-)/Hcy). Cells were analysed at 0 h, 72 h, and 120 h using the ImageXpress Pico automated cell imaging system to determine absolute cell numbers. The results for HeLa and HaCaT cells show a compensatory response to MetR via Hcy to 100% (Figure 3a,b). HNSCC cells, on the other hand, had difficulty compensating for MetR with Hcy, and no significant compensation was measured in FaDU, SCC9, and SCC25. Furthermore, only Detroit562 shows homocysteine-mediated compensation after 120 h, which does not correspond to the growth of the control. (Figure 3c–f). For completeness, the results of the compensation experiment in L929 cells are also shown (Figure 3g).

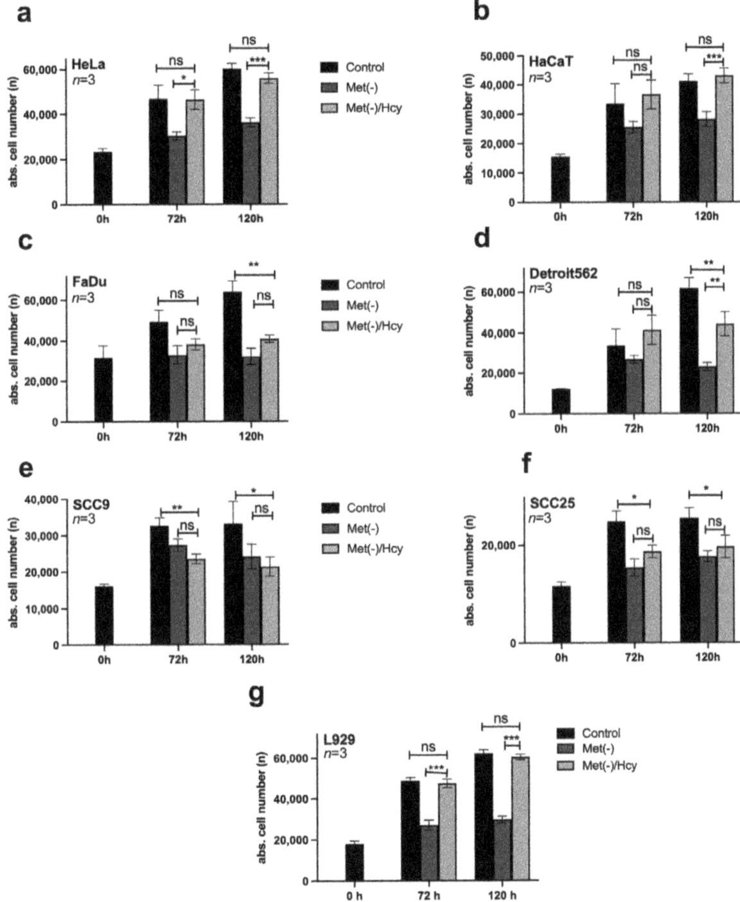

Figure 3. Homocysteine (Hcy)-based competition of MetR in different cell lines. (**a–g**) In each case, 10,000 cells/well were seeded on 96-well plates and stimulated in triplicate on the following day with control medium, Met(-) medium, or Met(-) medium to which 800 µM Hcy had been added (Met(-)/Hcy). Cells were analysed at 0 h, 72 h, and 120 h using the ImageXpress Pico automated cell imaging system to determine absolute cell counts. A summary of the results from three experimental replicates ($n = 3$) is shown in each case (ns; nonsignificant; * $p < 0.05$, ** $p < 0.01$; *** $p < 0.001$).

3.4. MetR Did Not Induce Significant Cell Death

In principle, the proliferation studies under MetR (Figure 1) do not directly prove that the reduced cell number is due to a purely antiproliferative mechanism and not to a form of cell death. For L929 cells, we have already shown in a previous publication that MetR is purely antiproliferative and does not induce significant cell death [23]. In this work, we analysed cell viability using the live/dead assay. As shown in Figure 1h, a significantly reduced cell number can already be detected after 48 h for most of the cell lines used here. For this reason, cell viability was analysed at the early time points of 6, 24, and 48 h, as the beginning stages of cell death should already be detectable during this period. It is necessary to analyse the early time points because at the later time points, dead cells can dissolve (apoptotic bodies or autophagy), or after death, adherent cells can sphere off, detach from the cell culture matrix, and go into suspension; thus, these cells are no longer clearly detectable via live/dead staining. The two dyes, calcein AM (CAM) and ethidium homodimer-III (EthD-III), were used for analysis by digital cell microscopy. CAM is a membrane-permeable dye. In living cells, the nonfluorescent CAM is converted into green, fluorescent calcein by means of acetoxymethyl ester hydrolysis by intracellular esterases. Thus, the majority of the cytoplasm is stained green, which, in addition to the transmitted light analysis, allows an additional optical analysis of the cell shape, since dying or dead cells usually change shape considerably and eventually sphere off. EthD-III is a red fluorescent dead cell stain for bacteria and mammalian cells. It is a cell membrane-impermeant nucleic acid dye that stains only dead cells with damaged cell membranes. Staurosporine (1 µM), a nonspecific inhibitor of various protein kinases (e.g., PKA and PKC) that can induce cell death, was used as a positive control [26]. Figure 4 shows the results for the time points 24 and 48 h for all cell lines examined. While staurosporine was able to induce increased cell death in all cell lines, the comparison between control and MetR shows neither an increase in cell death under amino acid restriction nor a significant change in cell shape, which would indicate that the cell is approaching cell death. The cells showed the same vitality as in the control.

3.5. Under MetR, Amino Acid Transporters Are Subsequently Upregulated Then Downregulated

Basically, amino acids are one of the most essential resources for cell growth and proliferation, and as already mentioned, they are responsible for the largest part of the cell mass [3]. Membrane-bound amino acid transporters (AAT), which allow the import and export of metabolites, are correspondingly important. The families of AATs are large and quite heterogeneous. The substrate specificity is just as varied as the transport mechanisms, which exploit all possibilities from UniProt to symport to antiport [27]. In this work, we focused on the gene regulation of four AATs that are very frequently analysed in the context of cancer. These are SLC7A5 (LAT1), SLC7A11 (Xc- or xCT system), SLC1A5 (ASCT2), and SLC43a2 (LAT4). SLC7A5 mediates the uptake of neutral AAs (leucine, isoleucine, phenylalanine, methionine, histidine, tryptophan, valine, and tyrosine) into cells in exchange for the efflux of intracellular substrates (AAs and/or glutamine). System Xc(-) is composed of a light-chain subunit xCT (also known as SLC7A11) and a heavy-chain subunit (CD98hc, also referred to as SLC3A2) and functions as a Na^+-independent transporter that mediates the exchange of extracellular cystine for intracellular glutamate. SLC1A5 serves as an obligatory exchanger that imports a sodium-coupled amino acid substrate into cells and exports another sodium-coupled amino acid substrate with 1:1 stoichiometry. It is the primary transporter for importing glutamine. Glutamine flux, which is dependent on the balance between the uptake of glutamine by SLC1A5 and its subsequent export by SLC7A5, leads to high intracellular availability of essential amino acids (EAAs) (overview of SLC7A5, SLC7A11, and SLC1A5 in [28]). The amino acid transport activity induced by SLC43a2 (LAT4) is sodium-, chloride- and pH-independent. The main amino acids transported are isoleucine, leucine, phenylalanine, and methionine [29]. This transporter is of particular importance because it is upregulated in neoplastic cells and thus enables methionine to be "stolen" from the environment to increase the growth of the tumour [30,31].

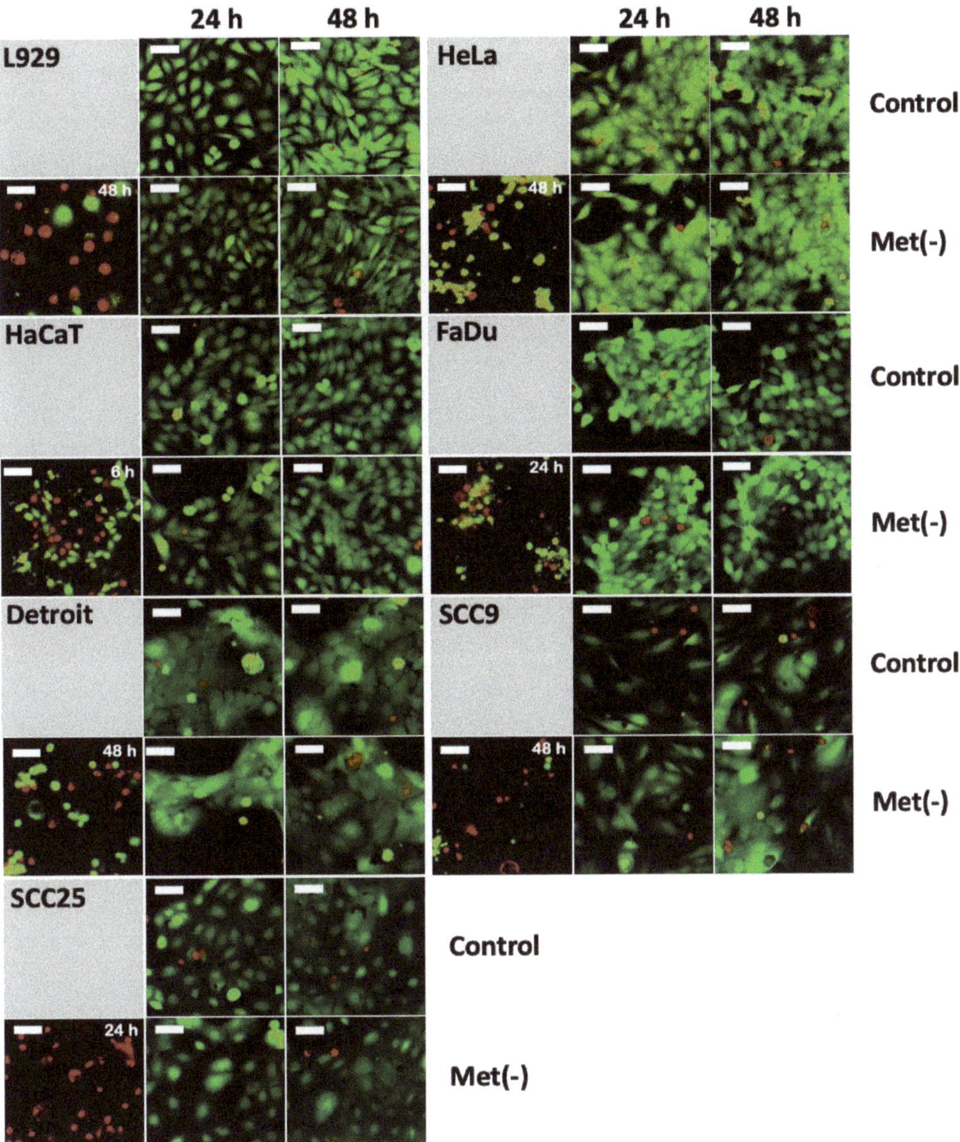

Figure 4. The cell lines L929, HeLa, HaCaT, Detroit562, FaDu, SCC9, and SCC25 were seeded at 10,000 cells/well on a 96-well plate. On the following day, the medium was removed, and the cells were stimulated with full or Met(-) medium. Cells stimulated with 1 µM staurosporine served as the dead control cells. After 6 h, 24 h, and 48 h, measurements were performed with the "EarlyTox Live/Dead Assay Kit" (Molecular Devices). Measurements were made using the "Cell Scoring: 3 Channels" program of the ImageXpress Pico Automated Cell Imaging System (Molecular Devices) with a 10× magnification. The cytoplasm of leaving cells was measured in the FITC channel with CAM, and dead cells were measured in Texas Red-(TRITC)-channel with EthD-III. The results for 24 h and 48 h are shown. The 6 h time point showed no cell death for the control or MetR group. The dead

cell control is on the left side for every cell line, and the time of the dead cell control measurement is indicated in the upper right corner. The size of the white bar corresponds to 52.44 μm. For each experimental condition, two wells were evaluated. The experiment was performed two times independently. The representative results of one experiment are shown. For better resolution, the figures are also added as a supplement (Suppl. Figure S1—Live/Dead Assay Figures).

The gene expression of the mentioned AAT was examined in the cell panel under MetR after 24 h and 72 h by RT–PCR. The experiment was repeated three times (V1–V3). Figure 5 shows the differential expression (x-fold) in comparison to the control at the same time point in the individual experiments. An overview of the numeric results is attached as a table in the supplement (Suppl. S2—RT_PCR_Results_Total).

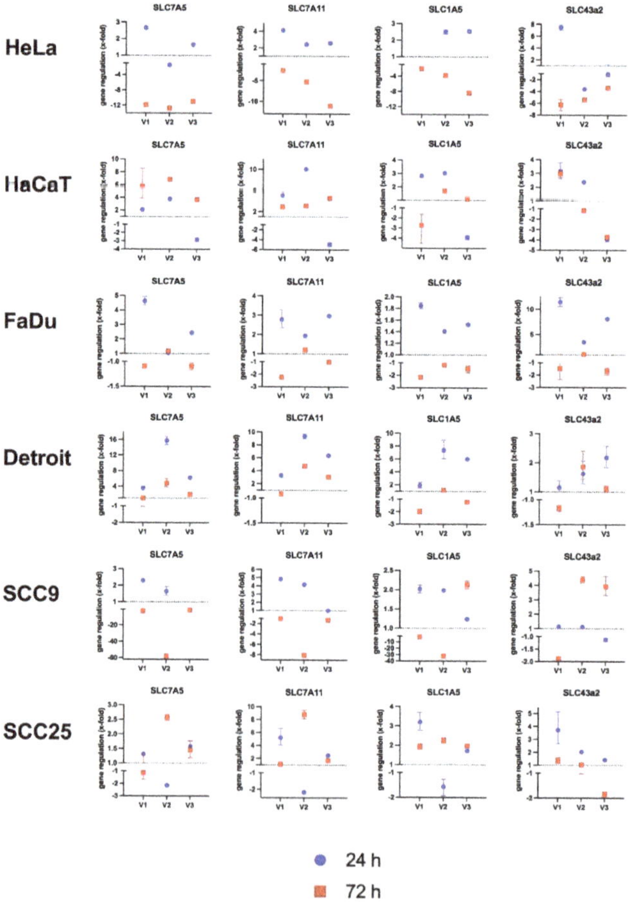

Figure 5. Analysis of the gene expression of different amino acid transporters. The cells were stimulated for 24 h or 72 h with or without methionine in the corresponding medium. The isolated RNA was transcribed into cDNA, and the relative expression was determined by RT–PCR using the $2^{-\Delta\Delta Ct}$ method. β-Actin was used as a standard. The results of the relative expression of the samples under MetR are shown. The dashed line represents an unchanged expression at a value of 1. The total of three independent experiments is shown individually, with the corresponding value of each experiment for 24 h and 72 h. Each PCR value was determined as a triplet. The range (error range) of each measured value is marked by error bars. An overview of the numeric results is attached as a table in the supplement (Suppl. S2—RT_PCR_Results_Total).

The results show no clearly significant, strictly reproducible results for the genes studied. However, in most cases, it can be observed that within the first 24 h, the receptors are upregulated, and the cells respond to the amino acid restriction with an increased expression compared to the control, while after 72 h, the expression then decreases again. This can be seen very well in the cell line HeLa for the receptor SLC7A11. For the cell line SSC9, on the other hand, the results for all analysed genes are very heterogeneous and do not show a clear trend.

3.6. The Efficacy of Cisplatin Is Only Marginally Affected by MetR

In the field of HNSCC therapy, the classic cytostatic drug cisplatin plays a very important role, as the range of adjuvant drug therapies has hardly improved in recent decades [32]. For this reason, we analysed the influence of MetR on the efficacy of cisplatin. Thus, cells were incubated for 72 h with a cisplatin log2 dilution series in full medium and under MetR, and subsequently evaluated using the ImageXpress Pico automated cell imaging system to determine absolute cell numbers. To compare the efficacy of cisplatin under different conditions, we used the IC_{50}, which is defined as the concentration at which the drug reaches its half-maximal efficacy. For this purpose, the absolute cell numbers must be converted into percentages. Therefore, the control value (without cisplatin) was set as 100%, and the lowest value in a measurement series was set as 0%. If one now places the two curves in a diagram, one can see whether there is a shift in the curve and, thus, a change in the IC_{50}. For objective representation, the results of the absolute cell numbers are shown for each cell line, as well as the normalised representation in percent to the right (Figure 6). If one compares the individual IC_{50} values with each other, one can see that the values are either almost identical or increase only slightly under MetR. As a trend, the curves are almost identical at the beginning and at the end, which means that the basic sensitivity of the cells to cisplatin changes only marginally. The slight "deterioration" of the IC50 can be explained simply by the mechanism of action of cisplatin. Cisplatin is incorporated into newly synthesised DNA during the initial phase of cell division and causes DNA cross-linking, which then leads to cell death via different mechanisms [33]. Cells under MetR, in which proliferation is inhibited and DNA synthesis comes to a standstill, accordingly show an increase in the IC_{50} of cisplatin. This may seem counterintuitive in the context of chemotherapy, but we show in the discussion that this is not the case precisely because of the phenomenon of "differential stress resistance."

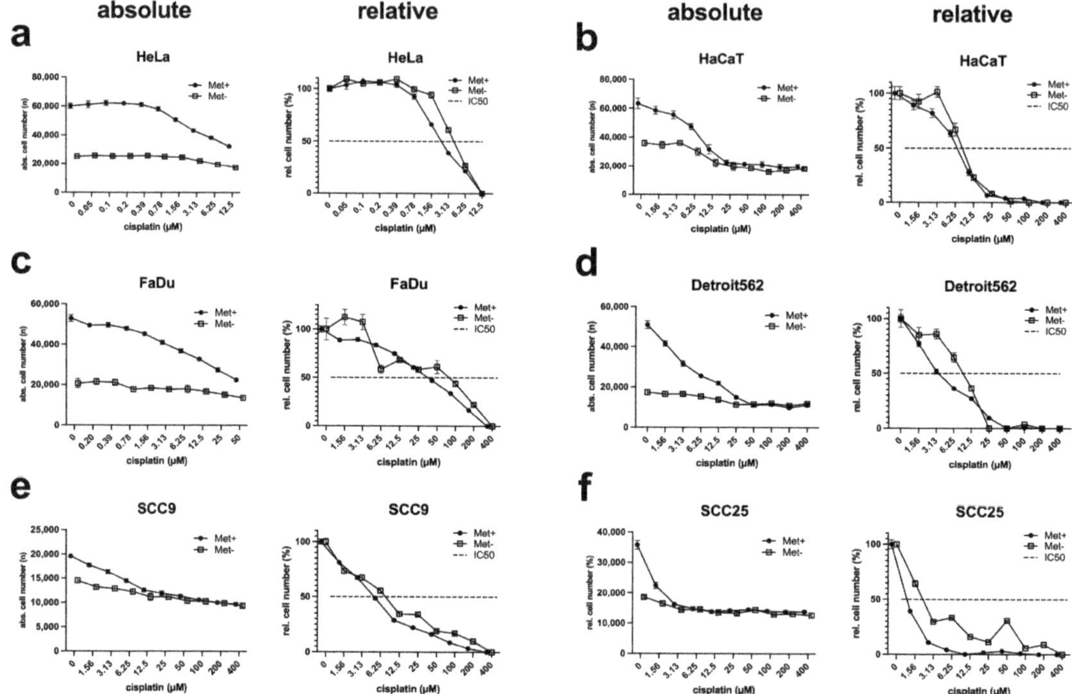

Figure 6. Analysis of the efficacy of cisplatin under MetR. (**a–f**) A total of 10,000 cells/well were seeded on 96-well plates. On the following day, cells were incubated in triplicate with control or Met(-) medium and a log2 dilution of cisplatin, with a starting concentration of 400 μM (12.5 μM for HeLa and 50 μM for FaDu). After 72 h, the cells were analysed using the ImageXpress Pico automated cell imaging system to determine absolute cell numbers. To compare the efficacy of cisplatin under different conditions, we used the IC_{50}, which is defined as the concentration at which the drug reaches its half-maximal efficacy. For this purpose, the absolute cell counts (shown on the left side for each cell line) must be converted into percentages (shown on the right side for each cell line). The respective control value (without cisplatin) was set as 100%, and the lowest value in a measurement series was set as 0%. If one now places the two curves in a diagram, one can see whether there is a shift in the curve and, thus, a change in the IC_{50}. For objective representation, the results of the absolute cell numbers are shown for each cell line, as well as the normalised representation in percent to the right. The dashed line indicates the inhibition of proliferation at 50% (IC_{50}). The experiments were carried out three times. The figures show a summary of the experiments (n = 3) in one diagram.

4. Discussion

In principle, different forms of restriction have great potential to prevent the development of age-associated diseases such as type II diabetes, cardiovascular diseases, and cancer if they are applied continuously. At the same time, they exert their effects through many common mechanisms at the molecular level, which, as already mentioned, are, above all, reduced proliferation, the induction of autophagy and the implementation of LEM. For this reason, restriction is, in principle, also an interesting approach in tumour therapy. The work of Levine et al. [20,34] in suppressing the development and progression of tumours in vivo in mice through protein restriction, as well as the successful use of different caloric restriction mimetics (CRMs), shows enormous potential.

In this paper, for the first time, we publish the effects of AAR on the proliferation of HNSCC cell lines. Regarding proliferation, the restriction of methionine had a significant effect on proliferation and the proliferation rate. Strikingly, the proliferation of the four

HNSCC cell lines FaDu, Detroit562, SCC9, and SCC25 reached a plateau, and restriction did not show the same efficiency in the other cell lines examined (Figure 1h). In essence, cancer cells are characterised by metabolic programming; in other words, they optimise their cellular metabolism to the conditions to enable growth and proliferation. One of the best-known mechanisms is the Warburg effect [35]. It is, therefore, not surprising that tumour cells can resist restriction to a certain extent. More importantly, all HNSCC cell lines showed a response to AAR within the first 72 h.

The results regarding Hcy compensation are also promising. All four HNSCC cell lines showed a lower ability or no ability to compensate for methionine with Hcy, which is in line with the consensus in the literature [21]. This may also be why the absolute number of HNSCC cells decreased slightly after a period of AAR. Of course, the complete absence of the amino acid methionine inevitably leads to cell death after a certain period of time. In contrast, cell death by MetR did not play a significant role and was not induced in the first 48 h in any cell line used in this publication, as shown by the live/dead assay (Figure 2). The decreasing cell numbers in some cell lines after 120 h can usually be attributed to the consumption of essential metabolites (e.g., glucose) in the medium, as the total cell number decreased both in the control and under MetR.

The most likely reason, however, for the lower influence of MetR on the proliferation rate in HNSCC cell lines is simply the time factor. A short impulse is probably not sufficient to have a lasting and significant influence on proliferation, which occurred with a delay and was not as efficient as in HaCaT or L929 cells, for example. We have already observed this phenomenon in many other experiments, which is why medium- and long-term experiments on AAR in HNSCC seem to be necessary. A very good approach here is the use of perfusion cultures, which make it possible to supply cells with fresh medium over a longer period permanently, and thus also investigate long-term effects on cells in culture. We used this method in another work to analyse the influence of MetR compared to glucose restriction over a longer time period on the murine cell line L929 [24]. A further argument for the necessity of long-term analyses in the form of perfusion culture is the differential expression results for the amino acid transporters. In the short term, the different amino acid transporters were upregulated after 24 h, but then generally downregulated after 72 h (Figure 5). Only an analysis over several days under constant conditions in the perfusion culture can show whether this is permanent.

However, the experiments clearly show that no extreme stress situation occurs due to MetR, but the cell can react or act specifically to deficits of energy and mass with its programmes developed during evolution. The experiments with cisplatin also prove that MetR in vitro has no direct negative influence on the effectiveness of the drug.

The fact that there is no more extreme stress situation can also be seen in the expression analyses of the amino acid transporters, which tend to be upregulated in the short term but then down-regulated again. However, transcriptome analyses, which analyse a large number of genes simultaneously, would be helpful to show the exact reaction of a cell to an amino acid restriction at the transcription level.

In the immediate context of cancer, the question arises to what extent, for example, MetR can influence the effectiveness of various drugs and what influence the restriction has on the growth of the tumour. Basically, there are two possible situations. First, MetR forces the cells in the entire organism into a LEM. This would result in an inhibition of proliferation, which would also affect the tumour. In this case, the growth of the tumour would at least be limited. The second situation appears counterproductive at first glance. LEM is only induced in healthy tissues, and proliferation is also only limited to healthy cells. The neoplastic cells ignore the signals and are further programmed to divide at a high rate. It is exactly this situation that creates an extreme, which is of enormous advantage for tumour therapy. This situation is called "differential stress resistance" (DSR). Cancer cells that express oncogenes and exhibit egocentric proliferative behaviour respond to certain cancer-promoting and growth-promoting factors. Moreover, cancer cells do not respond to protective signals generated by short-term fasting (SCR) or long-term nutrient restriction.

Hence, cells can be exposed to the following two different extreme situations: somatic cells may be protected, while cancer cells become increasingly vulnerable to attack [36].

A good example of this theory is the use of cisplatin in combination with fasting. Short-term starvation (STS) based on caloric and/or protein reduction protects normal cells while simultaneously sensitising malignant cells to high-dose chemotherapeutic drugs such as cisplatin in mice and possibly patients. The fasting-dependent protection of normal cells and sensitisation of malignant cells depend, in part, on reduced levels of IGF-1 and glucose [37].

We were able to show in this work for the first time that MetR represents a further promising alternative in HNSCC therapy. Furthermore, this approach is supported by the increased methionine dependence of HNSCC cell lines, which should enhance the efficiency of methionine-based therapy.

5. Conclusions

In this work, it was shown for the first time that methionine-based amino acid restriction represents a further promising alternative in HNSCC therapy. Furthermore, the use of this approach is supported by the increased methionine dependence of HNSCC cell lines, which should enhance the efficiency of methionine-based therapy.

Supplementary Materials: The following supporting information can be downloaded at: https://www.mdpi.com/article/10.3390/cimb45060289/s1 Figure S1. Live/Dead Assay Figures; Suppl. S2 RT_PCR_Results_Total.

Author Contributions: Conceptualisation, A.S.; methodology, A.S., E.R., A.C.W., S.H. (Sina Heinzelmann), A.F., P.C.W. and J.M.V.; software, S.H. (Stefan Hartmann); validation, M.E.-M., E.R., C.K., T.R., S.H. (Sina Heinzelmann), A.F., P.C.W. and A.C.W.; formal analysis, A.S.; statistical analysis, M.E.-M. and S.H. (Stefan Hartmann); investigation, A.C.W., E.R., C.K., J.M.V. and A.S.; data curation, A.S.; writing—original draft preparation, A.S. and M.E.-M.; writing—review and editing, S.H. and A.C.K.; visualisation, A.S.; supervision, A.S.; project administration, A.S. All authors have read and agreed to the published version of the manuscript.

Funding: This research received no external funding.

Institutional Review Board Statement: Not applicable.

Informed Consent Statement: Not applicable.

Data Availability Statement: For better resolution, the figures of the live/dead assays are added as a PowerPoint file (Suppl. Figure S1—Live/Dead Assay Figures). An overview of the numeric results of the PCR experiments is attached as a prism table in the supplement (Suppl. S2—RT_PCR_Results_Total).

Acknowledgments: This publication was supported by the Open Access Publication Fund of the University of Wuerzburg.

Conflicts of Interest: The authors declare no conflict of interest.

References

1. Kamangar, F.; Dores, G.M.; Anderson, W.F. Patterns of cancer incidence, mortality, and prevalence across five continents: Defining priorities to reduce cancer disparities in different geographic regions of the world. *J. Clin. Oncol.* **2006**, *24*, 2137–2150. [CrossRef] [PubMed]
2. Torre, L.A.; Bray, F.; Siegel, R.L.; Ferlay, J.; Lortet-Tieulent, J.; Jemal, A. Global cancer statistics, 2012. *CA Cancer J. Clin.* **2015**, *65*, 87–108. [CrossRef] [PubMed]
3. Hosios, A.M.; Hecht, V.C.; Danai, L.V.; Johnson, M.O.; Rathmell, J.C.; Steinhauser, M.L.; Manalis, S.R.; Vander Heiden, M.G. Amino acids rather than glucose account for the majority of cell mass in proliferating mammalian cells. *Dev. Cell* **2016**, *36*, 540–549. [CrossRef] [PubMed]
4. Gillespie, Z.E.; Pickering, J.; Eskiw, C.H. Better living through chemistry: Caloric restriction (CR) and CR mimetics alter genome function to promote increased health and lifespan. *Front. Genet.* **2016**, *7*, 142. [CrossRef]
5. Mirzaei, H.; Suarez, J.A.; Longo, V.D. Protein and amino acid restriction, aging and disease: From yeast to humans. *Trends Endocrinol. Metab. TEM* **2014**, *25*, 558–566. [CrossRef]

6. Ruckenstuhl, C.; Netzberger, C.; Entfellner, I.; Carmona-Gutierrez, D.; Kickenweiz, T.; Stekovic, S.; Gleixner, C.; Schmid, C.; Klug, L.; Sorgo, A.G.; et al. Lifespan extension by methionine restriction requires autophagy-dependent vacuolar acidification. *PLoS Genet.* **2014**, *10*, e1004347. [CrossRef]

7. Escobar, K.A.; Cole, N.H.; Mermier, C.M.; VanDusseldorp, T.A. Autophagy and aging: Maintaining the proteome through exercise and caloric restriction. *Aging Cell* **2019**, *18*, e12876. [CrossRef]

8. Shimokawa, I. Growth hormone and IGF-1 axis in aging and longevity. In *Healthy Ageing Longevity*; Rattan, S., Sharma, R., Eds.; Springer International Publishing: Cham, Switzerland, 2017; pp. 91–106.

9. Gongol, B.; Sari, I.; Bryant, T.; Rosete, G.; Marin, T. AMPK: An epigenetic landscape modulator. *Int. J. Mol. Sci.* **2018**, *19*, 3238. [CrossRef]

10. Longo, V.D.; Kennedy, B.K. Sirtuins in aging and age-related disease. *Cell* **2006**, *126*, 257–268. [CrossRef]

11. Goberdhan, D.C.I.; Wilson, C.; Harris, A.L. Amino acid sensing by mTORC1: Intracellular transporters mark the spot. *Cell Metab.* **2016**, *23*, 580–589. [CrossRef]

12. Lauinger, L.; Kaiser, P. Sensing and signaling of methionine metabolism. *Metabolites* **2021**, *11*, 83. [CrossRef] [PubMed]

13. Jeong, H.; Vacanti, N.M. Systemic vitamin intake impacting tissue proteomes. *Nutr. Metab.* **2020**, *17*, 73. [CrossRef] [PubMed]

14. Kim, J.; Guan, K.-L. mTOR as a central hub of nutrient signalling and cell growth. *Nat. Cell Biol.* **2019**, *21*, 63–71. [CrossRef] [PubMed]

15. Saxton, R.A.; Sabatini, D.M. mTOR Signaling in growth, metabolism, and disease. *Cell* **2017**, *168*, 960–976. [CrossRef]

16. Martínez, Y.; Li, X.; Liu, G.; Bin, P.; Yan, W.; Más, D.; Valdivié, M.; Hu, C.-A.A.; Ren, W.; Yin, Y. The role of methionine on metabolism, oxidative stress, and diseases. *Amino Acids* **2017**, *49*, 2091–2098. [CrossRef]

17. Cavuoto, P.; Fenech, M.F. A review of methionine dependency and the role of methionine restriction in cancer growth control and life-span extension. *Cancer Treat. Rev.* **2012**, *38*, 726–736. [CrossRef]

18. Sharma, B.; Singh, S.; Kanwar, S.S. L-methionase: A therapeutic enzyme to treat malignancies. *BioMed Res. Int.* **2014**, *2014*, 506287. [CrossRef]

19. Wang, Z.; Xie, Q.; Zhou, H.; Zhang, M.; Shen, J.; Ju, D. Amino acid degrading enzymes and autophagy in cancer therapy. *Front. Pharmacol.* **2021**, *11*, 582587. [CrossRef]

20. Levine, M.E.; Suarez, J.A.; Brandhorst, S.; Balasubramanian, P.; Cheng, C.-W.; Madia, F.; Fontana, L.; Mirisola, M.G.; Guevara-Aguirre, J.; Wan, J.; et al. Low protein intake is associated with a major reduction in IGF-1, cancer, and overall mortality in the 65 and younger but not older population. *Cell Metab.* **2014**, *19*, 407–417. [CrossRef]

21. Chaturvedi, S.; Hoffman, R.M.; Bertino, J.R. Exploiting methionine restriction for cancer treatment. *Biochem. Pharmacol.* **2018**, *154*, 170–173. [CrossRef]

22. Livak, K.J.; Schmittgen, T.D. Analysis of relative gene expression data using real-time quantitative PCR and the $2^{-\Delta\Delta CT}$ method. *Methods* **2001**, *25*, 402–408. [CrossRef] [PubMed]

23. Schmitz, W.; Ries, E.; Koderer, C.; Völter, M.F.; Wünsch, A.C.; El-Mesery, M.; Frackmann, K.; Kübler, A.C.; Linz, C.; Seher, A. Cysteine restriction in murine L929 fibroblasts as an alternative strategy to methionine restriction in cancer therapy. *Int. J. Mol. Sci.* **2021**, *22*, 11630. [CrossRef] [PubMed]

24. Volland, J.M.; Kaupp, J.; Schmitz, W.; Wünsch, A.C.; Balint, J.; Möllmann, M.; El-Mesery, M.; Frackmann, K.; Peter, L.; Hartmann, S.; et al. Mass spectrometric metabolic fingerprinting of 2-deoxy-D-Glucose (2-DG)-induced inhibition of glycolysis and comparative analysis of methionine restriction versus glucose restriction under perfusion culture in the murine L929 model system. *Int. J. Mol. Sci.* **2022**, *23*, 9220. [CrossRef] [PubMed]

25. Finkelstein, J.D. Methionine metabolism in mammals. *J. Nutr. Biochem.* **1990**, *1*, 228–237. [CrossRef] [PubMed]

26. Park, B.; Abdel-Azeem, A.; Al-Sanea, M.; Yoo, K.; Tae, J.; Lee, S. Staurosporine analogues from microbial and synthetic sources and their biological activities. *Curr. Med. Chem.* **2013**, *20*, 3872–3902. [CrossRef]

27. Kahya, U.; Köseer, A.S.; Dubrovska, A. Amino acid transporters on the guard of cell genome and epigenome. *Cancers* **2021**, *13*, 125. [CrossRef]

28. Yoshida, G.J. The harmonious interplay of amino acid and monocarboxylate transporters induces the robustness of cancer cells. *Metabolites* **2021**, *11*, 27. [CrossRef] [PubMed]

29. Bodoy, S.; Martín, L.; Zorzano, A.; Palacín, M.; Estévez, R.; Bertran, J. Identification of LAT4, a novel amino acid transporter with system L activity. *J. Biol. Chem.* **2005**, *280*, 12002–12011. [CrossRef]

30. Bian, Y.; Li, W.; Kremer, D.M.; Sajjakulnukit, P.; Li, S.; Crespo, J.; Nwosu, Z.C.; Zhang, L.; Czerwonka, A.; Pawłowska, A.; et al. Cancer SLC43A2 alters T cell methionine metabolism and histone methylation. *Nature* **2020**, *585*, 277–282. [CrossRef]

31. Gubser, P.M.; Kallies, A. Methio "mine"! Cancer cells steal methionine and impair CD8 T-cell function. *Immunol. Cell Biol.* **2020**, *98*, 623–625. [CrossRef]

32. Goel, B.; Tiwari, A.K.; Pandey, R.K.; Singh, A.P.; Kumar, S.; Sinha, A.; Jain, S.K.; Khattri, A. Therapeutic approaches for the treatment of head and neck squamous cell carcinoma-An update on clinical trials. *Transl. Oncol.* **2022**, *21*, 101426. [CrossRef] [PubMed]

33. Ghosh, S. Cisplatin: The first metal based anticancer drug. *Bioorg. Chem.* **2019**, *88*, 102925. [CrossRef] [PubMed]

34. Madeo, F.; Carmona-Gutierrez, D.; Hofer, S.J.; Kroemer, G. Caloric restriction mimetics against age-associated disease: Targets, mechanisms, and therapeutic potential. *Cell Metab.* **2019**, *29*, 592–610. [CrossRef] [PubMed]

35. Vander Heiden, M.G.; Cantley, L.C.; Thompson, C.B. Understanding the Warburg effect: The metabolic requirements of cell proliferation. *Science* **2009**, *324*, 1029–1033. [CrossRef]
36. Lee, C.; Longo, V.D. Fasting vs dietary restriction in cellular protection and cancer treatment: From model organisms to patients. *Oncogene* **2011**, *30*, 3305–3316. [CrossRef]
37. Brandhorst, S.; Wei, M.; Hwang, S.; Morgan, T.E.; Longo, V.D. Short-term calorie and protein restriction provide partial protection from chemotoxicity but do not delay glioma progression. *Exp. Gerontol.* **2013**, *48*, 1120–1128. [CrossRef]

current issues in molecular biology

MDPI

Article

Potential Diagnostic Value of Salivary Tumor Markers in Breast, Lung and Ovarian Cancer: A Preliminary Study

Lyudmila V. Bel'skaya [1,2,*], Elena A. Sarf [1], Alexandra I. Loginova [3,4], Dmitry M. Vyushkov [3,4] and En Djun Choi [5]

[1] Biochemistry Research Laboratory, Omsk State Pedagogical University, 14, Tukhachevsky Str., 644099 Omsk, Russia; nemcha@mail.ru
[2] Department of Biochemistry, Omsk State Medical University, 12, Lenina Str., 644099 Omsk, Russia
[3] Clinical Oncology Dispensary, 9/1, Zavertyayeva Str., 644013 Omsk, Russia; pai198585@mail.ru (A.I.L.); viushkov@mail.ru (D.M.V.)
[4] Department of Oncology, Omsk State Medical University, 12, Lenina Str., 644099 Omsk, Russia
[5] Clinic Lekar, 14/4, Presnensky Val Str., 107031 Moscow, Russia; drchoiworld@gmail.com
* Correspondence: belskaya@omgpu.ru or ludab2005@mail.ru

Abstract: The aim of the study was to determine the content of tumor markers for breast, lung and ovarian cancer in saliva, as well as for benign diseases of the corresponding organs and in the control group, and to evaluate their diagnostic significance. Strictly before the start of treatment, saliva samples were obtained and the concentrations of tumor markers (AFP, NSE, HE4, CA15-3, CA72-4, CA125 and CEA) were determined using an enzyme immunoassay (ELISA). CA125 and HE4 were simultaneously determined to be in the blood serum of patients with ovarian cancer. The concentrations of salivary CEA, NSE, CA15-3, CA72-4 and CA125 of the control group were significantly lower than in oncological diseases; however, these tumor markers also increased in saliva with benign diseases. The content of tumor markers depends on the stage of cancer, and the presence of lymph node metastasis; however, the identified patterns are statistically unreliable. The determination of HE4 and AFP in saliva was not informative. In general, the area of potential use of tumor markers in saliva is extremely narrow. Thus, CEA may be diagnostic for breast and lung cancer, but not for ovarian cancer. CA72-4 is most informative for ovarian mucinous carcinoma. None of the markers showed significant differences between malignant and non-malignant pathologies.

Keywords: saliva; tumor markers; breast cancer; ovarian cancer; lung cancer; CA125; CA15-3; CA72-4; HE4; CEA

check for updates

Citation: Bel'skaya, L.V.; Sarf, E.A.; Loginova, A.I.; Vyushkov, D.M.; Choi, E.D. Potential Diagnostic Value of Salivary Tumor Markers in Breast, Lung and Ovarian Cancer: A Preliminary Study. *Curr. Issues Mol. Biol.* **2023**, *45*, 5084–5098. https://doi.org/10.3390/cimb45060323

Academic Editor: Dumitru A. Iacobas

Received: 21 May 2023
Revised: 4 June 2023
Accepted: 9 June 2023
Published: 10 June 2023

1. Introduction

Saliva is an important bodily fluid, and interest in it as a diagnostic tool has increased in recent years [1–4]. Its main advantages are that saliva can be taken non-invasively and repeatedly without the discomfort associated with taking blood samples [5,6]. Saliva is already widely used in genetic testing [7] due to its better transport stability compared to blood [8]. Saliva contains various substances and biomarkers that can be used as indicators of health and disease, in particular for diagnosing cancer [9–14].

Several cancer biomarkers have been identified in saliva, such as increased levels of c-erbB-2 in the saliva of women with breast carcinoma compared to those in patients with benign diseases and healthy controls [15], increased levels of CA125 in ovarian and oral cancer [16,17], and increased salivary levels of cytokeratin fragment 19 (CYFRA 21-1) in patients with oral cancer [18,19]. Carcinoembryonic antigen (CEA) [20], carbohydrate antigen 15-3 (CA15-3) [21,22], α-fetoprotein (AFP) [23], and human epididymis protein 4 (HE4) [24] in saliva were also determined. Several studies have shown that salivary CA15-3 levels are useful in early breast cancer diagnosis and follow-up [25]. A number of studies combine the determination of several biomarkers in saliva, including CEA, CA125, and c-erbB-2 [26].

Despite numerous studies of tumor markers in saliva, a number of difficulties associated with the routine use of this method in clinical laboratory practice still need to be resolved. First, the content of tumor markers in saliva can differ significantly from their concentration in the blood; for example, the level of CA15-3 in the saliva is up to 10 times lower than in serum [15]. In this regard, it becomes necessary to establish separate criteria for the norm and pathology for saliva for each tumor marker. Secondly, the data on the content of tumor markers in saliva obtained by different authors differ significantly even in normal conditions, which often make it impossible to compare the results with each other. Thus, values obtained for CA125 differ by several times (137.12 ± 124.58 [27] and 319.27 ± 187.91 U/mL [28]), for CEA differ by order of magnitude (11.36 ± 13.94 [29], 77.34 ± 28.53 [30] and 188.0 ± 59.5 ng/mL [22]), etc. This, in turn, requires a more careful attitude to the methodological part of the research, and implies the obligatory analysis of the saliva of the control group in each individual laboratory.

In this work, we determined the content of tumor markers in saliva for breast, lung, and ovarian cancer, as well as for benign diseases of the corresponding organs and in the control group, and compared the results with the literature data. We compared the level of salivary tumor markers of various types of cancer within the same experiment with reagents from the same manufacturer and on the same equipment. The aim of the study was to evaluate the potential diagnostic value of salivary tumor markers.

2. Materials and Methods

2.1. Study Design and Description of Study Groups

The study included patients of the Omsk Clinical Oncology Center with histologically confirmed cancer (main group, Table 1), with benign diseases of the corresponding organs (comparison group), and healthy volunteers (control group). Inclusion in groups occurred in parallel. The inclusion criteria were considered: the age of patients 30–75 years, the absence of any treatment at the time of the study, including surgery, chemotherapy or radiation, histological verification of the diagnosis, the absence of signs of active infection (including purulent processes), and oral cavity sanitation.

Table 1. The structure of the study group.

Feature		Breast Cancer, n = 48	Lung Cancer, n = 34	Ovarian Cancer, n = 51
Age, years		56.7 [49.0; 64.0]	60.1 [55.3; 64.4]	54.6 [37.9; 61.6]
Clinical stage				
	In situ	4 (8.3%)	-	-
	Stage I	13 (27.1%)	4 (11.8%)	17 (33.4%)
	Stage II	19 (39.6%)	15 (44.1%)	5 (9.8%)
	Stage III	12 (25.0%)	7 (20.6%)	26 (50.9%)
	Stage IV	-	8 (23.5%)	3 (5.9%)
Lymph node status				
	pN_0	30 (62.5%)	18 (52.9%)	-
	pN_1	10 (28.1%)	4 (11.8%)	-
	$pN_2 + pN_3$	8 (16.6%)	12 (35.3%)	-
Metastasis status				
	pM_0	48 (100%)	26 (76.5%)	48 (94.1%)
	pM_1	-	8 (23.5%)	3 (5.9%)

2.1.1. Breast Cancer

The main group consisted of 48 patients with breast cancer (age 56.7 [49.0; 64.0] years), the comparison group consisted of 40 patients with fibroadenomas (age 46.1 [37.9; 56.7] years), the control group consisted of 32 healthy volunteers (age 45.9 [36.8; 57.8] years).

In all patients of the main group, invasive breast carcinoma of the following stages was histologically and cytological confirmed: in situ—4 (8.3%), stage I—13 (27.1%), stage IIa—15 (31.3%), stage IIb—4 (8.3%) and stage III—12 (25.0%). In 30 patients (pN_0—62.5%), there were no signs of regional lymph node metastases, in 10 patients (pN_1—28.1%) metastases were detected in the displaced axillary lymph nodes on the side of the lesion, in 8 patients more than 2 lymph nodes were affected ($pN_2 + N_3$—16.6%). Patients with distant metastasis were not included in the study. Breast tumors were classified according to the degree of tissue differentiation into highly differentiated (G1, n = 10), moderately differentiated (G2, n = 29) and poorly differentiated (G3, n = 9). In all cases, the status of HER2, estrogen and progesterone receptors was determined. In 28 patients (58.3%), HER2-negative status was confirmed, in 20 (41.7%)—HER2-positive; nine patients (18.8%) were confirmed ER-negative; 39 (81.2%) were ER-positive; PR-negative status was confirmed in 12 patients (25.0%), PR-positive in 36 (75.0%) patients. The values of the proliferative activity marker Ki-67 were less than 20% (Ki-67 low) in 22 patients (45.8%), and more than 20% (Ki-67 high) in 26 patients (54.2%). By molecular biological subtypes of breast cancer, the patients were distributed as follows: basal-like—4 (8.3%), luminal A-like—16 (33.3%), luminal B-like (HER2-negative)—8 (16.7%), luminal B-like (HER2-positive)—20 (41.7%).

In patients of the comparison group, the presence of fibroadenomas (single or multiple) of the mammary glands was confirmed.

2.1.2. Lung Cancer

The main group consisted of 34 patients with LC (age 60.1 [55.3; 64.4] years), the comparison group consisted of 11 patients with benign lung diseases (age 57.5 [53.2; 62.0] years), the control group consisted of 30 healthy volunteers (age 48.3 [39.1; 56.7] years). The main group included 25 men and 9 women; in the comparison and control groups, the same ratio of patients of both sexes (3 to 1) was maintained.

In all patients of the main group, lung cancer of the following stages was histologically confirmed: stage I—4 (11.8%), stage II—15 (44.1%), stage III—7 (20.6%) and stage VI—8 (23.5%). In eighteen patients, there were no signs of regional lymph node metastases (pN_0—52.9%), pN1 status was detected in four patients (11.8%), pN_2—in eleven patients (32.4%), pN_3—in one patient (2.9%), and eight patients had distant metastases (23.5%). In 12 patients histologically confirmed squamous cell lung cancer (35.3%), 15 patients confirmed adenocarcinoma (44.1%) and 6 patients confirmed small cell lung cancer (8.8%). In patients of the control group, the following lung pathologies were confirmed: tuberculoma—4, pneumofibrosis—3, inflammatory pseudotumor—2, pneumonia—2 people.

2.1.3. Ovarian Cancer

The main group consisted of 51 patients with ovarian cancer (age 54.6 [37.9; 61.6] years), the comparison group consisted of 16 patients with benign ovarian pathologies (age 46.7 [35.6; 61.3] years), the control group consisted of 22 healthy volunteers without ovarian pathologies (age 49.5 [37.7; 61.2] years).

In all patients of the main group, ovarian cancer of the following stages was histologically confirmed: stage Ia—11 (21.6%), stage Ic—6 (11.8%), stage II—5 (9.8%), stage IIIb—9 (17.6%), stage IIIc—17 (33.3%) and stage VI—3 (5.9%). According to the histological type, serous carcinoma—31 (60.8%), serous borderline tumor—5 (9.8%), endometrioid carcinoma—6 (11.8%), mucinous carcinoma—5 (9.8%) and granulosa cell tumor—4 (7.8%) were distinguished in the main group. In the comparison group, histologically confirmed serous cystadenoma—5 (31.3%), serous cystadenofibroma—3 (18.7%), mucinous cystadenoma—5 (31.3%) and endometrioid cyst—3 (18.7%).

2.2. Collection of Saliva and Determination of Tumor Markers

All participants had saliva sampling in the amount of 1 mL before the start of treatment. The saliva samples were collected in the morning on an empty stomach by spitting into sterile polypropylene tubes, centrifuged at $10,000 \times g$ for 10 min (CLb-16, Moscow, Russia). Cancer antigens (CA125, CA15-3, CA72-4), α-fetoprotein (AFP), carcinoembryonic antigen (CEA), human epididymis protein 4 (HE4), and neuron specific enolase (NSE) were analyzed using a commercially available kit (Vector Best, Novosibirsk, Russia), according to the instructions of the manufacturer without changes, including reagent volumes and incubation time. Reading was performed using Thermo Scientific Multiskan FC (Waltham, MA, USA). Values were calculated based on a standard curve plotted for the assay.

2.3. Determination of Tumor Markers in Blood Serum

For CA125 and HE4, a parallel determination of the content in blood serum was carried out according to the standard procedure described in Section 2.2.

2.4. Statistical Analysis

Statistical analysis of the obtained data was performed using the Statistica 13.0 (StatSoft, Tulsa, OK, USA) software by a non-parametric method using the Wilcoxon test in dependent groups, and the Mann–Whitney U-test in independent groups. The sample was described by calculating the median (Me) and the interquartile range in the form of the 25th and 75th percentiles [LQ; UQ]. Differences were considered statistically significant at $p < 0.05$.

3. Results

3.1. Determination of Tumor Markers in Saliva for Breast Cancer

Normally, the CEA level in saliva was 67.8 [59.6; 94.6] ng/mL, CEA content increases by 1.37 times in fibroadenomas (160.7 [139.2; 199.4] ng/mL) and by 1.79 times in breast cancer (189.2 [141.3; 215.4] ng/mL) (Figure 1A). For CA 15-3, the normal concentration in saliva was 4.52 [3.36; 5.08] U/mL, while its content increases equally in both fibroadenomas and breast cancer (13.5 [8.59; 17.6] and 11.10 [8.64; 20.2] U/mL, respectively) (Figure 1B). Differences with the control group were statistically significant in all cases, while differences between FA and breast cancer were not statistically confirmed.

For both markers, an increase in the concentration in saliva was shown depending on the stage of breast cancer (Figure 1C). For breast cancer in situ, the increase in the concentration of tumor markers was minimal, but the increase in the concentration of tumor markers at advanced stages of breast cancer was statistically significant. The concentration of tumor markers increases with the defeat of the lymph nodes: for CEA 212.8 [166.3; 220.6] ng/mL and for CA15-3 16.8 [11.4; 23.6] U/mL at pN_2 (Figure 1D). For highly differentiated tumors, the content of both markers is higher than for poorly differentiated ones (CEA—184.5 [155.0; 212.6] vs. 178.1 [159.7; 208.8] ng/mL; CA15-3—13.9 [7.85; 28.1] vs. 10.9 [8.64; 12.5] U/mL) (Figure 1E). It was shown that CEA and CA15-3 vary differently depending on the Ki-67 value (Figure 1F). Thus, the level of CEA is higher for a high marker of proliferative activity Ki-67 high; the level of CA 15-3 is higher for Ki-67 low. There were no significant differences in the level of tumor markers depending on the molecular biological subtype of breast cancer (Figure 1G). However, salivary marker levels have been shown to be lower for HER2-positive breast cancer than for HER2-negative breast cancer (Figure 1H). The differences were more pronounced for CA15-3 (8.00 [4.94; 10.3] vs. 11.4 [8.65; 21.0] U/mL) than for CEA (188.0 [159.7; 221.1] vs. 193.2 [127.7; 215.4] ng/mL). CEA was slightly increased for ER- and PR-positive tumors (Figure 1I), while no dependence on estrogen and progesterone receptors was found for CA15-3 (Figure 1J).

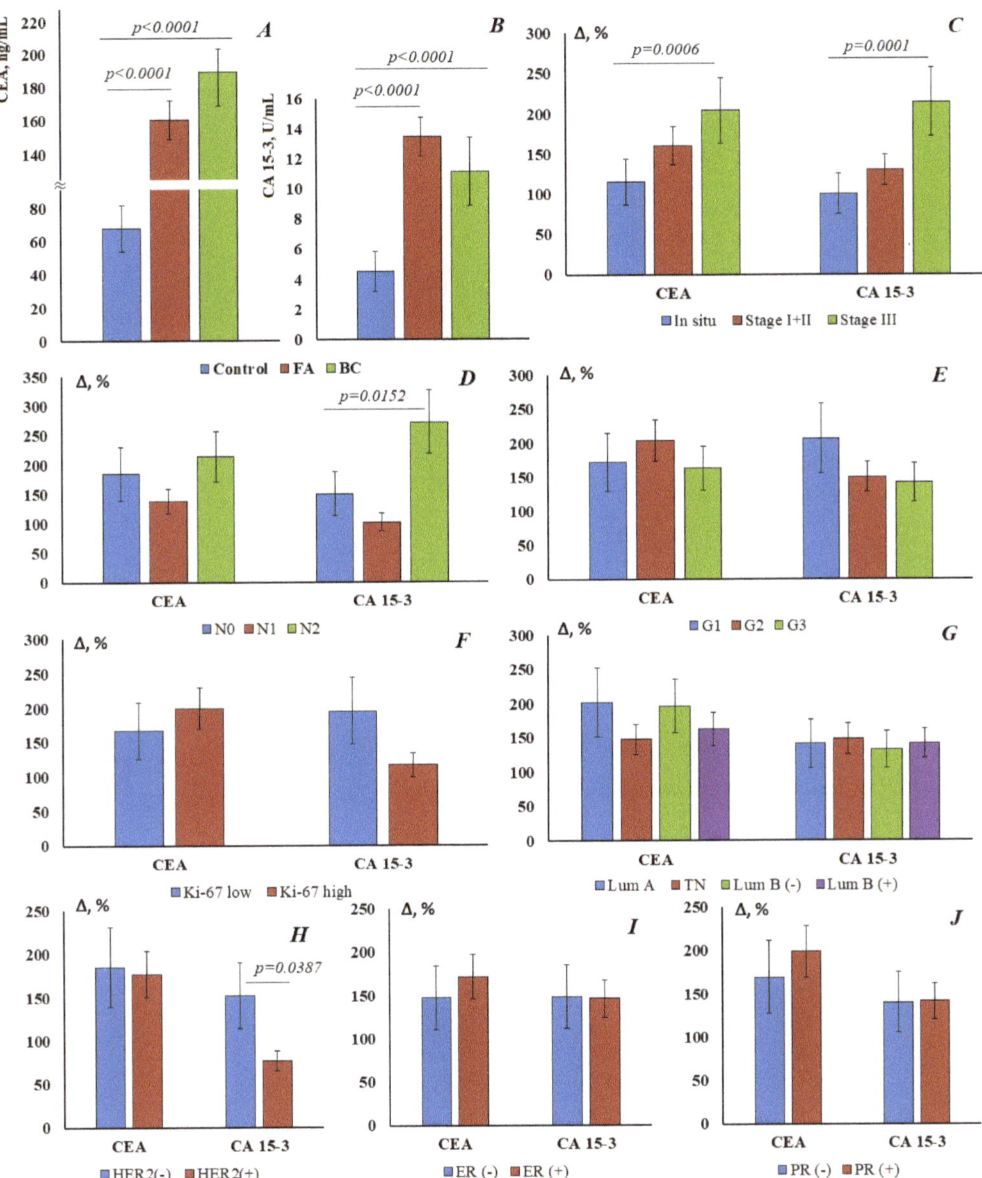

Figure 1. The content of CEA (**A**) and CA15-3 (**B**) in the saliva of the control group, with fibroadenomas and in breast cancer. The relative change in the level of tumor markers in saliva compared with the control group depending on the stage of breast cancer (**C**), the prevalence of regional metastasis (**D**), the degree of tumor differentiation (**E**), Ki-67 expression (**F**), molecular biological subtype (**G**), HER2 expression (**H**), estrogen receptor expression (**I**), and progesterone receptor expression (**J**). Differences with the control group were statistically significant in all cases ($p < 0.05$).

3.2. Determination of Tumor Markers in Saliva for Lung Cancer

In the control group, the CEA level was 72.3 [61.7; 98.1] ng/mL, while the level of CEA increases statistically significantly in the group with benign lung diseases (108.5 [95.2;

120.6] ng/mL) and in the group with lung cancer (103.4 [90.9; 110.4] ng/mL) (Figure 2A). Differences between BLD and lung cancer were not statistically confirmed (Figure 2A,D). CEA concentrations increase depending on the stage of the disease: from 98.5 [96.0; 101.1] ng/mL at pT_1 up to 108.9 [103.0; 109.7] ng/mL at pT_4 (Figure 2B). The most pronounced increase in CEA is for squamous cell lung cancer (107.3 [103.8; 112.4] ng/mL) (Figure 2C).

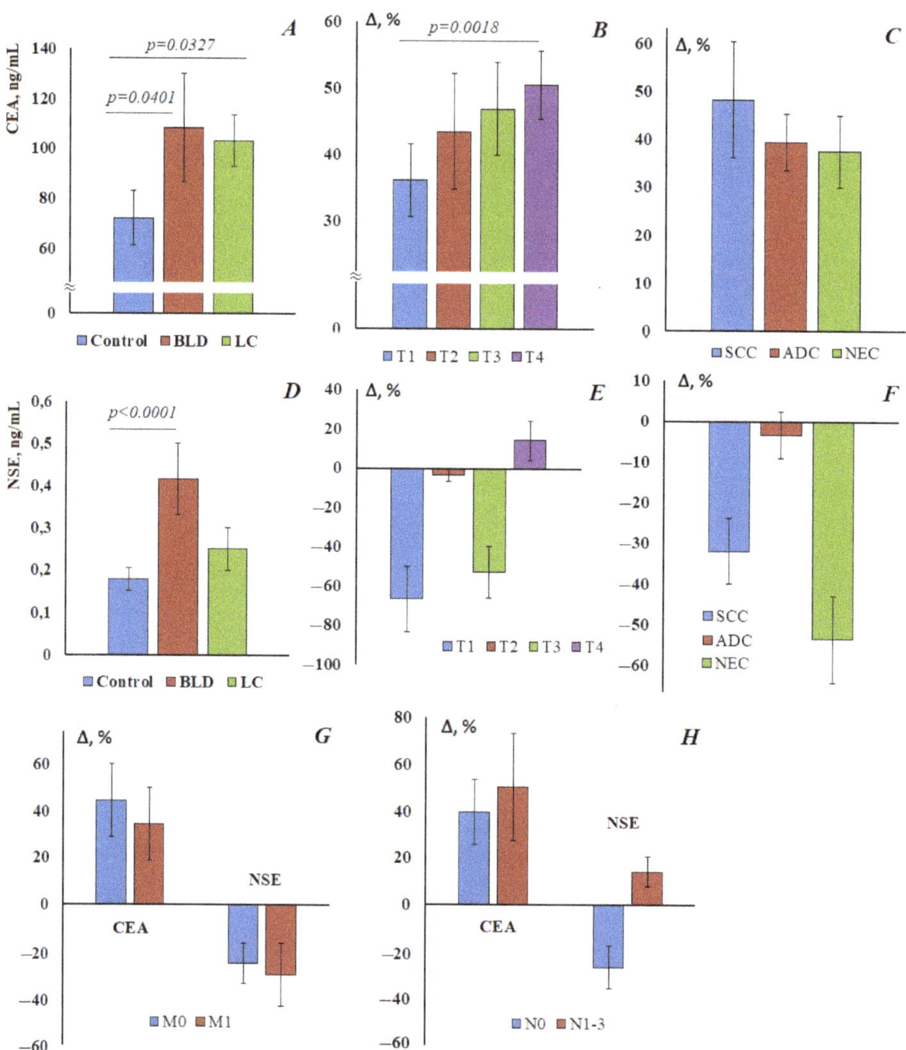

Figure 2. The content of CEA (**A**) and NSE (**D**) in the saliva of the control group, with benign lung diseases (BLD) and lung cancer (LC). Relative change in the level of tumor markers in saliva compared with the control group, depending on the stage of lung cancer: (**B**) CEA, (**E**) NSE; depending on the histological type of lung cancer: (**C**) CEA, (**F**) NSE; depending on the presence/absence of distant metastasis (**G**) and lymph node metastasis (**H**). SCC, squamous cell lung cancer; ADC, adenocarcinoma; NEC, neuroendocrine lung cancer. Differences with the control group were statistically significant in all cases ($p < 0.05$).

The salivary NSE concentration in the control group was 0.179 [0.104; 0.471] mIU/mL, NSE concentration increases in benign lung diseases (0.417 [0.164; 0.608] mIU/mL) and in the lung cancer group (0.252 [0.161; 0.388] mIU/mL) (Figure 2D); however, in this case, the differences are significant only for the group with benign lung diseases compared with the control. Comparison of the level of NSE with the control group at different stages of lung cancer showed that with pT_{1-3}, the concentration of NSE in saliva was lower than in healthy volunteers, and only with pT_4 was it slightly increased (Figure 2E). The maximum deviation in the concentration of NSE was observed for neuroendocrine tumors of the lung, but the concentration of NSE decreases (Figure 2F). There were no statistically significant differences in the level of tumor markers depending on the presence/absence of distant metastasis (Figure 2G). The values of both tumor markers increased in the presence of metastases in the lymph nodes (Figure 2H).

3.3. Determination of Tumor Markers in Saliva and Blood for Ovarian Cancer

The normal concentration of AFP in saliva was 0.572 [0.530; 0.711] IU/mL, significant changes in AFP concentration were not shown both in benign ovarian pathologies (0.625 [0.541; 0.715] IU/mL) and in ovarian cancer (0.562 [0.519; 0.668] IU/mL).

The CEA content in the saliva of the control group was 61.9 [53.7; 69.0] ng/mL, in benign ovarian diseases 63.1 [53.2; 73.6] ng/mL and in ovarian cancer 66.3 [55.4; 79.7] ng/mL (Figure 3A). No statistically significant differences were found between the groups. In addition, there were no differences in the content of CEA at different stages of ovarian cancer. When comparing subgroups with different histological types of ovarian cancer, it was shown that the content of CEA significantly increases compared with the control group only in the low-grade serous carcinoma group (Figure 3B).

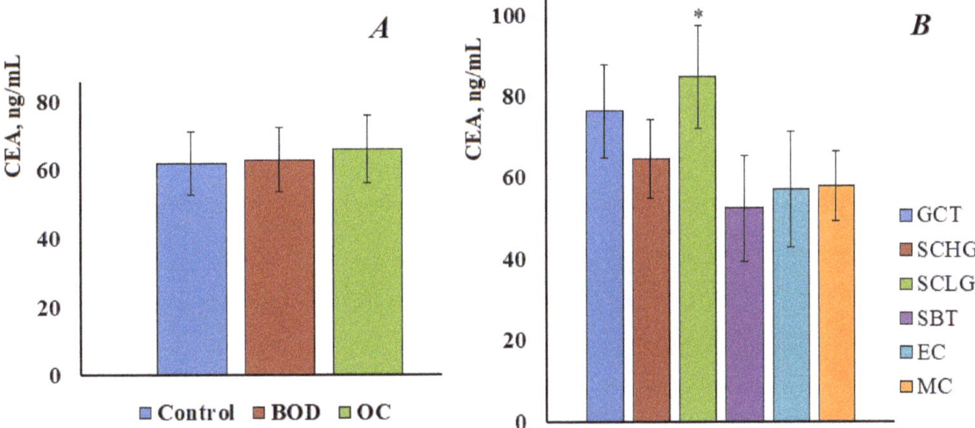

Figure 3. The content of CEA in the saliva of the control group, with benign ovarian diseases and ovarian cancer (**A**). CEA content in saliva depending on the histological type of ovarian cancer (**B**). GCT, granulose cell tumor, SCHG, high grade serous carcinoma, SCLG, low grade serous carcinoma, SBT, serous borderline tumor, EC, endometrioid carcinoma, MC, mucinous carcinoma. * differences with the control group are statistically significant, $p < 0.05$.

For CA 72-4, the content in the control group was 2.17 [1.21; 2.91] U/mL, with benign ovarian diseases was 2.52 [1.90; 4.46] U/mL, and in ovarian cancer was 3.16 [1.48; 4.82] U/mL (Figure 4A). Thus, there was a tendency to increase the level of CA 72-4, but it was not statistically confirmed. It was shown that the content of CA 72-4 increases at advanced stages of ovarian cancer; the differences between early and advanced stages were statistically significant (Figure 4B). A significant increase in the content of CA 72-4 was shown for ovarian mucinous carcinoma (Figure 4C), in this case, the differences both with the control group and with other histological types of ovarian cancer were statistically significant ($p < 0.0001$).

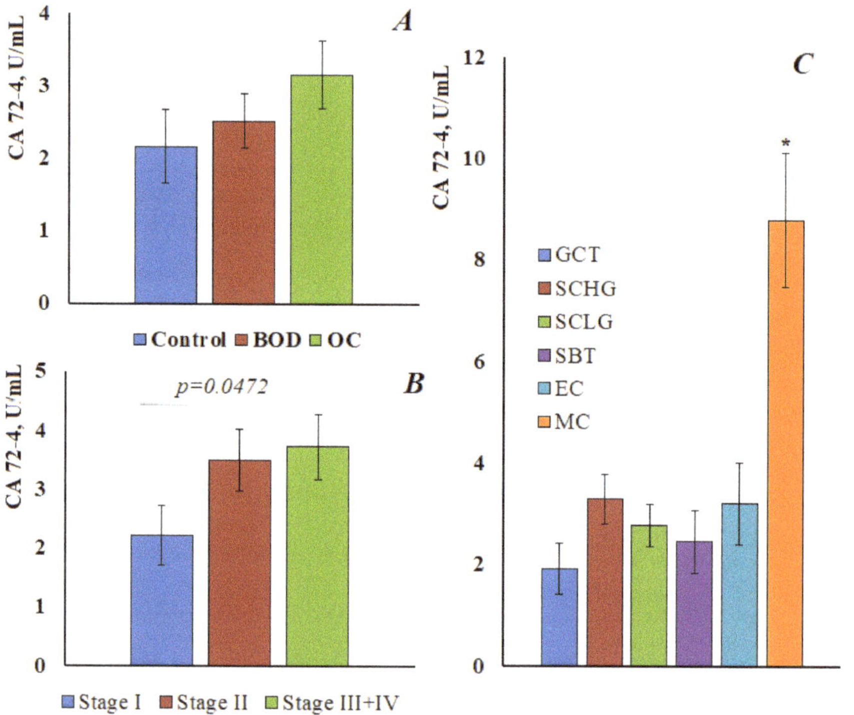

Figure 4. The content of CA 72-4 in the saliva of the control group, with benign ovarian diseases and ovarian cancer (**A**). The content of CA 72-4 in saliva depending on the stage (**B**) and the histological type of ovarian cancer (**C**). GCT, granulose cell tumor, SCHG, high grade serous carcinoma, SCLG, low grade serous carcinoma, SBT, serous borderline tumor, EC, endometrioid carcinoma, MC, mucinous carcinoma. * differences with the control group are statistically significant, $p < 0.05$.

For HE4 and CA125, a parallel determination was made both in saliva and in blood (Figure 5). It was found that the normal value for HE4 in saliva was 343 [302; 416] pmol/L (Figure 5A). The content of HE4 in saliva practically did not change in benign ovarian diseases and ovarian cancer, and did not depend on the stage of ovarian cancer (Figure 5C) or the histological subtype of the tumor (Figure 5E). For CA125 in saliva, a statistically significant increase was noted both in benign ovarian diseases (202.1 [142.9; 290.7] vs. 330.4 [198.4; 448.5] U/mL) and in ovarian cancer (376.5 [325.0; 509.5] U/mL) (Figure 5A). However, we also did not show differences in the level of CA125 in saliva depending on the stage of ovarian cancer (Figure 5C). A significant increase in the level of CA125 in saliva was noted for granulosa cell tumor (Figure 5E); but, due to the small sample size, this result can be considered as preliminary.

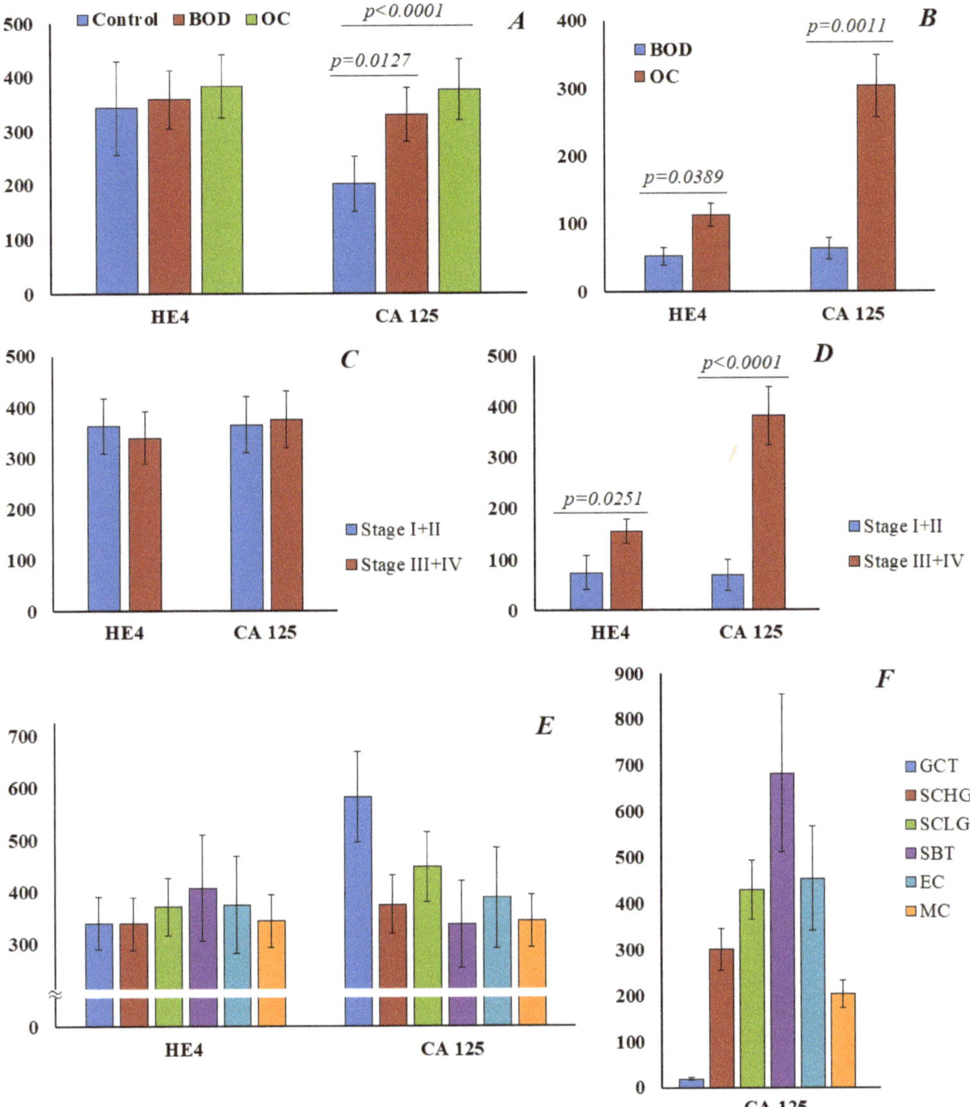

Figure 5. The content of HE4 and CA125 in saliva is normal, with benign ovarian diseases and cancer (**A**). The content of HE4 and CA125 in saliva in ovarian cancer, depending on the stage (**C**) and histological type (**E**). The content of HE4 and CA125 in the blood in benign ovarian diseases and ovarian cancer (**B**), and in different stages of ovarian cancer (**D**). The content of CA 125 in the blood in different histological types of ovarian cancer (**F**). GCT, granulose cell tumor, SCHG, high grade serous carcinoma, SCLG, low grade serous carcinoma, SBT, serous borderline tumor, EC, endometrioid carcinoma, MC, mucinous carcinoma.

Parallel analyses of HE4 and CA125 in the blood showed a statistically significant increase in the level of the marker in ovarian cancer compared with benign ovarian diseases (Figure 5B), and at advanced stages of ovarian cancer (Figure 5D). It was shown that the level of CA125 in saliva and blood in different histological types of ovarian cancer does not correlate with each other (Figure 5E,F). In general, the correlation between the content of HE4 and CA125 in saliva and blood in our study was not confirmed.

4. Discussion

CEA is a glycoprotein located on the cell surface, and widely used in clinical practice as an important routine auxiliary indicator for tumor diagnosis [31]. It is known that CEA is also found in the saliva of healthy people, but its concentrations turned out to be very low (0–3 ng/mL) [32]. In the works of other authors, CEA was determined in the saliva of healthy volunteers, but normal values varied within a wide range (11–188 ng/mL) [22,29,30]. In our study, we determined CEA in three groups of cancer patients and analyzed the control group in each case. It is shown that the normal value of CEA in saliva is 60–70 ng/mL. We have shown an increase in the concentration of CEA both in cancer and in benign diseases of the mammary glands and lungs, but not for ovarian diseases. This result is consistent with the fact that the determination of CEA in the blood for ovarian cancer is also not used. Both in the group of patients with breast cancer and lung cancer, an increase in CEA was shown, including in benign diseases, while the difference between benign and malignant pathologies is not statistically significant. We observed an increase in the concentration of CEA depending on the stage and metastasis in the lymph nodes. Previously, it was shown that the level of CEA positively correlates with tumor progression [30]. In their study, Zheng J. et al. found that salivary CEA levels in patients with OSCC correlated with clinical staging and lymph node metastases, so salivary CEA can be used as an indicator of OSCC severity and serve as a method for assessing OSCC staging and lymph node invasion. Brooks et al. [33] found a significant increase in salivary CEA concentrations in the breast cancer group compared with the control group, as in our study.

Carbohydrate antigen 15.3 (CA15-3), a 400 kDa glycoprotein from the MUC-1 family of mucins, is present at higher levels in the serum and saliva of breast cancer patients than in healthy women. It is used as a reference marker or "diagnostic gold standard" against which other markers of breast cancer are compared [34]. We have shown an increase in the concentration of CA15-3 in the saliva of patients with breast cancer compared with healthy controls, which is consistent with the literature data [22,25,35,36]. As expected from previous observations, elevated CA15-3 levels are more common in patients with advanced breast cancer than in patients with early breast cancer. The absence of a significant difference between the levels of CA15-3 in cancer and in the control group (9.2 ± 7.9 vs. 4.5 ± 2.7 U/mL) shown in the study by Farahani H. et al. could be attributed to the early stages of the disease, as the authors themselves substantiate [22]. However, we have shown a significant increase in the concentration of this marker both in benign diseases of the mammary glands and in breast cancer in situ. The same authors showed the existence of a positive significant correlation between the concentrations of CA15-3 in serum and saliva in healthy people. With cut-off values of 5.5 ng/mL for CA15-3 and 85 ng/mL for CEA, breast cancer can be diagnosed with a sensitivity of 80% and a specificity of 70–75% [22]. Assad et al. showed that the concentration of CA15-3 in the saliva of healthy volunteers is higher than in breast cancer (6.51 ± 7.18 vs. 4.73 ± 5.74 U/mL) [37]; a statistically significant increase in the concentration of CA15-3 was shown only at advanced stages, which is consistent with our results. The authors of this work compared the concentration of CA15-3 in saliva depending on the molecular biological subtype of breast cancer. It was noted that the maximum concentration of CA15-3 is characteristic of the Luminal B HER2+ subtype [37], which was not confirmed in our study (Figure 1G). We have shown for the first time that in HER2-negative tumors the concentration of CA15-3 in saliva was statistically significantly higher than in HER2-positive tumors, while the dependence on the status of estrogen and progesterone receptors was not shown. It is also interesting that the concentration of CA15-3 in saliva was higher at a low level of a marker of tumor proliferative activity, which showed the potential for evaluating the prognostic significance of CA15-3 in saliva [38], but this requires additional research.

Neuron-specific enolase (NSE) is a catalyst for glucose metabolism. NSE is predominantly synthesized in the brain, peripheral nerves, and neuroendocrine cells [39]. Elevated levels of NSE were found in various pathologies, such as traumatic brain injury, brain tumors, and small cell lung cancer [40]. Acute inflammation can also cause an increase in serum NSE levels [41]. NSE in saliva was determined in single studies, in particular in ischemic stroke [42]; the authors reported normal values of NSE in saliva of 2.2–3.5 µg/L, and an increase in the level of NSE in stroke up to 2.3–8 µg/L was noted. No works devoted to the determination of NSE in cancer have been found, so it was extremely difficult to compare our data with the literature.

Alpha-fetoprotein (AFP), a fetal serum protein, may be useful as a tumor marker for the detection of malignancies such as yolk sac tumors (YST) [43]. In addition to typical YSTs, ovarian epithelial carcinoma with elevated AFP levels can be easily misdiagnosed due to its infrequent occurrence with high AFP levels, especially in young women [44,45]. The determination of AFP in saliva has been described in a few studies, so in hepatocellular carcinoma, the level of AFP in saliva was 3552.6 ± 2829.9 ng/L, while the normal concentration was lower (18.1 ± 3.8 ng/L) [46]. López-Jornet P. et al. determined AFP in the norm and in breast cancer, the normal value was 72.2 (26.5–514.0) pg/mL, in breast cancer there was no increase in the concentration of AFP [24]. Our study also did not show an increase in AFP levels in ovarian cancer, but this fact was explained by the absence of patients with ovarian germ cell tumors in our sample.

CA72-4 is a human fetal epithelial surface glycoprotein used as a tumor marker for diagnosis and monitoring of gastric and ovarian cancer. CA72-4 can be considered as a marker of choice for monitoring patients with ovarian tumors of the mucinous type [47]. It has been shown that the level of CA72-4 may slightly increase during inflammatory processes [48]. Despite the fact that, according to some authors, this glycoprotein is absent in the saliva of healthy people [49], we determined its content in the saliva of patients in the control group at the level of 2.17 U/mL. We have shown a statistically significant increase in the concentration of CA72-4 in saliva at advanced stages of ovarian cancer and for mucinous carcinoma (Figure 4C), which is in good agreement with the literature data [49].

CA125 is a marker of serous ovarian carcinoma, and its increased value indicates the involvement of serous membranes in the process. Monitoring its concentration is important for evaluating the effectiveness of chemotherapy and surgery. In general, there is little data on salivary CA125 levels in the published literature to date. Geng X.F. et al. report elevated salivary CA125 levels in oral cancer patients compared with benign oral disease and controls [27]. Agha-Hosseini F. et al. reported that salivary and serum CA125 levels were significantly elevated in patients with untreated breast cancer compared with patients with treated breast cancer and controls [50]. Data on the concentration of CA125 in saliva in ovarian cancer are contradictory. Thus, Tay S.K. et al. showed no difference in salivary CA125 levels between ovarian cancer and controls [51]. Vuković A. et al. showed that patients with malignant ovarian tumors had significantly higher levels of CA125 in saliva and serum compared with patients with benign tumors; however, there was no significant correlation between salivary and serum CA125 [52]. In our study, we also did not confirm the existence of a correlation between saliva and blood in terms of CA125 levels. Zhang K-Y. et al. found that salivary CA125 levels were significantly higher than serum CA125 levels in both control and tumor groups, and also showed no significant correlation with serum-saliva CA125 levels [28]. CA125, with its high molecular weight (>200 kDa), is unlikely to diffuse into saliva. The authors suggest that salivary CA125 may be locally produced by the salivary glands and/or tumor tissue, rather than being derived from serum, and this may be the reason why there was no correlation between salivary and serum CA125 levels [28]. Plante et al. previously showed that the highest concentration of CA125 was found in whole saliva, while the level of CA125 was significantly lower in saliva from the parotid gland and even lower in blood serum [53]. Our study showed that the content of CA125 in saliva is normally higher than in serum, while in ovarian cancer, the concentrations in blood and saliva are comparable.

Other biomarkers have been developed to increase the specificity of ovarian cancer diagnosis, such as human epididymal protein 4 (HE4), a biomarker that is overexpressed in ovarian cancer [54]. In the only study, HE4 was detected in saliva in breast cancer, while its normal content was 117.9 ng/mL (265 pmol/L) and slightly increased in breast cancer to 154.9 ng/mL [24]. In our study, we showed that the normal content of HE4 is higher than that given in the literature, and was 343 pmol/L, and practically did not change in ovarian cancer and did not depend on the stage and histological subtype of the tumor (Figure 5A,C,E). It is known that HE4 is highly expressed in the epithelium of the oral cavity, the excretory ducts of the salivary glands, and the nasopharynx [55]. The physiological role of HE4 in the oral cavity is not completely clear; it is probably necessary for the normal functioning of the epithelium, but it is indicated that it supports the innate immune system of the respiratory tract and oral cavity [56]. Apparently, HE4 does not diffuse into saliva from serum, or its amount is less than its own content in saliva, which reduces or eliminates the possibility of using this saliva marker for diagnostic purposes. However, this hypothesis requires further research and verification.

The limitations of the study include the fact that we did not determine the level of Cyfra 21-1 for lung cancer, as well as the fact that not all tumor markers were subjected to parallel blood levels. The sample size was small, which does not allow a correct comparison of subgroups with each other.

5. Conclusions

The possibility of determining the content of tumor markers in saliva was shown, and the concentrations were established as normal. It was shown that the level of AFP and HE4 in the studied groups did not change, which is due to the characteristics of the sample in the case of AFP, and the probable features of the expression of HE4 in the epithelium of the oral cavity in the case of HE4. The concentrations of CEA, NSE, CA15-3, CA72-4 and CA125 in the saliva of the control group were significantly lower than in oncological diseases; however, these tumor markers also increased in saliva in benign diseases. The dependence of the content of tumor markers on the stage of cancer, and the presence of metastasis in the lymph nodes, was shown; however, in most cases, the revealed patterns are statistically unreliable. In general, the area of potential use of tumor markers in saliva is extremely narrow. Thus, CEA may be diagnostic for breast cancer and lung cancer, but not for ovarian cancer. CA72-4 is most informative for ovarian mucinous carcinoma. None of the markers showed significant differences between malignant and non-malignant pathologies. Although the potential diagnostic value of salivary tumor markers remains questionable, knowing whether proteins and tumor DNA are present in other fluids, including saliva, may contribute to understanding the biological behavior of the disease.

Author Contributions: Conceptualization, L.V.B. and E.D.C.; methodology, L.V.B.; software, A.I.L.; validation, D.M.V., L.V.B. and E.D.C.; formal analysis, E.A.S.; investigation, E.A.S.; resources, D.M.V.; data curation, A.I.L. and E.A.S.; writing—original draft preparation, E.A.S.; writing—review and editing, L.V.B.; visualization, A.I.L.; supervision, D.M.V.; project administration, E.D.C. All authors have read and agreed to the published version of the manuscript.

Funding: This research received no external funding.

Institutional Review Board Statement: The study was conducted in accordance with the Declaration of Helsinki, and approved by the Ethics Committee of Omsk Regional Clinical Oncological Dispensary (21 July 2016, protocol № 15).

Informed Consent Statement: Informed consent was obtained from all subjects involved in the study.

Data Availability Statement: All data and materials used in this study are available from the corresponding author and will be provided upon reasonable request.

Conflicts of Interest: The authors declare no conflict of interest.

References

1. Huang, Z.; Yang, X.; Huang, Y.; Tang, Z.; Chen, Y.; Liu, H.; Huang, M.; Qing, L.; Li, L.; Wang, Q.; et al. Saliva—A new opportunity for fluid biopsy. *Clin. Chem. Lab. Med.* **2023**, *61*, 4–32. [CrossRef] [PubMed]
2. Roblegg, E.; Coughran, A.; Sirjanim, D. Saliva: An all-rounder of our body. *Eur. J. Pharm. Biopharm.* **2019**, *142*, 133–141. [CrossRef] [PubMed]
3. Song, M.; Bai, H.; Zhang, P.; Zhou, X.; Ying, B. Promising applications of human-derived saliva biomarker testing in clinical diagnostics. *Int. J. Oral Sci.* **2023**, *15*, 2. [CrossRef] [PubMed]
4. Roi, A.; Rusu, L.C.; Roi, C.I.; Luca, R.E.; Boia, S.; Munteanu, R.I. A New Approach for the Diagnosis of Systemic and Oral Diseases Based on Salivary Biomolecules. *Hindawi Dis. Markers* **2019**, *2019*, 8761860. [CrossRef] [PubMed]
5. Khurshid, Z.; Warsi, I.; Moin, S.F.; Slowey, P.D.; Latif, M.; Zohaib, S.; Zafar, M.S. Biochemical analysis of oral fluids for disease detection. *Adv. Clin. Chem.* **2021**, *100*, 205–253. [PubMed]
6. Nunes, L.A.; Mussavira, S.; Bindhu, O.S. Clinical and diagnostic utility of saliva as a non-invasive diagnostic fluid: A systematic review. *Biochem. Med.* **2015**, *25*, 177–192. [CrossRef]
7. Meghnani, V.; Mohammed, N.; Giauque, C.; Nahire, R.; David, T. Performance characterization and validation of saliva as an alternative specimen source for detecting hereditary breast cancer mutations by next generation sequencing. *Int. J. Genom.* **2016**, *2016*, 2059041. [CrossRef]
8. Cascella, R.; Stocchi, L.; Strafella, C.; Mezzaroma, I.; Mannazzu, M.; Vullo, V.; Montella, F.; Parruti, G.; Borgiani, P.; Sangiuolo, F.; et al. Comparative analysis between saliva and buccal swabs as source of DNA: Lesson from HLA-B*57:01 testing. *Pharmacogenomics* **2015**, *16*, 1039–1046. [CrossRef]
9. Assad, D.X.; Mascarenhas, E.C.P.; de Lima, C.L.; de Toledo, I.P.; Chardin, H.; Combes, A.; Acevedo, A.C.; Guerra, E.N.S. Salivary metabolites to detect patients with cancer: A systematic review. *Int. J. Clin. Oncol.* **2020**, *25*, 1016–1036. [CrossRef]
10. Hasanzadeh, M.; Sharifi, S.; Dizaj, S.M.; Khalilov, R.; Ahmadian, E. Bioassay of saliva proteins: The best alternative for conventional methods in non-invasive diagnosis of cancer. *Int. J. Biol. Macromol.* **2019**, *124*, 1246–1255.
11. Kaczor-Urbanowicz, K.E.; Wei, F.; Rao, S.L.; Kim, J.; Shin, H.; Cheng, J.; Tu, M.; Wong, D.T.W.; Kim, Y. Clinical validity of saliva and novel technology for cancer detection. *BBA-Rev. Cancer* **2019**, *1872*, 49–59. [CrossRef] [PubMed]
12. de Lima, L.T.F.; Bark, J.M.; Rasheduzzaman, M.; Weeramange, C.E.; Punyadeera, C. Saliva as a matrix for measurement of cancer biomarkers. In *Clinical Aspects and Laboratory Determination. Cancer Biomarkers*; Elsevier: Amsterdam, The Netherlands, 2022; pp. 297–351.
13. Nijakowski, K.; Zdrojewski, J.; Nowak, M.; Gruszczyński, D.; Knoll, F.; Surdacka, A. Salivary Metabolomics for Systemic Cancer Diagnosis: A Systematic Review. *Metabolites* **2023**, *13*, 28. [CrossRef] [PubMed]
14. Eftekhari, A.; Dizaj, S.M.; Sharifi, S.; Salatin, S.; Khalilov, R.; Samiei, M.; Vahed, S.Z.; Ahmadian, E. Salivary biomarkers in cancer. *Adv. Clin. Chem.* **2022**, *110*, 171–192.
15. Streckfus, C.; Bigler, L.; Dellinger, T.; Dai, X.; Kingman, A.; Thigpen, J.T. The presence of soluble c-erbB-2 in saliva and serum among women with breast carcinoma: A preliminary study. *Clin. Cancer Res.* **2000**, *6*, 2363–2370. [PubMed]
16. Chen, D.X.; Schwartz, P.E.; Li, F.Q. Saliva and serum CA125 assays for detecting malignant ovarian tumors. *Obstet. Gynecol.* **1990**, *75*, 701–704. [PubMed]
17. Balan, J.J.; Rao, R.S.; Premalatha, B.R.; Patil, S. Analysis of tumor marker CA 125 in saliva of normal and oral squamous cell carcinoma patients: A comparative study. *J. Contemp. Dent. Pract.* **2012**, *13*, 671–675. [CrossRef]
18. Rajkumar, K.; Ramya, R.; Nandhini, G.; Rajashree, P.; Ramesh Kumar, A.; Nirmala Anandan, S. Salivary and serum level of CYFRA 21-1 in oral precancer and oral squamous cell carcinoma. *Oral Dis.* **2015**, *21*, 90–96. [CrossRef]
19. Jafari, M.; Hasanzadeh, M. Non-invasive bioassay of Cytokeratin Fragment 21.1 (Cyfra 21.1) protein in human saliva samples using immunoreaction method: An efficient platform for early-stage diagnosis of oral cancer based on biomedicine. *Biomed. Pharmacother.* **2020**, *131*, 110671. [CrossRef]
20. Joshi, S.; Kallappa, S.; Kumar, P.; Shukla, S.; Ghosh, R. Simple diagnosis of cancer by detecting CEA and CYFRA 21-1 in saliva using electronic sensors. *Sci. Rep.* **2022**, *12*, 15315. [CrossRef]
21. Liang, Y.-H.; Chang, C.-C.; Chen, C.-C.; Chu-Su, Y.; Lin, C.-W. Development of an Au/ZnO thin film surface plasmon resonance-based biosensor immunoassay for the detection of carbohydrate antigen 15-3 in human saliva. *Clin. Biochem.* **2012**, *45*, 1689–1693. [CrossRef]
22. Farahani, H.; Amri, J.; Alaee, M.; Mohaghegh, F.; Rafiee, M. Serum and Saliva Levels of Cancer Antigen 15-3, Carcinoembryonic Antigen, Estradiol, Vaspin, and Obestatin as Biomarkers for the Diagnosis of Breast Cancer in Postmenopausal Women. *Lab. Med.* **2020**, *51*, 620–627. [CrossRef] [PubMed]
23. Yio, X.Y.; Jiang, J.; Yin, F.Z.; Ruan, K.-H. Highly Sensitive Sandwich Enzyme Immunoassay for Alpha-Fetoprotein in Human Saliva. *Ann. Clin. Biochem.* **1992**, *29*, 519–522. [CrossRef] [PubMed]
24. López-Jornet, P.; Aznar, C.; Ceron, J.; Asta, T. Salivary biomarkers in breast cancer: A cross-sectional study. *Support. Care Cancer* **2021**, *29*, 889–896. [CrossRef] [PubMed]
25. Agha-Hosseini, F.; Mirzaii-Dizgah, I.; Rahimi, A. Correlation of serum and salivary CA15-3 levels in patients with breast cancer. *Med. Oral Patol. Oral Cir. Bucal* **2009**, *14*, e521-4. [CrossRef]

26. Jokerst, J.V.; Raamanathan, A.; Christodoulides, N.; Floriano, P.N.; Pollard, A.A.; Simmons, G.W.; Wong, J.; Gage, C.; Furmaga, W.B.; Redding, S.W.; et al. Nano-bio-chips for high performance multiplexed protein detection: Determinations of cancer biomarkers in serum and saliva using quantum dot bioconjugate labels. *Biosens. Bioelectron.* **2009**, *24*, 3622–3629. [CrossRef] [PubMed]

27. Geng, X.F.; Du, M.; Han, J.X.; Zhang, M.; Tang, X.F.; Xing, R.D. Saliva CA125 and TPS levels in patients with oral squamous cell carcinoma. *Int. J. Biol. Markers* **2013**, *28*, 216–220. [CrossRef]

28. Zhang, K.-Y.; Liu, C.-Y.; Hua, L.; Wang, S.-L.; Li, J. Clinical evaluation of salivary carbohydrate antigen 125 and leptin in controls and parotid tumours. *Oral Dis.* **2016**, *22*, 630–638. [CrossRef]

29. He, H.; Chen, G.; Zhou, L.; Liu, Y. A joint detection of CEA and CA-50 levels in saliva and serum of patients with tumors in oral region and salivary gland. *J. Cancer Res. Clin. Oncol.* **2009**, *135*, 1315–1321. [CrossRef]

30. Zheng, J.; Sun, L.; Yuan, W.; Xu, J.; Yu, X.; Wang, F.; Sun, L.; Zeng, Y. Clinical value of Naa10p and CEA levels in saliva and serum for diagnosis of oral squamous cell carcinoma. *J. Oral Pathol. Med.* **2018**, *47*, 830–835. [CrossRef]

31. Khosravi, N.; Bahrami, N.; Khosravi, A.; Abedini, A.; Kiani, A.; Sharifynia, S.; Gharaeeyan, P.; Seifi, S.; Mohamadnia, A. Expression of mammaglobin and carcinoembryonic antigen in peripheral blood of patients with breast cancer using real time polymerase chain reaction. *Open J. Clin. Diagn.* **2017**, *7*, 103–112. [CrossRef]

32. Rubins, J.B.; Dunitz, J.; Rubins, H.B.; Maddaus, M.A.; Niewoehner, D.E. Serum carcinoembryonic antigen as an adjunct to preoperative staging of lung cancer. *J. Torac. Cardiovasc. Surg.* **1998**, *116*, 412–416. [CrossRef] [PubMed]

33. Brooks, M.N.; Wang, J.; Li, Y.; Zhang, R.; Elashoff, D.; Wong, D.T. Salivary protein factors are elevated in breast cancer patients. *Mol. Med. Rep.* **2008**, *1*, 375–378. [CrossRef] [PubMed]

34. Streckfus, C.; Bigler, L. The use of soluble, salivary c-erbB-2 for the detection and post-operative follow-up of breast cancer in women: The results of a five-year translational research study. *Adv. Dent. Res.* **2005**, *18*, 17–24. [CrossRef]

35. Laidi, F.; Bouziane, A.; Lakhdar, A.; Khabouze, S.; Amrani, M.; Rhrab, B.; Zaoui, F. Significant correlation between salivary and serum CA15-3 in healthy women and breast cancer patients. *Asian Pac. J. Cancer Prev.* **2014**, *15*, 4659–4662. [CrossRef] [PubMed]

36. Zhang, S.J.; Hu, Y.; Qian, H.L.; Jiao, S.C.; Liu, Z.F.; Tao, H.T.; Han, L. Expression and significance of ER, PR, VEGF, CA15-3, CA125 and CEA in judging the prognosis of breast cancer. *Asian Pac. J. Cancer Prev.* **2013**, *14*, 3937–3940. [CrossRef]

37. Assad, D.X.; Mascarenhas, E.C.; Normando, A.C.; Chardin, H.; Barra, G.B.; Pratesi, R.; Nóbrega, Y.D.; Acevedo, A.; Guerra, E.N. Correlation between salivary and serum CA15-3 concentrations in patients with breast cancer. *Mol. Clin. Oncol.* **2020**, *13*, 155–161. [CrossRef]

38. Choi, S.B.; Park, J.M.; Ahn, J.H.; Go, J.; Kim, J.; Park, H.S.; Kim, S.I.; Park, B.-W.; Park, S. Ki-67 and breast cancer prognosis: Does it matter if Ki-67 level is examined using preoperative biopsy or postoperative specimen? *Breast Cancer Res. Treat.* **2022**, *192*, 343–352. [CrossRef]

39. Friedrich, R.E.; Davidoff, M.S.; Bartel-Friedrich, S. Expression of Neuron-specific Enolase in Irradiated Salivary Glands of the Rat: A Pilot Study. *Anticancer Res.* **2010**, *30*, 1569–1571.

40. Hergenroeder, G.W.; Redell, J.B.; Moore, A.N.; Dash, P.K. Biomarkers in the clinical diagnosis and management of traumatic brain injury. *Mol. Diagn. Ther.* **2008**, *12*, 345–358. [CrossRef]

41. Bayerl, C.; Lauk, J.; Moll, I.; Jung, E.G. Immunohistochemical characterization of HSP, alpha-MSH, Merkel cells and neuronal markers in acute UV dermatitis and acute contact dermatitis in vivo. *Inflamm. Res.* **1997**, *46*, 409–411. [CrossRef]

42. Al-Rawi, N.H.; Atiyah, K.M. Salivary neuron specific enolase: An indicator for neuronal damage in patients with ischemic stroke and stroke-prone patients. *Clin. Chem. Lab. Med.* **2009**, *47*, 1519–1524. [CrossRef] [PubMed]

43. Chen, J.; Wang, J.; Cao, D.; Yang, J.; Shen, K.; Huang, H.; Shi, X. Alpha-fetoprotein (AFP)-producing epithelial ovarian carcinoma (EOC): A retrospective study of 27 cases. *Arch. Gynecol. Obstet.* **2021**, *304*, 1043–1053. [CrossRef] [PubMed]

44. Meguro, S.; Yasuda, M. Alpha-fetoprotein-producing ovarian tumor in a postmenopausal woman with germ cell differentiation. *Ann. Diagn. Pathol.* **2013**, *17*, 140–144. [CrossRef]

45. Jha, S.; Sinha, J.; Bharti, S. Alpha Fetoprotein Secreting Mucinous Epithelial Ovarian Carcinoma in a Young Woman—A Rare Case Report and Review of Literature. *EJMO* **2021**, *5*, 80–84. [CrossRef]

46. You, X.Y.; Jiang, J.; Yin, F.Z. Preliminary observation on human saliva alpha-fetoprotein in patients with hepatocellular carcinoma. *Chin. Med. J.* **1993**, *106*, 179–182. [PubMed]

47. Buderath, P.; Kasimir-Bauer, S.; Aktas, B.; Rasch, J.; Kimmig, R.; Zeller, T.; Heubner, M. Evaluation of a novel ELISA for the tumorassociated antigen CA 72-4 in patients with ovarian cancer. *Future Sci. OA* **2016**, *2*, FSO145. [CrossRef]

48. Granato, T.; Midulla, C.; Longo, F.; Colaprisca, B.; Frati, L.; Anastasi, E. Role of HE4, CA72.4, and CA125 in monitoring ovarian cancer. *Tumor Biol.* **2012**, *33*, 1335–1339. [CrossRef]

49. Sergeeva, N.S.; Marshutina, N.V.; Solokhina, M.P.; Alentov, I.I.; Kaprin, A.D. Clinical significance of CA 72-4 as a serological tumor-associated marker. *Onkologiya. Zhurnal Im. P.A. Gertsena* **2019**, *8*, 120–125. (In Russian) [CrossRef]

50. Agha-Hosseini, F.; Mirzaii-Dizgah, I.; Rahimi, A.; Seilanian-Toosi, M. Correlation of serum and salivary CA125 levels in patients with breast cancer. *J. Contemp. Dent. Pract.* **2009**, *10*, E001-8.

51. Tay, S.K.; Chua, E.K. Correlation of serum, urinary and salivary CA 125 levels in patients with adnexal masses. *Ann. Acad. Med. Singap.* **1994**, *23*, 311–314.

52. Vuković, A.; Kuna, K.; Brzak, B.L.; Boras, V.V.; Šeparović, R.; Šekerija, M.; Šumilin, L.; Vidranski, V. The role of salivary and serum CA125 and routine blood tests in patients with ovarian malignancies. *Acta Clin. Croat.* **2021**, *60*, 55–62. [CrossRef] [PubMed]

53. Plante, M.; Wong, G.Y.; Nisselbaum, J.S.; Almadrones, L.; Hoskins, W.J.; Rubin, S.C. Relationship between saliva and serum CA125 in women with and without epithelial ovarian cancer. *Obstet. Gynecol.* **1993**, *81*, 989–992. [PubMed]
54. Dochez, V.; Caillon, H.; Vaucel, E.; Dimet, J.; Winer, N.; Ducarme, G. Biomarkers and algorithms for diagnosis of ovarian cancer: CA125, HE4, RMI and ROMA, a review. *J. Ovarian Res.* **2019**, *12*, 28. [CrossRef] [PubMed]
55. Bingle, L.; Cross, S.S.; High, A.S.; Wallace, W.A.; Rassl, D.; Yuan, G.; Hellstrom, I.; Campos, M.A.; Bingle, C.D.WFDC2 (HE4): A potential role in the innate immunity of the oral cavity and respiratory tract and the development of adenocarcinomas of the lung. *Respir. Res.* **2006**, *7*, 61. [CrossRef]
56. Karlsen, N.S.; Karlsen, M.A.; Høgdall, C.K.; Høgdall, E.V.S. HE4 Tissue Expression and Serum HE4 Levels in Healthy Individuals and Patients with Benign or Malignant Tumors: A Systematic Review. *Cancer Epidemiol. Biomark. Prev.* **2014**, *23*, 2285–2295. [CrossRef]

Review

Repurposing of the Drug Tezosentan for Cancer Therapy

Eduarda Ribeiro [1,2,3] and Nuno Vale [1,2,4,*]

1 OncoPharma Research Group, Center for Health Technology and Services Research (CINTESIS), Rua Dr. Plácido da Costa, 4200-450 Porto, Portugal; eduardaprr@gmail.com
2 CINTESIS@RISE, Faculty of Medicine, University of Porto, Alameda Professor Hernâni Monteiro, 4200-319 Porto, Portugal
3 Institute of Biomedical Sciences Abel Salazar (ICBAS), University of Porto, Rua de Jorge Viterbo Ferreira 228, 4050-313 Porto, Portugal
4 Department of Community Medicine, Health Information and Decision (MEDCIDS), Faculty of Medicine, University of Porto, Rua Doutor Plácido da Costa, 4200-450 Porto, Portugal
* Correspondence: nunovale@med.up.pt; Tel.: +351-220-426-537

Abstract: Tezosentan is a vasodilator drug that was originally developed to treat pulmonary arterial hypertension. It acts by inhibiting endothelin (ET) receptors, which are overexpressed in many types of cancer cells. Endothelin-1 (ET1) is a substance produced by the body that causes blood vessels to narrow. Tezosentan has affinity for both ET_A and ET_B receptors. By blocking the effects of ET1, tezosentan can help to dilate blood vessels, improve the blood flow, and reduce the workload on the heart. Tezosentan has been found to have anticancer properties due to its ability to target the ET receptors, which are involved in promoting cellular processes such as proliferation, survival, neovascularization, immune cell response, and drug resistance. This review intends to demonstrate the potential of this drug in the field of oncology. Drug repurposing can be an excellent way to improve the known profiles of first-line drugs and to solve several resistance problems of these same antineoplastic drugs.

Keywords: tezosentan; cancer; endothelin receptors; drug repurposing

Citation: Ribeiro, E.; Vale, N. Repurposing of the Drug Tezosentan for Cancer Therapy. *Curr. Issues Mol. Biol.* **2023**, *45*, 5118–5131. https://doi.org/10.3390/cimb45060325

Academic Editor: Dumitru A. Iacobas

Received: 28 April 2023
Revised: 2 June 2023
Accepted: 9 June 2023
Published: 11 June 2023

1. Introduction

Nowadays, cancer remains a significant health problem worldwide. In 2020 there were an estimated 19.3 million new cases of cancer and almost 10 million deaths from cancer [1]. Despite huge improvements, current anticancer pharmacological therapies are effective in a limited number of cancer cases. Tumors with a high mortality rate, targets not reachable by chemotherapy, and chemotherapy resistance, represent the current challenges of cancer treatments [2]. As the pharmaceutical productivity and drug efficacy in oncology seem to have reached a plateau, 'drug repurposing'—meaning the use of old drugs, already in clinical use, for a different therapeutic indication—is a promising and viable strategy to improve cancer therapy. Opportunities for drug repurposing are often based on occasional observations or on time-consuming pre-clinical drug screenings that are often not hypothesis-driven [3]. This approach is greatly beneficial because of the main benefits such as the reduced development timelines with an average saving of 5–7 years, high approval rates, lower development costs, and, as they are already approved drugs, these compounds have already been tested in humans, so comprehensive information exists on their pharmacology, dose, possible toxicity, and formulation [4,5]. A possible solution to tackle the complexity of treating different types of cancer is to focus on how a specific treatment functions at a molecular level. By understanding this, we can identify which types of cancer are more likely to respond to that treatment [4,6]. However, it is essential to note that the effects of drugs on cancer cells vary depending on the specific drug and cancer type being examined. This is important to consider, as the results observed in laboratory studies may not always translate to the same effects in humans.

One such approach is the use of vasodilators, drugs originally development to use in the management of hypertension, angina, preeclampsia, stroke, heart failure, chronic kidney disease, and myocardial infarction. Examples of vasodilators include drugs such as angiotensin-converting enzyme inhibitors, angiotensin receptor blockers, calcium channel blockers, nitrates, PDE5 inhibitors, and endothelin antagonists, among others. These drugs work through different mechanisms to relax the smooth muscles in blood vessels, thereby reducing blood pressure and improving blood flow throughout the body [7].

Tezosentan, a vasodilator drug originally developed to treat pulmonary arterial hypertension acts through inhibiting endothelin (ET) receptors, which are overexpressed in many types of cancer cells [8].

Therefore, additional research is necessary to fully comprehend the impact of tezosentan on cancer cells and determine their potential as a cancer treatment. In the following section, we review the current understanding of the mechanisms of action of tezosentan in cancer and its clinical applications as a novel anticancer agent.

2. Structure of Tezosentan and Mechanism of Action

Tezosentan is a small molecule that belongs to a class of drugs known as ET receptor antagonists, which work through blocking the effects of a hormone called endothelin. The chemical name of tezosentan is N-[(2R)-6-(2,6-dimethylphenyl)-5-[(3R)-3-methyl-2,3,4,5-tetrahydro-1,5-benzoxazepin-5-yl]pyridazin-3-yl]-4-methoxybenzamide, and its chemical formula is $C_{30}H_{31}N_5O_2$. Tezosentan belongs to the class of organic compounds known as pyridinylpyrimidines and is composed of three primary structural moieties (Figure 1), namely a pyridazine ring, a tetrahydrobenzoxazine ring, and a phenyl group. These moieties are connected through a linker, with the pyridazine ring being linked to the tetrahydrobenzoxazine ring and the phenyl group being attached to the pyridazine ring. Additionally, the phenyl group of the molecule is substituted with a methoxy group (-OCH$_3$).

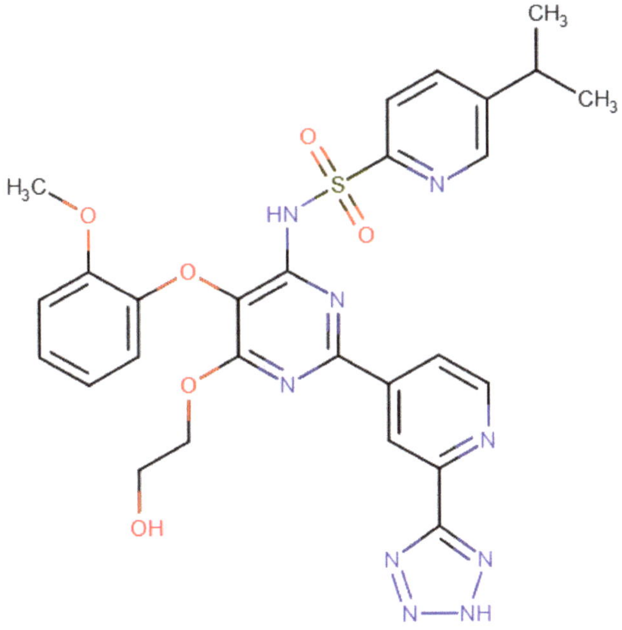

Figure 1. Chemical structure of the drug tezosentan (figure edited from DrugBank.com; accessed on 25 March 2023).

Endothelin-1 (ET1) is a naturally occurring substance that is produced by the body and has potent vasoconstrictive effects, meaning it causes blood vessels to narrow. ET1 is produced by the endothelial cells that line blood vessels, and acts on two types of receptors: endothelin type A (ET_A) and endothelin type B (ET_B). Tezosentan has affinity for both ET_A and ET_B receptors [9,10]. By blocking the effects of ET1, tezosentan can help to dilate blood vessels, improve the blood flow, and reduce the workload on the heart. This can help to alleviate symptoms of heart failure and improve overall cardiac function. The ET_A receptor is primarily found in vascular smooth muscle cells and is responsible for causing blood vessels to narrow and for promoting cell growth. The ET_B receptor is found in various tissues, including the brain, blood vessels, and heart, and can have different effects depending on where it is located. In blood vessels, it can cause both constriction and dilation, while in cardiac fibroblasts it is believed to play a role in fibrosis, which is the buildup of excess connective tissue that can impair heart function [11].

Tezosentan is rapidly distributed after intravenous administration, with a volume of distribution of approximately 3 L/kg. The drug has a linear pharmacokinetic profile over a wide range of doses. Tezosentan is metabolized primarily by the liver via the cytochrome P450 3A4 enzyme system, and its clearance is primarily via the biliary route. The drug has a half-life of approximately 30 min and is eliminated from the body within 24 h after administration [9,12].

3. Clinical Use against Cancer

3.1. Endothelin-1 Receptor

The ET1 signaling pathway can promote various cellular processes such as proliferation, survival, epithelial-to-mesenchymal transition, neovascularization, immune cell response, and drug resistance, but the effects may vary depending on the context (Figure 2) [13]. Several tumor cell lines have demonstrated increased levels of ET1 production, highlighting the targeting of endothelin receptors as an important approach for cancer therapy [14–18].

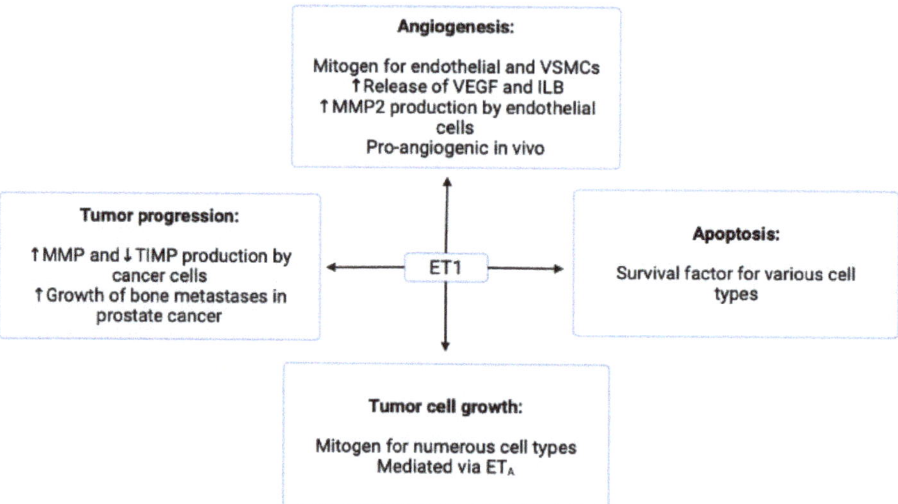

Figure 2. Actions of endothelin-1 in cancer.

The ET_B signaling pathway has been found to promote the migration and proliferation of microvascular endothelial cells [19]. ET_A signaling, on the other hand, induces angiogenesis, particularly through the induction of vascular endothelial growth factor [20,21]. ET1 triggers the activation of phospholipase C β (PLCβ), which increases intracellular calcium ion levels and activates protein kinase C. It also activates phosphoinositide 3-kinase (PI3K), c-Jun N-terminal kinase (JNK), extracellular-signal-regulated kinase/mitogen-activated protein kinase (ERK/MAPK), and epidermal growth factor receptor (EGFR) signaling (Figure 3). Additionally, ET1-induced signaling leads to the activation of FAK and paxillin. Blocking ET_A and ET_B receptors has shown promise in treating systemic pulmonary hypertension and chronic heart failure by inhibiting endothelial cell proliferation and survival [22]. In cancer therapy, inhibition of endothelin receptor signaling has shown potential.

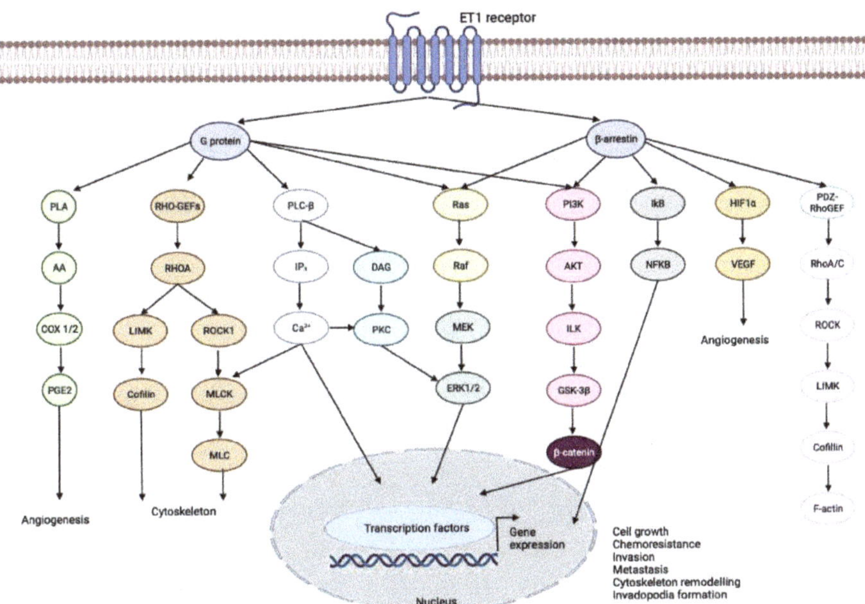

Figure 3. The ET1 signaling network. ET1, a signaling molecule involved in cancer, activates several pathways that contribute to various cellular processes. Upon binding to its receptor, it initiates a cascade of events involving G-protein coupled receptor activation and the activation of primary effectors. One of the pathways activated by ET1 is through the activation of PLCβ. This enzyme cleaves a molecule called phosphatidylinositol-4,5-bisphosphate (PtdIns(4,5)P2) into diacylglycerol (DAG) and inositol triphosphate (IP3). This leads to an increase in calcium levels and the activation of protein kinase C (PKC). Additionally, this pathway triggers the activation of members of the MAPK family, including ERK1/2, which are important for cellular signaling. ET1 activation also stimulates the Ras/Raf/MEK pathway, which converges on the activation of ERK1/2. This pathway plays a role in transmitting signals that regulate cell growth and proliferation. Furthermore, ET1 stimulation activates phospholipase A (PLA), resulting in the release of arachidonic acid (AA) and the activation of cyclooxygenase-1 (COX-1) and COX-2. These enzymes are involved in the production of prostaglandin E2 (PGE2). ET1 also activates PI3K, leading to the activation of AKT, integrin-linked kinase (ILK), and glycogen synthase kinase (GSK)-3β, which in turn stabilizes β-catenin. Importantly, ET1 can also activate ERK1/2 and PI3K/AKT/β-catenin signaling through the involvement of β-arrestin1. Through β-arrestin1, ET1 also activates nuclear factor-kB (NF-kB) signaling by inhibiting NF-kB inhibitor (IkB), leading to the dissociation and subsequent nuclear localization of active NF-kB. Additionally, ET1 activates PDZ-RhoGEF, leading to the activation of Rho-A and -C GTPases. This activation triggers Rho-dependent signaling events through RHO-associated coiled-coil containing protein kinase 1 (ROCK1), resulting in the activation of LIMK and the inhibition of cofilin. Collectively, these signaling pathways, orchestrated by ET-1R, promote cell growth, chemoresistance, angiogenesis, cytoskeleton remodeling, invadopodia formation, and metastasis. By understanding and targeting these pathways, it may be possible to develop therapeutic strategies to intervene in cancer progression.

Changes in the expression of ET1 and its associated receptors and signaling pathways can disrupt normal cellular processes and contribute to the development and progression of tumors. This can occur through both autocrine signaling (when cells produce ET1 to stimulate their own growth) and paracrine signaling (when cells produce ET1 to stimulate the growth of neighboring cells). These changes may be caused by genetic mutations or epigenetic modifications, which can lead to aberrant expression of ET1 and alterations in its downstream signaling pathways [13].

Due to the significant involvement of ET_A in the biology of tumor cells, there has been greater emphasis on the development of selective antagonists targeting ET_A compared to those targeting ET_B in the treatment of cancer. This is reflected in Table 1, which demonstrates a higher number of ET_A-selective antagonists developed for this purpose.

Table 1. Clinical trials involving ET1 receptors against types of cancer.

Drug	Target	Type of Cancer	Clinical Trial	Status	Identifier Trial Number (https://clinicaltrials.gov) Accessed on 25 April 2023
Atrasentan	ET_A	Prostate	Phase III	Completed (2021)	NCT00134056
		Prostate	Phase III	Completed (2006)	NCT00036543
		Prostate	Phase III	Completed (2011)	NCT00046943
		Prostate	Phase II	Completed (2010)	NCT00181558
		Prostate	Phase III	Completed (2007)	NCT00036556
		Prostate	Phase II	Completed (2006)	NCT00038662
		Kidney	Phase II	Completed (2010)	NCT00039429
		Ovarian	Phase II	Terminated (2012)	NCT00653328
		Prostate	Phase II/III	Completed (2007)	NCT00127478
		Brain	Phase I	Completed (2009)	NCT00017264
Zibotentan (ZD4054)	ET_A	Prostate	Phase III	Completed (2016)	NCT00554229
		Prostate	Phase III	Terminated (2012)	NCT00626548
		Prostate	Phase II	Terminated (2019)	NCT01119118
		Prostate	Phase III	Terminated (2012)	NCT00617669
		Prostate	NA	Unknown	NCT01168141
		Breast	Phase II	Withdrawn	NCT01134497
		Lung	Phase II	Completed (2012)	NCT00745875
		Prostate	Phase II	Completed (2013)	NCT00090363
		Prostate	Phase I	Completed (2013)	NCT00314782
		Ovarian	Phase II	Terminated (2012)	NCT00929162
		Prostate	Phase II	Completed (2012)	NCT00055471
		Colorectal	Phase II	Completed (2014)	NCT01205711
BQ788	ET_B	Melanoma	Phase I	Terminated (2015)	NCT02442466
Bosentan	ET_A/ET_B	Pancreatic	Phase I	Recruiting	NCT04158635
		Solid tumor	Phase I	Recruiting	NCT05072106
Macitentan	ET_A/ET_B	Glioblastoma	Phase I	Terminated (2018)	NCT02254954
		Glioblastoma	Phase I	Terminated (2016)	NCT01499251

NA—Not applicable.

Based on safety data on other ET1 receptor antagonists, bosentan has the ability to affect specific enzymes in the body, including CYP2C9, CYP2C19, and CYP3A4. These enzymes play a role in the breakdown of many contraceptives, which means that when bosentan is taken together with these contraceptives, there is a possibility of reduced effectiveness. Patients with moderate to severe liver problems should exercise caution when using bosentan because it can lead to increased levels of liver amino transferase. Additionally, bosentan has been found to have the potential to cause birth defects. Ambrisentan, when used in combination with drugs such as ciclosporin, ketoconazole, or omeprazole, which affect the activity of CYP enzymes, can potentially result in drug interactions. Similar caution is recommended for inhibitors or inducers of P-glycoprotein (P-gp), UDP-glucuronosyltransferase, and organic-anion-transporting polypeptide (OATP) when co-administered with ambrisentan. Ambrisentan is not recommended for patients with severe liver impairment due to the involvement of the liver and bile in its metabolism and elimination. Macitentan is not recommended for use in pregnant women due to the risk of causing birth defects, and in individuals with severe liver dysfunction or elevated liver enzyme levels [23]. Tezosentan should be avoided in patients with clinically significant renal failure [10]. With this, tezosentan seems to be the safest ET1 receptor inhibitor to be tested as an antineoplastic drug.

3.2. Endothelin Receptor Type A

Several types of cancer, including colorectal, ovarian, and prostate tumors, have been shown to have increased expression of ET_A receptors in malignant tissue, as demonstrated through immunohistochemistry and autoradiography [24–26]. In fact, ET_A receptor expression levels in prostate tumors have been found to correlate with both the Gleason score and the presence of metastasis [27].

Akhavan et al., found that blocking ET1 receptors with a selective ET_A receptor antagonist, atrasentan, inhibited the growth of prostate cancer cells in vitro and in vivo by inducing apoptosis and reducing angiogenesis [28]. The results of another study on LIM1215 and HT29 colorectal cancer cell lines have shown that ET1 can promote cell growth through the ET_A receptor. However, it is not yet clear whether this effect is due to a mitogenic stimulus, an antiapoptotic signal, or a combination of both. Nevertheless, the findings of previous studies that have shown that ET1 is produced by colorectal cancers, along with the results of this study, suggest that ET1 may act as a mitogen in colorectal cancer [29,30]. Therefore, targeting the ET_A receptor with antagonists may have therapeutic potential for the treatment of colorectal cancer.

There is some debate over the role of ET1, which acts as both an autocrine and paracrine cytokine. However, findings of Liu et al., on gastric cancer cell lines suggest that activating the $ET1/ET_A$ pathway contributes to cell proliferation, migration, and antiapoptosis. By acting mainly through ET_A, ET1 plays a crucial role in promoting the development of gastric cancer, indicating that inhibiting the $ET1/ET_A$ axis could potentially improve treatment outcomes [31]. Another study demonstrates that the drug ambrisentan, another ET_A receptor antagonist, inhibits in vitro cancer cell migration and invasion in COLO-357 metastatic pancreatic adenocarcinoma, OvCar3 ovarian carcinoma, MDA-MB-231 breast adenocarcinoma, and HL-60 promyelocytic leukemia. In vivo this drug has demonstrated its potential in reducing metastasis of a metastatic breast cancer into the lung and liver, thereby decreasing mortality [32]. Atrasentan (ABT-627), a specific antagonist of the endothelin-A receptor, has shown promising results in treating men with hormone-refractory metastatic prostate cancer. The safety profile of this drug aligns with its pharmacologic activity and makes it suitable for long-term, noncytolytic therapy. Further research is necessary to investigate its efficacy in earlier stages of prostate cancer and in other types of cancer in which the endothelin pathway may be involved [33].

Arabanian et al., demonstrate that ET_A is upregulated in leukemic cells that express high levels of *Hoxa9* and *Meis1*. Overexpression of ET_A has been shown to promote cell proliferation and enhance the repopulating capacity of *Hoxa9*-expressing cells. Moreover, cancer progression and metastasis rely heavily on the resistance to apoptosis, and leukemia-inducing cells require close contact with their niche to survive. This contact is supported and maintained by various surface molecules that modulate the homing capacity of leukemic cells and their ability to proliferate and remain quiescent [34]. Identification of these surface molecules can be helpful in targeting the residual leukemia-inducing cells after therapy. In this study, they investigated whether ET_A signaling contributes to protection against drug-induced apoptosis in leukemia cells [35]. They used daunorubicin, a common chemotherapeutic agent used in leukemia treatment, to induce apoptosis in the cells. The results suggest that upregulation of ET_A may contribute to chemo-resistance in leukemic cells in response to daunorubicin. ET_A may function as a signaling molecule that promotes cell growth and induces protection against cytotoxic effects. In conclusion, targeting ET_A, along with other chemotherapies, may have therapeutic benefits for leukemia by overcoming resistance to drug-induced apoptosis. ET_A inhibitors could potentially be used in the treatment of leukemia to enhance the effectiveness of chemotherapy. Further studies are needed to investigate the mechanism of ET_A function using additional chemotherapeutic agents and ET_A knockdown models as controls [36].

ATP-Binding Cassette Subfamily B Member 1 (ABCB1) Protein

Upregulation of transmembrane transporters such as ATP-binding cassette subfamily B member 1 (ABCB1 or P-glycoprotein) in tumor cells contributes to drug resistance. The ABCB1 protein is known for its involvement in multidrug resistance of tumor cells by preventing intracellular accumulation of cytotoxic drugs. It serves as a membrane efflux that extrudes the drugs from the cancer cells, reducing their effectiveness [37,38]. This is a major problem in cancer treatment, as drug resistance can limit the effectiveness of chemotherapy and ultimately lead to treatment failure.

Several studies have shown that ABCB1 overexpression is associated with poor prognosis in various types of cancer, including breast [39], lung [40], colon [41], and ovarian cancer [42–45]. In addition, ABCB1 inhibitors have been investigated as potential drugs to overcome drug resistance in cancer cells. In that study, Englinger *et al.*, found that a specific signaling pathway, called the $ET1/ET_A$ pathway, was overactive in a type of lung cancer cell called DMS114/NIN, and this contributed to the cells' resistance to a cancer drug called nintedanib. The overactive pathway led to increased levels of certain proteins (ABCB1, PKC, and NFκB), which are known to be involved in drug resistance. The researchers tested tezosentan, which blocks the ET_A protein, and found that it decreased the levels of ABCB1, PKC, and NFκB proteins in the cells. Importantly, this also led to a significant increase in the cells' sensitivity to nintedanib. In summary, blocking the ET_A pathway using tezosentan could be a promising strategy to overcome resistance to nintedanib in cancer cells [46].

3.3. Endothelin Receptor Type B

ET_B is overexpressed in bladder carcinoma [47], melanoma [48], small-cell lung cancer [49], vulvar cancer [50], clear-cell renal cell carcinoma [51], esophageal squamous cell carcinoma [52], and glioblastoma [53].

Methylation of the ET_B gene has also been found to be relatively increased in several cancer cell lines from the prostate, bladder, and colon, with a corresponding downregulation of transcription [54]. This suggests a potential mechanism for the reduced expression of ET_B receptors in malignant tissue.

One of the main signaling pathways activated by ET_B in cancer is the PI3K/AKT pathway, which is involved in cell growth and survival [55,56]. ET_B can also activate the MAPK/ERK pathway, which is involved in cell proliferation and differentiation [57]. In addition, ET_B can stimulate the expression of various genes that promote cancer cell invasion and metastasis [58].

Melanocytes use a surface receptor called ET_B to produce and transport melanin, and it has been found that this receptor works in synergy with many other factors involved in the melanin pathway [59]. However, studies have shown that using ET_B antagonists can slow the growth of melanoma cells and increase their programmed cell death, or apoptosis [60]. Research has also indicated that ET_B antagonists can prevent the progression of melanoma induced by ET and stabilize metastatic disease, making it an effective clinical approach for treating melanoma [61,62].

A study led by Lahav et al., indicates that the ET_B inhibitor BQ788 is a potent agent for inhibiting human melanoma tumor growth in a nude mouse model, with some tumors being completely halted. Although the underlying mechanism is not yet fully understood, experiments conducted in culture suggest that BQ788 may slow melanoma cell growth, induce differentiation, and ultimately lead to cell death. For example, some lines showed a slower growth rate, increased pigmentation, and flattened appearance, while others displayed dendritic morphology characteristic of differentiated melanocytes. The most sensitive cell lines showed a loss of viability accompanied by cytoplasmic vacuoles and increased TUNEL staining, indicative of apoptosis. Although further investigation is needed to determine whether BQ788 is the most effective ET_B antagonist for stopping melanoma growth in vivo, there is evidence that such drugs can be well tolerated. Future experiments should explore why systemic treatment with BQ788 results in variable responses in different animals, and whether the effects are reversible. It will also be important to determine whether ET_B antagonists can inhibit metastasis in addition to tumor growth [63].

Another study on melanoma has identified ET_B, which is associated with aggressive phenotypes, including the ability of melanoma cells to mimic blood vessels. By screening a variety of benign and malignant pigment cell lesions using immunohistochemistry and quantitative reverse transcription PCR analysis, the authors have found that ET_B is a marker of tumor progression in malignant melanoma and could be a clinically relevant target for the development of small molecules that can block ET. Furthermore, an orally active ET_B antagonist called A-192621 has displayed antitumor activity against established melanoma expressing ET_B, making it a promising candidate for targeted therapy. ET-1 and ET-3 are molecules that trigger several molecular effectors involved in melanoma progression, including cell–cell adhesion and communication molecules, tumor proteases, and integrins, through the activity of ET_B. By blocking this receptor with small molecules, melanoma growth can be inhibited, providing new possibilities for integrated treatments for this malignancy. This knowledge is particularly important given the known resistance of melanoma to current therapies [64].

Because of its role in promoting cancer cell growth and survival, ET_B has become a potential target for cancer therapy. The drugs that target ET_B work through blocking the activity of ET_B and inhibiting its downstream signaling pathways, thereby reducing cancer cell growth and invasion.

Nitric Oxide (NO) Release

When endothelin-1 binds to ETB receptors on endothelial cells, it triggers the synthesis of nitric oxide (NO) and the subsequent relaxation of smooth muscle [65]. Nitric oxide (NO), also known as nitrogen monoxide, is a naturally occurring gas that is soluble in water and acts as a free radical. Its presence is critical in several physiological and pathological processes. NO is produced by the oxidation of l-arginine through the catalytic activity of NO synthases (NOSs). This process requires nicotinamide adenine dinucleotide phosphate (NADPH) and O_2 as cosubstrates, and it results in the production of NO and l-citrulline as byproducts. NO's role in many bodily functions occurs primarily through a cyclic guanosine monophosphate (cGMP)-dependent pathway, leading to vasodilation, neurotransmission, inhibition of platelet aggregation, and smooth muscle relaxation. Another pathway for NO activity is cGMP-independent, and it involves the reaction of NO with molecular oxygen, superoxide (O2-), thiols, and transition metals such as zinc. Additionally, NO can directly modify proteins without enzyme involvement through nitration or nitrosylation. S-nitrosylation, a reversible modification, involves the signaling of cysteine thiol residues and regulates the function of several intracellular proteins [66].

NO plays a complex and context-dependent role in cancer, having both pro-tumor and antitumor effects (Figure 4). The effects of NO on cancer cells depend on its concentration and the stage of the cancer. At low concentrations, NO can activate pathways that promote cell growth and inhibit the immune response, leading to the development of cancer. At higher concentrations, NO can induce cell death and inhibit the growth and metastasis of cancer cells [67]. The role of NO in cancer is complex and depends on a variety of factors, including the stage of the cancer and the cellular response to stress. NO can alter the expression of genes involved in DNA repair and tumor suppression, as well as affect processes such as apoptosis and metastasis [68].

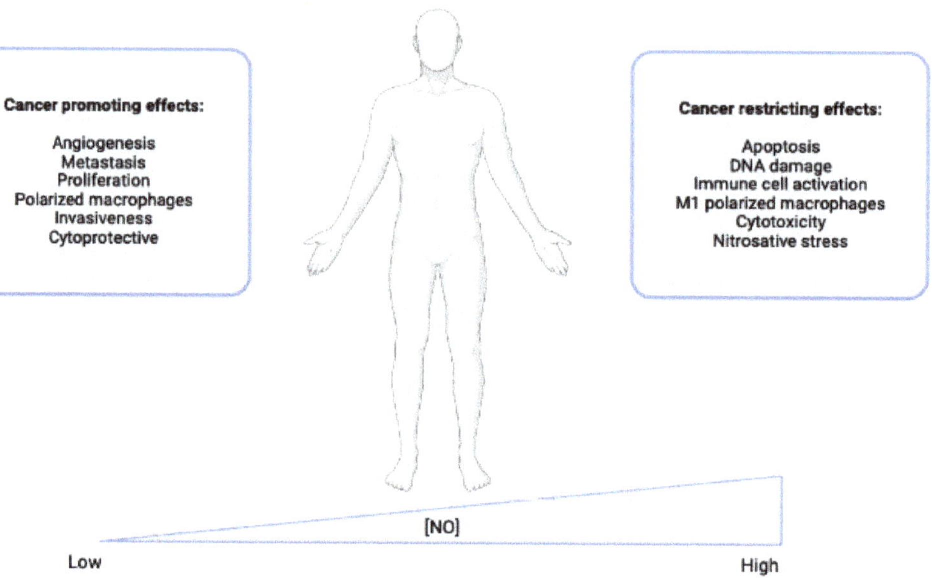

Figure 4. Concentration-dependent effects of NO in cancer. When NO is present in low concentrations, it can improve the molecular processes that maintain normal physiology. However, in already established cancers, low levels of NO may promote cancer progression by enhancing processes such as proliferation and angiogenesis, and the switch to immunologically suppressive immune cell types. In contrast, high levels of NO can induce DNA damage, activate p53, and cause nitrosative stress. While this may promote the development of cancer initially, in already established cancers, high levels of NO can actually activate immunity and improve the effectiveness of chemotherapy. Overall, the effects of NO on cancer depend on the stage of the cancer and the concentration of NO present. While low levels of NO can enhance cancer progression, high levels of NO can induce DNA damage and activate immunity to improve chemotherapeutic efficacy.

Studies have shown that NO can inhibit the growth of various cancer cells, including gastric [69], breast [70], prostate [71], and bladder cancer cells [72], as well as neural precursor cells [73]. GIT-27NO, a novel NO donor, has been shown to inhibit the growth of PC3 and LnCap prostate cancer cells in a concentration-dependent manner when xenografted into nude mice [74]. Saquinavir, a NO-derivative of the HIV protease inhibitor, has also been shown to induce apoptosis and production of proapoptotic BCL-2-interacting mediator of cell death (Bim) in PC3. In vivo studies have shown that Saq-NO inhibited PC3 xenotransplants to a greater extent than the parental compound [75].

One proposed mechanism for the anticancer effects of NO is through the upregulation of a pathway involving BRCA1, Chk1, p53, and p21. This pathway is involved in controlling the cell cycle and can cause cell cycle arrest and cell death in response to DNA damage or other types of cellular stress [76].

Additionally, a study has revealed that NO can increase the sensitivity of cancer cells to radiation, making them more vulnerable to destruction. In cancer radiotherapy, one of the challenges is that hypoxic cells in tumors, which have low levels of oxygen, are not very responsive to radiation. However, NO has been shown to sensitize hypoxic cells to radiation by increasing their oxygen levels through pathways that alter blood flow and oxygen intake by the cells [77]. The combination of NO and ionizing radiation can also induce apoptotic cell death by activating the p53 pathway. In one study, treating colorectal cancer cells with NO donors resulted in a significant increase in the radiosensitivity of the cancer cells [78,79]. Overall, NO has the potential to be a powerful tool in cancer treatment, particularly in combination with radiation therapy. By sensitizing hypoxic cells and inducing apoptotic cell death, NO can improve the effectiveness of cancer treatment and lead to better outcomes for patients.

3.4. METAP1 Methionyl Aminopeptidase 1

As said before, tezosentan belongs to the class of organic compounds known as pyridinylpyrimidines that selectively inhibit human methionyl aminopeptidase 1 (*METAP1*) [80,81].

METAP1 is a protein-coding gene that belongs to the family of metalloproteases. It is involved in the removal of the N-terminal methionine residue from newly synthesized proteins, a process known as N-terminal methionine excision (NME) [82]. *METAP1* is highly conserved across species and is expressed in a wide range of tissues, including the brain, liver, kidney, lung, and heart. It has been implicated in various cellular processes, such as protein synthesis, cell proliferation, and apoptosis [83,84]. Studies have shown that *METAP1* inhibitors have potential therapeutic benefits in the treatment of various diseases, including cancer [83,85] and malaria [86].

Frottin et al., propose that the assessment of *METAP1* levels and redox homeostasis imbalances can be advantageous in creating individualized anticancer tactics. These measurements provide vital information about the role of *METAP1* in cancer and the redox status of cancer cells. By identifying these characteristics in a patient's cancer cells, healthcare providers can determine the potential effectiveness of *METAP1* inhibitors and redox-targeted therapies in that specific individual. Therefore, *METAP1* levels and redox homeostasis imbalances can serve as innovative tools to develop personalized cancer treatments [83].

4. Conclusions

The repurposing of vasodilator drugs such as tezosentan for cancer therapy represents a promising approach to overcome the challenges of current anticancer pharmacological therapies. Tezosentan acts through inhibiting endothelin receptors, which are overexpressed in many types of cancer cells, and has been shown to have potential as a novel anticancer agent. Additionally, tezosentan has shown promising preclinical results in inhibiting cancer cell growth and inducing apoptosis, particularly in tumors with high expression of endothelin receptor type A. Further clinical studies are needed to confirm the efficacy and safety of tezosentan in cancer therapy, but its potential as a repurposed drug for cancer treatment is certainly worth exploring.

One of the aspects that can be developed in the future will be its potential as a new agent to be used in oncology compared to reference drugs if eventually used in combination. In drug combination models, using only repurposed drugs or including reference drugs, can be an excellent way to ensure additive or synergistic effects in a preclinical or even clinical context. Drug combination therapy, in which two or more drugs are used together, aims to increase the chances for better efficacy. Additionally, the efficacy will be observed by decreasing toxicity and reduced development of drug resistance compared to monotherapies. Being a repurposed drug, its clinical safety potential has already been duly proven and this aspect will be a stimulus for further studies involving the drug tezosentan.

Author Contributions: Conceptualization, N.V.; methodology E.R.; formal analysis, E.R. and N.V.; investigation, E.R.; writing—original draft preparation, E.R.; writing—review and editing, N.V.; supervision, N.V.; project administration, N.V.; funding acquisition, N.V. All authors have read and agreed to the published version of the manuscript.

Funding: This research was financed by *Fundo Europeu de Desenvolvimento Regional* (FEDER) funds through the COMPETE 2020 Operational Programme for Competitiveness and Internationalisation (POCI), Portugal 2020, and by Portuguese funds through *Fundação para a Ciência e a Tecnologia* (FCT) in the framework of the projects IF/00092/2014/CP1255/CT0004 and CHAIR in Onco-Innovation from the Faculty of Medicine, University of Porto (FMUP).

Institutional Review Board Statement: Not applicable.

Informed Consent Statement: Not applicable.

Data Availability Statement: Not applicable.

Acknowledgments: E.R. acknowledges CHAIR in Onco-Innovation/FMUP for funding her PhD grant.

Conflicts of Interest: The authors declare no conflict of interest.

References

1. Ferlay, J.; Colombet, M.; Soerjomataram, I.; Parkin, D.M.; Piñeros, M.; Znaor, A.; Bray, F. Cancer statistics for the year 2020: An overview. *Int. J. Cancer* **2021**, *149*, 778–789. [CrossRef]
2. Alfarouk, K.O.; Stock, C.M.; Taylor, S.; Walsh, M.; Muddathir, A.K.; Verduzco, D.; Bashir, A.H.; Mohammed, O.Y.; Elhassan, G.O.; Harguindey, S.; et al. Resistance to cancer chemotherapy: Failure in drug response from ADME to P-gp. *Cancer Cell. Int.* **2015**, *15*, 71. [CrossRef] [PubMed]
3. To, K.K.W.; Cho, W.C.S. Drug Repurposing for Cancer Therapy in the Era of Precision Medicine. *Curr. Mol. Pharmacol.* **2022**, *15*, 895–903. [CrossRef] [PubMed]
4. Khataniar, A.; Pathak, U.; Rajkhowa, S.; Jha, A.N. A Comprehensive Review of Drug Repurposing Strategies against Known Drug Targets of COVID-19. *COVID* **2022**, *2*, 148–167. [CrossRef]
5. Zhang, Z.; Zhou, L.; Xie, N.; Nice, E.C.; Zhang, T.; Cui, Y.; Huang, C. Overcoming cancer therapeutic bottleneck by drug repurposing. *Signal. Transduct. Target. Ther.* **2020**, *5*, 113. [CrossRef] [PubMed]
6. Sahragardjoonegani, B.; Beall, R.F.; Kesselheim, A.S.; Hollis, A. Repurposing existing drugs for new uses: A cohort study of the frequency of FDA-granted new indication exclusivities since 1997. *J. Pharm. Policy Pract.* **2021**, *14*, 3. [CrossRef]
7. Ribeiro, E.; Costa, B.; Vasques-Nóvoa, F.; Vale, N. In Vitro Drug Repurposing: Focus on Vasodilators. *Cells* **2023**, *12*, 671. [CrossRef]
8. Wang, R.; Dashwood, R.H. Endothelins and their receptors in cancer: Identification of therapeutic targets. *Pharmacol. Res.* **2011**, *63*, 519–524. [CrossRef] [PubMed]
9. Dingemanse, J.; Clozel, M.; van Giersbergen, P.L. Pharmacokinetics and pharmacodynamics of tezosentan, an intravenous dual endothelin receptor antagonist, following chronic infusion in healthy subjects. *Br. J. Clin. Pharmacol.* **2002**, *53*, 355–362. [CrossRef]
10. Cheng, J.W. Tezosentan in the management of decompensated heart failure. *Cardiol. Rev.* **2005**, *13*, 28–34. [CrossRef]
11. Schalcher, C.; Cotter, G.; Reisin, L.; Bertel, O.; Kobrin, I.; Guyene, T.T.; Kiowski, W. The dual endothelin receptor antagonist tezosentan acutely improves hemodynamic parameters in patients with advanced heart failure. *Am. Heart J.* **2001**, *142*, 340–349. [CrossRef]
12. Lebrec, D.; Bosch, J.; Jalan, R.; Dudley, F.J.; Jessic, R.; Moreau, R.; Garcia-Pagan, J.C.; Mookerjee, R.P.; Chiossi, E.; Van Giersbergen, P.L.; et al. Hemodynamics and pharmacokinetics of tezosentan, a dual endothelin receptor antagonist, in patients with cirrhosis. *Eur. J. Clin. Pharmacol.* **2012**, *68*, 533–541. [CrossRef]
13. Rosanò, L.; Spinella, F.; Bagnato, A. Endothelin 1 in cancer: Biological implications and therapeutic opportunities. *Nat. Rev. Cancer* **2013**, *13*, 637–651. [CrossRef] [PubMed]
14. Grant, K.; Loizidou, M.; Taylor, I. Endothelin-1: A multifunctional molecule in cancer. *Br. J. Cancer* **2003**, *88*, 163–166. [CrossRef] [PubMed]
15. Kusuhara, M.; Yamaguchi, K.; Nagasaki, K.; Hayashi, C.; Suzaki, A.; Hori, S.; Handa, S.; Nakamura, Y.; Abe, K. Production of endothelin in human cancer cell lines. *Cancer Res.* **1990**, *50*, 3257–3261. [PubMed]
16. Nakamuta, M.; Ohashi, M.; Tabata, S.; Tanabe, Y.; Goto, K.; Naruse, M.; Naruse, K.; Hiroshige, K.; Nawata, H. High plasma concentrations of endothelin-like immunoreactivities in patients with hepatocellular carcinoma. *Am. J. Gastroenterol.* **1993**, *88*, 248–252.
17. Nelson, J.B.; Hedican, S.P.; George, D.J.; Reddi, A.H.; Piantadosi, S.; Eisenberger, M.A.; Simons, J.W. Identification of endothelin-1 in the pathophysiology of metastatic adenocarcinoma of the prostate. *Nat. Med.* **1995**, *1*, 944–949. [CrossRef]
18. Ferrari-Bravo, A.; Franciosi, C.; Lissoni, P.; Fumagalli, L.; Uggeri, F. Effects of oncological surgery on endothelin-1 secretion in patients with operable gastric cancer. *Int. J. Biol. Mrk.* **2000**, *15*, 56–57. [CrossRef]
19. Morbidelli, L.; Orlando, C.; Maggi, C.A.; Ledda, F.; Ziche, M. Proliferation and migration of endothelial cells is promoted by endothelins via activation of ETB receptors. *Am. J. Physiol.* **1995**, *269*, H686–H695. [CrossRef]
20. Alberts, G.F.; Peifley, K.A.; Johns, A.; Kleha, J.F.; Winkles, J.A. Constitutive endothelin-1 overexpression promotes smooth muscle cell proliferation via an external autocrine loop. *J. Biol. Chem.* **1994**, *269*, 10112–10118. [CrossRef] [PubMed]
21. Salani, D.; Di Castro, V.; Nicotra, M.R.; Rosanò, L.; Tecce, R.; Venuti, A.; Natali, P.G.; Bagnato, A. Role of endothelin-1 in neovascularization of ovarian carcinoma. *Am. J. Pathol.* **2000**, *157*, 1537–1547. [CrossRef] [PubMed]
22. Bagnato, A.; Spinella, F.; Rosanò, L. Emerging role of the endothelin axis in ovarian tumor progression. *Endocr. Relat. Cancer* **2005**, *12*, 761–772. [CrossRef]
23. Rich, S.; McLaughlin, V.V. Endothelin Receptor Blockers in Cardiovascular Disease. *Circulation* **2003**, *108*, 2184–2190. [CrossRef] [PubMed]
24. Nelson, J.B.; Chan-Tack, K.; Hedican, S.P.; Magnuson, S.R.; Opgenorth, T.J.; Bova, G.S.; Simons, J.W. Endothelin-1 production and decreased endothelin B receptor expression in advanced prostate cancer. *Cancer Res.* **1996**, *56*, 663–668. [PubMed]

25. Bagnato, A.; Salani, D.; Di Castro, V.; Wu-Wong, J.R.; Tecce, R.; Nicotra, M.R.; Venuti, A.; Natali, P.G. Expression of endothelin 1 and endothelin A receptor in ovarian carcinoma: Evidence for an autocrine role in tumor growth. *Cancer Res.* **1999**, *59*, 720–727.

26. Ali, H.; Dashwood, M.; Dawas, K.; Loizidou, M.; Savage, F.; Taylor, I. Endothelin receptor expression in colorectal cancer. *J. Cardiovasc. Pharmacol.* **2000**, *36*, S69–S71. [CrossRef]

27. Gohji, K.; Kitazawa, S.; Tamada, H.; Katsuoka, Y.; Nakajima, M. Expression of endothelin receptor a associated with prostate cancer progression. *J. Urol.* **2001**, *165*, 1033–1036. [CrossRef]

28. Akhavan, A.; McHugh, K.H.; Guruli, G.; Bies, R.R.; Zamboni, W.C.; Strychor, S.A.; Nelson, J.B.; Pflug, B.R. Endothelin receptor A blockade enhances taxane effects in prostate cancer. *Neoplasia* **2006**, *8*, 725–732. [CrossRef]

29. Asham, E.; Shankar, A.; Loizidou, M.; Fredericks, S.; Miller, K.; Boulos, P.B.; Burnstock, G.; Taylor, I. Increased endothelin-1 in colorectal cancer and reduction of tumour growth by ET(A) receptor antagonism. *Br. J. Cancer* **2001**, *85*, 1759–1763. [CrossRef]

30. Shankar, A.; Loizidou, M.; Aliev, G.; Fredericks, S.; Holt, D.; Boulos, P.B.; Burnstock, G.; Taylor, I. Raised endothelin 1 levels in patients with colorectal liver metastases. *Br. J. Surg.* **1998**, *85*, 502–506. [CrossRef]

31. Liu, W.; Zhang, Q.; Zhang, Y.; Sun, L.; Xiao, H.; Luo, B. Epstein-Barr Virus Regulates Endothelin-1 Expression through the ERK/FOXO1 Pathway in EBV-Associated Gastric Cancer. *Microbiol. Spectr.* **2023**, *11*, e0089822. [CrossRef]

32. Kappes, L.; Amer, R.L.; Sommerlatte, S.; Bashir, G.; Plattfaut, C.; Gieseler, F.; Gemoll, T.; Busch, H.; Altahrawi, A.; Al-Sbiei, A.; et al. Ambrisentan, an endothelin receptor type A-selective antagonist, inhibits cancer cell migration, invasion, and metastasis. *Sci. Rep.* **2020**, *10*, 15931. [CrossRef] [PubMed]

33. Carducci, M.A.; Padley, R.J.; Breul, J.; Vogelzang, N.J.; Zonnenberg, B.A.; Daliani, D.D.; Schulman, C.C.; Nabulsi, A.A.; Humerick-house, R.A.; Weinberg, M.A.; et al. Effect of endothelin-A receptor blockade with atrasentan on tumor progression in men with hormone-refractory prostate cancer: A randomized, phase II, placebo-controlled trial. *J. Clin. Oncol.* **2003**, *21*, 679–689. [CrossRef]

34. Ayob, A.Z.; Ramasamy, T.S. Cancer stem cells as key drivers of tumour progression. *J. Biomed. Sci.* **2018**, *25*, 20. [CrossRef] [PubMed]

35. Yadav, L.; Puri, N.; Rastogi, V.; Satpute, P.; Sharma, V. Tumour Angiogenesis and Angiogenic Inhibitors: A Review. *J. Clin. Diagn. Res.* **2015**, *9*, Xe01–Xe05. [CrossRef]

36. Arabanian, L.S.; Johansson, P.; Staffas, A.; Nilsson, T.; Rouhi, A.; Fogelstrand, L.; Palmqvist, L. The endothelin receptor type A is a downstream target of Hoxa9 and Meis1 in acute myeloid leukemia. *Leuk. Res.* **2018**, *75*, 61–68. [CrossRef] [PubMed]

37. Ieiri, I. Functional significance of genetic polymorphisms in P-glycoprotein (MDR1, ABCB1) and breast cancer resistance protein (BCRP, ABCG2). *Drug. Metab. Pharmacokinet.* **2012**, *27*, 85–105. [CrossRef]

38. Fung, K.L.; Gottesman, M.M. A synonymous polymorphism in a common MDR1 (ABCB1) haplotype shapes protein function. *Biochim. Biophys. Acta* **2009**, *1794*, 860–871. [CrossRef]

39. Tulsyan, S.; Mittal, R.D.; Mittal, B. The effect of ABCB1 polymorphisms on the outcome of breast cancer treatment. *Pharmgenomics Pers. Med.* **2016**, *9*, 47–58. [CrossRef]

40. Zou, F.; Seike, M.; Noro, R.; Kunugi, S.; Kubota, K.; Gemma, A. Prognostic significance of ABCB1 in stage I lung adenocarcinoma. *Oncol. Lett.* **2017**, *14*, 313–321. [CrossRef]

41. Lei, Z.N.; Teng, Q.X.; Wu, Z.X.; Ping, F.F.; Song, P.; Wurpel, J.N.D.; Chen, Z.S. Overcoming multidrug resistance by knockout of ABCB1 gene using CRISPR/Cas9 system in SW620/Ad300 colorectal cancer cells. *MedComm (2020)* **2021**, *2*, 765–777. [CrossRef] [PubMed]

42. Raspollini, M.R.; Amunni, G.; Villanucci, A.; Boddi, V.; Taddei, G.L. Increased cyclooxygenase-2 (COX-2) and P-glycoprotein-170 (MDR1) expression is associated with chemotherapy resistance and poor prognosis. Analysis in ovarian carcinoma patients with low and high survival. *Int. J. Gynecol. Cancer* **2005**, *15*, 255–260. [CrossRef]

43. Yakirevich, E.; Sabo, E.; Naroditsky, I.; Sova, Y.; Lavie, O.; Resnick, M.B. Multidrug resistance-related phenotype and apoptosis-related protein expression in ovarian serous carcinomas. *Gynecol. Oncol.* **2006**, *100*, 152–159. [CrossRef] [PubMed]

44. Chen, H.; Hao, J.; Wang, L.; Li, Y. Coexpression of invasive markers (uPA, CD44) and multiple drug-resistance proteins (MDR1, MRP2) is correlated with epithelial ovarian cancer progression. *Br. J. Cancer* **2009**, *101*, 432–440. [CrossRef] [PubMed]

45. Sedláková, I.; Laco, J.; Caltová, K.; Červinka, M.; Tošner, J.; Řezáč, A.; Špaček, J. Clinical significance of the resistance proteins LRP, Pgp, MRP1, MRP3, and MRP5 in epithelial ovarian cancer. *Int. J. Gynecol. Cancer* **2015**, *25*, 236–243. [CrossRef] [PubMed]

46. Englinger, B.; Lötsch, D.; Pirker, C.; Mohr, T.; van Schoonhoven, S.; Boidol, B.; Lardeau, C.H.; Spitzwieser, M.; Szabó, P.; Heffeter, P.; et al. Acquired nintedanib resistance in FGFR1-driven small cell lung cancer: Role of endothelin-A receptor-activated ABCB1 expression. *Oncotarget* **2016**, *7*, 50161–50179. [CrossRef]

47. Eltze, E.; Wild, P.J.; Wülfing, C.; Zwarthoff, E.C.; Burger, M.; Stoehr, R.; Korsching, E.; Hartmann, A. Expression of the endothelin axis in noninvasive and superficially invasive bladder cancer: Relation to clinicopathologic and molecular prognostic parameters. *Eur. Urol.* **2009**, *56*, 837–845. [CrossRef]

48. Demunter, A.; De Wolf-Peeters, C.; Degreef, H.; Stas, M.; van den Oord, J.J. Expression of the endothelin-B receptor in pigment cell lesions of the skin. Evidence for its role as tumor progression marker in malignant melanoma. *Virchows Arch.* **2001**, *438*, 485–491. [CrossRef]

49. Blouquit-Laye, S.; Regnier, A.; Beauchet, A.; Zimmermann, U.; Devillier, P.; Chinet, T. Expression of endothelin receptor subtypes in bronchial tumors. *Oncol. Rep.* **2010**, *23*, 457–463. [CrossRef]

50. Eltze, E.; Bertolin, M.; Korsching, E.; Wülfing, P.; Maggino, T.; Lellé, R. Expression and prognostic relevance of endothelin-B receptor in vulvar cancer. *Oncol. Rep.* **2007**, *18*, 305–311. [CrossRef]

51. Wuttig, D.; Zastrow, S.; Füssel, S.; Toma, M.I.; Meinhardt, M.; Kalman, K.; Junker, K.; Sanjmyatav, J.; Boll, K.; Hackermüller, J.; et al. CD31, EDNRB and TSPAN7 are promising prognostic markers in clear-cell renal cell carcinoma revealed by genome-wide expression analyses of primary tumors and metastases. *Int. J. Cancer* **2012**, *131*, E693–E704. [CrossRef] [PubMed]

52. Tanaka, T.; Sho, M.; Takayama, T.; Wakatsuki, K.; Matsumoto, S.; Migita, K.; Ito, M.; Hamada, K.; Nakajima, Y. Endothelin B receptor expression correlates with tumour angiogenesis and prognosis in oesophageal squamous cell carcinoma. *Br. J. Cancer* **2014**, *110*, 1027–1033. [CrossRef] [PubMed]

53. Vasaikar, S.; Tsipras, G.; Landázuri, N.; Costa, H.; Wilhelmi, V.; Scicluna, P.; Cui, H.L.; Mohammad, A.A.; Davoudi, B.; Shang, M.; et al. Overexpression of endothelin B receptor in glioblastoma: A prognostic marker and therapeutic target? *BMC Cancer* **2018**, *18*, 154. [CrossRef]

54. Pao, M.M.; Tsutsumi, M.; Liang, G.; Uzvolgyi, E.; Gonzales, F.A.; Jones, P.A. The endothelin receptor B (EDNRB) promoter displays heterogeneous, site specific methylation patterns in normal and tumor cells. *Hum. Mol. Genet.* **2001**, *10*, 903–910. [CrossRef] [PubMed]

55. Halaka, M.; Hired, Z.A.; Rutledge, G.E.; Hedgepath, C.M.; Anderson, M.P.; St John, H.; Do, J.M.; Majmudar, P.R.; Walker, C.; Alawawdeh, A.; et al. Differences in Endothelin B Receptor Isoforms Expression and Function in Breast Cancer Cells. *J. Cancer* **2020**, *11*, 2688–2701. [CrossRef]

56. Wei, F.; Ge, Y.; Li, W.; Wang, X.; Chen, B. Role of endothelin receptor type B (EDNRB) in lung adenocarcinoma. *Thorac. Cancer* **2020**, *11*, 1885–1890. [CrossRef] [PubMed]

57. Chen, Q.W.; Edvinsson, L.; Xu, C.B. Role of ERK/MAPK in endothelin receptor signaling in human aortic smooth muscle cells. *BMC Cell. Biol.* **2009**, *10*, 52. [CrossRef] [PubMed]

58. Said, N.; Theodorescu, D. Permissive role of endothelin receptors in tumor metastasis. *Life Sci.* **2012**, *91*, 522–527. [CrossRef]

59. Fang, D.; Leishear, K.; Nguyen, T.K.; Finko, R.; Cai, K.; Fukunaga, M.; Li, L.; Brafford, P.A.; Kulp, A.N.; Xu, X.; et al. Defining the Conditions for the Generation of Melanocytes from Human Embryonic Stem Cells. *Stem Cells* **2006**, *24*, 1668–1677. [CrossRef]

60. Berger, Y.; Bernasconi, C.C.; Juillerat-Jeanneret, L. Targeting the endothelin axis in human melanoma: Combination of endothelin receptor antagonism and alkylating agents. *Exp. Biol. Med.* **2006**, *231*, 1111–1119.

61. Kefford, R.; Beith, J.M.; Van Hazel, G.A.; Millward, M.; Trotter, J.M.; Wyld, D.K.; Kusic, R.; Shreeniwas, R.; Morganti, A.; Ballmer, A.; et al. A Phase II study of bosentan, a dual endothelin receptor antagonist, as monotherapy in patients with stage IV metastatic melanoma. *Investig. New Drugs* **2007**, *25*, 247–252. [CrossRef] [PubMed]

62. Spinella, F.; Rosanò, L.; Di Castro, V.; Decandia, S.; Nicotra, M.R.; Natali, P.G.; Bagnato, A. Endothelin-1 and endothelin-3 promote invasive behavior via hypoxia-inducible factor-1alpha in human melanoma cells. *Cancer Res.* **2007**, *67*, 1725–1734. [CrossRef] [PubMed]

63. Lahav, R.; Heffner, G.; Patterson, P.H. An endothelin receptor B antagonist inhibits growth and induces cell death in human melanoma cells in vitro and in vivo. *Proc. Natl. Acad. Sci. USA* **1999**, *96*, 11496–11500. [CrossRef]

64. Bagnato, A.; Rosanò, L.; Spinella, F.; Di Castro, V.; Tecce, R.; Natali, P.G. Endothelin B receptor blockade inhibits dynamics of cell interactions and communications in melanoma cell progression. *Cancer Res.* **2004**, *64*, 1436–1443. [CrossRef] [PubMed]

65. Liu, S.; Premont, R.T.; Kontos, C.D.; Huang, J.; Rockey, D.C. Endothelin-1 activates endothelial cell nitric-oxide synthase via heterotrimeric G-protein betagamma subunit signaling to protein jinase B/Akt. *J. Biol. Chem.* **2003**, *278*, 49929–49935. [CrossRef] [PubMed]

66. Huerta, S. Nitric oxide for cancer therapy. *Future Sci. OA* **2015**, *1*, Fso44. [CrossRef]

67. Choudhari, S.K.; Chaudhary, M.; Bagde, S.; Gadbail, A.R.; Joshi, V. Nitric oxide and cancer: A review. *World J. Surg. Oncol.* **2013**, *11*, 118. [CrossRef]

68. Xu, W.; Liu, L.Z.; Loizidou, M.; Ahmed, M.; Charles, I.G. The role of nitric oxide in cancer. *Cell. Res.* **2002**, *12*, 311–320. [CrossRef]

69. Sang, J.; Chen, Y.; Tao, Y. Nitric oxide inhibits gastric cancer cell growth through the modulation of the Akt pathway. *Mol. Med. Rep.* **2011**, *4*, 1163–1167. [CrossRef]

70. Pervin, S.; Singh, R.; Chaudhuri, G. Nitric oxide-induced cytostasis and cell cycle arrest of a human breast cancer cell line (MDA-MB-231): Potential role of cyclin D1. *Proc. Natl. Acad. Sci. USA* **2001**, *98*, 3583–3588. [CrossRef]

71. Huguenin, S.; Fleury-Feith, J.; Kheuang, L.; Jaurand, M.C.; Bolla, M.; Riffaud, J.P.; Chopin, D.K.; Vacherot, F. Nitrosulindac (NCX 1102): A new nitric oxide-donating non-steroidal anti-inflammatory drug (NO-NSAID), inhibits proliferation and induces apoptosis in human prostatic epithelial cell lines. *Prostate* **2004**, *61*, 132–141. [CrossRef]

72. Huguenin, S.; Vacherot, F.; Kheuang, L.; Fleury-Feith, J.; Jaurand, M.C.; Bolla, M.; Riffaud, J.P.; Chopin, D.K. Antiproliferative effect of nitrosulindac (NCX 1102), a new nitric oxide-donating non-steroidal anti-inflammatory drug, on human bladder carcinoma cell lines. *Mol. Cancer Ther.* **2004**, *3*, 291–298. [CrossRef]

73. Peña-Altamira, E.; Petazzi, P.; Contestabile, A. Nitric oxide control of proliferation in nerve cells and in tumor cells of nervous origin. *Curr. Pharm. Des.* **2010**, *16*, 440–450. [CrossRef]

74. Donia, M.; Mijatovic, S.; Maksimovic-Ivanic, D.; Miljkovic, D.; Mangano, K.; Tumino, S.; Biondi, A.; Basile, F.; Al-Abed, Y.; Stosic-Grujicic, S.; et al. The novel NO-donating compound GIT-27NO inhibits in vivo growth of human prostate cancer cells and prevents murine immunoinflammatory hepatitis. *Eur. J. Pharmacol.* **2009**, *615*, 228–233. [CrossRef] [PubMed]

75. Donia, M.; Maksimovic-Ivanic, D.; Mijatovic, S.; Mojic, M.; Miljkovic, D.; Timotijevic, G.; Fagone, P.; Caponnetto, S.; Al-Abed, Y.; McCubrey, J.; et al. In vitro and in vivo anticancer action of Saquinavir-NO, a novel nitric oxide-derivative of the protease inhibitor saquinavir, on hormone resistant prostate cancer cells. *Cell. Cycle* **2011**, *10*, 492–499. [CrossRef] [PubMed]

76. Van de Wouwer, M.; Couzinié, C.; Serrano-Palero, M.; González-Fernández, O.; Galmés-Varela, C.; Menéndez-Antolí, P.; Grau, L.; Villalobo, A. Activation of the BRCA1/Chk1/p53/p21(Cip1/Waf1) pathway by nitric oxide and cell cycle arrest in human neuroblastoma NB69 cells. *Nitric Oxide* **2012**, *26*, 182–191. [CrossRef]
77. Sonveaux, P.; Jordan, B.F.; Gallez, B.; Feron, O. Nitric oxide delivery to cancer: Why and how? *Eur. J. Cancer* **2009**, *45*, 1352–1369. [CrossRef] [PubMed]
78. Wang, Z.; Cook, T.; Alber, S.; Liu, K.; Kovesdi, I.; Watkins, S.K.; Vodovotz, Y.; Billiar, T.R.; Blumberg, D. Adenoviral gene transfer of the human inducible nitric oxide synthase gene enhances the radiation response of human colorectal cancer associated with alterations in tumor vascularity. *Cancer Res.* **2004**, *64*, 1386–1395. [CrossRef]
79. Cook, T.; Wang, Z.; Alber, S.; Liu, K.; Watkins, S.C.; Vodovotz, Y.; Billiar, T.R.; Blumberg, D. Nitric oxide and ionizing radiation synergistically promote apoptosis and growth inhibition of cancer by activating p53. *Cancer Res.* **2004**, *64*, 8015–8021. [CrossRef]
80. Zhang, P.; Yang, X.; Zhang, F.; Gabelli, S.B.; Wang, R.; Zhang, Y.; Bhat, S.; Chen, X.; Furlani, M.; Amzel, L.M.; et al. Pyridinylpyrimidines selectively inhibit human methionine aminopeptidase-1. *Bioorg Med. Chem.* **2013**, *21*, 2600–2617. [CrossRef]
81. Hu, X.; Addlagatta, A.; Matthews, B.W.; Liu, J.O. Identification of pyridinylpyrimidines as inhibitors of human methionine aminopeptidases. *Angew. Chem. Int. Ed. Engl.* **2006**, *45*, 3772–3775. [CrossRef] [PubMed]
82. Jonckheere, V.; Fijałkowska, D.; Van Damme, P. Omics Assisted N-terminal Proteoform and Protein Expression Profiling On Methionine Aminopeptidase 1 (MetAP1) Deletion. *Mol. Cell. Proteom.* **2018**, *17*, 694–708. [CrossRef]
83. Frottin, F.; Bienvenut, W.V.; Bignon, J.; Jacquet, E.; Vaca Jacome, A.S.; Van Dorsselaer, A.; Cianferani, S.; Carapito, C.; Meinnel, T.; Giglione, C. MetAP1 and MetAP2 drive cell selectivity for a potent anti-cancer agent in synergy, by controlling glutathione redox state. *Oncotarget* **2016**, *7*, 63306–63323. [CrossRef] [PubMed]
84. Hu, X.; Addlagatta, A.; Lu, J.; Matthews, B.W.; Liu, J.O. Elucidation of the function of type 1 human methionine aminopeptidase during cell cycle progression. *Proc. Natl. Acad. Sci. USA* **2006**, *103*, 18148–18153. [CrossRef] [PubMed]
85. Mauriz, J.L.; Martín-Renedo, J.; García-Palomo, A.; Tuñón, M.J.; González-Gallego, J. Methionine aminopeptidases as potential targets for treatment of gastrointestinal cancers and other tumours. *Curr. Drug. Targets* **2010**, *11*, 1439–1457. [CrossRef] [PubMed]
86. Chen, X.; Chong, C.R.; Shi, L.; Yoshimoto, T.; Sullivan, D.J., Jr.; Liu, J.O. Inhibitors of Plasmodium falciparum methionine aminopeptidase 1b possess antimalarial activity. *Proc. Natl. Acad. Sci. USA* **2006**, *103*, 14548–14553. [CrossRef] [PubMed]

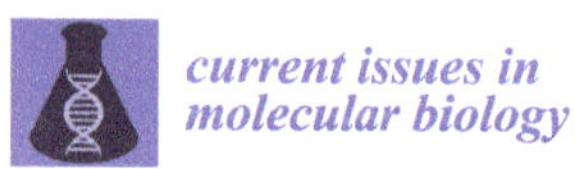

Review

Adenovirus as a Vector and Oncolytic Virus

Wataru Matsunaga [1] and Akinobu Gotoh [2],*

[1] Joint-Use Research Facilities, Hyogo Medical University, 1-1 Mukogawa, Nishinomiya 663-8501, Japan
[2] Department of Education for Medical Research Base, Hyogo Medical University, 1-1 Mukogawa, Nishinomiya 663-8501, Japan
* Correspondence: gotoh@hyo-med.ac.jp

Abstract: Adenoviral vectors, both oncolytic viruses and gene delivery vectors, are among the earliest approved and commercialised vectors for gene therapy. Adenoviruses have high cytotoxicity and immunogenicity. Therefore, lentiviruses or adeno-associated viruses as viral vectors and herpes simplex virus as an oncolytic virus have recently drawn attention. Thus, adenoviral vectors are often considered relatively obsolete. However, their high cargo limit and transduction efficiency are significant advantages over newer viral vectors. This review provides an overview of the new-generation adenoviral vectors. In addition, we describe the modification of the fiber knob region that enhances affinity of adenoviral vectors for cancer cells and the utilisation of cancer-cell specific promoters to suppress expression of unwanted transgenes in non-malignant tissues.

Keywords: adenovirus; adenoviral vector; oncolytic virus; midkine

Citation: Matsunaga, W.; Gotoh, A. Adenovirus as a Vector and Oncolytic Virus. *Curr. Issues Mol. Biol.* **2023**, *45*, 4826–4840. https://doi.org/10.3390/cimb45060307

Academic Editor: Dumitru A. Iacobas

Received: 4 May 2023
Revised: 30 May 2023
Accepted: 30 May 2023
Published: 2 June 2023

1. Introduction

Adenoviruses are DNA viruses with non-enveloped icosahedral particles that are approximately 90 nm in diameter. Currently, 88 types of human adenoviruses (1–88) have been identified and classified into seven serotypes (A–G) [1], many of which cause various diseases in humans. However, in 1993, gene transfer into neurons using adenoviral vectors was reported [2], whereafter adenoviruses began to attract attention as efficient vectors that could also transfer genes into non-dividing cells. Despite being one of the most common viral human pathogens, adenoviruses are also among the most widely used viral vectors in gene therapy research. Adenoviral vector use accounts for 15.5% of vector-based gene therapy clinical trials worldwide [3]. Moreover, adenoviruses have attracted attention because they have been used as a base material for viral vector vaccines against coronavirus disease 2019 [4].

Currently, 68.2% of clinical gene therapy trials are aimed at cancer treatment [3]. Virotherapy for cancer uses a genetically modified virus to introduce a therapeutic gene as a vector or uses it as an oncolytic virus. In recent years, viral immunotherapy has been widely studied.

Adenoviral vectors used in gene therapy research, including studies on virotherapy, are mostly based on adenovirus serotype 5 (Ad5). This is possibly because Ad5 was the first to be analysed for gene function, knowledge about Ad5 has already been accumulated, and high-titer viral particles can be produced easily. However, studies using lentiviruses and adeno-associated viruses (AAV) as subjects for virus therapy have recently increased, and when compared with these "newcomer" viral vectors, some researchers consider adenoviral vectors "obsolete".

In this article, we briefly describe the characteristics of adenoviruses as vectors or oncolytic viruses and the present status of studies using them.

2. Trends of Viral Vectors

Currently, the top five vectors used in gene therapy clinical trials worldwide are adenoviruses, retroviruses, plasmids, lentiviral vectors, and AAV vectors [3]. Adenoviral vectors are among the earliest vectors used in the history of gene therapy and, as of 2022, will account for the largest share of vectors used in gene therapy clinical trials worldwide (Table 1). Although adenoviruses still rank first in terms of the percentage of total cases in which they are used, the share of adenoviral vectors has declined over the past 10 years [5,6]. However, the use of lentiviral and AAV vectors has significantly increased. Comparing the number of cases from 2012 to 2017, lentiviral vectors ranked first and adenoviral vectors third, whereas those from 2017 to 2022 showed adenovirus vectors ranking the lowest among all vectors listed in Table 1. Considering this, adenoviruses appear to be "obsolete" vectors. However, they still offer several advantages. In addition, among the many serotypes of adenoviruses known, it is conceivable that a completely new adenoviral vector suitable for human clinical applications may be discovered.

Table 1. Changes in vectors used in gene therapy clinical trials over the past decade.

	Share in Clinical Trials until 2012 [5]		Share in Clinical Trials until 2017 [6]		Share in Clinical Trials until 2022 [3]		Increase 2012 to 17	Increase 2017 to 22
Adenovirus	23.3%	438	20.5%	547	15.5%	573	109	26
Retrovirus	19.7%	370	17.9%	478	14.6%	538	108	60
Plasmid DNA	18.3%	345	16.6%	442	13.1%	483	97	41
Lentivirus	2.9%	55	7.3%	196	9.9%	364	141	168
AAV	4.9%	92	7.6%	204	9.5%	350	112	146

AAV: adeno-associated virus.

Characteristics of the top five vectors used in gene therapy clinical trials are shown in Table 2.

Table 2. Characteristics of the top five gene vectors used in gene therapy clinical trials.

	Nondividing Cell	Cargo Limit	Duration of Gene Expression	Physical Containment	Immunogenicity of Vectors	Drawbacks	Safety
Adenovirus	Yes	7–36 kb	Transient *	P2	High	Strong cytotoxicity	Moderate
Retrovirus	No	8–9 kb	Stable	P2	Moderate	Risk of the carcinogenesis	Moderate
Plasmid DNA	Yes	limitless	Transient	-	-	Low transduction efficiency	High
Lentivirus	Yes	8–9 kb	Stable	P2	Moderate	Concerns about genes derived from HIV-1	High
AAV	Yes	4.7 kb	Potentially Stable	P1	Low	Difficult to purification, Small cargo limit	High

*: Third-generation adenoviral vectors are "stable" under certain conditions. AAV: adeno-associated virus.

Adenoviruses, which accounted for the largest share of clinical trials, are the main subject of this article and will be discussed in the next chapter. Below is a summary of the other viral vectors listed in Tables 1 and 2.

2.1. Retroviral Vectors

Retroviral vectors generally refer to vectors derived mainly from the Moloney murine leukaemia virus, a type of gammaretrovirus, and are distinguished from lentiviral vectors, which are retroviruses. Retroviral vectors are commonly used in gene therapy because of their low cytotoxicity. Long-term stable gene expression is also an advantage of retroviral vectors; however, their inability to transfer genes into non-dividing cells is a major drawback. Moreover, the potential problems of mutagenesis randomly affecting chromosomes and activating nearby proto-oncogenes to induce cancer is not negligible [7,8]. The retroviral vector cargo limit of 9–10 kb is moderately large but inferior to that of new-generation adenoviral vectors.

2.2. Lentiviral Vectors

Lentiviral vectors are a subtype of retroviral vectors developed based on HIV-1 and are capable of transducing genes into almost all mammalian cells, including non-dividing cells that cannot be transducted with retroviral vectors. Similar to retroviral vectors, lentiviral vectors can maintain stable gene expression. Lentiviral vectors have been widely used to study ex vivo cancer immunotherapy via T cells using chimeric antigen receptors [9]. As a result of improvements made to enhance safety, third-generation lentiviruses have acquired low immunogenicity and cytotoxicity. However, because lentiviral vectors are based on HIV-1, some HIV-1 genes are required for vector production, and safety concerns regarding their use in gene therapy remain. In addition, lentiviruses cannot be used as oncolytic viruses because they lack replication ability. Their cargo limit of 9–10 kb seems sufficient, but it is still inferior to that of new-generation adenoviruses.

2.3. Adeno-Associated Virus (AAV) Vectors

AAVs are single-stranded DNA parvoviruses that can replicate only in the presence of other "helper viruses", such as the adenovirus or human papillomavirus. Eleven AAV serotypes (1–11) have been identified [10], which differ in their tropism to the types of cells they infect. Therefore, AAV vectors are useful for the preferential transduction of specific cell types. AAV2 is the most well-characterised and most used AAV vector. Although AAV immunogenicity and high-dose AAV-vector-induced hepatotoxicity have been frequently reported [11], there are no known infections caused by AAV in humans, and its pathogenicity is thought to be almost negligible. Taken together, AAV vectors, which are safe and have a high transduction efficiency, are promising viral vectors. Studies on gene therapy using AAV vectors include the treatment of various intractable diseases, such as haemophilia [12] and Parkinson's disease [13]. Regarding cancer treatment, a genetically engineered AAV vector that selectively binds to the tumour antigen, Her2/neu, and can only infect cancer cells has been reported [14]. Moreover, Onasemnogene abeparvovec (Zolgensma), an AAV9-based gene therapy drug used to treat spinal muscular atrophy, has been marketed. This drug is very expensive but shows remarkable therapeutic efficacy [15,16].

However, AAV exhibits distinct drawbacks. The cargo limit of AAV (4.7–5 kb) is too small to accommodate the large sizes of therapeutic genes that may occur with the development of gene therapy in the future. AAV vectors allow for the stable expression of transducted genes; however, due to various factors, stable expression of transgenes is not always possible. Furthermore, purification of high-titer AAV vectors is relatively difficult.

3. Adenovirus as a Vector

The advantages of adenoviral vectors include a large cargo limit and high efficiency of infection. High cytotoxicity is a major drawback of adenoviral vectors; however, when used as oncolytic viruses, this trait proves to be powerful and advantageous. Transient gene expression is also a drawback, but this leads to better safety because gene expression would not last long when acting on unintended targets.

3.1. First-Generation Adenoviral Vectors

Adenoviral vectors are classified according to their genetic design, from first to third generation. First-generation adenoviral vectors were designed to delete the Early 1 (E1) and Early 3 (E3) regions (Figure 1). The E1 region encodes genetic information essential for adenovirus survival and replication [17]; however, this region is deleted in first-generation adenoviruses because it prevents uncontrolled adenoviral replication and can be replaced with the contents of packaging cells [18]. The E3 region encodes a protein that protects infected cells against the immune response [19]. It is often deleted in the first generation to increase the cargo capacity because it is not essential for replication. By deleting E1 and E3, 5–6 kbp of cargo capacity can be secured. First-generation adenoviral vectors are incapable of self-replication and require a packaging cell line expressing E1 protein for their production. Human embryonic kidney 293 cells are the most commonly used packaging cells [18].

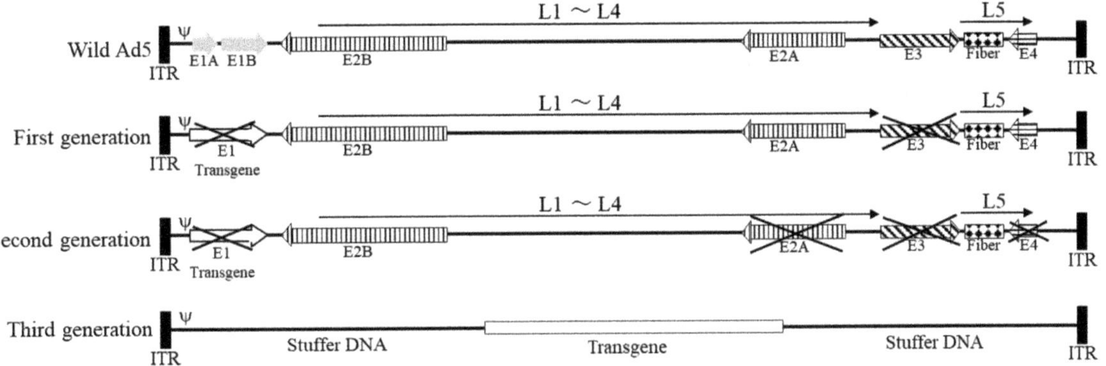

Figure 1. Genomic overview of each generation of adenoviral vectors. E1~E4, early regions 1–4; L1~L5, late regions 1–5; Ad5, adenovirus serotype 5; ITR, inverted terminal repeat; ψ, packaging signal.

3.2. Second-Generation Adenoviral Vectors

Because several human cells express E1A-like factors [20], even first-generation adenoviral vectors with deleted E1 regions can induce potent host immune responses and chronic cytotoxicity in transduced host cells [21]. A second-generation adenovirus with E2 and E4 deletions was developed to attenuate the host cell immune response to the vector (Figure 1). The E2 region contains genes required for adenovirus replication [22], and the E4 region encodes regulatory proteins for DNA transcription [23]. Deleting these regions provides approximately 14 kb free space and significantly reduces native adenovirus protein synthesis. Nevertheless, second-generation adenoviruses still induce host immune responses through adenoviral proteins expressed by the remaining genes, resulting in reduced transgene expression in target cells [24].

3.3. Third-Generation Adenoviral Vectors

Despite deletion of the E1–E4 regions, first- and second-generation adenoviral vectors still have considerable immunogenicity and cytotoxicity. In addition, at the beginning of the 21st century, genes exceeding the capacity limit of conventional adenoviral vectors were considered therapeutic genes; therefore, third-generation adenoviral vectors were developed [25,26]. Third-generation adenoviral vectors have their entire native adenoviral genome removed, except for the inverted terminal repeat and ψ packaging signal, thereby increasing the cargo limit to 36 kb. This third-generation feature, also called a "gutless adenovirus vector", allows high-level expression in host cells with no expression of viral proteins and little induction of an immune response [27].

Owing to the lack of genes required for self-assembly, a "helper adenovirus" carrying the genes required for replication must be co-introduced into the packaging cells when producing third-generation adenoviral vectors, which is also called a "helper-dependent vector". However, there are concerns that helper adenoviruses will contaminate the final product of third-generation adenoviruses and that homologous recombination between helper viruses and packaging cells will result in self-propagating adenoviruses. Therefore, genetic modifications of adenoviruses used as helper viruses or methods using non-adenoviruses as helpers have been investigated [28]; however, viral approaches have not resolved the concern of contamination. Nevertheless, a contamination-free third-generation adenoviral vector production method has been developed that uses plasmids encoding the necessary genes as "helpers" [28,29].

For third-generation adenoviruses, it is necessary to place other genes (stuffer DNA) in regions other than the target gene for efficient encapsidation [30]. However, the effect of stuffer DNA on vector gene transfer efficiency remains unclear. Some reports state that stuffer DNA affects transduction efficiency [31,32], whereas others claim that it does not [33].

3.4. Current Status of Adenoviral Vectors

At present (June 2023), Nadofaragene firadenovec (brand name: Adstiladrin, also known as rAd-IFNa/Syn3) is the only approved adenovirus vector for gene therapy. Nadofaragene firadenovec is an E1-deleted, replication-deficient adenoviral vector based on Ad5 that contains a human IFN-α2b gene [34,35]. It was approved in the United States for the treatment of non-muscle-invasive bladder cancer (NMIBC) in December 2022 [36]. Currently, most clinical trials using non-proliferating adenovirus vectors are vaccine studies for COVID19. Meanwhile, in cancer therapy research, trials using adenoviral thymidine kinase (ADV-Tk) to treat hepatocellular carcinoma are in phase III (NCT03313596) [37].

4. Adenovirus as an Oncolytic Virus

Surgery, radiation, and anticancer agents are the three pillars of cancer therapy, and cancer immunotherapy has recently been included as a standard cancer therapy. However, surgery places a heavy burden on the body and is not always feasible for all patients with cancer. Other standard treatments have side effects that significantly impair patients' quality of life (QOL). In this regard, oncolytic virus therapy has attracted attention as a therapeutic method that causes little deterioration in QOL.

The basic concept of an oncolytic virus is that it preferentially infects and proliferates in cancer cells, thereby lysing and destroying them. The virus released at that time also infects surrounding cancer cells, causing a chain of destruction in tumour tissues [38]. The origin of the concept of oncolytic viruses dates back to the discovery of viruses. In the late nineteenth and early twentieth centuries, when viruses began to be discovered, a relationship between temporary cancer remission and viral infections was reported [39–41].

In the 1920s, viral infection-induced oncolysis was confirmed in animal experiments. From the late 1940s to the 1960s, clinical studies of oncolytic virotherapy using wild-type viruses, such as yellow fever or West Nile viruses, were extensively conducted, but the results of these studies indicated a lack of safety or efficacy. Therefore, studies of oncolytic virus therapy were abandoned [39–41]. However, with the establishment of genetic engineering technology in the 1990s, highly safe viruses that replicate only in cancer cells have become a reality, and oncolytic virus therapy has once again become the focus of attention.

In 1997, the first oncolytic virus, named ONYX-015, was reported [42]. This oncolytic virus is an adenovirus lacking E1B 55-Kilodalton-Associated Protein, which inactivates the tumour suppressor gene p53 [43] and selectively infects and lyses human cancer cells in which the p53 gene is non-functional.

A clinical trial was conducted to treat head and neck cancer using ONYX-015 [44]. However, ONYX-015 must be injected directly into the tumour because it is highly toxic when administered intravenously. Therefore, it could only be applied to large tumours. Because it did not show the expected antitumour effect, the ONYX-015 project was aborted in 2000 [45]. However, the H101 strain (Oncorine), a genetically modified adenovirus based on ONYX-015, was approved by the Food and Drug Administration of the People's Republic of China in 2005 for treating head and neck cancer [43]. It is the first approved oncolytic virus worldwide, aside from Rigvir, which has unclear efficacy, and has been approved in the Republic of Latvia [46].

Recently, the herpes simplex virus seems to have attracted more attention than the adenovirus as an oncolytic virus, and T-VEC, G47Δ (Teserpaturev) [47,48] and C-REV (originally HF10) [49] have already been approved or are in clinical trials. Moreover, novel adenovirus-based oncolytic viruses have also been investigated. At present, OBP-301 (suratadenotureb) [50,51] is undergoing a phase II clinical trial [52] for the treatment of head and neck cancer (NCT04685499), and CG0070 Adenoviral vector is phase II/III (NCT01438112) [53].

Although oncolytic viruses have increased infection specificity for cancer cells, they are still capable of infecting non-malignant cells. Therefore, the development of new-generation oncolytic viruses requires further improvements to enhance their specificity. For adenovirus-based oncolytic viruses, replacement of the Ad5 fiber knob with other adenovirus serotypes has been shown to increase the infectivity of cancer cells. Initially, a vector was developed in which the adenovirus type 35 (Ad35) fiber knob region was introduced into Ad5. Currently, Ad3 and Ad37 fiber knobs are also being studied [54].

Ad5 is transduced into cells through the coxsackie and adenovirus receptor (CAR) [55]. However, CAR expression is generally low in cancer cells [56,57], although its upregulation has been reported in some of them [57]. Therefore, genetic modification of the fiber knob region, where the adenovirus binds to its receptor, is the main method used to resolve the reduced efficiency of CAR-mediated gene transfer to cancer cells.

Cluster of differentiation 46 (CD46), which acts as a receptor for serotype B adenoviruses [58], is expressed in almost every human cell [58,59]. Ad35, which belongs to serotype B, does not interact with CAR, and the adenovirus serotype 5 F35 (Ad5F35) vector, in which the fiber knob region of the Ad5 vector is replaced with that derived from Ad35, binds to CD46 (Figure 2). Thus, it is effective in cells lacking CAR [60,61]. Several groups have investigated the utility of Ad5F35 in the development of transduction efficacy [62–64]. In 2021, it was reported that Ad35 is an oncolytic virus. Ad35 was not inhibited by Ad5 antibodies, which many adults have, and showed a high antitumor effect [65]. However, the physical potency of Ad35 is less than one tenth of that of Ad5 [65], and considering the current efficiency of production/purification, Ad35 is likely not superior to Ad5F35.

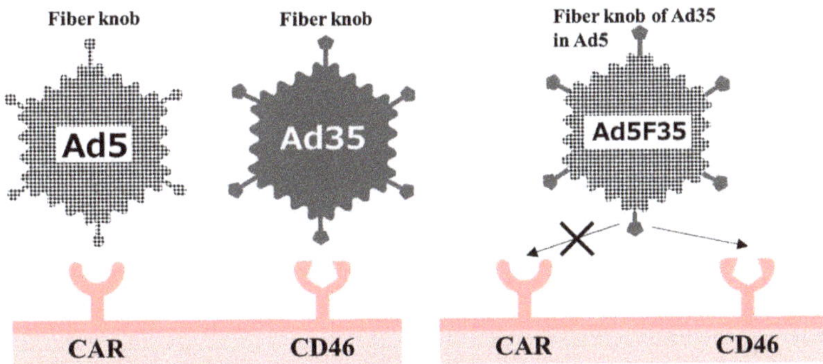

Figure 2. Concept of the chimeric adenoviral vector, adenovirus serotype 5 F35 (Ad5F35). Ad5 infects cells via the coxsackie and adenovirus receptor (CAR), but CAR is poorly expressed in cancer cells. Ad35 infects via ubiquitously expressed cluster of differentiation 46 (CD46). Ad5 and Ad35 have no affinity for CD46 and CAR, respectively. Therefore, when the fiber knob region of Ad5 is converted to that of Ad35, it is possible for adenoviral vectors to infect via CD46.

Onyx-015 is a genetically modified adenovirus engineered to grow only in p53-deficient cancer cells by deleting the E1B region [43,44]; however, this method is not effective for only propagating adenoviruses in cancer cells. Strategies using the promoters of tumour-specific genes are more effective. Inserting the promoter regions of genes that are highly active in cancer cells, such as human telomerase reverse transcriptase [66] and cyclooxygenase 2 [67], at positions that control the expression of genes involved in adenoviral replication allows the oncolytic virus to replicate only in cancer cells. In addition, especially for the development of adenovirus-based oncolytic viruses, because half of the population is immune to adenoviruses [68] and the effect of the virus is thought to decline in a short period, "arming" them with therapeutic genes to enhance antitumour efficacy is also important [53]. The therapeutic genes being investigated for addition into oncolytic viruses include tumour suppressor (p53 and p16), cytokine-inducible [69,70], and immune stimulator genes [71]. Adenovirus-based oncolytic viruses have the advantage of being theoretically capable of "heavy-arming" with multiple therapeutic genes owing to their greater capacity limitations compared with that of other oncolytic viruses.

Ad5-yCD/mutTKSR39rep-ADP is one of "heavy-arming" oncolytic viruses [72]. Yeast cytosine deaminase is catalyzes the deamination of the prodrug 5-fluorocytosine to form 5-fluorouracil [73]. Herpes simplex virus thymdine kinase is suicide gene in combination with ganciclovir [74]. ADP promotes the lysis and death of the host cell. At recent, the Ad5-yCD/mutTKSR39rep-ADP, second-generation, replicative adenoviral vector that containing a yeast cytosine deaminase (yCD), mutantSR39 herpes simplex virus thymdine kinase and adenovirus death protein (ADP) gene is in phase I trial (NCT03281382).

5. Use of Adenoviruses for Bladder Cancer

We have been researching virus therapy using adenoviruses to develop new treat-ment methods for urinary system cancer.

Cancer of the urinary system is intractable. Approximately 70% of bladder cancer cases are NMIBC, which often develops into recurrent tumors and progresses to a higher stage or grade [75]. Currently, Bacillus Calmette-Guérin therapy is considered the most effective treatment for NMIBC; however, its severe side effects cause a significant reduction in patient QOL [76]. Nadofaragene firadenovec has already been approved in the US for gene therapy of NMIBC [36], and NMIBC is one of the subjects of the clinical trial CG0070 [53]. Renal cell carcinoma (RCC) accounts for over 90% of all kidney cancer cases [77]. Chemotherapy and molecular immunotherapy are not remarkably effective for RCC [78,79] and surgery is the most common treatment. However, surgery is difficult in some cases. Therefore, we focused on adenoviruses to develop new treatments for these cancers, designed chimeric adenoviral vectors, and examined their efficacies.

The safety of first-generation adenoviral vectors has been confirmed in clinical trials; however, relatively high inflammation caused by adenoviruses remains a major concern. Moreover, the major problem with cancer virotherapy using adenoviral vectors is the limited efficacy of gene transfer or the oncolytic effect because of lower CAR expression in highly malignant cancer cells [54,55,80]. Ad5F35 infects via CD46, but CD46 is expressed in almost all human cells; thus, its side effects on non-malignant tissues should be prevented. The conditionally replicating adenovirus (CRAD) vector is an effective solution for reducing the side effects [81]. Therefore, we devised a method to control adenovirus vectors using the promoter regions of proteins abundantly expressed in cancer cells. Midkine is a heparin-binding growth factor induced by retinoic acid in embryonal carcinoma cells [82] and is involved in mitogenesis, angiogenesis, anti-apoptosis, fibrinolysis, and transfor-mation [83–87]. Midkine expression is enhanced in several cancers [88–90], whereas its expression in non-malignant cells is relatively limited. Thus, the promoter region [91] of the midkine gene (Figure 3A) can be used to control gene expression required for adenovirus replication to restrict the expression of introduced genes to cancer cells only. Therefore, we constructed an Ad5F35-Mkp-E1 CRAD vector encoding E1 under the control of a midkine promoter (Mkp; Figure 3A). The CRAD Ad5F35-Mkp-E1 vector, as an oncolytic virus, exhibited a significantly potent oncolytic effect on cancer cells neighbouring the primarily infected cells through secondary infection and replication (Figure 3B).

We examined the oncolytic effect of Ad5F35/Mkp-E1 in various cancer cell lines [92–96]. In bladder cancer cell lines (Figure 4) [92], midkine mRNA expression was observed in all experimental cell lines, with the lowest expression observed in UMUC-3 cells (Figure 4A). In 253J and KK47, CAR mRNA expression was considerably lower than that of CD46 mRNA (Figure 4A). In contrast, CAR mRNA expression in UMUC-3 was higher than that of CD46. In the 253J cell line, Ad5F35/MKp-E1 reduced cell viability in a PFU-dependent manner (Figure 4B), but no statistically significant difference was observed when compared to the antitumour effect of Ad5. Conversely, Ad5F35/Mkp-E1 had almost no effect on KK47 cells, even though the expression of CD46 mRNA was higher than that of CAR mRNA (Figure 4C). In UMUC3 cells, the expression of CAR mRNA was lower than that of CD46 (Figure 4A), and Ad5F35/Mkp-E1 was less effective in suppressing viability (Figure 4D).

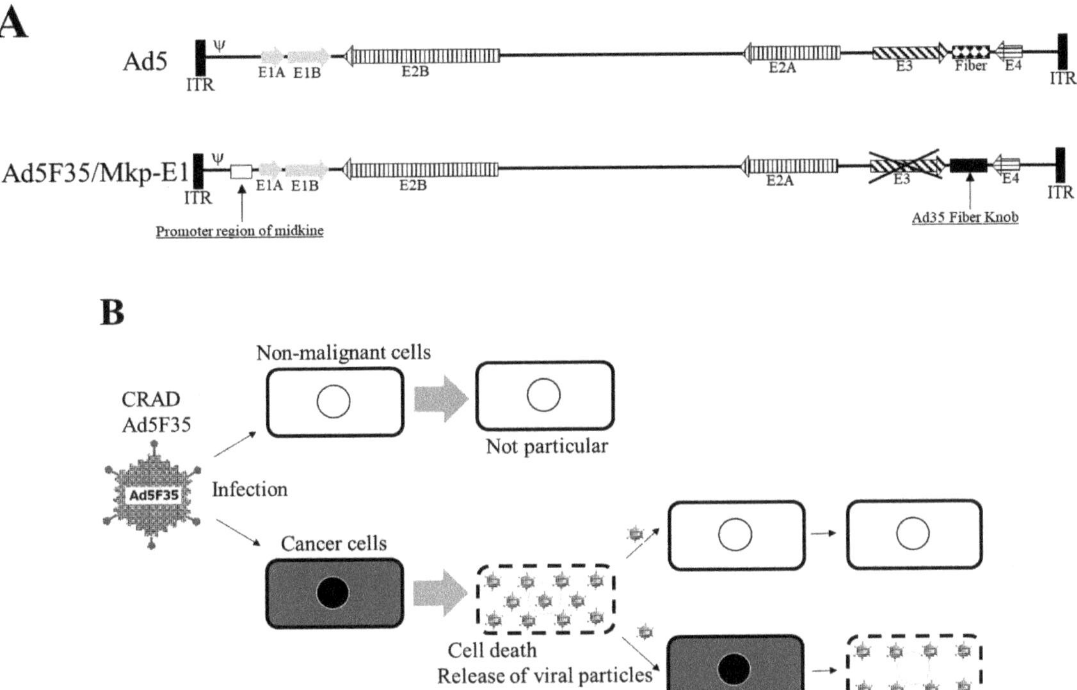

Figure 3. Combination of a chimeric adenovirus vector and midkine promoter. Because cluster of differentiation 46 (CD46) is also expressed in non-malignant cells, a safety device is required to prevent gene expression in these cells even if they are infected with an adenovirus serotype 5 F35 (AD5F35) vector. Safety is ensured by a conditionally replicating adenovirus (CRAD) that regulates genes necessary for adenovirus replication in the promoter region of midkine (**A**). The CRAD Ad5F35 vector as an oncolytic virus exhibits a remarkably potent oncolytic effect on cancer cells neighbouring primarily infected cells through secondary infection and replication (**B**). (**A**) is adapted from the data published by Nagaya et al., Anticancer Res, 2012 [95].

These results suggested that the oncolytic effect of Ad5F35/MKp-E1 on human bladder cancer cells depends not only on the expression level of CD46 mRNA, but also on some Ad5F35/MKp-E1-targeted apoptosis-related molecules.

Unlike in bladder cancer cell lines, only low CAR mRNA expression was observed in RCC cell lines (Figure 5). In contrast, CD46 mRNA expression was significantly higher than that of CAR mRNA (Figure 5A). When comparing antitumour effects, only Ad5F35/Mkp-E1 reduced cell viability in RCC cell lines; however, little or no effect was observed with Ad5/Mkp-E1 (Figure 5B–D). These results demonstrated the ineffectiveness of conventional adenoviruses against RCC and the efficacy of Ad5F35 [95].

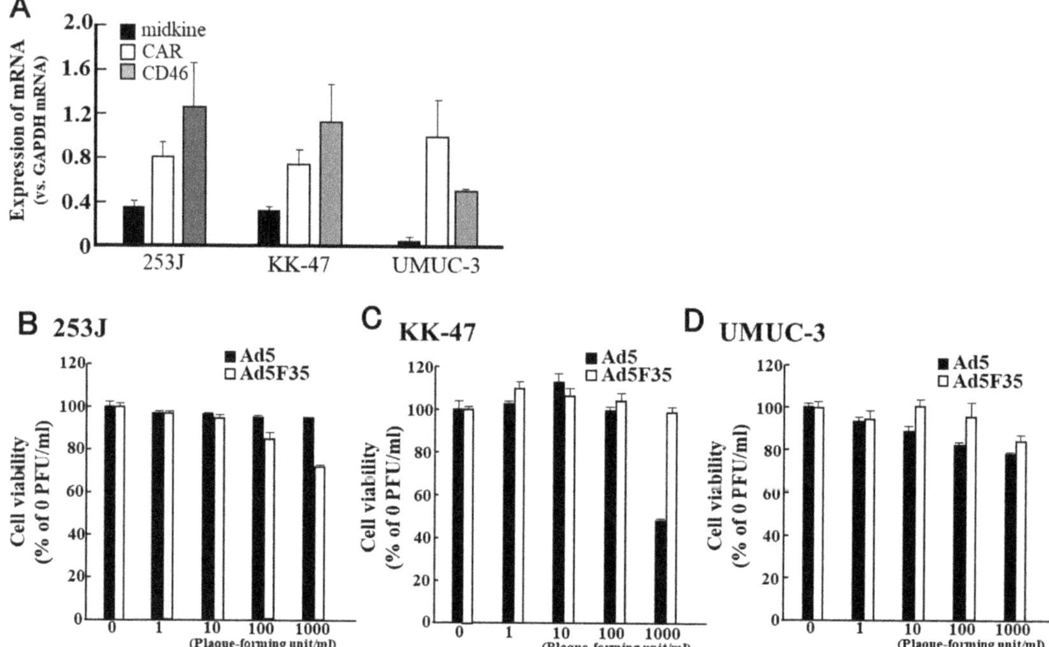

Figure 4. Expression of midkine, coxsackie and adenovirus receptor (CAR), and cluster of differentiation 46 (CD46) in human bladder cancer cell lines (**A**) and the oncolytic effects of adenovirus serotype 5 F35 (Ad5F35)/Mkp-E1 against bladder cancer cell lines (**B–D**). This figure is adapted from the data published by Gotoh et al., Urology, 2013 [92].

The combination of chimeric Ad5F35 and the Mkp can be applied for the development of oncolytic viruses and gene delivery vectors that express therapeutic genes in a cancer cell-specific manner.

Clustered regularly interspaced short palindromic repeat-associated protein 9 (CRISPR-Cas9) technology is a versatile gene-editing tool with excellent clinical potential in the field of cancer therapy [96,97]. However, off-target effects of the CRISPR-Cas9 system are a major concern for its clinical application [98]. Therefore, we constructed a chimeric Ad5F35 vector containing Cas9, which is regulated by Mkp, to restrict gene expression in cancer cells. The upstream regulatory region of the midkine gene was used as a promoter. The human codon-optimised *Streptococcus pyogenes* Cas9 gene (*hCas9*) was placed downstream of Mkp, and an enhanced green fluorescent protein gene inserted next to it as an expression marker (Figure 6A) [99]. In PNT1A cells, a non-malignant cell line, hCas9 was not expressed, even in cells infected with Ad5F35-Mkp-hCas9. On the other hand, in bladder cancer cell lines, hCas9 protein expression was confirmed after Ad5F35-Mkp-hCas9 infection (Figure 6B). CD46 and midkine expression in bladder cancer cell lines have already been confirmed [92].

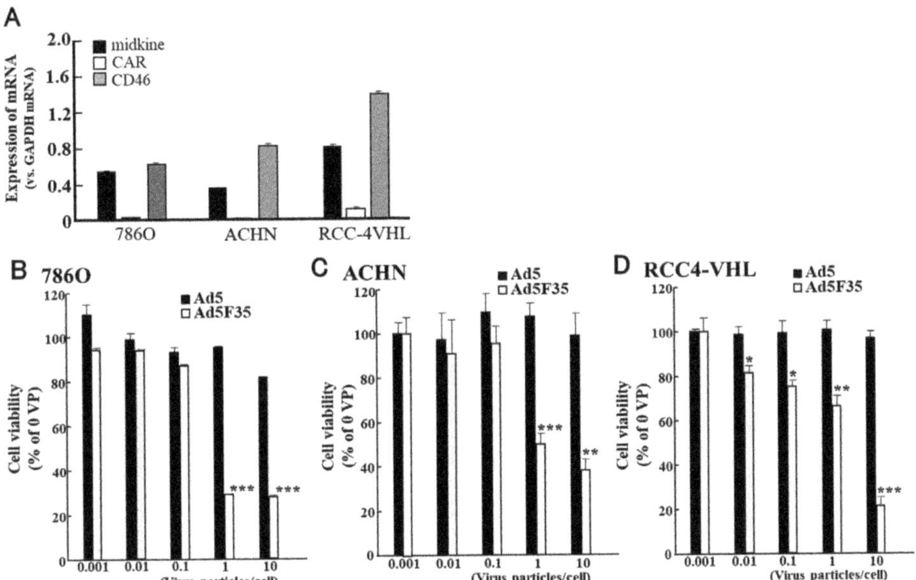

Figure 5. Coxsackie and adenovirus receptor (CAR) and cluster of differentiation 46 (CD46) expressions in renal cell carcinoma (RCC) cell lines (**A**) and the oncolytic effects of Ad5F35/Mkp-E1 against these cells (**B–D**). Only low CAR mRNA expression was found, and Ad5F35/Mkp-E1 reduced the cell viability of RCC cell lines. However, little or no effect was observed with Ad5/Mkp-E1 (**B–D**). This figure is adapted from the data published by Nagaya et al., Anticancer Res. 2012 [95]. $p < 0.05$: *, $p < 0.01$: **, $p < 0.001$: ***.

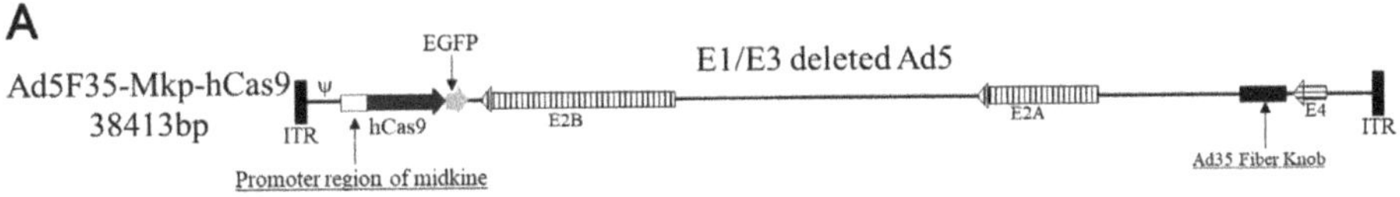

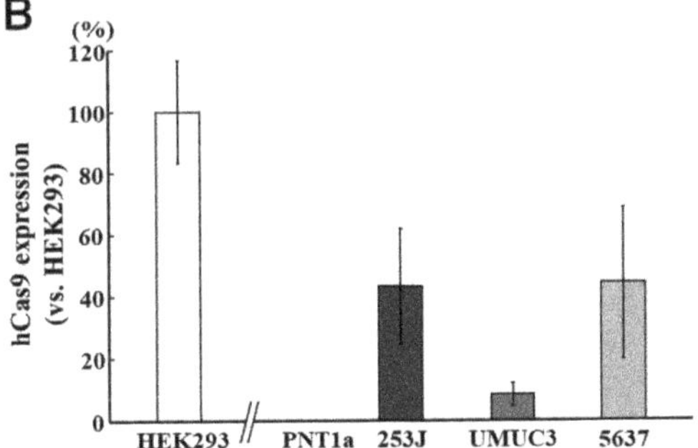

Figure 6. Human codon-optimised *Streptococcus pyogenes* Cas9 gene (*hCas9*) was placed downstream of the midkine promoter (Mkp) and the enhanced green fluorescent protein (*EGFP*) gene inserted next to it as an expression marker (**A**). In PNT1A, which is a non-malignant cell line, hCas9 expression was inhibited. In the bladder cancer cells lines, hCas9 protein expression was confirmed after Ad5F35-Mkp-hCas9 infection (**B**). This figure is adapted from the data published by Matsunaga et al., Anticancer Res. 2021 [99].

6. Discussion

Adenoviral vectors, including oncolytic viruses, are now considered relatively "old-fashioned", but they still have the largest share of vectors used in clinical trials. Their high gene transfer efficiency, ease of manufacturing, and large cargo capacity are major advantages over the "newcomer" vectors. Although their high cytotoxicity is a drawback, the use of adenoviruses as oncolytic viruses is advantageous. The disadvantages of adenoviruses include their high immunogenicity and transient gene expression. However, this is also advantageous because overgrowth of oncolytic adenoviral vectors and aberrant gene expression are prevented. In addition, because several types of viruses that infect humans have been discovered, adenoviruses that are more suitable as vectors than Ad5 are also expected to be discovered. Therefore, future studies on adenoviral vectors for viral therapy should be conducted.

However, Japan lags behind other developed countries when it comes to the development of gene therapies, including studies using viral vectors. Studies in Japan account for only 1.5% of the clinical gene therapy trials conducted worldwide [3].

In Japan, the government regulations for gene therapy studies are strict. When initiating clinical studies on gene therapy in Japan, the Clinical Research Act requires examination and reporting to the Ministry of Health, Labour, and Welfare. Moreover, a review by a committee stipulated in the "guidelines about the clinical study including gene therapies" that approval by the Minister of Health, Labour, and Welfare is also required. The Cartagena Protocol on Biosafety does not regulate medical studies. However, the Japanese law that corresponds to the protocol, the so-called "Cartagena Act", regulates medical studies, and clinical studies on gene therapy require approval from the minister in charge. These regulations involve complex and time-consuming procedures. We aimed to conduct clinical studies using adenoviruses in Japan. However, unless strict regulations are changed, the development of gene therapy studies will be hindered, regardless of whether adenovirus vectors are used. We hope that the Japanese government will promptly ease the restrictions on gene therapy studies.

Author Contributions: Writing—original draft preparation, W.M.; writing—review and editing, A.G. All authors have read and agreed to the published version of the manuscript.

Funding: This review received no external funding.

Institutional Review Board Statement: Not applicable.

Informed Consent Statement: Not applicable.

Data Availability Statement: No new data were created or analyzed in this study. Data sharing is not applicable to this article.

References

1. Dhingra, A.; Hage, E.; Ganzenmueller, T.; Böttcher, S.; Hofmann, J.; Hamprecht, K.; Obermeier, P.; Rath, B.; Hausmann, F.; Dobner, T.; et al. Molecular Evolution of Human Adenovirus (HAdV) Species C. *Sci. Rep.* **2019**, *9*, 1039. [CrossRef] [PubMed]
2. Akli, S.; Caillaud, C.; Vigne, E.; Stratford-Perricaudet, L.D.; Poenaru, L.; Perricaudet, M.; Kahn, A.; Peschanski, M.R. Transfer of a foreign gene into the brain using adenovirus vectors. *Nat. Genet.* **1993**, *3*, 224–228. [CrossRef]
3. Gene Therapy Clinical Trials Worldwide. 2022. Available online: https://a873679.fmphost.com/fmi/webd/GTCT (accessed on 28 February 2023).
4. Voysey, M.; Clemens, S.A.C.; Madhi, S.A.; Weckx, L.Y.; Folegatti, P.M.; Aley, P.K.; Angus, B.; Baillie, V.L.; Barnabas, S.L.; Bhorat, Q.E.; et al. Safety and efficacy of the ChAdOx1 nCoV-19 vaccine (AZD1222) against SARS-CoV-2: An interim analysis of four randomised controlled trials in Brazil, South Africa, and the UK. *Lancet* **2021**, *397*, 99–111. [CrossRef]
5. Ginn, S.L.; Alexander, I.E.; Edelstein, M.L.; Abedi, M.R.; Wixon, J. Gene therapy clinical trials worldwide to 2012—An update. *J. Gene Med.* **2013**, *15*, 65–77. [CrossRef] [PubMed]
6. Ginn, S.L.; Amaya, A.K.; Alexander, I.E.; Edelstein, M.; Abedi, M.R. Gene therapy clinical trials worldwide to 2017: An update. *J. Gene Med.* **2018**, *20*, e3015. [CrossRef] [PubMed]

7.   Hacein-Bey-Abina, S.; Garrigue, A.; Wang, G.P.; Soulier, J.; Lim, A.; Morillon, E.; Clappier, E.; Caccavelli, L.; Delabesse, E.; Beldjord, K.; et al. Insertional oncogenesis in 4 patients after retrovirus-mediated gene therapy of SCID-X1. *J. Clin. Investig.* **2008**, *118*, 3132–3142. [CrossRef]

8.   Howe, S.J.; Mansour, M.R.; Schwarzwaelder, K.; Bartholomae, C.; Hubank, M.; Kempski, H.; Brugman, M.H.; Pike-Overzet, K.; Chatters, S.J.; de Ridder, D.; et al. Insertional mutagenesis combined with acquired somatic mutations causes leukemogenesis following gene therapy of SCID-X1 patients. *J. Clin. Investig.* **2008**, *118*, 3143–3150. [CrossRef]

9.   Labbé, R.P.; Vessillier, S.; Rafiq, Q.A. Lentiviral Vectors for T Cell Engineering: Clinical Applications, Bioprocessing and Future Perspectives. *Viruses* **2021**, *13*, 1528. [CrossRef]

10.   Mori, S.; Wang, L.; Takeuchi, T.; Kanda, T. Two novel adeno-associated viruses from cynomolgus monkey: Pseudotyping characterization of capsid protein. *Virology* **2004**, *330*, 375–383. [CrossRef]

11.   Kishimoto, T.K.; Samulski, R.J. Addressing high dose AAV toxicity—'One and done' or 'slower and lower'? *Expert Opin. Biol. Ther.* **2022**, *22*, 1067–1071. [CrossRef]

12.   High, K.A.; Anguela, X.M. Adeno-associated viral vectors for the treatment of hemophilia. *Hum. Mol. Genet.* **2016**, *25*, R36–R41. [CrossRef] [PubMed]

13.   Fajardo-Serrano, A.; Rico, A.J.; Roda, E.; Honrubia, A.; Arrieta, S.; Ariznabarreta, G.; Chocarro, J.; Lorenzo-Ramos, E.; Pejenaute, A.; Vázquez, A.; et al. Adeno-Associated Viral Vectors as Versatile Tools for Parkinson's Research, Both for Disease Modeling Purposes and for Therapeutic Uses. *Int. J. Mol. Sci.* **2021**, *22*, 6389. [CrossRef] [PubMed]

14.   Reul, J.; Frisch, J.; Engeland, C.E.; Thalheimer, F.B.; Hartmann, J.; Ungerechts, G.; Buchholz, C.J. Tumor-Specific Delivery of Immune Checkpoint Inhibitors by Engineered AAV Vectors. *Front. Oncol.* **2019**, *9*, 52. [CrossRef] [PubMed]

15.   FDA Approves Innovative Gene Therapy to Treat Pediatric Patients with Spinal Muscular Atrophy, A Rare Disease and Leading Genetic Cause of Infant Mortality. 24 May 2019. Available online: https://www.fda.gov/news-events/press-announcements/fda-approves-innovative-gene-therapy-treat-pediatric-patients-spinal-muscular-atrophy-rare-disease (accessed on 19 January 2023).

16.   Day, J.W.; Finkel, R.S.; Chiriboga, C.A.; Connolly, A.M.; Crawford, T.O.; Darras, B.T.; Iannaccone, S.T.; Kuntz, N.L.; Peña, L.D.M.; Shieh, P.B.; et al. Onasemnogene abeparvovec gene therapy for symptomatic infantile-onset spinal muscular atrophy in patients with two copies of SMN2 (STR1VE): An open-label, single-arm, multicentre, phase 3 trial. *Lancet Neurol.* **2021**, *20*, 284–293. [CrossRef] [PubMed]

17.   King, C.R.; Zhang, A.; Tessier, T.M.; Gameiro, S.F.; Mymryk, J.S. Hacking the Cell: Network Intrusion and Exploitation by Adenovirus E1A. *mBio* **2018**, *9*, e00390-18. [CrossRef]

18.   Louis, N.; Evelegh, C.; Graham, F.L. Cloning and sequencing of the cellular-viral junctions from the human adenovirus type 5 transformed 293 cell line. *Virology* **1997**, *233*, 423–429. [CrossRef]

19.   Oliveira, E.R.A.; Bouvier, M. Immune evasion by adenoviruses: A window into host-virus adaptation. *FEBS Lett.* **2019**, *593*, 3496–3503. [CrossRef]

20.   Imperiale, M.J.; Kao, H.T.; Feldman, L.T.; Nevins, J.R.; Strickland, S. Common control of the heat shock gene and early adenovirus genes: Evidence for a cellular E1A-like activity. *Mol. Cell Biol.* **1984**, *4*, 867–874.

21.   Hartman, Z.C.; Appledorn, D.M.; Amalfitano, A. Adenovirus vector induced innate immune responses: Impact upon efficacy and toxicity in gene therapy and vaccine applications. *Virus Res.* **2008**, *132*, 1–14. [CrossRef]

22.   Saha, B.; Parks, R.J. Human adenovirus type 5 vectors deleted of early region 1 (E1) undergo limited expression of early replicative E2 proteins and DNA replication in non-permissive cells. *PLoS ONE* **2017**, *12*, e0181012. [CrossRef]

23.   Evans, J.D.; Hearing, P. Distinct roles of the adenovirus E4 ORF3 protein in viral DNA replication and inhibition of genome concatenation. *J. Virol.* **2003**, *77*, 5295–5304. [CrossRef] [PubMed]

24.   O'Neal, W.K.; Zhou, H.; Morral, N.; Aguilar-Cordova, E.; Pestaner, J.; Langston, C.; Mull, B.; Wang, Y.; Beaudet, A.L.; Lee, B. Toxicological comparison of E2a-deleted and first-generation adenoviral vectors expressing alpha1-antitrypsin after systemic delivery. *Hum. Gene Ther.* **1998**, *9*, 1587–1598. [CrossRef] [PubMed]

25.   Brunetti-Pierri, N.; Ng, P. Progress and prospects: Gene therapy for genetic diseases with helper-dependent adenoviral vectors. *Gene Ther.* **2008**, *15*, 553–560. [CrossRef] [PubMed]

26.   Ng, P.; Beauchamp, C.; Evelegh, C.; Parks, R.; Graham, F.L. Development of a FLP/frt system for generating helper-dependent adenoviral vectors. *Mol. Ther.* **2001**, *3*, 809–815. [CrossRef]

27.   Jozkowicz, A.; Dulak, J.; Nanobashvili, J.; Polterauer, P.; Prager, M.; Huk, I. Gutless adenoviral vectors—Promising tools for gene therapy. *Eur. Surg.* **2002**, *34*, 95–100. [CrossRef]

28.   Liu, J.; Seol, D.W. Helper virus-free gutless adenovirus (HF-GLAd): A new platform for gene therapy. *BMB Rep.* **2020**, *53*, 565–575. [CrossRef]

29.   Lee, D.; Liu, J.; Junn, H.J.; Lee, E.J.; Jeong, K.S.; Seol, D.W. No more helper adenovirus: Production of gutless adenovirus (GLAd) free of adenovirus and replication-competent adenovirus (RCA) contaminants. *Exp. Mol. Med.* **2019**, *51*, 1–18. [CrossRef]

30.   Parks, R.J.; Graham, F.L. A helper-dependent system for adenovirus vector production helps define a lower limit for efficient DNA packaging. *J. Virol.* **1997**, *71*, 3293–3329. [CrossRef]

31.   Parks, R.J.; Bramson, J.L.; Wan, Y.; Addison, C.L.; Graham, F.L. Effects of stuffer DNA on transgene expression from helper-dependent adenovirus vectors. *J. Virol.* **1999**, *73*, 8027–8034. [CrossRef]

32.   Ross, P.J.; Kennedy, M.A.; Parks, R.J. Host cell detection of noncoding stuffer DNA contained in helper-dependent adenovirus vectors leads to epigenetic repression of transgene expression. *J. Virol.* **2009**, *83*, 8409–8417. [CrossRef]

33. Schiedner, G.; Hertel, S.; Johnston, M.; Biermann, V.; Dries, V.; Kochanek, S. Variables affecting in vivo performance of high-capacity adenovirus vectors. *J. Virol.* **2002**, *76*, 1600–1609. [CrossRef]
34. Boorjian, S.A.; Alemozaffar, M.; Konety, B.R.; Shore, N.D.; Gomella, L.G.; Kamat, A.M.; Bivalacqua, T.J.; Montgomery, J.S.; Lerner, S.P.; Busby, J.E.; et al. Intravesical nadofaragene firadenovec gene therapy for BCG-unresponsive non-muscle-invasive bladder cancer: A single-arm, open-label, repeat-dose clinical trial. *Lancet Oncol.* **2021**, *1*, 107–117. [CrossRef]
35. Green, J.L.; Osterhout, R.E.; Klova, A.L.; Merkwirth, C.; McDonnell, S.R.P.; Zavareh, R.B.; Fuchs, B.C.; Kamal, A.; Jakobsen, J.S. Molecular characterization of type I IFN-induced cytotoxicity in bladder cancer cells reveals biomarkers of resistance. *Mol. Ther. Oncolytics* **2021**, *23*, 547–559. [CrossRef]
36. FDA Approves First Gene Therapy for the Treatment of High-Risk, Non-Muscle-Invasive Bladder Cancer. (6 December 2022). Available online: https://www.fda.gov/news-events/press-announcements/fda-approves-first-gene-therapy-treatment-high-risk-non-muscle-invasive-bladder-cancer (accessed on 22 May 2023).
37. Zhu, R.; Weng, D.; Lu, S.; Lin, D.; Wang, M.; Chen, D.; Lv, J.; Li, H.; Lv, F.; Xi, L.; et al. Double-Dose Adenovirus-Mediated Adjuvant Gene Therapy Improves Liver Transplantation Outcomes in Patients with Advanced Hepatocellular Carcinoma. *Hum. Gene Ther.* **2018**, *29*, 251–258. [CrossRef]
38. Kuruppu, D.; Tanabe, K.K. Viral oncolysis by herpes simplex virus and other viruses. *Cancer Biol. Ther.* **2005**, *4*, 524–531. [CrossRef]
39. Kelly, E.; Russell, S.J. History of oncolytic viruses: Genesis to genetic engineering. *Mol. Ther.* **2007**, *15*, 651–659. [CrossRef]
40. Eto, Y.; Nagai, K. Oncolytic Viruses as Cancer Therapeutic agents Current Status of Oncolytic Viruses. *J. Kyushu Univ. Health Welf.* **2017**, *18*, 49–53. (In Japanese)
41. Cao, G.D.; He, X.B.; Sun, Q.; Chen, S.; Wan, K.; Xu, X.; Feng, X.; Li, P.P.; Chen, B.; Xiong, M.M. The Oncolytic Virus in Cancer Diagnosis and Treatment. *Front. Oncol.* **2020**, *10*, 1786. [CrossRef]
42. Heise, C.; Sampson-Johannes, A.; Williams, A.; McCormick, F.; Von Hoff, D.D.; Kirn, D.H. ONYX-015, an E1B gene-attenuated adenovirus, causes tumor-specific cytolysis and antitumoral efficacy that can be augmented by standard chemotherapeutic agents. *Nat. Med.* **1997**, *3*, 639–645. [CrossRef]
43. Bischoff, J.R.; Kirn, D.H.; Williams, A.; Heise, C.; Horn, S.; Muna, M.; Ng, L.; Nye, J.A.; Sampson-Johannes, A.; Fattaey, A.; et al. An adenovirus mutant that replicates selectively in p53-deficient human tumor cells. *Science* **1996**, *274*, 373–376. [CrossRef]
44. Nemunaitis, J.; Ganly, I.; Khuri, F.; Arseneau, J.; Kuhn, J.; McCarty, T.; Landers, S.; Maples, P.; Romel, L.; Randlev, B.; et al. Selective replication and oncolysis in p53 mutant tumors with ONYX-015, an E1B-55kD gene-deleted adenovirus, in patients with advanced head and neck cancer: A phase II trial. *Cancer Res.* **2000**, *60*, 6359–6366.
45. Garber, K. China approves world's first oncolytic virus therapy for cancer treatment. *J. Nat. Cancer Inst.* **2006**, *98*, 298–300. [CrossRef]
46. Babiker, H.M.; Riaz, I.B.; Husnain, M.; Borad, M.J. Oncolytic virotherapy including Rigvir and standard therapies in malignant melanoma. *Oncolytic Virother.* **2017**, *6*, 11–18. [CrossRef]
47. Fukuhara, H.; Ino, Y.; Todo, T. Oncolytic virus therapy: A new era of cancer treatment at dawn. *Cancer Sci.* **2016**, *107*, 1373–1379. [CrossRef]
48. Press Release from Daiichi-Sankyo. 11 June 2021. Available online: https://www.daiichisankyo.co.jp/files/news/pressrelease/pdf/202106/20210611_J.pdf (accessed on 12 November 2022).
49. Esaki, S.; Goshima, F.; Ozaki, H.; Takano, G.; Hatano, Y.; Kawakita, D.; Ijichi, K.; Watanabe, T.; Sato, Y.; Murata, T.; et al. Oncolytic activity of HF10 in head and neck squamous cell carcinomas. *Cancer Gene Ther.* **2020**, *27*, 585–598. [CrossRef]
50. Kawashima, T.; Kagawa, S.; Kobayashi, N.; Shirakiya, Y.; Umeoka, T.; Teraishi, F.; Taki, M.; Kyo, S.; Tanaka, N.; Fujiwara, T. Telomerase-specific replication-selective virotherapy for human cancer. *Clin. Cancer Res.* **2004**, *10*, 285–292. [CrossRef]
51. Gohara, S.; Shinohara, K.; Yoshida, R.; Kariya, R.; Tazawa, H.; Hashimoto, M.; Inoue, J.; Kubo, R.; Nakashima, H.; Arita, H.; et al. An oncolytic virus as a promising candidate for the treatment of radioresistant oral squamous cell carcinoma. *Mol. Ther. Oncolytics* **2022**, *27*, 141–156. [CrossRef]
52. Phase 2 Study of OBP-301 (Telomelysin™) in Combination with Pembrolizumab and SBRT in Patients with HNSCC with Inoperable, Recurrent or Progressive Disease. 28 December 2020. Available online: https://clinicaltrials.gov/ct2/show/NCT046 85499 (accessed on 23 December 2022).
53. Packiam, V.T.; Lamm, D.L.; Barocas, D.A.; Trainer, A.; Fand, B.; Davis, R.L., 3rd; Clark, W.; Kroeger, M.; Dumbadze, I.; Chamie, K.; et al. An open label, single-arm, phase II multicenter study of the safety and efficacy of CG0070 oncolytic vector regimen in patients with BCG-unresponsive non-muscle-invasive bladder cancer: Interim results. *Urol. Oncol.* **2018**, *36*, 440–447. [CrossRef]
54. Zhao, Y.; Liu, Z.; Li, L.; Wu, J.; Zhang, H.; Zhang, H.; Lei, T.; Xu, B. Oncolytic Adenovirus: Prospects for Cancer Immunotherapy. *Front. Microbiol.* **2021**, *12*, 707290. [CrossRef]
55. Bergelson, J.M.; Cunningham, J.A.; Droguett, G.; Kurt-Jones, E.A.; Krithivas, A.; Hong, J.S.; Horwitz, M.S.; Crowell, R.L.; Finberg, R.W. Isolation of a common receptor for Coxsackie B viruses and adenoviruses 2 and 5. *Science* **1997**, *275*, 1320–1323. [CrossRef]
56. Okegawa, T.; Sayne, J.R.; Nutahara, K.; Pong, R.C.; Saboorian, H.; Kabbani, W.; Higashihara, E.; Hsieh, J.T. A histone deacetylase inhibitor enhances adenoviral infection of renal cancer cells. *J. Urol.* **2007**, *177*, 1148–1156. [CrossRef] [PubMed]
57. Hensen, L.C.M.; Hoeben, R.C.; Bots, S.T.F. Adenovirus Receptor Expression in Cancer and Its Multifaceted Role in Oncolytic Adenovirus Therapy. *Int. J. Mol. Sci.* **2020**, *21*, 6828. [CrossRef]

58. Gaggar, A.; Shayakhmetov, D.M.; Lieber, A. CD46 is a cellular receptor for group B adenoviruses. *Nat. Med.* **2003**, *9*, 1408–1412. [CrossRef] [PubMed]

59. Segerman, A.; Atkinson, J.P.; Marttila, M.; Dennerquist, V.; Wadell, G.; Arnberg, N. Adenovirus type 11 uses CD46 as a cellular receptor. *J. Virol.* **2003**, *77*, 9183–9191. [CrossRef] [PubMed]

60. Mizuguchi, H.; Hayakawa, T. Adenovirus vectors containing chimeric type 5 and type 35 fiber proteins exhibit altered and expanded tropism and increase the size limit of foreign genes. *Gene* **2002**, *285*, 69–77. [CrossRef]

61. Shayakhmetov, D.M.; Papayannopoulou, T.; Stamatoyannopoulos, G.; Lieber, A. Efficient gene transfer into human CD34(+) cells by a retargeted adenovirus vector. *J. Virol.* **2000**, *74*, 2567–2583. [CrossRef]

62. Toyoda, E.; Doi, R.; Kami, K.; Mori, T.; Ito, D.; Koizumi, M.; Kida, A.; Nagai, K.; Ito, T.; Masui, T.; et al. Adenovirus vectors with chimeric type 5 and 35 fiber proteins exhibit enhanced transfection of human pancreatic cancer cells. *Int. J. Oncol.* **2008**, *33*, 1141–1147.

63. Yu, L.; Shimozato, O.; Li, Q.; Kawamura, K.; Ma, G.; Namba, M.; Ogawa, T.; Kaiho, I.; Tagawa, M. Adenovirus type 5 substituted with type 11 or 35 fiber structure increases its infectivity to human cells enabling dual gene transfer in CD46-dependent and independent manners. *Anticancer Res.* **2007**, *27*, 2311–2316.

64. Ni, S.; Gaggar, A.; Di Paolo, N.; Li, Z.Y.; Liu, Y.; Strauss, R.; Sova, P.; Morihara, J.; Feng, Q.; Kiviat, N.; et al. Evaluation of adenovirus vectors containing serotype 35 fibers for tumor targeting. *Cancer Gene Ther.* **2006**, *13*, 1072–1081. [CrossRef]

65. Ono, R.; Takayama, K.; Sakurai, F.; Mizuguchi, H. Efficient antitumor effects of a novel oncolytic adenovirus fully composed of species B adenovirus serotype 35. *Mol. Ther. Oncolytics* **2021**, *20*, 399–409. [CrossRef]

66. Hemminki, O.; Diaconu, I.; Cerullo, V.; Pesonen, S.K.; Kanerva, A.; Joensuu, T.; Kairemo, K.; Laasonen, L.; Partanen, K.; Kangasniemi, L.; et al. Ad3-hTERT-E1A, a fully serotype 3 oncolytic adenovirus, in patients with chemotherapy refractory cancer. *Mol. Ther.* **2012**, *20*, 1821–1830. [CrossRef] [PubMed]

67. Bauerschmitz, G.J.; Guse, K.; Kanerva, A.; Menzel, A.; Herrmann, I.; Desmond, R.A.; Yamamoto, M.; Nettelbeck, D.M.; Hakkarainen, T.; Dall, P.; et al. Triple-targeted oncolytic adenoviruses featuring the cox2 promoter, E1A transcomplementation, and serotype chimerism for enhanced selectivity for ovarian cancer cells. *Mol. Ther.* **2006**, *14*, 164–174. [CrossRef] [PubMed]

68. Yu, B.; Zhou, Y.; Wu, H.; Wang, Z.; Zhan, Y.; Feng, X.; Geng, R.; Wu, Y.; Kong, W.; Yu, X. Seroprevalence of neutralizing antibodies to human adenovirus type 5 in healthy adults in China. *J. Med. Virol.* **2012**, *84*, 1408–1414. [CrossRef] [PubMed]

69. Havunen, R.; Siurala, M.; Sorsa, S.; Grönberg-Vähä-Koskela, S.; Behr, M.; Tähtinen, S.; Santos, J.M.; Karell, P.; Rusanen, J.; Nettelbeck, D.M.; et al. Oncolytic Adenoviruses Armed with Tumor Necrosis Factor Alpha and Interleukin-2 Enable Successful Adoptive Cell Therapy. *Mol. Ther. Oncolyt.* **2017**, *4*, 77–86. [CrossRef]

70. Santos, J.M.; Heiniö, C.; Cervera-Carrascon, V.; Quixabeira, D.C.A.; Siurala, M.; Havunen, R.; Butzow, R.; Zafar, S.; de Gruijl, T.; Lassus, H.; et al. Oncolytic adenovirus shapes the ovarian tumor microenvironment for potent tumor-infiltrating lymphocyte tumor reactivity. *J. Immunother. Cancer* **2020**, *8*, 000188. [CrossRef]

71. Kuryk, L.; Haavisto, E.; Garofalo, M.; Capasso, C.; Hirvinen, M.; Pesonen, S.; Ranki, T.; Vassilev, L.; Cerullo, V. Synergistic anti-tumor efficacy of immunogenic adenovirus ONCOS-102 (Ad5/3-D24-GM-CSF) and standard of care chemotherapy in preclinical mesothelioma model. *Int. J. Cancer* **2016**, *139*, 1883–1893. [CrossRef]

72. Barton, K.N.; Paielli, D.; Zhang, Y.; Koul, S.; Brown, S.L.; Lu, M.; Seely, J.; Kim, J.H.; Freytag, S.O. Second-generation replication-competent oncolytic adenovirus armed with improved suicide genes and ADP gene demonstrates greater efficacy without increased toxicity. *Mol. Ther.* **2006**, *13*, 347–356. [CrossRef]

73. Nyati, M.K.; Symon, Z.; Kievit, E.; Dornfeld, K.J.; Rynkiewicz, S.D.; Ross, B.D.; Rehemtulla, A.; Lawrence, T.S. The potential of 5-fluorocytosine/cytosine deaminase enzyme prodrug gene therapy in an intrahepatic colon cancer model. *Gene Ther.* **2002**, *9*, 844–849. [CrossRef]

74. Moolten, F.L. Tumor chemosensitivity conferred by inserted herpes thymidine kinase genes: Paradigm for a prospective cancer control strategy. *Cancer Res.* **1986**, *46*, 5276–5281.

75. Leopardo, D.; Cecere, S.C.; Di Napoli, M.; Cavaliere, C.; Pisano, C.; Striano, S.; Marra, L.; Menna, L.; Claudio, L.; Perdonà, S.; et al. Intravesical chemo-immunotherapy in non-muscle invasive bladder cancer. *Eur. Rev. Med. Pharmacol. Sci.* **2013**, *17*, 2145–2158.

76. Herr, H.W.; Morales, A. History of bacillus Calmette-Guerin and bladder cancer: An immunotherapy success story. *J. Urol.* **2008**, *179*, 53–56. [CrossRef] [PubMed]

77. Siegel, R.L.; Miller, K.D.; Jemal, A. Cancer Statistics, 2017. *CA Cancer J. Clin.* **2017**, *67*, 7–30. [CrossRef] [PubMed]

78. Chang, Y.; Lin, W.Y.; Chang, Y.C.; Huang, C.H.; Tzeng, H.E.; Abdul-Lattif, E.; Wang, T.H.; Tseng, T.H.; Kang, Y.N.; Chi, K.Y. The Association between Baseline Proton Pump Inhibitors, Immune Checkpoint Inhibitors, and Chemotherapy: A Systematic Review with Network Meta-Analysis. *Cancers* **2022**, *15*, 284. [CrossRef] [PubMed]

79. Yang, J.C.; Hughes, M.; Kammula, U.; Royal, R.; Sherry, R.M.; Topalian, S.L.; Suri, K.B.; Levy, C.; Allen, T.; Mavroukakis, S.; et al. Ipilimumab (anti-CTLA4 antibody) causes regression of metastatic renal cell cancer associated with enteritis and hypophysitis. *J. Immunother.* **2007**, *30*, 825–830. [CrossRef]

80. Guse, K.; Cerullo, V.; Hemminki, A. Oncolytic vaccinia virus for the treatment of cancer. *Expert Opin. Biol. Ther.* **2011**, *11*, 595–608. [CrossRef]

81. Mantwill, K.; Klein, F.G.; Wang, D.; Hindupur, S.V.; Ehrenfeld, M.; Holm, P.S.; Nawroth, R. Concepts in Oncolytic Adenovirus Therapy. *Int. J. Mol. Sci.* **2021**, *22*, 10522. [CrossRef]

82. Kadomatsu, K.; Huang, R.P.; Suganuma, T.; Murata, F.; Muramatsu, T. A retinoic acid responsive gene MK found in the teratocarcinoma system is expressed in spatially and temporally controlled manner during mouse embryogenesis. *J. Cell Biol.* **1990**, *110*, 607–616. [CrossRef]

83. Choudhuri, R.; Zhang, H.T.; Donnini, S.; Ziche, M.; Bicknell, R. An angiogenic role for the neurokines midkine and pleiotrophin in tumorigenesis. *Cancer Res.* **1997**, *57*, 1814–1819.

84. Kadomatsu, K.; Hagihara, M.; Akhter, S.; Fan, Q.W.; Muramatsu, H.; Muramatsu, T. Midkine induces the transformation of NIH3T3 cells. *Br. J. Cancer* **1997**, *75*, 354–359. [CrossRef]

85. Muramatsu, T. Structure and function of midkine as the basis of its pharmacological effects. *Br. J. Pharmacol.* **2014**, *171*, 814–826. [CrossRef]

86. Mashour, G.A.; Ratner, N.; Khan, G.A.; Wang, H.L.; Martuza, R.L.; Kurtz, A. The angiogenic factor midkine is aberrantly expressed in NF1-deficient Schwann cells and is a mitogen for neurofibroma-derived cells. *Oncogene* **2001**, *20*, 97–105. [CrossRef] [PubMed]

87. Tang, Y.; Kwiatkowski, D.J.; Henske, E.P. Midkine expression by stem-like tumor cells drives persistence to mTOR inhibition and an immune-suppressive microenvironment. *Nat. Commun.* **2022**, *13*, 5018. [CrossRef]

88. Ikematsu, S.; Yano, A.; Aridome, K.; Kikuchi, M.; Kumai, H.; Nagano, H.; Okamoto, K.; Oda, M.; Sakuma, S.; Aikou, T.; et al. Serum midkine levels are increased in patients with various types of carcinomas. *Br. J. Cancer* **2000**, *83*, 701–706. [CrossRef] [PubMed]

89. Kato, M.; Maeta, H.; Kato, S.; Shinozawa, T.; Terada, T. Immunohistochemical and in situ hybridization analyses of midkine expression in thyroid papillary carcinoma. *Mod. Pathol.* **2000**, *13*, 1060–1065. [CrossRef] [PubMed]

90. Gowhari Shabgah, A.; Ezzatifar, F.; Aravindhan, S.; Olegovna Zekiy, A.; Ahmadi, M.; Gheibihayat, S.M.; Gholizadeh Navashenaq, J. Shedding more light on the role of Midkine in hepatocellular carcinoma: New perspectives on diagnosis and therapy. *IUBMB Life* **2021**, *73*, 659–669. [CrossRef]

91. Uehara, K.; Matsubara, S.; Kadomatsu, K.; Tsutsui, J.; Muramatsu, T. Genomic structure of human midkine (MK), a retinoic acid-responsive growth/differentiation factor. *J. Biochem.* **1992**, *111*, 563–567. [CrossRef]

92. Gotoh, A.; Nagaya, H.; Kanno, T.; Tagawa, M.; Nishizaki, T. Fiber-substituted conditionally replicating adenovirus Ad5F35 induces oncolysis of human bladder cancer cells in in vitro analysis. *Urology* **2013**, *81*, 920.e7–920.e11. [CrossRef]

93. Kanno, T.; Gotoh, A.; Nakano, T.; Tagawa, M.; Nishizaki, T. Beneficial oncolytic effect of fiber-substituted conditionally replicating adenovirus on human lung cancer. *Anticancer Res.* **2012**, *32*, 4891–4895.

94. Gotoh, A.; Kanno, T.; Nagaya, H.; Nakano, T.; Tabata, C.; Fukuoka, K.; Tagawa, M.; Nishizaki, T. Gene therapy using adenovirus against malignant mesothelioma. *Anticancer Res.* **2012**, *32*, 3743–3747.

95. Nagaya, H.; Tagawa, M.; Hiwasa, K.; Terao, S.; Kanno, T.; Nishizaki, T.; Gotoh, A. Fiber-substituted conditionally replicating adenovirus for oncolysis of human renal carcinoma cells. *Anticancer Res.* **2012**, *32*, 2985–2989.

96. Yao, S.; He, Z.; Chen, C. CRISPR/Cas9-mediated genome editing of epigenetic factors for cancer therapy. *Hum. Gene Ther.* **2015**, *26*, 463–471. [CrossRef] [PubMed]

97. Akram, F.; Ul Haq, I.; Ahmed, Z.; Khan, H.; Ali, M.S. CRISPR-Cas9, a promising therapeutic tool for cancer therapy: A review. *Protein Pept. Lett.* **2020**, *27*, 931–944. [PubMed]

98. Fu, Y.; Foden, J.A.; Khayter, C.; Maeder, M.L.; Reyon, D.; Joung, J.K.; Sander, J.D. High-frequency off-target mutagenesis induced by CRISPR-Cas nucleases in human cells. *Nat. Biotechnol.* **2013**, *31*, 822–826. [CrossRef] [PubMed]

99. Matsunaga, W.; Hamada, K.; Tagawa, M.; Morinaga, T.; Gotoh, A. Cancer Cell-specific Transfection of hCas9 Gene Using Ad5F35 Vector. *Anticancer Res.* **2021**, *41*, 3731–3740. [CrossRef] [PubMed]

current issues in molecular biology

Case Report

The Case of an Endometrial Cancer Patient with Breast Cancer Who Has Achieved Long-Term Survival via Letrozole Monotherapy

Masako Ishikawa [1], Kentaro Nakayama [1,*], Sultana Razia [1], Hitomi Yamashita [1], Tomoka Ishibashi [1], Hikaru Haraga [1], Kosuke Kanno [1], Noriyoshi Ishikawa [2] and Satoru Kyo [1]

[1] Department of Obstetrics and Gynecology, Faculty of Medicine, Shimane University, Izumo 693-8501, Shimane, Japan
[2] Tokushukai Medical Corporation, Shonan Fujisawa Tokushukai Pathology Group, Fujisawa 251-0041, Kanagawa, Japan
* Correspondence: kn88@med.shimane-u.ac.jp; Tel.: +81-853-20-2268; Fax: +81-853-20-2264

Abstract: Herein, we present the successful treatment of a 92-year-old woman who experienced recurrent EC in the vaginal stump and para-aortic lymph nodes. The patient was first treated with paclitaxel and carboplatin for recurrent EC, which was abandoned after two cycles of chemotherapy because of G4 hematologic toxicity. Later, the patient was treated with letrozole for early-stage breast cancer, which was diagnosed simultaneously with EC recurrence. After four months of hormonal therapy, a partial response was observed not only in the lesions in the breast, but also those in the vaginal stump and para-aortic lymph nodes. She had no recurrence of breast cancer or EC, even after six years of treatment with letrozole-based hormonal therapy. Subsequent whole-exome sequencing using the genomic DNA isolated from the surgical specimen in the uterine tumor identified several genetic variants, including actionable mutations, such as *CTNNB1* (p.S37F), *PIK3R1* (p.M582Is_10), and *TP53* c.375 + 5G>T. These data suggest that the efficacy of letrozole is mediated by blocking the mammalian target of the rapamycin pathway. The findings of this study, substantiated via genetic analysis, suggest the possibility of long-term disease-free survival, even in elderly patients with recurrent EC, which was thought to be difficult to cure completely.

Keywords: breast cancer; endometrial cancer; letrozole; mammalian target of rapamycin mTOR pathway; whole-exome sequencing

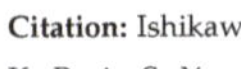

Citation: Ishikawa, M.; Nakayama, K.; Razia, S.; Yamashita, H.; Ishibashi, T.; Haraga, H.; Kanno, K.; Ishikawa, N.; Kyo, S. The Case of an Endometrial Cancer Patient with Breast Cancer Who Has Achieved Long-Term Survival via Letrozole Monotherapy. *Curr. Issues Mol. Biol.* **2023**, *45*, 2908–2916. https://doi.org/10.3390/cimb45040190

Academic Editor: Dumitru A. Iacobas

Received: 24 February 2023
Revised: 24 March 2023
Accepted: 25 March 2023
Published: 1 April 2023

1. Introduction

Endometrial cancer (EC) is a common gynecological malignancy in women, and its incidence is increasing in elderly women worldwide [1–3]. In 2020, an estimated 420,000 new cases of EC and 97,000 deaths were reported globally [4]. The risk of EC increases over time and across generations in some countries, including Japan [5]. Obesity and advanced age are currently considered major risk factors for EC development. Obesity is a major risk factor and is the main reason for rapidly rising incidence in first world countries [6]. Older age is associated with deeper myometrial invasion, higher tumor grade, and a more advanced stage [7–9] with subsequently higher recurrence rates.

The prognosis of recurrent EC is poor; the overall survival rate is reduced to 55% for pelvic recurrence and 17% for extrapelvic recurrence [10]. The treatment options for elderly women with advanced and recurrent EC include chemotherapy, hormonal therapy, radiation therapy, and combined treatment modalities. However, balancing comorbidities in elderly patients with treatment tolerance remains challenging for oncologists. In this study, we demonstrated the successful treatment of recurrent EC in an elderly patient aged 92 years, who was also diagnosed with concurrent breast cancer, using letrozole-based hormonal therapy for seven years. Subsequently, we performed whole-exome sequencing

(WES) using the uterine tumor samples to understand the mechanism underlying the efficacy of letrozole on EC. This case report outlines the successful treatment of recurrent EC and unfolds its genetic basis, suggesting the possibility of long-term disease-free survival, even in elderly patients with recurrent EC.

2. Case Presentation

An 84-year-old Japanese woman with a history of diabetes, hypertension, hyperlipidemia, and Alzheimer's disease was referred to our hospital with abnormal uterine bleeding. Her weight was 50.5 kg and her BMI was 24.3 (slightly obese). She was previously diagnosed with abnormal endometrial histology in a clinic and visited our hospital for the treatment of suspected endometrial carcinoma. An endometrial biopsy showed endometrioid carcinoma G3, and pelvic magnetic resonance imaging revealed a 67 mm mass lesion in the uterus (Figure 1).

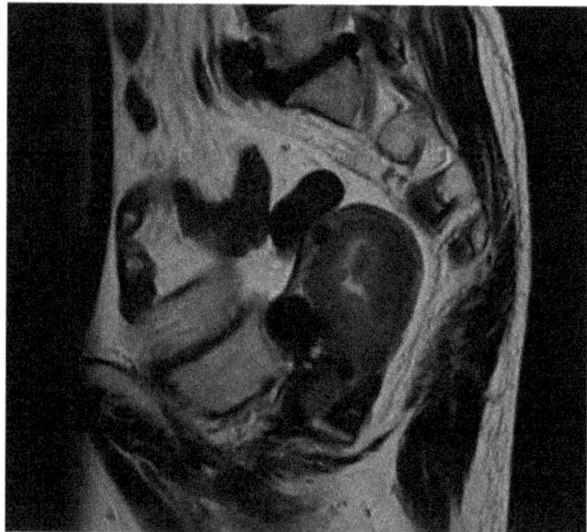

Figure 1. Magnetic resonance imaging of the pelvis. T2-weighted imaging revealed a thickened endometrial mass (50 × 73 × 43 mm).

The patient underwent a total abdominal hysterectomy and bilateral salpingo-oophorectomy. Lymph node dissection was not performed because the patient was too elderly to undergo this treatment. The pathological examination revealed endometrioid carcinoma G3; finally, the patient was diagnosed with stage 3A endometrial carcinoma according to FIGO 2018, and pT3apNxpMx according to TMN (myometrial invasion 100%) (Figure 2a,b). Immunohistochemistry showed 90% positivity for the estrogen receptor (ER) and 80% for the progesterone receptor (PgR). Strong and diffuse overexpression of p53 was identified, which indicated a *TP53* mutation (Supplementary Figure S1).

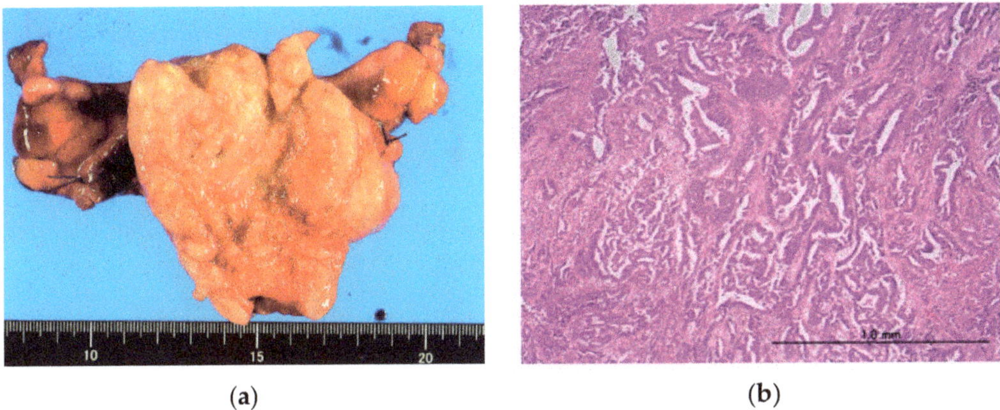

Figure 2. (**a**) Surgical findings showing a high volume tumor in the uterine corpus; (**b**) pathological diagnosis: endometrioid carcinoma G3.

Two months after surgery, the patient presented to our gynecological outpatient department with a small amount of vaginal bleeding that made us suspect a recurrence. A small tumor was observed in the vaginal stump. A tumor biopsy was performed, and endometrial carcinoma G3 was detected. Positron emission tomography (PET-CT) revealed metastatic regions in the vaginal stump (Figure 3a) and pelvic and para-aortic lymph nodes (Figure 3b).

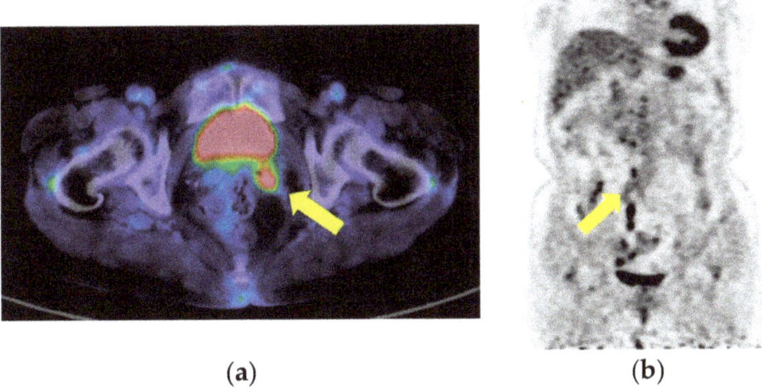

Figure 3. Positron emission tomography showing recurrence lesions in the (**a**) vaginal stump and (**b**) para-aortic lymph node. Yellow arrow point to the disease lesion.

Simultaneously, fluorodeoxyglucose uptake was detected in the patient's left breast A region (Figure 4a); however, lymph node and distant organ metastases were absent. The mediolateral oblique view of the screening mammogram revealed disordered construction in the U region of the mediolateral oblique (MLO) and in the I region of the craniocaudal (CC) (Figure 4b).

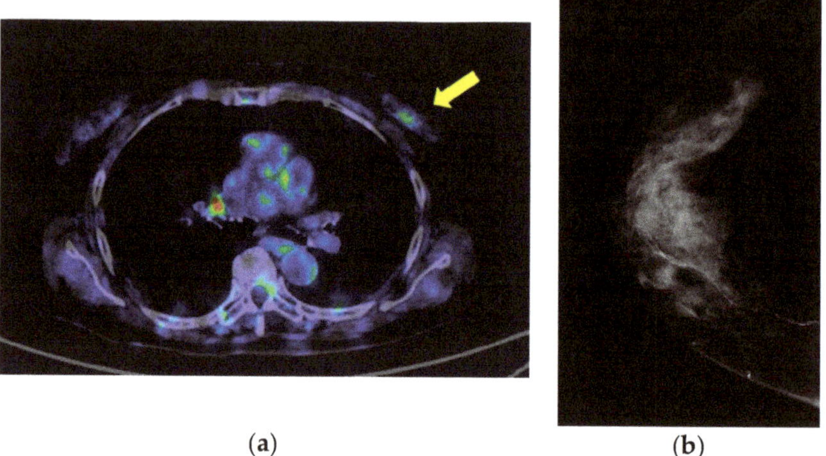

(a)   (b)

Figure 4. (**a**) Positron emission tomography showing the accumulation of fluorodeoxyglucose in a 2 cm mass lesion in the A region of the left breast; (**b**) mammography showing the U region of the MLO and the I region of the craniocaudal, indicating construction disorder.

Subsequently, a tumor biopsy was performed, which indicated the irregular dispersal of tumor cells in a densely fibrotic stroma. Immunohistochemical staining for the ER, PgR, human epidermal growth factor receptor 2 (HER2), and E-cadherin were positive (Supplementary Figure S2). The morphology, immunohistochemical findings, and clinical history supported the diagnosis of scirrhous type stage I invasive ductal carcinoma. Owing to her advanced age and recurrent EC, the patient did not undergo any surgery for breast cancer. Her doctor for breast cancer did not choose anti-HER2 therapy. The patient received 'paclitaxel and carboplatin' chemotherapy for EC and 'letrozole' hormone therapy for breast cancer. Unfortunately, after two cycles of chemotherapy, she developed a fever, diarrhea, and fatigue. Blood examination revealed a white blood cell (WBC) count of 1432/μL, a neutrophil count under 500/μL, and a C-reactive protein (CRP) level > 30, suggesting febrile neutropenia. Subsequently, chemotherapy was discontinued because of the patient's worsening conditions; nevertheless, the hormone therapy was continued.

After four months of hormonal therapy, regression of the disease lesion was observed not only in the patient's breast, but also in her vaginal stump and pelvic and para-aortic lymph nodes (Figure 5a,b). She has had no exacerbation of breast cancer or recurrent EC for seven years with hormonal therapy alone.

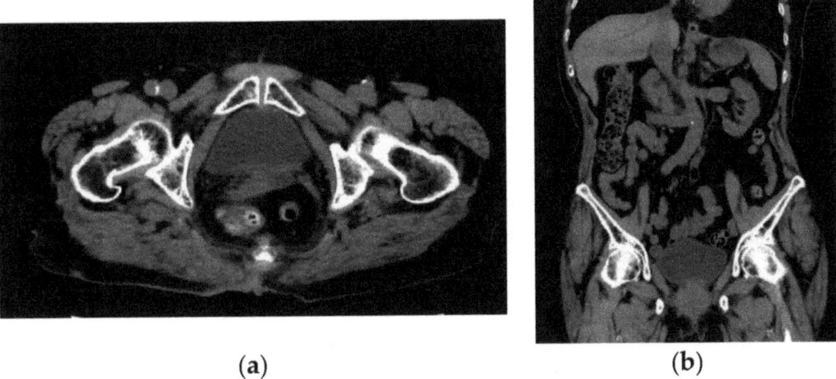

(a) **(b)**

Figure 5. Computed tomography showing reduced recurrent lesions of (**a**) the vaginal stump and (**b**) the para-aortic lymph node.

Based on the efficacy of letrozole (which is recommended for breast cancer) against recurrent EC, we hypothesized a unique gene profile in EC. To test this hypothesis, we performed WES using the genomic DNA isolated from the surgical specimen of the uterine tumor. The analysis identified a mutation rate of 10.7, indicating hypermutation; however, the microsatellite instability was stable. It detected actionable gene mutations, such as *CTNNB1* (p.S37F), *PIK3R1* (p.M582Is_10), and *TP53* c.375 + 5G>T (10.3%) in the tumor. The positions and allele frequencies of the variants identified are summarized in Table 1 and Supplementary Table S1. These variants were considered to have uncertain significance rather than being pathogenic. Moreover, we did not observe copy number alterations in oncogenes or tumor suppressor genes (Figure 6).

Table 1. Actionable variants in the uterine tumor.

Function	Actionable Gene Variants	Variant Allele Frequency (%)	Copy Number	CNV
OG	*CTNNB1* p.Ser37Phe (ClinVar: Pathogenic)	10.2	2.96	neutral
TSG	*PIK3R1* p.Met582Ile fs*10	45	1.9	UPD
TSG	*TP53* c.375+5G>T (ClinVar: Likely Pathogenic)	10.3	2.45	neutral

VAF, variant allele frequency; CN, copy number; OG, oncogene; TSG, tumor suppressor gene.

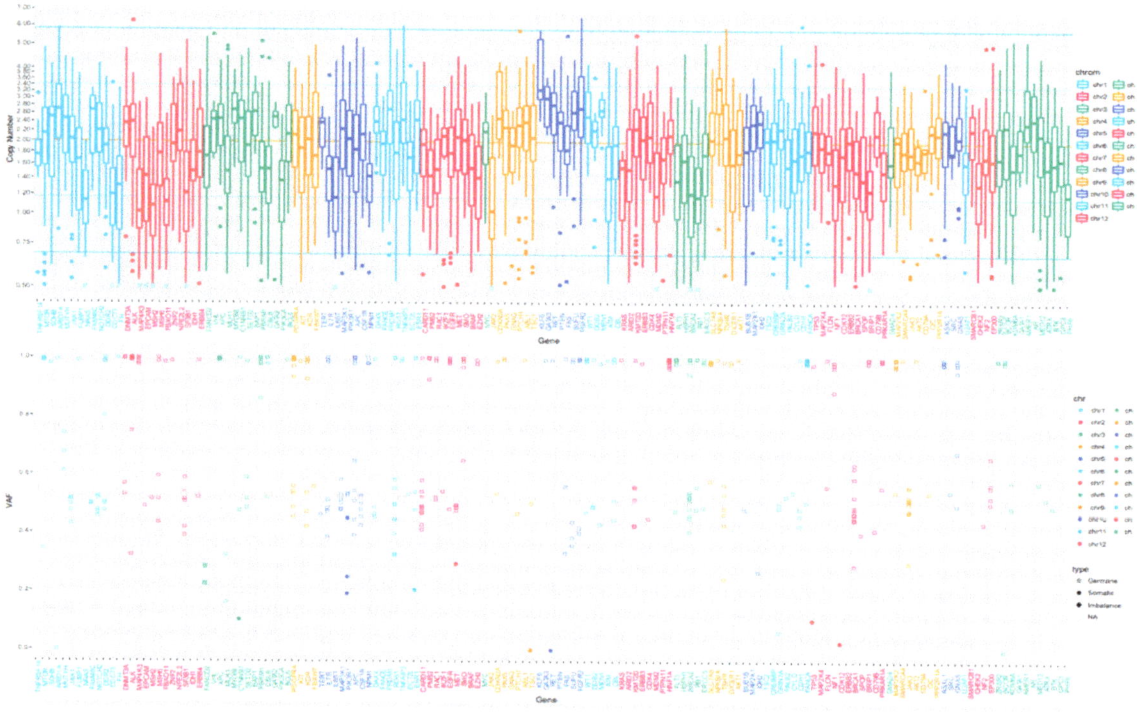

Figure 6. Copy number plots of the uterine tumor in the current case. The horizontal axis shows the chromosome location, and the vertical axis shows the gene copy number. The representative copy number plot appears normal.

3. Discussion

EC is the most common pelvic gynecological tumor in developed countries, and its incidence is increasing. According to different published studies, the mean age at diagnosis is 63 years, and 8–14% of patients diagnosed with EC are elderly [11,12]. Moreover, owing to their more aggressive tumor biology, less favorable clinicopathological features, and more advanced disease stage, elderly women are more likely to die from this disease than younger patients [13]. However, 20% of patients with advanced-stage disease experience recurrence [13–15]. The prognosis for these patients is poor, and treatment options are limited.

In the present case, the patient could not continue chemotherapy for recurrent EC because of worsening febrile neutropenia. Nevertheless, the aromatase inhibitor (letrozole, which is used for breast cancer) was remarkably effective against recurrent EC, and long-term survival was obtained in a state of progesterone receptor (PR) with tumor shrinkage. Subsequent WES to characterize the genetic profile underlying EC recurrence identified actionable gene mutations, such as *CTNNB1* (p.S37F), *PIK3R1* (p.M582Is_10), and *TP53* c.375 + 5G>T.

Recently, the phosphatidylinositide 3-kinase (PI3K)/AKT signaling pathway was found to be a major survival signal in cancer cells. Deregulation of the PI3K pathway is frequent in tumor cells, and can be caused by multiple changes affecting different signaling cascades. These changes include gene amplification, mutations, and alterations in gene expression. However, various changes in the PI3K pathway have been identified in different cancers. In addition, mutations of *PIK3R1* have been reported in breast cancer, colorectal cancer, and glioblastoma [16–20]. The overactivation of the PI3K/AKT/mammalian target of rapamycin (mTOR) pathway has recently been implicated in the pathogenesis of ECs [21]. Moreover, most endometrial tumors are hormonally driven; estrogen signaling through ERα acts as an oncogenic signal, and mTOR signaling is required for estrogen-induced tumors [22]. Therefore, synergistic antitumor effects may be achieved by combining PI3K/AKT/mTOR pathway inhibitors with agents that disrupt ER signaling. One study evaluated the potential of combining letrozole with RAD001 (everolimus) in two in vitro models of breast carcinoma (MCF7 and T47D) and found that estrogen-induced proliferation largely depended on mTOR signaling. Furthermore, RAD001, in combination with letrozole, has more profound effects on aromatase-mediated estrogen-induced proliferation in aromatase-expressing lines than either agent alone [23]. Another phase II trial of everolimus and letrozole in women with recurrent EC reported a high clinical benefit rate of 40%, a response rate of 32%, and a complete response rate of 20% [24]. Furthermore, the combination of everolimus and letrozole significantly improved progression-free survival in patients with breast cancer [25].

In the present case, the development of EC was suspected to involve mutations in the mTOR pathway. Therefore, it is considered that letrozole, which is a therapeutic drug for breast cancer, has been effective. To date, the patient survives with the EC lesion continuing to shrink. Thus far, it is clear that dual inhibition of ER signaling and the PI3K/AKT/mTOR pathway induces antitumor effects; however, in the present study, it was suggested that letrozole alone is effective against recurrent EC. The toxicity profile of letrozole was favorable. Therefore, letrozole can be used as a personalized medicine based on the genetic or molecular profiling of patients with recurrent EC. Furthermore, PIK3CA status could be used as a molecular marker to guide clinical decision-making regarding the use of letrozole in women with advanced EC. Thus, multiple gene panel tests can benefit elderly patients with useful and safe treatments.

In the current case, letrozole was used to suppress tumor growth in recurrent EC and improved the patient's disease-free survival beyond seven years without cytotoxic anticancer drugs. This indicates that letrozole affects a molecular biological inhibitory pathway not only in breast cancer, but also in metastatic lesions of EC. Genetic analysis of the tumor revealed mutations in *CTNNB1* and *PIK3R1*, the tumor suppressor genes located in the AKT/PI3K pathway. It has been shown that the loss *PIK3R1*'s function via mutation causes the AKT/PI3K pathway to turn around. Inhibition of the mTOR pathway using mTOR inhibitors, such as temsirolimus, has also been shown to control tumor growth [26]. Moreover, the AKT/PI3K pathway is associated with the ER pathway [27], and letrozole, an ER serine, plays an important role in preventing tumor growth. Taken together, we infer that the long-term suppression of tumor growth in the present case could have been achieved by blocking the AKT/PI3K pathway. Furthermore, tumor-like endometrial cancer (which is driven by unopposed estrogen from obesity) anti-estrogen therapy, such as megestrol acetate, medroxyprogesterone acetate, and aromatase inhibitors, might produce effectiveness for tumors. Additionally, the strategy of adding cdk4/6 inhibitor has already been noted in a recent study [28]. The fact that letrozole has shown efficacy in preventing tumor growth as a single agent may provide the possibility of long-term survival for recurrent endometrial uterine cancer in elderly patients, even though recurrence sites are difficult to control with conventional chemotherapy. However, confirming biological profiles with multiple gene panel testing is important when considering the effects of tumor suppression and patient treatment.

Supplementary Materials: The following supporting information can be downloaded at: https://www.mdpi.com/article/10.3390/cimb45040190/s1.

Author Contributions: M.I. contributed to the diagnosis and drafted the manuscript. K.N. made the diagnosis and advised regarding the manuscript's preparation. S.R. drafted the manuscript. H.Y. helped correct the patient's data. T.I. helped correct the patient's data. H.H. helped correct the patient's data. K.K. helped correct the patient's data. S.K. helped correct the patient's data. N.I. performed the pathological diagnosis. S.K. advised regarding the manuscript's preparation and revised the final manuscript. All authors have read and agreed to the published version of the manuscript.

Funding: This research received no external funding.

Institutional Review Board Statement: This case report protocol was approved by the institutional ethics and research review boards at Shimane University (IRB No. 20070305-1 and No. 20070305-2; approval date: 5 March 2007).

Informed Consent Statement: Informed consent was obtained from the subject involved in the study. Written informed consent was obtained from the patient to publish this case report.

Data Availability Statement: The datasets used or analyzed during the current study are available from the corresponding author upon reasonable request.

Conflicts of Interest: The authors declare no conflict of interest.

References

1. Poupon, C.; Bendifallah, S.; Ouldamer, L.; Canlorbe, G.; Raimond, E.; Hudry, N.; Coutant, C.; Graesslin, O.; Touboul, C.; Collinet, P.; et al. Management and survival of elderly and very elderly patients with of endometrial cancer: An age-stratified study of 1228 women from the FRANCOGYN Group. *Ann. Surg. Oncol.* **2017**, *24*, 1667–1676. [CrossRef]

2. Rauh-Hain, J.A.; Pepin, K.J.; Meyer, L.A.; Clemmer, J.T.; Lu, K.H.; Rice, L.W.; Uppal, S.; Schorge, J.O.; Del Carmen, M.G. Management for elderly women with advanced-stage, high-grade endometrial cancer. *Obstet. Gynecol.* **2015**, *126*, 1198–1206. [CrossRef]

3. Bourgin, C.; Saidani, M.; Poupon, C.; Cauchois, A.; Foucher, F.; Leveque, J.; Lavoue, V. Endometrial cancer in elderly women: Which disease, which surgical management? A systematic review of the literature. *Eur. J. Surg. Oncol.* **2016**, *42*, 166–175. [CrossRef] [PubMed]

4. Sung, H.; Ferlay, J.; Siegel, R.L.; Laversanne, M.; Soerjomataram, I.; Jemal, A.; Bray, F. Cancer Statistics 2020, GLOBOCAN estimates of Incidence and Mortality Worldwide for 36 cancers in 185 Countries. *CA Cancer J. Clin.* **2021**, *71*, 209–249. [CrossRef] [PubMed]

5. Lortet-Tieulent, J.; Ferlay, J.; Bray, F.; Jemal, A. International patterns and trends in endometrial cancer incidence, 1978–2013. *J. Natl. Cancer Inst.* **2018**, *110*, 354–361. [CrossRef]

6. Onstad, M.A.; Schmandt, R.E.; Lu, K.H. Addressing the Role of Obesity in Endometrial Cancer Risk, Prevention, and Treatment. *J. Clin. Oncol.* **2016**, *34*, 4225–4230. [CrossRef]

7. Mundt, A.J.; Waggoner, S.; Yamada, D.; Rotmensch, J.; Connell, P.P. Age as a prognostic factor for recurrence in patients with endometrial carcinoma. *Gynecol. Oncol.* **2000**, *79*, 79–85. [CrossRef]

8. Goodwin, J.S.; Samet, J.M.; Key, C.R.; Humble, C.; Kutvirt, D.; Hunt, C. Stage at diagnosis of cancer varies with the age of the patient. *J. Am. Geriatr. Soc.* **1986**, *34*, 20–26. [CrossRef] [PubMed]

9. Samet, J.; Hunt, W.C.; Key, C.; Humble, C.G.; Goodwin, J.S. Choice of cancer therapy varies with age of patient. *JAMA* **1986**, *255*, 3385–3390. [CrossRef]

10. Xu, Y.; Burmeister, C.; Hanna, R.K.; Munkarah, A.; Elshaikh, M.A. Predictors of survival after recurrence in women with early-stage endometrial carcinoma. *Int. J. Gynecol. Cancer* **2016**, *26*, 1137–1142. [CrossRef]

11. Ferlay, J.; Soerjomataram, I.; Dikshit, R.; Eser, S.; Mathers, C.; Rebelo, M.; Parkin, D.M.; Forman, D.; Bray, F. Cancer incidence and mortality worldwide: Sources, methods and major patterns in GLOBOCAN 2012. *Int. J. Cancer* **2015**, *136*, E359–E386. [CrossRef] [PubMed]

12. Torgeson, A.; Boothe, D.; Poppe, M.M.; Suneja, G.; Gaffney, D.K. Disparities in care for elderly women with endometrial cancer adversely effects survival. *Gynecol. Oncol.* **2017**, *147*, 320–328. [CrossRef]

13. Duska, L.; Shahrokni, A.; Powell, M. Treatment of Older Women with Endometrial Cancer: Improving Outcomes with Personalized Care. *Am. Soc. Clin. Oncol. Educ. Book* **2016**, *35*, 164–174. [CrossRef] [PubMed]

14. Van Der Putten, L.J.; Visser, N.; Van De Vijver, K.; Santacana, M.; Bronsert, P.; Bulten, J.; Hirschfeld, M.; Colas, E.; Gil-Moreno, A.; Garcia, A.; et al. L1CAM expression in endometrial carcinomas: An ENITEC collaboration study. *Br. J. Cancer* **2016**, *115*, 716–724. [CrossRef] [PubMed]

15. Trovik, J.; Wik, E.; Werner, H.M.; Krakstad, C.; Helland, H.; Vandenput, I.; Njolstad, T.S.; Stefansson, I.M.; Marcickiewicz, J.; Tingulstad, S.; et al. Hormone receptor loss in endometrial carcinoma curettage predicts lymph node metastasis and poor outcome in prospective multicentre trial. *Eur. J. Cancer* **2013**, *49*, 3431–3441. [CrossRef]
16. Creutzberg, C.L.; Nout, R.A.; Lybeert, M.L.; Wárlám-Rodenhuis, C.C.; Jobsen, J.J.; Mens, J.W.; Lutgens, L.C.; Pras, E.; van de Poll-Franse, L.V.; van Putten, W.L.; et al. Fifteen-year radiotherapy outcomes of the randomized PORTEC-1 trial for endometrial carcinoma. *Int. J. Radiat. Oncol. Biol. Phys.* **2011**, *81*, e631–e638. [CrossRef]
17. Velho, S.; Oliveira, C.; Ferreira, A.; Ferreira, A.C.; Suriano, G.; Schwartz, S.; Duval, A., Jr.; Carneiro, F.; Machado, J.C.; Hamelin, R.; et al. The prevalence of PIK3CA mutations in gastric and colon cancer. *Eur. J. Cancer* **2005**, *41*, 1649–1654. [CrossRef]
18. Lee, J.; van Hummelen, P.; Go, C.; Palescandolo, E.; Jang, J.; Park, H.Y.; Kang, S.Y.; Park, J.O.; Kang, W.K.; MacConaill, L.; et al. High-throughput mutation profiling identifies frequent somatic mutations in advanced gastric adenocarcinoma. *PLoS ONE* **2012**, *7*, e38892. [CrossRef]
19. Spoerke, J.M.; O'Brien, C.; Huw, L.; Koeppen, H.; Fridlyand, J.; Brachmann, R.K.; Haverty, P.M.; Pandita, A.; Mohan, S.; Sampath, D.; et al. Phosphoinositide 3-kinase (PI3K) pathway alterations are associated with histologic subtypes and are predictive of sensitivity to PI3K inhibitors in lung cancer preclinical models. *Clin. Cancer Res.* **2012**, *18*, 6771–6783. [CrossRef]
20. Gallia, G.L.; Rand, V.; Siu, I.M.; Eberhart, C.G.; James, C.D.; Marie, S.K.; Oba-Shinjo, S.M.; Carlotti, C.G.; Caballero, O.L.; Simpson, A.J.; et al. PIK3CA gene mutations in pediatric and adult glioblastoma multiforme. *Mol. Cancer Res.* **2006**, *4*, 709–714. [CrossRef]
21. Krakstad, C.; Birkeland, E.; Seidel, D.; Kusonmano, K.; Petersen, K.; Mjøs, S.; Hoivik, E.A.; Wik, E.; Halle, M.K.; Øyan, A.M.; et al. High-throughput mutation profiling of primary and metastatic endometrial cancers identifies KRAS, FGFR2 and PIK3CA to be frequently mutated. *PLoS ONE* **2012**, *7*, e52795. [CrossRef]
22. Kuo, K.T.; Mao, T.L.; Jones, S.; Veras, E.; Ayhan, A.; Wang, T.L.; Glas, R.; Slamon, D.; Velculescu, V.E.; Kuman, R.J.; et al. Frequent activating mutations of PIK3CA in ovarian clear cell carcinoma. *Am. J. Pathol.* **2009**, *174*, 1597–1601. [CrossRef]
23. Slomovitz, B.M.; Coleman, R.L. The PI3K/AKT/mTOR pathway as a therapeutic target in endometrial cancer. *Clin. Cancer Res.* **2012**, *18*, 5856–5864. [CrossRef] [PubMed]
24. Rodriguez, A.C.; Blanchard, Z.; Maurer, K.A.; Gertz, J. Estrogen Signaling in Endometrial Cancer: A Key Oncogenic Pathway with Several Open Questions. *Horm. Cancer* **2019**, *10*, 51–63. [CrossRef] [PubMed]
25. Boulay, A.; Rudloff, J.; Ye, J.; Zumstein-Mecker, S.; O'Reilly, T.; Evans, D.B.; Chen, S.; Lane, H.A. Dual inhibition of mTOR and estrogen receptor signaling in vitro induces cell death in models of breast cancer. *Clin. Cancer Res.* **2005**, *11*, 5319–5328. [CrossRef]
26. Slomovitz, B.M.; Jiang, Y.; Yates, M.S.; Soliman, P.T.; Johnston, T.; Nowakowski, M.; Levenback, C.; Zhang, Q.; Ring, K.; Munsell, M.F.; et al. Phase II study of everolimus and letrozole in patients with recurrent endometrial carcinoma. *J. Clin. Oncol.* **2015**, *33*, 930–936. [CrossRef]
27. Skriver, S.K.; Jensen, M.B.; Eriksen, J.O.; Ahlborn, L.B.; Knoop, A.S.; Rossing, M.; Ejlertsen, B.; Laenkholm, A.V. Induction of PIK3CA alterations during neoadjuvant letrozole may improve outcome in postmenopausal breast cancer patients. *Breast Cancer Res. Treat.* **2020**, *184*, 123–133. [CrossRef] [PubMed]
28. Konstantinopoulos, P.A.; Lee, E.K.; Xiong, N.; Krasner, C.; Campos, S.; Kolin, D.L.; Liu, J.F.; Horowitz, N.; Wright, A.A.; Bouberhan, S.; et al. A Phase II, Two-Stage Study of Letrozole and Abemaciclib in Estrogen Receptor–Positive Recurrent Endometrial Cancer. *J. Clin. Oncol.* **2023**, *41*, 599–608. [CrossRef]

MDPI

St. Alban-Anlage 66

4052 Basel

Switzerland

www.mdpi.com

Current Issues in Molecular Biology Editorial Office

E-mail: cimb@mdpi.com

www.mdpi.com/journal/cimb